To my friend Ron

BIRKHÄUSER

Progress in Mathematics
Volume 274

Lluís Puig

Frobenius Categories versus Brauer Blocks

The Grothendieck Group of the Frobenius Category of a Brauer Block

Birkhäuser
Basel · Boston · Berlin

Author:

Lluís Puig
CNRS, Institut de Mathématiques
de Jussieu
Université Denis Diderot (Paris VII)
175, Rue du Chevaleret
75013 Paris
France
e-mail: puig@math.jussieu.fr

2000 Mathematics Subject Classification 20C11

Library of Congress Control Number: 2009921943

Bibliographic information published by Die Deutsche Bibliothek.
Die Deutsche Bibliothek lists this publication in the Deutsche Nationalbibliografie;
detailed bibliographic data is available in the Internet at http://dnb.ddb.de

ISBN 978-3-7643-9997-9 Birkhäuser Verlag AG, Basel · Boston · Berlin

Basel · Boston · Berlin
P.O. Box 133, CH-4010 Basel, Switzerland
Part of Springer Science+Business Media
Printed on acid-free paper produced from chlorine-free pulp. TCF ∞
Printed in Germany

ISBN 978-3-7643-9997-9 e-ISBN 978-3-7643-9998-6

9 8 7 6 5 4 3 2 1 www.birkhauser.ch

Contents

Introduction

I1 More than one hundred years ago, Georg Frobenius [26] proved his remarkable theorem affirming that, *for a prime p and a finite group G, if the quotient of the normalizer by the centralizer of any p-subgroup of G is a p-group then, up to a normal subgroup of order prime to p, G is a p-group.* Of course, it would be an anachronism to pretend that Frobenius, when doing this theorem, was thinking the *category* — noted $\mathcal{F}_G$ in the sequel — where the objects are the p-subgroups of G and the morphisms are the group homomorphisms between them which are induced by the G-conjugation. Yet Frobenius' hypothesis is truly meaningful in this category.

I2 Fifty years ago, John Thompson [57] built his seminal proof of the nilpotency of the so-called *Frobenius kernel* of a *Frobenius group G* with arguments — at that time completely new — which might be rewritten in terms of $\mathcal{F}_G$; indeed, some time later, following these kind of arguments, George Glauberman [27] proved that, under some — rather strong — hypothesis on G, the normalizer N of a suitable nontrivial p-subgroup of G *controls fusion* in G, which amounts to saying that *the inclusion $N \subset G$ induces an equivalence of categories $\mathcal{F}_N \cong \mathcal{F}_G$*.

I3 Thus, when about forty years ago we start working on finite groups, these kind of results pushed us to introduce the *Frobenius category $\mathcal{F}_G$* in the language of Claude Chevalley's seminar. At the beginning, it was essentially a language to guide our research, as for instance in our refinement [35, Ch. II and III] of the *Alperin Fusion Theorem* [1]. Moreover, we quickly realized that, in order to have a better insight into the structure of G — for instance, to follow Thompson's arguments in Chapter IV of the so-called *Odd paper* [25] — we had to consider suitable *extensions* determined by G of this category, namely what in [35, Ch. VI] we call *"localités à épimorphismes"*.

I4 The next step in the gestation of the ideas contained in this book came thirty years ago, when Michel Broué talked us about his will of extending to the *Brauer blocks* our Frobenius categories for groups. As a matter of fact, we found that Richard Brauer already had partially realized this project in [9]; indeed, in this paper, for what Brauer calls a *block of characters* of G, he considers a new *order relation* refining the inclusion between a suitable set of subgroups of a *defect group* of the block, already proving a suitable generalization of the *Alperin Fusion Theorem* in the new context.

I5 Then, Michel Broué† reformulated the new inclusion introduced by Brauer in terms of *pairs* — which freed Broué from the choice of a *defect group* — formed by a p-subgroup P of G and a *block of characters* e (for short, a *block* in the sequel) of the centralizer $C_G(P)$ of P, and extended the new inclusion to *all* these pairs — ever since called *Brauer pairs*. At this point, we already had a *Frobenius category* for a block b of G, namely the category $\mathcal{F}_{(b,G)}$ where the objects are all the Brauer pairs (P,e) *containing* $(\{1\},b)$ — with respect to the new inclusion — and where, again, the morphisms are the group homomorphisms between the p-subgroups induced by the G-conjugation, up to Brauer-Broué's inclusion.

I6 The miracle happened again: the Frobenius Hypothesis on $\mathcal{F}_{(b,G)}$ — *all the automorphism groups of the objects are p-groups* — was meaningful too; in this situation — where b is called a *nilpotent block* — with Michel Broué we proved in [12] that there is a bijection between the set of ordinary *irreducible characters* of G in b and the set of ordinary *irreducible characters* of any defect group of b. This immediately implies that the *block algebra* over $\mathbb{C}$ is *Morita equivalent* to the group $\mathbb{C}$-algebra of the defect group of b; in other words, denoting by (P,e) a maximal Brauer pair containing $(\{1\},b)$ and identifying b with the corresponding central idempotent, it implies that *the categories of* $\mathbb{C}Gb$*- and* $\mathbb{C}P$*-modules are equivalent.*

I7 However, Broué guessed something more precise, namely that even over a complete discrete valuation ring $\mathcal{O}$, the categories of $\mathcal{O}Gb$- and $\mathcal{O}P$-modules were equivalent. In our joint work [12], we already proved that the centers $Z(\mathcal{O}Gb)$ and $Z(\mathcal{O}P)$ were isomorphic which, as Everett Dade pointed out to us, proved Broué's conjecture whenever P is Abelian. Soon after, following James Green's approach [29], we came to the idea of the *source algebra* of the block b — which is more than an $\mathcal{O}$-algebra: it is the $\mathcal{O}$-algebra $i(\mathcal{O}Gb)i$ endowed with the homomorphism mapping $u \in P$ on ui, for a primitive idempotent i of $(\mathcal{O}Gb)^P$ such that $\mathrm{Br}_P(i) \neq 0$ (cf. 1.12) — and succeeded in determining the source algebra of a *nilpotent block* [36], which in particular proved Broué's conjecture.

I8 As for the Frobenius Theorem mentioned above, it could be claimed that, strictly speaking, the category $\mathcal{F}_{(b,G)}$ is unnecessary to define a *nilpotent block* since we are just assuming that the quotient of the normalizer by the centralizer of any Brauer pair (Q,f) fulfilling $f\mathrm{Br}_Q(b) \neq 0$ (cf. 1.13) is a p-group. Yet, the existence of the *hyperfocal subalgebra* in the source algebra of a block b — proved in [49] ten years ago — involves $\mathcal{F}_{(b,G)}$ more seriously.

---

† A point of history. Before his stay at Chicago University where his joint paper with Jon Alperin [3] comes from, Michel Broué already had given a complete account of his reformulation in Chevalley's seminar.

I9 Indeed, it is well-known that the direct limit $\varinjlim \mathcal{F}_G$ is the maximal Abelian p-quotient of G [28, Ch. 7,Theorem 3.4]: pushing it further, from the Frobenius category $\mathcal{F}_G$ it is possible to compute, inside a Sylow p-subgroup, its intersection — called a *hyperfocal subgroup* of G — with the kernel of the maximal p-quotient of G. Mimicking this computation in $\mathcal{F}_{(b,G)}$, we can define the *hyperfocal subgroup* $H_{(b,G)}$ of the block b of G; the point is that *there exists an essentially unique P-stable subalgebra of the source algebra $i(kGb)i$ which intersects Pi in $H_{(b,G)}i$ and, together with Pi, generates the whole source algebra* [51, Theorems 14.7 and 15.10].

I10 However, the step which led us to seek for an abstract setting behind all these constructions was the discovery — twenty years ago — of the *localizer* of a *selfcentralizing* Brauer pair, together with the *localizing functor* over the category of *chains* of such Brauer pairs [44]. The *selfcentralizing* Brauer pairs (Q, f) are exactly those considered by Brauer in [9] and one of their possible definitions — justifying their name — is that $C_P(Q) \subset Q$ for any Brauer pair (P, e) containing (Q, f); in this case, it follows from [34] that there exists a suitable extension $L_G(Q, f)$ — the *localizer* of (Q, f) — of the quotient $N_G(Q, f)/C_G(Q)$ by the center $Z(Q)$. In the Frobenius category of a finite group G, these extensions were just the *automorphism group of the objects* in some of the *"localités à épimorphismes"* of G mentioned above.

I11 At that time there appeared the paper by Reinhard Knörr and Geoffrey Robinson [33] where they reformulated Jon Alperin's Conjecture [2] in terms of an alternating sum over a set of *chains* of Brauer pairs — we are more precise below. Thus, quite naturally we considered the localizers $L_G(\mathfrak{q})$ of *chains* $\mathfrak{q}$ — here *chain* stands for *totally ordered set throughout the inclusion* — of selfcentralizing Brauer pairs, and in [44] we prove that this correspondence can be extended to a *functor* from the suitable category of chains — where the morphisms are defined by the *set inclusion* and the *G-conjugation* — to the category of finite groups up to conjugation, namely to the *exterior quotient* $\widetilde{\mathfrak{Gr}}$ (cf. 1.3).

I12 Throughout all this work, it became clear that many arguments were *inner* arguments, in the sense that the blocks around it played no significant role, and we decided to look for a suitable abstract formulation. Actually, this was a reason to delay publication of [44] since we could hope to recover its contents in a more general setting, as we did. A key point of the endeavour to find such an axiomatic approach was the possibility to come back to Brauer's point of view in [9], namely to the subgroups of a defect group; indeed, as for $\mathcal{F}_G$, if (P, e) is a maximal Brauer pair containing $(\{1\}, b)$, then $\mathcal{F}_{(b,G)}$ is equivalent to the full subcategory over the set $\mathfrak{B}_{(P,e)}$ of Brauer pairs (Q, f) contained in (P, e) and *the correspondence mapping (Q, f) on Q is an order-preserving bijection between $\mathfrak{B}_{(P,e)}$ and the set of subgroups of P* [1, Theorem 3.4].

I13 In autumn 1990, we began to work in this direction when doing a series of lectures [44] at the MSRI, and in spring 1991, invited by James Green at Warwick University, we already could give a first definition of an abstract *Frobenius category*†. The starting point was obvious: from a finite p-group P, we had to consider a subcategory of the category of finite groups $\mathfrak{Gr}$ defined over the set of subgroups of P, the problem being to find suitable conditions tightening the situation enough to remain *near* the Frobenius categories of blocks of finite groups.

I14 That is to say, it was not difficult to imagine reasonable necessary conditions, yet they should allow us to go far toward mimicking the usual constructions in groups... till where? Although the definition did not change from spring 1991, we spent some time developing the machinery of normalizers, centralizers, quotients, the translation of our refinement of Alperin's Fusion Theorem... till we were able to state a reasonable criterion of *simplicity* — in the sense that, up to a normal subgroup of order prime to p, a finite group G is simple whenever the Frobenius category of G fulfills this abstract criterion (cf. 12.20). We wrote all this in a manuscript [46] which, in its first half, essentially covered chapters 2, 4, 5 and 12 below.

I15 In chapter 2, we state the conditions defining a *Frobenius P-category* — conditions admitting different equivalent forms — and show the existence of *normalizers* and *centralizers* of the subgroups of P, fulfilling the same conditions. The title of this chapter — *Frobenius P-categories: the first definition* — suggests that we have, at least, another definition; indeed, related with our effort for proving the existence of a *perfect locality* explained below, we found a quite different equivalent definition of a Frobenius P-category, stated in chapter 21. Although simpler from the formal point of view, this definition is farther from our main intuition and, for the moment, it takes second place.

I16 As announced in its title, the purpose of this book is not only to develop our abstract setting but to apply it to a better understanding of *Brauer blocks*. In chapter 3 we prove that the category $\mathcal{F}_{(b,G)}$ considered above (cf. I5) fulfills the conditions of a Frobenius P-category and, more generally, we illustrate in this case all the concepts introduced in 2.

I17 In chapter 4 we come back to the abstract Frobenius P-categories $\mathcal{F}$ to introduce the *selfcentralizing* objects, mimicking the definition of selfcentralizing Brauer pairs mentioned above (cf. I10); indeed, these objects play the most important role in the structure of $\mathcal{F}$ and, actually, the full subcategory $\mathcal{F}^{^{sc}}$ of $\mathcal{F}$ over the set of them determines $\mathcal{F}$, as we prove in this chapter. But, in many arguments, the selfcentralizing Brauer pairs (Q,f) play a

† At present, there is some confusion with the terminology employed around this concept. Although we hope that this book will contribute to fix the original one, we mention some alternative names in footnotes and they can be found in *italics* in the Index.

role simply because f is a *nilpotent block* of $C_G(Q)$ (cf. I6), and this condition is preserved in a quotient of G by a central p-subgroup Z, whereas the selfcentralizing condition need not. Analogously, we introduce the *nilcentralized* objects — we note $\mathcal{F}^{\mathrm{nc}}$ the full subcategory over the set of them; this is possible since we already have *centralizers* of the objects in $\mathcal{F}$ and the *nilpotency* of a Frobenius P-category makes sense *via* the Frobenius Hypothesis above.

I18 Chapters 5 and 6 further illustrate the importance of selfcentralizing objects. In chapter 5 we develop in $\mathcal{F}$ our refinement of the Alperin Fusion Theorem mentioned above, introducing the *essential* objects; actually, these objects already can be introduced in simpler structures called *divisible P-categories*, and then a suitable formulation of the Alperin Fusion Theorem becomes a necessary and "almost" sufficient condition to get a Frobenius P-category. In this case, the essential objects are selfcentralizing, proving again that $\mathcal{F}$ is completely determined by the subcategory $\mathcal{F}^{\mathrm{sc}}$. Moreover, in our approach we show that the Alperin Fusion Theorem concerns the *additive category* $\mathbb{Z}\mathcal{F}$ — where the morphisms between a pair of subgroups Q and R of P are the free $\mathbb{Z}$-modules over $\mathcal{F}(Q,R)$ and the composition is distributive — rather than $\mathcal{F}$ itself, and it is somehow related to the *projective resolution* of the *trivial contravariant* functor over $\mathcal{F}$; we believe that this relationship deserves more consideration.

I19 Chapter 6 exploits a remarkable feature of the *exterior quotient* $\tilde{\mathcal{F}}^{\mathrm{sc}}$ of $\mathcal{F}^{\mathrm{sc}}$ — the quotient of $\mathcal{F}^{\mathrm{sc}}$ by the inner automorphisms of the corresponding subgroups of P (cf. 1.3) — namely the fact that in this category any morphism is an *epimorphism*. This leads to a canonical *partition* of the set of morphisms between two objects relative to a morphism with the same origin (cf. Proposition 6.7); then, the existence of these partitions implies the existence of a *direct product* in the *additive cover* $\mathfrak{ac}(\tilde{\mathcal{F}}^{\mathrm{sc}})$ of $\tilde{\mathcal{F}}^{\mathrm{sc}}$ (cf. 6.2) — a construction introduced by Stefan Jackowski and James McClure in [32]. As a consequence, we get a *vanishing result for positive cohomology*, which is the key for the determination of the rank of the *Grothendieck groups associated with* $\mathcal{F}$ in chapter 14.

I20 As in chapter 3, in chapter 7 we analyze the *nilcentralized* — and the *selfcentralizing* — objects mentioned above (cf. I17), inside the category $\mathcal{F}_{(b,G)}$ associated with a block b of G. As a matter of fact, any nilcentralized Brauer pair (Q,f) fulfilling $f\mathrm{Br}_Q(b) \neq 0$ (cf. 1.13) appears associated with two meaningful *invariants*, namely with a central extension $\hat{\mathcal{F}}_{(b,G)}(Q)$ of the automorphism group

$$\mathcal{F}_{(b,G)}(Q) \cong N_G(Q,f)/C_G(Q) \qquad \text{I20.1}$$

by k^* — where k is a fixed algebraically closed field of characteristic p — and with a *Dade* $\bar{N}_P(Q)$*-algebra* S_Q — a simple k-algebra endowed with an action of $\bar{N}_P(Q) = N_P(Q)/Q$ which stabilizes a basis containing the unity element (cf. 1.20).

I21 As we explain below, from the point of view of Alperin's Conjecture [2] we were interested in the central extension $\hat{\mathcal{F}}_{(b,G)}(Q)$ rather than in the group $\mathcal{F}_{(b,G)}(Q)$ itself, and this raised a huge problem of coherence. Explicitly, for any *chain* $\mathfrak{q}$ (cf. I11) of nilcentralized Brauer pairs fulfilling $f\mathrm{Br}_Q(b) \neq 0$ (cf. 1.12), we can define $\mathcal{F}_{(b,G)}(\mathfrak{q})$ as the stabilizer of $\mathfrak{q}$ in $N_G(Q,f)/C_G(Q)$ for the maximal element (Q,f) in this chain, and then $\hat{\mathcal{F}}_{(b,G)}(\mathfrak{q})$ is the corresponding converse image. It was clear that the map sending $\mathfrak{q}$ to $\mathcal{F}_{(b,G)}(\mathfrak{q})$ might be extended to a functor from the corresponding category of chains to $\mathfrak{Gr}$; yet, from the point of view of Alperin's Conjecture, this functor was useless unless we were able to get a *lifting* of it sending $\mathfrak{q}$ to $\hat{\mathcal{F}}_{(b,G)}(\mathfrak{q})$.

I22 Since the eighties, we knew how to prove the existence of a lifting

$$\hat{\mathcal{F}}_{(b,G)}(\mathfrak{q}) \longrightarrow \hat{\mathcal{F}}_{(b,G)}(\mathfrak{r}) \qquad \text{I22.1}$$

for *any* morphism $\mathfrak{r} \to \mathfrak{q}$ between chains (cf. Theorem 7.16), and this was already announced in [44]; it is a consequence of a Splitting Theorem that Everett Dade announced in 1979 at the Santa Cruz Conference [20] and never published (cf. Theorem 9.21 below or [40]). But the huge problem is to prove that it is possible to do *coherent* choices in order to get a functor, a question that we only have solved when preparing this book.

I23 In the meanwhile, we tentatively have followed two strategies. In one of them, we tried to improve our proof in [40] of Dade's Splitting Theorem, showing that it was possible to make a choice once for ever — a *polarization* — in the set of *equivalence* classes of Dade P-algebras for any finite p-group P, which then we could apply to our problem. Although what we obtained in this direction did not solve the problem, we explain our result in chapter 9; in particular, it includes a proof of Dade's Splitting Theorem somewhat different and more detailed than in [40]. Previously, in chapter 8 we recall the main facts on Dade P-algebras we need in the sequel.

I24 Still since the eighties, we already knew that the existence of a $\mathcal{F}_{(b,G)}(\mathfrak{q})$-*stable* Dade P-algebra S *gluing* together all the Dade $\bar{N}_P(Q)$-algebras S_Q — namely, a Dade P-algebra S such that $\mathrm{Res}^P_Q(S)$ and $\mathrm{Res}_\varphi(S)$ are *equivalent* for any nilcentralized Brauer pair (Q,f) contained in (P,e) and any $\mathcal{F}_{(b,G)}$-morphism φ from (Q,f) to (P,e), and that S_Q is *equivalent* to the *Brauer quotient* $S(Q)$ (see 11.6) — easily would solve our problem. Indeed, replacing kGb by the P-algebra $S^\circ \otimes_k \mathrm{Res}^G_P(kGb)$, we could find inside the *same* extensions $\hat{\mathcal{F}}_{(b,G)}(\mathfrak{q})$ and $\hat{\mathcal{F}}_{(b,G)}(\mathfrak{r})$ — up to *noncanonical* isomorphisms — and then this P-algebra provided coherent *extension group homomorphisms*

$$\hat{\mathcal{F}}_{(b,G)}(\mathfrak{q}) \longrightarrow \hat{\mathcal{F}}_{(b,G)}(\mathfrak{r}) \qquad \text{I24.1}$$

(see Proposition 11.8 and Theorem 11.10 for the precise arguments).

I25 Thus, the other strategy was to look for such a Dade P-algebra. One way to find S was from a conjectural *gluing* result on Dade P-algebras, which is indeed true when P is Abelian [45]; but presently, after the complete classification of the Dade P-algebras in [17] and [6], such a general result seems impossible. In chapter 10 we state a result in this direction which should be combined with a suitable *polarization*... that does not exist. Another way to find S could be from the P-algebra $\mathrm{Res}_P^G(kGb)$ itself and, although we have no construction to propose, we cannot close this possibility.

I26 Actually, the existence of such a Dade P-algebra S also would prove the existence of a suitable k^*-extension $\hat{\mathcal{F}}^{^{\mathrm{nc}}}$ of $\mathcal{F}^{^{\mathrm{nc}}}$ — a question raised by Markus Linckelmann in the 2002 Durham Symposium — which naturally appears inside $S^\circ \otimes_k \mathrm{Res}_P^G(kGb)$. It is this fact that gives a solution to our problem: from the Dade $\bar{N}_P(Q)$-algebras S_Q themselves, we are able to construct k^*-extensions of suitable subcategories of $\mathcal{F}^{^{\mathrm{nc}}}$ and the point is that these k^*-extensions are quite independent of our choice of Dade algebras; although we do not get a complete k^*-extension $\hat{\mathcal{F}}^{^{\mathrm{nc}}}$, we get a complete coherent choice for the *extension group homomorphisms* I24.1. All this is explained in chapter 11.

I27 In chapters 12 and 13, we pursue the development of the abstract Frobenius P-categories. In chapter 12 we discuss the existence of *quotients* of a Frobenius P-category $\mathcal{F}$ by suitable subgroups of P. If $\mathcal{F} = \mathcal{F}_G$ is the Frobenius category of a finite group G, any Sylow p-subgroup of a normal subgroup of G naturally determines one of those quotients of $\mathcal{F}$; although the converse is not true, using the Classification of Finite Simple Groups, it is not difficult to check that, provided P is not Abelian, there are not so many exceptions. But the main purpose of this chapter is to determine the *p'-quotients* of $\mathcal{F}$ — the quotients reduced to a finite group of order prime to p — since we need them to state the *simplicity* criterion (cf. 12.20) and to talk about *solvability* in Frobenius P-categories. As a matter of fact, all the p'-quotients of $\mathcal{F}$ have a *kernel* which is also a Frobenius P-category, and there is a smallest such Frobenius P-category (cf. Corollary 12.17).

I28 In chapter 13 we discuss the *p-quotients* of $\mathcal{F}$; as implicitly mentioned in I9 above, it is still possible to define the *hyperfocal subgroup* $H_{\mathcal{F}}$ of P in $\mathcal{F}$; once again, $P/H_{\mathcal{F}}$ is indeed the maximal *p-quotient* of $\mathcal{F}$ and there exists a suitable Frobenius $H_{\mathcal{F}}$-category $\mathcal{F}^{\mathfrak{h}}$ — called the *hyperfocal* subcategory of $\mathcal{F}$. Then, the contents of these two chapters allow us to state a definition of *solvability* for Frobenius P-categories, and one main point is that *a solvable Frobenius P-category is necessarily the Frobenius category of a p-solvable finite group*, which responds positively to our requirement of remaining near the Frobenius categories of finite groups (cf. I13). This result does not appear till chapter 19 since its proof needs the existence of the *localizer* proved in chapter 18.

I29 Coming back to the Frobenius category $\mathcal{F}_{(b,G)}$ of a block (b,G), we already know from chapter 11 that there is a functor from the *proper category of* $(\mathcal{F}_{(b,G)})^{\mathrm{nc}}$*-chains* (cf. A2.8) to the category k^*-$\mathfrak{Gr}$ of the *central* k^**-extensions of finite groups* sending a $(\mathcal{F}_{(b,G)})^{\mathrm{nc}}$*-chain* $\mathfrak{q}$ to the k^*-extension $\hat{\mathcal{F}}_{(b,G)}(\mathfrak{q})$ above (cf. I21). On the other hand, the *modular Grothendieck group* evidently defines a *contravariant* functor from k^*-$\mathfrak{Gr}$ to the category of free $\mathbb{Z}$-modules. A main fact stated in this book is that *the inverse limit of the composition of both functors* — called the *Grothendieck group* of $\mathcal{F}_{(b,G)}$ (cf. 14.3) — is a free $\mathbb{Z}$-module tightly related with Alperin's Conjecture.

I30 Let us be more explicit. Recall that, following Alperin [2], a *weight* of the block b is a pair (Q,χ) formed by a p-subgroup Q of G and by an irreducible character χ of $N_G(Q)$, associated with $\mathrm{Br}_Q(b)$ (cf. 1.13), which comes from a modular irreducible character $\bar\chi$ in a block of $\bar N_G(Q) = N_G(Q)/Q$ of *defect zero* (cf. 1.17). Then, Alperin's Conjecture affirms that *the number of G-conjugacy classes of weights of the block b coincides with the number* $|\mathrm{Irr}_k(G,b)|$ *of modular irreducible characters in* b.

I31 But, for a weight (Q,χ), an $\bar N_G(Q)$-module M affording $\bar\chi$ is simple and projective (cf. 1.17), and therefore its restriction to $\bar C_G(Q)$ is semisimple and projective too, determining a set of blocks $\bar f$ of $\bar C_G(Q)$ of *defect zero* (cf. 1.17). That is to say, lifting $\bar f$ to a block f in $Z\big(kC_G(Q)\big)$, (Q,f) is a *selfcentralizing Brauer pair* (cf I10 and Corollary 7.3) and, up to G-conjugation, may assume it is contained in (P,e); moreover, it is not difficult to prove that *the set of weights determining the same selfcentralizing Brauer pair (Q,f) bijectively correspond with the set of isomorphism classes of $k_*\hat{\mathcal{F}}_{(b,G)}(Q)$-modules, both simple and projective* [53, Theorem 3.7]. Consequently, denoting by $\mathrm{IrPr}_k\big(\hat{\mathcal{F}}_{(b,G)}(Q)\big)$ the corresponding set of modular characters, Alperin's Conjecture affirms

$$|\mathrm{Irr}_k(G,b)| = \sum_Q \big|\mathrm{IrPr}_k\big(\hat{\mathcal{F}}_{(b,G)}(Q)\big)\big| \qquad \text{I31.1}$$

where Q runs over a set of representatives for the $\mathcal{F}_{(b,G)}$-isomorphism classes of $\mathcal{F}_{(b,G)}$-selfcentralizing objects.

I32 At this point, the arguments of Reinhard Knörr and Geoffrey Robinson in [33] prove, in the language above, that Alperin's Conjecture is equivalent to the equality

$$|\mathrm{Irr}_k(G,b)| = \sum_{\mathfrak{q}} (-1)^{|\mathfrak{q}|-1}\big|\mathrm{Irr}_k\big(\hat{\mathcal{F}}_{(b,G)}(\mathfrak{q})\big)\big| \qquad \text{I32.1}$$

where $\mathfrak{q}$ runs over a set of representatives for the $\mathcal{F}_{(b,G)}$-isomorphism classes of $(\mathcal{F}_{(b,G)})^{\mathrm{sc}}$*-chains* (cf. I11). But, in chapter 14 we prove that *the $\mathbb{Z}$-rank of the Grothendieck group of $\mathcal{F}_{(b,G)}$ coincides with the right-hand member of this* *equality* — actually, the statement makes sense for any Frobenius P-category $\mathcal{F}$ endowed with an analogous functor from the category of nilcentralized $\mathcal{F}$-chains to k^*-$\mathfrak{Gr}$, and we prove that equality in this general context. Thus, Alperin's Conjecture is equivalent to the assertion that *the $\mathbb{Z}$-ranks of the Grothendieck groups of the block (b, G) and the category $\mathcal{F}_{(b,G)}$ coincide.*

I33 Everyone understands that the sentence *"an alternating sum of ranks which coincides with the rank of an inverse limit"* necessarily suggests the possible existence of a *differential complex* with a unique nonzero *cohomology group* at degree zero. But, it has to be noticed that our *contravariant* functor mapping an $(\mathcal{F}_{(b,G)})^{^{\mathrm{sc}}}$*-chain* $\mathfrak{q}$ on the *Grothendieck group* of $\hat{\mathcal{F}}_{(b,G)}(\mathfrak{q})$ does *not* come from a *contravariant* functor defined over the category $(\hat{\mathcal{F}}_{(b,G)})^{^{\mathrm{sc}}}$; in other words, we are not dealing with the usual *cohomology groups* of $(\hat{\mathcal{F}}_{(b,G)})^{^{\mathrm{sc}}}$. Moreover, our sum runs over a set of representatives for a set of suitable $\mathcal{F}_{(b,G)}$-isomorphism classes and this fact has to be integrated in our hypothetical differential complex.

I34 All these remarks forced us to enlarge the usual construction of the cohomology groups of a category $\mathfrak{C}$ in order to include our situation; the new cohomology groups we consider need not fulfill the *long exact sequence* condition (cf. A3.11.4), but they are useful for our purposes. We explain our point of view — which possibly has been already employed in other situations — in the Appendix. We have adopted the language of the 2-categories since, when constructing the cohomology groups of $\mathfrak{C}$, we dislike expressions such as *"consider a sequence of n $\mathfrak{C}$-morphisms which can be composed"*, that we replace by *"consider a functor from Δ_n to $\mathfrak{C}$"*; but, if the *simplex* Δ_n becomes a *category* then the *simplicial category* $\mathbf{\Delta}$ becomes a 2-*category*...

I35 Let us come back to our discussion on Alperin's Conjecture. Of course, two free $\mathbb{Z}$-modules with the same $\mathbb{Z}$-rank are isomorphic, and therefore Alperin's Conjecture is also equivalent to the assertion that *the Grothendieck groups of the block (b, G) and the category $\mathcal{F}_{(b,G)}$ are isomorphic.* But, the stabilizer $\mathrm{Out}(G)_b$ of b in the group of *outer automorphisms* of G has a natural action over both Grothendieck groups and then an obvious question arises: *is there an $\mathrm{Out}(G)_b$-stable isomorphism between the Grothendieck groups of the block (b, G) and the category $\mathcal{F}_{(b,G)}$?* Actually, even a positive answer for a suitable scalar extension of the Grothendieck groups would be welcome.

I36 A good indication towards a positive answer to this question is that, up to a suitable scalar extension, both Grothendieck groups have the *same* behaviour throughout the restriction to the normal subgroups, provided we can "follow" the subcategory $(\mathcal{F}_{(b,G)})^{^{\mathrm{nc}}}$ in the normal subgroups; we expose our reduction results in chapter 15. Here there appears a significant difference between our method and the method which consists of restricting any irreducible character in the block b individually; indeed, in the second one, we are forced to apply the so-called *Clifford Theory* [31, Ch. V] which involves the unknown *Clifford extensions* of the stabilizers, which makes any tentative induction enormously difficult.

I37 In chapter 16 we develop a strategy toward reducing a possible positive answer to the question above, to a positive *verification* of the same question "around" the noncommutative simple groups; by "around" we mean that, for any noncommutative simple group S, we have to consider the central k^*-extensions of suitable subgroups of $\mathrm{Aut}(S)$ containing S. Our strategy itself already needs the Classification of Simple Groups since it quotes some facts which are only known from this classification, as for instance the *solvability* of $\mathrm{Out}(S)$. The precise result is stated in Theorem 16.45.

I38 The last part of this book deals again with an abstract Frobenius P-category $\mathcal{F}$. In the second half of our manuscript [46], we investigated to what extent we still got *localizers* together with the *localizing functor* (cf. I10) in our abstract setting — the main purpose and the crucial test in building it. The answer had been "almost" positive — as we explain in 18.5, it remained to prove that some 1-*cocycle* was a 1-*coboundary*, which now is done — and, since it was not reasonable to foresee a finite group as a possible direct limit of the *localizing functor*, we considered the possibility of the existence of a *topological space* as, roughly speaking, a direct limit of the functor defined by the *classifying spaces* of the *localizers*.

I39 In 1994, we proposed this idea to Dave Benson, who already had constructed a *topological space* [5] from a configuration considered by Ron Solomon when discussing finite simple groups with the same Sylow 2-subgroups P as the third Conway's group [56]; actually, as it could be expected, Solomon's configuration is nothing but a Frobenius P-category [13]. After a while, Benson raised the question of the existence of an *extension* $\mathcal{L}^{^{sc}}$ of the full subcategory $\mathcal{F}^{^{sc}}$ mimicking a suitable *"localité à épimorphismes"* (cf. I3), namely having the *localizer* as the automorphism group of any selfcentralizing object.

I40 After Benson's publication [5], Carles Broto, Ran Levi and Bob Oliver became interested in our manuscript [46]† in order to prove that the topological space coming from the category $\mathcal{L}^{^{sc}}$ guessed by Benson had good enough properties to be a *"classifying space"* of $\mathcal{F}$; in [13] they proved the "good properties" of $\mathcal{L}^{^{sc}}$ but did not succeed in proving its existence and its uniqueness, just giving some sufficient conditions.

† A point of history. In december 1999, coming back from Wuhan, we found an e-mail sent by Bob Oliver asking us for a copy of our manuscript. We personally gave him a copy on the basis of a possible collaboration. Only in October 2000, did we learn that Carles Broto and Ran Levi were not only interested but already deeply engaged in our manuscript.

I41 It has to be understood that the existence of a suitable extension of $\mathcal{F}^{^{\mathrm{sc}}}$, or even of $\mathcal{F}$, already supplies *localizers* and *localizing functors* — namely, the *automorphism groups* of the objects in such an extension and the *automorphism group functor* over the corresponding category of chains (cf. Proposition A2.10). Thus, in chapter 17 we systematically consider extensions of $\mathcal{F}$ — called *$\mathcal{F}$-localities* since they generalize our point of view in [35] — which we would like to be *determined* by $\mathcal{F}$ and P; for some precise meaning of the word *determined*, this condition imposes a biggest possibility that we call *perfect $\mathcal{F}$-locality* and corresponds to the category expected by Benson. In particular, we prove that if $\mathcal{F}$ holds a perfect $\mathcal{F}$-locality $\mathcal{L}$, then any quotient $\bar{\mathcal{F}}$ of $\mathcal{F}$ as considered in chapter 12 holds a perfect $\bar{\mathcal{F}}$-locality $\bar{\mathcal{L}}$ defined as a quotient of $\mathcal{L}$; of course, had we the existence and uniqueness of a perfect $\mathcal{F}$-locality such a result would be redundant.

I42 In this case, chapter 18 on the *localizers* would be somewhat redundant too, since the *localizers* we announce would be nothing but the automorphism groups of the objects in this category, as we said above. But, the *localizer* of a selfcentralizing object admits a *direct* group-theoretical characterization, given in [46] and coming from [34], which deserves to be stated. Moreover, as we mention above, this result is useful to prove, in chapter 19, that a solvable Frobenius P-category is necessarily the Frobenius category of a p-solvable finite group. In chapter 18 we also prove the existence and the uniqueness of the *localizing functor* $\mathfrak{loc}_{\mathcal{F}}$ mentioned above (cf. I10), together with some kind of "universality" of it, which is quite useful in chapter 23.

I43 In chapter 20 we prove that the existence of a perfect $\mathcal{F}^{^{\mathrm{sc}}}$-locality $\mathcal{L}^{^{\mathrm{sc}}}$ forces the existence of a perfect $\mathcal{F}$-locality $\mathcal{L}$ by a direct *necessarily unique* construction of $\mathcal{L}$ from $\mathcal{L}^{^{\mathrm{sc}}}$; we obviously proceed by induction, but cannot avoid the distinction between the *fully centralized* subgroups of P (cf. 2.6) and the others, as we cannot avoid the distinction between normal and ordinary inclusions. All this generates a long proof even if it is nothing but routine. Does there exist a general result guaranteeing that some kind of properties of $\mathcal{F}^{^{\mathrm{sc}}}$ can be extended to $\mathcal{F}$?

I44 In chapter 21 we expose the second definition of a Frobenius P-category $\mathcal{F}$, which leads to the *basic $\mathcal{F}$-locality*. This equivalent definition comes from an original contribution of Broto, Levi and Oliver† to the behaviour of a Frobenius P-category $\mathcal{F}$, namely the existence of a suitable $P\times P$-set Ω, where P acts freely on the left and on the right, which has some precise *$\mathcal{F}$-stable* property and $P\times P$-orbits determined by $\mathcal{F}$ [13, Proposition 5.5] — that we call *$\mathcal{F}$-basic*. Roughly speaking, Ω keeps some properties — which can be stated in terms of the Frobenius P-category $\mathcal{F}_G$ — of the action of a Sylow p-subgroup P on a finite group G, by left and right multiplication.

† They credit Markus Linckelmann and Peter Webb for the original idea.

I45 But, the point is that we can define a *basic* $P \times P$-set Ω independently of any Frobenius P-category, as a $P \times P$-set with free actions on the left and on the right, fulfilling suitable extreme equalities — actually, the conditions are so simple that to give further details amounts to stating our definition here! Then, any *basic* $P \times P$-set Ω supplies a Frobenius P-category $\mathcal{F}^{\Omega}$ and, by the Broto-Levi-Oliver result mentioned above, any Frobenius P-category comes from a *basic* $P \times P$-set; of course, a Frobenius P-category $\mathcal{F}$ may come from two different basic $P \times P$-sets Ω and Ω', but then we can construct a third basic $P \times P$-set Ω'' containing both Ω and Ω', and still fulfilling $\mathcal{F}^{\Omega''} = \mathcal{F}$.

I46 In particular, for any Frobenius P-category $\mathcal{F}$ and any $\mathcal{F}$-*basic* $P \times P$-*set* Ω — a basic $P \times P$-set such that $\mathcal{F}^{\Omega} = \mathcal{F}$ — we consider the group G of permutations σ of Ω which centralize the action of P on the right; then, by the action on the left, P becomes a subgroup of G and, by the very definition of a basic $P \times P$-set, for any subgroup Q of P we have

$$N_G(Q)/C_G(Q) \cong \mathcal{F}(Q) \qquad \text{I46.1.}$$

The elementary but careful work we do in chapter 22 consists of determining all the centralizers $C_G(Q)$ and the inclusions between them.

I47 Naturally, these centralizers contain full *symmetric groups* coming from the possible mutually isomorphic $P \times P$-orbits of Ω, but fortunately the minimal normal subgroups in the centralizers containing these *symmetric groups* form a *"localité"* in the old sense of [35] and therefore they determine an $\mathcal{F}$-*locality*. Although this $\mathcal{F}$-locality — called the *basic* $\mathcal{F}$-*locality* $\mathcal{L}^{\mathrm{b}}$ — is far from being *perfect*, it is *canonically* associated with $\mathcal{F}$ in the sense that it does not depend on the choice of Ω provided it is "big enough".

I48 The "universal" property of the *localizing functor* proved in chapter 18 guarantees that, if a perfect $\mathcal{F}$-locality does exist, it should be related to the *basic* $\mathcal{F}$-*locality* $\mathcal{L}^{\mathrm{b}}$ — at least over the set of $\mathcal{F}$-selfcentralizing subgroups of P — because of the rich structure of the *additive cover* $\mathfrak{ac}(\tilde{\mathcal{F}}^{\mathrm{sc}})$ of the *exterior* quotient $\tilde{\mathcal{F}}^{\mathrm{sc}}$ of $\mathcal{F}^{\mathrm{sc}}$ developed in chapter 6. The problem with the relationship between the *basic* and the possible *perfect* $\mathcal{F}^{\mathrm{sc}}$-localities is the *thickness* of the first one — in the sense that the *kernel* of the canonical functor $\mathcal{L}^{\mathrm{b,sc}} \to \mathcal{F}^{\mathrm{sc}}$ is too big. In chapter 23 we start by showing that the full subcategory $\mathcal{L}^{\mathrm{b,sc}}$ of $\mathcal{L}^{\mathrm{b}}$ over the set of $\mathcal{F}$-selfcentralizing subgroups of P admits a quotient — the *polycentral* $\mathcal{F}^{\mathrm{sc}}$-*locality* $\mathcal{L}^{\mathrm{c,sc}}$ — already *narrowing* $\mathcal{L}^{\mathrm{b,sc}}$.

I49 As a matter of fact, the kernel of the corresponding canonical functor $\mathcal{L}^{\mathrm{c,sc}} \to \mathcal{F}^{\mathrm{sc}}$ admits a quite general description — we discuss it in our Appendix from A2.14 to A2.17 — in terms of *representations* and *semidirect products*. In our situation in chapter 23, this formulation leads to a vanishing result for *stable* positive cohomology groups — a type of nonstandard cohomology groups introduced in our Appendix (cf. I34 and A3.8) — *narrowing* twice the *polycentral* $\mathcal{F}^{sc}$*-locality* $\mathcal{L}^{c,sc}$ till we reach the *reduced* $\mathcal{F}^{sc}$*-locality* $\mathcal{L}^{r,sc}$. Moreover, in chapter 23 we explain our difficulties with finding a perfect $\mathcal{F}^{sc}$-sublocality of $\mathcal{L}^{r,sc}$ and give a (strongly!) sufficient condition to overcome them.

I50 Finally, in order to prove in chapter 24 that any perfect $\mathcal{F}^{sc}$-locality $\mathcal{L}^{sc}$ is contained in the *reduced* $\mathcal{F}^{sc}$*-locality*, we have to exhibit a suitable *basic* $P \times P$*-set* Ω. Where to find such a $P \times P$-set? The answer comes from the fact that any morphism in $\mathcal{L}^{sc}$ is an *epimorphism* — actually, it is a *monomorphism* too — and, in particular, all the arguments on $\mathfrak{ac}(\tilde{\mathcal{F}}^{sc})$ in chapter 6 can be repeated in $\mathfrak{ac}(\mathcal{L}^{sc})$; namely, the category $\mathfrak{ac}(\mathcal{L}^{sc})$ admits a *direct product*, allowing us to consider *the direct product of* P *by* P which, being an $\mathfrak{ac}(\mathcal{L}^{sc})$-object, involves some finite set Ω (cf. 6.2): this is the set we are looking for.

I51 A last remark. The reader may ask himself whether or not it is possible to define an *ordinary Grothendieck group* for the Frobenius category $\mathcal{F}_{(b,G)}$ of a block (b,G); namely, to carry out an analogous construction with the Grothendieck groups obtained from the categories of representations over a field of characteristic zero, opening the possibility of dealing with Dade's Conjecture [21]. Firstly note that, without any extra effort, the direct sum of suitable scalar extensions of the *Grothendieck group* of the Frobenius categories $\mathcal{F}_{(g,C_G(u))}$ when (g,u) runs over a set of representatives for the set of G-conjugacy classes of *Brauer* (b,G)*-elements* [11] provides a satisfactory definition for the ordinary Grothendieck group of $\mathcal{F}_{(b,G)}$.

I52 Does it coincide with the *inverse limit* of the composition of the k^**-localizing functor* $\widehat{\mathfrak{loc}}_{\mathcal{F}_{(b,G)}}$ with the corresponding *ordinary Grothendieck group* functor? — here $\widehat{\mathfrak{loc}}_{\mathcal{F}_{(b,G)}}$ denotes the *pull-back* of the *localizing functor* $\mathfrak{loc}_{\mathcal{F}_{(b,G)}}$ (cf. I42) and the functor $\hat{\mathcal{F}}_{(b,G)}(\bullet)$ in I29 above. But, even if the answer was in the affirmative, in order to deal with Dade's Conjecture some extra idea would be necessary to fit the *defect* of ordinary irreducible characters considered in [21] inside the functorial framework.

Paris, October 2007

Chapter 1

General notation and quoted results

1.1 Since the title of this book, the word *category* appears often; but, we employ this concept in a very restricted sense. Mostly we bound ourself to consider categories where the isomorphism classes of objects form a *set*, as for instance the categories of finite groups — noted $\mathfrak{Gr}$ — or of finitely generated Abelian groups — noted $\mathfrak{Ab}$ — and the category of finite sets — noted $\aleph$. Although in the Appendix we expose all the results on categories and on their cohomology groups we need in the book, in this chapter we list some common terminology and standard notation, often employed without reference in the sequel.

1.2 When we consider a new category $\mathfrak{C}$, we simply assume that the objects form a *set* — often called a *small category* — and, if X and Y are $\mathfrak{C}$-objects, we denote by $\mathfrak{C}(Y,X)$ the set of $\mathfrak{C}$-morphisms *from* X *to* Y, an order somewhat helpful when composing morphisms; also, we write $\mathfrak{C}(X)$ instead of $\mathfrak{C}(X,X)$; we call *opposite category* of $\mathfrak{C}$, and denote by $\mathfrak{C}^\circ$, the category with the *same* objects and with the morphisms

$$\mathfrak{C}^\circ(Y,X) = \mathfrak{C}(X,Y) \qquad 1.2.1.$$

As usual, we call *contravariant* functors from $\mathfrak{C}$ to $\mathfrak{Ab}$ the functors $\mathfrak{a}$ from $\mathfrak{C}^\circ$ to $\mathfrak{Ab}$. For any $n \in \mathbb{N}$, let us denote by $\mathfrak{a}^n$ the composition of $\mathfrak{a}$ with the functor $\mathfrak{Ab} \to \mathfrak{Ab}$ determined by the n-th power.

1.3 Very often we deal with categories where it makes sense to talk about the *inner* automorphisms of the objects — the category $\mathfrak{Gr}$ for instance; more precisely, an *interior structure* in a category $\mathfrak{C}$ is a correspondence sending any $\mathfrak{C}$-object X to a subgroup $\mathfrak{I}(X)$ of $\mathfrak{C}(X)$ in such a way that, for any $\mathfrak{C}$-object Y and any $f \in \mathfrak{C}(Y,X)$, we have

$$f \circ \mathfrak{I}(X) \subset \mathfrak{I}(Y) \circ f \qquad 1.3.1.$$

In this case, we call *exterior quotient* of $\mathfrak{C}$, and denote by $\tilde{\mathfrak{C}}$, the *quotient category* defined by

$$\tilde{\mathfrak{C}}(Y,X) = \mathfrak{I}(Y)\backslash\mathfrak{C}(Y,X) \qquad 1.3.2$$

for any pair of $\mathfrak{C}$-objects X and Y; usually, we denote by $\tilde{f}$ the $\tilde{\mathfrak{C}}$-morphism determined by f.

1.4 A *subcategory* $\mathfrak{D}$ of a category $\mathfrak{C}$ is a category where the set of objects is contained in the set of $\mathfrak{C}$-objects and, for any pair of $\mathfrak{D}$-objects X and Y, the set of $\mathfrak{D}$-morphisms $\mathfrak{D}(Y,X)$ from X to Y is contained in $\mathfrak{C}(Y,X)$, the composition in $\mathfrak{D}$ being induced by the composition in $\mathfrak{C}$; we say that $\mathfrak{D}$ is a *full subcategory* of $\mathfrak{C}$ whenever $\mathfrak{D}(Y,X) = \mathfrak{C}(Y,X)$ for any pair of $\mathfrak{D}$-objects X and Y.

1.5 In particular, we denote by $\mathfrak{C}^*$ and $\mathfrak{C}^o$ the subcategories of $\mathfrak{C}$ with the same set of objects as $\mathfrak{C}$ and only with the $\mathfrak{C}$-isomorphisms and with the $\mathfrak{C}$-identities respectively. If $\mathfrak{D}$ and $\mathfrak{D}'$ are subcategories of $\mathfrak{C}$ then we can define the *intersection* $\mathfrak{D} \cap \mathfrak{D}'$ in the obvious way. Note that a category where the *isomorphism classes* of objects form a set is *equivalent* to a full subcategory where the *objects* form a set, and therefore it is easy to translate results from the one to the other. If $\mathfrak{A}$ and $\mathfrak{B}$ are two small categories, we denote by $\mathfrak{A} \times \mathfrak{B}$ their *direct product* where the objects are the pairs (A, B) formed by an $\mathfrak{A}$-object A and a $\mathfrak{B}$-object B, and where the set of morphisms from (A, B) to (A', B') is the direct product

$$\mathfrak{A}(A', A) \times \mathfrak{B}(B', B) \qquad 1.5.1.$$

Moreover, we denote by $\mathfrak{Fct}(\mathfrak{A}, \mathfrak{B})$ the small category of *functors* from $\mathfrak{A}$ to $\mathfrak{B}$ and, if $\mathfrak{f}, \mathfrak{g} \in \mathfrak{Fct}(\mathfrak{A}, \mathfrak{B})$, by $\mathfrak{Nat}(\mathfrak{f}, \mathfrak{g})$ and $\mathfrak{Nat}(\mathfrak{f}, \mathfrak{g})^*$ the respective sets of *natural maps* and *natural isomorphisms* from $\mathfrak{f}$ to $\mathfrak{g}$.

1.6 Any functor $\mathfrak{f} : \mathfrak{B} \to \mathfrak{C}$ holds the *inverse image* throughout $\mathfrak{f}$

$$\mathfrak{f}^* : \mathfrak{Fct}(\mathfrak{C}^\circ, \mathfrak{Ab}) \longrightarrow \mathfrak{Fct}(\mathfrak{B}^\circ, \mathfrak{Ab}) \qquad 1.6.1,$$

which is a functor mapping any *contravariant* functor $\mathfrak{b} : \mathfrak{C} \to \mathfrak{Ab}$ on $\mathfrak{b} \circ \mathfrak{f}$. Recall that $\mathfrak{f}^*$ admits a *right adjoin* called the *direct image* throughout $\mathfrak{f}$

$$\mathfrak{f}_* : \mathfrak{Fct}(\mathfrak{B}^\circ, \mathfrak{Ab}) \longrightarrow \mathfrak{Fct}(\mathfrak{C}^\circ, \mathfrak{Ab}) \qquad 1.6.2,$$

namely a functor (cf. A2.11 below) such that the evident functors

$$\mathfrak{Ab} \longleftarrow \mathfrak{Fct}(\mathfrak{B}^\circ, \mathfrak{Ab}) \times \mathfrak{Fct}(\mathfrak{C}^\circ, \mathfrak{Ab})^\circ \longrightarrow \mathfrak{Ab} \qquad 1.6.3$$

respectively mapping $\mathfrak{a} \in \mathfrak{Fct}(\mathfrak{B}^\circ, \mathfrak{Ab})$ and $\mathfrak{b} \in \mathfrak{Fct}(\mathfrak{C}^\circ, \mathfrak{Ab})$ on $\mathfrak{Nat}\big(\mathfrak{f}^*(\mathfrak{b}), \mathfrak{a}\big)$ and $\mathfrak{Nat}\big(\mathfrak{b}, \mathfrak{f}_*(\mathfrak{a})\big)$ are *naturally isomorphic*.

1.7 For any $\mathfrak{C}$-object X we denote by $\mathfrak{C}_X$ the new category where the objects are the $\mathfrak{C}$-morphisms $f : Y \to X$ from any $\mathfrak{C}$-object Y to X and where the morphisms from $f : Y \to X$ to another $\mathfrak{C}_X$-object $f' : Y' \to X$ are the $\mathfrak{C}$-morphisms $g : Y \to Y'$ fulfilling $f = f' \circ g$, the composition being induced by the composition in $\mathfrak{C}$. Note that we have an obvious *forgetful functor* from $\mathfrak{C}_X$ to $\mathfrak{C}$. Moreover,

$$\mathfrak{C}^X = \big((\mathfrak{C}^\circ)_X\big)^\circ \qquad 1.7.1$$

is the analogous category where the objects are the $\mathfrak{C}$-morphisms $f : X \to Y$ from X to any $\mathfrak{C}$-object Y and the morphisms from $f : X \to Y$ to another $\mathfrak{C}^X$-object $f' : X \to Y'$ are the $\mathfrak{C}$-morphisms $g : Y \to Y'$ fulfilling $f' = g \circ f$.

1.8 Throughout this book p is a fixed prime number. The *Frobenius category* associated with a finite group G — noted $\mathcal{F}_G$ — is the category where the objects are the p-subgroups of G and the morphisms are the group homomorphisms between them which are induced by the inclusion and by the G-conjugation (cf. I1); actually, if P is a Sylow p-subgroup of G, it follows from the well-known Sylow Theorem(s) that $\mathcal{F}_G$ is equivalent to the full subcategory over the set of subgroups of P — that we still denote by $\mathcal{F}_G$. More explicitly, if Q and R are p-subgroups of G and $x \in G$ fulfills $R \subset Q^x$, we denote by

$$\kappa_{Q,R}(x) : R \longrightarrow Q \qquad 1.8.1$$

(or by $\kappa_Q(x)$ if $R = Q$) the group homomorphism induced by the conjugation by x.

1.9 As usual, we call *order* of G and denote by $|G|$ the cardinal of a finite group G; we call *p'-group* any finite group of order prime to p. If H is a subgroup of G, we denote by $\iota_H^G : H \to G$ the *inclusion map* and we set $|G : H| = |G|/|H|$; occasionally, we write 1 instead of $\{1\}$ for the trivial subgroup of G. Besides the standard notation $N_G(H)$ and $C_G(H)$ for the *normalizer* and the *centralizer* of H in G, for any subgroup X of the group $\mathrm{Aut}(H)$ of automorphisms of H, we call *X-normalizer* of H in G, and denote by $N_G^X(H)$, the converse image of X in $N_G(H)$; note that

$$N_G^{\{\mathrm{id}_H\}}(H) = C_G(H) \qquad 1.9.1;$$

moreover, it is handy to employ the notation

$$\bar N_G^X(H) = N_G^X(H)/\big(H \cap N_G^X(H)\big) \qquad 1.9.2.$$

If $C_G(H) \subset H$, we say that H is a *selfcentralizing subgroup* of G.

1.10 Throughout this book k is an algebraically closed field of characteristic p. A *Brauer block*, or a *block* for short, is a pair (b, G) formed by a finite group G and a primitive idempotent b of the center $Z(kG)$ of the group algebra kG; sometimes we also say that *b is a block of G*, but note that the same central idempotent may be called a *block* with respect to several groups. Then, a *Brauer G-pair* is a pair (P, e) formed by a p-subgroup P of G and a primitive idempotent e of $Z\big(kC_G(P)\big)$ or, equivalently, a block of $C_G(P)$; it is well-known that the image $\bar e$ of e in $Z\big(k\bar C_G(P)\big)$ remains a primitive idempotent [50, Corollary 2.13]. Note that, for any subgroup H of G containing $P{\cdot}C_G(P)$, (P, e) is also a Brauer H-pair.

1.11 In order to define the *inclusion* between Brauer G-pairs, it is better to adopt the more general point of view of the G-algebras, that we also need in this book. Following Green [29], a *G-algebra* is a k-algebra A of finite dimension endowed with a G-action; we say that A is *primitive* if the unity element is primitive in A^G. A G-algebra homomorphism from A to another

G-algebra A' is a *not necessarily unitary* algebra homomorphism $f: A \to A'$ compatible with the G-actions; we say that f is an *embedding* whenever

$$\mathrm{Ker}(f) = \{0\} \quad \text{and} \quad \mathrm{Im}(f) = f(1_A)A'f(1_A) \qquad 1.11.1.$$

We say that f is a *strict semicovering* — we only need this definition in Lemma 15.49 below — if f is *unitary*, $\mathrm{Ker}(f)$ is contained in the *radical* $J(A)$ of A and, for any p-subgroup P of G, $f\big(J(A^P)\big)$ is contained in $J(A'^P)$ and $f(i)$ is primitive in A'^P for any primitive idempotent i in A^P [34, §3]. For any group homomorphism $\varphi: G' \to G$, we denote by $\mathrm{Res}_\varphi(A)$ the same algebra A endowed with the obvious G'-action.

1.12 Recall that for any subgroup H of G, a *point* α of H on A is an $(A^H)^*$-conjugacy class of primitive idempotents of A^H and the pair H_α is a *pointed group* on A [37, 1.1]; we denote by $\mathcal{P}_A(H)$ the set of points of H on A. For any $i \in \alpha$, iAi has an evident structure of H-algebra and we denote by A_α one of these mutually $(A^H)^*$-conjugate H-algebras, and by $A(H_\alpha)$ the *simple quotient* of A^H determined by α. A second pointed group K_β on A is *contained* in H_α if $K \subset H$ and, for any $i \in \alpha$, there is $j \in \beta$ such that [37, 1.1]

$$ij = j = ji \qquad 1.12.1;$$

then, it is quite clear that the $(A^K)^*$-conjugation induces K-algebra embeddings

$$f_\beta^\alpha : A_\beta \longrightarrow \mathrm{Res}_K^H(A_\alpha) \qquad 1.12.2.$$

We set $\mathcal{P}(A) = \mathcal{P}_A(\{1\})$ and call *points of* A the points of $\{1\}$ on A; recall that, for any two-sided ideal $\mathfrak{a}$ of A, *the image in* $\bar{A} = A/\mathfrak{a}$ *of a point of* A *not contained in* $\mathfrak{a}$ *is a point of* $\bar{A}$ and that this correspondence determines an injective map [51, Proposition 3.23]

$$\mathcal{P}(\bar{A}) \longrightarrow \mathcal{P}(A) \qquad 1.12.3.$$

1.13 Following Broué, we consider the *Brauer quotient* and the *Brauer* (algebra) *homomorphism* [11, 1.2]

$$\mathrm{Br}_P^A : A^P \longrightarrow A(P) = A^P \Big/ \sum_Q A_Q^P \qquad 1.13.1,$$

where Q runs over the set of proper subgroups of P, and call *local* any point γ of P on A not contained in $\mathrm{Ker}(\mathrm{Br}_P^A)$ [37, 1.1]; we denote by $\mathcal{LP}_A(P)$ the set of local points of P on A. Recall that *a local pointed group* P_γ *contained in* H_α *is maximal if and only if* $\mathrm{Br}_P(\alpha) \subset A(P_\gamma)_P^{N_H(P_\gamma)}$ [37, Proposition 1.3] and then *the* P*-algebra* A_γ — called a *source algebra* of A_α — *is Morita* *equivalent to* A_α [51, 6.10]; moreover, *the maximal local pointed groups* P_γ *contained in* H_α — called the *defect pointed groups* of H_α — *are mutually* H*-conjugate* [37, Theorem 1.2].

1.14 Let us say that A is a *p-permutation G-algebra* if a Sylow p-subgroup of G stabilizes a basis of A [11, 1.1]. In this case, choosing a point α of G on A, we call *Brauer* (α, G)*-pair* any pair (P, e_A) formed by a p-subgroup P of G such that $\mathrm{Br}_P^A(\alpha) \neq \{0\}$ and by a primitive idempotent e_A of the center $Z\big(A(P)\big)$ of $A(P)$ such that

$$e_A \mathrm{Br}_P^A(\alpha) \neq \{0\} \qquad 1.14.1;$$

note that any local pointed group Q_δ on A *contained* in G_α determines the Brauer (α, G)-pair (Q, f_A) fulfilling $f_A \mathrm{Br}_Q^A(\delta) \neq \{0\}$.

1.15 Then, it follows from Theorem 1.8 in [11] that *the inclusion between the local pointed groups on A induces an inclusion between the Brauer* (α, G)*-pairs*; explicitly, if (P, e_A) and (Q, f_A) are two Brauer (α, G)-pairs then we have

$$(Q, f_A) \subset (P, e_A) \qquad 1.15.1$$

whenever there are local pointed groups P_γ and Q_δ on A fulfilling

$$Q_\delta \subset P_\gamma \subset G_\alpha \quad , \quad f_A \mathrm{Br}_Q^A(\delta) \neq \{0\} \quad \text{and} \quad e_A \mathrm{Br}_P^A(\gamma) \neq \{0\} \qquad 1.15.2.$$

Actually, according to the same result, for any p-subgroup P of G, any primitive idempotent e_A of $Z\big(A(P)\big)$ fulfilling $e_A \mathrm{Br}_P^A(\alpha) \neq \{0\}$ and any subgroup Q of P, there is a unique primitive idempotent f_A of $Z\big(A(Q)\big)$ fulfilling

$$e_A \mathrm{Br}_P^A(\alpha) \neq \{0\} \quad \text{and} \quad (Q, f_A) \subset (P, e_A) \qquad 1.15.3.$$

Once again, *the maximal Brauer* (α, G)*-pairs are pairwise* G*-conjugate* [11, Theorem 1.14].

1.16 Obviously, the *group algebra kG* is a p-permutation G-algebra and, for any primitive idempotent b of $Z(kG)$, $\alpha = \{b\}$ is a point of G on A; for any p-subgroup P of G, the Brauer homomorphism $\mathrm{Br}_P = \mathrm{Br}_P^{kG}$ induces a k-algebra isomorphism [42, 2.8.4]

$$kC_G(P) \cong (kG)(P) \qquad 1.16.1;$$

thus, in a Brauer $(\{b\}, G)$-pair (P, e) — simply called a *Brauer* (b, G)*-pair* — e is nothing but a block of $C_G(P)$ such that $e\mathrm{Br}_P(b) \neq 0$, and we have an *inclusion relation* between them; note that $(\{1\}, b)$ is a Brauer (b, G)-pair and that the condition $e\mathrm{Br}_P(b) \neq 0$ is equivalent to

$$(\{1\}, b) \subset (P, e) \qquad 1.16.2.$$

Then, the *Brauer First Main Theorem* affirms that (P,e) *is maximal if and only if the algebra* $k\bar{C}_G(P)\bar{e}$ *is simple and the quotient* $N_G(P,e)/P{\cdot}C_G(P)$ *is a* p'*-group* [51, Theorem 10.14]. We still call *Brauer* G*-pair* a Brauer (b,G)-pair for some block b of G.

1.17 Recall that if (P,e) is a maximal Brauer (b,G)-pair then, in Brauer terms, P is a *defect group* of the block b and if $|P| = p^d$, d is the *defect* of b. In particular, the sentence b *has defect zero* means that $P = \{1\}$ which, by the very definition of *defect pointed group*, is clearly equivalent to the condition $b \in (kG)_1^G$; moreover, according to *Higman's Criterion* [51, Theorem 5.12], this condition is equivalent to *any* kGb*-module is projective*, which amounts to saying that kGb *is simple*.

1.18 More generally, we say that the block b is *nilpotent* whenever the quotients $N_G(Q,f)/C_G(Q)$ are p-groups for all the Brauer (b,G)-pairs (Q,f) [12, Definition 1.1]; by the main result in [41], *the block* b *is nilpotent if and only if, for a maximal local pointed group* P_γ *on* kGb, P *stabilizes a simple unitary subalgebra* S *of* $(kGb)_\gamma$ *fulfilling*

$$(kGb)_\gamma = SP \cong S \otimes_k kP \qquad 1.18.1$$

where we denote by SP the obvious k-algebra $\bigoplus_{u\in P} Su$ and, for the isomorphism in the right, we consider the lifting to a group homomorphism $P \to S^*$ of the P-action on S [41, statement (1.8.1)]. Note that, although in [41] we argue over a complete discrete valuation ring $\mathcal{O}$ of characteristic zero, by the classification of the Dade P-algebras in [17] and [6], S can be lifted to $\mathcal{O}$ and the comments in [41, 7.7-7.9] apply, allowing us to argue either in characteristic 0 or in characteristic p indifferently.

1.19 Always in the case of the *group algebra*, for any p-subgroup P of G and any subgroup H of $N_G(P)$ containing $P{\cdot}C_G(P)$, we have

$$\mathrm{Br}_P\big((kG)^H\big) = (kG)(P)^H \qquad 1.19.1$$

and therefore *any block* e *of* $C_G(P)$ *determines a unique point* β *of* H *on* kG (cf. 1.12.3) *such that* H_β *contains* P_γ *for a local point* γ *of* P *on* kG *fulfilling* [41, Lemma 3.9]

$$e\mathrm{Br}_P(\gamma) \neq \{0\} \qquad 1.19.3.$$

Moreover, if Q is a subgroup of P such that $C_G(Q) \subset H$, then the blocks of $C_G(Q) = C_H(Q)$ determined by (P,e) from G and from H coincide [11, Theorem 1.8]. Note that if P is normal in G then the kernel of the obvious k-algebra homomorphism $kG \to k(G/P)$ is contained in the *radical* $J(kG)$ and contains $\mathrm{Ker}(\mathrm{Br}_P)$; thus, in this case, isomorphism 1.16.1 implies that *any point of* P *on* kG *is local* and that *any block of* G *is contained in* $Z\big(kC_G(P)\big)$.

1.20 Let P be a finite p-group; a *Dade P-algebra* is a p-permutation P-algebra S which is a simple k-algebra and fulfills $S(P) \neq \{0\}$ [43, 1.3]; since k is algebraically closed and $\mathrm{Aut}_k(S) \cong S^*/k^*$, we have $S \cong \mathrm{End}_k(M)$ for a suitable kP-module M that in [19] Everett Dade calls *capped endopermutation kP-module.* Since

$$\big(S(Q)\big)\big(\bar{N}_P(Q)\big) \cong S\big(N_P(Q)\big) \qquad 1.20.1$$

for any subgroup Q of P [11, Proposition 1.5], $\mathrm{Res}^P_Q(S)$ is a Dade Q-algebra; moreover, according to 1.8 in [43], the *Brauer quotient* $S(Q)$ is a Dade $\bar{N}_P(Q)$-algebra.

1.21 In particular Q has a unique *local point* on S or, equivalently, $\mathrm{Res}^P_Q(M)$ has a unique isomorphism class of direct kQ-summands of *vertex* Q. We say that two Dade P-algebras

$$S \cong \mathrm{End}_k(M) \quad \text{and} \quad S' \cong \mathrm{End}_k(M') \qquad 1.21.1$$

are *similar* if M and M' have the same isomorphism class of direct kP-summands of vertex P or, equivalently, if S can be *embedded* (cf. 1.11) in the *tensor product* $\mathrm{End}(N) \otimes_k S'$ for a suitable kP-module N with a P-stable basis [43, 1.5 and 2.5.1]; we denote by $\mathcal{D}_k(P)$ the set of *similarity classes.* On the other hand, the *tensor product* $S \otimes_k S'$ is a Dade P-algebra since we have [41, Proposition 5.6]

$$(S \otimes_k S')(P) \cong S(P) \otimes_k S'(P) \qquad 1.21.2,$$

and it induces a *group structure* on $\mathcal{D}_k(P)$ — called the *Dade group* of P — where the opposite P-algebra S° determines the inverse of the class of S since

$$S^\circ \otimes_k S \cong \mathrm{End}_k(S) \qquad 1.21.3$$

is clearly *similar* to the trivial P-algebra k.

1.22 Now, for any subgroup Q of P, it is quite clear that the restriction induces a group homomorphism

$$\mathrm{res}^P_Q : \mathcal{D}_k(P) \longrightarrow \mathcal{D}_k(Q) \qquad 1.22.1.$$

If Q is normal in P then the *Brauer quotient functor* $\mathfrak{Br}^P_Q$ from the category of P-algebras to the category of P/Q-algebras, mapping a P-algebra A on the P/Q-algebra $A(Q)$, preserves the *similarity* and determines a group homomorphism

$$\widetilde{\mathfrak{Br}}{}^P_Q : \mathcal{D}_k(P) \longrightarrow \mathcal{D}_k(P/Q) \qquad 1.22.2.$$

Moreover, for a second normal subgroup R of P contained in Q, we have the obvious *transitivity* [11, Proposition 1.5]

$$\widetilde{\mathfrak{Br}}_{Q/R}^{P/R} \circ \widetilde{\mathfrak{Br}}_R^P = \widetilde{\mathfrak{Br}}_Q^P \qquad 1.22.3.$$

1.23 As announced in the Introduction (cf. I20), we have to deal with central extensions $\hat{G}$ of finite groups G by k^*; however, note that $\hat{G}$ always contains a *finite* subgroup G' covering G [42, Lemma 5.5]. Precisely, we call *k^*-group* a group X endowed with an injective group homomorphism $\theta : k^* \to Z(X)$ [42, 5], and *k^*-quotient* of (X, θ) the group $X/\theta(k^*)$; we denote by X° the k^*-group formed by X and by the composition of θ with the automorphism $k^* \cong k^*$ mapping $\lambda \in k^*$ on λ^{-1}; we say that a k^*-group is *finite* whenever its k^*-quotient is finite. Usually, we denote by $\hat{G}$ a k^*-group and by G its k^*-quotient, and we write $\lambda \cdot \hat{x}$ for the product of $\hat{x} \in \hat{G}$ by the image of $\lambda \in k^*$ in $\hat{G}$.

1.24 If $\hat{G}'$ is a second k^*-group, we denote by $\hat{G} \hat{\times} \hat{G}'$ the quotient of the direct product $\hat{G} \times \hat{G}'$ by the image of the *inverse* diagonal in $k^* \times k^*$, which has an obvious structure of k^*-group with k^*-quotient $G \times G'$; moreover, if $G = G'$ then we denote by $\hat{G} * \hat{G}'$ the k^*-group obtained from the converse image of $\Delta(G) \subset G \times G$ in $\hat{G} \hat{\times} \hat{G}'$, which is nothing but the so-called *sum* of both central extensions of G by k^*; in particular, we have a canonical k^*-group isomorphism

$$\hat{G} * \hat{G}^\circ \cong k^* \times G \qquad 1.24.1.$$

A k^*-group homomorphism $\varphi : \hat{G} \to \hat{G}'$ is a group homomorphism which preserves the k^*-multiplication.

1.25 Note that, for any k-algebra A of finite dimension, the group A^* of invertible elements has a canonical k^*-group structure. If S is a simple algebra then $\mathrm{Aut}_k(S)$ coincides with the k^*-quotient of S^*; in particular, any finite group G acting on S determines — by *pull-back* — a k^*-group $\hat{G}$ of k^*-quotient G, together with a k^*-group homomorphism [42, 5.7]

$$\rho : \hat{G} \longrightarrow S^* \qquad 1.25.1.$$

It is clear that the inclusion $k^* \subset k$ determines a k-algebra homomorphism from the *group algebra* kk^* of k^* to k and that, for any finite k^*-group $\hat{G}$, *the group algebra* $k\hat{G}$ of the *group* $\hat{G}$ is also a kk^*-algebra; we call *k^*-group algebra* of $\hat{G}$ the algebra

$$k_*\hat{G} = k \otimes_{kk^*} k\hat{G} \qquad 1.25.2$$

which has dimension $|G|$. We also call *block of* $\hat{G}$ any primitive idempotent b of $Z(k_*\hat{G})$, and denote by $\mathrm{Irr}_k(\hat{G}, b)$ the canonical $\mathbb{Z}$-basis — formed by the isomorphism classes of simple $k_*\hat{G}b$-modules — of the *Grothendieck group* of the category of finitely generated $k_*\hat{G}b$-modules.

1.26 In particular, for any local pointed group P_γ on kG — we are avoiding unnecessary generality — $\mathrm{Br}_P(\gamma)$ is a $(kG)(P)^*$-conjugacy class of primitive idempotents in $(kG)(P)$ and therefore it determines a *simple quotient* $(kG)(P_\gamma)$ of this k-algebra; thus, the action of $\bar{N}_G(P_\gamma)$ on $(kG)(P_\gamma)$ determines a k^*-group $\hat{\bar{N}}_G(P_\gamma)$ [42, 6.2]; but, the *Brauer homomorphism* Br_P induces a group homomorphism

$$\bar{C}_G(P) \longrightarrow (kG)(P_\gamma)^* \qquad 1.26.1$$

and therefore an $N_G(P_\gamma)$-stable *group* homomorphism

$$\bar{C}_G(P) \longrightarrow \hat{\bar{N}}_G(P_\gamma) \qquad 1.26.2;$$

consequently, setting

$$E_G(P_\gamma) = N_G(P_\gamma)/C_G(P) \quad \text{and} \quad \tilde{E}_G(P_\gamma) = \bar{N}_G(P_\gamma)/\bar{C}_G(P) \qquad 1.26.3,$$

we obtain the k^*-group $\hat{\tilde{E}}_G(P_\gamma) = \hat{\bar{N}}_G(P_\gamma)/\bar{C}_G(P)$†.

1.27 Let $\hat{G}$ be a finite k^*-group; recall that we call *$\hat{G}$-interior algebra* [42, 5.10] any k-algebra A of finite dimension endowed with a k^*-group homomorphism

$$\rho : \hat{G} \longrightarrow A^* \qquad 1.27.1$$

and, as usual, we write $\hat{x}{\cdot}a$ and $a{\cdot}\hat{x}$ instead of $\rho(\hat{x})a$ and $a\rho(\hat{x})$ for any $\hat{x} \in \hat{G}$ and any $a \in A$. As above, a $\hat{G}$-interior algebra homomorphism from A to another $\hat{G}$-interior algebra A' is a *not necessarily unitary* algebra homomorphism $f : A \to A'$ this time fulfilling

$$f(\hat{x}{\cdot}a) = \hat{x}{\cdot}f(a) \quad \text{and} \quad f(a{\cdot}\hat{x}) = f(a){\cdot}\hat{x} \qquad 1.27.2;$$

we say that f is an *embedding* if it fulfills condition 1.11.1. For a k^*-group homomorphism $\varphi : \hat{G}' \to \hat{G}$, we denote by $\mathrm{Res}_\varphi(A)$ the $\hat{G}'$-interior algebra defined by $\rho \circ \varphi$.

1.28 The conjugation in A induces an action of the k^*-quotient G of $\hat{G}$ on A, so that A becomes an ordinary G-algebra; thus, all the *pointed group language* developed above applies to $\hat{G}$-interior algebras; if $\hat{H}$ is a k^*-subgroup of $\hat{G}$ and α a point of H on A, we call *pointed k^*-group* on A the pair $\hat{H}_\alpha$, and note that now A_α becomes an *$\hat{H}$-interior algebra*. More generally, if $\hat{H}_\alpha$ and $\hat{K}_\beta$ are two pointed k^*-groups on A, we say that an injective k^*-group homomorphism $\hat{\varphi} : \hat{K} \to \hat{H}$ is an *A-fusion from $\hat{K}_\beta$ to $\hat{H}_\alpha$* whenever there is a $\hat{K}$-interior algebra *embedding*

$$f_{\hat{\varphi}} : A_\beta \longrightarrow \mathrm{Res}^{\hat{H}}_{\hat{K}}(A_\alpha) \qquad 1.28.1$$

† Note that we slightly modify our usual notation introduced in [39].

such that the inclusion $A_\beta \subset A$ and the composition of $f_{\hat\varphi}$ with the inclusion $A_\alpha \subset A$ are A^*-conjugate; we denote by $F_A(\hat K_\beta, \hat H_\alpha)$ the set of them [39, Definition 2.5] and by $\tilde F_A(\hat K_\beta, \hat H_\alpha)$ its quotient by the action of H†.

1.29 As usual, we write $F_A(\hat H_\alpha)$ instead of $F_A(\hat H_\alpha, \hat H_\alpha)$; choosing $i \in \alpha$ and setting $A_\alpha = iAi$, it follows from Corollary 2.13 in [39] that we have a group homomorphism

$$F_A(\hat H_\alpha) \longrightarrow N_{A_\alpha^*}(\hat H{\cdot}i)/(A_\alpha^H)^* \qquad 1.29.1$$

and then we consider the k^*-group $\hat F_A(\hat H_\alpha)$ defined by the *pull-back*

$$\begin{array}{ccc} F_A(\hat H_\alpha) & \longrightarrow & N_{A_\alpha^*}(\hat H{\cdot}i)/(A_\alpha^H)^* \\ \uparrow & & \uparrow \\ \hat F_A(\hat H_\alpha) & \longrightarrow & N_{A_\alpha^*}(\hat H{\cdot}i)/\big(i + J(A_\alpha^H)\big) \end{array} \qquad 1.29.2.$$

It is clear that the *group algebra* kG is a $(k^*\times)G$-interior algebra and recall that, for any local pointed group P_γ on kG, we have a *canonical* k^*-group isomorphism [42, Proposition 6.12]

$$\hat E_G(P_\gamma)^\circ \cong \hat F_{kG}(P_\gamma) \qquad 1.29.3.$$

1.30 As usual, we denote by $\mathrm{Res}_{\hat H}^{\hat G}(A)$ the corresponding $\hat H$-interior algebra. Conversely, for any $\hat H$-interior algebra B, we consider the *induced $\hat G$-interior algebra*

$$\mathrm{Ind}_{\hat H}^{\hat G}(B) = k_*\hat G \otimes_{k_*\hat H} B \otimes_{k_*\hat H} k_*\hat G \qquad 1.30.1,$$

where the distributive product is defined by the *formula*

$$(\hat x \otimes b \otimes \hat y)(\hat x' \otimes b' \otimes \hat y') = \begin{cases} \hat x \otimes b.\hat y\hat x'.b' \otimes \hat y' & \text{if } \hat y\hat x' \in \hat H \\ 0 & \text{otherwise} \end{cases} \qquad 1.30.2$$

for any $\hat x, \hat y, \hat x', \hat y' \in \hat G$ and any $b, b' \in B$, and where the structural homomorphism

$$\hat G \longrightarrow \mathrm{Ind}_{\hat H}^{\hat G}(B) \qquad 1.30.3$$

maps $\hat x \in \hat G$ on $\sum_{\hat y} \hat x\hat y \otimes 1_B \otimes \hat y^{-1}$, $\hat y \in \hat G$ running over a set of representatives for $\hat G/\hat H$. Recall that if I is a $\hat G$-orbit of pairwise orthogonal idempotents of A such that $1_A = \sum_{i\in I} i$ and $\hat H$ is the stabilizer in $\hat G$ of $i \in I$, we have a $\hat G$-interior algebra isomorphism [53, Proposition 2.3]

$$A \cong \mathrm{Ind}_{\hat H}^{\hat G}(iAi) \qquad 1.30.4.$$

† Note that we slightly modify our usual notation introduced in [39].

1.31 Our general notation mainly concerns *group theory* — our standard reference being [28] — and *homological algebra* — our standard reference being [18]. In particular, if G is a finite group, recall that $\mathbf{O}^p(G)$, $\mathbf{O}^{p'}(G)$, $\mathbf{O}_p(G)$ and $\mathbf{O}_{p'}(G)$ respectively denote the minimal or the maximal normal subgroups of G with their index or their order being a power of p or prime to p; note that this notation still makes sense for a *finite* k^*-group $\hat{G}$ except that $\mathbf{O}_p(\hat{G})$ remains a p-group. For any pair of subgroups H and K of G, we denote by $T_G(K,H)$ the set of $x \in G$ fulfilling $xKx^{-1} \subset H$. If P is a finite p-group, $\Phi(P)$ denotes the *Frattini subgroup*, namely the intersection of all the maximal subgroups of P, and $\Omega_1(P)$ denotes the subgroup of P generated by all the elements of order p.

1.32 We denote by $\mathfrak{ab} : \mathfrak{Gr} \to \mathfrak{Ab}$ the functor mapping G on its maximal Abelian quotient $G/[G,G]$ and by $\mathfrak{ab}^\circ$ the *contravariant* $\mathfrak{Ab}$-valued functor defined by the *transfer* [28, Ch. 7, Theorem 3.2] from the subcategory of $\mathfrak{Gr}$ formed by all injective group homomorphisms. For any finite set Ω, we denote by $\mathfrak{S}_\Omega$ the group of permutations of Ω and we also set $\mathfrak{S}_n = \mathfrak{S}_\Omega$ whenever $n = |\Omega|$; if G is a subgroup of $\mathfrak{S}_\Omega$ and H a group, $H \wr G$ denotes the *wreath product* (cf. 15.1 in [31, Ch. I]), namely the obvious semidirect product

$$H \wr G = \Big(\prod_{\omega \in \Omega} H\Big) \rtimes G \qquad 1.32.1.$$

Finally, we denote by μ the so-called *Möbius function* mapping any *ordered finite set* X on $\sum_Y (-1)^{|Y|}$ where Y runs over the set of *totally ordered subsets* of X, setting $\mu_X(x)$ for the value of μ over the set of $y \in X$ strictly smaller than $x \in X$.

Chapter 2

Frobenius P-categories: the first definition

2.1 Let P be a finite p-group. In this chapter we introduce the main objects of this book, namely the *Frobenius P-categories*; actually, in chapter 21 below, we give an equivalent — somewhat *easier* — definition, but this first definition is more directly related to the blocks, as we show in the next chapter.

2.2 Let us call *P-category* any subcategory $\mathcal{F}$ (cf. 1.4) of the category of finite groups $\mathfrak{Gr}$ (cf. 1.1) containing the Frobenius category $\mathcal{F}_P$ of P (cf. 1.8), where the objects are the $\mathcal{F}_P$-objects — the subgroups of P — and where all the homomorphisms are injective. Note that the *intersection* $\mathcal{F} \cap \mathcal{F}'$ (cf. 1.5) of two P-categories is a P-category too, and that there is a *unique maximal* P-category, namely the P-category containing *all* injective group homomorphisms between the subgroups of P.

2.3 We say that a P-category $\mathcal{F}$ is *divisible* † whenever it fulfills:

2.3.1 *If Q, R and T are subgroups of P, for any $\varphi \in \mathcal{F}(Q,R)$ and any group homomorphism $\psi\colon T \to R$ the composition $\varphi \circ \psi$ belongs to $\mathcal{F}(Q,T)$ (if and) only if $\psi \in \mathcal{F}(R,T)$.*

Or, equivalently, whenever for any subgroup Q of P, the category $(\mathcal{F})_Q$ (cf. 1.7) is a *full* subcategory of $(\mathfrak{Gr})_Q$ (cf 1.4). Note that the maximal P-category is divisible, and that the intersection of two divisible P-categories is divisible too. Actually, all the P-categories we will consider are divisible.

2.4 Note that $\mathcal{F}$ is divisible if and only if, for any $\varphi \in \mathcal{F}(Q,R)$ which is a group isomorphism, the inverse φ^{-1} is also an $\mathcal{F}$-morphism; moreover, in this case, if Q' and R' are respective subgroups of Q and R, the restriction of any $\varphi \in \mathcal{F}(Q,R)$ fulfilling $\varphi(R') \subset Q'$ belongs to $\mathcal{F}(Q',R')$. In particular, a divisible P-category $\mathcal{F}$ is determined by the sets $\mathcal{F}(P,Q)$ where Q runs over the set of all the subgroups of P. Conversely, if $\mathcal{H}(Q) \subset \mathrm{Hom}(Q,P)$ is a set of injective homomorphisms containing $\mathcal{F}_P(P,Q)$ for any subgroup Q of P and this family fulfills the following condition

2.4.1 *for any pair of subgroups Q and R of P and any $\theta \in \mathrm{Hom}(Q,R)$ fulfilling $\mathcal{H}(Q) \cap \big(\mathcal{H}(R) \circ \theta\big) \neq \emptyset$, we have $\mathcal{H}(R) \circ \theta \subset \mathcal{H}(Q)$,*

there exists an evident divisible P-category $\mathcal{F}'$ fulfilling $\mathcal{F}'(P,Q) = \mathcal{H}(Q)$ for any $Q \in \mathfrak{X}$.

---

† A divisible P-category is called a *fusion system over P* in [13].

2.5 In a finite group G with a Sylow p-subgroup S, it is obvious that the centralizer or the normalizer in S of a subgroup Q of S need not be a Sylow p-subgroup of the centralizer or the normalizer of Q in G, and in our abstract setting we will determine when they are so. For our purpose, it is handy to introduce the following notation: if Q, R and T are subgroups of P and $Q \subset T$, any injective group homomorphism $\psi : T \to R$ determines a group isomorphism $\mathrm{Aut}(Q) \cong \mathrm{Aut}\big(\psi(Q)\big)$ and we simply denote by ${}^{\psi}K$ and ${}^{\psi}\chi$ the images of $K \subset \mathrm{Aut}(Q)$ and $\chi \in \mathrm{Aut}(Q)$ respectively.

2.6 Let $\mathcal{F}$ be a divisible P-category, Q a subgroup of P and K a subgroup of $\mathrm{Aut}(Q)$; it is quite clear that, for any $\mathcal{F}$-morphism $\psi : Q{\cdot}N_P^K(Q) \to P$ (cf. 1.9), we have

$$\psi\big(N_P^K(Q)\big) \subset N_P^{\psi K}\big(\psi(Q)\big) \qquad 2.6.1;$$

then, we say that Q is *fully K-normalized* in $\mathcal{F}$ whenever it fulfills

2.6.2 *For any $\psi \in \mathcal{F}\big(P, Q{\cdot}N_P^K(Q)\big)$, we have $\psi\big(N_P^K(Q)\big) = N_P^{\psi K}\big(\psi(Q)\big)$.*

If $K = \{\mathrm{id}_Q\}$ or $K = \mathrm{Aut}(Q)$, we respectively say that Q is *fully centralized* or *fully normalized* in $\mathcal{F}$; note that K, $K{\cdot}\mathcal{F}_Q(Q)$ and $K \cap \mathcal{F}(Q)$ play the same role in this condition, so that we always may assume that

$$\mathcal{F}_Q(Q) \subset K \subset \mathcal{F}(Q) \qquad 2.6.3.$$

Proposition 2.7 *With the notation above, let R be a subgroup of $Q{\cdot}N_P^K(Q)$ containing Q and assume that an $\mathcal{F}$-morphism $\psi : R \to P$ fulfills*

2.7.1 *For any $\psi' \in \mathcal{F}(P,R)$ we have $\big|N_P^{\psi' K}\big(\psi'(Q)\big)\big| \le \big|N_P^{\psi K}\big(\psi(Q)\big)\big|$.*

Then $\psi(Q)$ is fully ${}^{\psi}K$-normalized in $\mathcal{F}$. In particular there is $\varphi \in \mathcal{F}(P,Q)$ such that $\varphi(Q)$ is both fully centralized and fully ${}^{\varphi}K$-normalized in $\mathcal{F}$.

Proof: Set $Q' = \psi(Q)$; since $\psi(R)$ is contained in $Q'{\cdot}N_P^{\psi K}(Q')$ and $\mathcal{F}$ is divisible, any $\mathcal{F}$-morphism $\xi' : Q'{\cdot}N_P^{\varphi K}(Q') \to P$ determines an $\mathcal{F}$-morphism $\psi' : R \to P$ mapping $v \in R$ on $\xi'\big(\psi(v)\big)$ and therefore simultaneously we have

$$\big|N_P^{\psi' K}\big(\psi'(Q)\big)\big| \le |N_P^{\psi K}(Q')| \quad \text{and} \quad \xi'\big(N_P^{\psi K}(Q')\big) \subset N_P^{\psi' K}\big(\psi'(Q)\big) \qquad 2.7.2,$$

so that we get the equality $\xi'\big(N_P^{\psi K}(Q')\big) = N_P^{\psi' K}\big(\xi'(Q')\big)$.

In particular, it is clear that there is an $\mathcal{F}$-morphism $\varphi : Q \to P$ such that $Q'' = \varphi(Q)$ is fully centralized in $\mathcal{F}$, and then it is still clear that we can find an $\mathcal{F}$-morphism $\xi'' : Q''{\cdot}C_P(Q'') \to P$ such that $\xi''(Q'')$ is fully ${}^{\xi''}({}^{\varphi}K)$-normalized in $\mathcal{F}$; but $\xi''(Q'')$ is fully centralized too, as it is easily checked.

2.8 We are ready to state our main definition; our conditions below mimic suitable formulations of the Sylow Theorem(s), as we show in the next chapter. We say that a P-category $\mathcal{F}$ is a *Frobenius P-category* or a *Frobenius category over P*† if it is *divisible* and fulfills the following two conditions:

2.8.1 *The group $\mathcal{F}_P(P)$ of inner automorphisms of P is a Sylow p-subgroup of $\mathcal{F}(P)$.*

2.8.2 *For any subgroup Q of P, any subgroup K of $\mathrm{Aut}(Q)$ and any $\mathcal{F}$-morphism $\varphi : Q \to P$ such that $\varphi(Q)$ is fully ${}^{\varphi}K$-normalized in $\mathcal{F}$, there are an $\mathcal{F}$-morphism $\psi : Q{\cdot}N_P^K(Q) \to P$ and $\chi \in K$ such that $\psi(u) = \varphi\big(\chi(u)\big)$ for any $u \in Q$.*

2.9 Actually, in condition 2.8.2 we may assume that $\mathcal{F}_Q(Q) \subset K \subset \mathcal{F}(Q)$ (cf. 2.6.3) and that $Q' = \varphi(Q)$ is *fully centralized* too. Indeed, by Proposition 2.7, there is $\bar\varphi \in \mathcal{F}(P, Q)$ such that $\bar Q = \bar\varphi(Q)$ is both fully centralized and fully ${}^{\bar\varphi}K$-normalized in $\mathcal{F}$; then, setting $K' = {}^{\varphi}K$ and denoting by $\varphi' : Q' \cong \bar Q$ the $\mathcal{F}$-isomorphism fulfilling $\varphi'\big(\varphi(u)\big) = \bar\varphi(u)$ for any $u \in Q$, assume that there are $\mathcal{F}$-morphisms

$$\bar\psi : Q{\cdot}N_P^K(Q) \longrightarrow P \quad \text{and} \quad \psi' : Q'{\cdot}N_P^{K'}(Q') \longrightarrow P \qquad 2.9.1,$$

and elements $\bar\chi \in K$ and $\chi' \in K'$ such that we have

$$\bar\psi(u) = \bar\varphi\big(\bar\chi(u)\big) \quad \text{and} \quad \psi'(u') = \varphi'\big(\chi'(u')\big) \qquad 2.9.2$$

for any $u \in Q$ and any $u' \in Q'$; since $\mathcal{F}$ is divisible and we have

$$\psi'\big(N_P^{K'}(Q')\big) = N_P^{{}^{\psi'}K'}(\bar Q) \quad \text{and} \quad {}^{\psi'}K' = {}^{\bar\varphi}K \qquad 2.9.3,$$

we finally get an $\mathcal{F}$-isomorphism

$$\psi'^* : \bar Q{\cdot}N_P^{{}^{\bar\varphi}K}(\bar Q) \cong Q'{\cdot}N_P^{K'}(Q') \qquad 2.9.4$$

such that $w' = \psi'^*\big(\psi'(w')\big)$ for any $w' \in Q'{\cdot}N_P^{K'}(Q')$; thus, the $\mathcal{F}$-morphism $\psi : Q{\cdot}N_P^K(Q) \to P$ mapping $w \in Q{\cdot}N_P^K(Q)$ on $\psi'^*\big(\bar\psi(w)\big)$ maps $u \in Q$ on

$$\psi'^*\Big(\bar\varphi\big(\bar\chi(u)\big)\Big) = \psi'^*\Big(\varphi'\Big(\varphi\big(\bar\chi(u)\big)\Big)\Big) = \chi'^{-1}\Big(\varphi\big(\bar\chi(u)\big)\Big) = \varphi\big(\chi(u)\big) \quad 2.9.5$$

for a suitable $\chi \in K$.

2.10 Moreover, condition 2.8.2 implies that

2.10.1 *For any subgroup Q of P, any $\mathcal{F}$-morphism $\varphi : Q \to P$ such that $\varphi(Q)$ is fully centralized in $\mathcal{F}$, and any subgroup R of $N_P(Q)$ such that $Q \subset R$ and ${}^{\varphi}\mathcal{F}_R(Q) \subset \mathcal{F}_P\big(\varphi(Q)\big)$, there is an $\mathcal{F}$-morphism $\rho : R \to P$ extending φ.*

Indeed, first of all note that if Q is a subgroup of P fully centralized in $\mathcal{F}$ and R is a subgroup of $N_P(Q)$ containing Q then we have $N_P^{\mathcal{F}_R(Q)}(Q) = R{\cdot}C_P(Q)$ and, for any $\mathcal{F}$-morphism $\eta : R{\cdot}C_P(Q) \to P$, we get

$$\eta\big(R{\cdot}C_P(Q)\big) = \eta(R){\cdot}C_P\big(\eta(Q)\big) = N_P^{{}^{\eta}\mathcal{F}_R(Q)}\big(\eta(Q)\big) \qquad 2.10.2,$$

† Called a *saturated fusion system over P* in [13].

so that Q is also fully $\mathcal{F}_R(Q)$-normalized in $\mathcal{F}$; hence, in condition 2.10.1 the subgroup $\varphi(Q)$ is fully $\mathcal{F}_{R'}(Q)$-normalized in $\mathcal{F}$ where R' is the converse image of ${}^{\varphi}\mathcal{F}_R(Q)$ in $N_P\big(\varphi(Q)\big)$; then, condition 2.8.2 implies the existence of $\psi \in \mathcal{F}(P,R)$ and $\chi \in \mathcal{F}_R(Q)$ such that $\psi(u) = \varphi\big(\chi(u)\big)$ for any $u \in Q$, and it suffices to choose $w \in R$ lifting χ and to define $\rho \in \mathcal{F}(P,R)$ by $\rho(v) = \psi(v^w)$ for any $v \in R$. Conversely, condition 2.8.2 can be replaced by condition 2.11.1 below†.

Proposition 2.11 *Let $\mathcal{F}$ be a divisible P-category such that $\mathcal{F}_P(P)$ is a Sylow p-subgroup of $\mathcal{F}(P)$. Then, $\mathcal{F}$ is a Frobenius P-category if and only if it fulfills the condition*

2.11.1 *For any subgroup Q of P, any $\mathcal{F}$-morphism $\varphi\colon Q \to P$ such that $\varphi(Q)$ is both fully centralized and fully normalized in $\mathcal{F}$, and any subgroup R of $N_P(Q)$ such that $Q \subset R$ and ${}^{\varphi}\mathcal{F}_R(Q) \subset \mathcal{F}_P\big(\varphi(Q)\big)$, there is an $\mathcal{F}$-morphism $\rho\colon R \to P$ extending φ.*

In this case, for any subgroups Q of P and K of $\mathrm{Aut}(Q)$, the following statements are equivalent:

2.11.2 *The subgroup Q is fully K-normalized in $\mathcal{F}$.*

2.11.3 *For any $\varphi \in \mathcal{F}(P,Q)$, we have $\big|N_P^{{}^{\varphi}K}\big(\varphi(Q)\big)\big| \le |N_P^K(Q)|$.*

2.11.4 *The subgroup Q is fully centralized in $\mathcal{F}$ and $K \cap \mathcal{F}_P(Q)$ is a Sylow p-subgroup of $K \cap \mathcal{F}(Q)$.*

Proof: We already know that statement 2.11.3 implies statement 2.11.2. Assume that $\mathcal{F}$ fulfills statement 2.11.1 and let us prove that condition 2.8.2 holds; actually, it is easily checked from Proposition 2.7 that $\mathcal{F}$ fulfills statement 2.10.1 too. Let Q be a subgroup of P, K a subgroup of $\mathrm{Aut}(Q)$ and $\varphi\colon Q \to P$ an $\mathcal{F}$-morphism such that $Q' = \varphi(Q)$ is fully centralized and fully ${}^{\varphi}K$-normalized in $\mathcal{F}$; by Lemma 2.12 below, setting $K' = {}^{\varphi}K$ we know that $K' \cap \mathcal{F}_P(Q')$ is a Sylow p-subgroup of $K' \cap \mathcal{F}(Q')$ and therefore, for a suitable $\chi \in K$, we have

$$ {}^{\varphi\circ\chi}\big(K \cap \mathcal{F}_P(Q)\big) \subset K' \cap \mathcal{F}_P(Q') \qquad 2.11.5. $$

Moreover, choose $\varphi' \in \mathcal{F}(P,Q')$ such that $Q'' = \varphi'(Q')$ is fully centralized and fully normalized in $\mathcal{F}$ (cf. Proposition 2.7); once again, up to a modification of our choice, we may assume that ${}^{\varphi'}K' \cap \mathcal{F}_P(Q'')$ is a Sylow p-subgroup of ${}^{\varphi'}K' \cap \mathcal{F}(Q'')$. Then, it follows from statement 2.11.1 that φ' can be extended to an $\mathcal{F}$-morphism $\rho'\colon Q'{\cdot}N_P^{K'}(Q') \to P$ and that there is an $\mathcal{F}$-morphism $\rho''\colon Q{\cdot}N_P^K(Q) \to P$ fulfilling $\rho''(u) = \varphi'\big((\varphi\circ\chi)(u)\big)$ for any

† In [13] Broto, Levi and Oliver show that condition 2.8.2 can be replaced by condition 2.10.1 and Radu Stancu has noticed that the same is true by replacing "fully centralized" by "fully normalized".

$u \in Q$; but, since Q' is fully K'-normalized in $\mathcal{F}$, we have

$$\rho'\big(N_P^{K'}(Q')\big) = N_P^{\varphi' K'}(Q'') \supset \rho''\big(Q{\cdot}N_P^K(Q)\big) \qquad 2.11.6;$$

consequently, since $\mathcal{F}$ is divisible, the $\mathcal{F}$-morphism $\rho : Q{\cdot}N_P^K(Q) \to P$ fulfilling $\rho'\big(\rho(w)\big) = \rho''(w)$ for any element $w \in Q{\cdot}N_P^K(Q)$ extends $\varphi \circ \chi$.

Now, assume that $\mathcal{F}$ is a Frobenius P-category; we firstly prove that statement 2.11.2 implies statement 2.11.4; indeed, since $\mathcal{F}$ is divisible, any $\mathcal{F}$-morphism $\psi : Q{\cdot}C_P(Q) \to P$ determines $\varphi' \in \mathcal{F}\big(P, \psi(Q)\big)$ such that $\varphi'\big(\psi(u)\big) = u$ for any $u \in Q$ (cf. 2.3) and therefore, setting $Q' = \psi(Q)$ and $K' = {}^{\psi}K$, by condition 2.8.2 there are an $\mathcal{F}$-morphism $\xi' : N_P^{K'}(Q') \to P$ and an element $\chi' \in K'$ such that we have $\xi'(u') = \varphi'\big(\chi'(u')\big)$ for any $u' \in Q'$; in particular, we have $\xi'\big(C_P(Q')\big) \subset C_P(Q)$ and therefore $\psi\big(C_P(Q)\big) = C_P(Q')$; that is to say, Q is fully centralized in $\mathcal{F}$ and then statement 2.11.4 follows from Lemma 2.12 below.

Furthermore, if Q is a subgroup of P and $\varphi : Q \to P$ an $\mathcal{F}$-morphism such that $Q' = \varphi(Q)$ is fully centralized and fully normalized in $\mathcal{F}$ then, for any subgroup R of $N_P(Q)$ fulfilling $Q \subset R$ and ${}^{\varphi}\mathcal{F}_R(Q) \subset \mathcal{F}_P(R')$, denoting by R' the converse image of ${}^{\varphi}\mathcal{F}_R(Q)$ in $N_P(Q')$ so that ${}^{\varphi}\mathcal{F}_R(Q) = \mathcal{F}_{R'}(Q')$, it is easily checked that Q' is also fully $\mathcal{F}_{R'}(Q')$-normalized in $\mathcal{F}$; hence, condition 2.8.2 implies the existence of $\psi \in \mathcal{F}(P, R)$ and $\chi \in \mathcal{F}_R(Q)$ such that $\psi(u) = \varphi\big(\chi(u)\big)$ for any $u \in Q$; thus, in order to prove statement 2.11.1, it suffices to choose $w \in R$ lifting χ and to define $\rho \in \mathcal{F}(P, R)$ by $\rho(v) = \psi(v^w)$ for any $v \in R$.

Finally, assume that statement 2.11.4 holds; it follows from the divisibility of $\mathcal{F}$ and from condition 2.8.2 that, if $\varphi \in \mathcal{F}(P, Q)$ and we set $Q' = \varphi(Q)$, there is an $\mathcal{F}$-morphism $\psi' : Q'{\cdot}C_P(Q') \to P$ such that $\psi'\big(\varphi(u)\big) = u$ for any $u \in Q$, and therefore we have $\psi'\big(C_P(Q')\big) \subset C_P(Q)$; hence, we get

$$\begin{aligned} \big|N_P^{\varphi K}(Q')\big| &= |C_P(Q')|{\cdot}|{}^{\varphi}K \cap \mathcal{F}_P(Q')| \\ &\le |C_P(Q)|{\cdot}|K \cap \mathcal{F}(Q)|_p = |N_P^K(Q)| \end{aligned} \qquad 2.11.7$$

which proves statement 2.11.3 (cf. Proposition 2.7). We are done.

Lemma 2.12 *Let $\mathcal{F}$ be a divisible P-category such that $\mathcal{F}_P(P)$ is a Sylow p-subgroup of $\mathcal{F}(P)$, and $\mathfrak{X}$ a nonempty set of subgroups of P such that if $Q \in \mathfrak{X}$ then any subgroup R of P fulfilling $\mathcal{F}(R, Q) \neq \emptyset$ belongs to $\mathfrak{X}$. Assume that for any subgroup $Q \in \mathfrak{X}$, any $\mathcal{F}$-morphism $\varphi : Q \to P$ such that $\varphi(Q)$ is fully centralized in $\mathcal{F}$, and any subgroup R of $N_P(Q)$ fulfilling $Q \subset R$ and ${}^{\varphi}\mathcal{F}_R(Q) \subset \mathcal{F}_P\big(\varphi(Q)\big)$, there is an $\mathcal{F}$-morphism $\rho : R \to P$ extending φ . Then, for any $Q \in \mathfrak{X}$ and any subgroup K of* $\mathrm{Aut}(Q)$ *such that Q is fully centralized and fully K-normalized in $\mathcal{F}$, $K \cap \mathcal{F}_P(Q)$ is a Sylow p-subgroup of $K \cap \mathcal{F}(Q)$.*

Proof: We may assume that $Q \neq P$ and argue by induction on $|P\!:\!Q|$; set $R = N_P^K(Q)$. In the case where $K = \mathrm{Aut}(Q)$, denoting by J the set of automorphisms of R stabilizing Q, it is clear that $N_P^J(R) = R$ and therefore, since Q is fully normalized in $\mathcal{F}$, R is fully J-normalized in $\mathcal{F}$ so that, according to the induction hypothesis, $J \cap \mathcal{F}_P(R) = \mathcal{F}_R(R)$ is a Sylow p-subgroup of $J \cap \mathcal{F}(R)$. But, since Q is fully centralized in $\mathcal{F}$, it follows from our hypothesis that any element of $N_{\mathcal{F}(Q)}\big(\mathcal{F}_R(Q)\big)$ can by lifted to $J \cap \mathcal{F}(R)$; consequently, $\mathcal{F}_R(Q)$ is a Sylow p-subgroup of $N_{\mathcal{F}(Q)}\big(\mathcal{F}_R(Q)\big)$, so it is a Sylow p-subgroup of $\mathcal{F}(Q)$.

In the general case, choose an $\mathcal{F}$-morphism $\varphi' : Q{\cdot}C_P(Q) \to P$ such that $Q' = \varphi'(Q)$ is fully normalized and fully centralized in $\mathcal{F}$ (cf. Proposition 2.7); thus, by the above argument, $\mathcal{F}_P(Q')$ is a Sylow p-subgroup of $\mathcal{F}(Q')$ and therefore it contains a Sylow p-subgroup of ${}^{\tau\circ\varphi'}K$ for a suitable $\tau \in \mathcal{F}(Q')$. That is to say, up to a modification of our choice of φ', we may assume that ${}^{\varphi'}K \cap \mathcal{F}_P(Q')$ is a Sylow p-subgroup of ${}^{\varphi'}K \cap \mathcal{F}(Q')$ containing ${}^{\varphi'}\mathcal{F}_P(Q)$; in this case, according to our hypothesis, φ' can be extended to an $\mathcal{F}$-morphism $\rho' : Q{\cdot}R \to P$ and moreover, if Q is fully K-normalized in $\mathcal{F}$, we actually have $\rho'(R) = N_P^{{}^{\varphi'}K}(Q')$, so that we get

$$ {}^{\varphi'}\big(K \cap \mathcal{F}_P(Q)\big) = {}^{\varphi'}K \cap \mathcal{F}_P(Q') \qquad 2.12.1; $$

hence, since we also have ${}^{\varphi'}\big(K\cap\mathcal{F}(Q)\big) = {}^{\varphi'}K\cap\mathcal{F}(Q')$, $K\cap\mathcal{F}_P(Q)$ is a Sylow p-subgroup of $K \cap \mathcal{F}(Q)$. We are done.

Corollary 2.13 *Let $\mathcal{F}$ be a divisible P-category such that $\mathcal{F}_P(P)$ is a Sylow p-subgroup of $\mathcal{F}(P)$. Then, $\mathcal{F}$ is a Frobenius P-category if and only if it fulfills the following two conditions:*

2.13.1 *For any pair of $\mathcal{F}$-isomorphic subgroups Q and Q' of P fully normalized and fully centralized in $\mathcal{F}$, there is an $\mathcal{F}$-isomorphism $N_P(Q) \cong N_P(Q')$ mapping Q onto Q'.*

2.13.2 *For any subgroup Q of P fully normalized and fully centralized in $\mathcal{F}$ and any subgroup R of $N_P(Q)$ containing $Q{\cdot}C_P(Q)$†, denoting by $\mathcal{F}(R)_Q$ the stabilizer of Q in $\mathcal{F}(R)$, the group homomorphism $\mathcal{F}(R)_Q \to N_{\mathcal{F}(Q)}\big(\mathcal{F}_R(Q)\big)$ induced by the restriction is surjective.*

Moreover, in this case, for any pair of subgroups Q and Q' of P, and any subgroups K of $\mathrm{Aut}(Q)$ and K' of $\mathrm{Aut}(Q')$, if Q and Q' are respectively fully K- and K'-normalized and there is an $\mathcal{F}$-isomorphism $\varphi : Q \cong Q'$ such that $K' = {}^{\varphi}K$, then there is an $\mathcal{F}$-isomorphism $Q{\cdot}N_P^K(Q) \cong Q'{\cdot}N_P^{K'}(Q')$ mapping Q onto Q' and K onto K'.

† In [52, Corollary 2.14] the hypothesis that R contains the centralizer of Q in P has been forgotten.

Proof: Assume that $\mathcal{F}$ is a Frobenius P-category; if Q is a subgroup of P, K a subgroup of $\mathrm{Aut}(Q)$ and $\varphi : Q \to P$ an $\mathcal{F}$-morphism, setting $Q' = \varphi(Q)$ and $K' = {}^{\varphi}K$, it follows from condition 2.8.2 above that, if Q' is fully K'-normalized in $\mathcal{F}$, then there is an $\mathcal{F}$-morphism $\psi : Q{\cdot}N_P^K(Q) \to P$ extending $\varphi \circ \chi$ for a suitable $\chi \in \mathcal{F}(Q) \cap K$, and therefore, if Q is fully K-normalized in $\mathcal{F}$, we actually have $\psi\big(Q{\cdot}N_P^K(Q)\big) = Q'{\cdot}N_P^{K'}(Q')$ (cf. 2.6.2); this proves condition 2.13.1 and the last statement. Moreover, in condition 2.13.2, the surjectivity of the homomorphism $\mathcal{F}(R)_Q \to N_{\mathcal{F}(Q)}\big(\mathcal{F}_R(Q)\big)$ follows from Proposition 2.11; indeed, if $\sigma \in \mathcal{F}(Q)$ normalizes $\mathcal{F}_R(Q)$, it follows from condition 2.11.1 that there is $\rho \in \mathcal{F}(P,R)$ extending σ, which forces $\rho(R) = R$ since R contains $Q{\cdot}C_P(Q)$.

Conversely, assume that $\mathcal{F}$ fulfills the two conditions above and let Q be a subgroup of P fully normalized and fully centralized in $\mathcal{F}$; first of all, we claim that $\mathcal{F}_P(Q)$ is a Sylow p-subgroup of $\mathcal{F}(Q)$. Indeed, let $\xi : N_P(Q) \to P$ be an $\mathcal{F}$-morphism such that $N' = \xi\big(N_P(Q)\big)$ is fully normalized in $\mathcal{F}$ (cf. Proposition 2.7); arguing by induction on $|P : Q|$, we may assume that $\mathcal{F}_P(N')$ is already a Sylow p-subgroup of $\mathcal{F}(N')$ and therefore, up to a modification of our choice of ξ, we still may assume that $\mathcal{F}_P(N')_{Q'}$ is a Sylow p-subgroup of $\mathcal{F}(N')_{Q'}$.

On the other hand, since Q is fully centralized in $\mathcal{F}$, N' contains $C_P(Q')$. Then, it follows from condition 2.13.2 that the image of $\mathcal{F}_P(N')_{Q'}$ is a Sylow p-subgroup of $N_{\mathcal{F}(Q')}\big(\mathcal{F}_{N'}(Q')\big)$; but, this image is contained in $\mathcal{F}_{N'}(Q')$; consequently, $\mathcal{F}_{N'}(Q')$ is a Sylow p-subgroup of its normalizer in $\mathcal{F}(Q')$, so that it is a Sylow p-subgroup of $\mathcal{F}(Q')$ and therefore $\mathcal{F}_P(Q)$ is a Sylow p-subgroup of $\mathcal{F}(Q)$.

Now, let Q be a subgroup of P, $\varphi : Q \to P$ an $\mathcal{F}$-morphism such that $Q' = \varphi(Q)$ is fully normalized and fully centralized in $\mathcal{F}$, and R a subgroup of $N_P(Q)$ such that $Q \subset R$ and ${}^{\varphi}\mathcal{F}_R(Q) \subset \mathcal{F}_P(Q')$; according to Proposition 2.7, there is $\psi \in \mathcal{F}\big(P, N_P(Q)\big)$ such that $Q'' = \psi(Q)$ is fully normalized and fully centralized in $\mathcal{F}$, and then, by condition 2.13.1, there is an $\mathcal{F}$-isomorphism $\zeta : N_P(Q'') \cong N_P(Q')$ such that $\zeta(Q'') = Q'$; in particular, since $\mathcal{F}$ is divisible, there is $\sigma' \in \mathcal{F}(Q')$ fulfilling $\varphi(u) = \sigma'\Big(\zeta\big(\psi(u)\big)\Big)$ for any $u \in Q$. That is to say, setting $R' = \zeta\big(\psi(R)\big)$, the p-groups

$$\mathcal{F}_{R'}(Q') \quad \text{and} \quad \sigma' \circ \mathcal{F}_{R'}(Q') \circ \sigma'^{-1} = {}^{\varphi}\mathcal{F}_R(Q) \qquad 2.13.3$$

are contained in $\mathcal{F}_P(Q')$ which is a Sylow p-subgroup of $\mathcal{F}(Q')$.

At this point, it suffices to prove that there is $\theta' \in \mathcal{F}(P, R')$ fulfilling $\theta'(Q') = Q'$ and $\theta'(u') = \sigma'(u')$ for any $u' \in Q'$; indeed, in this case the $\mathcal{F}$-morphism $R \to P$ sending $v \in R$ to $\theta'\Big(\zeta\big(\psi(v)\big)\Big)$ extends φ and the corollary follows from Proposition 2.11. We apply the *Alperin Fusion Theorem* (cf. Theorem 2.6 in [28, Ch. 7] or chapter 5 below) to the group $\mathcal{F}(Q')$ and argue by induction on the *length* of the decomposition of σ' in Alperin's statement (see 5.14 below).

That is to say, we may assume that $\sigma' = \tau' \circ \sigma''$ where $\sigma'' \in \mathcal{F}(Q')$ already fulfills $\sigma''(u') = \theta''(u')$ for some $\theta'' \in \mathcal{F}(P,R')$ and any $u' \in Q'$, and where $\tau' \in \mathcal{F}(Q')$ normalizes $\mathcal{F}_{T'}(Q')$ for some subgroup T' of $N_P(Q')$ containing $\theta''(R')\cdot C_P(Q')$; then, it follows from the induction hypothesis and condition 2.13.2 that τ' can be lifted to some $\rho' \in \mathcal{F}(T')_{Q'}$, so that we have $\rho'\big(\theta''(u')\big) = \sigma'(u')$ for any $u' \in Q'$; since $\theta''(R') \subset T'$, we are done.

2.14 The first successful test for Frobenius P-categories — in order to mimic finite groups and, more generally, blocks — is the existence of *normalizers* and *centralizers* of the objects. Let $\mathcal{F}$ be a divisible P-category, Q a subgroup of P and K a subgroup of $\mathrm{Aut}(Q)$; assume that Q is fully K-normalized in $\mathcal{F}$; the K-*normalizer* — or the *centralizer* if $K = \{1\}$ and the *normalizer* if $K = \mathrm{Aut}(Q)$ — *of* Q *in* $\mathcal{F}$ is the $N_P^K(Q)$-subcategory $N_{\mathcal{F}}^K(Q)$ where, for any pair of subgroups R and T of $N_P^K(Q)$, the set of morphisms from T to R is the set of $\varphi \in \mathcal{F}(R,T)$ fulfilling the condition

2.14.1 *There are an* $\mathcal{F}$*-morphism* $\psi : Q\cdot T \to Q\cdot R$ *and an element* $\chi \in K$ *such that* $\chi(u) = \psi(u)$ *for any* $u \in Q$ *and that* $\psi(v) = \varphi(v)$ *for any* $v \in T$.

2.15 It is quite clear that $N_{\mathcal{F}}^K(Q)$ is an $N_P^K(Q)$-category. Note that, since $\mathcal{F}$ is divisible, it is easy to check that the isomorphism $T \cong \varphi(T)$ determined by $\varphi \in \mathcal{F}(R,T)$ belongs to $\big(N_{\mathcal{F}}^K(Q)\big)(R,T)$ and therefore $N_{\mathcal{F}}^K(Q)$ is divisible too; actually, $\big(N_{\mathcal{F}}^K(Q)\big)(R,T)$ also coincides with the set of group homomorphisms $\varphi : T \to R$ fulfilling condition 2.14.1. Moreover, if Q' is a subgroup of P , K' is a subgroup of $\mathrm{Aut}(Q')$, Q' is fully K'-normalized in $\mathcal{F}$ and there is an $\mathcal{F}$-isomorphism $Q\cdot N_P^K(Q) \cong Q'\cdot N_P^{K'}(Q')$ mapping Q onto Q' and K onto K' , from the divisibility condition it is straightforward to prove that such an $\mathcal{F}$-isomorphism induces an equivalence of categories between $N_{\mathcal{F}}^K(Q)$ and $N_{\mathcal{F}}^{K'}(Q')$.

Proposition 2.16 *Let* $\mathcal{F}$ *be a Frobenius* P*-category,* Q *a subgroup of* P *and* K *a subgroup of* $\mathrm{Aut}(Q)$ *. If* Q *is fully* K*-normalized in* $\mathcal{F}$ *then* $N_{\mathcal{F}}^K(Q)$ *is a Frobenius* $N_P^K(Q)$*-category.*

Proof: Set $\mathcal{F}' = N_{\mathcal{F}}^K(Q)$ and $P' = N_P^K(Q)$; since P' is obviously fully normalized in $\mathcal{F}'$, denoting by K' the subgroup of automorphisms of $Q\cdot P'$ which stabilize Q and P' , and act on Q *via* elements of K , it follows from Lemma 2.17 below that $Q\cdot P'$ is fully K'-normalized in $\mathcal{F}$ and then, it follows from Proposition 2.11 that $K' \cap \mathcal{F}_P(Q\cdot P')$ is a Sylow p-subgroup of $K' \cap \mathcal{F}(Q\cdot P')$; but, by the very definition of $\mathcal{F}'$ (cf. 2.14.1), the restriction to P' determines a surjective homomorphism $K' \cap \mathcal{F}(Q\cdot P') \to \mathcal{F}'(P')$ mapping $K' \cap \mathcal{F}_P(Q\cdot P')$ onto $\mathcal{F}_{P'}(P')$, so that $\mathcal{F}'$ fulfills condition 2.8.1.

Let R be a subgroup of P' , J a subgroup of $\mathrm{Aut}(R)$ and $\varphi : R \to P'$ an $\mathcal{F}'$-morphism such that $\varphi(R)$ is fully ${}^{\varphi}J$-normalized in $\mathcal{F}'$, and assume that

$\psi : Q{\cdot}R \to Q{\cdot}P'$ is an $\mathcal{F}$-morphism and χ an element of K fulfilling

$$\psi(v) = \varphi(v) \quad \text{and} \quad \psi(u) = \chi(u) \qquad 2.16.1$$

for any $v \in R$ and any $u \in Q$ (cf. 2.14.1); set $T = Q{\cdot}R$ and denote by J' the subgroup of automorphisms of T which stabilize Q and R, and act on them *via* elements of K and J respectively.

According to Lemma 2.17 below, $\psi(T)$ is fully ${}^{\psi}J'$-normalized in $\mathcal{F}$ and therefore, it follows from condition 2.8.2 that there are an $\mathcal{F}$-morphism $\zeta : T{\cdot}N_P^M(T) \to P$ and an element $\mu \in J'$ such that $\zeta(w) = \psi\big(\mu(w)\big)$ for any $w \in T$; in particular, for any $u \in Q$ we get $\zeta(u) = \chi\big(\mu(u)\big)$ and therefore the action of ζ on Q determines an element of K; thus, $\zeta\big(R{\cdot}N_P^{J'}(T)\big)$ also normalizes Q and acts on it *via* a subgroup of K. Consequently, since $N_P^{J'}(T) = N_{P'}^J(R)$ (cf. Lemma 2.17 below) and the action of μ on R determines an element λ of J, the restriction of ζ over $R{\cdot}N_{P'}^J(R)$ determines an $\mathcal{F}'$-morphism $R{\cdot}N_{P'}^J(R) \to P'$ and, for any $v \in R$, we have

$$\zeta(v) = \psi\big(\mu(v)\big) = \varphi\big(\lambda(v)\big) \qquad 2.16.2.$$

This proves that $\mathcal{F}'$ fulfills condition 2.8.2 too. We are done.

Lemma 2.17 *Let $\mathcal{F}$ be a Frobenius P-category, Q a subgroup of P and K a subgroup of* $\mathrm{Aut}(Q)$. *Assume that Q is fully K-normalized in $\mathcal{F}$. Let R be a subgroup of $N_P^K(Q)$ and J a subgroup of* $\mathrm{Aut}(R)$, *and denote by I the subgroup of automorphisms of $Q{\cdot}R$ which stabilize Q and R, and act on them via elements of K and J respectively. Then, we have*

$$N_P^I(Q{\cdot}R) = N_P^J(R) \cap N_P^K(Q) \qquad 2.17.1$$

and if R is fully J-normalized in $N_{\mathcal{F}}^K(Q)$ then $Q{\cdot}R$ is fully I-normalized in $\mathcal{F}$.

Proof: Set $\mathcal{F}' = N_{\mathcal{F}}^K(Q)$, $P' = N_P^K(Q)$ and $T = Q{\cdot}R$; firstly, the equality $N_P^I(T) = N_{P'}^J(R)$ is easily checked and needs no hypothesis on $\mathcal{F}$. Secondly, for any $\mathcal{F}$-morphism $\psi : T{\cdot}N_P^I(T) \to P$, set $Q' = \psi(Q)$ and consider the $\mathcal{F}$-morphism $Q' \to P$ obtained from the composition of the inclusion map $\iota_Q^P : Q \to P$ and the inverse of the isomorphism $Q \cong Q'$ determined by ψ; since Q is fully K-normalized in $\mathcal{F}$, it follows from condition 2.8.2 that there are an $\mathcal{F}$-morphism $\zeta : Q'{\cdot}N_P^{\psi K}(Q') \to P$ and an element $\chi \in K$ such that $\zeta\big(\psi(u)\big) = \chi(u)$ for any $u \in Q$; in particular, we have $\zeta\big(N_P^{\psi K}(Q')\big) \subset P'$ and, since $\psi(Q{\cdot}P')$ is contained in $Q'{\cdot}N_P^{\psi K}(Q')$ and $\mathcal{F}$ is divisible, the homomorphism

$$\eta : T{\cdot}N_P^I(T) = Q{\cdot}\big(R{\cdot}N_{P'}^J(R)\big) \longrightarrow Q{\cdot}P' \qquad 2.17.2$$

mapping $w \in T{\cdot}N_P^I(T)$ on $\zeta\big(\psi(w)\big)$ belongs to $\mathcal{F}\big(Q{\cdot}P', T{\cdot}N_P^I(T)\big)$; moreover, since $\psi(R) \subset N_P^{\psi K}\big(\psi(Q)\big)$, it determines an $\mathcal{F}'$-morphism $R{\cdot}N_{P'}^J(R) \to P'$ (cf. 2.14.1).

Consequently, if we assume that R is fully J-normalized in $\mathcal{F}'$, we get $\eta\big(N_{P'}^J(R)\big) = N_{P'}^{\eta J}\big(\eta(R)\big)$; but, we already have $N_P^I(T) = N_{P'}^J(R)$ and, according to the same equality applied to $\eta(Q) = Q$, ${}^{\eta}K = K$, $\eta(R)$ and ${}^{\eta}J$, we still have $N_P^{\eta I}\big(\eta(T)\big) = N_{P'}^{\eta J}\big(\eta(R)\big)$, so that we get

$$\zeta\Big(\psi\big(N_P^I(T)\big)\Big) = \eta\big(N_P^I(T)\big) = N_P^{\eta I}\big(\eta(T)\big) \supset \zeta\Big(N_P^{\psi I}\big(\psi(T)\big)\Big) \qquad 2.17.3,$$

which forces $\psi\big(N_P^I(T)\big) = N_P^{\psi I}\big(\psi(T)\big)$. We are done.

2.18 It is clear that, in a divisible P-category $\mathcal{F}$, we can iterate the *normalizer construction* until to define the *normalizer* of an $\mathcal{F}$*-chain* — $\mathfrak{C}$*-chains* are introduced in A2.8 for any small category $\mathfrak{C}$ — provided it fulfills a suitable *iterated fully normalized* condition. Explicitly, recall that for any $n \in \mathbb{N}$, an $\mathcal{F}$*-chain* — more precisely called a $(n,\mathcal{F})$*-chain* — is a functor

$$\mathfrak{q} : \Delta_n \longrightarrow \mathcal{F} \qquad 2.18.1$$

from the category Δ_n formed by the objects $0 \le i \le n$ and the morphisms $0 \le j \le i \le n$, with the obvious composition (cf. A2.2); then, arguing by induction on n , we say that $\mathfrak{q}$ is *fully normalized* in $\mathcal{F}$ if $\mathfrak{q}(n)$ is fully normalized in $\mathcal{F}$ and moreover, setting $P' = N_P\big(\mathfrak{q}(n)\big)$ and $\mathcal{F}' = N_{\mathcal{F}}\big(\mathfrak{q}(n)\big)$, in the case where $n \ge 1$ the $\mathcal{F}'$-chain $\mathfrak{q}' : \Delta_{n-1} \to \mathcal{F}'$ mapping $0 \le i \le n-1$ on the image of $\mathfrak{q}(i \bullet n)$, and the Δ_{n-1}-morphisms on the corresponding inclusion maps, is *fully normalized* in $\mathcal{F}'$. Note that, by Proposition 2.7, any $\mathcal{F}$-chain admits a $\mathfrak{ch}(\mathcal{F})$-isomorphic $\mathcal{F}$-chain fully normalized in $\mathcal{F}$.

2.19 Assume that $\mathfrak{q}$ is an $(n,\mathcal{F})$-chain fully normalized in $\mathcal{F}$. In the case where $n \ge 1$, we inductively define

$$N_P(\mathfrak{q}) = N_{P'}(\mathfrak{q}') \quad \text{and} \quad N_{\mathcal{F}}(\mathfrak{q}) = N_{\mathcal{F}'}(\mathfrak{q}') \qquad 2.19.1,$$

and it follows from Proposition 2.16 that if $\mathcal{F}$ is a Frobenius P-category then $N_{\mathcal{F}}(\mathfrak{q})$ is a Frobenius $N_P(\mathfrak{q})$-category. Actually, if $\mathcal{F}$ is a Frobenius P-category, denoting by $\mathcal{F}(\mathfrak{q})$ the group of $\mathfrak{ch}(\mathcal{F})$-automorphisms of $\mathfrak{q}$ — identified with the stabilizer in $\mathcal{F}\big(\mathfrak{q}(n)\big)$ of all the subgroups $\mathrm{Im}\big(\mathfrak{q}(i \bullet n)\big)$ when i runs over Δ_n (cf. A2.8) — it follows from Lemma 2.17 that $\mathfrak{q}(n)$ is fully $\mathcal{F}(\mathfrak{q})$-normalized in $\mathcal{F}$ and that we have

$$N_P(\mathfrak{q}) = N_P^{\mathcal{F}(\mathfrak{q})}\big(\mathfrak{q}(n)\big) \quad \text{and} \quad N_{\mathcal{F}}(\mathfrak{q}) = N_{\mathcal{F}}^{\mathcal{F}(\mathfrak{q})}\big(\mathfrak{q}(n)\big) \qquad 2.19.2.$$

These comments are useful in chapter 18 for the construction of the *localizing functor*.

2.20 But in chapter 5, in order to exhibit a sufficient condition to get a Frobenius P-category from a suitable formulation of our version of the Alperin Fusion Theorem, we are also interested on the so-called *normal* $\mathcal{F}$-chains, which can be reduced from the bottom. Let us say that an $\mathcal{F}$-chain $\mathfrak{q}:\Delta_n \to \mathcal{F}$ is *normal* if the image of $\mathfrak{q}(i\bullet n)$ is normal in $\mathfrak{q}(n)$ for any $i \in \Delta_n$; then, arguing by induction on n, we say that $\mathfrak{q}$ is *fully conormalized* in $\mathcal{F}$ if the image Q of $\mathfrak{q}(0 \bullet n)$ is fully normalized in $\mathcal{F}$ and, setting $P' = N_P(Q)$ and $\mathcal{F}' = N_{\mathcal{F}}(Q)$, in the case where $n \geq 1$ the $\mathcal{F}'$-chain $\mathfrak{q}':\Delta_{n-1} \to \mathcal{F}'$ mapping $0 \leq i \leq n-1$ on the image of $\mathfrak{q}(i+1\bullet n)$, and the Δ_{n-1}-morphisms on the corresponding inclusions, is *fully conormalized* in $\mathcal{F}'$. Note that, by Proposition 2.7, any $\mathcal{F}$-chain admits a $\mathfrak{ch}(\mathcal{F})$-isomorphic $\mathcal{F}$-chain fully conormalized in $\mathcal{F}$. Once again, if $\mathfrak{q}$ is a normal $(n,\mathcal{F})$-chain fully conormalized in $\mathcal{F}$, in the case where $n \geq 1$ we inductively define

$$N_P(\mathfrak{q}) = N_{P'}(\mathfrak{q}') \quad \text{and} \quad N_{\mathcal{F}}(\mathfrak{q}) = N_{\mathcal{F}'}(\mathfrak{q}') \qquad 2.20.1;$$

if $\mathfrak{q}$ is also fully normalized in $\mathcal{F}$, it is not difficult to check from condition 2.14.1 that both definitions coincide.

Corollary 2.21 *Let $\mathcal{F}$ be a Frobenius P-category. For any normal $\mathcal{F}$-chain $\mathfrak{q}:\Delta_n \to \mathcal{F}$ there is an $\mathcal{F}$-morphism $\varphi:\mathfrak{q}(n) \to P$ such that the image of $\varphi\circ\mathfrak{q}(i\bullet n)$ is fully centralized in $\mathcal{F}$ for any $i \in \Delta_n$. Moreover, any subgroup Q of P having a selfcentralizing subgroup R fully centralized in $\mathcal{F}$ is also fully centralized in $\mathcal{F}$.*

Proof: We may assume that $n \geq 1$ and that $\mathfrak{q}$ is fully conormalized in $\mathcal{F}$, and argue by induction on n; with the notation in 2.20 above, we already know that the image T of $\mathfrak{q}(0\bullet n)$ is fully centralized in $\mathcal{F}$ since it is fully normalized (cf. Proposition 2.11); then, it follows from Proposition 2.16 that $\mathcal{F}' = N_{\mathcal{F}}(T)$ is a Frobenius P'-category, and from the induction hypothesis applied to the normal $\mathcal{F}'$-chain $\mathfrak{q}'$ that there is an $\mathcal{F}'$-morphism $\varphi':\mathfrak{q}'(n-1) \to P'$ such that the image of $\varphi'\circ\mathfrak{q}'(i\bullet n-1)$ is fully centralized in $\mathcal{F}'$ for any $i \in \Delta_{n-1}$; since T is fully normalized in $\mathcal{F}$ and normal in the image of $\varphi'\circ\mathfrak{q}(i+1\bullet n)$, it follows from Lemma 2.17 that this image is also fully centralized in $\mathcal{F}$.

In order to prove the last statement arguing by induction on $|Q:R|$, we clearly may assume that Q normalizes R; in this case, we already know that there is an $\mathcal{F}$-morphism $\varphi:Q \to P$ such that $\varphi(Q)$ and $\varphi(R)$ are both fully centralized in $\mathcal{F}$. But, since ${}^{\varphi}\mathcal{F}_{Q\cdot C_P(R)}(R) = \mathcal{F}_{\varphi(Q)}\big(\varphi(R)\big)$, it follows from condition 2.10 that there is an $\mathcal{F}$-morphism $\rho:Q\cdot C_P(R) \to P$ extending φ; moreover, since $C_Q(R) \subset R$ (cf. 1.9), we necessarily have $\rho(Q) = \varphi(Q)$ and therefore, since $\varphi(R)$ is fully centralized in $\mathcal{F}$, we get

$$\begin{aligned}\rho\big(C_P(Q)\big) &= \rho\big(C_{C_P(R)}(Q)\big) = C_{\rho(C_P(R))}\big(\rho(Q)\big)\\ &= C_{C_P(\varphi(R))}\big(\varphi(Q)\big) = C_P\big(\varphi(Q)\big)\end{aligned} \qquad 2.21.1.$$

Consequently, since $\varphi(Q)$ is fully centralized in $\mathcal{F}$, Q is fully centralized too (cf. Proposition 2.7). We are done.

Chapter 3

The Frobenius P-category of a block

3.1 Let G be a finite group, k an algebraically closed field of characteristic p, b a primitive idempotent of the center $Z(kG)$ of the group algebra kG — so that (b, G) is a *block*, or b is a *block of* G (cf. 1.10) — and (P, e) a *maximal Brauer* (b, G)*-pair* (cf. 1.15). In this chapter, we describe the Frobenius P-category $\mathcal{F}_{(b,G)}$ associated with (b, G); although strictly speaking it depends on the choice of (P, e), it will be immediately clear that the different choices determine equivalent categories and that this dependence is ultimately unessential.

3.2 Let Q and R be a pair of subgroups of P; recall that there are unique blocks f and g of the respective centralizers $C_G(Q)$ and $C_G(R)$ fulfilling (cf. 1.15)

$$(Q, f) \subset (P, e) \quad \text{and} \quad (R, g) \subset (P, e) \qquad 3.2.1;$$

the uniqueness of f and g allows us to define $\mathcal{F}_{(b,G)}(Q, R)$ as the set of group homomorphisms $\varphi : R \to Q$ such that there is $x \in G$ fulfilling

$$(R, g) \subset (Q, f)^x \quad \text{and} \quad \varphi(v) = xvx^{-1} \text{ for any } v \in R \qquad 3.2.2;$$

the transitivity of the *inclusion* between Brauer (b, G)-pairs (cf. 1.15) guarantees that the sets $\mathcal{F}_{(b,G)}(Q, R)$ define a subcategory $\mathcal{F}_{(b,G)}$ of $\mathfrak{Gr}$ over the set of subgroups of P and it is clear that all these group homomorphisms are injective. Moreover, for any $u \in P$, we still have $(Q, f)^u \subset (P, e)$ and therefore if $R \subset Q^u$ then we get $(R, g) \subset (Q, f)^u$ (cf. 1.15.3); hence, we have

$$\mathcal{F}_P(Q, R) \subset \mathcal{F}_{(b,G)}(Q, R) \qquad 3.2.3,$$

which proves that $\mathcal{F}_{(b,G)}$ is a P-category. Note that if b is the so-called *principal block of* G then $\mathcal{F}_{(b,G)}$ coincides with $\mathcal{F}_G$ (cf. Theorem 3.13 in [3]).

Proposition 3.3 *The P-category $\mathcal{F}_{(b,G)}$ is divisible.*

Proof: Let Q, R and T be subgroups of P, φ an element of $\mathcal{F}_{(b,G)}(Q, R)$ and ψ an element of $\mathcal{F}_{(b,G)}(Q, T)$ fulfilling $\psi(T) \subset \varphi(R)$, so that there is a group homomorphism $\eta : T \to R$ such that $\psi = \varphi \circ \eta$; thus, denoting by f, g and h the respective blocks of the centralizers $C_G(Q)$, $C_G(R)$ and $C_G(T)$ such that (P, e) contains (Q, f), (R, g) and (T, h), there are $x, y \in G$ fulfilling

$$\begin{aligned} &(R, g)^{x^{-1}} \subset (Q, f) \quad \text{and} \quad \varphi(v) = xvx^{-1} \text{ for any } v \in R \\ &(T, h)^{y^{-1}} \subset (Q, f) \quad \text{and} \quad \psi(w) = ywy^{-1} \text{ for any } w \in T \end{aligned} \qquad 3.3.1;$$

consequently, we get $x\eta(w)x^{-1} = ywy^{-1}$ for any $w \in T$, so that $T^{y^{-1}} \subset R^{x^{-1}}$ and therefore, by the uniqueness of the blocks $g^{x^{-1}}$ and $h^{y^{-1}}$, we still get $(T,h)^{y^{-1}} \subset (R,g)^{x^{-1}}$ (cf. 1.15.3), so that η and $x^{-1}y$ fulfill condition 3.2.2; thus, η belongs to $\mathcal{F}_{(b,G)}(R,T)$.

3.4 Let (Q,f) be a Brauer (b,G)-pair contained in (P,e) and K a subgroup of $\mathrm{Aut}(Q)$; since the converse image $N_G^K(Q,f)$ of K in $N_G(Q,f)$ contains $C_G(Q)$, it is clear that f is also a block of $N_G^K(Q,f)$ (cf. 1.19); moreover, the uniqueness of f implies that $N_P(Q)$ — and *a fortiori* $N_P^K(Q)$ — fixes f. On the other hand, since $C_G\big(Q{\cdot}N_P^K(Q)\big)$ is a normal subgroup of

$$C_{N_G^K(Q,f)}\big(N_P^K(Q)\big) = C_G\big(N_P^K(Q)\big) \cap N_G^K(Q,f) \qquad 3.4.1,$$

the block g of $C_G\big(Q{\cdot}N_P^K(Q)\big)$ such that (P,e) contains $(Q{\cdot}N_P^K(Q),g)$ determines a block g^K of $C_{N_G^K(Q,f)}\big(N_P^K(Q)\big)$, so that $(N_P^K(Q),g^K)$ is a Brauer $\big(f,N_G^K(Q,f)\big)$-pair (cf. 1.16).

Proposition 3.5 *With the notation above, Q is fully K-normalized in $\mathcal{F}_{(b,G)}$ if and only if $(N_P^K(Q),g^K)$ is a maximal Brauer $\big(f,N_G^K(Q,f)\big)$-pair.*

Proof: We know that for any $\mathcal{F}_{(b,G)}$-morphism $\psi\colon Q{\cdot}N_P^K(Q) \to P$ there is $y \in G$ fulfilling

$$(Q{\cdot}N_P^K(Q),g)^y \subset (P,e) \quad \text{and} \quad \psi(v) = v^y \text{ for any } v \in Q{\cdot}N_P^K(Q) \qquad 3.5.1;$$

hence, setting $K_y = {}^{\psi}K$ and denoting by g_y the block of $C_G\big(Q^y{\cdot}N_P^{K_y}(Q^y)\big)$ such that (P,e) contains $(Q^y{\cdot}N_P^{K_y}(Q^y),g_y)$, we have (cf. 1.15.3)

$$(Q{\cdot}N_P^K(Q),g)^y \subset (Q^y{\cdot}N_P^{K_y}(Q^y),g_y) \qquad 3.5.2;$$

moreover, it is clear that f^y is a block of $N_G^{K_y}(Q^y)$ and that this inclusion of Brauer (b,G)-pairs implies the inclusion

$$(N_P^K(Q),g^K)^y \subset (N_P^{K_y}(Q^y),(g_y)^K) \qquad 3.5.3$$

of Brauer $\big(f^y,N_G^{K_y}(Q^y,f^y)\big)$-pairs (cf. 1.15) where, as above, $(g_y)^K$ denotes the block of $C_{N_G^{K_y}(Q^y,f^y)}\big(N_P^{K_y}(Q^y)\big)$ determined by g_y. But, if $(N_P^K(Q),g^K)$ is a maximal Brauer $\big(f,N_G^K(Q,f)\big)$-pair then it is clear that $(N_P^K(Q),g^K)^y$ is a maximal Brauer $\big(f^y,N_G^{K_y}(Q^y,f^y)\big)$-pair too. In particular, in this case we get

$$\psi\big(N_P^K(Q)\big) = N_P^K(Q)^y = N_P^{K_y}(Q^y) = N_P^{{}^{\psi}K}\big(\psi(Q)\big) \qquad 3.5.4,$$

which proves that Q is fully K-normalized in $\mathcal{F}_{(b,G)}$ (cf. 2.6.2).

Conversely, assume that Q is fully K-normalized in $\mathcal{F}_{(b,G)}$; any maximal Brauer $(f, N_G^K(Q,f))$-pair (T, h^K) containing $(N_P^K(Q), g^K)$ is determined by a Brauer (b,G)-pair $(Q{\cdot}T, h)$ which contains $(Q{\cdot}N_P^K(Q), g)$ (cf. 3.4) and therefore there is $y \in G$ such that $(Q{\cdot}T, h)^y \subset (P, e)$. Thus, we have the inclusions

$$(Q{\cdot}N_P^K(Q), g)^y \subset (Q{\cdot}T, h)^y \subset (P, e) \qquad 3.5.5$$

and, in particular, the group homomorphism $\psi : Q{\cdot}N_P^K(Q) \to P$ mapping $v \in Q{\cdot}N_P^K(Q)$ on v^y belongs to $\mathcal{F}_{(b,G)}\big(P, Q{\cdot}N_P^K(Q)\big)$; hence, since Q is fully K-normalized in $\mathcal{F}_{(b,G)}$, we get

$$N_P^K(Q)^y = \psi\big(N_P^K(Q)\big) = N_P^{\psi K}\big(\psi(Q)\big) = N_P^{K_y}(Q^y) \qquad 3.5.6.$$

But, it is clear that $T^y \subset P \cap N_G^{K_y}(Q^y, f^y)$ which, with equality 3.5.6, forces the equality $(N_P^K(Q), g^K) = (T, h^K)$. We are done.

Corollary 3.6 *With the notation above, assume that Q is fully K-normalized in $\mathcal{F}_{(b,G)}$. Then, the K-normalizer $N_{\mathcal{F}_{(b,G)}}^K(Q)$ of Q in $\mathcal{F}_{(b,G)}$ coincides with the $N_P^K(Q)$-category $\mathcal{F}_{(f,N_G^K(Q,f))}$ associated with the block f of $N_G^K(Q,f)$.*

Proof: We know that $(N_P^K(Q), g^K)$ is a maximal Brauer $(f, N_G^K(Q,f))$-pair and therefore that $\mathcal{F}_{(f,N_G^K(Q,f))}$ is a $N_P^K(Q)$-category; moreover, if (T, h^K) is a Brauer $(f, N_G^K(Q,f))$-pair contained in $(N_P^K(Q), g^K)$ and $\psi : T \to N_P^K(Q)$ is an $\mathcal{F}_{(f,N_G^K(Q,f))}$-morphism, there is $n \in N_G^K(Q,f)$ fulfilling

$$(T, h^K) \subset (N_P^K(Q), g^K)^n \quad \text{and} \quad \psi(w) = nwn^{-1} \text{ for any } w \in T \qquad 3.6.1.$$

On the other hand, we know that the block h of $C_G(Q{\cdot}T)$ such that the Brauer (b,G)-pair $(Q{\cdot}T, h)$ is contained in $(Q{\cdot}N_P^K(Q), g)$, determines the block h^K (cf. 1.15). In particular, we have $h^K = \sum_{z\in Z} h^z$ for a suitable subset $Z \subset C_G(T) \cap N_G^K(Q,f)$ and, similarly, it follows from the inclusion in 3.6.1 that there is $z \in Z$ such that we have the inclusion of (b,G)-pairs

$$(Q{\cdot}T, h^z) \subset (Q{\cdot}N_P^K(Q), g)^n \qquad 3.6.2;$$

hence, the group homomorphism $\xi : Q{\cdot}T \to Q{\cdot}N_P^K(Q)$ mapping $u \in Q{\cdot}T$ on $nz^{-1}uzn^{-1}$ is an $\mathcal{F}_{(b,G)}$-morphism; finally, since ξ extends ψ and the restriction of ξ to Q induces an element of K, ψ is an $N_{\mathcal{F}_{(b,G)}}^K(Q)$-morphism (cf. 2.14.1).

Conversely, if $\psi : T \to N_P^K(Q)$ is an $N_{\mathcal{F}_{(b,G)}}^K(Q)$-morphism, we know that ψ can be extended to an $\mathcal{F}_{(b,G)}$-morphism $\xi : Q{\cdot}T \to Q{\cdot}N_P^K(Q)$ such that its restriction to Q induces an element of K (cf. 2.14.1); thus, with the notation above, there is $x \in G$ fulfilling

$$(Q{\cdot}T, h) \subset (Q{\cdot}N_P^K(Q), g)^x \quad \text{and} \quad \xi(u) = xux^{-1} \text{ for any } u \in Q{\cdot}T \qquad 3.6.3;$$

in particular, x normalizes Q and induces on it an element of K; moreover, we have $(Q,f) \subset (P,e)^x$ and therefore $f = f^x$ (cf. 1.15.3). Then, with the notation above, $(N_P^K(Q), g^K)^x$ is a maximal Brauer $(f, N_G^K(Q,f))$-pair, whereas (T, h^K) is a Brauer $(f, N_G^K(Q,f))$-pair and we still have (cf. 1.15)

$$(T, h^K) \subset (N_P^K(Q), g^K)^x \qquad 3.6.4;$$

now, inclusion 3.6.4 shows that ψ is a $\mathcal{F}_{(f,N_G^K(Q,f))}$-morphism too.

Consequently, for any subgroup T of $N_P^K(Q)$ we have obtained

$$\big(N^K_{\mathcal{F}_{(b,G)}}(Q)\big)\big(N_P^K(Q),T\big) = \mathcal{F}_{(f,N_G^K(Q,f))}\big(N_P^K(Q),T\big) \qquad 3.6.5$$

and, since the $N_P^K(Q)$-categories $N^K_{\mathcal{F}_{(b,G)}}(Q)$ and $\mathcal{F}_{(f,N_G^K(Q,f))}$ are both divisible (cf. 2.14 and Proposition 3.3), they coincide (cf. 2.4).

Theorem 3.7 *$\mathcal{F}_{(b,G)}$ is a Frobenius P-category.*

Proof: By Proposition 3.3, we already know that $\mathcal{F}_{(b,G)}$ is divisible and, according to condition 3.2.2, we have

$$\mathcal{F}_{(b,G)}(P) \cong N_G(P,e)/C_G(P) \qquad 3.7.1;$$

thus, since p does not divide $|N_G(P,e)/P{\cdot}C_G(P)|$ (cf. 1.16), $\mathcal{F}_{(b,G)}$ fulfills condition 2.8.1.

Let Q be a subgroup of P, K a subgroup of $\mathrm{Aut}(Q)$ containing $\mathcal{F}_Q(Q)$ and $\varphi: Q \to P$ an $\mathcal{F}_{(b,G)}$-morphism such that $\varphi(Q)$ is fully ${}^{\varphi}K$-normalized in $\mathcal{F}_{(b,G)}$; in particular, denoting by f the block of $C_G(Q)$ such that (P,e) contains (Q,f), there is $x \in G$ fulfilling (cf. 3.2.2)

$$(Q,f)^x \subset (P,e) \quad \text{and} \quad \varphi(u) = u^x \text{ for any } u \in Q \qquad 3.7.2.$$

As above, denote by g the block of $C_G\big(N_P^K(Q)\big)$ such that (P,e) contains the Brauer (b,G)-pair $(N_P^K(Q), g)$.

Moreover, setting $Q' = \varphi(Q)$, $K' = {}^{\varphi}K$ and $f' = f^x$ in kG and denoting by g' the block of $C_G\big(N_P^{K'}(Q')\big)$ such that (P,e) contains the Brauer (b,G)-pair $(N_P^{K'}(Q'), g')$ (cf. 1.15), it follows from Proposition 3.5 that $(N_P^{K'}(Q'), g')$ is a maximal Brauer $(f', N_G^{K'}(Q',f'))$-pair; thus, since $(N_P^K(Q), g)^x$ is also a Brauer $(f', N_G^{K'}(Q',f'))$-pair, there is $n' \in N_G^{K'}(Q',f')$ such that (cf. 1.15)

$$(N_P^K(Q), g)^{xn'} \subset (N_P^{K'}(Q'), g') \qquad 3.7.3.$$

But, this inclusion is an inclusion of Brauer (b,G)-pairs too (cf. 1.19). Consequently, (P,e) contains the Brauer (b,G)-pair $(N_P^K(Q),g)^{xn'}$ and therefore the group homomorphism $\psi : N_P^K(Q) \to P$ mapping $v \in N_P^K(Q)$ on $v^{xn'}$ is an $\mathcal{F}_{(b,G)}$-morphism; moreover, denoting by $\chi : Q \cong Q$ the group automorphism mapping $u \in Q$ on $u^{xn'x^{-1}}$, it is clear that χ belongs to K and that we have $\psi(u) = \varphi\big(\chi(u)\big)$ for any $u \in Q$. So, $\mathcal{F}_{(b,G)}$ fulfills condition 2.8.2 too. We are done.

3.8 As a matter of fact, a similar argument covers a more general situation, namely the case of the *p-permutation G-algebras* (cf. 1.14) introduced in [11]. Precisely, let A be a G-algebra over k such that a Sylow p-subgroup of G stabilizes a basis of A; consider a point α of G on A (cf. 1.12) and the Brauer (α,G)-pairs (cf. 1.14). Recall that the inclusion between the local pointed groups on A (cf. 1.12) contained in G_α induces an inclusion between the Brauer (α,G)-pairs (cf. 1.15), and that all the maximal Brauer (α,G)-pairs are G-conjugate (cf. Theorem 1.14 in [11]); denote by (P,e_A) one of them and let us describe the P-category $\mathcal{F}_{(\alpha,G)}$ associated with A and α.

3.9 Let Q and R be a pair of subgroups of P; once again, recall that there are unique primitive idempotents f_A of $Z\big(A(Q)\big)$ and g_A of $Z\big(A(R)\big)$ fulfilling (cf. 1.15)

$$(Q,f_A) \subset (P,e_A) \quad \text{and} \quad (R,g_A) \subset (P,e_A) \qquad 3.9.1;$$

the uniqueness of f_A and g_A allows us to define $\mathcal{F}_{(\alpha,G)}(Q,R)$ as the set of group homomorphisms $\varphi : R \to Q$ such that there is $x \in G$ fulfilling

$$(R,g_A) \subset (Q,f_A)^x \quad \text{and} \quad \varphi(v) = xvx^{-1} \text{ for any } v \in R \qquad 3.9.2;$$

the transitivity of the *inclusion* between Brauer (α,G)-pairs guarantees that the sets $\mathcal{F}_{(\alpha,G)}(Q,R)$ define a subcategory $\mathcal{F}_{(\alpha,G)}$ of $\mathfrak{Gr}$ over the set of subgroups of P and it is clear that all these group homomorphisms are injective. Moreover, for any $u \in P$, we still have $(Q,f_A)^u \subset (P,e_A)$ and therefore if $R \subset Q^u$ then we get $(R,g_A) \subset (Q,f_A)^u$ (cf. 1.15); hence, we still have

$$\mathcal{F}_P(Q,R) \subset \mathcal{F}_{(\alpha,G)}(Q,R) \qquad 3.9.3$$

which proves that $\mathcal{F}_{(\alpha,G)}$ is a P-category. As above, it is easily proved that

3.9.4 *$\mathcal{F}_{(\alpha,G)}$ is a divisible P-category.*

Remark 3.10 As in the particular case of the Frobenius P-category $\mathcal{F}_{(b,G)}$, this divisible P-category $\mathcal{F}_{(\alpha,G)}$ depends on the choice of the maximal Brauer (α,G)-pair (P,e_A); although this dependence is unessential, it is sometimes more handy to consider the equivalent category — noted $\mathcal{F}^{\mathfrak{e}}_{(\alpha,G)}$ — where the objects are *all* the Brauer (α,G)-pairs, avoiding the choices of (P,e_A) and of a *functorial section* $\mathcal{F}^{\mathfrak{e}}_{(\alpha,G)} \to \mathcal{F}_{(\alpha,G)}$. This point of view will be very useful in chapter 11.

Proposition 3.11 *Let Q be a subgroup of P, K a subgroup of $\mathrm{Aut}(Q)$ containing $\mathcal{F}_Q(Q)$ and n_A the primitive idempotent of the center of $A\big(N_P^K(Q)\big)$ such that $(N_P^K(Q), n_A) \subset (P, e_A)$. If Q is fully K-normalized in $\mathcal{F}_{(\alpha,G)}$ then there is a point β of $N_G^K(Q, f_A)$ on A such that $N_G^K(Q, f_A)_\beta \subset G_\alpha$ and $(N_P^K(Q), n_A)$ is a maximal Brauer $(\beta, N_G^K(Q, f_A))$-pair.*

Proof: Let (T, m_A) be a Brauer (α, G)-pair which is maximal fulfilling

$$T \subset N_G^K(Q, f_A) \quad \text{and} \quad (N_P^K(Q), n_A) \subset (T, m_A) \qquad 3.11.1;$$

since $m_A \mathrm{Br}_T(\alpha) \neq \{0\}$, we can consider a point β of $N_G^K(Q, f_A)$ on A such that (cf. 1.12)

$$N_G^K(Q, f_A)_\beta \subset G_\alpha \quad \text{and} \quad m_A \mathrm{Br}_T(\beta) \neq \{0\} \qquad 3.11.2;$$

then, we still have $n_A \mathrm{Br}_{N_P^K(Q)}(\beta) \neq \{0\}$ and $f_A \mathrm{Br}_Q(\beta) \neq \{0\}$, and therefore (T, m_A), $(N_P^K(Q), n_A)$ and (Q, f_A) are also Brauer $(\beta, N_G^K(Q, f_A))$-pairs.

Moreover, we already know that we have $(T, m_A)^y \subset (P, e_A)$ for some $y \in G$ (cf. 1.15) and therefore we still have the inclusion

$$(N_P^K(Q), n_A)^y \subset (T, m_A)^y \subset (P, e_A) \qquad 3.11.3;$$

in particular, the group homomorphism $\psi : N_P^K(Q) \to P$ mapping $v \in N_P^K(Q)$ on v^y belongs to $\mathcal{F}_{(\alpha,G)}\big(P, N_P^K(Q)\big)$; hence, since we assume that Q is fully K-normalized in $\mathcal{F}_{(\alpha,G)}$, setting $K_y = {}^\psi K$ we get

$$N_P^K(Q)^y = \psi\big(N_P^K(Q)\big) = N_P^{{}^\psi K}\big(\psi(Q)\big) = N_P^{K_y}(Q^y) \qquad 3.11.4.$$

But, it is clear that $T^y \subset N_P^{K_y}(Q^y)$ which, with equality 3.11.4, forces the equality $(N_P^K(Q), n_A) = (T, m_A)$; in particular, $(N_P^K(Q), n_A)$ is a maximal Brauer $(\beta, N_G^K(Q, f_A))$-pair. We are done.

3.12 A specially interesting case where we find the above situation concerns the group algebra kG but considered as an H-algebra with respect to some subgroup H of G. That is to say, in 3.8 above, we replace G by a subgroup H and consider $A = kGb$ as an H-algebra; then, for any point β of H on kGb, choosing a maximal Brauer (b, G)-pair (Q, f) such that $f\mathrm{Br}_Q(\beta) \neq \{0\}$, we get the divisible Q-category $\mathcal{F}_{(\beta,H)}$ and, from the point of view of Remark 3.10, it is quite clear that we simply have an *inclusion* of categories

$$\mathcal{F}^{\mathfrak{c}}_{(\beta,H)} \subset \mathcal{F}^{\mathfrak{c}}_{(b,G)} \qquad 3.12.1.$$

Nevertheless, any element $x \in G$ such that $(Q, f) \subset (P, e)^x$ determines an *inclusion* functor from $\mathcal{F}_{(\beta,H)}$ to $\mathcal{F}_{(b,G)}$ noted

$$\mathfrak{i}^b_\beta : \mathcal{F}_{(\beta,H)} \longrightarrow \mathcal{F}_{(b,G)} \qquad 3.12.2,$$

and, for another element $x' \in G$ such that $(Q,f) \subset (P,e)^{x'}$, $x'x^{-1}$ determines a natural isomorphism between both inclusion functors. All this framework will be widely employed in chapter 11.

3.13 A last remark; whenever $H = N_G^K(Q,f)$ where (Q,f) is a Brauer (b,G)-pair and K a subgroup of $\mathrm{Aut}(Q)$, assuming that Q is fully K-normalized in $\mathcal{F}_{(b,G)}$ and denoting by ν the point of $N_G^K(Q,f)$ on kGb determined by f (cf. 1.19), we have

$$\mathcal{F}_{(\nu,N_G^K(Q,f))} = N^K_{\mathcal{F}_{(b,G)}}(Q) = \mathcal{F}_{(f,N_G^K(Q,f))} \qquad 3.13.1.$$

Chapter 4

Nilcentralized, selfcentralizing and intersected objects in Frobenius P-categories

4.1 Let P be a finite p-group and $\mathcal{F}$ a divisible P-category such that $\mathcal{F}_P(P)$ is a Sylow p-subgroup of $\mathcal{F}(P)$. As a matter of fact, in order to check whether or not $\mathcal{F}$ is a Frobenius P-category, we only need to check condition 2.11.1 above over a restricted set of subgroups of P — the so-called *$\mathcal{F}$-intersected subgroups* — that we introduce in 4.11 below.

4.2 They appear inside a larger family of subgroups of P — the *$\mathcal{F}$-selfcentralizing* subgroups (cf. 4.8 below) — which has an interest on its own: denoting by $\mathcal{F}^{^{sc}}$ the full subcategory of $\mathcal{F}$ over this family, all the morphisms in its *exterior quotient* $\tilde{\mathcal{F}}^{^{sc}}$ (cf. 1.3) are *epimorphisms* (see Proposition 4.6 below). In chapter 6 we exhibit more interesting features of this exterior quotient, and give a meaningful interpretation of the *$\mathcal{F}$-intersected* subgroups (see Proposition 6.16 below).

4.3 Actually, in Proposition 4.6 the key property of the $\mathcal{F}$-selfcentralizing subgroups holds for a larger set of subgroups, namely for the *$\mathcal{F}$-nilcentralized* subgroups of P; we say that a subgroup Q of P is *$\mathcal{F}$-nilcentralized* if, for any $\varphi \in \mathcal{F}(P,Q)$ such that $Q' = \varphi(Q)$ is fully centralized in $\mathcal{F}$, the $C_P(Q')$-categories $C_{\mathcal{F}}(Q')$ (cf. 2.14) and $\mathcal{F}_{C_P(Q')}$ (cf. 1.8) coincide — loosely speaking, this amounts to saying that Q has a *nilpotent centralizer* in $\mathcal{F}$, which motivates the terminology.

4.4 Note that, by 2.15 above, if $\mathcal{F}$ is a Frobenius P-category and this condition is fulfilled for some $\varphi \in \mathcal{F}(P,Q)$, it is fulfilled for any. As a matter of fact, this condition is preserved by the *central quotients of $\mathcal{F}$* (cf. Proposition 12.3 below and Theorem 3.4 in [28, Ch. 5]), whereas it need not be the case for the condition defining the $\mathcal{F}$-selfcentralizing subgroups (cf. 4.8 below).

Proposition 4.5 *Assume that $\mathcal{F}$ is a Frobenius P-category. A subgroup Q of P which contains an $\mathcal{F}$-nilcentralized subgroup R is $\mathcal{F}$-nilcentralized too. If moreover R is fully centralized in $\mathcal{F}$, then Q is fully centralized in $\mathcal{F}$ too.*

Proof: Arguing by induction on $|Q\!:\!R|$, we may assume that R is normal in Q; then, by Proposition 2.7, there is an $\mathcal{F}$-morphism $\xi : Q{\cdot}C_P(Q) \to P$ such that $\xi(R)$ is both fully centralized and fully normalized in $\mathcal{F}$; hence, up to replacing R and Q by $\xi(R)$ and $\xi(Q)$, we may assume that R is fully centralized in $\mathcal{F}$.

Choose $\varphi \in \mathcal{F}(P,Q)$ such that $Q' = \varphi(Q)$ is fully centralized in $\mathcal{F}$ (cf. Proposition 2.7) and set $R' = \varphi(R)$; then, since we assume that R is fully centralized in $\mathcal{F}$ and we clearly have

$$^{\varphi}\mathcal{F}_Q(R) = \mathcal{F}_{Q'\cdot C_P(R')}(R') \qquad 4.5.1,$$

according to statement 2.10.1 there is an $\mathcal{F}$-morphism $\psi : Q'\cdot C_P(R') \to P$ such that $\psi(\varphi(v)) = v$ for any $v \in R$; moreover, setting $U = \mathcal{F}_Q(R)$, we already know that R is also fully U-normalized in $\mathcal{F}$ (cf. Proposition 2.11) and we clearly have $N_P^U(R) = Q\cdot C_P(R)$, so that we can consider the Frobenius $Q\cdot C_P(R)$-category $N_{\mathcal{F}}^U(R)$ (cf. Proposition 2.16).

In particular, the composition of φ with the restriction of ψ to Q' determines an element of $\big(N_{\mathcal{F}}^U(R)\big)\big(Q\cdot C_P(R), Q\big)$; but, we are assuming that $C_{\mathcal{F}}(R) = \mathcal{F}_{C_P(R)}$ and therefore we also have $N_{\mathcal{F}}^U(R) = \mathcal{F}_{Q\cdot C_P(R)}$, as it is easily checked; consequently, up to a modification of our choice of ψ, we may assume that $\psi(\varphi(u)) = u$ for any $u \in Q$ and then we have

$$\psi\big(C_P(Q')\big) \subset C_P(Q) \qquad 4.5.2,$$

which forces the equality (cf. 2.6.2) and proves that Q is fully centralized too. Since it is now clear that $C_{\mathcal{F}}(Q)$ is a subcategory of $C_{\mathcal{F}}(R) = \mathcal{F}_{C_P(R)}$, it is then straightforward to prove that $C_{\mathcal{F}}(Q) = \mathcal{F}_{C_P(Q)}$. We are done.

Proposition 4.6 *Assume that $\mathcal{F}$ is a Frobenius P-category. Let Q be a subgroup of P and R an $\mathcal{F}$-nilcentralized subgroup of Q. If $\varphi, \varphi' \in \mathcal{F}(P,Q)$ fulfill $\varphi(v) = \varphi'(v)$ for any $v \in R$ and the subgroup $R' = \varphi(R) = \varphi'(R)$ is fully centralized in $\mathcal{F}$, then there is $u \in C_P(R')$ fulfilling $\varphi'(v) = \varphi(v)^u$ for any $v \in Q$.*

Proof: We argue by induction on $|Q:R|$ and may assume that $R \neq Q$; moreover, from the divisibility of $\mathcal{F}$, up to replacing Q and R by $\varphi'(Q)$ and $\varphi'(R)$ we may assume that φ' is the inclusion map and that R is fully centralized in $\mathcal{F}$. Set $N = N_Q(R)$ and $U = \mathcal{F}_P(R)$ so that $N_P^U(R) = N_P(R)$; then, according to Proposition 2.11, R is also fully U-normalized in $\mathcal{F}$ and, since R is $\mathcal{F}$-nilcentralized, it is not difficult to see that $N_{\mathcal{F}}^U(R) = \mathcal{F}_{N_P(R)}$ (cf. Theorem 3.2 in [28, Ch. 5]).

In this situation, since $\varphi(v) = v$ for any $v \in R$, the restriction of φ to N determines an element of $\big(N_{\mathcal{F}}^U(R)\big)\big(N_P(R), N\big)$ and therefore there is $w \in N_P(R)$ such that $\varphi(v)^w = v$ for any $v \in N$; but, by Proposition 4.5, N is $\mathcal{F}$-nilcentralized and fully centralized in $\mathcal{F}$ too; consequently, it follows from the induction hypothesis that there is an element $v \in C_P(N) \subset C_P(R)$ such that $\varphi(u)^{wv} = u$ for any $u \in Q$. We are done.

Corollary 4.7 *Assume that $\mathcal{F}$ is a Frobenius P-category. Let Q be a subgroup of P and R an $\mathcal{F}$-nilcentralized subgroup of Q fully centralized in $\mathcal{F}$; denote by $\mathcal{F}(Q)_R$ the stabilizer of R in $\mathcal{F}(Q)$. Then, the kernel of the restriction homomorphism $\mathcal{F}(Q)_R \to \mathcal{F}(R)$ coincides with $\mathcal{F}_{C_P(R)}(Q)$.*

Proof: If $\sigma \in \mathcal{F}(Q)$ fulfills $\sigma(v) = v$ for any $v \in R$, then Proposition 4.6 states that σ belongs to $\mathcal{F}_{C_P(R)}(Q)$. We are done.

4.8 We say that a subgroup Q of P is $\mathcal{F}$-*selfcentralizing*† if we have

$$C_P\big(\varphi(Q)\big) \subset \varphi(Q) \qquad 4.8.1$$

for any $\varphi \in \mathcal{F}(P,Q)$; then, Q is clearly fully centralized in $\mathcal{F}$ and $\mathcal{F}$-nilcentralized; moreover, any subgroup R of P such that $\mathcal{F}(R,Q) \neq \emptyset$ is $\mathcal{F}$-selfcentralizing too. Recall that $\tilde{\mathcal{F}}$ denotes the *exterior quotient* of $\mathcal{F}$ (cf. 1.3).

Corollary 4.9 *Assume that $\mathcal{F}$ is a Frobenius P-category. For any $\mathcal{F}$-selfcentralizing subgroups Q, R and T of P and any $\mathcal{F}$-morphism $\varphi : Q \to R$, the map $\tilde{\mathcal{F}}(T,R) \to \tilde{\mathcal{F}}(T,Q)$ determined by the composition with $\tilde{\varphi}$ is injective. In particular, any $\tilde{\mathcal{F}}$-morphism from an $\mathcal{F}$-selfcentralizing subgroup of P is an epimorphism.*

Proof: If two elements $\tilde{\psi}, \tilde{\psi}' \in \tilde{\mathcal{F}}(T,R)$ fulfill $\tilde{\psi} \circ \tilde{\varphi} = \tilde{\psi}' \circ \tilde{\varphi}$, we may choose representatives ψ of $\tilde{\psi}$ and ψ' of $\tilde{\psi}'$ such that $\psi \circ \varphi = \psi' \circ \varphi$, and then it follows from Proposition 4.6 that there is $z \in Z(Q)$ fulfilling

$$\psi'(v) = \psi(v^z) = \psi(v)^{\psi(z)} \qquad 4.9.1$$

for any $v \in R$, so that $\tilde{\psi}' = \tilde{\psi}$. We are done.

4.10 Note that, for any subgroup Q of P fully centralized in $\mathcal{F}$, $Q{\cdot}C_P(Q)$ is clearly $\mathcal{F}$-selfcentralizing; in particular, a subgroup Q of P fully centralized in $\mathcal{F}$ is $\mathcal{F}$-selfcentralizing if and only if we have $C_P(Q) = Z(Q)$, namely if Q is *selfcentralizing* in P (cf. 1.9). If $\mathcal{F}$ is a Frobenius P-category, a subgroup Q of P fully normalized in $\mathcal{F}$ is also fully centralized (cf. Proposition 2.11) and therefore it is $\mathcal{F}$-selfcentralizing if and only if it is selfcentralizing in P; in this case, by statement 2.10.1 and Proposition 4.6, if R is a subgroup of $N_P(Q)$ containing Q then any $\varphi \in \mathcal{F}(P,Q)$ such that ${}^{\varphi}\mathcal{F}_R(Q) \subset \mathcal{F}_P\big(\varphi(Q)\big)$ can be extended to R in a unique way up to conjugation by $Z(Q)$.

4.11 Actually, there is a maximal subgroup R of $N_P(Q)$ fulfilling the inclusion ${}^{\varphi}\mathcal{F}_R(Q) \subset \mathcal{F}_P\big(\varphi(Q)\big)$ for any $\varphi \in \mathcal{F}(P,Q)$; namely, for any subgroup Q of P fully centralized in $\mathcal{F}$, we denote by $\mathcal{I}_{\mathcal{F}}(Q)$ the converse image in $N_P(Q)$ of the intersection

$$\bar{\mathcal{I}}_{\mathcal{F}}(Q) = \bigcap_{\varphi \in \mathcal{F}(P,Q)} {}^{\varphi^*}\mathcal{F}_P\big(\varphi(Q)\big) \qquad 4.11.1,$$

where $\varphi^* : \varphi(Q) \cong Q$ denotes the inverse of the isomorphism induced by φ, and we say that Q is an $\mathcal{F}$-*intersected subgroup* of P whenever $\mathcal{I}_{\mathcal{F}}(Q) = Q$.

† Called an $\mathcal{F}$-*centric subgroup* in [13]

Note that $\mathcal{I}_{\mathcal{F}}(Q)$ is an $\mathcal{F}$-intersected subgroup of P, and that an $\mathcal{F}$-intersected subgroup of P is $\mathcal{F}$-selfcentralizing. We have the following criterion†.

Theorem 4.12 *A divisible P-category $\mathcal{F}$ is a Frobenius P-category if and only if the following conditions hold:*

4.12.1 *$\mathcal{F}_P(P)$ is a Sylow p-subgroup of $\mathcal{F}(P)$.*

4.12.2 *If Q is an $\mathcal{F}$-intersected subgroup of P, R is a subgroup of $N_P(Q)$ containing Q and $\varphi: Q \to P$ is an $\mathcal{F}$-morphism fulfilling ${}^{\varphi}\mathcal{F}_R(Q) \subset \mathcal{F}_P\big(\varphi(Q)\big)$ then there is an $\mathcal{F}$-morphism $\psi: R \to P$ extending φ.*

4.12.3 *Any divisible P-category $\mathcal{F}'$ fulfilling $\mathcal{F}'(P,Q) \supset \mathcal{F}(P,Q)$ for every $\mathcal{F}$-intersected subgroup Q of P contains $\mathcal{F}$.*

Proof: Conditions 4.12.1 and 4.12.2 are clearly necessary. In a Frobenius P-category $\mathcal{F}'$, condition 4.12.3 is necessary too since any $\mathcal{F}'$-morphism to P from a subgroup Q of P fully centralized in $\mathcal{F}'$ can be extended to $\mathcal{I}_{\mathcal{F}'}(Q)$ (cf. statement 2.10.1) which is an $\mathcal{F}'$-intersected subgroup (cf. 4.11).

For any pair Q and Q' of $\mathcal{F}$-isomorphic subgroups of P, consider the set $\mathcal{F}'(Q',Q)$ of elements $\varphi \in \mathcal{F}(Q',Q)$ such that there are subgroups U and U' of P, $\mathcal{F}$-isomorphic to Q and Q', which are both fully centralized and fully normalized in $\mathcal{F}$ and admit $\mathcal{F}$-morphisms

$$N_P(Q) \xrightarrow{\lambda} N_P(U) \quad , \quad \mathcal{I}_{\mathcal{F}}(U) \xrightarrow{\sigma} \mathcal{I}_{\mathcal{F}}(U') \quad , \quad N_P(U') \xleftarrow{\lambda'} N_P(Q') \qquad 4.12.4$$

fulfilling

$$\lambda(Q) = U \quad , \quad \sigma(U) = U' \quad , \quad U' = \lambda'(Q') \qquad 4.12.5$$

and $\lambda'\big(\varphi(u)\big) = \sigma\big(\lambda(u)\big)$ for any $u \in Q$.

Note that $\mathcal{F}_P(Q',Q) \subset \mathcal{F}'(Q',Q)$; indeed, if φ is the conjugation by some $u \in P$ then, choosing an $\mathcal{F}$-morphism $\lambda: N_P(Q) \to P$ such that $U = \lambda(Q)$ is fully centralized and fully normalized in $\mathcal{F}$ (cf. Proposition 2.7), it is clear that $U' = uUu^{-1} = {}^{u}\lambda(Q')$ is also fully centralized and fully normalized in $\mathcal{F}$, and it suffices to consider $\lambda' = {}^{u}\lambda$ and the isomorphism $\mathcal{I}_{\mathcal{F}}(U) \cong \mathcal{I}_{\mathcal{F}}(U')$ determined by the conjugation by u.

On the other hand, if Q is an $\mathcal{F}$-intersected subgroup of P then we have $\mathcal{F}'(Q',Q) = \mathcal{F}(Q',Q)$; indeed, it suffices to choose $\mathcal{F}$-morphisms

$$\zeta : N_P(Q) \longrightarrow P \quad \text{and} \quad \zeta' : N_P(Q') \longrightarrow P \qquad 4.12.6$$

such that the subgroups $U = \zeta(Q)$ and $U' = \zeta'(Q')$ are both fully normalized and fully centralized in $\mathcal{F}$ (cf. Proposition 2.7); in this case, we have the equalities $\mathcal{I}_{\mathcal{F}}(U) = U$ and $\mathcal{I}_{\mathcal{F}}(U') = U'$, and therefore the existence of an $\mathcal{F}$-morphism $\sigma: U \to U'$ fulfilling the conditions above is clear.

† The interested reader will see that the hypothesis (*) in [15, Theorem 2.2, 331-339] forces the set of subgroups $\mathcal{H}$ considered there to include *all* the $\mathcal{F}$-intersected subgroups. Our proof here already appeared in [46] — quoted with the number 15 in [15] — except that in our old notes we considered all the $\mathcal{F}$-selfcentralizing subgroups, and $U{\cdot}C_P(U)$ instead of $\mathcal{I}_{\mathcal{F}}(U)$.

From now on, we assume that $\mathcal{F}$ fulfills the conditions above. More generally, for any pair Q and R of subgroups of P we set

$$\mathcal{F}'(Q,R) = \bigcup_{\varphi\in\mathcal{F}(Q,R)} \iota_{\varphi(R)}^{Q}\circ\mathcal{F}'\big(\varphi(R),R\big) \qquad 4.12.7,$$

where $\iota_{\varphi(R)}^{Q}$ denotes the corresponding inclusion map (cf. 1.9), and we will prove that $\mathcal{F}'$ is a Frobenius P-category, so that $\mathcal{F}'=\mathcal{F}$ by condition 4.12.3. Since $\mathcal{F}'(P,P)=\mathcal{F}(P)$, it suffices to prove that $\mathcal{F}'$ is a category and fulfills conditions 2.3.1 and 2.8.2. First of all, with the notation above we claim that if φ is an element of $\mathcal{F}'(Q',Q)$ then φ^{-1} belongs to $\mathcal{F}'(Q,Q')$; indeed, since U is fully centralized in $\mathcal{F}$, we have $\sigma\big(U{\cdot}C_P(U)\big)=U'{\cdot}C_P(U')$ and therefore we still have $\sigma\big(\mathcal{I}_{\mathcal{F}}(U)\big)=\mathcal{I}_{\mathcal{F}}(U')$ (cf. 4.11.1), so that it suffices to consider the triple $(\lambda',\sigma^{-1},\lambda)$ of $\mathcal{F}$-morphisms.

Let Q'' be a third subgroup of P, $\mathcal{F}$-isomorphic to Q and Q', and consider a homomorphism $\varphi'\in\mathcal{F}'(Q'',Q')$; in order to prove that $\varphi'\circ\varphi$ belongs to $\mathcal{F}'(Q'',Q)$, it follows from our argument above that we may assume Q, Q' and Q'' are *not* $\mathcal{F}$-intersected groups. *Mutatis mutandis* we have subgroups V and V' of P, $\mathcal{F}$-isomorphic to Q' and Q'', which are both fully centralized and fully normalized in $\mathcal{F}$, and admit $\mathcal{F}$-morphisms

$$N_P(Q')\stackrel{\mu}{\longrightarrow}N_P(V)\ ,\ \mathcal{I}_{\mathcal{F}}(V)\stackrel{\tau}{\longrightarrow}\mathcal{I}_{\mathcal{F}}(V')\ ,\ N_P(V')\stackrel{\mu'}{\longleftarrow}N_P(Q'') \quad 4.12.8$$

fulfilling

$$\mu(Q')=V\quad,\quad\tau(V)=V'\quad,\quad V'=\mu'(Q'') \qquad 4.12.9$$

and $\mu'\big(\varphi'(u')\big)=\tau\big(\mu(u')\big)$ for any $u'\in Q'$; in particular, denoting by λ'^* the inverse of the group isomorphism $N_P(Q')\cong\lambda'\big(N_P(Q')\big)$ induced by λ', we have the $\mathcal{F}$-morphism

$$\mu\circ\lambda'^*:\lambda'\big(N_P(Q')\big)\longrightarrow N_P(V) \qquad 4.12.10$$

which induces an $\mathcal{F}$-isomorphism $\theta:U'\cong V$.

But, arguing by induction on $|P:Q|$, we may assume that the set $\mathfrak{X}$ of all the subgroups of P of smaller index fulfills the hypothesis in Lemma 4.13 below; then, since $\lambda'\big(N_P(Q')\big)$ belongs to $\mathfrak{X}$, since $\mu\circ\lambda'^*$ extends θ and since V is also fully $\bar{\mathcal{I}}_{\mathcal{F}}(V)$-normalized in $\mathcal{F}$ (cf. 2.10), it follows from this lemma applied to U', to $\bar{\mathcal{I}}_{\mathcal{F}}(U')$ and to $\iota_V^P\circ\theta$ that θ can be extended to some $\mathcal{F}$-morphism $\rho:\mathcal{I}_{\mathcal{F}}(U')\to\mathcal{I}_{\mathcal{F}}(V)$†; then, the existence of U, V', λ, μ' and the $\mathcal{F}$-morphism $\mathcal{I}_{\mathcal{F}}(U)\to\mathcal{I}_{\mathcal{F}}(V')$ mapping $u\in U$ on $\tau\Big(\rho\big(\sigma(u)\big)\Big)$ proves that $\varphi'\circ\varphi$ belongs to $\mathcal{F}'(Q'',Q)$.

† This argument has been scratched in the proof of [52, Theorem 3.8].

Let R and R' be subgroups of P respectively containing Q and Q', and assume that $\psi \in \mathcal{F}'(R', R)$ fulfills $\psi(Q) = Q'$; we claim that the $\mathcal{F}$-isomorphism $\varphi: Q \cong Q'$ induced by ψ (cf. 2.4) belongs to $\mathcal{F}'(Q', Q)$. We may assume that Q and Q' are *not* $\mathcal{F}$-intersected groups, that we have $|R:Q| \neq 1$ and that R and R' respectively normalize Q and Q'. We argue by induction on $|R:Q|$ and we already know that there are $\zeta \in \mathcal{F}(P, R)$ and $\zeta' \in \mathcal{F}(P, R')$ such that $V = \zeta(Q)$ and $V' = \zeta'(Q')$ are both fully centralized and fully normalized in $\mathcal{F}$ (cf. Proposition 2.7).

Once again, since R and R' belong to $\mathfrak{X}$, it follows from Lemma 4.13 below, applied to Q, to $\mathrm{Aut}(Q)$ and to $\iota_Q^P \circ \varphi$, that there are $\mathcal{F}$-morphisms

$$\nu : N_P(Q) \longrightarrow N_P(V) \quad \text{and} \quad \nu' : N_P(Q') \longrightarrow N_P(V') \qquad 4.12.11$$

fulfilling $\nu(Q) = V$ and $\nu'(Q') = V'$; moreover, it is clear that the $\mathcal{F}$-isomorphism $\omega: \nu(R) \cong \nu'(R')$ defined by $\omega\big(\nu(v)\big) = \nu'\big(\psi(v)\big)$ for any $v \in R$ maps V onto V' and, since V' is fully centralized in $\mathcal{F}$, it is also fully $\bar{\mathcal{I}}_{\mathcal{F}}(V')$-normalized (cf. 2.10); then, it follows from Lemma 4.13 below, applied to V, to $\bar{\mathcal{I}}_{\mathcal{F}}(V)$ and to the restriction of $\iota^P_{\nu'(R')} \circ \omega$ to V, that there is an $\mathcal{F}$-isomorphism $\eta: \mathcal{I}_{\mathcal{F}}(V) \cong \mathcal{I}_{\mathcal{F}}(V')$ fulfilling $\eta\big(\nu(u)\big) = \nu'\big(\psi(u)\big)$ for any $u \in Q$; now, the existence of V, V', ν, ν' and η proves that φ belongs to $\mathcal{F}'(Q', Q)$.

In conclusion, if R and T are subgroups of P, φ is an element of $\mathcal{F}'(R, Q)$ and ψ an element of $\mathcal{F}'(T, R)$, we claim that $\psi \circ \varphi$ belongs to $\mathcal{F}'(T, Q)$; indeed, setting $Q' = \varphi(Q)$ and $Q'' = \psi(Q')$, and denoting by $\varphi_*: Q \cong Q'$ and $\psi_*: Q' \cong Q''$ the corresponding $\mathcal{F}$-isomorphisms, it follows from our definition that φ_* belongs to $\mathcal{F}'(Q', Q)$ and, by the arguments above, we already know that ψ_* and $\psi_* \circ \varphi_*$ respectively belong to $\mathcal{F}'(Q'', Q')$ and to $\mathcal{F}'(Q'', Q)$; hence, $\psi \circ \varphi$ belongs to $\mathcal{F}'(T, Q)$ (cf. definition 4.12.7).

It remains to prove that $\mathcal{F}'$ fulfills condition 2.8.2; let K be a subgroup of $\mathrm{Aut}(Q)$ containing $\mathcal{F}_Q(Q)$ and $\varphi \in \mathcal{F}'(P, Q)$ such that $Q' = \varphi(Q)$ is fully K'-normalized in $\mathcal{F}'$ where we set $K' = {}^{\varphi}K$; actually, since $\bar{\mathcal{I}}_{\mathcal{F}}(Q')$ is normal in $\mathcal{F}(Q')$, setting $\hat{K}' = \bar{\mathcal{I}}_{\mathcal{F}}(Q')\cdot\big(K' \cap \mathcal{F}(Q')\big)$ it is easily checked that Q' is also fully $\hat{K}'$-normalized in $\mathcal{F}$ (cf. 2.10) and therefore we may assume that K also contains $\bar{\mathcal{I}}(Q)$. Firstly assume that Q is an $\mathcal{F}$-intersected group; since $\mathcal{F}_P(Q') \cap K'$ is a Sylow p-subgroup of $\mathcal{F}(Q') \cap K'$ (cf. Lemma 2.12 and the induction hypothesis), there is $\chi \in \mathcal{F}(Q) \cap K$ such that

$$^{\varphi\circ\chi}\big(\mathcal{F}_P(Q) \cap K\big) \subset \mathcal{F}_P(Q') \cap K' \qquad 4.12.12;$$

then, according to condition 4.12.2, there is an $\mathcal{F}$-morphism $\zeta: N_P^K(Q) \to P$ extending $\varphi \circ \chi$; in this case†, we are done.

† This case has been forgotten in the proof of [52, Theorem 3.8].

Otherwise, since the isomorphism $\varphi_* : Q \cong Q'$ induced by φ belongs to $\mathcal{F}'(Q', Q)$, as in 4.12.4 we have subgroups U and U' of P, $\mathcal{F}$-isomorphic to Q and Q', which are both fully centralized and fully normalized in $\mathcal{F}$, and admit $\mathcal{F}$-morphisms

$$N_P(Q) \xrightarrow{\lambda} N_P(U) \ , \ \mathcal{I}_{\mathcal{F}}(U) \xrightarrow{\sigma} \mathcal{I}_{\mathcal{F}}(U') \ , \ N_P(U') \xleftarrow{\lambda'} N_P(Q') \qquad 4.12.13$$

fulfilling equalities 4.12.5 and $\lambda'\big(\varphi(u)\big) = \sigma\big(\lambda(u)\big)$ for any $u \in Q$; set

$$R = \mathcal{I}_{\mathcal{F}}(U) \quad \text{and} \quad R' = \mathcal{I}_{\mathcal{F}}(U') \qquad 4.12.14$$

and denote by λ'^* the inverse of the $\mathcal{F}$-isomorphism $N_P(Q') \cong \lambda'\big(N_P(Q')\big)$ induced by λ'.

Since U is fully centralized in $\mathcal{F}$ and we have ${}^\sigma\bar{\mathcal{I}}_{\mathcal{F}}(U) = \bar{\mathcal{I}}_{\mathcal{F}}(U')$ (cf. definition 4.11.1), we still have $\sigma(R) = R'$ (cf. 2.6.2); moreover, since Q' is fully K'-normalized in $\mathcal{F}'$, we get (cf. 2.6.2)

$$\lambda'\big(N_P^{K'}(Q')\big) = N_P^{\lambda' K'}(U') \supset C_P(U') \qquad 4.12.15$$

and therefore we still get

$$R' = \mathcal{I}_{\mathcal{F}}(U') \subset \lambda'\Big(N_P\big(\lambda'^*(U')\big)\Big) = \lambda'\big(N_P(Q')\big) \qquad 4.12.16;$$

hence, there is an $\mathcal{F}$-morphism $\psi : R \to P$ fulfilling $\psi(v) = \lambda'^*\big(\sigma(v)\big)$ for any $v \in R$. Finally, since R belongs to $\mathfrak{X}$, since $U \triangleleft R$ and $\mathcal{F}_R(U) \subset {}^\lambda K$, and since $\psi(U) = Q'$ is fully ${}^\psi({}^\lambda K)$-normalized in $\mathcal{F}'$, it follows from Lemma 4.13 below applied to U, to ${}^\lambda K$ and to the restriction to U of ψ that there are an $\mathcal{F}$-morphism $\xi : N_P^{\lambda K}(U) \to P$ and an element $\chi \in K$ such that, for any $u \in Q$, we have

$$\xi\big(\lambda(u)\big) = \psi\Big(\lambda\big(\chi(u)\big)\Big) = \lambda'^*\Big(\sigma\Big(\lambda\big(\chi(u)\big)\Big)\Big) = \varphi\big(\chi(u)\big) \qquad 4.12.17.$$

We are done.

Lemma 4.13. *Let $\mathfrak{X}$ be a nonempty set of subgroups Q of P such that any subgroup T of P fulfilling $\mathcal{F}(T, Q) \neq \emptyset$ belongs to $\mathfrak{X}$, and that, for any subgroup K of* Aut(Q) *and any $\mathcal{F}$-morphism $\varphi : Q \to P$ such that $\varphi(Q)$ is fully ${}^\varphi K$-normalized in $\mathcal{F}$, the following condition holds:*

4.13.1. *There are an $\mathcal{F}$-morphism $\zeta : Q \cdot N_P^K(Q) \to P$ and an element $\chi \in K$ such that ζ extends $\varphi \circ \chi$.*

A subgroup R of P, a subgroup J of Aut(R) *and an $\mathcal{F}$-morphism $\psi : R \to P$ such that $\psi(R)$ is fully ${}^\psi J$-normalized in $\mathcal{F}$ fulfill condition* 4.13.1 *provided there are $Q \in \mathfrak{X}$ having R as a normal subgroup and stabilizing J, and an $\mathcal{F}$-morphism $\eta : Q \to P$ extending ψ.*

Proof: Let Q be an element of $\mathfrak{X}$ such that $R \subset Q \subset N_P(R)$, J a Q-stable subgroup of $\mathrm{Aut}(R)$ such that $\psi(R)$ is fully ${}^{\psi}J$-normalized in $\mathcal{F}$ and $\eta : Q \to P$ an $\mathcal{F}$-morphism extending ψ; we argue by induction on $|N_P(R) : Q|$ and may assume that $N_P^J(R) \not\subset Q$ or, equivalently, that $Q \neq Q{\cdot}N_P^J(R)$. Denote by K the converse image of $J{\cdot}\mathcal{F}_Q(R)$ in the stabilizer $\mathrm{Aut}(Q)_R$ of R in $\mathrm{Aut}(Q)$.

Choose an $\mathcal{F}$-morphism $\zeta : N_P^K(Q) \to P$ in such a way that $\zeta(R)$ is fully ${}^{\zeta}\big(J{\cdot}\mathcal{F}_Q(R)\big)$-normalized and $\zeta(Q)$ is fully ${}^{\zeta}K$-normalized in $\mathcal{F}$. This is possible since, applying Proposition 2.7 to $Q \subset N_P^K(Q)$, there is an $\mathcal{F}$-morphism $\zeta' : N_P^K(Q) \to P$ such that $\bar{Q} = \zeta'(Q)$ is fully ${}^{\zeta'}K$-normalized in $\mathcal{F}$; then, setting $\bar{J} = {}^{\zeta'}\big(J{\cdot}\mathcal{F}_Q(R)\big)$ and $\bar{K} = {}^{\zeta'}K$, and applying Proposition 2.7 to

$$\bar{R} = \zeta'(R) \subset N_P^{\bar{K}}(\bar{Q}) \subset N_P^{\bar{J}}(\bar{R}) \qquad 4.13.2,$$

there is an $\mathcal{F}$-morphism $\zeta'' : N_P^{\bar{K}}(\bar{Q}) \to P$ such that $\zeta''(\bar{R})$ is fully ${}^{\zeta''}\bar{J}$-normalized in $\mathcal{F}$; since we have (cf. 2.6.2)

$$\zeta''\big(N_P^{\bar{K}}(\bar{Q})\big) = N_P^{{}^{\zeta''}\bar{K}}\big(\zeta''(\bar{Q})\big) \qquad 4.13.3,$$

$\zeta''(\bar{Q})$ is also fully ${}^{\zeta''}\bar{K}$-normalized in $\mathcal{F}$.

Set $R' = \psi(R)$, $Q' = \eta(Q)$ and $J' = {}^{\psi}J$; it follows from condition 4.13.1 applied to Q', to $K' = {}^{\eta}K$ and to the group homomorphism $Q' \to P$ determined by η and ζ that there are an $\mathcal{F}$-morphism $\xi : N_P^{K'}(Q') \to P$ and an element $\chi \in K$ fulfilling $\xi\big(\eta(u)\big) = \zeta\big(\chi(u)\big)$ for any $u \in Q$; in particular, we have $\xi(R') = \zeta(R)$ since $\chi(R) = R$, and we set $R'' = \xi(R') = \zeta(R)$ and $J'' = {}^{\zeta}J$.

On the other hand, we still have

$$Q \neq N_P(Q) \cap Q{\cdot}N_P^J(R) = N_P^K(Q) \quad \text{and} \quad Q' \neq N_P^{K'}(Q') \qquad 4.13.4$$

and therefore, since R'' is $J''{\cdot}{}^{\zeta}\mathcal{F}_Q(R)$-fully normalized in $\mathcal{F}$, it follows from the induction hypothesis applied to R, to $N_P^K(Q)$ and to the restriction of ζ, and further to R', to $N_P^{K'}(Q')$ and to the restriction of ξ, that there are $\mathcal{F}$-morphisms $\alpha : R{\cdot}N_P^J(R) \to P$ and $\alpha' : R'{\cdot}N_P^{J'}(R') \to P$, and elements θ and θ' of J such that, for any $v \in R$, we have

$$\alpha(v) = \zeta\big(\theta(v)\big) \quad \text{and} \quad \alpha'\big(\psi(v)\big) = \xi\Big(\psi\big(\theta'(v)\big)\Big) = \zeta\Big(\chi\big(\theta'(v)\big)\Big) \qquad 4.13.5.$$

Moreover, since we are assuming that R' is fully J'-normalized in $\mathcal{F}$, we get (cf. 2.6.2)

$$\alpha\big(R{\cdot}N_P^J(R)\big) \subset R''{\cdot}N_P^{J''}(R'') = \alpha'\big(R'{\cdot}N_P^{J'}(R')\big) \qquad 4.13.6$$

and therefore, denoting by $\omega: R{\cdot}N_P^J(R) \to R'{\cdot}N_P^{J'}(R')$ the $\mathcal{F}$-morphism fulfilling $\alpha' \circ \omega = \alpha$ (cf. 2.3.1), for any $v \in R$ we still get

$$\zeta\big(\theta(v)\big) = \alpha'\big(\omega(v)\big) = \zeta\Big(\chi\Big(\big(\theta' \circ \psi^*\big)\big(\omega(v)\big)\Big)\Big) \qquad 4.13.7$$

where $\psi^* \in \mathcal{F}(R, R')$ is the inverse of the isomorphism $R \cong R'$ determined by ψ; thus, since ζ is injective and the image of χ in $\mathrm{Aut}(R)$ is contained in $J{\cdot}\mathcal{F}_Q(R)$, there are $\theta'' \in J$ and $u \in Q$ such that from equality 4.13.7 we obtain (cf. 1.8)

$$\theta(v) = \big({}^{\chi}\theta' \circ \theta'' \circ \kappa_R(u) \circ \psi^*\big)\big(\omega(v)\big) \qquad 4.13.8.$$

Finally, since η extends ψ, for any $v \in R$ we still obtain (cf. 1.8)

$$\psi\big((\theta''^{-1} \circ {}^{\chi}\theta'^{-1} \circ \theta)(v)\big) = \big(\kappa_{P, R'\cdot N_P^{J'}(R')}\big(\eta(u)\big) \circ \omega\big)(v) \qquad 4.13.9$$

and, since $\theta''^{-1} \circ {}^{\chi}\theta'^{-1} \circ \theta$ belongs to J and $\kappa_{P, R'\cdot N_P^{J'}(R')}\big(\eta(u)\big) \circ \omega$ is an $\mathcal{F}$-morphism from $R{\cdot}N_P^J(R)$ to P, equality 4.13.9 proves that R, J and ψ fulfill condition 4.13.1. We are done.

Chapter 5

Alperin fusions in Frobenius P-categories

5.1 Let P be a finite p-group and $\mathcal{F}$ a divisible P-category such that $\mathcal{F}_P(P)$ is a Sylow p-subgroup of $\mathcal{F}(P)$. In this chapter, we prove that the framework in [50, Appendix] can be translated to this abstract setting and that a suitable reformulation of the main statement in that Appendix is then *equivalent* to saying that $\mathcal{F}$ is a *Frobenius P-category*. As we explain in the Introduction (cf. I3), the origin of the concepts and the results below goes back to [35], where we formulate the first systematic treatment of the *Alperin Fusion Theorem* [1].

5.2 It is well-known that the Alperin Fusion Theorem [1] can be applied to compute the *cohomology groups* of a finite group G from the *cohomology groups* of its p-subgroups (cf. §2 in [35, Ch. III]) — precisely, they can be computed from the corresponding *contravariant* functors from $\mathcal{F}_G$ to the category of finitely generated Abelian groups $\mathfrak{Ab}$. But, as a matter of fact, when dealing with *contravariant* functors $\mathfrak{a}$ from $\mathcal{F}$ to $\mathfrak{Ab}$ (cf. 1.2), it is handy to consider the category $\mathbb{Z}\mathcal{F}$ still defined over the set of the subgroups of P where, for any pair of subgroups Q and R of P, the set of morphisms from R to Q is the *free $\mathbb{Z}$-module* $\mathbb{Z}\mathcal{F}(Q,R)$ over $\mathcal{F}(Q,R)$, with the *distributive* composition extending the composition in $\mathcal{F}$; indeed, it is clear that $\mathfrak{a}$ can be extended to a *contravariant* functor

$$\mathbb{Z}\mathfrak{a} : \mathbb{Z}\mathcal{F} \longrightarrow \mathfrak{Ab} \qquad 5.2.1.$$

5.3 Then, for any two different elements $\varphi, \varphi' \in \mathcal{F}(Q,R)$, we call *$\mathcal{F}$-dimorphism* from R to Q the difference $\varphi' - \varphi$ in the $\mathbb{Z}$-module $\mathbb{Z}\mathcal{F}(Q,R)$; moreover, let us call *Alperin $\mathcal{F}$-fusions* the $\mathcal{F}$-dimorphisms with $Q = P$; it is clear that the set of $\mathcal{F}$-dimorphisms is stable by left and right composition with $\mathcal{F}$-morphisms; note that, for any $\varphi \in \mathcal{F}(Q,R)$, the family $\{\varphi' - \varphi\}_{\varphi' \in \mathcal{F}(Q,R) - \{\varphi\}}$ is a $\mathbb{Z}$-basis of the kernel of the evident *augmentation* $\mathbb{Z}$-linear map

$$\varepsilon_{Q,R} : \mathbb{Z}\mathcal{F}(Q,R) \longrightarrow \mathbb{Z} \qquad 5.3.1$$

sending any $\varphi \in \mathcal{F}(Q,R)$ to 1. The next elementary lemma relates any decomposition of an $\mathcal{F}$-dimorphism with the *partially defined linear combinations* introduced in [35, Ch. III]. Moreover, note that equalities 5.4.2 below coincide with Alperin's original formulation in the case where $Q = P$, φ is the inclusion map $\iota_R^P : R \to P$, and for any $i \in I$, we have $\mu_i = \iota_{Q_i}^P$, $Q_i = R_i$ and $\varphi_i = \mathrm{id}_{R_i}$ (cf. 2 in [28, Ch. 7]).

Lemma 5.4 *With the notation above, let $\{Q_i\}_{i\in I}$ and $\{R_i\}_{i\in I}$ be finite families of subgroups of P and, for any $i \in I$, $\varphi'_i - \varphi_i$ an $\mathcal{F}$-dimorphism from R_i to Q_i and $\mu_i : Q_i \to Q$ and $\nu_i : R \to R_i$ two $\mathcal{F}$-morphisms. We have*

$$\varphi' - \varphi = \sum_{i\in I} \mu_i \circ (\varphi'_i - \varphi_i) \circ \nu_i \qquad 5.4.1$$

if and only if there are $n \in \mathbb{N}$ and an injective map $\sigma : \Delta_n \to I$ fulfilling

$$\begin{aligned} \varphi &= \mu_{\sigma(0)} \circ \varphi_{\sigma(0)} \circ \nu_{\sigma(0)} \\ \mu_{\sigma(\ell-1)} \circ \varphi'_{\sigma(\ell-1)} \circ \nu_{\sigma(\ell-1)} &= \mu_{\sigma(\ell)} \circ \varphi_{\sigma(\ell)} \circ \nu_{\sigma(\ell)} \text{ for any } 1 \le \ell \le n \qquad 5.4.2 \\ \mu_{\sigma(n)} \circ \varphi'_{\sigma(n)} \circ \nu_{\sigma(n)} &= \varphi' . \end{aligned}$$

Proof: It is clear that equalities 5.4.2 imply equality 5.4.1. Conversely, equality 5.4.1 is obviously equivalent to

$$\varphi' + \sum_{i\in I} \mu_i \circ \varphi_i \circ \nu_i = \varphi + \sum_{i\in I} \mu_i \circ \varphi'_i \circ \nu_i \qquad 5.4.3$$

and therefore, since $\varphi' \neq \varphi$, there are $i', i'' \in I$ (the possibility $i' = i''$ is not excluded!) and a bijection $\pi : I - \{i''\} \cong I - \{i'\}$ such that

$$\begin{aligned} \mu_{i'} \circ \varphi_{i'} \circ \nu_{i'} &= \varphi \\ \mu_i \circ \varphi'_i \circ \nu_i &= \mu_{\pi(i)} \circ \varphi_{\pi(i)} \circ \nu_{\pi(i)} \text{ for any } i \in I - \{i'\} \qquad 5.4.4; \\ \varphi' &= \mu_{i''} \circ \varphi'_{i''} \circ \nu_{i''} \end{aligned}$$

then, we inductively define σ setting $\sigma(0) = i'$ and $\sigma(\ell + 1) = \pi\big(\sigma(\ell)\big)$ for any $\ell \in \mathbb{N}$ such that $\sigma(\ell)$ is already defined and different from i'', and we denote by $n \in \mathbb{N}$ the maximal ℓ where σ is defined (so that $\sigma(n) = i''$). It remains to prove that σ is injective; arguing by contradiction, assume that there are $0 \le \ell < \ell' \le n$ such that $\sigma(\ell) = \sigma(\ell')$; that is to say, denoting by $\hat\pi$ the permutation of I extending π, we have $\hat\pi^{\ell}(i') = \hat\pi^{\ell'}(i')$ and therefore we get $i' = \hat\pi^{\ell'-\ell}(i')$, a contradiction.

5.5 On the other hand, by the so-called *Yoneda's Lemma* (cf. [30, §1]), the *contravariant* functor $\mathfrak{h}_{\mathcal{F}} : \mathcal{F} \to \mathfrak{Ab}$ mapping any subgroup Q of P on $\mathbb{Z}\mathcal{F}(P, Q)$ and any morphism $\alpha : R \to Q$ in $\mathcal{F}$ on the group homomorphism $\mathfrak{h}_{\mathcal{F}}(Q) \to \mathfrak{h}_{\mathcal{F}}(R)$ defined by the composition with α, is *projective* in the category of functors $\mathfrak{Fct}(\mathcal{F}^\circ, \mathfrak{Ab})$; moreover, still denoting by $\mathbb{Z} : \mathcal{F} \to \mathfrak{Ab}$ the trivial *contravariant* functor mapping all the $\mathcal{F}$-objects on $\mathbb{Z}$ and all the $\mathcal{F}$-morphisms on $\mathrm{id}_{\mathbb{Z}}$, the family of augmentation maps $\varepsilon_{P,Q}$, when Q runs over the set of subgroups of P, define a surjective *natural map*

$$\varepsilon_{\mathcal{F}} : \mathfrak{h}_{\mathcal{F}} \longrightarrow \mathbb{Z} \qquad 5.5.1$$

and therefore the kernel $\mathfrak{w}_\mathcal{F} = \mathrm{Ker}(\varepsilon_\mathcal{F})$ is the *Heller translate* of the trivial functor $\mathbb{Z}$; in particular, for any *contravariant* functor $\mathfrak{a}: \mathcal{F} \to \mathfrak{Ab}$, we have the exact sequence (cf. §4 and §8 of Ch. V and §1 of Ch. VI in [18])

$$0 \longrightarrow \mathfrak{Nat}(\mathbb{Z}, \mathfrak{a}) \longrightarrow \mathfrak{Nat}(\mathfrak{h}_\mathcal{F}, \mathfrak{a}) \longrightarrow \mathfrak{Nat}(\mathfrak{w}_\mathcal{F}, \mathfrak{a}) \longrightarrow \mathbb{H}^1(\mathcal{F}, \mathfrak{a}) \longrightarrow 0 \quad 5.5.2$$

since the projectivity of $\mathfrak{h}_\mathcal{F}$ forces $\mathbf{Ext}^i_\mathcal{F}(\mathfrak{h}_\mathcal{F}, \mathfrak{a}) = \{0\}$ for any $i \geq 1$ and we have $\mathbf{Ext}^1_\mathcal{F}(\mathbb{Z}, \mathfrak{a}) = \mathbb{H}^1(\mathcal{F}, \mathfrak{a})$ (cf. A3.6.3 and A3.9.2).

5.6 Now, if $\mathfrak{a}: \mathcal{F} \to \mathfrak{Ab}$ is a *contravariant* functor, let us say that a family $\mathcal{S} = \{S_Q\}_Q$ of subsets $S_Q \subset \mathfrak{a}(Q)$, where Q runs over the set of proper subgroups of P, is a *generator family* of $\mathfrak{a}$ whenever, for any proper subgroup Q of P, we have

$$\mathfrak{a}(Q) = \sum_R \sum_{\varphi \in \mathcal{F}(R,Q)} \sum_{a \in S_R} \mathbb{Z}\cdot\big(\mathfrak{a}(\varphi)\big)(a) \quad 5.6.1,$$

where R runs over the set of subgroups of P (such that $|R| \geq |Q|$). Thus, according to Lemma 5.4, in the Frobenius category $\mathcal{F}_G$ associated with a finite group G, it is clear that the genuine purpose of the Alperin Fusion Theorem (cf. [1]) is to describe suitable *generator families of the contravariant functor*

$$\mathfrak{w}_{\mathcal{F}_G} : \mathcal{F}_G \longrightarrow \mathfrak{Ab} \quad 5.6.2.$$

5.7 With the analogous purpose, in the divisible P-category $\mathcal{F}$ let us set $\mathfrak{r}_\mathcal{F}(P) = \mathfrak{w}_\mathcal{F}(P)$ and, for any proper subgroup Q of P, define

$$\mathfrak{r}_\mathcal{F}(Q) = \sum_R \mathfrak{w}_\mathcal{F}(R) \circ \mathbb{Z}\mathcal{F}(R,Q) \quad 5.7.1$$

where R runs over the set of subgroups of P such that $|R| > |Q|$. Note that, since there is $\psi \in \mathcal{F}(P,R)$ such that $\psi(R)$ is fully normalized in $\mathcal{F}$ (cf. Proposition 2.7) and since the inverse of the isomorphism $R \cong \psi(R)$ determined by ψ belongs to $\mathcal{F}\big(R, \psi(R)\big)$ (cf. 2.4), in definition 5.7.1 it suffices to restrict the sum to the subgroups R which are fully normalized in $\mathcal{F}$; moreover, if Q' is a subgroup of P and $\theta \in \mathcal{F}(Q,Q')$ is an isomorphism then we clearly have

$$\mathfrak{r}_\mathcal{F}(Q) \circ \theta = \mathfrak{r}_\mathcal{F}(Q') \quad 5.7.2.$$

We say that Q is *$\mathcal{F}$-essential* when $\mathfrak{r}_\mathcal{F}(Q) \neq \mathfrak{w}_\mathcal{F}(Q)$ and call *$\mathcal{F}$-irreducible* the elements in $\mathfrak{w}_\mathcal{F}(Q) - \mathfrak{r}_\mathcal{F}(Q)$.

5.8 Coherently, the elements of $\mathfrak{r}_\mathcal{F}(Q)$ are called *$\mathcal{F}$-reducible*; actually, any element of $\mathfrak{r}_\mathcal{F}(Q)$ is a sum of a family of $\mathcal{F}$-reducible *Alperin $\mathcal{F}$-fusions* from Q. Considering the canonical map

$$\mathfrak{h}_\mathcal{F}(Q) \longrightarrow \mathfrak{h}_\mathcal{F}(Q)/\mathfrak{r}_\mathcal{F}(Q) = \bar{\mathfrak{h}}_\mathcal{F}(Q) \quad 5.8.1,$$

it is clear that $\mathcal{F}(Q)$ acts on the image $\overline{\mathcal{F}(P,Q)}$ of $\mathcal{F}(P,Q)$ in $\mathfrak{h}_\mathcal{F}(Q)/\mathfrak{r}_\mathcal{F}(Q)$ by composition on the left; we denote by $\mathcal{F}(Q)_{\bar\varphi}$ the stabilizer of the image $\bar\varphi$ of $\varphi \in \mathcal{F}(P,Q)$ and, according to equality 5.7.2, we clearly have

$$^\theta\mathcal{F}(Q')_{\overline{\varphi\circ\theta}} = \mathcal{F}(Q)_{\bar\varphi} \qquad 5.8.2$$

for any $\mathcal{F}$-isomorphism $\theta : Q' \cong Q$. Note that the correspondence mapping Q on $\mathfrak{r}_\mathcal{F}(Q)$ defines a subfunctor $\mathfrak{r}_\mathcal{F}$ of $\mathfrak{w}_\mathcal{F}$. Moreover, if P' is a subgroup of P and $\mathcal{F}'$ a divisible P'-subcategory of $\mathcal{F}$ such that $\mathcal{F}_{P'}(P')$ is a Sylow p-subgroup of $\mathcal{F}'(P')$ then, for any *proper* subgroup Q' of P', it is clear that $\iota_{P'}^P \circ \mathfrak{r}_{\mathcal{F}'}(Q')$ is contained in $\mathfrak{r}_\mathcal{F}(Q')$ and therefore the inclusion $P' \subset P$ induces a $\mathbb{Z}$-module homomorphism

$$\bar{\mathfrak{h}}_{\mathcal{F}'}(Q') \longrightarrow \bar{\mathfrak{h}}_\mathcal{F}(Q') \qquad 5.8.3$$

sending $\overline{\mathcal{F}'(P',Q')}$ to $\overline{\mathcal{F}(P,Q')}$.

Proposition 5.9 *Let $\mathcal{S} = \{S_Q\}_Q$ be a generator family of $\mathfrak{w}_\mathcal{F}$, where Q runs over the set of proper subgroups of P. The family formed by the $\mathcal{F}$-irreducible elements of S_Q, where Q runs over the set of proper subgroups of P, is also a generator family of $\mathfrak{w}_\mathcal{F}$. Moreover, for any $\mathcal{F}$-essential subgroup Q of P, there is $\varphi \in \mathcal{F}(P,Q)$ such that $S_{\varphi(Q)}$ contains an $\mathcal{F}$-irreducible element of $\mathfrak{w}_\mathcal{F}(\varphi(Q))$.*

Proof: Let $\omega \in \mathfrak{w}_\mathcal{F}(Q)$ be $\mathcal{F}$-irreducible and, setting $S_P = \mathfrak{w}_\mathcal{F}(P)$, assume that we have

$$\omega = \sum_R \sum_{\sigma\in S_R} \sigma \circ \alpha_{R,\sigma} \qquad 5.9.1$$

for suitable $\alpha_{R,\sigma} \in \mathbb{Z}\mathcal{F}(R,Q)$, where R runs over the set of subgroups of P; then, necessarily there are a suitable subgroup R of P such that $|R| = |Q|$ and an $\mathcal{F}$-irreducible element $\sigma \in S_R$ such that $0 \neq \alpha_{R,\sigma}$; in particular, we have $R = \varphi(Q)$ for some $\varphi \in \mathcal{F}(P,Q)$.

On the other hand, if $\tau \in S_Q$ is an $\mathcal{F}$-reducible element then either $Q = P$ or we have

$$\tau = \sum_R \sum_{\theta\in\mathfrak{w}_\mathcal{F}(R)} \theta \circ \beta_{R,\theta} \qquad 5.9.2$$

for suitable $\beta_{R,\theta} \in \mathbb{Z}\mathcal{F}(R,Q)$ where R runs over the set of subgroups of P such that $|R| > |Q|$; in the second case, considering an $\mathcal{S}$-decomposition of any $\theta \in \mathfrak{w}_\mathcal{F}(R)$, we still have

$$\tau = \sum_R \sum_{\sigma\in S_R} \sigma \circ \gamma_{R,\sigma} \qquad 5.9.3$$

for suitable $\gamma_{R,\sigma} \in \mathbb{Z}\mathcal{F}(R,Q)$ where R runs over the set of subgroups of P such that $|R| > |Q|$; so that the new family where we replace S_Q by $S_Q - \{\tau\}$ is a generator family of $\mathfrak{w}_\mathcal{F}$ too. We are done.

5.10 Let Q be an $\mathcal{F}$-essential subgroup of P; it is clear that $\mathcal{F}_Q(Q)$ acts trivially on $\overline{\mathcal{F}(P,Q)}$ and, if $\mathcal{F}$ is a Frobenius P-category, we prove below that the action of $\tilde{\mathcal{F}}(Q) = \mathcal{F}(Q)/\mathcal{F}_Q(Q)$ on $\overline{\mathcal{F}(P,Q)}$ is actually transitive and that any nontrivial p-subgroup fixes a unique element in $\overline{\mathcal{F}(P,Q)}$. Then, since we also prove that Q is $\mathcal{F}$-selfcentralizing, it is actually an $\mathcal{F}$-*intersected subgroup* (cf. 4.11).

Theorem 5.11 *Assume that $\mathcal{F}$ is a Frobenius P-category. A subgroup Q of P is $\mathcal{F}$-essential if and only if it fulfills the two conditions*

5.11.1 *Q is $\mathcal{F}$-selfcentralizing.*

5.11.2 *$\tilde{\mathcal{F}}(Q)$ has a proper subgroup $\tilde{M}$ such that p divides $|\tilde{M}|$ and does not divide $|\tilde{M} \cap \tilde{M}^{\tilde{\sigma}}|$ for any $\tilde{\sigma} \in \tilde{\mathcal{F}}(Q) - \tilde{M}$.*

In this case, the groups $\tilde{\mathcal{F}}(Q)_{\bar{\varphi}}$, when φ runs over $\mathcal{F}(P,Q)$, coincide with the minimal proper subgroups $\tilde{M}$ of $\tilde{\mathcal{F}}(Q)$ in condition 5.11.2, *and they contain Sylow p-subgroups of $\tilde{\mathcal{F}}(Q)$. Moreover, $\overline{\mathcal{F}(P,Q)}$ is a $\mathbb{Z}$-basis of $\overline{\mathfrak{h}_{\mathcal{F}}(Q)}$ and $\tilde{\mathcal{F}}(Q)$ acts transitively on this set.*

Proof: Let $\varphi: Q \to P$ be an $\mathcal{F}$-morphism such that $Q' = \varphi(Q)$ is fully normalized in $\mathcal{F}$ (cf. Proposition 2.7); for another $\mathcal{F}$-morphism $\varphi' : Q \to P$, set $R' = N_P\big(\varphi'(Q)\big)$ and consider the isomorphism $\varphi'(Q) \cong Q'$ determined by φ and φ'; since Q' is also fully centralized in $\mathcal{F}$ (cf. Proposition 2.11), it follows from statement 2.10.1 that there are an $\mathcal{F}$-morphism $\rho' : R' \to P$ and an element $\sigma \in \mathcal{F}(Q)$ such that $\rho'\big(\varphi'(u)\big) = \varphi\big(\sigma(u)\big)$ for any $u \in Q$; consequently, denoting by $\psi' : Q \to R'$ the group homomorphism determined by φ', we get (cf. 1.9)

$$\varphi' - \varphi \circ \sigma = (\iota_{R'}^P - \rho') \circ \psi' \qquad 5.11.3.$$

In particular, assuming that Q is $\mathcal{F}$-essential, φ' and $\varphi\circ\sigma$ have the same image in $\overline{\mathfrak{h}_{\mathcal{F}}(Q)}$, so that the $\mathbb{Z}$-linear map

$$\mathbb{Z}\mathcal{F}(Q) \longrightarrow \overline{\mathfrak{h}_{\mathcal{F}}(Q)} \qquad 5.11.4$$

sending $\tau \in \mathcal{F}(Q)$ to the class of $\varphi\circ\tau$ is surjective; thus, $\mathcal{F}(Q)$ acts transitively on the image of $\mathcal{F}(P,Q)$; moreover, from the very definition of $\mathcal{F}(Q)_{\bar{\varphi}}$, we get the factorization

$$\mathbb{Z}\big(\mathcal{F}(Q)/\mathcal{F}(Q)_{\bar{\varphi}}\big) \longrightarrow \overline{\mathfrak{h}_{\mathcal{F}}(Q)} \qquad 5.11.5,$$

where the left term denotes the free $\mathbb{Z}$-module over $\mathcal{F}(Q)/\mathcal{F}(Q)_{\bar{\varphi}}$.

Furthermore, if we assume that an element $\sigma \in \mathcal{F}(Q)$ does not fix $\bar{\varphi}$ then the Alperin $\mathcal{F}$-fusion $\varphi - \varphi\circ\sigma$ is $\mathcal{F}$-irreducible; but, setting $U = Q'{\cdot}C_P(Q')$

and considering the element ${}^{\varphi}\sigma$ of $\mathcal{F}(Q')$ determined by σ, it follows from statement 2.10.1 that there is $\rho \in \mathcal{F}(U)$ such that, for any $u \in Q$, we have $\rho\big(\varphi(u)\big) = \varphi\big(\sigma(u)\big)$; hence, denoting by $\psi : Q \to U$ the group homomorphism determined by φ, we get

$$\varphi - \varphi \circ \sigma = \iota_U^P \circ (\mathrm{id}_U - \rho) \circ \psi \qquad 5.11.6$$

which forces the equality $Q' = \varphi(Q) = U$; since $\rho'\Big(C_P\big(\varphi'(Q)\big)\Big)$ is contained in $C_P(Q')$, it is clear that Q is $\mathcal{F}$-selfcentralizing, fulfilling condition 5.11.1.

Set $R = N_P(Q')$; according to Proposition 2.11, $\mathcal{F}_R(Q')$ is a Sylow p-subgroup of $\mathcal{F}(Q')$, which does not coincide with $\mathcal{F}_{Q'}(Q')$ since $Q' \neq P$ (cf. 5.7); moreover, if $v \in R$ and ν is the image of v in $\mathcal{F}(Q)$ by the isomorphism determined by φ, it is easily checked that

$$\varphi - \varphi \circ \nu = \big(\mathrm{id}_P - \kappa_P(v)\big) \circ \varphi \qquad 5.11.7$$

where $\kappa_P(v)$ is the image of v in $\mathcal{F}(P)$ (cf. 1.8), so that ν belongs to $\mathcal{F}(Q)_{\bar{\varphi}}$; that is to say, ${}^{\varphi}\mathcal{F}(Q)_{\bar{\varphi}}$ contains $\mathcal{F}_R(Q')$ and, in particular, $\mathcal{F}(Q)_{\bar{\varphi}}$ contains $\mathcal{F}_Q(Q)$ and p divides $|\tilde{\mathcal{F}}(Q)_{\bar{\varphi}}|$.

Now, consider the intersection

$$F = \mathcal{F}(Q)_{\bar{\varphi}'} \cap \mathcal{F}(Q)_{\bar{\varphi}} \qquad 5.11.8$$

and, assuming that p divides $|F/\mathcal{F}_Q(Q)|$, choose a p-subgroup V of F strictly containing $\mathcal{F}_Q(Q)$; thus, since $\mathcal{F}(Q)_{\bar{\varphi}'} = \big(\mathcal{F}(Q)_{\bar{\varphi}}\big)^{\sigma}$ and $\mathcal{F}_R(Q')$ is a Sylow p-subgroup of ${}^{\varphi}\big(\mathcal{F}(Q)_{\bar{\varphi}}\big)$, there are $\tau \in \mathcal{F}(Q)_{\bar{\varphi}}$ and $\tau' \in \mathcal{F}(Q)_{\bar{\varphi}'}$ such that

$${}^{\varphi\circ\tau}V \subset \mathcal{F}_R(Q') \supset {}^{\varphi\circ\sigma\circ\tau'}V \qquad 5.11.9$$

and therefore, since we already have $C_P(Q') = Z(Q')$, it follows from statement 2.10.1 that, denoting by T the converse image of ${}^{\varphi\circ\tau}V$ in P, there is an $\mathcal{F}$-morphism $\zeta : T \to R$ fulfilling $\zeta\big(\varphi(u)\big) = \varphi\big((\sigma\circ\tau'\circ\tau^{-1})(u)\big)$ for any $u \in Q$.

In conclusion, denoting by $\xi : T \to P$ the inclusion map ι_T^P, by $\xi' : T \to P$ the composition of ζ with the corresponding inclusion map, and by η the $\mathcal{F}$-morphism from Q to T determined by $\varphi\circ\tau$ and by the inclusion $\varphi(Q) \subset T$, we have

$$\xi \circ \eta = \varphi \circ \tau \quad \text{and} \quad \xi' \circ \eta = \varphi \circ \sigma \circ \tau' \qquad 5.11.10$$

and therefore the Alperin $\mathcal{F}$-fusions

$$\varphi - \xi \circ \eta \quad , \quad (\xi - \xi') \circ \eta \quad \text{and} \quad \xi' \circ \eta - \varphi \circ \sigma \qquad 5.11.11$$

are $\mathcal{F}$-reducible, so that we get

$$\mathcal{F}(Q)_{\bar{\varphi}} = \mathcal{F}(Q)_{\overline{\varphi\circ\sigma}} = \mathcal{F}(Q)_{\bar{\varphi}'} \qquad 5.11.12.$$

Conversely, assume that Q fulfills conditions 5.11.1 and 5.11.2, and denote by $\tilde{M}$ a proper subgroup of $\tilde{\mathcal{F}}(Q)$ as in condition 5.11.2 and by M its converse image in $\mathcal{F}(Q)$; in particular, for any p-subgroup V of M strictly containing $\mathcal{F}_Q(Q)$, M contains $N_{\mathcal{F}(Q)}(V)$; hence, a Sylow p-subgroup of M is also a Sylow p-subgroup of $\mathcal{F}(Q)$ and we have $N_{\mathcal{F}(Q)}(M) = M$. Consequently, with the notation above, there is $\sigma' \in \mathcal{F}(Q)$ such that ${}^{\varphi'\circ\sigma'}M$ contains the image of R' in $\mathcal{F}\big(\varphi'(Q)\big)$; but, since $C_P\big(\varphi'(Q)\big) = \varphi'\big(Z(Q)\big)$, this image strictly contains $\mathcal{F}_{\varphi'(Q)}\big(\varphi'(Q)\big)$ and therefore such elements σ' determine a unique class in $\mathcal{F}(Q)/M$.

Consider the map $\mathcal{F}(P,Q) \to \mathcal{F}(Q)/M$ sending φ' to the class of σ'; we claim that the corresponding $\mathbb{Z}$-linear map

$$\mathfrak{h}_{\mathcal{F}}(Q) \longrightarrow \mathbb{Z}\big(\mathcal{F}(Q)/M\big) \qquad 5.11.13$$

annihilates $\mathfrak{r}_{\mathcal{F}}(Q)$. According to definition 5.8.1, it suffices to prove that, for any subgroup T of P such that $|T| > |Q|$, any $\xi, \xi' \in \mathcal{F}(P,T)$ and any $\eta \in \mathcal{F}(T,Q)$, this map annihilates $(\xi' - \xi)\circ\eta$; moreover, it is clear that we may assume that $\eta(Q)$ is normal in T. Since M contains a Sylow p-subgroup of $\mathcal{F}(Q)$, for suitable $\tau, \tau' \in \mathcal{F}(Q)$ the groups ${}^{\xi\circ\eta\circ\tau}M$ and ${}^{\xi'\circ\eta\circ\tau'}M$ contain the respective images of $\xi(T)$ in $\mathcal{F}\big((\xi\circ\eta)(Q)\big)$ and of $\xi'(T)$ in $\mathcal{F}\big((\xi'\circ\eta)(Q)\big)$; consequently, the image of T in $\mathcal{F}\big(\eta(Q)\big)$ is contained in the intersection ${}^{\eta\circ\tau}M \cap {}^{\eta\circ\tau'}M = {}^{\eta}({}^{\tau}M \cap {}^{\tau'}M)$ and contains $\mathcal{F}_{\eta(Q)}\big(\eta(Q)\big)$ strictly by condition 5.11.1, so that we have ${}^{\tau}M = {}^{\tau'}M$ which forces τ and τ' to be in the same class since $N_{\mathcal{F}(Q)}(M) = M$.

In conclusion, Q is $\mathcal{F}$-essential and, from the $\mathbb{Z}$-linear maps 5.11.5 and 5.11.13, we get the composed $\mathbb{Z}$-linear map

$$\mathbb{Z}\big(\mathcal{F}(Q)/\mathcal{F}(Q)_{\bar{\varphi}}\big) \longrightarrow \overline{\mathfrak{h}_{\mathcal{F}}(Q)} \longrightarrow \mathbb{Z}\big(\mathcal{F}(Q)/M\big) \qquad 5.11.14$$

sending $\sigma\mathcal{F}(Q)_{\bar{\varphi}}$ to σM; in particular, this proves that $\mathcal{F}(Q)_{\bar{\varphi}}$ is contained in M and, applying it to the choice $M = \mathcal{F}(Q)_{\bar{\varphi}}$, that the $\mathbb{Z}$-linear map 5.11.5 is injective too. We are done.

5.12 If $\mathcal{F}$ is a Frobenius P-category, the $\mathcal{F}$-essential subgroups Q of P behave as the *“sous-groupes C-essentiels”* in [35, Ch. II and Ch. III], namely we have

5.12.1 *A proper subgroup $\tilde{M}$ of $\tilde{\mathcal{F}}(Q)$ contains $\tilde{\mathcal{F}}(Q)_{\bar{\varphi}}$ for some $\varphi \in \mathcal{F}(P,Q)$ if and only if p divides $|\tilde{M}|$ and does not divide $|\tilde{M}\cap\tilde{M}^{\tilde{\sigma}}|$ for $\tilde{\sigma} \in \tilde{\mathcal{F}}(Q) - \tilde{M}$. In particular, if Q is fully normalized in $\mathcal{F}$ then we have*

$$\tilde{\mathcal{F}}(Q)_{\overline{\iota_Q^P}} = \big\langle N_{\tilde{\mathcal{F}}(Q)}(\tilde{R}) \mid \tilde{R} \textit{ is a nontrivial subgroup of } \tilde{\mathcal{F}}_P(Q)\big\rangle\,.$$

Indeed, if $\tilde{M}$ contains $\tilde{\mathcal{F}}(Q)_{\bar{\varphi}}$ then, since the index $|\tilde{M}:\tilde{\mathcal{F}}(Q)_{\bar{\varphi}}|$ is prime to p and any nontrivial p-subgroup of $\tilde{\mathcal{F}}(Q)$ fixes a unique element in $\overline{\mathcal{F}(P,Q)}$, it is clear that any nontrivial p-subgroup of $\tilde{M}$ is not contained in $\tilde{M}^{\tilde{\sigma}}$ for any $\tilde{\sigma}\in\tilde{\mathcal{F}}(Q)-\tilde{M}$; the converse follows from Theorem 5.11. Moreover, it is quite clear that $N_{\tilde{\mathcal{F}}(Q)}(\tilde{R})$ fixes the unique element of $\overline{\mathcal{F}(P,Q)}$ fixed by $\tilde{R}$ for any nontrivial subgroup $\tilde{R}$ of $\tilde{\mathcal{F}}_P(Q)$; conversely, if Q is fully normalized in $\mathcal{F}$, $\mathcal{F}_P(Q)$ is a Sylow p-subgroup of $\mathcal{F}(Q)$ (cf. Proposition 2.11) and then it is not difficult to check that the right-hand member of the equality above fulfills the above condition. Similarly, let us translate Proposition 6 in [35, Ch. II] to our present context.

Corollary 5.13 *Assume that $\mathcal{F}$ is a Frobenius P-category and let Q be an $\mathcal{F}$-essential subgroup of P. Then, the set of normal subgroups X of $\tilde{\mathcal{F}}(Q)$ such that p divides $|X|$ has a unique minimal element $X_{\tilde{\mathcal{F}}}(Q)$. In particular, the quotient $X_{\tilde{\mathcal{F}}}(Q)/\mathrm{O}_{p'}\big(X_{\tilde{\mathcal{F}}}(Q)\big)$ is simple and, denoting by $X_{\mathcal{F}}(Q)$ the converse image of $X_{\tilde{\mathcal{F}}}(Q)$ in $\mathcal{F}(Q)$, we have*

$$\mathcal{F}(Q)=\mathbf{O}^p\big(X_{\mathcal{F}}(Q)\big)\cdot\mathcal{F}(Q)_{\overline{\iota_Q^P}} \qquad 5.13.1.$$

Proof: Arguing by contradiction, we may assume that there are two normal subgroups X and X' of $\tilde{\mathcal{F}}(Q)$ such that p divides $|X|$ and $|X'|$ but does not divide $|X\cap X'|$; in particular, if $\tilde{R}$ is a Sylow p-subgroup of $\tilde{\mathcal{F}}(Q)_{\overline{\iota_Q^P}}$, the intersections $\tilde{T}=X\cap\tilde{R}$ and $\tilde{T}'=X'\cap\tilde{R}$ are nontrivial and we have

$$[\tilde{T},\tilde{T}']\subset X\cap X'\cap\tilde{R}=\{1\} \qquad 5.13.2,$$

so that $\tilde{T}\cdot\tilde{T}'$ contains a noncyclic subgroup of order p^2 ; in this case, it is well-known that (cf. Theorem 3.16 in [28, Ch. 5])

$$X\cap X'=\langle C_{X\cap X'}(\tilde{\tau})\mid\tilde{\tau}\in\tilde{T}\cdot\tilde{T}'-\{1\}\rangle \qquad 5.13.3.$$

But, since $\tilde{T}\cdot\tilde{T}'\subset\tilde{R}\subset\tilde{\mathcal{F}}(Q)_{\overline{\iota_Q^P}}$ and $\tilde{\mathcal{F}}(Q)_{\overline{\iota_Q^P}}$ fulfills condition 5.11.2, this stabilizer contains $C_{\tilde{\mathcal{F}}(Q)}(\tilde{\tau})$ for any $\tilde{\tau}\in\tilde{T}\cdot\tilde{T}'-\{1\}$. Hence, $\tilde{\mathcal{F}}(Q)_{\rho_Q(\varphi)}$ contains $X\cap X'$.

Moreover, the images of X and X' in the quotient $\tilde{\mathcal{F}}(Q)/(X\cap X')$ centralize each other and, in particular, the image of $\tilde{T}$ centralizes the image of X', so that X' normalizes $(X\cap X')\cdot\tilde{T}$; hence, since $\tilde{T}$ is a Sylow p-subgroup of this product, the *Frattini argument* shows that $X'=(X\cap X').N_{X'}(\tilde{T})$; but, as above, $\tilde{\mathcal{F}}(Q)_{\overline{\iota_Q^P}}$ contains $N_{\tilde{\mathcal{F}}(Q)}(\tilde{T})$; consequently, $\tilde{\mathcal{F}}(Q)_{\overline{\iota_Q^P}}$ contains X', a contradiction. Finally, equality 5.13.1 still follows from the *Frattini argument*. We are done.

Corollary 5.14 *Assume that $\mathcal{F}$ is a Frobenius P-category. Let $\mathcal{E}$ be an $\mathcal{F}(P)$-stable set of $\mathcal{F}$-essential subgroups of P containing at least a representative for each $\mathcal{F}$-isomorphism class. For any subgroup Q of P and any $\varphi \in \mathcal{F}(P,Q)$, there are $\sigma \in \mathcal{F}(P)$, a finite family $\{Q_i\}_{i\in I}$ of elements of $\mathcal{E}$ and, for any $i \in I$, a p'-element $\sigma_i \in X_{\mathcal{F}}(Q_i)$ not fixing $\overline{\iota_{Q_i}^P}$ and an $\mathcal{F}$-morphism $\nu_i : Q \to Q_i$ fulfilling*

$$\varphi = \sigma \circ \iota_Q^P + \sum_{i\in I} \iota_{Q_i}^P \circ (\sigma_i - \mathrm{id}_{Q_i}) \circ \nu_i \qquad 5.14.1.$$

Proof: Setting $\mathcal{E}' = \mathcal{E} \cup \{P\}$ and $X_{\mathcal{F}}(P) = \mathcal{F}(P)$, we firstly prove that, for any ψ, $\psi' \in \mathcal{F}(P,Q)$, there are a finite family $\{Q_j\}_{j\in J}$ of elements of $\mathcal{E}'$ and, for any $j \in J$, a p'-element $\eta_j \in X_{\mathcal{F}}(Q_j)$ not fixing $\overline{\iota_{Q_j}^P}$ and an $\mathcal{F}$-morphism $\mu_j : Q \to Q_j$ fulfilling

$$\psi - \psi' = \sum_{j\in J} \iota_{Q_j}^P \circ (\eta_j - \mathrm{id}_{Q_j}) \circ \mu_j \qquad 5.14.2.$$

It is clear that, arguing by induction on $|P:Q|$, we may assume that $Q \neq P$ and that $\psi - \psi'$ is $\mathcal{F}$-irreducible; but, in this case, Q is $\mathcal{F}$-essential and therefore we have an $\mathcal{F}$-isomorphism $\theta : Q \cong Q'$ for some $Q' \in \mathcal{E}$; then, it follows from Theorem 5.11 and equality 5.13.1 that there are elements $\tau, \tau' \in \mathbf{O}^p\big(X_{\mathcal{F}}(Q')\big)$ such that the Alperin $\mathcal{F}$-fusions

$$\rho = \psi \circ \theta^{-1} - \iota_{Q'}^P \circ \tau \qquad \text{and} \qquad \rho' = \psi' \circ \theta^{-1} - \iota_{Q'}^P \circ \tau' \qquad 5.14.3$$

are $\mathcal{F}$-reducible; thus, setting

$$\begin{aligned} \delta_\tau &= \iota_{Q'}^P \circ (\tau - \mathrm{id}_{Q'}) \circ \theta \\ \delta_{\tau'} &= \iota_{Q'}^P \circ (\tau'^{-1} - \mathrm{id}_{Q'}) \circ (\tau' \circ \theta) \end{aligned} \qquad 5.14.4,$$

we already get the following decomposition

$$\psi - \psi' = (\rho - \rho') \circ \theta + \delta_\tau + \delta_{\tau'} \qquad 5.14.5;$$

moreover, for any decomposition $\eta = \eta' \circ \eta''$ in $\mathbf{O}^p\big(X_{\mathcal{F}}(Q')\big)$, we still have

$$\eta - \mathrm{id}_{Q'} = (\eta' - \mathrm{id}_{Q'}) \circ \eta'' + (\eta'' - \mathrm{id}_{Q'}) \qquad 5.14.6;$$

hence, since τ and τ' can be decomposed as suitable products of p'-elements of $X_{\mathcal{F}}(Q')$, it suffices to apply again the induction hypothesis.

Now, we set $\psi = \varphi$ and $\psi' = \iota_Q^P$ and argue by induction on $|J|$; we may assume that $J \neq \emptyset$ and, according to Lemma 5.4, there is $j \in J$ such that $\varphi = \iota_{Q_j}^P \circ \eta_j \circ \mu_j$ and then, according to the induction hypothesis, the $\mathcal{F}$-morphism $\varphi' = \iota_{Q_j}^P \circ \mu_j$ admits the announced decomposition 5.14.1. Thus, if Q_j belongs to $\mathcal{E}$ then the equality

$$\varphi = \varphi' + \iota_{Q_j}^P \circ (\eta_j - \mathrm{id}_{Q_j}) \circ \mu_j \qquad 5.14.7$$

gives the announced decomposition for φ. If $Q_j = P$ then $\iota_{Q_j}^P = \mathrm{id}_P$, η_j belongs to $\mathcal{F}(P)$ and we have $\varphi = \eta_j \circ \mu_j$ and $\varphi' = \mu_j$; in this case, it is easy to check that the equality $\varphi = \eta_j \circ \varphi'$ still gives the announced decomposition for φ. We are done.

5.15 With the hypothesis and the notation of the corollary above, it is handy to introduce the $\mathcal{E}$-*length* of φ for inductive purposes: it is the smallest integer $\ell_{\mathcal{E}}(\varphi)$ such that we have a decomposition 5.14.1 with $|I| = \ell_{\mathcal{E}}(\varphi)$. Note that if $\ell_{\mathcal{E}}(\varphi) \geq 1$ then there are $R \in \mathcal{E}$, $\eta \in \mathcal{F}(R, Q)$ and a p'-element $\tau \in X_{\mathcal{F}}(R)$ not fixing $\overline{\iota_R^P}$ such that we have

$$\varphi = \iota_R^P \circ \tau \circ \eta \quad \text{and} \quad \ell_{\mathcal{E}}(\iota_R^P \circ \eta) = \ell_{\mathcal{E}}(\varphi) - 1 \qquad 5.15.1.$$

When $\mathcal{E}$ is the set of all the $\mathcal{F}$-essential subgroups of P fully normalized in $\mathcal{F}$, we simply write $\ell(\varphi)$ and call it the *length* of φ.

5.16 To what extent does the behaviour of the $\mathcal{F}$-essential subgroups of P characterize the Frobenius P-categories? In order to give some answer to this question, let us consider the *normal* $\mathcal{F}$-*chains* (cf. 2.20); recall that, for any $n \in \mathbb{N}$, an $\mathcal{F}$-*chain* — more precisely called $(n, \mathcal{F})$-*chain* — is a functor

$$\mathfrak{q} : \Delta_n \longrightarrow \mathcal{F} \qquad 5.16.1$$

from the category Δ_n formed by the objects $0 \leq i \leq n$ and the morphisms $0 \leq j \leq i \leq n$, with the obvious composition (cf. A2.2), and we say that $\mathfrak{q}$ is *normal* if the image of $\mathfrak{q}(i \bullet n)$ is normal in $\mathfrak{q}(n)$ for any $i \in \Delta_n$. Let us denote by $\mathcal{F}(\mathfrak{q})$ the group of $\mathfrak{ch}(\mathcal{F})$-automorphisms of $\mathfrak{q}$ — identified with the stabilizer in $\mathcal{F}\big(\mathfrak{q}(n)\big)$ of the image of $\mathfrak{q}(i \bullet n)$ for any $i \in \Delta_n$ (cf. 2.19) — and, for any subgroup Q of P containing $\mathfrak{q}(n)$, we set

$$\mathcal{F}_Q(\mathfrak{q}) = \mathcal{F}(\mathfrak{q}) \cap \mathcal{F}_Q\big(\mathfrak{q}(n)\big) \quad \text{and} \quad \tilde{\mathcal{F}}(\mathfrak{q}) = \mathcal{F}(\mathfrak{q})/\mathcal{F}_{\mathfrak{q}(n)}(\mathfrak{q}) \qquad 5.16.2.$$

5.17 Let $\mathfrak{q}\colon \Delta_n \to \mathcal{F}$ be a normal $\mathcal{F}$-chain; let us say that a normal $\mathcal{F}$-chain $\mathfrak{r}\colon \Delta_m \to \mathcal{F}$ *extends* $\mathfrak{q}$ if $n < m$, $\mathfrak{q}(n) \not\cong \mathfrak{r}(m)$, and there is a $\mathfrak{ch}^*(\mathcal{F})$-morphism (cf. A2.8)

$$(\nu, \delta) : (\mathfrak{q}, \Delta_n) \longrightarrow (\mathfrak{r}, \Delta_m) \qquad 5.17.1$$

such that $\delta(i) = i$ for any $i \in \Delta_n$; in this case, it is clear that the $\mathcal{F}$-morphism $\mathfrak{r}(n \bullet m)$ induces a group homomorphism

$$\rho_{\mathfrak{q}}^{\mathfrak{r}} : \mathcal{F}(\mathfrak{r}) \longrightarrow \mathcal{F}(\mathfrak{q}) \qquad 5.17.2;$$

note that, if $\mathcal{F}$ is a Frobenius P-category, there exists a normal $\mathcal{F}$-chain $\mathfrak{r}$ extending $\mathfrak{q}$ if and only if either $\mathfrak{q}(n)$ is not $\mathcal{F}$-selfcentralizing or p divides $|\tilde{\mathcal{F}}(\mathfrak{q})|$

(cf. Propositions 2.7 and 2.11). In general, we say that $\mathfrak{q}$ is *$\mathcal{F}$-maximal* if it cannot be extended; thus, if $n \geq 1$ and $\mathfrak{q}$ is *fully conormalized* in $\mathcal{F}$ (cf. 2.20), $\mathfrak{q}$ is $\mathcal{F}$-maximal if and only if we have $\mathfrak{q}(n) = N_P(\mathfrak{q} \circ \delta_n^{n-1})$ where as usual $\delta_n^{n-1} : \Delta_{n-1} \to \Delta_n$ is the injective order-preserving map which does not cover n in Δ_n (cf. A3.1).

5.18 Similarly, we say that a normal $\mathcal{F}$-chain is *$\mathcal{F}$-essential* if it is $\mathfrak{ch}(\mathcal{F})$-isomorphic to a normal $\mathcal{F}$-chain $\mathfrak{q} : \Delta_n \to \mathcal{F}$ *fully conormalized* in $\mathcal{F}$ (cf. 2.20) such that $\mathfrak{q}(n)$ is $N_{\mathcal{F}}(\mathfrak{q} \circ \delta_n^{n-1})$-essential where for $n = 0$ we set $N_{\mathcal{F}}(\mathfrak{q} \circ \delta_0^{-1}) = \mathcal{F}$. In particular, since we have

$$\big(N_{\mathcal{F}}(\mathfrak{q} \circ \delta_n^{n-1})\big)\big(\mathfrak{q}(n)\big) = \mathcal{F}(\mathfrak{q}) \qquad 5.18.1,$$

if $\mathcal{F}$ is a Frobenius P-category then it follows from Proposition 2.16 and Theorem 5.11 that $\mathfrak{q}$ is $\mathcal{F}$-essential if and only if we have $C_P\big(\mathfrak{q}(n)\big) \subset \mathfrak{q}(n)$ and $\tilde{\mathcal{F}}(\mathfrak{q})$ has a proper subgroup $\tilde{M}$ such that p divides $|\tilde{M}|$ and does not divide $|\tilde{M} \cap \tilde{M}^{\tilde{\sigma}}|$ for any $\tilde{\sigma} \in \tilde{\mathcal{F}}(\mathfrak{q}) - \tilde{M}$; otherwise, as in 5.12 above, it is easily checked that we have

$$\mathcal{F}(\mathfrak{q}) = \big\langle N_{\mathcal{F}(\mathfrak{q})}\big(\mathcal{F}_R(\mathfrak{q})\big) \mid R \big\rangle \qquad 5.18.2$$

where R runs over the set of subgroups of $\mathcal{F}_P(\mathfrak{q})$ strictly containing $\mathcal{F}_{\mathfrak{q}(n)}(\mathfrak{q})$.

Proposition 5.19 *Assume that $\mathcal{F}$ is a Frobenius P-category. With the notation above, for any normal $\mathcal{F}$-chain $\mathfrak{q} : \Delta_n \to \mathcal{F}$ neither $\mathcal{F}$-maximal nor $\mathcal{F}$-essential, we have*

$$\mathcal{F}(\mathfrak{q}) = \big\langle \rho_{\mathfrak{q}}^{\mathfrak{r}}\big(\mathcal{F}(\mathfrak{r})\big) \mid \mathfrak{r} \big\rangle \qquad 5.19.1$$

where $\mathfrak{r}$ runs over the set of normal $\mathcal{F}$-chains which extend $\mathfrak{q}$ and are either $\mathcal{F}$-maximal or $\mathcal{F}$-essential.

Proof: We may assume that $\mathfrak{q}$ is conormalized in $\mathcal{F}$ (cf. 2.20) and argue by induction on $|P : \mathfrak{q}(n)|$ and on n; thus, up to replacing $\mathcal{F}$ by $N_{\mathcal{F}}\big(\mathfrak{q} \circ \delta_n^{n-1}\big)$ if $n \geq 1$, we may assume that $n = 0$ where $Q = \mathfrak{q}(0)$ is neither equal to P nor $\mathcal{F}$-essential; thus, since $N_{\mathcal{F}}(Q)$ is a Frobenius $N_P(Q)$-category (cf. Proposition 2.16), it follows from Theorem 5.11 that Q is also not $N_{\mathcal{F}}(Q)$-essential. Then, according to Corollary 5.14, for any $\sigma \in \mathcal{F}(Q)$ there are a finite family $\{R_i\}_{i \in I}$ of subgroups of $N_P(Q)$ which are fully normalized in $N_{\mathcal{F}}(Q)$ and contain Q strictly, and two families of elements $\tau_i \in \big(N_{\mathcal{F}}(Q)\big)(R_i)$ and $\nu_i \in \big(N_{\mathcal{F}}(Q)\big)(R_i, Q)$, where i runs over I, such that

$$\iota_Q^{N_P(Q)} \circ \sigma - \iota_Q^{N_P(Q)} = \sum_{i \in I} \iota_{R_i}^{N_P(Q)} \circ (\tau_i - \mathrm{id}_{R_i}) \circ \nu_i \qquad 5.19.2.$$

But, for any $i \in I$, denoting by $\mathfrak{r}_i : \Delta_1 \to \mathcal{F}$ the normal $\mathcal{F}$-chain obviously induced by the inclusion $Q \triangleleft R_i$, it is clear that $\big(N_{\mathcal{F}}(Q)\big)(R_i) = \mathcal{F}(\mathfrak{r}_i)$; moreover, assuming that our choice makes $|I|$ minimal, it follows from Lemma 5.4

and equality 5.19.2 that, for a suitable total order in I, σ coincides with the composition of the family of group homomorphisms $\{\rho_Q^{\mathfrak{r}_i}(\tau_i)\}_{i\in I}$, where we denote by Q the corresponding $(0,\mathcal{F})$-chain. Hence, we have obtained

$$\mathcal{F}(Q) = \big\langle\, \rho_Q^{\mathfrak{r}}\big(\mathcal{F}(\mathfrak{r})\big) \mid \mathfrak{r}\,\big\rangle \qquad 5.19.3$$

where $\mathfrak{r}: \Delta_1 \to \mathcal{F}$ runs over the set of normal $(1,\mathcal{F})$ -chains fulfilling $\mathfrak{r}(0) = Q$ and $\mathfrak{r}(1) \not\cong Q$. Finally, by our induction hypothesis, all these normal $\mathcal{F}$-chains which are neither $\mathcal{F}$-maximal nor $\mathcal{F}$-essential fulfill equality 5.19.1. We are done.

5.20 Let Q be a subgroup of P fully normalized and fully centralized in $\mathcal{F}$; for any normal $N_{\mathcal{F}}(Q)$-chain $\mathfrak{q}: \Delta_n \to N_{\mathcal{F}}(Q)$ such that $\mathfrak{q}(0)$ contains $Q{\cdot}C_P(Q)$, denote by $\mathcal{F}_{\mathfrak{q}}(Q)$ the corresponding normal $\mathcal{F}_{\mathcal{F}(Q)}$-chain mapping $i \in \Delta_n$ on $\mathcal{F}_{\mathfrak{q}(i)}(Q)$; as a matter of fact, if $\mathcal{F}_{\mathfrak{q}}(Q)$ is neither $\mathcal{F}_{\mathcal{F}(Q)}$-maximal nor $\mathcal{F}_{\mathcal{F}(Q)}$-essential, Proposition 5.19 still may supply a *generator set* for the group $N_{\mathcal{F}(Q)}\big(\mathcal{F}_{\mathfrak{q}}(Q)\big)$.

5.21 Indeed, considering the semidirect product $L = Q \rtimes \mathcal{F}(Q)$ and denoting by $Q \rtimes \mathcal{F}_{\mathfrak{q}}(Q)$ the corresponding normal $\mathcal{F}_L$-chain mapping $i \in \Delta_n$ on $Q \rtimes \mathcal{F}_{\mathfrak{q}(i)}(Q)$, we have an evident *surjective* group homomorphism

$$\mathcal{F}_L\big(Q \rtimes \mathcal{F}_{\mathfrak{q}}(Q)\big) \longrightarrow N_{\mathcal{F}(Q)}\big(\mathcal{F}_{\mathfrak{q}}(Q)\big) \qquad 5.21.1.$$

But, since $\mathcal{F}_L$ is a Frobenius $Q \rtimes \mathcal{F}_P(Q)$-category (cf. 3.2 and Theorem 3.7), it follows from Proposition 5.19 that

$$\mathcal{F}_L\big(Q \rtimes \mathcal{F}_{\mathfrak{q}}(Q)\big) = \big\langle\, \rho_{Q\rtimes\mathcal{F}_{\mathfrak{q}}(Q)}^{\mathfrak{r}^L}\big(\mathcal{F}_L(\mathfrak{r}^L)\big) \mid \mathfrak{r}^L\,\big\rangle \qquad 5.21.2$$

where $\mathfrak{r}^L$ runs over the set of normal $\mathcal{F}_L$-chains which extend $Q \rtimes \mathcal{F}_{\mathfrak{q}}(Q)$ and are either $\mathcal{F}_L$-maximal or $\mathcal{F}_L$-essential; moreover, it is quite clear that $\mathfrak{r}^L = Q \rtimes \mathcal{F}_{\mathfrak{r}}(Q)$ where $\mathfrak{r}$ is a normal $N_{\mathcal{F}}(Q)$-chain extending $\mathfrak{q}$ and that we have (cf. 5.16.2)

$$\tilde{\mathcal{F}}_L\big(Q \rtimes \mathcal{F}_{\mathfrak{q}}(Q)\big) \cong N_{\mathcal{F}(Q)}\big(\mathcal{F}_{\mathfrak{q}}(Q)\big)/\mathcal{F}_{\mathfrak{q}(n)}(Q) = \bar{N}_{\mathcal{F}(Q)}\big(\mathcal{F}_{\mathfrak{q}}(Q)\big) \qquad 5.21.3.$$

Consequently, the image in $\mathcal{F}(Q)$ of equality 5.21.2 yields

$$N_{\mathcal{F}(Q)}\big(\mathcal{F}_{\mathfrak{q}}(Q)\big) = \big\langle N_{\mathcal{F}(Q)}\big(\mathcal{F}_{\mathfrak{r}}(Q)\big) \mid \mathfrak{r}\,\big\rangle \qquad 5.21.4$$

where $\mathfrak{r}$ runs over the set of normal $N_{\mathcal{F}}(Q)$-chains extending $\mathfrak{q}$ such that $\bar{N}_{\mathcal{F}(Q)}\big(\mathcal{F}_{\mathfrak{r}}(Q)\big)$ either is a p'-group or has a proper subgroup $\bar{M}$ such that p divides $|\bar{M}|$ and does not divide $|\bar{M} \cap \bar{M}^{\bar{\sigma}}|$ for any $\bar{\sigma} \in \bar{N}_{\mathcal{F}(Q)}\big(\mathcal{F}_{\mathfrak{q}}(Q)\big) - \bar{M}$. We are ready to state our answer to the question above.

Theorem 5.22 *$\mathcal{F}$ is a Frobenius P-category if and only if it fulfills the following two conditions*

5.22.1 *For any $\mathcal{F}$-essential subgroup Q of P, $\mathcal{F}(Q)$ is transitive on $\overline{\mathcal{F}(P,Q)}$.*

5.22.2 *For any subgroup Q of P fully normalized and fully centralized in $\mathcal{F}$, and any normal $N_{\mathcal{F}}(Q)$-chain $\mathfrak{q}:\Delta_n \to N_{\mathcal{F}}(Q)$ such that $Q{\cdot}C_P(Q) \subset \mathfrak{q}(0)$ and that $\bar{N}_{\mathcal{F}(Q)}\big(\mathcal{F}_{\mathfrak{q}}(Q)\big)$ either is a p'-group or has a proper subgroup $\bar{M}$ such that p divides $|\bar{M}|$ and does not divide $|\bar{M}\cap\bar{M}^{\bar{\sigma}}|$ for any element $\bar{\sigma}$ in $\bar{N}_{\mathcal{F}(Q)}\big(\mathcal{F}_{\mathfrak{q}}(Q)\big) - \bar{M}$, the restriction induces a surjective group homomorphism*

$$\big(N_{\mathcal{F}}(Q)\big)(\mathfrak{q}) \longrightarrow N_{\mathcal{F}(Q)}\big(\mathcal{F}_{\mathfrak{q}}(Q)\big) \qquad 5.22.3.$$

Proof: If $\mathcal{F}$ is a Frobenius P-category then, it follows from Theorem 5.11 that $\mathcal{F}(Q)$ acts transitively on $\overline{\mathcal{F}(P,Q)}$, and from Corollary 2.13 and Proposition 2.16 that the restriction induces a surjective group homomorphism

$$\big(N_{\mathcal{F}}(Q)\big)\big(\mathfrak{q}(n)\big) \longrightarrow N_{\mathcal{F}(Q)}\big(\mathcal{F}_{\mathfrak{q}(n)}(Q)\big) \qquad 5.22.4;$$

moreover, since $Q{\cdot}C_P(Q) \subset \mathfrak{q}(0)$, any element of the end term which normalizes $\mathcal{F}_{\mathfrak{q}(i)}(Q)$ comes from an element of the origin which stabilizes $\mathfrak{q}(i)$ for any $i \in \Delta_n$.

Conversely, we will prove that the two conditions above imply that $\mathcal{F}$ fulfills both conditions in Corollary 2.13. Let Q and Q' be $\mathcal{F}$-isomorphic proper subgroups of P both fully normalized and fully centralized in $\mathcal{F}$, and $\varphi: Q \to P$ an $\mathcal{F}$-morphism such that $\varphi(Q) = Q'$; then, according to condition 5.22.1, we actually have $\bar{\varphi} = \overline{\iota_Q^P \circ \sigma}$ for some $\sigma \in \mathcal{F}(Q)$; that is to say, the difference $\varphi - \iota_Q^P \circ \sigma$ is $\mathcal{F}$-reducible and therefore we have (cf. 5.8)

$$\varphi - \iota_Q^P \circ \sigma = \sum_R \sum_{\theta \in \mathfrak{w}_{\mathcal{F}}(R)} \theta \circ \alpha_{R,\theta} \qquad 5.22.5$$

for suitable $\alpha_{R,\theta} \in \mathbb{Z}\mathcal{F}(R,Q)$, where R runs over the set of subgroups of P such that $|R| > |Q|$.

Consequently, it follows from 5.3 that we still have

$$\varphi - \iota_Q^P \circ \sigma = \sum_{j\in J} (\psi'_j - \psi_j) \circ \mu_j \qquad 5.22.6$$

where J is a finite set and, for any $j \in J$, we have $\psi_j, \psi'_j \in \mathcal{F}(P, R_j)$ and $\mu_j \in \mathcal{F}(R_j, Q)$ for a suitable subgroup R_j of P such that $|R_j| > |Q|$; more precisely, applying again condition 5.22.1 and arguing by induction on $|P:Q|$, we actually get

$$\varphi - \iota_Q^P \circ \sigma = \sum_{i\in I} \iota_{U_i}^P \circ (\tau_i - \mathrm{id}_{U_i}) \circ \nu_i \qquad 5.22.7$$

where I is a finite set and, for any $i \in I$, τ_i is an element of $\mathcal{F}(U_i)$ and ν_i an element of $\mathcal{F}(U_i, Q)$ for a suitable subgroup U_i of P such that $|U_i| > |Q|$. Then, it follows from Lemma 5.4 that, for a suitable ℓ, we can identify Δ_ℓ with a subset of I in such a way that, setting $Q_{-1} = Q$ and $Q_i = \tau_i(Q_{i-1})$ for any $i \in \Delta_\ell$, we have $Q_\ell = Q'$ and, denoting by $\varphi_i : Q_{i-1} \cong Q_i$ the $\mathcal{F}$-isomorphism induced by τ_i, the composition of all these isomorphisms coincides with the isomorphism $Q \cong Q'$ induced by $\varphi \circ \sigma^{-1}$. Moreover, note that U_i contains Q_i and Q_{i-1} for any $i \in \Delta_\ell$.

For any $i \in \Delta_\ell$, choose $\eta_i \in \mathcal{F}(P, N_P(Q_i))$ such that $Q'_i = \eta_i(Q_i)$ is fully normalized in $\mathcal{F}$ (cf. Proposition 2.7) and, setting $Q'_{-1} = Q$ and $\eta_{-1} = \iota^P_{N_P(Q)}$, denote by $\varphi'_i : Q'_{i-1} \cong Q'_i$ the $\mathcal{F}$-morphism mapping $\eta_{i-1}(u)$ on $\eta_i\big(\varphi_i(u)\big)$ for any $u \in Q_{i-1}$; we still may assume that

$$Q'_\ell = Q' \quad \text{and} \quad \eta_\ell = \iota^P_{N_P(Q')} \qquad 5.22.8.$$

Arguing by induction on $|P:Q|$, for any $i \in \Delta_\ell$ we claim that we can apply Lemma 4.13 to Q'_{i-1}, to $\mathrm{Aut}(Q'_{i-1})$ and to φ'_i; indeed, from the induction hypothesis we may assume that all the subgroups of P of smaller index fulfill condition 4.13.1; moreover, it is clear that Q'_{i-1} is a proper normal subgroup of $\eta_{i-1}\big(N_{U_i}(Q_{i-1})\big)$ and this group clearly stabilizes $\mathrm{Aut}(Q'_{i-1})$; finally, the $\mathcal{F}$-morphism

$$\eta_{i-1}\big(N_{U_i}(Q_{i-1})\big) \longrightarrow \eta_i\big(N_{U_i}(Q_i)\big) \qquad 5.22.9$$

mapping $\eta_{i-1}(v)$ on $\eta_i\big(\tau_i(v)\big)$ for any $v \in N_{U_i}(Q_{i-1})$ clearly extends φ'_i.

Hence, since Q'_i is fully normalized in $\mathcal{F}$, it follows from this lemma that thereis an $\mathcal{F}$-morphism

$$\zeta'_{i-1} : N_P(Q'_{i-1}) \longrightarrow P \qquad 5.22.10$$

extending $\chi'_i \circ \varphi'_i$ for some $\chi'_i \in \mathcal{F}(Q'_i)$; moreover, since Q'_{i-1} is fully normalized in $\mathcal{F}$, we actually get

$$\zeta'_{i-1}\big(N_P(Q'_{i-1})\big) = N_P(Q'_i) \qquad 5.22.11,$$

so that ζ'_{i-1} induces an $\mathcal{F}$-isomorphism $\xi'_i : N_P(Q'_{i-1}) \cong N_P(Q'_i)$. Finally, the composition of all these $\mathcal{F}$-isomorphisms when i runs over Δ_ℓ yields an $\mathcal{F}$-isomorphism $N_P(Q) \cong N_P(Q')$, proving condition 2.13.1.

Let Q be a subgroup of P fully normalized and fully centralized in $\mathcal{F}$ and R a subgroup of $N_P(Q)$ containing $Q{\cdot}C_P(Q)$; in order to prove condition 2.13.2, we may assume that $Q \neq R$; then, it follows from equality 5.21.4 that

$$N_{\mathcal{F}(Q)}\big(\mathcal{F}_R(Q)\big) = \big\langle N_{\mathcal{F}(Q)}\big(\mathcal{F}_{\mathfrak{r}}(Q)\big) \mid \mathfrak{r}\,\big\rangle \qquad 5.22.12$$

where $\mathfrak{r}$ runs over the set of normal $N_{\mathcal{F}}(Q)$-chains such that $\mathfrak{r}(0) = R$ and that $\bar{N}_{\mathcal{F}(Q)}\big(\mathcal{F}_{\mathfrak{r}}(Q)\big)$ either is a p'-group or has a proper subgroup $\bar{M}$ such that p

divides $|\bar{M}|$ and does not divide $|\bar{M} \cap \bar{M}^{\bar{\sigma}}|$ for any $\bar{\sigma} \in \bar{N}_{\mathcal{F}(Q)}\big(\mathcal{F}_{\mathfrak{r}}(Q)\big) - \bar{M}$. But, according to condition 5.22.2, for such an $\mathfrak{r}$ the restriction induces a surjective group homomorphism

$$\big(N_{\mathcal{F}}(Q)\big)(\mathfrak{r}) \longrightarrow N_{\mathcal{F}(Q)}\big(\mathcal{F}_{\mathfrak{r}}(Q)\big) \qquad 5.22.13;$$

moreover, it is clear that the restriction still induces the following commutative diagram

$$\begin{array}{ccc} \big(N_{\mathcal{F}}(Q)\big)(\mathfrak{r}) & \longrightarrow & N_{\mathcal{F}(Q)}\big(\mathcal{F}_{\mathfrak{r}}(Q)\big) \\ \downarrow & & \cap \\ \mathcal{F}(R)_Q & \longrightarrow & N_{\mathcal{F}(Q)}\big(\mathcal{F}_R(Q)\big) \end{array} \qquad 5.22.14.$$

Finally, equality 5.22.12 and the surjectivity of the top arrow in all these diagrams force the surjectivity of the bottom arrow. We are done.

Chapter 6

Exterior quotient of a Frobenius P-category over the selfcentralizing objects

6.1 Let P be a finite p-group and $\mathcal{F}$ a Frobenius P-category. Denote by $\mathcal{F}^{^{sc}}$ the full subcategory of $\mathcal{F}$ over the set of all the $\mathcal{F}$-selfcentralizing subgroups of P (cf. 4.8) and consider the *exterior quotient* $\tilde{\mathcal{F}}^{^{sc}}$† of $\mathcal{F}^{^{sc}}$ (cf. 1.3); that is to say, for any pair of $\mathcal{F}$-selfcentralizing subgroups Q and R of P, $\tilde{\mathcal{F}}(Q,R)$ is the the set of Q-conjugacy classes in $\mathcal{F}(Q,R)$. Although Corollary 5.14 supplies a suitable decomposition for any morphism in $\tilde{\mathcal{F}}^{^{sc}}$, Proposition 6.7 below leads to a more precise description of the structure of this category inside its *additive cover* (cf. A2.7)

$$\mathfrak{ac}(\tilde{\mathcal{F}}^{^{sc}}) = \mathfrak{pr}\big((\tilde{\mathcal{F}}^{^{sc}})^\circ\big)^\circ \qquad 6.1.1.$$

6.2 Recall that the $\mathfrak{ac}(\tilde{\mathcal{F}}^{^{sc}})$-objects are the finite sequences $\{Q_i\}_{i\in I}$ — denoted by $Q = \bigoplus_{i\in I} Q_i$ — of $\mathcal{F}$-selfcentralizing subgroups Q_i of P, and an $\mathfrak{ac}(\tilde{\mathcal{F}}^{^{sc}})$-morphism from another object $R = \bigoplus_{j\in J} R_j$ to $Q = \bigoplus_{i\in I} Q_i$ is a pair $(\tilde\alpha, f)$ formed by a map $f: J \to I$ and a family $\tilde\alpha = \{\tilde\alpha_j\}_{j\in J}$ of $\tilde{\mathcal{F}}^{^{sc}}$-morphisms $\tilde\alpha_j: R_j \to Q_{f(j)}$. The composition of $(\tilde\alpha, f)$ with another $\mathfrak{ac}(\tilde{\mathcal{F}}^{^{sc}})$-morphism

$$(\tilde\beta, g) : T = \bigoplus_{\ell\in L} T_\ell \longrightarrow R \qquad 6.2.1$$

where $\tilde\beta = \{\tilde\beta_\ell\}_{\ell\in L}$, is the pair $\big((\tilde\alpha * g)\circ\tilde\beta, f\circ g\big)$ (cf. A2.6.3) where $(\tilde\alpha * g)\circ\tilde\beta$ is the family $\{\tilde\alpha_{g(\ell)}\circ\tilde\beta_\ell\}_{\ell\in L}$ of composed morphisms

$$\tilde\alpha_{g(\ell)}\circ\tilde\beta_\ell : T_\ell \longrightarrow R_{g(\ell)} \longrightarrow Q_{(f\circ g)(\ell)} \qquad 6.2.2.$$

6.3 Actually, with no extra effort, we not only will describe the special structure of this category but, for any finite group K, the structure of the *category of K-objects of* $\mathfrak{ac}(\tilde{\mathcal{F}}^{^{sc}})$. Precisely, for any category $\mathfrak{C}$, let us call *K-object of* $\mathfrak{C}$ any pair (ρ, C) formed by an object C in $\mathfrak{C}$ and a group homomorphism $\rho: K \to \mathfrak{C}(C)$; then, the *category of K-objects of* $\mathfrak{C}$ is the category — denoted by ${}^K\mathfrak{C}$ — where the objects are the K-objects of $\mathfrak{C}$ and where the morphisms between two K-objects (ρ, C) and (ρ', C') of $\mathfrak{C}$ are the $\mathfrak{C}$-morphisms $f: C \to C'$ such that $\rho'(x)\circ f = f\circ\rho(x)$ for any $x\in K$. In other words,

6.3.1 *${}^K\mathfrak{C}$ is the category of functors to $\mathfrak{C}$ from the category with a unique object having K as the monoid of endomorphisms.*

---

† Called the *centric orbit category* in [13], whereas their *orbit category* is our $\tilde{\mathcal{F}}$ (cf. 1.3).

Note that, for any $z \in Z(K)$ and any K-object (ρ, C) of $\mathfrak{C}$, $\rho(z)$ is a ${}^K\mathfrak{C}$-automorphism of (ρ, C) and therefore the correspondence sending (ρ, C) to $\rho\big(Z(K)\big)$ defines an *inner automorphism structure* in ${}^K\mathfrak{C}$, so that ${}^K\mathfrak{C}$ admits an *exterior quotient* — denoted by ${}^K\widetilde{\mathfrak{C}}$ (cf. 1.3).

6.4 It follows from Corollary 4.9 that, for any triple of $\mathcal{F}$-selfcentralizing subgroups Q, R and T of P, any $\tilde{\mathcal{F}}$-morphism $\tilde\alpha : Q \to R$ induces an injective map from $\tilde{\mathcal{F}}(T,R)$ to $\tilde{\mathcal{F}}(T,Q)$ and we will consider the elements of $\tilde{\mathcal{F}}(T,Q)$ which, even *partially*, cannot be extended *via* $\tilde\alpha$; precisely, we set

$$\tilde{\mathcal{F}}(T,Q)_{\tilde\alpha} = \tilde{\mathcal{F}}(T,Q) - \bigcup_{\tilde\theta'} \tilde{\mathcal{F}}(T,Q') \circ \tilde\theta' \qquad 6.4.1,$$

where $\tilde\theta'$ runs over the set of $\tilde{\mathcal{F}}$-nonisomorphisms $\tilde\theta' : Q \to Q'$ from Q — the set of *nonfinal* $(\tilde{\mathcal{F}}^\circ)_Q$-objects (cf. 1.7) — fulfilling $\tilde\alpha' \circ \tilde\theta' = \tilde\alpha$ for some $\tilde\alpha' \in \tilde{\mathcal{F}}(R,Q')$ which is then unique, and we simply say that $\tilde\theta'$ *divides* $\tilde\alpha$ setting $\tilde\alpha' = \tilde\alpha/\tilde\theta'$; thus, if $\tilde\alpha$ is an isomorphism, we have $\tilde{\mathcal{F}}(T,Q)_{\tilde\alpha} = \tilde{\mathcal{F}}(T,Q)$. Note that the existence of $\tilde\alpha'$ is equivalent to the existence of a subgroup of R which is $\mathcal{F}$-isomorphic to Q' and contains $\alpha(Q)$ for a representative $\alpha \in \tilde\alpha$.

6.5 Actually, an element $\tilde\beta \in \tilde{\mathcal{F}}(T,Q)$ which can be extended to Q' *via* $\tilde\theta'$, *a fortiori* it can be extended to $N_{Q'}\big(\theta'(Q)\big)$ for a representative $\theta' \in \tilde\theta'$; hence, it follows from condition 2.10.1 that $\tilde\beta$ belongs to $\tilde{\mathcal{F}}(T,Q)_{\tilde\alpha}$ if and only if, for some representative $\beta \in \tilde\beta$, we have

$${}^{\alpha^*}\mathcal{F}_R\big(\alpha(Q)\big) \cap {}^{\beta^*}\mathcal{F}_T\big(\beta(Q)\big) = \mathcal{F}_Q(Q) \qquad 6.5.1$$

where $\alpha^* : \alpha(Q) \cong Q$ and $\beta^* : \beta(Q) \cong Q$ denote the inverse of the isomorphisms respectively induced by α and β (cf. 2.4) — which is a *symmetric* condition. That is to say, with the same notation we have

6.5.2 $\tilde\beta \in \tilde{\mathcal{F}}(T,Q)_{\tilde\alpha}$ *is equivalent to* $\tilde\alpha \in \tilde{\mathcal{F}}(R,Q)_{\tilde\beta}$.

6.6 Note that, if $\tilde{\mathcal{F}}(P,Q)_{\tilde\alpha} = \tilde{\mathcal{F}}(P,Q)$ then, since $\tilde\iota_R^P \circ \tilde\alpha$ obviously belongs to $\tilde{\mathcal{F}}(P,Q)$, $\tilde\alpha$ belongs to $\mathcal{F}(R,Q)_{\tilde\iota_R^P \circ \tilde\alpha}$ which forces $\tilde\alpha$ to be an isomorphism; conversely, by the very definition of the $\mathcal{F}$*-intersected subgroups* of P in 4.11, it follows from equality 6.5.1 that

6.6.1 *If* $\mathcal{F}(P,Q)_{\tilde\iota_Q^P} \neq \emptyset$ *then* Q *is an* $\mathcal{F}$*-intersected subgroup of* P.

Moreover, the quotient

$$\bar N_R\big(\alpha(Q)\big) \cong {}^{\alpha^*}\tilde{\mathcal{F}}_R\big(\alpha(Q)\big) \qquad 6.6.2$$

clearly acts on $\tilde{\mathcal{F}}(T,Q)_{\tilde\alpha}$ by composition on the right and whenever we have $\tilde\beta \circ \tilde\alpha^* \circ \tilde\kappa_{\alpha(Q)}(v) \circ \tilde\alpha = \tilde\beta$ (cf. 1.8) for some $\tilde\beta \in \tilde{\mathcal{F}}(T,Q)_{\tilde\alpha}$ and some $v \in R$, we still have

$$\alpha^* \circ \kappa_{\alpha(Q)}(v) \circ \alpha = \beta^* \circ \kappa_{\beta(Q)}(w) \circ \beta \qquad 6.6.3$$

for a suitable $w \in T$, so that from equality 6.5.1 we get

6.6.4 *$\bar{N}_R(\alpha(Q))$ acts freely on $\tilde{\mathcal{F}}(T,Q)_{\tilde{\alpha}}$. In particular, if $\tilde{\alpha}$ is not an $\tilde{\mathcal{F}}$-isomorphism then p divides $|\tilde{\mathcal{F}}(T,Q)_{\tilde{\alpha}}|$.*

Proposition 6.7 *For any triple of $\mathcal{F}$-selfcentralizing subgroups Q, R and T of P and any $\tilde{\alpha} \in \tilde{\mathcal{F}}(R,Q)$, we have*

$$\tilde{\mathcal{F}}(T,Q) = \bigsqcup_{\tilde{\theta}'} \tilde{\mathcal{F}}(T,Q')_{\tilde{\alpha}/\tilde{\theta}'} \circ \tilde{\theta}' \qquad 6.7.1$$

where $\tilde{\theta}' : Q \to Q'$ runs over a set of representatives for the isomorphism classes of $(\tilde{\mathcal{F}}^\circ)_Q$-objects dividing $\tilde{\alpha}$. In particular, we have

$$|\tilde{\mathcal{F}}(T,Q)| \equiv |\tilde{\mathcal{F}}(T)| \pmod{p} \qquad 6.7.2.$$

Proof: It is quite clear that arguing by induction on $|T|/|Q|$ we get

$$\tilde{\mathcal{F}}(T,Q) = \bigcup_{\tilde{\theta}'} \tilde{\mathcal{F}}(T,Q')_{\tilde{\alpha}/\tilde{\theta}'} \circ \tilde{\theta}' \qquad 6.7.3$$

where $\tilde{\theta}'$ runs over the set of $\tilde{\mathcal{F}}$-morphisms $\tilde{\theta}' : Q \to Q'$ from Q dividing $\tilde{\alpha}$; hence, it suffices to prove that, whenever for another such an $\tilde{\mathcal{F}}$-morphism $\tilde{\theta}'' : Q \to Q''$ we have

$$\big(\tilde{\mathcal{F}}(T,Q')_{\tilde{\alpha}/\tilde{\theta}'} \circ \tilde{\theta}'\big) \cap \big(\tilde{\mathcal{F}}(T,Q'')_{\tilde{\alpha}/\tilde{\theta}''} \circ \tilde{\theta}''\big) \neq \emptyset \qquad 6.7.4,$$

there is an $\tilde{\mathcal{F}}$-isomorphism $\tilde{\eta} : Q' \cong Q''$ fulfilling $\tilde{\eta} \circ \tilde{\theta}' = \tilde{\theta}''$ and, in particular, we have

$$\tilde{\mathcal{F}}(T,Q')_{\tilde{\alpha}/\tilde{\theta}'} \circ \tilde{\theta}' = \tilde{\mathcal{F}}(T,Q'')_{\tilde{\alpha}/\tilde{\theta}''} \circ \tilde{\theta}'' \qquad 6.7.5.$$

We argue by induction on $|R|/|Q|$ and may assume that $|R| \neq |Q|$, that Q' and Q'' are subgroups of R containing $\hat{Q} = \alpha(Q)$ for some $\alpha \in \tilde{\alpha}$, and that the respective homomorphisms $\theta' : Q \to Q'$ and $\theta'' : Q \to Q''$ determined by α are representatives of $\tilde{\theta}'$ and $\tilde{\theta}''$, so that

$$\tilde{\alpha}/\tilde{\theta}' = \tilde{\iota}_{Q'}^R \quad \text{and} \quad \tilde{\alpha}/\tilde{\theta}'' = \tilde{\iota}_{Q''}^R \qquad 6.7.6;$$

then, if $\tilde{\beta}' \in \tilde{\mathcal{F}}(T,Q')_{\tilde{\alpha}/\tilde{\theta}'}$ and $\tilde{\beta}'' \in \tilde{\mathcal{F}}(T,Q'')_{\tilde{\alpha}/\tilde{\theta}''}$ fulfill $\tilde{\beta}' \circ \tilde{\theta}' = \tilde{\beta}'' \circ \tilde{\theta}''$, choosing $\beta' \in \tilde{\beta}'$ and $\beta'' \in \tilde{\beta}''$ such that $\beta' \circ \theta' = \beta'' \circ \theta''$, and denoting by $\beta \in \mathcal{F}(T,\hat{Q})$ the element determined by the restriction of $\beta' \circ \theta' = \beta'' \circ \theta''$, we have

$${}^{\beta}\mathcal{F}_{Q'}(\hat{Q}) \subset \mathcal{F}_T\big(\beta(\hat{Q})\big) \supset {}^{\beta}\mathcal{F}_{Q''}(\hat{Q}) \qquad 6.7.7.$$

Hence, setting $N' = N_{Q'}(\hat Q)$, $N'' = N_{Q''}(\hat Q)$ and $N = \langle N', N''\rangle$, we still have

$$^{\beta}\mathcal{F}_N(\hat Q) \subset \mathcal{F}_T\big(\beta(\hat Q)\big) \qquad 6.7.8$$

and therefore, denoting by $\nu : Q \to N$ the homomorphism induced by α, it follows from condition 2.10.1 that there is $\tilde\xi \in \tilde{\mathcal{F}}(T,N)$ such that

$$\tilde\xi \circ \tilde\nu = \tilde\beta' \circ \tilde\theta' = \tilde\beta'' \circ \tilde\theta'' \qquad 6.7.9;$$

moreover, according to equality 6.7.3, there is an $\tilde{\mathcal{F}}$-morphism $\tilde\theta''' : N \to Q'''$ dividing $\tilde\iota_N^R$ such that $\tilde\xi = \tilde\beta''' \circ \tilde\theta'''$ for some $\tilde\beta''' \in \tilde{\mathcal{F}}(T,Q''')_{\tilde\iota_N^R/\tilde\theta'''}$.

Consequently, respectively denoting by $\nu' : Q \to N'$ and $\nu'' : Q \to N''$ the homomorphisms induced by α, we get

$$\begin{aligned}\tilde\beta' \circ \tilde\iota_{N'}^{Q'} \circ \tilde\nu' &= \tilde\beta' \circ \tilde\theta' = \tilde\beta''' \circ \tilde\theta''' \circ \tilde\nu = \tilde\beta''' \circ \tilde\theta''' \circ \tilde\iota_{N'}^{N} \circ \tilde\nu' \\ \tilde\beta'' \circ \tilde\iota_{N''}^{Q''} \circ \tilde\nu'' &= \tilde\beta'' \circ \tilde\theta'' = \tilde\beta''' \circ \tilde\theta''' \circ \tilde\nu = \tilde\beta''' \circ \tilde\theta''' \circ \tilde\iota_{N''}^{N} \circ \tilde\nu''\end{aligned} \qquad 6.7.10$$

and therefore, according to Corollary 4.9, we still get

$$\tilde\beta' \circ \tilde\iota_{N'}^{Q'} = \tilde\beta''' \circ \tilde\theta''' \circ \tilde\iota_{N'}^{N} \quad\text{and}\quad \tilde\beta'' \circ \tilde\iota_{N''}^{Q''} = \tilde\beta''' \circ \tilde\theta''' \circ \tilde\iota_{N''}^{N} \qquad 6.7.11;$$

now, it follows from the induction hypothesis that there are $\tilde{\mathcal{F}}$-isomorphisms $\tilde\eta' : Q''' \cong Q'$ and $\tilde\eta'' : Q''' \cong Q''$ fulfilling

$$\tilde\eta' \circ \tilde\theta''' \circ \tilde\iota_{N'}^{N} = \tilde\iota_{N'}^{Q'} \quad\text{and}\quad \tilde\eta'' \circ \tilde\theta''' \circ \tilde\iota_{N''}^{N} = \tilde\iota_{N''}^{Q''} \qquad 6.7.12;$$

hence, setting $\tilde\eta = \tilde\eta'' \circ \tilde\eta'^{-1}$, we obtain

$$\begin{aligned}\tilde\eta \circ \tilde\theta' &= \tilde\eta'' \circ \tilde\eta'^{-1} \circ \tilde\iota_{N'}^{Q'} \circ \tilde\nu' = \tilde\eta'' \circ \tilde\theta''' \circ \tilde\iota_{N'}^{N} \circ \tilde\nu' \\ &= \tilde\eta'' \circ \tilde\theta''' \circ \tilde\iota_{N''}^{N} \circ \tilde\nu'' = \tilde\iota_{N''}^{Q''} \circ \tilde\nu'' = \tilde\theta''\end{aligned} \qquad 6.7.13$$

and, in particular, we still obtain

$$\tilde{\mathcal{F}}(T,Q')_{\tilde\alpha/\tilde\theta'} \circ \tilde\theta' = \tilde{\mathcal{F}}(T,Q''')_{\tilde\iota_N^R/\tilde\theta'''} \circ \tilde\theta''' \circ \tilde\nu = \tilde{\mathcal{F}}(T,Q'')_{\tilde\alpha/\tilde\theta''} \circ \tilde\theta'' \qquad 6.7.14.$$

The last statement follows from statement 6.6.4 and equality 6.7.1.

6.8 The decomposition 6.7.1 allows us to consider in $\mathfrak{ac}(\tilde{\mathcal{F}}^{sc})$ (cf. 6.2) the *exterior intersection* of two $\mathcal{F}$-selfcentralizing subgroups of P, which is just a *direct product* from the categorical point of view. Although, loosely speaking, the category of functors from any category to a category with a *direct product* inherits a direct product, as announced above we directly discuss the categories $^{K}\mathfrak{ac}(\tilde{\mathcal{F}}^{sc})$ of K-objects of $\mathfrak{ac}(\tilde{\mathcal{F}}^{sc})$ (cf. 6.3), fixing the notation. Actually, for finite p'-groups K, we are mainly interested in the cohomological properties of some functors from these categories.

6.9 Explicitly, fix a finite group K and consider the category ${}^K\tilde{\mathcal{F}}$ of K-objects of $\tilde{\mathcal{F}}$ — namely the pairs $(\rho, Q) = Q^\rho$ formed by a subgroup Q of P and by a group homomorphism $\rho: K \to \tilde{\mathcal{F}}(Q)$ (cf. 6.3) — and the full subcategory ${}^K(\tilde{\mathcal{F}}^{^{\rm sc}})$ over the K-objects where Q is $\mathcal{F}$-selfcentralizing. First of all, if R and T are two $\mathcal{F}$-selfcentralizing subgroups of P, we consider the set $\mathfrak{T}_{R,T}$ of *triples* $(\tilde{\alpha}, Q, \tilde{\beta})$ where Q is an $\mathcal{F}$-selfcentralizing subgroup of P and we have

$$\tilde{\alpha} \in \tilde{\mathcal{F}}(R, Q)_{\tilde{\beta}} \quad \text{and} \quad \tilde{\beta} \in \tilde{\mathcal{F}}(T, Q)_{\tilde{\alpha}} \tag{6.9.1;}$$

we say that two triples $(\tilde{\alpha}, Q, \tilde{\beta})$ and $(\tilde{\alpha}', Q', \tilde{\beta}')$ are *equivalent* if there is an $\tilde{\mathcal{F}}$-isomorphism $\tilde{\theta}: Q \cong Q'$ fulfilling

$$\tilde{\alpha}' \circ \tilde{\theta} = \tilde{\alpha} \quad \text{and} \quad \tilde{\beta}' \circ \tilde{\theta} = \tilde{\beta} \tag{6.9.2;}$$

we denote by $\widetilde{\mathfrak{T}}_{R,T}$ the set of equivalence classes of such triples.

6.10 Moreover, if two triples $(\tilde{\alpha}, Q, \tilde{\beta})$ and $(\tilde{\alpha}', Q', \tilde{\beta}')$ are equivalent, an $\tilde{\mathcal{F}}$-isomorphism $\tilde{\theta}: Q \cong Q'$ fulfilling $\tilde{\alpha}' \circ \tilde{\theta} = \tilde{\alpha}$ and $\tilde{\beta}' \circ \tilde{\theta} = \tilde{\beta}$ is unique. Indeed, we may assume that the triples coincide with each other and, choosing $\alpha \in \tilde{\alpha}$, $\beta \in \tilde{\beta}$ and $\theta \in \tilde{\theta}$, it is easily checked that θ belongs to both ${}^{\alpha^*}\mathcal{F}_R\big(\alpha(Q)\big)$ and ${}^{\beta^*}\mathcal{F}_T\big(\beta(Q)\big)$, and therefore it belongs to $\mathcal{F}_Q(Q)$ (cf. equality 6.5.1), so that $\tilde{\theta}$ is the trivial element in $\tilde{\mathcal{F}}(Q)$.

6.11 On the other hand, if R^σ and T^τ are two K-objects of $\tilde{\mathcal{F}}^{^{\rm sc}}$, it is clear that, for any triple $(\tilde{\alpha}, Q, \tilde{\beta})$ in $\mathfrak{T}_{R,T}$ and any $x \in K$, the triple $\big(\sigma(x) \circ \tilde{\alpha}, Q, \tau(x) \circ \tilde{\beta}\big)$ still belongs to $\mathfrak{T}_{R,T}$; that is to say, σ and τ induce an action of K on $\mathfrak{T}_{R,T}$ preserving the equivalence relation, and therefore they induce an action of K on $\widetilde{\mathfrak{T}}_{R,T}$; explicitly, if $x \in K$ maps the equivalence class of $t = (\tilde{\alpha}, Q, \tilde{\beta})$ on the equivalence class of ${}^x t = ({}^x\tilde{\alpha}, {}^x Q, {}^x\tilde{\beta})$, the triples $\big(\sigma(x) \circ \tilde{\alpha}, Q, \tau(x) \circ \tilde{\beta}\big)$ and $({}^x\tilde{\alpha}, {}^x Q, {}^x\tilde{\beta})$ are equivalent and therefore there is a unique $\rho_{t,x} \in \tilde{\mathcal{F}}({}^x Q, Q)$ fulfilling

$${}^x\tilde{\alpha} \circ \rho_{t,x} = \sigma(x) \circ \tilde{\alpha} \quad \text{and} \quad {}^x\tilde{\beta} \circ \rho_{t,x} = \tau(x) \circ \tilde{\beta} \tag{6.11.1;}$$

Consequently, for a set of representatives $\check{\mathfrak{T}}_{R,T}$ of $\widetilde{\mathfrak{T}}_{R,T}$ in $\mathfrak{T}_{R,T}$, we have obtained a group homomorphism

$$\rho : K \longrightarrow \big(\mathfrak{ac}(\mathcal{F}^{^{\rm sc}})\big)\Big(\bigoplus_{(\tilde{\alpha},Q,\tilde{\beta})\in\check{\mathfrak{T}}_{R,T}} Q\Big) \tag{6.11.2}$$

or, equivalently, a K-object $\big(\bigoplus_{(\tilde{\alpha},Q,\tilde{\beta})\in\check{\mathfrak{T}}_{R,T}} Q\big)^\rho$ of $\mathfrak{ac}(\mathcal{F}^{^{\rm sc}})$, and finally we define the *exterior intersection* of R^σ and T^τ by

$$R^\sigma \mathbin{\tilde{\cap}} T^\tau = \Big(\bigoplus_{(\tilde{\alpha},Q,\tilde{\beta})\in\check{\mathfrak{T}}_{R,T}} Q\Big)^\rho \tag{6.11.3;}$$

moreover, it follows from equalities 6.11.1 that the two families of $\mathcal{F}$-morphisms $\tilde\alpha : Q \to R$ and $\tilde\beta : Q \to T$ when $(\tilde\alpha, Q, \tilde\beta)$ runs over $\check{\mathfrak{T}}_{R,T}$ determine two *canonical* ${}^K\mathfrak{ac}(\mathcal{F}^{^{sc}})$-morphisms

$$R^\sigma \longleftarrow R^\sigma \mathbin{\tilde\cap} T^\tau \longrightarrow T^\tau \qquad 6.11.4.$$

6.12 Note that, for another choice of the set of representatives, we get an isomorphic K-object of $\mathfrak{ac}(\mathcal{F}^{^{sc}})$ and a *unique* ${}^K\mathfrak{ac}(\mathcal{F}^{^{sc}})$-isomorphism compatible with the canonical morphisms; actually, we may assume that any representative $(\tilde\alpha, Q, \tilde\beta) \in \check{\mathfrak{T}}_{R,T}$ fulfills $Q \subset R$ and $\tilde\alpha = \tilde\iota_Q^R$. Moreover, in the case where there are $\tilde\gamma \in \tilde{\mathcal{F}}(P,R)$ and $\tilde\delta \in \tilde{\mathcal{F}}(P,T)$ fulfilling $\tilde\gamma \circ \tilde\alpha = \tilde\delta \circ \tilde\beta$, choosing respective representatives α, β, γ and δ of $\tilde\alpha$, $\tilde\beta$, $\tilde\gamma$ and $\tilde\delta$ fulfilling $\gamma \circ \alpha = \delta \circ \beta$, it follows from equality 6.5.1 that

$$N_{\gamma(R)}\big((\gamma\circ\alpha)(Q)\big) \cap N_{\delta(T)}\big((\gamma\circ\alpha)(Q)\big) = (\gamma\circ\alpha)(Q) \qquad 6.12.1$$

and therefore we get $\gamma(R) \cap \delta(T) = (\gamma\circ\alpha)(Q)$, which motivates our terminology.

6.13 In our next result we prove the *functorial* nature of the *exterior intersection* defined above. It is then easy by "distributivity" to extend this *exterior intersection* to all the category ${}^K\mathfrak{ac}(\tilde{\mathcal{F}}^{^{sc}})$; indeed, considering now two K-objects of the category $\mathfrak{ac}(\tilde{\mathcal{F}}^{^{sc}})$

$$R^\sigma = \Big(\bigoplus_{i\in I} R_i\Big)^\sigma \quad\text{and}\quad T^\tau = \Big(\bigoplus_{j\in J} T_j\Big)^\tau \qquad 6.13.1,$$

the actions of K on I and on J determine an action of K on $I \times J$ and, moreover, if $x \in K$ maps $(i,j) \in I \times J$ on (i',j') then we have $\mathcal{F}$-morphisms $\sigma_i^{i'}(x) : R_i \to R_{i'}$ and $\tau_j^{j'}(x) : T_j \to T_{j'}$; thus, by the functoriality proved below, we get an $\mathfrak{ac}(\tilde{\mathcal{F}}^{^{sc}})$-morphism

$$\rho_{(i,j)}^{(i',j')}(x) : R_i \mathbin{\tilde\cap} T_j \longrightarrow R_{i'} \mathbin{\tilde\cap} T_{j'} \qquad 6.13.2$$

and therefore we still get an action ρ of K on the object $\bigoplus_{(i,j)\in I\times J} R_i \mathbin{\tilde\cap} T_j$ in $\mathfrak{ac}(\tilde{\mathcal{F}}^{^{sc}})$; then, we define

$$R^\sigma \mathbin{\tilde\cap} T^\tau = \Big(\bigoplus_{(i,j)\in I\times J} R_i \mathbin{\tilde\cap} T_j\Big)^\rho \qquad 6.13.3;$$

Proposition 6.14 *For any finite group K, the category ${}^K\mathfrak{ac}(\tilde{\mathcal{F}}^{^{sc}})$ admits a distributive direct product given by the exterior intersection. Moreover, for any homomorphism $\kappa : K \to K'$ between finite groups, the restriction functor from ${}^{K'}\mathfrak{ac}(\tilde{\mathcal{F}}^{^{sc}})$ to ${}^K\mathfrak{ac}(\tilde{\mathcal{F}}^{^{sc}})$ preserves the direct products.*

Proof: With the notation above, in order to discuss the functorial nature of the exterior intersection, consider three $\mathcal{F}$-selfcentralizing subgroups R, T and U of P and two morphisms $\tilde{\psi} \in \tilde{\mathcal{F}}(R,U)$ and $\tilde{\eta} \in \tilde{\mathcal{F}}(T,U)$; it follows from Proposition 6.7 that $\tilde{\eta}$ determines an isomorphism class of $(\tilde{\mathcal{F}}^\circ)_U$-objects $\tilde{\theta}' : U \to U'$ dividing $\tilde{\psi}$ such that, setting $\tilde{\psi}' = \tilde{\psi}/\tilde{\theta}'$, we have $\tilde{\eta} = \tilde{\eta}' \circ \tilde{\theta}'$ for a suitable $\tilde{\eta}' \in \tilde{\mathcal{F}}(T,U)_{\tilde{\psi}'}$ and, once again, $\tilde{\eta}'$ is uniquely determined.

That is to say, according to statement 6.5.2, the pair $(\tilde{\psi}, \tilde{\eta})$ determines an equivalence class of triples in $\widetilde{\mathfrak{X}}_{R,T}$ and, once we have chosen a set of representatives $\check{\mathfrak{X}}_{R,T}$, it determines a unique triple $(\tilde{\psi}', U', \tilde{\eta}')$ and a unique morphism $\tilde{\theta}' : U \to U'$ fulfilling $\tilde{\psi} = \tilde{\psi}' \circ \tilde{\theta}'$ and $\tilde{\eta} = \tilde{\eta}' \circ \tilde{\theta}'$, so that the canonical map

$$\bigsqcup_{(\tilde{\alpha},Q,\tilde{\beta}) \in \check{\mathfrak{X}}_{R,T}} \tilde{\mathcal{F}}(Q,U) \longrightarrow \tilde{\mathcal{F}}(R,U) \times \tilde{\mathcal{F}}(T,U) \qquad 6.14.1$$

sending $\tilde{\varphi} \in \tilde{\mathcal{F}}(Q,U)$ to $(\tilde{\alpha} \circ \tilde{\varphi}, \tilde{\beta} \circ \tilde{\varphi})$ is bijective.

In particular, considering two $\tilde{\mathcal{F}}$-morphisms $\tilde{\zeta} : R \to R'$ and $\tilde{\xi} : T \to T'$ and a triple $(\tilde{\alpha}, Q, \tilde{\beta}) \in \check{\mathfrak{X}}_{R,T}$, we have the morphisms

$$\tilde{\zeta} \circ \tilde{\alpha} : Q \longrightarrow R' \quad \text{and} \quad \tilde{\xi} \circ \tilde{\beta} : Q \longrightarrow T' \qquad 6.14.2$$

and therefore we obtain a triple $(\tilde{\alpha}', Q', \tilde{\beta}')$ in $\check{\mathfrak{X}}_{R',T'}$ and an $\tilde{\mathcal{F}}$-morphism $\tilde{\theta} : Q \to Q'$ fulfilling

$$\tilde{\zeta} \circ \tilde{\alpha} = \tilde{\alpha}' \circ \tilde{\theta} \quad \text{and} \quad \tilde{\xi} \circ \tilde{\beta} = \tilde{\beta}' \circ \tilde{\theta} \qquad 6.14.3.$$

Thus, we have obtained a map

$$\widetilde{\mathfrak{t}}_{\tilde{\zeta},\tilde{\xi}} : \widetilde{\mathfrak{X}}_{R,T} \longrightarrow \widetilde{\mathfrak{X}}_{R',T'} \qquad 6.14.4$$

and it is not difficult to check that, for two other $\tilde{\mathcal{F}}$-morphisms $\tilde{\zeta}' : R' \to R''$ and $\tilde{\xi}' : T' \to T''$, we get

$$\widetilde{\mathfrak{t}}_{\tilde{\zeta}',\tilde{\xi}'} \circ \widetilde{\mathfrak{t}}_{\tilde{\zeta},\tilde{\xi}} = \widetilde{\mathfrak{t}}_{\tilde{\zeta}' \circ \tilde{\zeta},\tilde{\xi}' \circ \tilde{\xi}} \qquad 6.14.5.$$

Moreover, if we consider respective K-actions σ, σ', τ and τ' on R, R', T and T', and assume that $\tilde{\zeta}$ and $\tilde{\xi}$ are ${}^K\tilde{\mathcal{F}}$-morphisms, it follows from equality 6.14.5 that the map $\widetilde{\mathfrak{t}}_{\tilde{\zeta},\tilde{\xi}}$ is compatible with the actions of K on the sets $\widetilde{\mathfrak{X}}_{R,T}$ and $\widetilde{\mathfrak{X}}_{R',T'}$; once we have chosen sets of representatives $\check{\mathfrak{X}}_{R,T}$ and $\check{\mathfrak{X}}_{R',T'}$ for $\widetilde{\mathfrak{X}}_{R,T}$ and $\widetilde{\mathfrak{X}}_{R',T'}$ respectively, this map induces a new map

$\check{\mathfrak{t}}_{\tilde\zeta,\tilde\xi} : \check{\mathfrak{T}}_{R,T} \to \check{\mathfrak{T}}_{R',T'}$ and it is easily checked from equalities 6.11.1 that the morphism $\tilde\theta : Q \to Q'$ above becomes a ${}^K\tilde{\mathcal{F}}$-morphism; hence, from definition 6.12.2, we obtain a ${}^K\mathfrak{ac}(\tilde{\mathcal{F}}^{^{\mathrm{sc}}})$-morphism

$$\tilde\zeta \,\tilde\Cap\, \tilde\xi : R^\sigma \,\tilde\Cap\, T^\tau \longrightarrow R'^{\sigma'} \,\tilde\Cap\, T'^{\tau'} \qquad 6.14.6.$$

Finally, always from equality 6.14.5, it is not difficult to check that, for two other ${}^K\tilde{\mathcal{F}}$-morphisms $\tilde\zeta' : R'^{\sigma'} \to R''^{\sigma''}$ and $\tilde\xi' : T'^{\tau'} \to T''^{\tau''}$, we get

$$(\tilde\zeta' \,\tilde\Cap\, \tilde\xi') \circ (\tilde\zeta \,\tilde\Cap\, \tilde\xi) = (\tilde\zeta' \circ \tilde\zeta) \,\tilde\Cap\, (\tilde\xi' \circ \tilde\xi) \qquad 6.14.7.$$

More generally, if we now consider four objects in the category ${}^K\mathfrak{ac}(\tilde{\mathcal{F}}^{^{\mathrm{sc}}})$

$$\begin{aligned} R^\sigma &= \Big(\bigoplus_{i\in I} R_i\Big)^\sigma \quad , \qquad R'^{\sigma'} = \Big(\bigoplus_{i'\in I'} R_{i'}\Big)^{\sigma'} \\ T^\tau &= \Big(\bigoplus_{j\in J} T_j\Big)^\tau \quad \text{and} \quad T'^{\tau'} = \Big(\bigoplus_{j'\in J'} T'_{j'}\Big)^{\tau'} \end{aligned} \qquad 6.14.8$$

and two ${}^K\mathfrak{ac}(\tilde{\mathcal{F}}^{^{\mathrm{sc}}})$-morphisms

$$(\tilde\zeta, f) : R^\sigma \longrightarrow R'^{\sigma'} \quad \text{and} \quad (\tilde\xi, g) : T^\tau \longrightarrow T'^{\tau'} \qquad 6.14.9,$$

where $f : I \to I'$ and $g : J \to J'$ are K-compatible maps and moreover $\tilde\zeta$ and $\tilde\xi$ are K-compatible families of $\tilde{\mathcal{F}}$-morphisms $\tilde\zeta_i : R_i \to R'_{f(i)}$ and $\tilde\xi_j : T_j \to T'_{g(j)}$, i and j respectively running over I and J, then we clearly have a K-compatible family of $\mathfrak{ac}(\tilde{\mathcal{F}}^{^{\mathrm{sc}}})$-morphisms

$$\tilde\zeta_i \,\tilde\Cap\, \tilde\xi_j : R_i \,\tilde\Cap\, T_j \longrightarrow R'_{f(i)} \,\tilde\Cap\, T'_{g(j)} \qquad 6.14.10$$

which define a ${}^K\mathfrak{ac}(\tilde{\mathcal{F}}^{^{\mathrm{sc}}})$-morphism

$$(\tilde\zeta, f) \,\tilde\Cap\, (\tilde\xi, g) : R^\sigma \,\tilde\Cap\, T^\tau \longrightarrow R'^{\sigma'} \,\tilde\Cap\, T'^{\tau'} \qquad 6.14.11.$$

Finally, it is not difficult (but painful!) to check that bijections 6.14.1 imply the bijections

$$\begin{aligned} &\big({}^K\mathfrak{ac}(\tilde{\mathcal{F}}^{^{\mathrm{sc}}})\big)(R^\sigma \,\tilde\Cap\, T^\tau, U^\omega) \\ &\qquad \cong \big({}^K\mathfrak{ac}(\tilde{\mathcal{F}}^{^{\mathrm{sc}}})\big)(R^\sigma, U^\omega) \times \big({}^K\mathfrak{ac}(\tilde{\mathcal{F}}^{^{\mathrm{sc}}})\big)(T^\tau, U^\omega) \end{aligned} \qquad 6.14.12$$

for any ${}^K\mathfrak{ac}(\tilde{\mathcal{F}}^{^{\mathrm{sc}}})$-object U^ω, which proves that the exterior intersection is a *direct product* in this category. The proof of the last statement is straightforward.

6.15 As any *direct product*, the *exterior intersection* is associative and commutative (from the categorical point of view!). Moreover, the exterior intersection $\tilde{\bigcap}_{i\in I} R_i$ of any finite family $\mathcal{R} = \{R_i\}_{i\in I}$ of $\mathcal{F}$-selfcentralizing subgroups of P depends on the following set of indices; first of all, for any $\mathcal{F}$-selfcentralizing subgroup Q of P, let us consider the families $\tilde{\alpha} = \{\tilde{\alpha}_i\}_{i\in I}$ with $\tilde{\alpha}_i \in \tilde{\mathcal{F}}(R_i, Q)$ fulfilling

$$\bigcap_{i\in I} {}^{(\alpha_i)^*}\mathcal{F}_{R_i}\big(\alpha_i(Q)\big) = \mathcal{F}_Q(Q) \qquad 6.15.1$$

where $\alpha_i \in \tilde{\alpha}_i$ and $(\alpha_i)^* : \alpha_i(Q) \cong Q$ denotes the inverse of the isomorphism determined by α_i — called the $\mathcal{R}$-*intersectional families at* Q; note that, for any $\tilde{\mathcal{F}}$-isomorphism $\tilde{\theta} : Q' \cong Q$, $\{\tilde{\alpha}_i \circ \tilde{\theta}\}_{i\in I}$ is an $\mathcal{R}$-intersectional family at Q'. On the other hand, in order to state their relationship with the $\mathcal{F}$-*intersected* subgroups of P (cf. 4.11), we write $\tilde{\bigcap}^{|I|} P = \tilde{\bigcap}_{i\in I} R_i$ whenever $R_i = P$ for any $i \in I$.

Proposition 6.16 *For any finite family* $\mathcal{R} = \{R_i\}_{i\in I}$ *of* $\mathcal{F}$*-selfcentralizing subgroups of* P *we have*

$$\tilde{\bigcap}_{i\in I} R_i \cong \bigoplus_{(Q,\tilde{\alpha})} Q \qquad 6.16.1$$

where $(Q, \tilde{\alpha})$ *runs over a set of representatives for the* $\tilde{\mathcal{F}}$*-isomorphism classes of pairs formed by an* $\mathcal{F}$*-selfcentralizing subgroup* Q *of* P *and an* $\mathcal{R}$*-intersectional family* $\tilde{\alpha}$ *at* Q, *and where the structural morphism* $\bigoplus_{(Q,\tilde{\alpha})} Q \to R_i$ *is determined by* $\tilde{\alpha}_i : Q \to R_i$ *for any* $i \in I$. *In particular,* Q *is an* $\mathcal{F}$*-intersected subgroup of* P *if and only if it is a direct summand of* $\tilde{\bigcap}^{n} P$ *for some* $n \geq 1$.

Proof: If $|I| = 2$, the $\mathfrak{ac}(\tilde{\mathcal{F}}^{\mathrm{sc}})$-isomorphism 6.16.1 follows from the very definition of the *exterior intersection*; otherwise, it suffices to argue by induction on $|I|$ and to apply the distributivity and equality 6.5.1.

In particular, from its very definition, an $\mathcal{F}$-intersected subgroup Q of P is a direct summand of $\tilde{\bigcap}^{|\tilde{\mathcal{F}}(P,Q)|} P$; conversely, it follows from equality 6.15.1 that, for any $n \geq 1$, any direct summand Q of $\tilde{\bigcap}^{n} P$ is an $\mathcal{F}$-intersected subgroup of P.

Corollary 6.17 *Denoting by* $\tilde{\mathcal{F}}^{\mathrm{sci}}$ *the full subcategory of* $\tilde{\mathcal{F}}^{\mathrm{sc}}$ *over the set of* $\mathcal{F}$*-intersected subgroups of* P, *the subcategory* $\mathfrak{ac}(\tilde{\mathcal{F}}^{\mathrm{sci}})$ *of* $\mathfrak{ac}(\tilde{\mathcal{F}}^{\mathrm{sc}})$ *is closed with respect to the exterior intersection.*

Proof: If Q and R are $\mathcal{F}$-intersected subgroups of P then, by Proposition 6.16, they appear as *direct summands* in $\tilde{\bigcap}^{n} P$ and $\tilde{\bigcap}^{m} P$ for suitable $n, m \in \mathbb{N} - \{0\}$ and therefore $Q \tilde{\cap} R$ appears in $\tilde{\bigcap}^{n+m} P$ by the distributivity of the exterior intersection (cf. Proposition 6.14); consequently, $Q \tilde{\cap} R$ is an object in $\mathfrak{ac}(\tilde{\mathcal{F}}^{\mathrm{sci}})$ by the same proposition.

6.18 As a matter of fact, for any K-object Q^ρ of $\mathfrak{ac}(\tilde{\mathcal{F}}^{\mathrm{sc}})$, the category $\big({}^K\mathfrak{ac}(\tilde{\mathcal{F}}^{\mathrm{sc}})\big)_{Q^\rho}$ (cf. 1.7) still admits a *direct product* or, equivalently, ${}^K\mathfrak{ac}(\tilde{\mathcal{F}}^{\mathrm{sc}})$ admits *pull-backs*; in order to show it, let us introduce the *relative exterior intersection*. For any triple of $\mathcal{F}$-selfcentralizing subgroups R, R' and T and any pair of $\tilde{\mathcal{F}}$-morphisms $R \xrightarrow{\tilde{\beta}} T \xleftarrow{\tilde{\beta}'} R'$, denote by $\mathfrak{T}_{\tilde{\beta},\tilde{\beta}'}$ the set of triples $(\tilde{\alpha}, Q, \tilde{\alpha}') \in \mathfrak{T}_{R,R'}$ fulfilling $\tilde{\beta} \circ \tilde{\alpha} = \tilde{\beta}' \circ \tilde{\alpha}'$ (cf. 6.12); it is clear that $\mathfrak{T}_{\tilde{\beta},\tilde{\beta}'}$ is a union of equivalence classes in $\mathfrak{T}_{R,R'}$; moreover, considering ${}^K\tilde{\mathcal{F}}$-objects R^σ, $R'^{\sigma'}$ and T^τ, and ${}^K\tilde{\mathcal{F}}$-morphisms

$$R^\sigma \xrightarrow{\tilde{\beta}} T^\tau \xleftarrow{\tilde{\beta}'} R'^{\sigma'} \qquad 6.18.1,$$

it is quite clear that K stabilizes $\tilde{\mathfrak{T}}_{\tilde{\beta},\tilde{\beta}'}$ and therefore, in the set of representatives $\check{\mathfrak{T}}_{\tilde{\beta},\tilde{\beta}'}$ for $\tilde{\mathfrak{T}}_{\tilde{\beta},\tilde{\beta}'}$, the action ρ of K on $R \tilde{\cap} R'$ determined by σ and σ' induces an action $\rho_{\tilde{\beta},\tilde{\beta}'}$ on the subfamily indexed by $\check{\mathfrak{T}}_{\tilde{\beta},\tilde{\beta}'}$. Then, we define

$$R^\sigma \tilde{\cap}_{T^\tau} R'^{\sigma'} = R^\sigma {}_{\tilde{\beta}}\tilde{\cap}_{\tilde{\beta}'} R'^{\sigma'} = \Big(\bigoplus_{(\tilde{\alpha},Q,\tilde{\alpha}') \in \check{\mathfrak{T}}_{\tilde{\beta},\tilde{\beta}'}} Q\Big)^{\rho_{\tilde{\beta},\tilde{\beta}'}} \qquad 6.18.2$$

endowed with the $\mathfrak{ac}(\tilde{\mathcal{F}}^{\mathrm{sc}})$-morphisms $R^\sigma \leftarrow R^\sigma \tilde{\cap}_{T^\tau} R'^{\sigma'} \rightarrow R'^{\sigma'}$ determined by the two families of $\mathcal{F}$-morphisms $\tilde{\alpha}$ and $\tilde{\alpha}'$ when $(\tilde{\alpha}, Q, \tilde{\alpha}')$ runs over $\check{\mathfrak{T}}_{\tilde{\beta},\tilde{\beta}'}$. As in 6.12 above, $R \tilde{\cap}_T R'$ admits the following explicit description.

Proposition 6.19 *With the notation above, choosing representatives β of $\tilde{\beta}$ and β' of $\tilde{\beta}'$, we have*

$$R \tilde{\cap}_T R' \cong \bigoplus_u \beta(R)^u \cap \beta'(R') \qquad 6.19.1,$$

where $u \in T$ runs over the set of elements such that $\beta(R)^u \cap \beta'(R')$ is $\mathcal{F}$-selfcentralizing in a set of representatives for $\beta(R)\backslash T/\beta'(R')$, and we consider the $\tilde{\mathcal{F}}$-morphisms from $\beta(R)^u \cap \beta'(R')$ to R and to R' determined by β, u and β'.

Proof: With the notation above, for any triple $(\tilde{\alpha}, Q, \tilde{\alpha}') \in \mathfrak{T}_{\tilde{\beta},\tilde{\beta}'}$ it follows from 6.12 that

$$\beta(R)^u \cap \beta'(R') = (\beta^u \circ \alpha)(Q) \qquad 6.19.2,$$

where $\alpha \in \tilde{\alpha}$, $\alpha' \in \tilde{\alpha}'$ and $u \in T$ fulfill $\beta' \circ \alpha' = \beta^u \circ \alpha$. Conversely, for any $w \in T$ such that $\beta(R)^w \cap \beta'(R')$ is $\mathcal{F}$-selfcentralizing, we already know that the respective $\tilde{\mathcal{F}}$-morphisms $\tilde{\gamma}$ and $\tilde{\gamma}'$ from $\beta(R)^w \cap \beta'(R')$ to R and to R' determined by β, w and β' induce an $\mathfrak{ac}(\tilde{\mathcal{F}}^{\mathrm{sc}})$-morphism to $R \tilde{\cap} R'$, and therefore they determine a *triple* which clearly belongs to $\mathfrak{T}_{\tilde{\beta},\tilde{\beta}'}$.

Consequently, by the argument above, there are a suitable element $u \in T$ and an $\tilde{\mathcal{F}}$-morphism (cf. 6.19.2)

$$\tilde{\theta} : \beta(R)^w \cap \beta'(R') \longrightarrow \beta(R)^u \cap \beta'(R') \qquad 6.19.3$$

compatible with the $\tilde{\mathcal{F}}$-morphisms determined by β, w, u and β'; actually, since $\tilde{\gamma} \circ \tilde{\theta}$ coincides with the $\tilde{\mathcal{F}}$-morphism determined by the inclusion, we may assume that

$$\beta(R)^w \cap \beta'(R') \subset \beta(R)^u \cap \beta'(R') \qquad 6.19.4$$

and that $\tilde{\theta}$ is determined by this inclusion; hence, there is $v \in R$ fulfilling

$$\beta(v)wtw^{-1}\beta(v)^{-1} = utu^{-1} \qquad 6.19.5$$

for any $t \in \beta(R)^w \cap \beta'(R')$, and therefore the element $u^{-1}\beta(v)w \in T$ centralizes $\beta(R)^w \cap \beta'(R')$; since $\beta(R)^w \cap \beta'(R')$ is $\mathcal{F}$-selfcentralizing, we actually have $u^{-1}\beta(v)w = \beta'(v')$ for some $v' \in R'$, so that u and w determine the same double class in $\beta(R)\backslash T/\beta'(R')$ and therefore we have

$$\beta(R)^w \cap \beta'(R') = \beta(R)^u \cap \beta'(R') \qquad 6.19.6.$$

We are done.

6.20 As above, we can extend the *relative exterior intersection* to the K-objects of $\mathfrak{ac}(\tilde{\mathcal{F}}^{sc})$; indeed, consider three ${}^K\mathfrak{ac}(\tilde{\mathcal{F}}^{sc})$-objects

$$R^\sigma = \Big(\bigoplus_{i\in I} R_i\Big)^\sigma \quad , \quad R'^{\sigma'} = \Big(\bigoplus_{i'\in I'} R_{i'}\Big)^{\sigma'} \quad \text{and} \quad T^\tau = \Big(\bigoplus_{j\in J} T_j\Big)^\tau \qquad 6.20.1$$

and two ${}^K\mathfrak{ac}(\tilde{\mathcal{F}}^{sc})$-morphisms

$$(\tilde{\beta}, f) : R^\sigma \longrightarrow T^\tau \quad \text{and} \quad (\tilde{\beta}', f') : R'^{\sigma'} \longrightarrow T^\tau \qquad 6.20.2,$$

where $f : I \to J$ and $f' : I' \to J$ are K-compatible maps and $\tilde{\beta}$ and $\tilde{\beta}'$ are K-compatible families of $\tilde{\mathcal{F}}$-morphisms $\tilde{\beta}_i : R_i \to T_{f(i)}$ and $\tilde{\beta}'_{i'} : R'_{i'} \to T_{f'(i')}$, i and i' respectively running over I and I'. Then, for any pair $(i, i') \in I \times I'$ such that $f(i) = f'(i')$, we already have defined $R_i \,{}_{\tilde{\beta}_i}\!\tilde{\cap}_{\tilde{\beta}'_{i'}} R'_{i'}$ in the category $\mathfrak{ac}(\tilde{\mathcal{F}}^{sc})$ and it is clear that K stabilizes the set of such pairs, namely the *pull-back* $I \times_J I'$; moreover, it follows from Proposition 6.19 that, for any $x \in K$ and any $(i, i') \in I \times_J I'$, $\sigma(x)$ and $\sigma'(x)$ induce an $\mathfrak{ac}(\tilde{\mathcal{F}}^{sc})$-morphism

$$R_i \,{}_{\tilde{\beta}_i}\!\tilde{\cap}_{\tilde{\beta}'_{i'}} R'_{i'} \longrightarrow R_{(\sigma(x))(i)} \,{}_{\tilde{\beta}_{(\sigma(x))(i)}}\!\tilde{\cap}_{\tilde{\beta}'_{(\sigma'(x))(i')}} R'_{(\sigma'(x))(i')} \qquad 6.20.3.$$

That is to say, we get an action ρ of K over the $\mathfrak{ac}(\tilde{\mathcal{F}}^{sc})$-object

$$R\,{}_{(\tilde\beta,f)}\widetilde{\cap}_{(\tilde\beta',f')}\,R' = \bigoplus_{(i,i')\in I\times_J I'} R_i\,{}_{\tilde\beta_i}\widetilde{\cap}_{\tilde\beta'_{i'}}\,R'_{i'} \qquad 6.20.4$$

and then we define the *relative exterior intersection* in ${}^K\mathfrak{ac}(\tilde{\mathcal{F}}^{sc})$ by

$$R^\sigma\,\widetilde{\cap}_{T^\tau}\,R'^{\sigma'} = R^\sigma\,{}_{(\tilde\beta,f)}\widetilde{\cap}_{(\tilde\beta',f')}\,R'^{\sigma'} = \Big(\bigoplus_{(i,i')\in I\times_J I'} R_i\,{}_{\tilde\beta_i}\widetilde{\cap}_{\tilde\beta'_{i'}}\,R'_{i'}\Big)^\rho \qquad 6.20.5$$

endowed with the evident structural ${}^K\mathfrak{ac}(\tilde{\mathcal{F}}^{sc})$-morphisms to R^σ and $R'^{\sigma'}$.

Proposition 6.21 *For any finite group K, the category ${}^K\mathfrak{ac}(\tilde{\mathcal{F}}^{sc})$ admits distributive pull-backs given by the relative exterior intersection. Moreover, for any homomorphism $\kappa : K \to K'$ between finite groups, the restriction functor from ${}^{K'}\mathfrak{ac}(\tilde{\mathcal{F}}^{sc})$ to ${}^K\mathfrak{ac}(\tilde{\mathcal{F}}^{sc})$ preserves the pull-backs.*

Proof: With the notation above, let $U^\omega = \big(\bigoplus_{\ell\in L} U_\ell\big)^\omega$ be a fourth K-object of $\mathfrak{ac}(\tilde{\mathcal{F}}^{sc})$ and

$$(\tilde\gamma, g) : U^\omega \longrightarrow R^\sigma \quad \text{and} \quad (\tilde\gamma', g') : U^\omega \longrightarrow R'^{\sigma'} \qquad 6.21.1$$

two ${}^K\mathfrak{ac}(\tilde{\mathcal{F}}^{sc})$-morphisms fulfilling

$$(\tilde\beta, f)\circ(\tilde\gamma, g) = (\tilde\beta', f')\circ(\tilde\gamma', g') \qquad 6.21.2;$$

in particular, the K-compatible maps $g : L \to I$ and $g' : L \to I'$ also fulfill $f\circ g = f'\circ g'$ and thus they determine a K-compatible map $h : L \to I\times_J I'$.

Moreover, $\tilde\gamma$ and $\tilde\gamma'$ are K-compatible families of $\tilde{\mathcal{F}}$-morphisms

$$\tilde\gamma_\ell : U_\ell \longrightarrow R_{g(\ell)} \quad \text{and} \quad \tilde\gamma'_\ell : U_\ell \longrightarrow R'_{g'(\ell)} \qquad 6.21.3,$$

where ℓ runs over L, and equality 6.21.2 yields $\tilde\beta_{g(\ell)}\circ\tilde\gamma_\ell = \tilde\beta'_{g'(\ell)}\circ\tilde\gamma'_\ell$; then, it easily follows from Proposition 6.19 that there is a unique $\mathfrak{ac}(\tilde{\mathcal{F}}^{sc})$-morphism

$$\tilde\varepsilon_\ell : U_\ell \longrightarrow R_{g(\ell)}\,{}_{\tilde\beta_{g(\ell)}}\widetilde{\cap}_{\tilde\beta'_{g'(\ell)}}\,R'_{g'(\ell)} \qquad 6.21.4$$

compatible with $\tilde\gamma$, $\tilde\gamma'$ and the structural $\mathfrak{ac}(\tilde{\mathcal{F}}^{sc})$-morphisms, the uniqueness always being a consequence of Corollary 4.9; it is this uniqueness which guarantees that we still get a K-compatible family of $\mathfrak{ac}(\tilde{\mathcal{F}}^{sc})$-morphisms and therefore an ${}^K\mathfrak{ac}(\tilde{\mathcal{F}}^{sc})$-morphism

$$(\tilde\varepsilon, h) : U^\omega \longrightarrow R^\sigma\,\widetilde{\cap}_{T^\tau}\,R'^{\sigma'} \qquad 6.21.5.$$

All the remaining verifications and the last statement are straightforward.

6.22 From now on, we assume that K *is a* p'*-group*. Let $\mathcal{O}$ be the *unramified complete discrete valuation ring* of characteristic zero admitting our field k (cf. 1.10) as residue class field, and respectively denote by $\mathcal{O}$-$\mathfrak{mod}$ and k-$\mathfrak{mod}$ the categories of finitely generated $\mathcal{O}$- and k-modules. We are interested in the vanishing cohomological properties of some *contravariant* functors to $\mathcal{O}$-$\mathfrak{mod}$ from full subcategories $\mathfrak{F}$ of ${}^K\mathfrak{ac}(\tilde{\mathcal{F}}^{^{sc}})$ *closed by direct sums, direct summands, direct products and pull-backs.*

6.23 More explicitly, it is clear that any K-object of $\mathfrak{ac}(\tilde{\mathcal{F}}^{^{sc}})$ is the *direct sum* of K-objects $Q^\rho = (\bigoplus_{i\in I} Q_i)^\rho$ such that K is transitive on I — called the *indecomposable* K*-objects*; note that, if $R^\sigma = \left(\bigoplus_{j\in J} R_j\right)^\sigma$ is another indecomposable K-object and $(\tilde{\alpha}, f): R^\sigma \to Q^\rho$ is an ${}^K\mathfrak{ac}(\tilde{\mathcal{F}}^{^{sc}})$-morphism, the map $f: J \to I$ is surjective and $|I|$ divise $|J|$. We will consider the full subcategories $\mathfrak{F}$ of ${}^K\mathfrak{ac}(\tilde{\mathcal{F}}^{^{sc}})$ over sets of indecomposable K-objects such that the subcategory $\mathfrak{ac}(\mathfrak{F})$ is closed by *direct products* and *pull-backs*. Recall that any *contravariant* functor $\mathfrak{m}$ from $\mathfrak{F}$ to $\mathcal{O}$-$\mathfrak{mod}$ can be uniquely extended to an *additive contravariant* functor

$$\mathfrak{ac}(\mathfrak{m}) : \mathfrak{ac}(\mathfrak{F}) \longrightarrow \mathcal{O}\text{-}\mathfrak{mod} \qquad 6.23.1$$

mapping the direct sums of K-objects on the direct sum of their images, and the $\mathfrak{ac}(\mathfrak{F})$-morphisms on the corresponding $\mathcal{O}$-module homomorphisms between the direct sums (cf. A4.10).

6.24 In [32], Stephan Jackowski and James McClure show a general result on vanishing cohomology, which applied to $\mathfrak{ac}(\mathfrak{F})$ and $\mathfrak{ac}(\mathfrak{m})$ would depend on the existence of a so-called *Mackey complement* for $\mathfrak{ac}(\mathfrak{m})$ — in order to get a *Mackey functor* from $\mathfrak{ac}(\mathfrak{F})$, a concept introduced by Andreas Dress in [22]. A careful analysis of their proof shows that the point is just the existence of a *natural section* of some *adjoinness natural map*, which *a priori* could be easier to reach; this is the case in the proof our next result. This result is the key in chapter 14 for determining the $\mathcal{O}$-rank of the so-called *local Grothendieck group.*

6.25 Let us consider the object $\bigoplus_{x\in K} P$ of the category $\mathfrak{ac}(\tilde{\mathcal{F}}^{^{sc}})$, endowed with the K-action π defined by the *regular* action of K on itself and by the identity on P between the corresponding terms, so that $\left(\bigoplus_{x\in K} P\right)^\pi$ is an indecomposable K-object of $\mathfrak{ac}(\tilde{\mathcal{F}}^{^{sc}})$. Note that, since $|Z(K)|$ is invertible in $\mathcal{O}$, if $\mathfrak{F}$ is a full subcategory of ${}^K\mathfrak{ac}(\tilde{\mathcal{F}}^{^{sc}})$ and $\mathfrak{m}: \mathfrak{F} \to \mathcal{O}$-$\mathfrak{mod}$ is a *contravariant* functor which factorizes *via* the image $\tilde{\mathfrak{F}}$ of $\mathfrak{F}$ in the *exterior quotient* ${}^{\tilde{K}}\mathfrak{ac}(\tilde{\mathcal{F}}^{^{sc}})$ (cf. 6.3) throughout a *contravariant* functor $\tilde{\mathfrak{m}}: \tilde{\mathfrak{F}} \to \mathcal{O}$-$\mathfrak{mod}$, it follows from Proposition A4.13 that, for any $n \in \mathbb{N}$, we have

$$\mathbb{H}^n(\tilde{\mathfrak{F}}, \tilde{\mathfrak{m}}) = \mathbb{H}^n(\mathfrak{F}, \mathfrak{m}) \qquad 6.25.1$$

since, considering the subcategory $\mathfrak{Z}$ of $\mathfrak{F}$ formed by the same objects and by the automorphisms of the $\mathfrak{F}$-objects induced by $Z(K)$, $\mathbb{H}^n(\tilde{\mathfrak{F}}, \tilde{\mathfrak{m}})$ clearly coincides with the $\mathfrak{Z}$-*stable* n-*cohomology group* of $\mathfrak{F}$ over $\mathfrak{m}$ (cf. A3.18).

Theorem 6.26 *Assume that K is a finite p'-group. Let $\mathfrak{F}$ be a full subcategory of ${}^K\mathfrak{ac}(\tilde{\mathcal{F}}^{sc})$ over indecomposable K-objects, including $\big(\bigoplus_{x\in K} P\big)^\pi$, such that the subcategory $\mathfrak{ac}(\mathfrak{F})$ is closed by direct products and pull-backs, and $\mathfrak{m} : \mathfrak{F} \to \mathcal{O}\text{-}\mathfrak{mod}$ a contravariant functor which maps any morphism on an isomorphism. Then, we have $\mathbb{H}^n(\mathfrak{F}, \mathfrak{m}) = \{0\}$ for any $n \geq 1$.*

Proof: First of all, we prove the theorem assuming that $p{\cdot}\mathfrak{m} = 0$ or, equivalently, that $\mathfrak{m}$ is a *contravariant* functor from $\mathfrak{F}$ to k-$\mathfrak{mod}$. Moreover, it follows from Proposition A4.11 that for any $n \geq 1$ we have

$$\mathbb{H}^n(\mathfrak{F}, \mathfrak{m}) \cong \mathbb{H}^n\big(\mathfrak{ac}(\mathfrak{F}), \mathfrak{ac}(\mathfrak{m})\big) \qquad 6.26.1,$$

so that it suffices to prove that $\mathbb{H}^n\big(\mathfrak{ac}(\mathfrak{F}), \mathfrak{ac}(\mathfrak{m})\big) = \{0\}$ for any $n \geq 1$.

Set $S = \bigoplus_{x\in K} P$; it follows from Proposition 6.14 that the *exterior intersection with* S^π (cf. definition 6.13.3) defines a functor

$$\mathfrak{int}_{S^\pi} : \mathfrak{ac}(\mathfrak{F}) \longrightarrow \mathfrak{ac}(\mathfrak{F}) \qquad 6.26.2;$$

then, the existence of the structural $\mathfrak{ac}(\mathfrak{F})$-morphism $Q^\rho \,\tilde{\cap}\, S^\pi \to S^\pi$ for any $\mathfrak{ac}(\mathfrak{F})$-object Q^ρ, shows that $\mathfrak{int}_{S^\pi}$ factorizes throughout the evident *forgetful* functor (cf. 1.7)

$$\mathfrak{fg}_{S^\pi} : \mathfrak{ac}(\mathfrak{F})_{S^\pi} \longrightarrow \mathfrak{ac}(\mathfrak{F}) \qquad 6.26.3;$$

explicitly, it suffices to consider the functor $\mathfrak{ac}(\mathfrak{F}) \to \mathfrak{ac}(\mathfrak{F})_{S^\pi}$ mapping any $\mathfrak{ac}(\mathfrak{F})$-object Q^ρ on the structural $\mathfrak{ac}(\mathfrak{F})$-morphism above and any $\mathfrak{ac}(\mathfrak{F})$-morphism $\tilde{\alpha} : R^\sigma \to Q^\rho$ on $\tilde{\alpha} \,\tilde{\cap}\, \widetilde{\mathrm{id}}_{S^\pi}$ (cf. Proposition 6.14). But, since the category $\mathfrak{ac}(\mathfrak{F})_{S^\pi}$ has the final object $\widetilde{\mathrm{id}}_{S^\pi} : S^\pi \to S^\pi$, it follows from Corollary A4.8 that for any $n \geq 1$ we have

$$\mathbb{H}^n\big(\mathfrak{ac}(\mathfrak{F})_{S^\pi}, \mathfrak{ac}(\mathfrak{m}) \circ \mathfrak{fg}_{S^\pi}\big) = \{0\} \qquad 6.26.4$$

and therefore, we still have (cf. A3.10.4).

$$\mathbb{H}^n\big(\mathfrak{ac}(\mathfrak{F}), \mathfrak{ac}(\mathfrak{m}) \circ \mathfrak{int}_{S^\pi}\big) = \{0\} \qquad 6.26.5.$$

Moreover, the existence of the structural morphism $\tilde{\omega}_{Q^\rho} : Q^\rho \,\tilde{\cap}\, S^\pi \to Q^\rho$ for any $\mathfrak{ac}(\mathfrak{F})$-object Q^ρ, shows the existence of a natural map

$$\omega : \mathfrak{int}_{S^\pi} \longrightarrow \mathfrak{id}_{\mathfrak{ac}(\mathfrak{F})} \qquad 6.26.6$$

sending Q^ρ to $\tilde{\omega}_{Q^\rho}$; thus, in order to prove that $\mathbb{H}^n\big(\mathfrak{ac}(\mathfrak{F}), \mathfrak{ac}(\mathfrak{m})\big) = \{0\}$, it suffices to prove that the natural map $\mathfrak{ac}(\mathfrak{m}) * \omega$ admits a *natural section*

$$\theta : \mathfrak{ac}(\mathfrak{m}) \circ \mathfrak{int}_{S^\pi} \longrightarrow \mathfrak{ac}(\mathfrak{m}) \qquad 6.26.7,$$

so that $\mathfrak{ac}(\mathfrak{m})$ becomes a direct summand of $\mathfrak{ac}(\mathfrak{m}) \circ \mathfrak{int}_{S^\pi}$.

Explicitly, for any $\mathfrak{F}$-object $Q^\rho = (\bigoplus_{i\in I} Q_i)^\rho$, we have (cf. 6.13)

$$Q^\rho \mathbin{\widetilde{\cap}} S^\pi = \Big(\bigoplus_{(i,x)\in I\times K}\ \bigoplus_{(\tilde\tau_i,T_i,\tilde\iota_{T_i}^P)\in\check{\mathfrak{T}}_{Q_i,P}} T_i\Big)^{\hat\rho} \qquad 6.26.8$$

for a set of representatives $\check{\mathfrak{T}}_{Q_i,P}$ of $\tilde{\mathfrak{T}}_{Q_i,P}$ in $\mathfrak{T}_{Q_i,P}$ and a suitable action $\hat\rho$ of K on the $\mathfrak{ac}(\tilde{\mathcal{F}}^{sc})$-object $Q \mathbin{\widetilde{\cap}} S$; in particular, K acts freely on the disjoint union

$$\hat I = \bigsqcup_{(i,x)\in I\times K} \check{\mathfrak{T}}_{Q_i,P} \qquad 6.26.9$$

and let us denote by $\hat I/K$ the set of K-orbits on $\hat I$ and, for any $O \in \hat I/K$, by $(T_O)^{\hat\rho_O}$ the corresponding indecomposable "direct summand" of $Q^\rho \mathbin{\widetilde{\cap}} S^\pi$ and by $\tilde\tau_O$ the composition

$$\tilde\tau_O : (T_O)^{\hat\rho_O} \longrightarrow Q^\rho \mathbin{\widetilde{\cap}} S^\pi \longrightarrow Q^\rho \qquad 6.26.10$$

of the structural $\mathfrak{ac}(\mathfrak{F})$-morphism with $\tilde\omega_{Q^\rho}$. Moreover, we denote by $\hat I^\circ/K$ the set of "special" orbits $O \in \hat I/K$ where the $\tilde{\mathcal{F}}^{sc}$-morphisms determining $\tilde\tau_O$ are isomorphisms; note that, according to Proposition 6.14, we have a canonical bijection

$$\hat I^\circ/K \cong \bigsqcup_{i\in I} \tilde{\mathcal{F}}(P, Q_i) \qquad 6.26.11.$$

Then, we consider the homomorphism

$$\begin{array}{rccc} \theta_{Q^\rho}: & \big(\mathfrak{ac}(\mathfrak{m})\big)(Q^\rho \mathbin{\widetilde{\cap}} S^\pi) & \longrightarrow & \mathfrak{m}(Q^\rho) \\ & \| & & \\ & \prod_{O\in\hat I/K} \mathfrak{m}\big((T_O)^{\hat\rho_O}\big) & & \end{array} \qquad 6.26.12$$

sending an element $m = (m_O)_{O\in\hat I/K}$ of this product to

$$\theta_{Q^\rho}(m) = |\hat I^\circ/K|^{-1}\cdot \sum_{O\in\hat I^\circ/K} \mathfrak{m}(\tilde\tau_O)^{-1}(m_O) \qquad 6.26.13,$$

so that we clearly have

$$\theta_{Q^\rho} \circ \big(\mathfrak{ac}(\mathfrak{m}) * \omega\big)_{Q^\rho} = \mathrm{id}_{\mathfrak{m}(Q^\rho)} \qquad 6.26.14$$

By the *distributivity* of the *exterior intersection* (cf. 6.13), we easily can extend this correspondence to all the $\mathfrak{ac}(\mathfrak{F})$-objects and then we claim that the extended correspondence is a natural map from $\mathfrak{ac}(\mathfrak{m}) \circ \mathfrak{int}_{S^\pi}$ to $\mathfrak{ac}(\mathfrak{m})$; once again, it suffices to consider an $\mathfrak{F}$-morphism $\tilde\alpha : R^\sigma \to Q^\rho$ and to prove the commutativity of the following diagram

$$\begin{array}{rcl} \big(\mathfrak{ac}(\mathfrak{m})\big)(Q^\rho \mathbin{\widetilde{\cap}} S^\pi) & \xrightarrow{\theta_{Q^\rho}} & \mathfrak{m}(Q^\rho) \\ {\scriptstyle (\mathfrak{ac}(\mathfrak{m}))(\tilde\alpha \mathbin{\widetilde{\cap}} \mathrm{id}_{S^\pi})} \downarrow & & \downarrow {\scriptstyle \mathfrak{m}(\tilde\alpha)} \\ \big(\mathfrak{ac}(\mathfrak{m})\big)(R^\sigma \mathbin{\widetilde{\cap}} S^\pi) & \xrightarrow{\theta_{R^\sigma}} & \mathfrak{m}(R^\sigma) \end{array} \qquad 6.26.15.$$

Explicitly, if $R^\sigma = (\bigoplus_{j\in J} R_j)^\sigma$ then $\tilde\alpha$ is given by a K-compatible map $f\colon J \to I$ and by a K-compatible family of $\tilde{\mathcal{F}}$-morphisms $\tilde\alpha_j\colon R_j \to Q_{f(j)}$ where j runs over J, and as above we have

$$R^\sigma \mathbin{\tilde\cap} S^\pi = \Big(\bigoplus_{(j,x)\in J\times K}\ \bigoplus_{(\tilde v_j, U_j, \tilde\iota^P_{U_j})\in\check{\mathfrak{T}}_{R_j,P}} U_j\Big)^{\hat\sigma} \qquad 6.26.16$$

for a set of representatives $\check{\mathfrak{T}}_{R_j,P}$ of $\tilde{\mathfrak{T}}_{R_j,P}$ in $\mathfrak{T}_{R_j,P}$ and a suitable action $\hat\sigma$ of K on the $\mathfrak{ac}(\tilde{\mathcal{F}}^{sc})$-object $R\mathbin{\tilde\cap} S$; again, we set

$$\hat J = \bigsqcup_{(j,x)\in J\times K} \check{\mathfrak{T}}_{R_j,P} \qquad 6.26.17$$

and denote by $\hat J/K$ the set of K-orbits on $\hat J$, by $\hat J^\circ/K$ the set of "special" K-orbits on $\hat J$ and, for any $O\in\hat J/K$, by $(U_O)^{\hat\sigma_O}$ the corresponding indecomposable "direct summand" of $R^\sigma\mathbin{\tilde\cap} S^\pi$ and by $\tilde v_O\colon (U_O)^{\hat\sigma_O}\to R^\sigma$ the analogous composition 6.26.10; moreover, it is clear that the map f and the family $\{\tilde\alpha_j\}_{j\in J}$ determine a K-compatible map $\hat f\colon \hat J\to\hat I$ and, for any $O\in\hat J/K$, an $\mathfrak{F}$-morphism

$$\tilde\alpha_O : (U_O)^{\hat\sigma_O} \longrightarrow (T_{\hat f(O)})^{\hat\rho_{\hat f(O)}} \qquad 6.26.18.$$

It is easily checked from Propositions 6.14 and 6.21 that

$$R^\sigma\ {}_{\tilde\alpha}{\tilde\cap}_{\tilde\omega_{Q^\rho}}\,(Q^\rho\mathbin{\tilde\cap} S^\pi) \cong R^\sigma\mathbin{\tilde\cap} S^\pi \qquad 6.26.19$$

and, by the *distributivity*, we may assume that the exterior intersection $R^\sigma\mathbin{\tilde\cap} S^\pi$ coincides with

$$\Big(\bigoplus_{(j,x)\in J\times K}\ \bigoplus_{(\tilde\tau_{f(j)}, T_{f(j)}, \tilde\iota^P_{T_{f(j)}})\in\check{\mathfrak{T}}_{Q_{f(j)},P}} R_j\ {}_{\tilde\alpha_j}{\tilde\cap}_{\tilde\tau_{f(j)}}\, T_{f(j)}\Big)^{\hat\sigma} \qquad 6.26.20.$$

Then, for any $(j,x)\in J\times K$ and any $\mathfrak{t}_{f(j)} = (\tilde\tau_{f(j)}, T_{f(j)}, \tilde\iota^P_{T_{f(j)}})\in\check{\mathfrak{T}}_{Q_{f(j)},P}$, choosing $\tau_{f(j)}\in\tilde\tau_{f(j)}$, $\alpha_{f(j)}\in\tilde\alpha_{f(j)}$ and a set of representatives $W_{(j,x,\mathfrak{t}_{f(j)})}$ in $Q_{f(j)}$ for the set of double classes $\tau_{f(j)}(T_{f(j)})\backslash Q_{f(j)}/\alpha_j(R)$, and denoting by $W^{sc}_{(j,x,\mathfrak{t}_{f(j)})}$ the set of $w\in W_{(j,x,\mathfrak{t}_{f(j)})}$ such that (cf. 1.8)

$$U_w = \big(\kappa_{Q_{f(j)}}(w)\circ\alpha_j\big)^{-1}\big(\tau_{f(j)}(T_{f(j)})\big) \qquad 6.26.21$$

remains $\mathcal{F}$-selfcentralizing, the $\mathfrak{ac}(\tilde{\mathcal{F}}^{sc})$-morphism

$$\tilde\alpha\mathbin{\tilde\cap}\mathrm{id}_{S^\pi} : R^\sigma\mathbin{\tilde\cap} S^\pi \longrightarrow Q^\rho\mathbin{\tilde\cap} S^\pi \qquad 6.26.22$$

is the "direct sum", over the set of triples $(j, x, \mathfrak{t}_{f(j)})$ above, of the $\mathfrak{ac}(\tilde{\mathcal{F}}^{sc})$-morphisms

$$\bigoplus_{w\in W^{sc}_{(j,x,\mathfrak{t}_{f(j)})}} U_w \longrightarrow T_{f(j)} \qquad 6.26.23$$

defined by the $\tilde{\mathcal{F}}^{sc}$-morphisms $\tilde{\beta}_{(j,x,\mathfrak{t}_{f(j)},w)} : U_w \to T_{f(j)}$ determined by the compositions $\kappa_{Q_{f(j)}}(w) \circ \alpha_j$ (cf. 1.8).

Moreover, K acts on all the situation and we denote by $O_{(j,x,\mathfrak{t}_{f(j)},w)}$ the K-orbit — which actually is *regular* — of $(j, x, \mathfrak{t}_{f(j)}, w) \in \hat{J}$, by $O_{(f(j),x,\mathfrak{t}_{f(j)})}$ the image in $\hat{I}$ *via* $\hat{f}$ of $O_{(j,x,\mathfrak{t}_{f(j)},w)}$ — which actually does not depend on w — and by $\tilde{\beta}_{O_{(j,x,\mathfrak{t}_{f(j)},w)}}$ and $\tilde{v}_{O_{(j,x,\mathfrak{t}_{f(j)},w)}}$ the $\mathfrak{F}$-morphisms

$$\begin{aligned} &(U_{O_{(j,x,\mathfrak{t}_{f(j)},w)}})^{\hat{\sigma}_{O_{(j,x,\mathfrak{t}_{f(j)},w)}}} \longrightarrow (T_{O_{(f(j),x,\mathfrak{t}_{f(j)})}})^{\hat{\rho}_{O_{(f(j),x,\mathfrak{t}_{f(j)})}}} \\ &(U_{O_{(j,x,\mathfrak{t}_{f(j)},w)}})^{\hat{\sigma}_{O_{(j,x,\mathfrak{t}_{f(j)},w)}}} \longrightarrow R^{\sigma} \end{aligned} \qquad 6.26.24$$

respectively determined by the K-orbits of $\tilde{\beta}_{(j,x,\mathfrak{t}_{f(j)},w)}$ and $\tilde{\iota}^{R_j}_{U_w}$; then, the naturality of $\mathfrak{ac}(\mathfrak{m}) * \omega$ forces

$$\begin{aligned} &\mathfrak{m}(\tilde{v}_{O_{(j,x,\mathfrak{t}_{f(j)},w)}}) \circ \mathfrak{m}(\tilde{\alpha}) \\ &\qquad = \mathfrak{m}(\tilde{\beta}_{O_{(j,x,\mathfrak{t}_{f(j)},w)}}) \circ \mathfrak{m}(\tilde{\tau}_{O_{(f(j),x,\mathfrak{t}_{f(j)})}}) \end{aligned} \qquad 6.26.25,$$

so that we still have

$$\begin{aligned} &\mathfrak{m}(\tilde{\alpha}) \circ \mathfrak{m}(\tilde{\tau}_{O_{(f(j),x,\mathfrak{t}_{f(j)})}})^{-1} \\ &\qquad = \mathfrak{m}(\tilde{v}_{O_{(j,x,\mathfrak{t}_{f(j)}),w)}})^{-1} \circ \mathfrak{m}(\tilde{\beta}_{O_{(j,x,\mathfrak{t}_{f(j)}),w)}}) \end{aligned} \qquad 6.26.26$$

and therefore the right member of this equality does not depend on w.

Now, we are ready to prove the commutativity of the diagram 6.29.15; according to our definition, the composition $\mathfrak{m}(\tilde{\alpha}) \circ \theta_{Q^{\rho}}$ sends the element

$$m = (m_O)_{O\in\hat{I}/K} = (m_{(i,x,\mathfrak{t}_i)})_{(i,x)\in I\times K\,,\mathfrak{t}_i\in\tilde{\mathfrak{T}}_{Q_i,P}} \qquad 6.26.27,$$

where $m_O \in \mathfrak{m}(T_O)$ and $m_{(i,x,\mathfrak{t}_i)} \in \mathfrak{m}(T_i)$ if $\mathfrak{t}_i = (\tilde{\tau}_i, T_i, \tilde{\iota}^P_{T_i})$, to the sum

$$|\hat{I}^{\circ}/K|^{-1}\cdot \sum_{O\in\hat{I}^{\circ}/K} (\mathfrak{m}(\tilde{\alpha}))\big(\mathfrak{m}(\tilde{\tau}_O)^{-1}(m_O)\big) \qquad 6.26.28,$$

whereas we have

$$\begin{aligned}&\big((\mathfrak{ac}(\mathfrak{m}))(\tilde\alpha\,\tilde\cap\,\mathrm{id}_{S^\pi})\big)(m)\\ &\quad=\sum_j\sum_{\mathfrak{t}_{f(j)}}\sum_w\big(\mathfrak{m}(\tilde\beta_{O_{(j,1,\mathfrak{t}_{f(j)},w)}})\big)(m_{O_{(f(j),1,\mathfrak{t}_{f(j)})}})\end{aligned}\qquad 6.26.29,$$

where j runs over J, $\mathfrak{t}_{f(j)}$ over $\check{\mathfrak{T}}_{Q_{f(j)},P}$ and w over $W^{\mathrm{sc}}_{(j,1,\mathfrak{t}_{f(j)})}$, and therefore, denoting by $W^{R_j}_{(j,1,\mathfrak{t}_{f(j)})}$ the set of $w\in W^{\mathrm{sc}}_{(j,1,\mathfrak{t}_{f(j)})}$ such that $U_w=R_j$, it follows from our definition of θ_{R^σ} and from equality 6.26.25 that

$$\begin{aligned}&\big(\theta_{R^\sigma}\circ(\mathfrak{ac}(\mathfrak{m}))(\tilde\alpha\,\tilde\cap\,\mathrm{id}_{S^\pi})\big)(m)\\ &=\sum_j\sum_{\mathfrak{t}_{f(j)}}\frac{|W^{R_j}_{(j,1,\mathfrak{t}_{f(j)})}|}{|\hat J^\circ/K|}\cdot\mathfrak{m}(\tilde\alpha)\big(\mathfrak{m}(\tilde\tau_{O_{(f(j),1,\mathfrak{t}_{f(j)})}})^{-1}(m_{O_{(f(j),1,\mathfrak{t}_{f(j)})}})\big)\end{aligned}\qquad 6.26.30,$$

where j runs over J and $\mathfrak{t}_{f(j)}$ over $\check{\mathfrak{T}}_{Q_{f(j)},P}$.

On the other hand, note that if $O_{(f(j),1,\mathfrak{t}_{f(j)})}$ belongs to $\hat I^\circ/K$ then we have $|W^{R_j}_{(j,1,\mathfrak{t}_{f(j)})}|=1$; moreover, we already know that $\tilde\alpha_j$ induces an injective map from $\tilde{\mathcal{F}}(P,Q_{f(j)})$ to $\tilde{\mathcal{F}}(P,R_j)$ (cf. Corollary 4.9) and it is clear that $|f^{-1}(i)|=|J|/|I|$ for any $i\in I$. Hence, since according to bijection 6.26.11 we have (cf. 6.7.2)

$$|\hat I^\circ/K|\equiv|I||\tilde{\mathcal{F}}(P)|\quad\text{and}\quad|\hat J^\circ/K|\equiv|J||\tilde{\mathcal{F}}(P)|\pmod p\qquad 6.26.31,$$

the sum of all these terms in the second member of equality 6.26.30 coincides with the sum 6.26.28 above.

Consequently, in order to show the commutativity of diagram 6.26.14, it suffices to prove that, for any $j\in J$ and any $\mathfrak{t}_{f(j)}=(\tilde\tau_{f(j)},T_{f(j)},\tilde\iota^P_{T_{f(j)}})$ in $\check{\mathfrak{T}}_{Q_{f(j)},P}$ such that $\tilde\tau_{f(j)}:T_{f(j)}\to Q_{f(j)}$ is *not* and isomorphism, p divides $|W^{R_j}_{(j,1,\mathfrak{t}_{f(j)})}|$; but, it is clear that $W^{R_j}_{(j,1,\mathfrak{t}_{f(j)})}$ is a set of representatives for the quotient set

$$\tau_{f(j)}(T_{f(j)})\backslash\mathcal{T}_{Q_{f(j)}}\big(\tau_{f(j)}(T_{f(j)}),\alpha_j(R_j)\big)\qquad 6.26.32$$

and that the nontrivial p-group $\bar N_{Q_{f(j)}}\big(\tau_{f(j)}(T_{f(j)})\big)$ acts freely on this set. This completes the proof of the naturality of θ and therefore the proof of the theorem in the case where $p\cdot\mathfrak{m}=0$.

In the general case, there is a subfunctor $\mathfrak{m}^{\mathrm{tor}}:\mathfrak{F}\to\mathcal{O}\text{-}\mathfrak{mod}$ mapping any $\mathfrak{F}$-object Q^ρ on the *torsion* $\mathcal{O}$-submodule of $\mathfrak{m}(Q^\rho)$ and then we have the quotient functor

$$\mathfrak{m}/\mathfrak{m}^{\mathrm{tor}}:\mathfrak{F}\longrightarrow\mathcal{O}\text{-}\mathfrak{mod}\qquad 6.26.33$$

which maps any object on a *free* $\mathcal{O}$-module; consequently, since we have exact sequences (cf. A3.11.4)

$$\mathbb{H}^n(\mathfrak{F}, \mathfrak{m}^{\mathrm{tor}}) \longrightarrow \mathbb{H}^n(\mathfrak{F}, \mathfrak{m}) \longrightarrow \mathbb{H}^n(\mathfrak{F}, \mathfrak{m}/\mathfrak{m}^{\mathrm{tor}}) \qquad 6.26.34,$$

we already may assume that either $\mathfrak{m} = \mathfrak{m}^{\mathrm{tor}}$ or $\mathfrak{m}^{\mathrm{tor}} = 0$.

In the first case we have $p^\ell \cdot \mathfrak{m} = 0$ for some $\ell \in \mathbb{N}$ and, considering the exact sequences (cf. A3.11.4)

$$\mathbb{H}^n(\mathfrak{F}, p\cdot\mathfrak{m}) \longrightarrow \mathbb{H}^n(\mathfrak{F}, \mathfrak{m}) \longrightarrow \mathbb{H}^n(\mathfrak{F}, \mathfrak{m}/p\cdot\mathfrak{m}) = \{0\} \qquad 6.26.35,$$

it suffices to argue by induction on ℓ. In the second case, if c_0 is an n-*cocycle*, we already have proved that

$$c_0 \equiv d_{n-1}(a_0) \pmod{p} \qquad 6.26.36$$

for a suitable $(n-1)$-*cochain* a_0, so that we have $c_0 - d_{n-1}(a_0) = p\cdot c_1$ for a suitable n-*cocycle* c_1 since we are dealing with free $\mathcal{O}$-modules; thus, we inductively can define n-*cocycles* c_i and $(n-1)$-*cochains* a_i fulfilling

$$c_i \equiv d_{n-1}(a_i) \pmod{p} \quad \text{and} \quad c_i - d_{n-1}(a_i) = p\cdot c_{i+1} \qquad 6.26.37$$

and then, according to the completeness of $\mathcal{O}$, it is quite clear that

$$c_0 = d_{n-1}\Big(\sum_{i\in\mathbb{N}} p^i\cdot a_i\Big) \qquad 6.26.38.$$

We are done.

Chapter 7

Nilcentralized and selfcentralizing Brauer pairs in blocks

7.1 As in chapter 3, let G be a finite group, k an algebraically closed field of characteristic p, b a block of G and (P, e) a maximal Brauer (b, G)-pair, and denote by $\mathcal{F}_{(b,G)}$ the corresponding Frobenius P-category defined there. In this chapter, we give alternative descriptions of the $\mathcal{F}_{(b,G)}$-nilcentralized and the $\mathcal{F}_{(b,G)}$-selfcentralizing subgroups Q of P in terms of the blocks in their associated Brauer (b, G)-pairs. As a matter of fact, in this context the group $\mathcal{F}_{(b,G)}(Q)$ can be *canonically* lifted to a k^**-group* $\hat{\mathcal{F}}_{(b,G)}(Q)$ (cf. 1.23) and there exists a canonical *Dade* $\bar{N}_P(Q)$*-algebra* S_Q (cf. 1.20).

Proposition 7.2 *Let Q be a subgroup of P and f the block of $C_G(Q)$ such that (P, e) contains (Q, f). Then, Q is $\mathcal{F}_{(b,G)}$-nilcentralized if and only if f is a nilpotent block.*

Proof: By Proposition 2.7, there is $\varphi \in \mathcal{F}_{(b,G)}(P, Q)$ such that $Q' = \varphi(Q)$ is fully centralized in $\mathcal{F}_{(b,G)}$; in particular, there is $x \in G$ such that $Q' = Q^x$ and that, setting $f' = f^x$, (P, e) contains (Q', f') (cf. 1.15) and therefore we may assume that Q is fully centralized in $\mathcal{F}_{(b,G)}$. In this case, by Proposition 3.5, $C_P(Q)$ is a defect subgroup of the block f of $C_G(Q)$ and, by Proposition 2.16, the *centralizer* $C_{\mathcal{F}_{(b,G)}}(Q)$ is a Frobenius $C_P(Q)$-category; moreover, by Corollary 3.6, the Frobenius $C_P(Q)$-categories $C_{\mathcal{F}_{(b,G)}}(Q)$ and $\mathcal{F}_{(f,C_G(Q))}$ coincide. Thus, if Q *is* $\mathcal{F}_{(b,G)}$*-nilcentralized* then the $C_P(Q)$-categories $\mathcal{F}_{C_P(Q)}$ and $\mathcal{F}_{(f,C_G(Q))}$ coincide, so that f is a nilpotent block (cf. 1.18).

Conversely, if f is a nilpotent block (cf. 1.18), it follows from statements (1.7.2) and (1.9.2) in [41] that the $C_P(Q)$-categories $\mathcal{F}_{(f,C_G(Q))}$ and $\mathcal{F}_{C_P(Q)}$ coincide, so that $C_{\mathcal{F}_{(b,G)}}(Q) = \mathcal{F}_{C_P(Q)}$ and Q is $\mathcal{F}_{(b,G)}$-nilcentralized (cf. 4.3).

Corollary 7.3 *With the notation above, denote by $\bar{f}$ the image of f in $k\bar{C}_G(Q)$. Then, Q is $\mathcal{F}_{(b,G)}$-selfcentralizing if and only if $\bar{f}$ has defect zero.*

Proof: If Q is $\mathcal{F}_{(b,G)}$-selfcentralizing then it is $\mathcal{F}_{(b,G)}$-nilcentralized, so that f is a nilpotent block and $C_P(Q) = Z(Q)$ is a defect subgroup of f; hence, $\bar{f}$ has defect zero (cf. 1.17). Conversely, if $\bar{f}$ has defect zero then $Z(Q)$ is a defect subgroup of f (cf. 1.17 and 1.19); but, as above, we may assume that Q is fully centralized in $\mathcal{F}_{(b,G)}$ and then, by Proposition 3.5, $C_P(Q)$ is a defect subgroup of the block f of $C_G(Q)$; hence, we have $C_P(Q) = Z(Q)$ and therefore Q is $\mathcal{F}_{(b,G)}$-selfcentralizing (cf. 4.10).

7.4 Coherently, we say that a Brauer (b,G)-pair (Q,f) is *nilcentralized* if f is a nilpotent block of $C_G(Q)$ (cf 1.17) and that it is *selfcentralizing* if $\bar{f}$ is a block of defect zero of $\bar{C}_G(Q) = C_G(Q)/Z(Q)$ (cf. 1.17). Assume that (Q,f) is nilcentralized; then, the quotient $kC_G(Q)f/J\big(kC_G(Q)f\big)$ is a simple k-algebra (cf. (1.9.1) in [41]) and therefore the action of the normalizer $N_G(Q,f)$ on this k-algebra determines a k^**-group* $\hat{N}_G(Q,f)$ (cf. 1.24) together with a k-algebra homomorphism (cf. 1.25)

$$k_*\hat{N}_G(Q,f) \longrightarrow kC_G(Q)f/J\big(kC_G(Q)f\big) \qquad 7.4.1;$$

moreover, since we have an obvious $N_G(Q,f)$-stable k-algebra homomorphism

$$kC_G(Q) \longrightarrow kC_G(Q)f/J\big(kC_G(Q)f\big) \qquad 7.4.2,$$

we still have a $N_G(Q,f)$-stable group homomorphism $C_G(Q) \to \hat{N}_G(Q,f)$ lifting the inclusion $C_G(Q) \subset N_G(Q,f)$ and therefore, assuming that (Q,f) is contained in (P,e), we get an exact sequence

$$1 \longrightarrow C_G(Q) \longrightarrow \hat{N}_G(Q,f) \longrightarrow \hat{\mathcal{F}}_{(b,G)}(Q) \longrightarrow 1 \qquad 7.4.3$$

for a suitable k^*-group $\hat{\mathcal{F}}_{(b,G)}(Q)$ lifting $\mathcal{F}_{(b,G)}(Q)$.

Proposition 7.5 *Let Q and Q' be $\mathcal{F}$-isomorphic $\mathcal{F}_{(b,G)}$-nilcentralized subgroups of P. Any $\mathcal{F}$-isomorphism $\varphi\colon Q \cong Q'$ determines a unique k^*-isomorphism $\hat{\mathcal{F}}_{(b,G)}(Q) \cong \hat{\mathcal{F}}_{(b,G)}(Q')$ induced by some $x \in G$ such that $Q' = Q^x$, $(Q,f)^x \subset (P,e)$ and $\varphi(u) = u^x$ for any $u \in Q$.*

Proof: By condition 3.2.2, there is such an $x \in G$ and, setting $f' = f^x$, it is clear that x induces isomorphisms compatible with the respective actions

$$\begin{gathered} N_G(Q,f) \cong N_G(Q',f') \\ kC_G(Q)f/J\big(kC_G(Q)f\big) \cong kC_G(Q')f'/J\big(kC_G(Q')f'\big) \end{gathered} \qquad 7.5.1;$$

hence, x induces a k^*-isomorphism $\hat{N}_G(Q,f) \cong \hat{N}_G(Q',f')$ and a commutative diagram

$$\begin{array}{ccccccccc} 1 & \longrightarrow & C_G(Q) & \longrightarrow & \hat{N}_G(Q,f) & \longrightarrow & \hat{\mathcal{F}}_{(b,G)}(Q) & \longrightarrow & 1 \\ & & \wr\| & & \wr\| & & \wr\| & & \\ 1 & \longrightarrow & C_G(Q') & \longrightarrow & \hat{N}_G(Q',f') & \longrightarrow & \hat{\mathcal{F}}_{(b,G)}(Q') & \longrightarrow & 1 \end{array} \qquad 7.5.2;$$

moreover, if $x' \in G$ is a second element fulfilling the same condition, the difference $x'x^{-1}$ centralizes Q and therefore it induces the identity map on $\hat{\mathcal{F}}_{(b,G)}(Q)$.

7.6 Recall that a maximal Brauer $(f, Q{\cdot}C_G(Q))$-pair is also a Brauer (b, G)-pair (cf. 1.10) and thus we may assume that (P, e) contains a maximal Brauer $(f, Q{\cdot}C_G(Q))$-pair (cf. 1.15). Denote by n the block of $C_G\big(N_P(Q)\big)$ such that (P, e) contains the Brauer (b, G)-pair $(N_P(Q), n)$ (cf. 1.15) and always assume that (Q, f) is nilcentralized. Then, since f is also a block of $C_G(Q){\cdot}N_P(Q)$ (cf. 1.19), it is easily checked that $(N_P(Q), n)$ is also a maximal Brauer $(f, C_G(Q){\cdot}N_P(Q))$-pair (cf. 1.16) and, according to Proposition 6.5 in [34], the block $(f, C_G(Q){\cdot}N_P(Q))$ is nilpotent too.

7.7 Now, it follows from the Main Theorem in [41] that, denoting by ν the local point of $N_P(Q)$ determined by n on the $N_P(Q)$-algebra $kC_G(Q)f$ (cf. 1.19), the quotient

$$S_Q = \big(kC_G(Q)f\big)_\nu \Big/ J\Big(\big(kC_G(Q)f\big)_\nu\Big) \qquad 7.7.1$$

is a *primitive Dade* $\bar{N}_P(Q)$*-algebra* (cf. 1.11 and 1.20) which, in particular, determines an element s_Q in the *Dade group* $\mathcal{D}_k\big(\bar{N}_P(Q)\big)$ (cf. 1.21); freely using our results in chapter 8 below — namely, some statement depending on results of [17] — we can be more precise on the nature of s_Q, but we will not need this precision in the sequel. Let us denote by $\mathbb{C}$ the field of *complex numbers*, by κ the *conjugation automorphism* of $\mathbb{C}$ and by $\kappa_G : \mathbb{C}G \cong (\mathbb{C}G)^\circ$ the κ*-anti-isomorphism* mapping any $x \in G$ on x^{-1}.

Theorem 7.8 *With the notation above,* s_Q *is a torsion element in the Dade group* $\mathcal{D}_k\big(\bar{N}_P(Q)\big)$.

Proof: Let $\mathcal{O}$ be an unramified complete discrete valuation ring of characteristic zero lifting k; without loss of generality, we may assume that k is the algebraic closure of the prime field and, in particular, that $\mathcal{O}$ is contained in $\mathbb{C}$; let us denote by $\mathcal{O}^\kappa$ the conjugate of $\mathcal{O}$, and set $\mathcal{R} = \mathcal{O} \cap \mathcal{O}^\kappa$ so that $\mathcal{R}$ is a *Dedekind ring*.

Set $T = N_P(Q)$ and $H = C_G(Q){\cdot}T$, and as above denote by n the block of $C_G(T)$ such that (P, e) contains (T, n). Let $\check{f}$ be the unique idempotent lifting f to $Z(\mathcal{O}H)$; since the coefficients of $\check{f}$ in the canonical $\mathcal{O}$-basis of $\mathcal{O}H$ are contained in the extension of $\mathbb{Z}_{(p)}$ by the p'-roots of unity (cf. [8]), $\check{f}$ still belongs to $Z(\mathcal{R}H)$; then, it is easily checked that $\check{f}^\kappa$ remains a primitive idempotent in $Z(\mathcal{O}H)$ — $\mathcal{O}$*-block* of H in the sequel — so that its image f^κ in $Z(kH)$ is again a block of H. Similarly, the Brauer (f^κ, H)-pairs are the pairs (R, g^κ) where (R, g) runs over the set of Brauer (f, H)-pairs and therefore f^κ is nilpotent too. In particular, (T, n^κ) is a maximal Brauer (f^κ, H)-pair.

It is well-known (cf. Lemma 3.13 in [12] or 7.11 in [41]) that the unique simple kHf-module comes from an ordinary absolutely irreducible character

χ in the $\mathcal{O}$-block $\check{f}$ and then, denoting by i_χ the corresponding primitive idempotent in $Z(\mathbb{C}H)$, the image of the canonical $\mathcal{O}$-algebra homomorphism

$$\rho_\chi : \mathcal{O}H\check{f} \longrightarrow \mathbb{C}Hi_\chi \qquad 7.8.1$$

is isomorphic to the matrix algebra $\mathfrak{M}_{\chi(1)}(\mathcal{O})$; but, by the argument above, $\check{f}$ is also an $\mathcal{O}^\kappa$-block of H and χ still belongs to this $\mathcal{O}^\kappa$-block; thus, denoting by $\rho_\chi^\kappa : \mathcal{O}^\kappa H\check{f} \longrightarrow \mathbb{C}Hi_\chi$ the corresponding $\mathcal{O}^\kappa$-algebra homomorphism and setting

$$\mathcal{M} = \mathrm{Ker}(\rho_\chi) \cap \mathrm{Ker}(\rho_\chi^\kappa) \qquad 7.8.2,$$

$\mathcal{M}$ is an ideal of $\mathcal{R}H\check{f}$ such that $\overline{\mathcal{R}H\check{f}} = \mathcal{R}H\check{f}/\mathcal{M}$ maps into $\mathbb{C}Hi_\chi$ and we have

$$\mathcal{O} \otimes_{\mathcal{R}} \overline{\mathcal{R}H\check{f}} \cong \mathfrak{M}_{\chi(1)}(\mathcal{O}) \quad \text{and} \quad \mathcal{O}^\kappa \otimes_{\mathcal{R}} \overline{\mathcal{R}H\check{f}} \cong \mathfrak{M}_{\chi(1)}(\mathcal{O}^\kappa) \qquad 7.8.3.$$

From now on, we consider $\overline{\mathcal{R}H\check{f}}$ as an $\mathcal{R}H$-module by the multiplication on the left; then, denoting by k_κ the field k endowed with the composed homomorphism

$$\mathcal{R} \xrightarrow{\kappa} \mathcal{R} \subset \mathcal{O} \longrightarrow k \qquad 7.8.4,$$

it follows from the very definition of $\mathcal{M}$ that the kH-modules $k \otimes_{\mathcal{R}} \overline{\mathcal{R}H\check{f}}$ and $k_\kappa \otimes_{\mathcal{R}} \overline{\mathcal{R}H\check{f}}$ are just the respective $\chi(1)$-multiples of the simple kHf- and kHf^κ-modules. More precisely, considering the unique idempotent $\check{n}$ lifting n to $Z\big(\mathcal{O}C_G(T)\big)$, we restrict ourselves to the $\mathcal{R}T$-submodule $\check{n}\overline{\mathcal{R}H\check{f}}$ which is a direct summand of $\mathrm{Res}_T^H(\overline{\mathcal{R}H\check{f}})$. In particular, the *endopermutation* kT-module V involved in the Dade T-algebra S_Q (cf. 1.20 and 7.7.1) is an indecomposable direct summand of the kT-module $k \otimes_{\mathcal{R}} \check{n}\overline{\mathcal{R}H\check{f}}$ and any indecomposable direct summand of vertex T of this kT-module is isomorphic to V (cf. 1.21).

It is clear that $\kappa_H(\mathcal{R}H) = (\mathcal{R}H)^\circ$ and in particular κ_H induces an $\mathcal{R}H$-module κ*-isomorphism* $\mathcal{R}H \cong (\mathcal{R}H)^\circ$; it is well-known that $\kappa_H(\check{f}) = \check{f}$, $\kappa_H(i_\chi) = i_\chi$ and $\kappa_H(\check{n}) = \check{n}$, and thus it is easily checked that the action of κ_H on $\mathbb{C}Hi_\chi$ stabilizes the image of $\check{n}\overline{\mathcal{R}H\check{f}}$; in particular, $\check{n}\overline{\mathcal{R}H\check{f}}$ is a $\langle\kappa_T\rangle$-stable $\mathcal{R}T$-module. At the same time, it is clear that the $\mathcal{R}$-algebras $\mathcal{R}H\check{f}^\kappa$ and $(\mathcal{R}H\check{f})^\circ$ are isomorphic, so that we have a kH-module isomorphism

$$k_\kappa \otimes_{\mathcal{R}} \overline{\mathcal{R}H\check{f}} \cong (k \otimes_{\mathcal{R}} \overline{\mathcal{R}H\check{f}})^* \qquad 7.8.5;$$

in particular, the k-dual kT-module V^* is an indecomposable direct summand of the kT-module $k_\kappa \otimes_{\mathcal{R}} \check{n}\overline{\mathcal{R}H\check{f}}$ and any indecomposable direct summand of vertex T of this kT-module is isomorphic to V^*.

On the other hand, it follows from Propositions 8.11 and 8.18, and Corollary 8.22 below that, for a suitable $n \in \mathbb{N}$, we have

$$p^n \cdot s_Q = \sum_U z_U \cdot \mathrm{ten}_U^T(r_1^U) \qquad 7.8.6,$$

where U runs over the set of nontrivial subgroups of T and, for such an U, z_U is an integer. Moreover, for any nontrivial subgroup U of T, mimicking the definition of r_1^U (cf. 8.17.2) consider the $\mathcal{R}U$-module

$$N_U = \bigotimes_R \mathrm{Res}_{\pi_R^U}\Big(\bigotimes^{\mu(U/R)} \mathcal{I}\big(\mathcal{R}(U/R)\big)\Big) \qquad 7.8.7,$$

where R runs over the set of subgroups of U containing $\Phi(U)$ (cf. 1.31) and, for such an R, $\pi_R^U : U \to U/R$ denotes the canonical map, $\mathcal{I}\big(\mathcal{R}(U/R)\big)$ the *augmentation* ideal of $\mathcal{R}(U/R)$, $\mu(U/R)$ the value of the *Möbius function* (cf. 1.32) over the set of proper subgroups of U strictly containing R, and $\bigotimes^{\mu(U/R)}$ the $\mu(U/R)$-th tensor power over $\mathcal{R}$, with the conventional rule that the 0-th and the (-1)-th tensor powers of an $\mathcal{R}$-torsion-free $\mathcal{R}(U/R)$-module are $\mathcal{R}$ and the $\mathcal{R}$-dual of this module respectively. Note that N_U is $\mathcal{R}$-torsion-free and $\langle \kappa_U \rangle$-stable as $\mathcal{R}U$-module.

Consequently, the $\mathcal{R}T$-module

$$N = \bigotimes_U \mathrm{Ten}_U^T\big(\bigotimes^{z_U} N_U\big) \qquad 7.8.8,$$

where U runs over the set of nontrivial subgroups of T, is $\mathcal{R}$-torsion-free and $\langle \kappa_T \rangle$-stable as $\mathcal{R}T$-module, and the scalar extensions $k \otimes_{\mathcal{R}} N$ and $k_\kappa \otimes_{\mathcal{R}} N$ are mutually isomorphic *endopermutation* kT-modules. Moreover, according to equality 7.8.6, the tensor product

$$(k \otimes_{\mathcal{R}} N)^* \otimes_k \big(k \otimes_{\mathcal{R}} \bigotimes^{p^n} \overline{\check{n}\mathcal{R}H\check{f}}\big) \cong k \otimes_{\mathcal{R}} \big(N^* \otimes_{\mathcal{R}} \bigotimes^{p^n} \overline{\check{n}\mathcal{R}H\check{f}}\,\big) \qquad 7.8.9$$

admits the trivial kT-module k as a direct summand.

Now, it follows from Lemma 7.10 below applied to the $\mathcal{R}$-torsion-free $\langle \kappa_T \rangle$-stable $\mathcal{R}T$-module $N^* \otimes_{\mathcal{R}} \bigotimes^{p^n} \overline{\check{n}\mathcal{R}H\check{f}}$ that the tensor product

$$(k_\kappa \otimes_{\mathcal{R}} N)^* \otimes_k \big(k_\kappa \otimes_{\mathcal{R}} \bigotimes^{p^n} \overline{\check{n}\mathcal{R}H\check{f}}\big) \cong k_\kappa \otimes_{\mathcal{R}} \big(N^* \otimes_{\mathcal{R}} \bigotimes^{p^n} \overline{\check{n}\mathcal{R}H\check{f}}\,\big) \quad 7.8.10$$

also admits the trivial kT-module k as a direct summand and therefore the class of the Dade T-algebra $\mathrm{End}_k(V^*)$ coincides with the class of the Dade T-algebra

$$\mathrm{End}_k(k \otimes_{\mathcal{R}} N) \cong \mathrm{End}_k(k_\kappa \otimes_{\mathcal{R}} N) \qquad 7.8.11,$$

which by construction coincides with $p^n \cdot s_Q$; that is to say, we get $2p^n \cdot s_Q = 0$. We are done.

Remark 7.9 By 8.16 and Corollary 8.22 below, the torsion subgroup of $\mathcal{D}_k\big(\bar{N}_P(Q)\big)$ is a 2-group and, if $p \neq 2$, it is elementary Abelian.

Lemma 7.10 *With the notation above, if M is an $\mathcal{R}$-torsion-free $\langle \kappa_T \rangle$-stable $\mathcal{R}T$-module then the trivial kT-module k is a direct summand of $k \otimes_{\mathcal{R}} M$ if and only if it is a direct summand of $k_\kappa \otimes_{\mathcal{R}} M$.*

Proof: Set $\bar{M} = M/p{\cdot}M$ and, following Green, consider the $\langle \kappa \rangle$-stable $\mathcal{R}$-module $\bar{M}[T]$ defined by the exact sequence

$$0 \longrightarrow [T, \bar{M}]^T \longrightarrow \bar{M}^T \longrightarrow \bar{M}[T] \longrightarrow 0 \qquad 7.10.1;$$

since $\mathcal{R}/p{\cdot}\mathcal{R} \cong k \times k_\kappa$, it is clear that we still have the exact sequences

$$\begin{aligned} 0 \longrightarrow [T, k \otimes_{\mathcal{R}} \bar{M}]^T \longrightarrow (k \otimes_{\mathcal{R}} \bar{M})^T \longrightarrow k \otimes_{\mathcal{R}} \bar{M}[T] \longrightarrow 0 \\ 0 \longrightarrow [T, k_\kappa \otimes_{\mathcal{R}} \bar{M}]^T \longrightarrow (k_\kappa \otimes_{\mathcal{R}} \bar{M})^T \longrightarrow k_\kappa \otimes_{\mathcal{R}} \bar{M}[T] \longrightarrow 0 \end{aligned} \qquad 7.10.2,$$

so that, considering the analogous over k of definition 7.10.1, we get

$$(k \otimes_{\mathcal{R}} M)[T] \cong k \otimes_{\mathcal{R}} \bar{M}[T] \quad \text{and} \quad (k_\kappa \otimes_{\mathcal{R}} M)[T] \cong k_\kappa \otimes_{\mathcal{R}} \bar{M}[T] \qquad 7.10.3;$$

but, if $k \otimes_{\mathcal{R}} M$ admits the trivial kT-module k as a direct summand, we have

$$(k \otimes_{\mathcal{R}} M)[T] \neq \{0\} \qquad 7.10.4;$$

consequently, we have $\bar{M}[T] \neq \{0\}$ and, since $\bar{M}[T]$ is a $\langle \kappa \rangle$-stable $\mathcal{R}$-module, it follows from the right isomorphism in 7.10.3 that we still have

$$(k_\kappa \otimes_{\mathcal{R}} M)[T] \neq \{0\} \qquad 7.10.5.$$

Then, for any element $1 \otimes n$ in $(k_\kappa \otimes_{\mathcal{R}} \bar{M})^T - [T, k_\kappa \otimes_{\mathcal{R}} \bar{M}]$, it is quite clear that $k{\cdot}(1 \otimes n)$ is a trivial direct summand in the kT-module $k_\kappa \otimes_{\mathcal{R}} \bar{M}$. We are done.

7.11 With the notation in 7.6 and 7.7, it follows from Theorem 6.6 in [34] that we have an $N_P(Q)$-algebra isomorphism (cf. 1.18.1)

$$\big(kC_G(Q)f\big)_\nu \cong S_Q C_P(Q) \qquad 7.11.1$$

and, considering the unique group homomorphism $C_P(Q) \to (S_Q)^*$ lifting the action of $C_P(Q)$ on S_Q, as usual we get a new $N_P(Q)$-algebra isomorphism (cf. (1.8.1) in [41])

$$\big(kC_G(Q)f\big)_\nu \cong S_Q \otimes_k kC_P(Q) \qquad 7.11.2.$$

Recall that, if Q' is a subgroup of P fully normalized in $\mathcal{F}_{(b,G)}$ and $\mathcal{F}_{(b,G)}$-isomorphic to Q, there is an $\mathcal{F}_{(b,G)}$-morphism $N_P(Q) \to N_P(Q')$ mapping Q onto Q' (cf. condition 2.8.2 and Theorem 3.7).

Proposition 7.12 *Let Q and Q' be $\mathcal{F}_{(b,G)}$-isomorphic $\mathcal{F}_{(b,G)}$-nilcentralized subgroups of P and assume that Q' is fully normalized in $\mathcal{F}_{(b,G)}$. Then, for any $\mathcal{F}_{(b,G)}$-morphism $\zeta : N_P(Q) \to N_P(Q')$ mapping Q onto Q', we have an $\bar{N}_P(Q)$-algebra embedding $S_Q \to \mathrm{Res}_{\bar{\zeta}}(S_{Q'})$ and the equality $s_Q = \mathrm{res}_{\bar{\zeta}}(s_{Q'})$, where*

$$\mathrm{res}_{\bar{\zeta}} : \mathcal{D}_k\big(\bar{N}_P(Q')\big) \longrightarrow \mathcal{D}_k\big(\bar{N}_P(Q)\big) \qquad 7.12.1$$

is the restriction map determined the homomorphism $\bar{\zeta} : \bar{N}_P(Q) \to \bar{N}_P(Q')$ induced by ζ.

Proof: Respectively denote by n and n' the blocks of $C_G\big(N_P(Q)\big)$ and of $C_G\big(N_P(Q')\big)$ such that (P,e) contains $(N_P(Q),n)$ and $(N_P(Q'),n')$ (cf. 1.15); by condition 3.2.2, there is $x \in G$ such that $(N_P(Q),n)^x \subset (N_P(Q'),n')$, $Q^x = Q'$ and $u^x = \zeta(u)$ for any $u \in N_P(Q)$; moreover, setting $f' = f^x$ and denoting by $\xi : N_G(Q,f) \cong N_G(Q',f')$ the isomorphism determined by x, it is clear that x induces an $N_G(Q,f)$-algebra isomorphism

$$kC_G(Q)f \cong \mathrm{Res}_{\,\xi}\big(kC_G(Q')f'\big) \qquad 7.12.2;$$

in particular, denoting by ν and ν' the respective local points of $N_P(Q)$ and $N_P(Q')$ determined by n and n' on the $N_P(Q)$- and $N_P(Q')$-algebras $kC_G(Q)f$ and $kC_G(Q')f'$ (cf. 1.19), we get an $\bar N_P(Q)$-algebra embedding

$$\big(kC_G(Q)f\big)_\nu \longrightarrow \mathrm{Res}_{\,\zeta}\big(kC_G(Q')f'\big)_{\nu'} \qquad 7.12.3.$$

Consequently, we get the announced $\bar N_P(Q)$-algebra embedding and equality (cf. 1.21).

7.13 Let R be an $\mathcal{F}_{(b,G)}$-nilcentralized subgroup of Q; in the Dade groups $\mathcal{D}_k\big(\bar N_P(Q)\big)$ and $\mathcal{D}_k\big(\bar N_P(R)\big)$ we have the respective elements s_Q and s_R; then, setting $N_{Q,R} = N_P(Q) \cap N_P(R)$, we have evident homomorphisms from the quotient $\bar N_{Q,R} = N_{Q,R}/R$ to $\bar N_P(Q)$ and to $\bar N_P(R)$, and we may ask ourselves for the relationship between the restrictions of s_Q and s_R to $\mathcal{D}_K(\bar N_{Q,R})$; if R is normal in Q, we can give an easy answer in terms of the homomorphism induced by the *Brauer quotient functor* (cf. 1.22)

$$\widetilde{\mathfrak{Br}}_{\bar Q}^{\bar N_{Q,R}} : \mathcal{D}_k(\bar N_{Q,R}) \longrightarrow \mathcal{D}_k(\bar N_{Q,R}/\bar Q) \subset \mathcal{D}_k(\bar N_{Q,R}) \qquad 7.13.1,$$

where we denote by $\bar Q$ the image of Q in $\bar N_{Q,R}$ and identify $\mathcal{D}_k(\bar N_{Q,R}/\bar Q)$ with its image in $\mathcal{D}_k(\bar N_{Q,R})$ by the restriction homomorphism.

Theorem 7.14 *Let Q and R be $\mathcal{F}_{(b,G)}$-nilcentralized subgroups of P such that Q contains and normalizes R, denote by $\bar N_{Q,R}$ and $\bar Q$ the respective images of $N_{Q,R} = N_P(Q) \cap N_P(R)$ and Q in $\bar N_P(R)$, and set $\bar{\bar N}_{Q,R} = \bar N_{Q,R}/\bar Q$. Then, we have*

$$\mathrm{res}^{\bar N_P(Q)}_{\bar{\bar N}_{Q,R}}(s_Q) = \widetilde{\mathfrak{Br}}_{\bar Q}^{\bar N_{Q,R}}\big(\mathrm{res}^{\bar N_P(R)}_{\bar N_{Q,R}}(s_R)\big) \qquad 7.14.1.$$

Proof: Denote by f and g the respective blocks of $C_G(Q)$ and $C_G(R)$ such that (P,e) contains (Q,f) and (R,g); in particular, we have $(R,g) \triangleleft (Q,f)$ (cf. 1.15) and therefore we get $f\,\mathrm{Br}_Q^{kC_G(R)}(g) = f$ (cf. Theorem 1.8 in [11]). As above, respectively denote by n and m the blocks of $C_G\big(N_P(Q)\big)$ and $C_G(N_{Q,R})$ such that (P,e) contains $(N_P(Q),n)$ and $(N_{Q,R},m)$ (cf. 1.15).

Since we have $C_G(N_{Q,R}) \subset C_G(Q) \subset C_G(R)$, $(N_{Q,R}, m)$ is also a Brauer $(f, N_{Q,R}{\cdot}C_G(Q))$-pair and a Brauer $(g, N_{Q,R}{\cdot}C_G(R))$-pair, and therefore m respectively determines local points ν of $\bar{\bar{N}}_{Q,R}$ and μ of $\bar{N}_{Q,R}$ on the $\bar{\bar{N}}_{Q,R}$- and $\bar{N}_{Q,R}$-algebras $\mathrm{Res}^{\bar{N}_P(Q)}_{\bar{\bar{N}}_{Q,R}}(kC_G(Q)f)$ and $\mathrm{Res}^{\bar{N}_P(R)}_{\bar{N}_{Q,R}}(kC_G(R)g)$ (cf. 1.19), so that we have $\bar{\bar{N}}_{Q,R}$- and $\bar{N}_{Q,R}$-algebra isomorphisms (cf. 7.11.2)

$$\begin{aligned}(kC_G(Q)f)_\nu &\cong (S_Q)_\nu \otimes_k kC_P(Q)\\ (kC_G(R)g)_\mu &\cong (S_R)_\mu \otimes_k kC_P(R)\end{aligned} \qquad 7.14.2,$$

where we identify ν and μ with the respective unique local points of $\bar{\bar{N}}_{Q,R}$ on S_Q and of $\bar{N}_{Q,R}$ on S_R (cf. Corollary 5.8 in [41]).

Moreover, it follows from the existence of $\bar{N}_{Q,R}$-stable bases that we have an $\bar{\bar{N}}_{Q,R}$-algebra isomorphism (cf. Proposition 5.6 in [41])

$$(kC_G(R)g)_\mu(\bar{Q}) \cong (S_R)_\mu(\bar{Q}) \otimes_k kC_P(Q) \qquad 7.14.3$$

and that $\mathrm{Br}^{S_R}_{\bar{Q}}(\mu)$ is the local point of $\bar{\bar{N}}_{Q,R}$ on $S_R(Q)$ (cf. Corollary 5.8 in [41]), so that

$$(S_R)_\mu(\bar{Q}) \cong S_R(Q)_{\mathrm{Br}^{S_R}_{\bar{Q}}(\mu)} \qquad 7.14.4.$$

Since the Dade $\bar{\bar{N}}_{Q,R}$- and $\bar{N}_{Q,R}$-algebras $(S_Q)_\nu$ and $(S_R)_\mu$ respectively determine $\mathrm{res}^{\bar{N}_P(Q)}_{\bar{\bar{N}}_{Q,R}}(s_Q)$ in $\mathcal{D}_k(\bar{\bar{N}}_{Q,R})$ and $\mathrm{res}^{\bar{N}_P(R)}_{\bar{N}_{Q,R}}(s_R)$ in $\mathcal{D}_k(\bar{N}_{Q,R})$ (cf. 1.22), equality 7.14.1 follows from isomorphism 7.14.4 above and Proposition 7.12.

7.15 A fundamental result on Dade P-algebras (cf. 9.2) allows us to answer the analogous of the question above on the k^*-groups $\hat{\mathcal{F}}_{(b,G)}(Q)$ introduced in 7.4. As above, let R be an $\mathcal{F}_{(b,G)}$-nilcentralized subgroup of Q and denote by $\mathcal{F}_{(b,G)}(Q)_R$ the stabilizer of R in $\mathcal{F}_{(b,G)}(Q)$; it follows from the divisibility of $\mathcal{F}_{(b,G)}$ (cf. Proposition 3.3) that the restriction induces a group homomorphism

$$\mathcal{F}_{(b,G)}(Q)_R \longrightarrow \mathcal{F}_{(b,G)}(R) \qquad 7.15.1;$$

but the k^*-group $\hat{\mathcal{F}}_{(b,G)}(Q)$ above determines a k^*-subgroup $\hat{\mathcal{F}}_{(b,G)}(Q)_R$; thus, we may ask for the relationship between $\hat{\mathcal{F}}_{(b,G)}(Q)_R$ and $\hat{\mathcal{F}}_{(b,G)}(R)$ and, in the next result, the answer is that they agree. Actually, in chapter 11 this result will appear as a consequence of a more general framework.

Theorem 7.16 *If Q and R are $\mathcal{F}_{(b,G)}$-nilcentralized subgroups of P such that Q contains R, the restriction group homomorphism $\mathcal{F}_{(b,G)}(Q)_R \to \mathcal{F}_{(b,G)}(R)$ can be lifted to a k^*-group homomorphism*

$$\hat{\mathcal{F}}_{(b,G)}(Q)_R \longrightarrow \hat{\mathcal{F}}_{(b,G)}(R) \qquad 7.16.1$$

and the kernel is a p-group.

Proof: The last statement follows from Corollary 4.7. Arguing by induction on $|Q:R|$ and setting $N = N_Q(R)$, if R is not normal in Q then we may assume that we have k^*-group homomorphisms

$$\hat{\mathcal{F}}_{(b,G)}(Q)_N \longrightarrow \hat{\mathcal{F}}_{(b,G)}(N) \quad \text{and} \quad \hat{\mathcal{F}}_{(b,G)}(N)_R \longrightarrow \hat{\mathcal{F}}_{(b,G)}(R) \qquad 7.16.2$$

lifting the corresponding restriction homomorphisms; then, since $\mathcal{F}_{(b,G)}(Q)_R$ is contained in $\mathcal{F}_{(b,G)}(Q)_N$ and moreover its image in $\mathcal{F}_{(b,G)}(N)$ is still contained in $\mathcal{F}_{(b,G)}(N)_R$, we get the announced lifting by composition.

In the sequel, we assume that Q normalizes R; then, by Proposition 2.7, we may assume that R is fully normalized in $\mathcal{F}_{(b,G)}$. Denote by f and g the respective blocks of $C_G(Q)$ and $C_G(R)$ such that (P,e) contains (Q,f) and (R,g) (cf. 1.15), set $A = kGb$ and consider the local points δ of Q and ε of R on A respectively determined by f and g (cf. 1.13 and 1.18)). The existence of the lifting 7.16.1 will follow from the coincidence of the k^*-group structures on the intersection $N_G(Q,f) \cap N_G(R,g)$, coming from $\hat{N}_G(Q,f)$ and from $\hat{N}_G(R,g)$; actually, it suffices to prove this coincidence over any p'-subgroup K of this intersection or, equivalently, to prove that the central extension of K by k^* which is the *difference* of the respective extensions coming from $\hat{N}_G(Q,f)$ and $\hat{N}_G(R,g)$ is split.

We borrow our notation from section 10 in [42]; more precisely, let K be a p'-subgroup of $\mathcal{F}_{(b,G)}(Q)_R$ and denote by $\widehat{K}^\delta$ and by $\widehat{K}^\varepsilon$ the respective extensions coming from $\hat{\mathcal{F}}_{(b,G)}(Q)_R$ and from $\hat{\mathcal{F}}_{(b,G)}(R)$ *via* the restriction map $\mathcal{F}_{(b,G)}(Q)_R \to \mathcal{F}_{(b,G)}(R)$, and by $\widehat{K}^{\varepsilon^\circ \delta}$ their *difference*; then, it follows from Proposition 10.13 in [42] that we have a k^*-group homomorphism

$$\rho : \widehat{K}^{\varepsilon^\circ \delta} \longrightarrow A_\delta(R_\varepsilon)^* \qquad 7.16.3;$$

moreover, since Q fixes g (cf. 1.15), $\bar{Q} = Q/R$ acts on the simple k-algebra $A_\delta(R_\varepsilon)$ (cf. 1.18); thus, we have a unique group homomorphism $\bar{Q} \to A_\delta(R_\varepsilon)^*$ lifting this action and then it is easily checked that $\rho(\widehat{K}^{\varepsilon^\circ \delta})$ normalizes the image of $\bar{Q}$.

On the other hand, we know that $A(R) \cong kC_G(R)\mathrm{Br}_R(b)$ (cf. 1.19) and that $\mathrm{Br}_R(\delta)$ still remains a local point of $\bar{Q}$ on $A(R)$; then, according to 7.11.2, it is not difficult to get the $\bar{Q}$-algebra isomorphisms

$$A_\delta(R) \cong A(R)_{\mathrm{Br}_R(\delta)} \cong (S_R)_{\mathrm{Br}_R(\delta)} \otimes_k C_P(R) \qquad 7.16.4,$$

where S_R denotes the Dade $\bar{N}_P(R)$-algebra introduced in 7.7.1 above and where we identify $\mathrm{Br}_R(\delta)$ with the unique local point of $\bar{Q}$ on S_R (cf. Corollary 5.8 in [41]). Consequently, we finally get a $\bar{Q}$-algebra isomorphism

$$A_\delta(R_\varepsilon) \cong (S_R)_{\mathrm{Br}_R(\delta)} \qquad 7.16.5.$$

But, it follows from the splitting theorem for Dade $\bar{Q}$-algebras (cf. Theorem 9.21 below or [40]) that we have a k^*-group homomorphism

$$\omega_{\bar{Q},S_R} : \hat{F}_{S_R}(\bar{Q}) \longrightarrow k^* \tag{7.16.6}$$

and, by the very definition of the k^*-group $\hat{F}_{S_R}(\bar{Q})$ (cf. 1.29), it is clear that ρ determines a k^*-group homomorphism $\widehat{K}^{\varepsilon^\circ\delta} \rightarrow \hat{F}_{S_R}(\bar{Q})$. We are done.

Chapter 8

Decompositions for Dade P-algebras

8.1 Let P be a nontrivial finite p-group and k an algebraically closed field of characteristic p. In the previous chapter we have seen that the *Dade P-algebras* over k (cf. 1.20) appear in a natural way when dealing with blocks. In order to handle them easily, in this chapter we show that any Dade P-algebra over k raised to some p^n-th tensor power admits — up to *similarity* (cf. 1.21) — a canonical decomposition in terms of the tensor product of suitable *tensor induced* Dade Q-algebras over k from the subgroups Q of P.

8.2 More generally, if G is a finite group, H a subgroup of G and B an H-algebra over k (cf. 1.11), recall that, following Serre, the *tensor induction* of B from H to G is the G-algebra (cf. A2.2 in [48])

$$\mathrm{Ten}_H^G(B) = \bigotimes_{C\in G/H} (kC \otimes_{kH} B) \qquad 8.2.1,$$

where kC denotes the k-vector space over the class $C \subset G$, endowed with the (right) kH-module structure determined by the multiplication on the right, where we consider B as a kH-module *via* the action of H, and where the product and the action of G are defined by the *formulæ*

$$\Big(\bigotimes_{C\in G/H} (x_C \otimes a_C)\Big)\Big(\bigotimes_{C\in G/H} (x_C \otimes a'_C)\Big) = \bigotimes_{C\in G/H} (x_C \otimes a_C a'_C)$$
$$\Big(\bigotimes_{C\in G/H} (x_C \otimes a_C)\Big)^x = \bigotimes_{C\in G/H} (x^{-1}x_C \otimes a_C) \qquad 8.2.2,$$

for any $x_C \in C$, any $a_C, a'_C \in B$ and any $x \in G$.

8.3 Note that we have an injective H-algebra homomorphism

$$^B\mathrm{t}_H^G : B \longrightarrow \mathrm{Res}_H^G\big(\mathrm{Ten}_H^G(B)\big) \qquad 8.3.1$$

mapping $a \in B$ on

$$^B\mathrm{t}_H^G(a) = \bigotimes_{C\in G/H} (x_C \otimes a_C) \qquad 8.3.2$$

where $x_H = 1$, $a_H = a$ and $a_C = 1$ for any $C \in G/H - \{H\}$; we write $\mathrm{t}_H^G(a)$ instead of $^B\mathrm{t}_H^G(a)$ for short, and note that, for any $x \in G - H$, the elements $\mathrm{t}_H^G(a)$ and $\mathrm{t}_H^G(a)^x$ centralize each other. Thus, for any subgroup K of G, any $a \in B^{K\cap H}$ and any set of representatives $X \subset K$ for the set of classes $(K\cap H)\backslash K$, the product

$$\mathrm{Nr}^K_{K\cap H}\big(\mathrm{t}^G_H(a)\big) = \prod_{x\in X} \mathrm{t}^G_H(a)^x \qquad 8.3.3$$

does not depend neither on the choice of X nor on the choice of a *total order* on X, and it is clearly an element of $\mathrm{Ten}^G_H(B)^K$.

8.4 All this construction is clearly *functorial* on B; that is to say, any H-algebra homomorphism $g: B \to B'$ determines a G-algebra homomorphism

$$\mathrm{Ten}^G_H(g) : \mathrm{Ten}^G_H(B) \longrightarrow \mathrm{Ten}^G_H(B') \qquad 8.4.1,$$

this correspondence is compatible with the composition of homomorphisms and the H-algebra homomorphisms ${}^B\mathrm{t}^G_H$ are *natural* on B. Moreover, we have a canonical P-algebra isomorphism

$$\mathrm{Ten}^G_H(B\otimes_k B') \cong \mathrm{Ten}^G_H(B)\otimes_k \mathrm{Ten}^G_H(B') \qquad 8.4.2$$

compatible with ${}^{B\otimes_k B'}\mathrm{t}^G_H$ and ${}^B\mathrm{t}^G_H \otimes {}^{B'}\mathrm{t}^G_H$. Furthermore, for any subgroup L of G containing H, we have a *natural* G-algebra isomorphism

$$\mathrm{Ten}^G_L\big(\mathrm{Ten}^L_H(B)\big) \cong \mathrm{Ten}^G_H(B) \qquad 8.4.3.$$

On the other hand, as in the *ordinary induction*, we have an evident *Mackey decomposition*; namely, for any subgroup K of G, there is a *natural* K-algebra isomorphism

$$\mathrm{Res}^G_K\big(\mathrm{Ten}^G_H(B)\big) \cong \bigotimes_{x\in X} \mathrm{Ten}^K_{K\cap H^x}\big(\mathrm{Res}_{\kappa_x}(B)\big) \qquad 8.4.4,$$

where $X \subset G$ is a set of representatives for $H\backslash G/K$ and, for any $x \in X$, $\kappa_x : K\cap H^x \to H$ is the group homomorphism induced by the conjugation by x.

8.5 Let us come back to the the finite p-groups; for any P-algebra A over k and any $\sigma \in \mathrm{Aut}(k)$, we set ${}^\sigma\!A = \mathrm{Res}_\sigma(A)$; moreover, we denote by $\phi : k \cong k$ the *Frobenius automorphism* of k; as a matter of fact, the Frobenius automorphism appears when computing the *Brauer quotients* (cf. 1.13) of a tensor induction†.

† When proving Proposition 8.6 in the eighties we were misled on the behaviour of the scalar coefficients via homomorphism 8.6.1 — we just forgot to check it — and it was Serge Bouc, ten years later, who mention to us that this behaviour involves the Frobenius automorphism. His independent version of Proposition 8.6 for Q-algebras B with a Q-stable basis appears in [6].

Proposition 8.6 *Let Q and R be subgroups of P and B a Q-algebra. Assume that R is normal in P, set $\bar{P} = P/R$, $|R| = p^{v_R}$ and $|Q \cap R| = p^{v_{Q\cap R}}$, denote by $\bar{Q}$ the image of Q in $\bar{P}$, and consider $B(Q\cap R)$ as a $\bar{Q}$-algebra. We have a unique $\bar{P}$-algebra homomorphism*

$$\mathrm{Ten}_{\bar{Q}}^{\bar{P}}\big({}^{\phi^{v_{Q\cap R}}}B(Q\cap R)\big) \longrightarrow {}^{\phi^{v_R}}\big(\mathrm{Ten}_Q^P(B)\big)(R) \qquad 8.6.1$$

mapping $\mathrm{t}_{\bar{Q}}^{\bar{P}}\big(\mathrm{Br}_{Q\cap R}(a)\big)$ *on* $\mathrm{Br}_R\Big(\mathrm{Nr}_{Q\cap R}^{R}\big(\mathrm{t}_Q^P(a)\big)\Big)$ *for any $a \in B^{Q\cap R}$, which is an isomorphism whenever Q stabilizes a basis of B.*

Proof: Firstly assume that R is contained in Q; on the one hand, in that case the inclusion $B^R \subset B$ induces a $\bar{P}$-algebra homomorphism

$$\mathrm{Ten}_{\bar{Q}}^{\bar{P}}(B^R) \cong \mathrm{Ten}_Q^P(B^R) \longrightarrow \mathrm{Ten}_Q^P(B)^R \longrightarrow \big(\mathrm{Ten}_Q^P(B)\big)(R) \qquad 8.6.2$$

mapping $\mathrm{t}_{\bar{Q}}^{\bar{P}}(a)$ on $\mathrm{Br}_R\big(\mathrm{t}_Q^P(a)\big)$ for any $a \in B^R$; on the other hand, it is easily checked that the kernel of the canonical map

$$\mathrm{Ten}_{\bar{Q}}^{\bar{P}}(\mathrm{Br}_R^B) : \mathrm{Ten}_{\bar{Q}}^{\bar{P}}(B^R) \longrightarrow \mathrm{Ten}_{\bar{Q}}^{\bar{P}}\big(B(R)\big) \qquad 8.6.3$$

is generated by $\{\mathrm{t}_{\bar{Q}}^{\bar{P}}\big(\mathrm{Ker}(\mathrm{Br}_R^B)\big)^u\}_{u\in P}$. But, if $a = \mathrm{Tr}_T^R(c)$ for a proper subgroup T of R and an element $c \in B^T$, we have $\mathrm{t}_Q^P(a) = \mathrm{Tr}_T^R\big(\mathrm{t}_Q^P(c)\big)$ since $R \subset Q$; consequently, homomorphism 8.6.2 determines a $\bar{P}$-algebra homomorphism

$$\mathrm{Ten}_{\bar{Q}}^{\bar{P}}\big(B(R)\big) \longrightarrow \big(\mathrm{Ten}_Q^P(B)\big)(R) \qquad 8.6.4$$

mapping $\mathrm{t}_{\bar{Q}}^{\bar{P}}\big(\mathrm{Br}_R(a)\big)$ on $\mathrm{Br}_R\big(\mathrm{t}_Q^P(a)\big)$ for any $a \in B^R$. Moreover, starting from a Q-stable basis of B, it is not difficult to obtain bases of both ends and to check that this homomorphism induces a bijection between them.

By isomorphism 8.4.3, now it suffices to discuss the case where $P = Q{\cdot}R$; setting $T = Q \cap R$, isomorphism 8.4.4 yields the R-algebra isomorphism

$$\mathrm{Res}_R^P\big(\mathrm{Ten}_Q^P(B)\big) \cong \mathrm{Ten}_T^R\big(\mathrm{Res}_T^Q(B)\big) \qquad 8.6.5$$

and, up to the obvious identification, we claim that the map

$${}^{\phi^{v_T}}B^T \longrightarrow {}^{\phi^{v_R}}\big(\mathrm{Ten}_Q^P(B)\big)(R) \qquad 8.6.6$$

sending $a \in B^T$ to $\mathrm{Br}_R\Big(\mathrm{Nr}_T^R\big(\mathrm{t}_Q^P(a)\big)\Big)$ is a $\bar{Q}$-algebra homomorphism.

Indeed, for any $a, a' \in B^T$, we have

$$\mathrm{t}_Q^P(a + a') = \mathrm{t}_Q^P(a) + \mathrm{t}_Q^P(a') \qquad 8.6.7$$

and, choosing a set of representatives $X \subset R$ for R/T, we get

$$\begin{aligned}\mathrm{Nr}_T^R\big(\mathrm{t}_Q^P(a+a')\big) &= \bigotimes_{u\in X} u\otimes(a+a') \\ &= \sum_{Y\subset X}\Big(\bigotimes_{u\in Y} u\otimes a\Big)\otimes\Big(\bigotimes_{u'\in X-Y} u'\otimes a'\Big)\end{aligned} \qquad 8.6.8;$$

but, the tensor products

$$a_Y = \bigotimes_{u\in Y} u\otimes a \quad \text{and} \quad a'_{X-Y} = \bigotimes_{u'\in X-Y} u'\otimes a' \qquad 8.6.9$$

only depend on the image of Y in R/T and for any $w \in R$ we have (cf. 8.2.2)

$$(a_Y \otimes a'_{X-Y})^w = a_{w^{-1}\cdot Y} \otimes a'_{w^{-1}\cdot X - w^{-1}\cdot Y} \qquad 8.6.10;$$

moreover, the stabilizer in R of this image by left multiplication is a proper subgroup of R unless we have $Y = \emptyset$ or $Y = X$; hence, by the very definition of the Brauer quotient (cf. 1.13) we finally get

$$\begin{aligned}&\mathrm{Br}_R\Big(\mathrm{Nr}_T^R\big(\mathrm{t}_Q^P(a+a')\big)\Big) \\ &\qquad = \mathrm{Br}_R\Big(\mathrm{Nr}_T^R\big(\mathrm{t}_Q^P(a)\big)\Big) + \mathrm{Br}_R\Big(\mathrm{Nr}_T^R\big(\mathrm{t}_Q^P(a')\big)\Big)\end{aligned} \qquad 8.6.11.$$

Furthermore, it is quite clear that

$$\mathrm{Nr}_T^R\big(\mathrm{t}_Q^P(aa')\big) = \mathrm{Nr}_T^R\big(\mathrm{t}_Q^P(a)\big)\mathrm{Nr}_T^R\big(\mathrm{t}_Q^P(a')\big) \qquad 8.6.12$$

and that, for any $\lambda \in k^*$, we get

$$\mathrm{Nr}_T^R\big(\mathrm{t}_Q^P(\lambda\cdot a)\big) = \lambda^{p^{v_R - v_T}}\cdot\mathrm{Nr}_T^R\big(\mathrm{t}_Q^P(a)\big) \qquad 8.6.13.$$

Finally, for any $w \in Q$, we obtain (cf. 8.2.2)

$$\begin{aligned}\mathrm{Nr}_T^R\big(\mathrm{t}_Q^P(a^w)\big) &= \bigotimes_{u\in X} u\otimes a^w \\ &= \Big(\bigotimes_{u\in X} w\cdot u\cdot w^{-1}\otimes a\Big)^w = \mathrm{Nr}_T^R\big(\mathrm{t}_Q^P(a)\big)^w\end{aligned} \qquad 8.6.14,$$

which completes the proof of the claim.

On the other hand, if we have $a = \mathrm{Tr}_U^T(c)$ for a proper subgroup U of T and an element $c \in B^U$, considering the canonical map $R/U \to R/T$, we get

$$\mathrm{Nr}_T^R\big(\mathrm{t}_Q^P(a)\big) = \bigotimes_{u\in X}\Big(u\otimes\sum_w {}^w c\Big) = \sum_s\Big(\bigotimes_{u\in X} u\otimes {}^{u^{-1}s(u)}c\Big) \qquad 8.6.15,$$

where w runs over a set of representatives for T/U in T and $s : R/T \to R/U$ runs over the sections of the canonical map, once we have identified R/T with X and R/U with some set of representatives Y so that $u^{-1}s(u)$ belongs to T for any $u \in X$.

It is clear that R acts on R/T and R/U by the multiplication on the left, which is compatible with the canonical map; hence, R still acts on the set of sections of the canonical map and, for any $v \in R$, we have (cf. 8.2.2)

$$\Big(\bigotimes_{u\in X} u \otimes {}^{u^{-1}s(u)}c\Big)^v = \bigotimes_{u\in X} v^{-1}u \otimes {}^{u^{-1}s(u)}c = \bigotimes_{u'\in X} u' \otimes {}^{u'^{-1}s^v(u')}c \qquad 8.6.16;$$

consequently, choosing a set of representatives $\mathcal{S}$ for the set of R-orbits in the set of sections of the canonical map and, for any $s \in \mathcal{S}$, denoting by R_s the stabilizer of s in R, equality 8.6.15 becomes

$$\mathrm{Nr}_T^R\big(\mathrm{t}_Q^P(a)\big) = \sum_{s\in\mathcal{S}} \mathrm{Tr}_{R_s}^R\Big(\bigotimes_{u\in X} u \otimes {}^{u^{-1}s(u)}c\Big) \qquad 8.6.17$$

which proves that $\mathrm{Br}_R\Big(\mathrm{Nr}_T^R\big(\mathrm{t}_Q^P(a)\big)\Big) = 0$ since R fixes no section.

In conclusion, homomorphism 8.6.6 induces a $\bar{P}$-algebra homomorphism

$${}^{\phi^{vT}}B(T) \longrightarrow {}^{\phi^{vR}}\big(\mathrm{Ten}_Q^P(B)\big)(R) \qquad 8.6.18.$$

As above, starting from a Q-stable basis of B, it is not difficult to obtain bases of both ends and to check that this homomorphism induces a bijection between them. We are done.

8.7 In particular, if Q is a subgroup of P and S a Dade Q-algebra, it is quite clear that P stabilizes a basis of $\mathrm{Ten}_Q^P(S)$ and that $\mathrm{Ten}_Q^P(S)$ is a simple k-algebra, and it is easily checked from Proposition 8.6 that

$$\big(\mathrm{Ten}_Q^P(S)\big)(P) \neq \{0\} \qquad 8.7.1;$$

that is to say, $\mathrm{Ten}_Q^P(S)$ is a Dade P-algebra. Clearly, for any kQ-module N, we have an analogous definition for the *tensor induction* $\mathrm{Ten}_Q^P(N)$, and a P-algebra isomorphism

$$\mathrm{Ten}_Q^P\big(\mathrm{End}_k(N)\big) \cong \mathrm{End}_k\big(\mathrm{Ten}_Q^P(N)\big) \qquad 8.7.2;$$

then, if N has a Q-stable basis, $\mathrm{Ten}_Q^P(N)$ has a P-stable basis too. Thus, according to isomorphism 8.4.2 above, for any Dade Q-algebra S' *similar* to S (cf. 1.21), $\mathrm{Ten}_Q^P(S')$ is also similar to $\mathrm{Ten}_Q^P(S)$. That is to say, the tensor induction from Q to P determines a map between the Dade groups (cf. 1.21)

$$\mathrm{ten}_Q^P : \mathcal{D}_k(Q) \longrightarrow \mathcal{D}_k(P) \tag{8.7.3}$$

which, according again to isomorphism 8.4.2, is a group homomorphism.

8.8 These group homomorphisms inherit the properties of the tensor induction; namely, for any pair of subgroups Q and R of P, we have the *Mackey formula*

$$\mathrm{res}_R^P \circ \mathrm{ten}_Q^P = \sum_{u\in X} \mathrm{ten}_{R\cap Q^u}^R \circ \mathrm{res}_{\kappa_u} \qquad 8.8.1,$$

where $X \subset P$ is a set of representatives for $Q\backslash P/R$ and, for any $u \in X$, $\kappa_u : R \cap Q^u \to Q$ is the group homomorphism induced by u (cf. 1.8.1); also, whenever Q contains R we get (cf. 8.4.3)

$$\mathrm{ten}_Q^P \circ \mathrm{ten}_R^Q = \mathrm{ten}_R^P \qquad 8.8.2.$$

Moreover, if R is normal in P, it follows from Proposition 8.6 that, respectively denoting by v_R and $v_{Q\cap R}$ the p-valuations of R and $Q \cap R$, we have (cf. 1.22)

$$\mathrm{ten}_{\bar{Q}}^{\bar{P}} \circ {}^{\phi^{v_{Q\cap R}}}\widetilde{\mathfrak{Br}}_{Q\cap R}^Q = {}^{\phi^{v_R}}\widetilde{\mathfrak{Br}}_R^P \circ \mathrm{ten}_Q^P \qquad 8.8.3,$$

where we also denote by ϕ the automorphism of $\mathcal{D}_k(\bar{P})$ determined by the Frobenius automorphism (cf. 8.5).

8.9 As we did in [42, §3] with the ordinary induction here, for the tensor induction, also we consider the *residual Dade groups*

$$\mathcal{R}^\circ\mathcal{D}_k(P) = \bigcap_Q \mathrm{Ker}(\mathrm{res}_Q^P) \quad \text{and} \quad \mathcal{R}\mathcal{D}_k(P) = \mathcal{D}_k(P)\Big/\sum_Q \mathrm{Im}(\mathrm{ten}_Q^P) \tag{8.9.1}$$

where Q runs over the set of proper subgroups of P. We denote by

$$\mathrm{td}_P^P : \mathcal{R}^\circ\mathcal{D}_k(P) \longrightarrow \mathcal{D}_k(P) \quad \text{and} \quad \mathrm{rd}_P^P : \mathcal{D}_k(P) \longrightarrow \mathcal{R}\mathcal{D}_k(P) \tag{8.9.2}$$

the canonical maps and, more generally, for any subgroup Q of P, we set

$$\mathrm{td}_Q^P = \mathrm{ten}_Q^P \circ \mathrm{td}_Q^Q \quad \text{and} \quad \mathrm{rd}_Q^P = \mathrm{rd}_Q^Q \circ \mathrm{res}_Q^P \qquad 8.9.3.$$

8.10 These two families of maps determine two group homomorphisms

$$\Big(\bigoplus_Q \mathcal{R}^\circ\mathcal{D}_k(Q)\Big)^P \xrightarrow{\mathrm{td}^P} \mathcal{D}_k(P) \quad \text{and} \quad \mathcal{D}_k(P) \xrightarrow{\mathrm{td}^P} \Big(\prod_Q \mathcal{R}\mathcal{D}_k(Q)\Big)^P \qquad 8.10.1,$$

where Q runs over the set of subgroups of P, which actually have *p-torsion* kernels and cokernels; namely, denoting by $\mathbb{Z}_p$ the ring $\mathbb{Z}$ localized over the set $\{p^n\}_{n\in\mathbb{N}}$ of powers of p, and setting

$$\mathbb{Z}_p \otimes_{\mathbb{Z}} \mathcal{D}_k(P) = \mathcal{D}_k(P)_p \qquad 8.10.2,$$

the corresponding homomorphisms are isomorphisms, as we show below. Denote by s^u the obvious image of $s \in \mathcal{D}_k(Q)_p$ in $\mathcal{D}_k(Q^u)_p$ for any $u \in P$.

Proposition 8.11 *The maps* rd^P *and* td^P *induce* $\mathbb{Z}_p$*-module isomorphisms*

$$\Big(\bigoplus_Q \mathcal{R}^\circ\mathcal{D}_k(Q)_p\Big)^P \cong \mathcal{D}_k(P)_p \cong \Big(\prod_Q \mathcal{R}\mathcal{D}_k(Q)_p\Big)^P \qquad 8.11.1$$

where Q runs over the set of subgroups of P. In particular, for any subgroup Q of P, rd_Q^Q *induces an isomorphism* $\mathcal{R}^\circ\mathcal{D}_k(Q)_p \cong \mathcal{R}\mathcal{D}_k(Q)_p$ *and, denoting by*

$$\varpi_Q^P : \mathcal{D}_k(P)_p \longrightarrow \mathcal{R}^\circ\mathcal{D}_k(Q)_p \subset \mathcal{D}_k(Q)_p \qquad 8.11.2$$

the composition of rd_Q^P *with the inverse of this isomorphism, for any element* $s \in \mathcal{D}_k(P)_p$ *we have*

$$s = \sum_Q \frac{|Q|}{|P|}\cdot \mathrm{ten}_Q^P\big(\varpi_Q^P(s)\big) \qquad 8.11.3,$$

where Q runs over the set of all the subgroups of P.

Proof: First of all, if Q and R are subgroups of P and s is an element of $\mathcal{R}^\circ\mathcal{D}_k(Q)$, it follows from the Mackey formula (cf. 8.8.1) that

$$\mathrm{rd}_R^P\big(\mathrm{ten}_Q^P(s)\big) = \sum_u \mathrm{rd}_R^R(s^u) \qquad 8.11.4,$$

where $u \in P$ runs over a set of representatives for the set of classes

$$\{u \in P \mid Q^u = R\}/R \subset Q\backslash P/R \qquad 8.11.5.$$

Now, we claim that it suffices to prove the isomorphism

$$\mathcal{R}^\circ\mathcal{D}_k(P)_p \cong \mathcal{R}\mathcal{D}_k(P)_p \qquad 8.11.6.$$

Indeed, on the one hand it follows from equality 8.11.4 that for any element $\sum_Q s_Q$ of $\big(\bigoplus_Q \mathcal{R}^\circ\mathcal{D}_k(Q)_p\big)^P$, where $s_Q \in \mathcal{R}^\circ\mathcal{D}_k(Q)_p$, we get

$$\big((\mathrm{rd}^{\mathrm{P}})_p \circ (\mathrm{td}^P)_p\big)\big(\sum_Q s_Q\big) = \big(|P/Q|\cdot(\mathrm{rd}_Q^Q)_p(s_Q)\big)_Q \qquad 8.11.7$$

where Q runs over the set of subgroups of P, so that $(\mathrm{td}^P)_p$ is injective and $(\mathrm{rd}^P)_p$ is surjective. On the other hand, once isomorphism 8.11.6 is proved, by applying this isomorphism to any subgroup Q of P , we get

$$\mathcal{D}_k(Q)_p = \mathcal{R}^\circ\mathcal{D}_k(Q)_p + \sum_R \mathrm{ten}_R^Q\big(\mathcal{D}_k(R)_p\big) \qquad 8.11.8,$$

where R runs over the set of proper subgroups of Q, and therefore, arguing by induction on $|Q|$, we still get

$$\mathcal{D}_k(P)_p = \sum_Q \mathrm{ten}_Q^P\big(\mathcal{R}^\circ\mathcal{D}_k(Q)_p\big) \qquad 8.11.9,$$

where Q runs over the set of subgroups of P; moreover, if $s = \sum_Q \mathrm{ten}_Q^P(s_Q)$ in $\mathcal{D}_k(P)_p$, where $s_Q \in \mathcal{R}^\circ\mathcal{D}_k(Q)_p$, we still have

$$s = \frac{1}{|P|}\cdot\sum_Q \mathrm{ten}_Q^P\Big(\sum_{u\in P}(s_{Q^u})^{u^{-1}}\Big) \qquad 8.11.10,$$

so that s belongs to the image of $(\mathrm{td}^P)_p$, proving the claim.

Let s be an element of the kernel of the map

$$(\mathrm{rd}_P^P)_p \circ (\mathrm{td}_P^P)_p : \mathcal{R}^\circ\mathcal{D}_k(P)_p \to \mathcal{R}\mathcal{D}_k(P)_p \qquad 8.11.11;$$

in particular, s belongs to $\sum_Q \mathrm{ten}_Q^P\big(\mathcal{D}_k(Q)_p\big)$ where Q runs over the set of proper subgroups of P; arguing by induction on $|Q|$ and applying again equality 8.11.10, we get $s = \sum_R \mathrm{ten}_R^P(s_R)$ where R runs over the set of proper subgroups of P and, for such an R, s_R belongs to $\mathcal{R}^\circ\mathcal{D}_k(R)_p$ and fulfills $(s_R)^u = s_{R^u}$ for any $u \in P$; hence, by equality 8.11.4, we get

$$0 = (\mathrm{rd}_Q^P)_p(s) = |\bar N_P(Q)|\cdot(\mathrm{rd}_Q^Q)_p(s_Q) \qquad 8.11.12$$

and therefore, applying again the induction hypothesis, we obtain $s_Q = 0$.

In order to prove the surjectivity of the map 8.11.11, it suffices to prove that, for any $s \in \mathcal{D}_k(P)_p$, $(\mathrm{rd}_P^P)_p(s)$ belongs to $(\mathrm{rd}_P^P)_p\big(\mathcal{R}^\circ\mathcal{D}_k(P)_p\big)$. Let $\mathfrak{Q}$ be the set of subgroups Q of P such that $\mathrm{res}_Q^P(s) \neq 0$; we argue by induction on $|\mathfrak{Q}|$ and may assume that $|\mathfrak{Q}| > 1$; let Q be a minimal element of $\mathfrak{Q}$ and set

$$s' = s - \frac{1}{|\bar N_P(Q)|}\cdot\mathrm{ten}_Q^P\big(\mathrm{res}_Q^P(s)\big) \qquad 8.11.13;$$

since $Q \neq P$, we have $(\mathrm{rd}_P^P)_p(s') = (\mathrm{rd}_P^P)_p(s)$; moreover, once again by the Mackey formula, for any subgroup R of P we get

$$\mathrm{res}_R^P(s') = \mathrm{res}_R^P(s) - \frac{1}{|\bar N_P(Q)|}\cdot\sum_u \mathrm{ten}_{R\cap Q^u}^R\big(\mathrm{res}_{R\cap Q^u}^P(s)\big) \qquad 8.11.14,$$

where $u \in P$ runs over a set of representatives for $Q\backslash P/R$, and therefore, by the minimality of Q, it is clear that $\mathrm{res}^P_R(s') \neq 0$ implies the existence of $u \in P$ such that $Q^u \subset R$ and $Q^u \neq R$; hence, by our induction hypothesis, $(\mathrm{rd}^P_P)_p(s')$ belongs to $(\mathrm{rd}^P_P)_p\big(\mathcal{R}^\circ\mathcal{D}_k(P)_p\big)$. We are done.

Corollary 8.12 *For any normal subgroup N of P, any subgroup Q of P containing N and any $\bar{s} \in \mathcal{D}_k(P/N)_p$, setting $\bar{P} = P/N$ and $\bar{Q} = Q/N$, and denoting by $\pi^P_N : P \to \bar{P}$ and $\pi^Q_N : Q \to \bar{Q}$ the canonical maps, we have*

$$\mathrm{res}_{\pi^Q_N}\big(\varpi^{\bar{P}}_{\bar{Q}}(\bar{s})\big) = \sum_R \frac{|R|}{|Q|}\cdot\mathrm{ten}^Q_R\Big(\varpi^P_R\big(\mathrm{res}_{\pi^P_N}(\bar{s})\big)\Big) \qquad 8.12.1$$

where R runs over the set of subgroups of Q such that $R{\cdot}N = Q$.

Proof: It is clear that, for any subgroup R of Q such that $R{\cdot}N \neq Q$, we have

$$\mathrm{res}^Q_R\Big(\mathrm{res}_{\pi^Q_N}\big(\mathcal{R}^\circ\mathcal{D}_k(\bar{Q})\big)\Big) = \{0\} \qquad 8.12.2$$

and therefore we still have (cf. Proposition 8.11)

$$\mathrm{res}_{\pi^Q_N}\big(\mathcal{R}^\circ\mathcal{D}_k(\bar{Q})\big) \subset \bigoplus_R \mathrm{ten}^Q_R\big(\mathcal{R}^\circ\mathcal{D}_k(R)\big) \qquad 8.12.3,$$

where R runs over the set of subgroups of Q such that $R{\cdot}N = Q$. Thus, setting $s = \mathrm{res}_{\pi^P_N}(\bar{s})$, since (cf. 8.11.3)

$$\bar{s} = \sum_{\bar{Q}} \frac{|\bar{Q}|}{|\bar{P}|}\cdot\mathrm{ten}^{\bar{P}}_{\bar{Q}}\big(\varpi^{\bar{P}}_{\bar{Q}}(\bar{s})\big) \quad \text{and} \quad s = \sum_R \frac{|R|}{|P|}\cdot\mathrm{ten}^P_R\big(\varpi^P_R(s)\big) \qquad 8.12.4,$$

where $\bar{Q}$ and R respectively run over the sets of subgroups of $\bar{P}$ and P, equality 8.12.1 follows from Proposition 8.11 and inclusion 8.12.3 above.

Corollary 8.13 *For any pair of subgroups Q and R of P such that $R \subset \Phi(Q)$ and Q normalizes R, setting $N = N_P(R)$, $\bar{N} = \bar{N}_P(R)$ and $\bar{Q} = Q/R$ we have*

$$\varpi^{\bar{N}}_{\bar{Q}} \circ \widetilde{\mathfrak{Br}}^N_R = \widetilde{\mathfrak{Br}}^Q_R \circ \varpi^N_Q \qquad 8.13.1.$$

Proof: From equality 8.11.3 applied to Q, for any $s \in \mathcal{D}_k(Q)$ we get

$$\begin{aligned} {}^{\phi^{vR}}\widetilde{\mathfrak{Br}}^N_R(s) &= \sum_T \frac{|T|}{|N|}\cdot{}^{\phi^{vR}}\widetilde{\mathfrak{Br}}^N_R\Big(\mathrm{ten}^N_T\big(\varpi^N_T(s)\big)\Big) \\ &= \sum_T \frac{|T|}{|N|}\cdot\mathrm{ten}^{\bar{N}}_{\bar{T}}\Big(\mathrm{res}_{\theta_{T,R}}\Big({}^{\phi^{vT\cap R}}\widetilde{\mathfrak{Br}}^T_{T\cap R}\big(\varpi^N_T(s)\big)\Big)\Big) \end{aligned} \qquad 8.13.2,$$

where T runs over the set of subgroups of N and, for such a T, we denote by $\bar{T}$ its image in $\bar{N}$ and by $\theta_{T,R}$ the canonical isomorphism $\bar{T} \cong T/(T\cap R)$; now, since $\bar{T} = \bar{Q}$ forces $T = Q$, equality 8.13.1 follows from equality 8.11.3 applied to $\bar{Q}$.

8.14 Proposition 8.11 above states that, for any Dade P-algebra S, a suitable p^n-th tensor power of S admits, up to similarity, a *canonical decomposition* as a tensor product of the tensor induction from the subgroups Q of P of a Dade Q-algebra in $\mathcal{R}^\circ\mathcal{D}_k(Q)$; thus, we are reduced to the description of $\mathcal{R}\mathcal{D}_k(P)_p$ which, as a matter of fact, is very simple since it has at most $\mathbb{Z}_p$-rank one. Unfortunately, we have no direct proof of this fact and we need the strong classification results in [19], [17] and [6].

8.15 In the case where P is Abelian, for any subgroup Q of P and any Dade P-algebra S, $S(Q)$ is a Dade P/Q-algebra and therefore, denoting by $\pi_Q^P : P \to P/Q$ the canonical map, $\mathrm{Res}_{\pi_Q^P}\big(S(Q)\big)$ is a Dade P-algebra. Consequently, we get a group endomorphism of $\mathcal{D}_k(P)$

$$\beta_Q^P = \mathrm{res}_{\pi_Q^P} \circ \widetilde{\mathfrak{Br}}_Q^P : \mathcal{D}_k(P) \longrightarrow \mathcal{D}_k(P/Q) \longrightarrow \mathcal{D}_k(P) \qquad 8.15.1$$

which is clearly an *idempotent* endomorphism; moreover, for another subgroup R of P, it is easily checked that (cf. equality 1.22.3)

$$\beta_Q^P \circ \beta_R^P = \beta_{Q\cdot R}^P = \beta_R^P \circ \beta_Q^P \qquad 8.15.2;$$

hence, inductively defining a family $\{\alpha_Q^P\}_Q$ of endomorphisms of $\mathcal{D}_k(P)$ by

$$\beta_Q^P = \sum_R \alpha_R^P \qquad 8.15.3,$$

where R runs over the set of subgroups of P containing Q, it is quite easy to check from equality 8.15.2 that $\{\alpha_Q^P\}_Q$ is a *pairwise orthogonal idempotent decomposition* of the identity; thus, we get

$$\begin{gathered}\mathcal{D}_k(P) = \bigoplus_Q \alpha_Q^P\big(\mathcal{D}_k(P)\big) \\ \alpha_Q^P\big(\mathcal{D}_k(P)\big) \cong \alpha_1^{P/Q}\big(\mathcal{D}_k(P/Q)\big)\end{gathered} \qquad 8.15.4,$$

where Q runs over the set of subgroups of P.

8.16 Recall that, since we have a kP-module isomorphism

$$\mathrm{End}_k\big(J(kP)\big) \cong k \oplus kP^{|P|-2} \qquad 8.16.1,$$

$\mathrm{End}_k\big(J(kP)\big)$ is a Dade P-algebra (cf. 1.20) and, for any nontrivial subgroup Q of P, we have

$$\Big(\mathrm{End}_k\big(J(kP)\big)\Big)(Q) \cong k \qquad 8.16.2;$$

denote by j_1^P the element of $\mathcal{D}_k(P)$ determined by this Dade P-algebra and, for any subgroup Q of P, note that $\mathrm{res}_Q^P(j_1^P) = j_1^Q$. When P is Abelian, it is quite clear that j_1^P belongs to $\alpha_1^P\big(\mathcal{D}_k(P)\big)$ and Dade's Classification in [19] states that this group is generated by this element; more precisely, he proves:

8.16.3 $\mathcal{D}_k(\mathbb{Z}/2\mathbb{Z}) \cong \{0\}$.

8.16.4 *If P is cyclic and $|P| \geq 3$ then $\alpha_1^P\big(\mathcal{D}_k(P)\big) = \mathbb{Z}{\cdot}j_1^P \cong \mathbb{Z}/2\mathbb{Z}$.*

8.16.5 *If P is Abelian noncyclic then $\alpha_1^P\big(\mathcal{D}_k(P)\big) = \mathbb{Z}{\cdot}j_1^P \cong \mathbb{Z}$.*

8.17 In this case, we set $j_Q^P = \mathrm{res}_{\pi_Q^P}(j_1^{P/Q})$ for any subgroup Q of P and, by the equality and the isomorphisms in 8.15.4, we get

$$\mathcal{D}_k(P) = \bigoplus_Q \mathbb{Z}{\cdot}j_Q^P \qquad 8.17.1$$

where Q runs over the set of proper subgroups of P. Moreover, denoting by $\mu(P)$ the value of the Möbius function (cf. 1.32) on the *ordered* set of proper nontrivial subgroups of P, in $\mathcal{D}_k(P)_p$ we still consider the following element

$$r_1^P = \sum_Q \mu(P/Q){\cdot}j_Q^P \qquad 8.17.2,$$

where Q runs over the set of proper subgroups of P.

Proposition 8.18 *If P is elementary Abelian we have*

$$\mathcal{R}^\circ\mathcal{D}_k(P)_p = \mathbb{Z}_p{\cdot}r_1^P \qquad 8.18.1.$$

In particular $\mathcal{R}^\circ\mathcal{D}_k(P)_p \cong \mathbb{Z}_p$ unless $|P| = p$ where either $p \neq 2$ and we have $\mathcal{R}^\circ\mathcal{D}_k(P)_p \cong \mathbb{Z}/2\mathbb{Z}$ or $p = 2$ and we have $\mathcal{R}^\circ\mathcal{D}_k(P)_p = \{0\}$.

Proof: For any proper subgroup R of P, it is quite clear that

$$\mathrm{res}_R^P(r_1^P) = \sum_Q \mu(P/Q){\cdot}j_{Q\cap R}^R = \sum_T \Big(\sum_U \mu(P/U)\Big){\cdot}j_T^R \qquad 8.18.2,$$

where Q runs over the set of (proper) subgroups of P, T runs over the set of (proper) subgroups of R and, for such a T, U runs over the set of subgroups of P containing T and fulfilling $U \cap R = T$, which implies that the sum $\sum_U \mu(P/U)$ is equal to zero.

Moreover, since $\mu(P)$ is invertible in $\mathbb{Z}_p$ and we have $\mathbb{Z}_p{\cdot}r_1^P \cong \mathbb{Z}_p{\cdot}j_1^P$ by statements 8.16.3, 8.16.4 and 8.16.5, it follows from equality 8.17.1 that

$$\mathcal{D}_k(P)_p = \mathbb{Z}_p{\cdot}r_1^P \oplus \Big(\bigoplus_Q \mathbb{Z}_p{\cdot}j_Q^P\Big) \qquad 8.18.3$$

where Q runs over the set of nontrivial (proper) subgroups of P; but, since for any nontrivial (proper) subgroup Q of P and any complement R of Q in P we have $\mathrm{res}^P_R(j^P_Q) = j^R_1$, and since Q is clearly determined by the set of its complements in P, the restrictions to the proper subgroups R of P map injectively the direct summand $\bigoplus_Q \mathbb{Z}_p{\cdot}j^P_Q$ above into $\bigoplus_R \mathcal{D}_k(R)_p$, where R runs over the set of proper subgroups of P. We are done.

8.19 In the general case, set $\bar{P}^{\Phi} = P/\Phi(P)$ and denote by $\pi^P_\Phi : P \to \bar{P}^{\Phi}$ the canonical map; since, for any proper subgroup Q of P, $\pi^P_\Phi(Q)$ is also a proper subgroup of $\bar{P}^{\Phi}$, we have

$$\mathrm{res}_{\pi^P_\Phi}\big(\mathcal{R}^\circ\mathcal{D}_k(\bar{P}^{\Phi})\big) \subset \mathcal{R}^\circ\mathcal{D}_k(P) \qquad 8.19.1;$$

moreover, since the Brauer homomorphism is compatible with the restriction, it is quite clear that

$$\widetilde{\mathfrak{Br}}^P_{\Phi(P)}\big(\mathcal{R}^\circ\mathcal{D}_k(P)\big) \subset \mathcal{R}^\circ\mathcal{D}_k(\bar{P}^{\Phi}) \qquad 8.19.2$$

and, since $\widetilde{\mathfrak{Br}}^P_{\Phi(P)} \circ \mathrm{res}_{\pi^P_\Phi} = \mathrm{id}_{\mathcal{D}_k(\bar{P}^{\Phi})}$, we get the equality

$$\widetilde{\mathfrak{Br}}^P_{\Phi(P)}\big(\mathcal{R}^\circ\mathcal{D}_k(P)\big) = \mathcal{R}^\circ\mathcal{D}_k(\bar{P}^{\Phi}) \qquad 8.19.3.$$

8.20 Actually, we claim that $\widetilde{\mathfrak{Br}}^P_{\Phi(P)}$ induces an isomorphism between $\mathcal{R}^\circ\mathcal{D}_k(P)$ and $\mathcal{R}^\circ\mathcal{D}_k(\bar{P}^{\Phi})$ and now it suffices to prove the injectivity; that is to say, it suffices to prove that, for any $s \in \mathcal{R}^\circ\mathcal{D}_k(P)$ such that $\widetilde{\mathfrak{Br}}^P_{\Phi(P)}(s) = 0$, we have $s = 0$; moreover, since for any normal subgroup Q of P, we have (cf. equality 1.22.3)

$$\widetilde{\mathfrak{Br}}^{P/Q}_{\Phi(P/Q)} \circ \widetilde{\mathfrak{Br}}^P_Q = \widetilde{\mathfrak{Br}}^{\bar{P}^{\Phi}}_{\pi^P_\Phi(Q)} \circ \widetilde{\mathfrak{Br}}^P_{\Phi(P)} \qquad 8.20.1,$$

arguing by induction on $|P|$ we may assume that s belongs to the subgroup

$$\bigcap_Q \mathrm{Ker}(\widetilde{\mathfrak{Br}}^P_Q) \cap \mathcal{R}^\circ\mathcal{D}_k(P) = \bigcap_R \mathrm{Ker}(\widetilde{\mathfrak{Br}}^{N_P(R)}_R) \cap \mathcal{R}^\circ\mathcal{D}_k(P) \qquad 8.20.2,$$

where Q runs over the set of nontrivial normal subgroups of P and R over the set of all the nontrivial ones.

8.21 As a matter of fact, if P is not elementary Abelian, this subgroup coincides with the kernel of homomorphism (2.2.1) in [43] where we prove that this kernel is *finite*, which already implies that

$$\mathbb{Q} \otimes_{\mathbb{Z}} \mathcal{R}^\circ\mathcal{D}_k(P) \cong \mathbb{Q} \otimes_{\mathbb{Z}} \mathcal{R}^\circ\mathcal{D}_k(\bar{P}^{\Phi}) \qquad 8.21.1;$$

the main ingredient for proving this finiteness was Carlson's main theorem in [16] and, pushing further all the machinery Jon Carlson developed since, it is proved in [17] that

8.21.2 *If P is not elementary Abelian, intersection* 8.20.2 *is trivial unless $p = 2$ and then it is a 2-group.*

Thus, the argument above and statements 8.16.5 and 8.21.2 yield:

Corollary 8.22 *With the notation above, $\widetilde{\mathfrak{Br}}^P_{\Phi(P)}$ induces an isomorphism*

$$\mathcal{R}^\circ\mathcal{D}_k(P)_p \cong \mathcal{R}^\circ\mathcal{D}_k(\overline{P}^\Phi)_p \qquad 8.22.1.$$

Thus, if P is not cyclic then $\mathcal{R}^\circ\mathcal{D}_k(P)_p \cong \mathbb{Z}_p$. Moreover, if $p \neq 2$ then it also induces an isomorphism $\mathcal{R}^\circ\mathcal{D}_k(P) \cong \mathcal{R}^\circ\mathcal{D}_k(\overline{P}^\Phi)$ and $\mathcal{D}_k(P)$ maps into $\mathcal{D}_k(P)_p$.

Chapter 9

Polarizations for Dade P-algebras

9.1 Let P be a finite p-group, k an algebraically closed field of characteristic p, S a Dade P-algebra over k and s the class of S in the *Dade group* $\mathcal{D}_k(P)$ (cf. 1.21); recall that S can be considered as a *P-interior algebra* (cf. 1.20 and 1.27), that P has a unique *local point* γ^S on S (cf. 1.20) and that the group of *S-fusions* from P_{γ^S} to P_{γ^S} (cf. 1.28) — simply noted $F_S(P)$ — coincides with the stabilizer $\mathrm{Aut}(P)_s$ of s in $\mathrm{Aut}(P)$ (cf. Proposition 2.18 in [39]). Choosing $i \in \gamma^S$ and denoting by P_i the canonical lifting to $(iSi)^*$ of the action of P on iSi, it follows from the very definition of the S-fusions that we have a group homomorphism (cf. homomorphism 1.29.1)

$$F_S(P) \longrightarrow N_{(iSi)^*}(P_i)/(iS^Pi)^* \qquad 9.1.1$$

and then, we define the *k^*-group* $\hat{F}_S(P)$ (cf. 1.29) by the *pull-back*

$$\begin{array}{ccc} F_S(P) & \longrightarrow & N_{(jSj)^*}(P_i)/(jS^Pj)^* \\ \uparrow & & \uparrow \\ \hat{F}_S(P) & \longrightarrow & N_{(iSi)^*}(P_i)/\big(i + J(iS^Pi)\big) \end{array} \qquad 9.1.2.$$

9.2 Although we only have a poor reference [40], it is known that *this k^*-group is split* or, equivalently, that there is a k^*-group homomorphism

$$\omega_{P,S} : \hat{F}_S(P) \longrightarrow k^* \qquad 9.2.1$$

and the main purpose of this chapter is to discuss the possibility of making *coherent choices* for $\omega_{P,S}$; by the way we give a complete proof of the *existence* of such a k^*-group homomorphism, short cutting the reference [40].

9.3 In order to formulate what we mean by *coherent choices*, let us consider the following category $\mathfrak{D}_k$. The *$\mathfrak{D}_k$-objects* are the pairs (P, S) formed by a finite p-group P and by a Dade P-algebra S over k; for a second $\mathfrak{D}_k$-object (P', S'), the *$\mathfrak{D}_k$-morphisms* from (P, S) to (P', S') are the pairs (π, f) formed by a surjective group homomorphism $\pi : P \to P'$ such that $\mathrm{Ker}(\pi)$ is $F_S(P)$-stable, and by a P-interior algebra *embedding* (cf. 1.11)

$$f : \mathrm{Res}_\pi(S') \longrightarrow S \qquad 9.3.1.$$

9.4 Then, for a second $\mathfrak{D}_k$-morphism (π', f') from (P', S') to a third $\mathfrak{D}_k$-object (P'', S''), embedding 9.3.1 forces (cf. Proposition 2.14 in [39])

$$F_S(P) = F_{\mathrm{Res}_\pi(S')}(P) \qquad 9.4.1$$

and, since $F_S(P)$ stabilizes $\mathrm{Ker}(\pi)$, it follows from the very definition of the S'-fusions that the action of $F_S(P)$ on $P/\mathrm{Ker}(\pi) \cong P'$ is contained in $F_{S'}(P')$, so that we get a group homomorphism

$$F_{\mathfrak{f}}(\pi) : F_S(P) \longrightarrow F_{S'}(P') \qquad 9.4.2;$$

hence, $F_S(P)$ stabilizes $\mathrm{Ker}(\sigma')$ and thus, since

$$\mathrm{Ker}(\sigma' \circ \sigma) = \sigma^{-1}\big(\mathrm{Ker}(\sigma')\big) \qquad 9.4.3,$$

$F_S(P)$ stabilizes $\mathrm{Ker}(\sigma' \circ \sigma)$ too. Consequently, the pair $\big(\pi' \circ \pi, f \circ \mathrm{Res}_\pi(f')\big)$ is a $\mathfrak{D}_k$-morphism from (P,S) to (P'',S'') and we define the composition in $\mathfrak{D}_k$ by

$$(\pi', f') \circ (\pi, f) = \big(\pi' \circ \pi, f \circ \mathrm{Res}_\pi(f')\big) \qquad 9.4.4,$$

which clearly fulfills the associative condition.

9.5 Now, we have a functor $\mathfrak{f} : \mathfrak{D}_k \to \mathfrak{Gr}$ from $\mathfrak{D}_k$ to the category of finite groups $\mathfrak{Gr}$ mapping any $\mathfrak{D}_k$-object (P,S) on the group $F_S(P)$ and any $\mathfrak{D}_k$-morphism $(\pi, f) : (P,S) \to (P',S')$ on the group homomorphism $F_{\mathfrak{f}}(\pi)$ above; actually, it follows from Proposition 6.8 in [42] that embedding 9.3.1 determines a k^*-group homomorphism

$$\hat{F}_{\mathfrak{f}}(\pi) : \hat{F}_S(P) \longrightarrow \hat{F}_{S'}(P') \qquad 9.5.1$$

lifting $F_{\mathfrak{f}}(\pi)$; consequently, $\mathfrak{f}$ can be lifted to a functor

$$\hat{\mathfrak{f}} : \mathfrak{D}_k \longrightarrow k^*\text{-}\mathfrak{Gr} \qquad 9.5.2,$$

where k^*-$\mathfrak{Gr}$ is the category of k^*-groups with finite k^*-quotients (cf. 1.23), mapping (P,S) on $\hat{F}_S(P)$ and (π, f) on $\hat{F}_{\mathfrak{f}}(\pi)$; note that if S is a P-algebra with trivial P-action then we have $F_S(P) = \mathrm{Aut}(P)$ and the pull-back 9.1.2 determines a k^*-group isomorphism

$$\hat{F}_S(P) \cong k^* \times \mathrm{Aut}(P) \qquad 9.5.3.$$

We are interested in the natural maps ω — called the *polarizations* — from this functor to the *trivial* one — namely, to the functor mapping (P,S) on k^* and (π, f) on id_{k^*} — such that if S is a P-algebra with trivial P-action then $\omega_{P,S}$ coincides with the first projection in isomorphism 9.5.3.

9.6 While we shall prove the existence of a polarization, as we show in the following example for $p \neq 2$ it is not possible to get a polarization *compatible with the tensor product*, namely a polarization ω fulfilling

9.6.1 *If S and S′ are Dade P-algebras, the following diagram is commutative*

$$\begin{array}{ccc} \hat{F}_S(P) \mathbin{\hat\cap} \hat{F}_{S'}(P) & \longrightarrow & \hat{F}_{S\otimes_k S'}(P) \\ {\scriptstyle \omega_{P,S} \hat{\times} \omega_{P,S'}} \searrow & & \swarrow {\scriptstyle \omega_{P,S\otimes_k S'}} \\ & k^* & \end{array}$$

where $\hat{F}_S(P) \mathbin{\hat\cap} \hat{F}_{S'}(P)$ is the converse image in $\hat{F}_S(P) \hat{\times} \hat{F}_{S'}(P)$ (cf. 1.24) of the *diagonal image* of $F_S(P) \cap F_{S'}(P)$ in $F_S(P) \times F_{S'}(P)$, and where the top k^*-group homomorphism is given by Proposition 5.11 in [41] (see 9.15 below for more detail).

Example 9.7 Assume that $|P| = p \neq 2$ and set $S = \mathrm{End}_k\big(J(kP)\big)$ (cf. 8.16); then, it is quite clear that

$$F_S(P) = \mathrm{Aut}(P) \cong \mathbb{Z}/(p-1)\mathbb{Z} \qquad 9.7.1,$$

and it is well-known that the kP-module $J(kP)$ is *selfdual* or, equivalently, that there is an *embedding* (cf. 1.10) from k to $S \otimes_k S$. Explicitly, choosing a nontrivial element $u \in P$, we have an obvious kP-module isomorphism

$$J(kP) \cong kP/k{\cdot}(1-u)^{p-1} \qquad 9.7.2$$

mapping $1-u$ on the class $\bar{1}$ of 1 and the scalar product in kP — defined by the product and by the k-linear form $\pi_1 : kP \to k$ mapping $1 \in P$ on $1 \in k$ and $v \in P - \{1\}$ on 0 — induces a kP-module isomorphism

$$kP/k{\cdot}(1-u)^{p-1} \cong \big(J(kP)\big)^* \qquad 9.7.3.$$

Since isomorphism 9.7.2 maps $(1-u)^m$ on $\overline{(1-u)^{m-1}}$ for any $m \geq 1$, then isomorphism 9.7.3 determines a scalar product in $J(kP)$ which, for any $n \geq 1$, sends $\big((1-u)^n, (1-u)^m\big)$ to

$$\pi_1\big((1-u)^{n+m-1}\big) = \begin{cases} 1 & \text{if } n+m \leq p \\ 0 & \text{if } n+m > p \end{cases} \qquad 9.7.4;$$

in particular, the family $\{(1-u)^m - (1-u)^{m+1}\}_{0<m<p}$ is the *dual* basis of $\{(1-u)^n\}_{0<n<p}$. Consequently, the kP-module isomorphism

$$S \cong J(kP) \otimes_k J(kP) \qquad 9.7.5$$

determined by isomorphisms 9.7.2 and 9.7.3 maps $\mathrm{id}_{J(kP)}$ on

$$r = \sum_{0<n<p} (1-u)^n \otimes \big((1-u)^{p-n} - (1-u)^{p-n+1}\big) \qquad 9.7.6.$$

Thus, since we clearly have

$$S = \mathrm{Ker}(\mathrm{tr}_{J(kP)}) \oplus k\cdot\mathrm{id}_{J(kP)} \qquad 9.7.7,$$

$k\cdot r$ is a trivial direct summand of the kP-module $M = J(kP) \otimes_k J(kP)$.

On the other hand, $\mathrm{Aut}(P)$ acts naturally on kP, $J(kP)$ and M, and, for any $\sigma \in \mathrm{Aut}(P)$, denoting by $\lambda_\sigma \in k^*$ the element of the prime subfield fulfilling $\sigma(u) = u^{\check\lambda_\sigma}$ for the lifting $\check\lambda_\sigma \in \Delta_{p-1}$ of λ_σ, we have

$$\sigma(1-u) = 1 - u^{\check\lambda_\sigma} = \Big(\sum_{h=0}^{\check\lambda_\sigma - 1} u^h\Big)(1-u) \qquad 9.7.8$$

and therefore we still have

$$\sigma(r) = \sum_{0<n<p} (1-u^{\check\lambda_\sigma})^n \otimes \big((1-u^{\check\lambda_\sigma})^{p-n} - (1-u^{\check\lambda_\sigma})^{p-n+1}\big) \qquad 9.7.9$$

which the scalar product maps on

$$\begin{aligned}\sum_{0<n<p} \pi_1\big((1-u^{\check\lambda_\sigma})^{p-1} - (1-u^{\check\lambda_\sigma})^p\big) &= \sum_{0<n<p} \pi_1\big((1-u^{\check\lambda_\sigma})^{p-1}\big)\\ &= \sum_{0<n<p} \pi_1\Big(\Big(\sum_{h=0}^{\check\lambda_\sigma-1} u^h\Big)^{p-1}(1-u)^{p-1}\Big) \qquad 9.7.10\\ &= \sum_{0<n<p} \pi_1\big((\lambda_\sigma)^{p-1}\cdot(1-u)^{p-1}\big)\\ &= -(\lambda_\sigma)^{-1}\end{aligned}$$

since $u(1-u)^{p-1} = (1-u)^{p-1}$ and $\pi_1\big((1-u)^{p-1}\big) = 1$.

In conclusion, it is clear that P still fixes the following element of M

$$t = \sum_{\sigma\in\mathrm{Aut}(P)} \lambda_\sigma\cdot\sigma(r) \qquad 9.7.11,$$

and it is easily checked that the scalar product maps t on 1, so that $k\cdot t$ is also a trivial direct summand the kP-module M, and that $\sigma(t) = (\lambda_\sigma)^{-1}\cdot t$ for any $\sigma \in \mathrm{Aut}(P)$.

Now, the $\mathfrak{D}_k$-morphism $(P, S\otimes_k S) \to (\{1\}, k)$ defined by the embedding

$$k \longrightarrow S\otimes_k S = \mathrm{End}_k(M) \qquad 9.7.12$$

induced by $k\cdot t$ determines a k^*-group homomorphism (cf. 9.5.1)

$$\hat F_{S\otimes_k S}(P) \longrightarrow k^* \qquad 9.7.13$$

which determines a k^*-group isomorphism

$$\hat{F}_{S\otimes_k S}(P) \cong k^* \times \mathrm{Aut}(P) \qquad 9.7.14;$$

moreover, since $\omega_{P,k} = \mathrm{id}_{k^*}$, homomorphism 9.7.13 has to coincide with $\omega_{P,S\otimes_k S}$ for any polarization ω. But, the analysis above shows that the action of $\mathrm{Aut}(P)$ on $J(kP)$ induces a group homomorphism

$$\mathrm{Aut}(P) \longrightarrow \hat{F}_S(P) \hat{\cap} \hat{F}_S(P) \longrightarrow \hat{F}_{S\otimes_k S}(P) \cong k^* \times \mathrm{Aut}(P) \qquad 9.7.15$$

mapping $\sigma \in \mathrm{Aut}(P)$ on $\big((\lambda_\sigma)^{-1}, \sigma\big)$; thus, denoting by $\hat{\sigma}$ the image of σ in $\hat{F}_S(P) \hat{\cap} \hat{F}_S(P)$, in the diagram in 9.6.1 above, the images of $\hat{\sigma}$ in k^* coincide with $(\lambda_\sigma)^{-1}$ and with $\omega_{P,S}(\hat{\sigma})^2$ which *cannot coincide* over any generator of $\mathrm{Aut}(P)$.

9.8 Before going further, let us consider a second compatibility condition involving the *Brauer quotient functor* (cf. 1.22); this condition depends on a *canonical k^*-group homomorphism* defined as follows. As above, let P be a finite p-group, S a Dade P-algebra and Q a normal subgroup of P; set $T = \mathrm{Res}^P_Q(S)$ and $\bar{P} = P/Q$, and denote by $\hat{F}_S(P)_Q$ the stabilizer of Q in $\hat{F}_S(P)$; if $\varphi \in F_S(P)_Q$ and $i \in S^P$ is a primitive idempotent such that $\mathrm{Br}^S_P(i) \neq 0$, by the very definition of the S-fusions (cf. 1.28) there is an element $a \in S^*$ fulfilling $i^a = i$ and $ai{\cdot}u = \varphi(u){\cdot}ai$ for any $u \in P$; then, setting

$$\overline{ai}^S = a\big(i + J(iS^P i)\big) \qquad 9.8.1,$$

it follows from its very definition, that $\hat{F}_S(P)$ contains $(\varphi, \overline{ai}^S)$.

9.9 In particular, a acts on iSi and, since $\varphi(Q) = Q$, a still acts on $(iSi)(Q)$; since this k-algebra is simple, this action can be lifted to an element of $(iSi)(Q)^*$ and therefore there is an element $c \in (S^Q)^*$ such that $i^c = i$ and that ac^{-1} acts trivially on $(iSi)(Q)$; then, choosing a primitive idempotent j in S^Q such that $ij = j = ji$ and $\mathrm{Br}^S_Q(j) \neq 0$, we have

$$\mathrm{Br}^S_Q(j)^{ac^{-1}} = \mathrm{Br}^S_Q(j) \qquad 9.9.1$$

and therefore we can modify our choice of c in such a way that $j^{ac^{-1}} = j$; in this case, since $c \in (S^Q)^*$, for any $u \in Q$ we have

$$ac^{-1}j{\cdot}u = \varphi(u){\cdot}ac^{-1}j \qquad 9.9.2$$

which proves that the restriction ψ of φ to Q belongs to $F_T(Q)$ and, setting

$$\overline{ac^{-1}j}^T = ac^{-1}\big(j + J(jT^Q j)\big) \qquad 9.9.3,$$

that $(\psi, \overline{ac^{-1}j}^T)$ belongs to $\hat{F}_T(Q)$.

9.10 Finally, since $\mathrm{Br}_Q\big(S^P\big) = S(Q)^{\bar P}$ and $\big(S(Q)\big)(\bar P) \cong S(P)$ (cf. isomorphism 1.20.1), $\mathrm{Br}_Q(i)$ is a primitive idempotent of $S(Q)^{\bar P}$ and we have $\mathrm{Br}_{\bar P}^{S(Q)}\big(\mathrm{Br}_Q(i)\big) \neq 0$; moreover, denoting by $\widehat{\mathrm{Br}}_Q^*(u{\cdot}i)$ the *canonical lifting* of $u{\cdot}i$ to $(iSi)(Q)^*$ for any $u \in P$, we get

$$\widehat{\mathrm{Br}}_Q^*(u{\cdot}i) = \widehat{\mathrm{Br}}_Q^*\big(\varphi(u){\cdot}i\big)^a = \widehat{\mathrm{Br}}_Q^*\big(\varphi(u){\cdot}i\big)^c \qquad 9.10.1,$$

which proves that the automorphism $\bar\varphi$ of $\bar P$ determined by φ belongs to $F_{S(Q)}(\bar P)$ and, setting

$$\overline{\mathrm{Br}_Q(ci)}^{S(Q)} = \mathrm{Br}_Q(c)\Big(\mathrm{Br}_Q(i) + J\big(\mathrm{Br}_Q(i)S(Q)^P\mathrm{Br}_Q(i)\big)\Big) \qquad 9.10.2,$$

that $(\bar\varphi, \overline{\mathrm{Br}_Q(ci)}^{S(Q)})$ belongs to $\hat F_{S(Q)}(\bar P)$ †.

Proposition 9.11 *With the notation above, identify $\hat F_T(Q)$ and $\hat F_{S(Q)}(\bar P)$ with their images in $\hat F_T(Q) \hat\times \hat F_{S(Q)}(\bar P)$. There is a unique k^*-group homomorphism*

$$\Delta_{P,S,Q} : \hat F_S(P)_Q \longrightarrow \hat F_T(Q) \hat\times \hat F_{S(Q)}(\bar P) \qquad 9.11.1$$

defined by

$$\Delta_{P,S,Q}(\varphi, \overline{ai}^S) = (\psi, \overline{ac^{-1}j}^T){\cdot}(\bar\varphi, \overline{\mathrm{Br}_Q(ci)}^{S(Q)}) \qquad 9.11.2.$$

Proof: First of all, it is quite clear that this map does not depend on the choices of i and j ; moreover, note that we only can modify our choice of a by an element $z \in (S^P)^*$ such that $i^z = i$, so that $\mathrm{Br}_P(zi) = \lambda{\cdot}\mathrm{Br}_P(i)$ for some $\lambda \in k^*$; then, we only can modify our choice of c by an element $x \in (S^Q)^*$ such that $i^x = i$, that $\mathrm{Br}_Q(z^{-1}xi) = \mu{\cdot}\mathrm{Br}_Q(i)$ for some $\mu \in k^*$ and that $j^{z^{-1}x} = j$; these modifications simply multiply $\overline{ai}^S$ by λ and $\overline{ac^{-1}j}^B$ by μ, and then it is easily checked that $\overline{\mathrm{Br}_Q(ci)}^{S(Q)}$ becomes multiplied by $\lambda\mu^{-1}$. Thus, the map $\Delta_{P,S,Q}$ does not depend on our choices.

For a second $\varphi' \in F_S(P)_Q$, choosing a' and c' as above and denoting by ψ' the restriction of φ' to Q, it is clear that aa' and cc' are a suitable choice for $\varphi \circ \varphi'$, and that

$$\begin{gathered}(\overline{ai}^S)(\overline{a'i}^S) = \overline{aa'i}^S \\ \overline{\mathrm{Br}_Q(ci)}^{S(Q)}\,\overline{\mathrm{Br}_Q(c'i)}^{S(Q)} = \overline{\mathrm{Br}_Q(cc'i)}^{S(Q)}\end{gathered} \qquad 9.11.3;$$

† All these arguments can be suitably extended to any P-interior algebra as it is shown by Proposition 10.3 in [42].

moreover, we have

$$ac^{-1}ja'c'^{-1}j = ac^{-1}a'c'^{-1}j = aa'(cc')^{-1}[c^{-1}, a'c'^{-1}]j \qquad 9.11.4$$

and, since $a'c'^{-1}$ acts trivially on $(iSi)(Q)$, we still have

$$\mathrm{Br}_Q\big([c^{-1}, a'c'^{-1}]i\big) = \mathrm{Br}_Q(i) \qquad 9.11.5,$$

so that we get

$$(\overline{ac^{-1}j}^{\,T})(\overline{a'c'^{-1}j}^{\,T}) = \overline{aa'(cc')^{-1}j}^{\,T} \qquad 9.11.6.$$

Hence, the map $\Delta_{P,S,Q}$ is a k^*-group homomorphism. We are done.

Corollary 9.12 *With the notation above, for any $\mathfrak{D}_k$-morphism*

$$(\pi, \tilde{f}) : (P, S) \longrightarrow (P', S') \qquad 9.12.1,$$

setting $Q' = \pi(Q)$, $T' = \mathrm{Res}^{P'}_{Q'}(S')$, $\tilde{g} = \mathrm{Res}^{P}_{Q}(\tilde{f})$ and $\bar{P}' = P'/Q'$, and denoting by $\rho: Q \to Q'$ and by $\bar{\pi} : \bar{P} \to \bar{P}'$ the group homomorphisms induced by π, we have the commutative diagram

$$\begin{array}{ccc} \hat{F}_S(P)_Q & \xrightarrow{\Delta_{P,S,Q}} & \hat{F}_T(Q) \hat{\times} \hat{F}_{S(Q)}(\bar{P}) \\ {\scriptstyle \hat{F}_{\tilde{f}}(\pi)} \downarrow & & \downarrow {\scriptstyle \hat{F}_{\tilde{g}}(\rho) \hat{\times} \hat{F}_{\tilde{f}(Q)}(\bar{\pi})} \\ \hat{F}_{S'}(P')_{Q'} & \xrightarrow{\Delta_{P',S',Q'}} & \hat{F}_{T'}(Q') \hat{\times} \hat{F}_{S'(Q')}(\bar{P}') \end{array} \qquad 9.12.2.$$

Proof: With the notation above, for a representative $f : \mathrm{Res}_\pi(S') \to S$ of $\tilde{f}$, we may assume that $i = f(i')$ for a suitable $i' \in S'^{P'}$, and that a and c centralize $f(1_{S'})$; then, it is clear that the elements $a', c', j' \in S'$ such that $f(a') = af(1_{S'})$, $f(c') = cf(1_{S'})$ and $f(j') = j$ fulfill the corresponding conditions with respect to the images φ' of φ, ψ' of ψ and $\bar{\varphi}'$ of $\bar{\varphi}$; hence, since

$$\begin{aligned} \hat{F}_{\tilde{f}}(\pi)(\varphi, \overline{ai}^{\,S}) &= (\varphi', \overline{a'i'}^{\,S'}) \\ \hat{F}_{\tilde{g}}(\rho)(\psi, \overline{ac^{-1}j}^{\,T}) &= (\psi', \overline{a'c'^{-1}j'}^{\,T'}) \\ \hat{F}_{\tilde{f}(Q)}(\bar{\pi})(\bar{\varphi}, \overline{\mathrm{Br}_Q(ci)}^{\,S(Q)}) &= (\bar{\varphi}', \overline{\mathrm{Br}_{Q'}(c'i')}^{\,S'(Q')}) \end{aligned} \qquad 9.12.3,$$

we are done.

9.13 When Q runs over the set of normal subgroups of P, the k^*-group homomorphisms $\Delta_{P,S,Q}$ may be seen as a kind of *coproduct* for the functor $\hat{f}$ above, and we need the *coassociativity* and the *distributivity* with respect to the tensor product. Firstly, we make the *coassociativity* explicit; let R be a second normal subgroup of P contained in Q, set $U = \mathrm{Res}^P_R(S)$, denote by $\hat{F}_S(P)_{Q,R}$ the stabilizer of Q and R in $\hat{F}_S(P)$, and identify $\big(S(R)\big)(Q/R)$ with $S(Q)$ (cf. isomorphism 1.20.1).

Proposition 9.14 *With the notation above, we have the following commutative diagram:*

$$\begin{array}{ccc} & \hat{F}_T(Q)_R \hat{\times} \hat{F}_{S(Q)}(P/Q) & \\ {\scriptstyle\Delta_{P,S,Q}} \nearrow & & \searrow {\scriptstyle\Delta_{Q,T,R} \hat{\times}\, \mathrm{id}_{\hat{F}_{S(Q)}(P/Q)}} \\ \hat{F}_S(P)_{Q,R} & & \hat{F}_U(R) \hat{\times} \hat{F}_{T(R)}(Q/R) \hat{\times} \hat{F}_{S(Q)}(P/Q) \\ {\scriptstyle\Delta_{P,S,R}} \searrow & & \nearrow {\scriptstyle \mathrm{id}_{\hat{F}_U(R)} \hat{\times} \Delta_{P/R,S(R),Q/R}} \\ & \hat{F}_U(R) \hat{\times} \hat{F}_{S(R)}(P/R)_{Q/R} & \end{array} \qquad 9.14.1.$$

Proof: With the notation above, assume that $\varphi(R) = R$ and denote by η the restriction of φ to R; then, $(\varphi, \overline{ai}^S)$ and $(\psi,\ \overline{ac^{-1}j}^T)$ respectively belong to $\hat{F}_S(P)_{Q,R}$ and to $\hat{F}_T(Q)_R$, and, setting $\alpha = \Delta_{P,S,Q}$, we already know that

$$\alpha(\varphi, \overline{ai}^S) = (\psi, \overline{ac^{-1}j}^T)\cdot(\bar{\varphi}^Q, \overline{\mathrm{Br}_Q(ci)}^{S(Q)}) \qquad 9.14.2,$$

where $\bar{\varphi}^Q$ denotes the automorphism of P/Q induced by φ and where we identify $\hat{F}_T(Q)_R$ and $\hat{F}_{S(Q)}(P/Q)$ with their images in $\hat{F}_T(Q)_R \hat{\times} \hat{F}_{S(Q)}(P/Q)$.

Moreover, since $\varphi(R) = R$, it is clear that $ac^{-1}i$ normalizes $(iSi)^R$ and acts on $(iSi)(R)$, so that there is $e \in (S^R)^*$ fulfilling $i^e = i$ and such that $ac^{-1}e^{-1}i$ acts trivially on $(iSi)(R)$; in particular, choosing a primitive idempotent ℓ in S^R such that $\mathrm{Br}_R(\ell) \neq 0$ and $\ell j = \ell = j\ell$, as above we can modify our choice of e in such a way that $j^{e^{-1}} = j$ and then that $\ell^{ac^{-1}e^{-1}} = \ell$. In this situation, for any $v \in R$, we still have

$$ac^{-1}e^{-1}\ell\cdot v = \varphi(v)\cdot ac^{-1}e^{-1}\ell \qquad 9.14.3,$$

so that, setting

$$\overline{ac^{-1}e^{-1}\ell}^U = ac^{-1}e^{-1}\big(\ell + J(\ell U^R \ell)\big) \qquad 9.14.4,$$

the pair $(\eta, \overline{ac^{-1}e^{-1}\ell}^U)$ belongs to $\hat{F}_U(R)$.

As above, denoting by $\widehat{\mathrm{Br}}^*_R(u\cdot i)$ the canonical lifting to $(iSi)(R)^*$ of $u\cdot i$ for any $u \in P$, we get

$$\widehat{\mathrm{Br}}^*_R(u\cdot i) = \widehat{\mathrm{Br}}^*_R\big(\varphi(u)\cdot i\big)^{ai} = \widehat{\mathrm{Br}}^*_R\big(\varphi(u)\cdot i\big)^{eci} \qquad 9.14.5$$

and therefore, setting

$$\overline{\mathrm{Br}_R(eci)}^{S(R)} = \mathrm{Br}_R(ec)\Big(\mathrm{Br}_R(i) + J\big(\mathrm{Br}_R(i)S(R)^P\mathrm{Br}_R(i)\big)\Big) \qquad 9.14.6$$

and denoting by $\bar{\varphi}^R$ the automorphism of P/R induced by φ, the pair $(\bar{\varphi}^R, \overline{\mathrm{Br}_R(eci)}^{S(R)})$ belongs to $\hat{F}_{S(R)}(P/R)_{Q/R}$. Thus, setting $\beta = \Delta_{P,S,R}$, similarly we have

$$\beta(\varphi, \overline{ai}^A) = (\eta, \overline{ac^{-1}e^{-1}\ell}^U)\cdot(\bar{\varphi}^R, \overline{\mathrm{Br}_R(eci)}^{S(R)}) \qquad 9.14.7.$$

On the other hand, since ac^{-1} and e^{-1} centralize j, and $ac^{-1}e^{-1}j$ acts trivially on $(jRj)(Q)$, setting $\gamma = \Delta_{Q,T,R} \hat{\times} \mathrm{id}_{\hat{F}_{S(Q)}(P/Q)}$ we still have

$$\begin{aligned}&\gamma\big(\alpha(\varphi, \overline{ai}^A)\big)\\ &\quad = (\eta, \overline{ac^{-1}e^{-1}\ell}^U)\cdot(\bar{\psi}^R, \overline{\mathrm{Br}_R(ej)}^{T(R)})\cdot(\bar{\varphi}^Q, \overline{\mathrm{Br}_Q(ci)}^{S(Q)})\end{aligned} \qquad 9.14.8$$

where $\bar{\psi}^R$ denotes the automorphism of Q/R induced by ψ and, as above, we also identify $\hat{F}_U(R)$, $\hat{F}_{T(R)}(Q/R)$ and $\hat{F}_{S(Q)}(P/Q)$ with their images in

$$\hat{F}_U(R) \hat{\times} \hat{F}_{T(R)}(Q/R) \hat{\times} \hat{F}_{S(Q)}(P/Q) \qquad 9.14.9.$$

Finally, we know that $\mathrm{Br}_R(i)$ is a primitive idempotent of $S(R)^{P/R}$ such that (cf. equality 1.22.3)

$$\mathrm{Br}_{P/R}^{S(R)}\big(\mathrm{Br}_R(i)\big) \neq 0 \qquad 9.14.10,$$

that $\mathrm{Br}_R(eci)$ normalizes $(iSi)(R)^{Q/R}$ (cf. equality 9.14.5), that $\mathrm{Br}_R(e)$ belongs to $\big(S(R)^{Q/R}\big)^*$ and that, since ac^{-1} acts trivially on $(iSi)(Q)$ (cf. 9.9) and $ac^{-1}e^{-1}i$ acts trivially on $(iSi)(R)$, $\mathrm{Br}_R(ei)$ still acts trivially on (cf. isomorphism 1.24.1)

$$\big((iSi)(R)\big)(Q/R) \cong (iSi)(Q) \qquad 9.14.11;$$

consequently, setting $\delta = \mathrm{id}_{\hat{F}_U(R)} \hat{\times} \Delta_{P/R,S(R),Q/R}$, we still get

$$\begin{aligned}&\delta\big(\beta(\varphi, \overline{ai}^A)\big)\\ &\quad = (\eta, \overline{ac^{-1}e^{-1}\ell}^U)\cdot(\bar{\psi}^R, \overline{\mathrm{Br}_R(ej)}^{T(R)})\cdot(\bar{\varphi}^Q \overline{\mathrm{Br}_Q(ci)}^{S(Q)})\end{aligned} \qquad 9.14.12.$$

We are done.

9.15 For the *distributivity*, let S' be another Dade P-algebra and, as above, set $T' = \mathrm{Res}^P_Q(S')$ and $\bar{P} = P/Q$; moreover, identify $(S\otimes_k S')(Q)$ with $S(Q)\otimes_k S'(Q)$ (cf. 1.21.2). It follows from Propositions 5.9 and 5.11 in [41] that

$$F_S(P)\cap F_{S'}(P)\subset F_{S\otimes_k S'}(P) \qquad 9.15.1$$

and that, denoting by $\hat{F}_S(P)\mathbin{\hat\cap}\hat{F}_{S'}(P)$ the converse image in $\hat{F}_S(P)\mathbin{\hat\times}\hat{F}_{S'}(P)$ of the *diagonal* image of $F_S(P)\cap F_{S'}(P)$ in $F_S(P)\times F_{S'}(P)$, this inclusion lifts to a *natural* k^*-group homomorphism

$$\hat{\nu}_{P,S,S'}:\hat{F}_S(P)\mathbin{\hat\cap}\hat{F}_{S'}(P)\longrightarrow \hat{F}_{S\otimes_k S'}(P) \qquad 9.15.2$$

which, with the notation above and with analogous choices i' and a' for S', maps the element $(\varphi,\overline{ai}^{\,S})\cdot(\varphi,\overline{a'i'}^{\,S'})$, that clearly belongs to $\hat{F}_S(P)\mathbin{\hat\cap}\hat{F}_{S'}(P)$, on

$$\hat{\nu}_{P,S,S'}\big((\varphi,\overline{ai}^{\,S})\cdot(\varphi,\overline{a'i'}^{\,S'})\big)=(\varphi,\overline{(a\otimes a')a''i''}^{\,S\otimes_k S'}) \qquad 9.15.3$$

where i'' is a primitive idempotent in $(S\otimes_k S')^P-\mathrm{Ker}(\mathrm{Br}_P^{S\otimes_k S'})$ such that

$$(i\otimes i')i''=i''=i''(i\otimes i') \qquad 9.15.4,$$

so that we have

$$\mathrm{Br}_P^{S\otimes_k S'}(i'')=\mathrm{Br}_P^S(i)\otimes\mathrm{Br}_P^{S'}(i') \qquad 9.15.5,$$

and where a'' is an element of $1+\mathrm{Ker}(\mathrm{Br}_P^{S\otimes_k S'})$ fulfilling $i''^{(a\otimes a')a''}=i''$ (cf. isomorphism 1.21.2).

Proposition 9.16 *With the notation above, let S' be a Dade P-algebra and set $T'=\mathrm{Res}^P_Q(S')$, $\alpha=\Delta_{P,S\otimes_k S',Q}$ and $\beta=\Delta_{P,S,Q}\mathbin{\hat\times}\Delta_{P,S',Q}$. Then, we have the following commutative diagram:*

$$\begin{array}{ccc}
\hat{F}_{S\otimes_k S'}(P)_Q & \stackrel{\alpha}{\longrightarrow} & \hat{F}_{T\otimes_k T'}(Q)\mathbin{\hat\times}\hat{F}_{(S\otimes_k S')(Q)}(\bar{P}) \\
\hat{\nu}_{P,S,S'}\uparrow & & \uparrow \hat{\nu}_{Q,T,T'}\mathbin{\hat\times}\hat{\nu}_{\bar{P},S(Q),S'(Q)} \\
\hat{F}_S(P)_Q\mathbin{\hat\cap}\hat{F}_{S'}(P)_Q & \stackrel{\beta}{\longrightarrow} & \big(\hat{F}_T(Q)\mathbin{\hat\cap}\hat{F}_{T'}(Q)\big)\mathbin{\hat\times}\big(\hat{F}_{S(Q)}(\bar{P})\mathbin{\hat\cap}\hat{F}_{S'(Q)}(\bar{P})\big)
\end{array} \qquad 9.16.1.$$

Proof: With the notation above, we consider the pairs $(\varphi,\overline{ai}^{\,S})$, $(\psi,\overline{ac^{-1}j}^{\,T})$ and $(\bar\varphi,\overline{\mathrm{Br}_Q(ci)}^{\,S(Q)})$, which respectively belong to $\hat{F}_S(P)_Q$, to $\hat{F}_T(Q)$ and to $\hat{F}_{S(Q)}(\bar{P})$; *mutatis mutandis*, assuming that $\varphi\in F_S(P)_Q\cap F_{S'}(P)_Q$, for S' we choose i', a', j' and c' as above, and thus the pairs $(\varphi,\overline{a'i'}^{\,S'})$, $(\psi,\overline{a'c'^{-1}j'}^{\,T'})$ and $(\bar\varphi,\overline{\mathrm{Br}_Q(c'i')}^{\,S'(Q)})$ belong to $\hat{F}_{S'}(Q)$, to $\hat{F}_{T'}(R)$ and to

$\hat{F}_{S'(Q)}(\bar{P})$ respectively. Consequently, β maps $(\varphi, \overline{ai}^{S})\cdot(\varphi, \overline{a'i'}^{S'})$ on the element

$$(\psi, \overline{ac^{-1}j}^{T})\cdot(\bar{\varphi}, \overline{\mathrm{Br}_Q(ci)}^{S(Q)})\cdot(\psi, \overline{a'c'^{-1}j'}^{T'})\cdot(\bar{\varphi}, \overline{\mathrm{Br}_Q(c'i')}^{S'(Q)}) \qquad 9.16.2.$$

On the other hand, choosing $i'' \in (S \otimes_k S')^P$ and $a'' \in 1 + \mathrm{Ker}(\mathrm{Br}_P^{S\otimes_k S'})$ as above, by equality 9.15.5 we have

$$\begin{aligned} \nu_{P,S,S'}\big((\varphi, \overline{ai}^{S})\cdot(\varphi, \overline{a'i'}^{S'})\big) &= (\varphi, \overline{(a \otimes a')a''i''}^{S\otimes_k S'}) \\ &= (\varphi, \overline{ai \otimes a'i'}^{S\otimes_k S'}) \end{aligned} \qquad 9.16.3.$$

Moreover, $ac^{-1} \otimes a'c'^{-1}$ acts trivially on $(iSi \otimes_k i'S'i')(Q)$ and *a fortiori* on $\big(i''(S\otimes_k S')i''\big)(Q)$; then, choosing a primitive idempotent j'' in $(S\otimes_k S')^Q$ such that

$$\mathrm{Br}_Q(j'') \neq 0 \quad \text{and} \quad i''j'' = j'' = j''i'' \qquad 9.16.4,$$

we may find $c'' \in 1 + \mathrm{Ker}(\mathrm{Br}_Q^{S\otimes_k S'})$ fulfilling

$$i''^{(c^{-1}\otimes c'^{-1})c''} = i'' \quad \text{and} \quad j''^{(ac^{-1}\otimes a'c'^{-1})c''} = j'' \qquad 9.16.5;$$

thus, α maps $(\varphi, \overline{ai \otimes a'i'}^{S\otimes_k S'})$ on the element

$$(\psi, \overline{(ac^{-1} \otimes a'c'^{-1})c''j''}^{T\otimes_k T'})\cdot(\bar{\varphi}, \overline{\mathrm{Br}_Q(c''^{-1}(c \otimes c')i'')}^{(S\otimes S')(Q)}) \qquad 9.16.6.$$

As above, it is easily checked that

$$\overline{\mathrm{Br}_Q(c''^{-1}(c \otimes c')i'')}^{(S\otimes_k S')(Q)} = \overline{\mathrm{Br}_Q(ci \otimes c'i')}^{(S\otimes_k S')(Q)} \qquad 9.16.7$$

and it is clear that

$$\begin{aligned} \nu_{P,S(Q),S'(Q)}\big((\bar{\varphi}, \overline{\mathrm{Br}_Q(ci)}^{S(Q)})\cdot(\bar{\varphi}, \overline{\mathrm{Br}_Q(c'i')}^{S'(Q)})\big) & \\ = (\bar{\varphi}, \overline{\mathrm{Br}_Q(ci \otimes c'i')}^{(S\otimes_k S')(Q)}) & \end{aligned} \qquad 9.16.8.$$

On the other hand, since $\mathrm{Br}_Q^T(j) \otimes \mathrm{Br}_Q^{T'}(j')$ and $\mathrm{Br}_Q^{T\otimes T'}(j'')$ are both primitive in $(T \otimes T')(Q)$, they are mutually conjugate and in $\hat{F}_{T\otimes_k T'}(Q)$ up to suitable identifications we get

$$(\psi, \overline{(ac^{-1} \otimes a'c'^{-1})c''j''}^{T\otimes_k T'}) = (\psi, \overline{ac^{-1}j \otimes a'c'^{-1}j'}^{T\otimes_k T'}) \qquad 9.16.9;$$

similarly, it is quite clear that $\nu_{Q,T,T'}$ maps $(\psi, \overline{ac^{-1}j}^{T})\cdot(\psi, \overline{a'c'^{-1}j'}^{T'})$ on this element. In conclusion, $\nu_{Q,T,T'} \hat{\times} \nu_{\bar{P},S(Q),S'(Q)}$ maps the element 9.16.2 on the element 9.16.6. We are done.

9.17 Now, we are ready to discuss our second condition. As in the case of the tensor product, if $p \neq 2$ there are *no* polarizations ω *compatible* with the kind of *coproduct* above and, more generally, we consider a *relative condition* with respect to a full subcategory $\mathfrak{E}_k$ of $\mathfrak{D}_k$; we say that a polarization ω is $\mathfrak{E}_k$*-compatible with the Brauer quotient functor* whenever it fulfills the condition

9.17.1 *If (P,S) is a $\mathfrak{E}_k$-object and Q a normal subgroup of P, the following diagram is commutative*

$$\begin{array}{ccc} \hat{F}_S(P)_Q & \xrightarrow{\Delta_{P,S,Q}} & \hat{F}_T(Q)\,\hat{\times}\,\hat{F}_{S(Q)}(\bar{P}) \\ {\scriptstyle \omega_{P,S}}\searrow & & \swarrow{\scriptstyle \omega_{Q,T}\,\hat{\times}\,\omega_{\bar{P},S(Q)}} \\ & k^* & \end{array}$$

where we set $T = \mathrm{Res}^P_Q(S)$ and $\bar{P} = P/Q$.

Similarly, we say that a polarization is $\mathfrak{E}_k$*-compatible with the tensor product* whenever it fulfills the following condition

9.17.2 *If (P,S) and (P,S') are $\mathfrak{E}_k$-objects the following diagram is commutative*

$$\begin{array}{ccc} \hat{F}_S(P)\,\hat{\cap}\,\hat{F}_{S'}(P) & \longrightarrow & \hat{F}_{S\otimes_k S'}(P) \\ {\scriptstyle \omega_{P,S}\,\hat{\times}\,\omega_{P,S'}}\searrow & & \swarrow{\scriptstyle \omega_{P,S\otimes_k S'}} \\ & k^* & \end{array}$$

For any $n \in \mathbb{N}$, let us denote by $n\mathfrak{D}_k$ the full subcategory of $\mathfrak{D}_k$ over the objects (P,S) such that the class of S belongs to $n{\cdot}\mathcal{D}_k(P)$; we have to consider the full subcategories $2\mathfrak{D}_k$ and $2(p-1)\mathfrak{D}_k$.

9.18 In order to define a polarization ω, we start by discussing the elementary Abelian case. In this case, by 8.13.4 and 8.15 above, any element $s \in \mathcal{D}_k(P)$ admits a unique decomposition

$$s = \sum_Q z_Q{\cdot}\mathrm{res}_{\pi^P_Q}(j_1^{P/Q}) \qquad 9.18.1,$$

where Q runs over the set of subgroups of P of index at least 3 and, for any Q, z_Q is an element of $\mathbb{Z}$, which belongs to $\{0,1\}$ if $|P/Q| = p$; consequently, any Dade P-algebra S in the class s can be *embedded* (cf. 1.11) in the tensor product (cf. 1.21)

$$\mathrm{End}_k(M) \otimes_k \Big(\bigotimes_Q \Big(\otimes^{z_Q}\mathrm{End}_k\big(J(k(P/Q))\big)\Big)\Big) \qquad 9.18.2,$$

where Q runs over the set of subgroups of P of index at least 3, where M is a kP-module admitting a P-stable basis with some fixed element and where

$\bigotimes^{z_Q}$ is the z_Q-th tensor power over k, with the conventional rule that if $z_Q = 0$ then we replace the corresponding tensor factor by k and if $z_Q < 0$ by $\bigotimes^{-z_Q}\mathrm{End}_k\big(J(k(P/Q))\big)^\circ$.

9.19 If i is a primitive idempotent of S^P such that $\mathrm{Br}_P^S(i) \neq 0$ then we have $\hat{F}_{iSi}(P) \cong \hat{F}_S(P)$ (cf. Proposition 2.14 in [39]) and the Dade P-algebra iSi can be *embedded* (cf. 1.11) in the above tensor factor (cf. 2.5.1 in [43])

$$T = \bigotimes_Q \Big(\bigotimes^{z_Q}\mathrm{End}_k\big(J(k(P/Q))\big)\Big) \qquad 9.19.1,$$

so that we may assume that $M \cong k$. Moreover, we know that (cf. Propositions 2.14 and 2.18 in [39])

$$F_S(P) = F_T(P) = \mathrm{Aut}(P)_s \qquad 9.19.2$$

and therefore, by the uniqueness of decomposition 9.18.1, $F_S(P)$ coincides with the stabilizer of the family $\{z_Q\}_Q$; but, for any subgroup Q of P, the stabilizer $\mathrm{Aut}(P)_Q$ of Q in $\mathrm{Aut}(P)$ acts on $J\big(k(P/Q)\big)$ and actually $J\big(k(P/Q)\big)$ becomes a $k\big(P \rtimes \mathrm{Aut}(P)_Q\big)$-module; consequently, the tensor product

$$N = \bigotimes_Q \big(\bigotimes^{z_Q} J(k(P/Q))\big) \qquad 9.19.3,$$

where Q runs over the set of subgroups of P of index at least 3, becomes a $k\big(P \rtimes F_S(P)\big)$-module. Since $T \cong \mathrm{End}_k(N)$, denoting by P_N the image of P in T^*, we get a group homomorphism $F_S(P) \to N_{T^*}(P_N)$ which determines a k^*-group homomorphism

$$\omega_{P,S} : \hat{F}_S(P) \longrightarrow k^* \qquad 9.19.4.$$

mapping the corresponding image of $F_S(P)$ on $1 \in k$. Note that, if $s = 0$, we have $iSi \cong k$ and homomorphism 9.19.4 agrees with the first projection in isomorphism 9.5.3.

9.20 Arguing by induction on $|P|$, we will extend this correspondence to all the pairs (P, S), where P is a finite p-group and S a Dade P-algebra. We may assume that P is not elementary Abelian; then, setting $Q = \Phi(P)$ and $\bar{P} = P/Q$, and denoting by $\pi : P \to \bar{P}$ the canonical map, it is clear that $F_S(P)$ stabilizes Q and acts on $\bar{P}$; thus, setting $T = \mathrm{Res}_Q^P(S)$, the k^*-group homomorphisms $\omega_{Q,T}$ and $\omega_{\bar{P},S(Q)}$ are already defined and we define $\omega_{P,S}$ by the composition (cf. Proposition 9.11)

$$\hat{F}_S(P) \xrightarrow{\Delta_{P,S,Q}} \hat{F}_T(Q) \hat{\times} \hat{F}_{S(Q)}(\bar{P}) \xrightarrow{\omega_{Q,T}\hat{\times}\omega_{\bar{P},S(Q)}} k^* \qquad 9.20.1.$$

Theorem 9.21 *With the notation above, ω is a polarization $2\mathfrak{D}_k$-compatible with the tensor product and $2(p-1)\mathfrak{D}_k$-compatible with the Brauer quotient functor.*

Proof: Let $(\pi, f) : (P, S) \to (P', S')$ be a $\mathfrak{D}_k$-morphism and firstly assume that P is elementary Abelian; thus, P' is elementary Abelian too and, denoting by $s' \in \mathcal{D}_k(P')$ the class of S', we have a unique decomposition of s'

$$s' = \sum_{Q'} z_{Q'} {\cdot} \mathrm{res}_{\pi^{P'}_{Q'}}(j_1^{P'/Q'}) \qquad 9.21.1$$

where Q' runs over the set of subgroups of P' of index at least 3 and, for such a Q', $z_{Q'}$ is an element of $\mathbb{Z}$, which belongs to $\{0,1\}$ if $|P'/Q'| = p$ (cf. 9.18); then, we get

$$s = \mathrm{res}_\pi(s') = \sum_{Q'} z_{Q'} {\cdot} \mathrm{res}_{\pi^{P}_{\pi^{-1}(Q')}}(j_1^{P/\pi^{-1}(Q')}) \qquad 9.21.2,$$

where Q' runs again over the set of subgroups of P' of index at least 3, and it is the unique decomposition of the class $s \in \mathcal{D}_k(P)$ of S; consequently, from the very definition of ω it is easily checked that

$$\omega_{P,S} = \omega_{P',S'} \circ \hat{F}_f(\pi) \qquad 9.21.3.$$

Otherwise, the surjectivity of π forces $\pi\big(\Phi(P)\big) = \Phi(P')$ and therefore, arguing by induction on $|P|$, setting $Q = \Phi(P)$, $Q' = \Phi(P')$, $\bar{P} = P/Q$, $\bar{P}' = P'/Q'$, $T = \mathrm{Res}^P_Q(S)$, $T' = \mathrm{Res}^{P'}_{Q'}(S')$ and $g = \mathrm{res}^P_Q(f)$, and denoting by $\rho : Q \to Q'$ and by $\bar{\pi} : \bar{P} \to \bar{P}'$ the group homomorphisms induced by π, we already have

$$\omega_{Q,T} = \omega_{Q',T'} \circ \hat{F}_g(\rho) \quad \text{and} \quad \omega_{\bar{P},S(Q)} = \omega_{\bar{P}',S'(Q')} \circ \hat{F}_{f(Q)}(\bar{\pi}) \qquad 9.21.4.$$

On the other hand, from Corollary 9.12 we get the following commutative diagram

$$\begin{array}{ccc} \hat{F}_S(P) & \xrightarrow{\Delta_{P,S,Q}} & \hat{F}_T(Q) \hat{\times} \hat{F}_{S(Q)}(\bar{P}) \\ {\scriptstyle \hat{F}_f(\pi)} \downarrow & & \downarrow {\scriptstyle \hat{F}_g(\rho) \hat{\times} \hat{F}_{f(Q)}(\bar{\pi})} \\ \hat{F}_{S'}(P') & \xrightarrow{\Delta_{P',S',Q'}} & \hat{F}_{T'}(Q') \hat{\times} \hat{F}_{S'(Q')}(\bar{P}') \end{array} \qquad 9.21.5.$$

Consequently, from the definition above, from equalities 9.21.4 and from the commutativity of this diagram, in this case we still get

$$\omega_{P,S} = \omega_{P',S'} \circ \hat{F}_f(\pi) \qquad 9.21.6.$$

This already proves that ω is a polarization.

Now, let S and S' be Dade P-algebras and denote by s and s' their respective classes in $\mathcal{D}_k(P)$; first of all, assume that P is elementary Abelian, so that we have unique decompositions

$$s = \sum_Q z_Q{\cdot}\mathrm{res}_{\pi_Q^P}(j_1^{P/Q}) \quad \text{and} \quad s' = \sum_Q z'_Q{\cdot}\mathrm{res}_{\pi_Q^P}(j_1^{P/Q}) \qquad 9.21.7,$$

where Q runs over the set of subgroups of P of index at least 3 and, for any Q, z_Q and z'_Q are elements of $\mathbb{Z}$, which belong to $\{0,1\}$ if $|P/Q| = p$ (cf. 9.18). Hence, for such a Q, the coefficient of $\mathrm{res}_{\pi_Q^P}(j_1^{P/Q})$ in the unique decomposition of $s+s'$ coincides with $z_Q + z'_Q$ unless $|P/Q| = p$ and we have $z_Q = 1 = z'_Q$; in this case this coefficient is zero, whereas the factor

$$J\big(k(P/Q)\big) \otimes_k J\big(k(P/Q)\big) \qquad 9.21.8$$

appears in the tensor product of the corresponding decompositions 9.19.3; moreover, according to example 9.7, the natural action of $P \rtimes \mathrm{Aut}(P)_Q$ on this factor stabilizes a subspace J_Q of dimension 1 where $\mathrm{Aut}(P)_Q$ acts *via* the inverse of the natural monomorphism $\theta_{P/Q} : \mathrm{Aut}(P/Q) \to k^*$.

Consequently, denoting by $\big(F_S(P) \cap F_{S'}(P)\big)_Q$ the stabilizer of Q in $F_S(P) \cap F_{S'}(P)$, the composition of $\theta_{P/Q}$ with the canonical homomorphism

$$\big(F_S(P) \cap F_{S'}(P)\big)_Q \longrightarrow \mathrm{Aut}(P/Q) \qquad 9.21.9$$

determines, *via* the *transfer homomorphism* from $\big(F_S(P) \cap F_{S'}(P)\big)_Q$ to $F_S(P) \cap F_{S'}(P)$ (cf. 1.32) a group homomorphism

$$\zeta_O : F_S(P) \cap F_{S'}(P) \longrightarrow k^* \qquad 9.21.10$$

which describes the action of $F_S(P) \cap F_{S'}(P)$ on $\bigotimes_{Q' \in O} J_{Q'}$ where O is the $F_S(P) \cap F_{S'}(P)$-orbit of Q.

In conclusion, it is easy to check that, for any $\hat\varphi$ in $\hat F_S(P) \mathbin{\hat\cap} \hat F_{S'}(P)$, we get

$$(\omega_{P,S\otimes_k S'} \circ \hat\nu_{P,S,S'})(\hat\varphi) = \Big(\prod_O \zeta_O(\varphi)\Big)(\omega_{P,S} \mathbin{\hat\times} \omega_{P,S'})(\hat\varphi) \qquad 9.21.11$$

where φ is the image of $\hat\varphi$ in $F_S(P) \cap F_{S'}(P)$ and O runs over the set of orbits of $F_S(P) \cap F_{S'}(P)$ on the set of subgroups Q of P such that $|P/Q| = p$ and $z_Q = 1 = z'_Q$. In particular, if s and s' belong to $2\mathcal{D}_k(P)$ the corresponding diagram in 9.17.2 is commutative.

Assume that P is not elementary Abelian and that (P,S) and (P,S') are actually $2\mathfrak{D}_k$-objects; setting $Q = \Phi(P)$ and $\bar P = P/Q$, and denoting by $\pi : P \to \bar{P}$ the canonical map, it follows from Proposition 9.16 that the following diagram is commutative

$$\begin{array}{ccc} \hat{F}_{S\otimes_k S'}(P) & \stackrel{\alpha}{\longrightarrow} & \hat{F}_{T\otimes_k T'}(Q) \hat{\times} \hat{F}_{(S\otimes_k S')(Q)}(\bar{P}) \\ \hat{\nu}_{P,S,S'} \uparrow & & \uparrow \hat{\nu}_{Q,T,T'} \hat{\times} \hat{\nu}_{\bar{P},S(Q),S'(Q)} \\ \hat{F}_S(P) \hat{\cap} \hat{F}_{S'}(P) & \stackrel{\beta}{\longrightarrow} & (\hat{F}_T(Q) \hat{\cap} \hat{F}_{T'}(Q)) \hat{\times} (\hat{F}_{S(Q)}(\bar{P}) \hat{\cap} \hat{F}_{S'(Q)}(\bar{P})) \end{array} \qquad 9.21.12,$$

where we set $\alpha = \Delta_{P,S\otimes_k S',Q}$, $\beta = \Delta_{P,S,Q} \hat{\times} \Delta_{P,S',Q}$, $T = \mathrm{Res}^P_Q(S)$ and $T' = \mathrm{Res}^P_Q(S')$, so that (Q,T) and (Q,T') are $2\mathfrak{D}_k$-objects too; thus, arguing by induction on $|P|$, we already have the commutative diagram

$$\begin{array}{ccc} \hat{F}_T(Q) \hat{\cap} \hat{F}_{T'}(Q) & \xrightarrow{\hat{\nu}_{Q,T,T'}} & \hat{F}_{T\otimes_k T'}(Q) \\ \omega_{Q,T} \hat{\times} \omega_{Q,T'} \searrow & & \swarrow \omega_{Q,T\otimes_k T'} \\ & k^* & \end{array} \qquad 9.21.13$$

Similarly, $(\bar{P}, S(Q))$ and $(\bar{P}, S'(Q))$ are also $2\mathfrak{D}_k$-objects and therefore, arguing by induction on $|P|$, we still have the commutative diagram

$$\begin{array}{ccc} \hat{F}_{S(Q)}(\bar{P}) \hat{\cap} \hat{F}_{S'(Q)}(\bar{P}) & \xrightarrow{\hat{\nu}_{\bar{P},S(Q),S'(Q)}} & \hat{F}_{S(Q)\otimes_k S'(Q)}(\bar{P}) \\ \omega_{\bar{P},S(Q)} \hat{\times} \omega_{\bar{P},S'(Q)} \searrow & & \swarrow \omega_{\bar{P},S(Q)\otimes_k S'(Q)} \\ & k^* & \end{array} \qquad 9.21.14.$$

Now, the k^*-*product* of diagrams 9.21.13 and 9.21.14, together with diagram 9.21.12, prove the commutativity of the diagram in 9.17.2.

Finally, we prove that ω is $2(p-1)\mathfrak{D}_k$-compatible with the *Brauer quotient functor*. Consider S and s as above, assume that s belongs to $2(p-1)\cdot\mathcal{D}_k(P)$, let R be a normal subgroup of P, set $U = \mathrm{Res}^P_R(S)$ and $\bar{P} = P/R$, and denote by $\pi : P \to \bar{P}$ the canonical map; first of all, assume that P is elementary Abelian, so that we have a unique decomposition

$$s = \sum_Q 2(p-1) z_Q \cdot \mathrm{res}_{\pi^P_Q}(j_1^{P/Q}) \qquad 9.21.15$$

where Q runs over the set of subgroups of P of index at least p^2 and, for such a Q, z_Q is an element of $\mathbb{Z}$ (cf. 9.18); the group $F_S(P)_R$ acts on this set of subgroups and obviously we still have

$$s = \sum_O 2(p-1) z_O \cdot s_O \qquad 9.21.16$$

where O runs over the set of $F_S(P)_R$-orbits and, for such an $F_S(P)_R$-orbit, we set $s_O = \sum_{Q\in O} \mathrm{res}_{\pi_Q^P}(j_1^{P/Q})$ and $z_O = z_Q$ for $Q \in O$.

Actually, we may assume that $S \cong \bigotimes_O \left(\bigotimes^{2(p-1)z_O} S_O\right)$ where O runs over the set of $F_S(P)_R$-orbits and, for such an orbit, we set

$$S_O = \bigotimes_{Q\in O} \mathrm{End}_k\big(J(k(P/Q))\big) \qquad 9.21.17.$$

But, it follows from Lemma 1.17 in [34] that we have

$$F_S(P)_R = \bigcap_O F_{S_O}(P)_R \qquad 9.21.18,$$

where O runs over the set of $F_S(P)_R$-orbits, and then, iterating our definition in 9.15 of $\hat{\cap}$, it follows from Proposition 5.11 in [41] that we still have

$$\hat{F}_S(P)_R = \widehat{\bigcap_O}\big(\hat{\cap}^{2(p-1)z_O} \hat{F}_{S_O}(P)_R\big) \qquad 9.21.19$$

where O runs again over the set of $F_S(P)_R$-orbits and where, for such an orbit, if $z_O = 0$ we set

$$\hat{\cap}^0 \hat{F}_{S_O}(P)_R = k^* \times \mathrm{Aut}(P)_R \qquad 9.21.20$$

whereas if $z_O < 0$ we define (cf. 1.24)

$$\hat{\cap}^{2(p-1)z_O} \hat{F}_{S_O}(P)_R = \hat{\cap}^{-2(p-1)z_O} \big(\hat{F}_{S_O}(P)_R\big)^\circ \qquad 9.21.21.$$

Thus, since we already know that ω is $2\mathfrak{D}_k$-compatible with the *tensor product*, it suffices to discuss the case of $S_O \otimes S_O = \bigotimes^2 S_O$ for any orbit O; clearly, either $R \subset Q$ for any $Q \in O$ or $R \not\subset Q$ for any $Q \in O$. In the first case, the tensor product

$$\bigotimes^2 U_O = \mathrm{Res}_R^P\big(\bigotimes^2 S_O\big) \qquad 9.21.22$$

is a trivial R-algebra and therefore, by the very definition, $\omega_{R,\otimes^2 U_O}$ is the first projection in 9.5.3; moreover, in this case we have $J(k(\bar{P}/\bar{Q})) \cong J(k(P/Q))$ for any $Q \in O$, so that we clearly have

$$\bigotimes^2 S_O = \mathrm{Res}_\pi\big(\bigotimes^2 S_O(R)\big) \qquad 9.21.23$$

and therefore we get

$$\omega_{P,\otimes^2 S_O} = \omega_{\bar{P},\otimes^2 S_O(R)} \circ \hat{F}_{\widetilde{\mathrm{id}}_{\otimes^2 S_O}}(\pi) \qquad 9.21.24;$$

hence, in this case, for any $\hat{\varphi} \in \hat{F}_{S_O}(P)_R$ we simply get

$$\omega_{P,\otimes^2 S_O}(\hat{\varphi}) = \big(\omega_{R,\otimes^2 U_O} \hat{\times}\, \omega_{\bar{P},\otimes^2 S_O(R)}\big)\big(\Delta_{P,\otimes^2 S_O,R}(\hat{\varphi})\big) \qquad 9.21.25.$$

In the second case, we have $\bigotimes^2 S_O(R) \cong k$ (cf. isomorphism 8.16.2) and therefore $\bigotimes^2 S_O(R)$ is a trivial $\bar{P}$-algebra; moreover, for any $Q \in O$, we have a $k\big(R \rtimes F_{S_O}(P)_{Q,R}\big)$-module isomorphism

$$J\big(k(P/Q)\big) \cong J\big(k(R/(Q\cap R))\big) \oplus k\big((P - Q{\cdot}R)/Q\big) \qquad 9.21.26$$

and therefore the splitting homomorphisms defined in 9.19 for $\bigotimes^2 S_O$ and $\bigotimes^2 U_O$ agree *provided* $|R/Q\cap R| > p$; if $|R/Q\cap R| = p \neq 2$ then we know from Example 9.7 that the image of $F_{S_O}(P)_{Q,R}$ in $\mathrm{Aut}\big(R/(Q{\cap}R)\big)$ is mapped injectively into $\bigotimes^2 U_O$ but, since

$$\big|\mathrm{Aut}\big(R/(Q\cap R)\big)\big| = p-1 \qquad 9.21.27,$$

it is finally mapped trivially in $\bigotimes^{2(p-1)} U_O$. Consequently, for all the orbits O, setting $T_O = \otimes^{2(p-1)} S_O$ and $V_O = \otimes^{2(p-1)} U_O$ we have the commutative diagram

$$\begin{array}{ccc}
\hat{F}_{T_O}(P)_R & \xrightarrow{\Delta_{P,T_O,R}} & \hat{F}_{V_O}(R) \,\hat{\times}\, \hat{F}_{T_O(R)}(\bar{P}) \\
{\scriptstyle \omega_{P,T_O}} \searrow & & \swarrow {\scriptstyle \omega_{R,V_O} \,\hat{\times}\, \omega_{\bar{P},T_O(R)}} \\
 & k^* &
\end{array} \qquad 9.21.28.$$

When P is not elementary Abelian, we argue by induction on $|P|$; if R contains $\Phi(P)$, setting $Q = \Phi(P)$, $T = \mathrm{Res}^P_Q(S)$ and $U = \mathrm{Res}^P_R(S)$, it follows from Proposition 9.14 that we have the commutative diagram

$$\begin{array}{ccc}
 & \hat{F}_T(Q) \,\hat{\times}\, \hat{F}_{S(Q)}(P/Q)_{R/Q} & \\
{\scriptstyle \Delta_{P,S,Q}} \nearrow & & \searrow {\scriptstyle \mathrm{id}_{\hat{F}_T(Q)} \,\hat{\times}\, \Delta_{P/Q,S(Q),R/Q}} \\
\hat{F}_S(P)_R & & \hat{F}_T(Q) \,\hat{\times}\, \hat{F}_{U(Q)}(R/Q) \,\hat{\times}\, \hat{F}_{S(R)}(P/R) \\
{\scriptstyle \Delta_{P,S,R}} \searrow & & \nearrow {\scriptstyle \Delta_{R,U,Q} \,\hat{\times}\, \mathrm{id}_{\hat{F}_{S(R)}(P/R)}} \\
 & \hat{F}_U(R)_Q \,\hat{\times}\, \hat{F}_{S(R)}(P/R) &
\end{array} \qquad 9.21.29$$

and we already know that, for any $\hat{\psi} \in \hat{F}_U(R)_Q$ and any $\hat{\hat{\varphi}} \in \hat{F}_{S(Q)}(P/Q)_{R/Q}$, we have

$$\begin{gathered}
(\omega_{Q,T} \,\hat{\times}\, \omega_{R/Q,U(Q)})\big(\Delta_{R,U,Q}(\hat{\psi})\big) = \omega_{R,U}(\hat{\psi}) \\
(\omega_{R/Q,U(Q)} \,\hat{\times}\, \omega_{P/R,S(R)})\big(\Delta_{P/Q,S(Q),R/Q}(\hat{\hat{\varphi}})\big) = \omega_{P/Q,S(Q)}(\hat{\hat{\varphi}})
\end{gathered} \qquad 9.21.30.$$

Consequently, if $\hat{\varphi} \in \hat{F}_S(P)_R$ and we have

$$\Delta_{P,S,Q}(\hat{\varphi}) = \hat{\psi}{\cdot}\hat{\hat{\varphi}} \quad \text{and} \quad \Delta_{P,S,R}(\hat{\varphi}) = \hat{\hat{\psi}}{\cdot}\hat{\varphi} \qquad 9.21.31,$$

where $\hat{\psi} \in \hat{F}_T(Q)$, $\hat{\bar{\bar{\varphi}}} \in \hat{F}_{S(Q)}(P/Q)_{R/Q}$, $\hat{\bar{\psi}} \in \hat{F}_U(R)_Q$ and $\hat{\bar{\varphi}} \in \hat{F}_{S(R)}(P/R)$, then we get (cf. 9.20.1, 9.21.29 and 9.21.30)

$$\begin{aligned}(\omega_{R,U} \hat{\times} \omega_{P/R,S(R)})\big(\Delta_{P,S,R}(\hat{\varphi})\big) &= \omega_{R,U}(\hat{\psi})\,\omega_{P/R,S(R)}(\hat{\bar{\bar{\varphi}}})\\ &= (\omega_{Q,T}\hat{\times}\omega_{R/Q,U(Q)})\big(\Delta_{R,U,Q}(\hat{\psi})\big)\omega_{P/R,S(R)}(\hat{\bar{\bar{\varphi}}})\\ &= (\omega_{Q,T}\hat{\times}\omega_{R/Q,U(Q)}\hat{\times}\omega_{P/R,S(R)})\big(\Delta_{R,U,Q}(\hat{\psi})\cdot\hat{\bar{\bar{\varphi}}}\big)\\ &= (\omega_{Q,T}\hat{\times}\omega_{R/Q,U(Q)}\hat{\times}\omega_{P/R,S(R)})\big(\hat{\bar{\psi}}\cdot\Delta_{P/Q,S(Q),R/Q}(\hat{\bar{\varphi}})\big)\\ &= \omega_{Q,T}(\hat{\bar{\psi}})(\omega_{R/Q,U(Q)}\hat{\times}\omega_{P/R,S(R)})\big(\Delta_{P/Q,S(Q),R/Q}(\hat{\bar{\varphi}})\big)\\ &= \omega_{Q,T}(\hat{\bar{\psi}})\omega_{P/Q,S(Q)}(\hat{\bar{\varphi}}) = \omega_{P,S}(\hat{\varphi})\end{aligned} \qquad 9.21.32.$$

Finally, if R does not contain $\Phi(P)$, we set $Q = \Phi(P)\cdot R$ and, as above, we have again by Proposition 9.14 the following commutative diagram

$$\begin{array}{ccc} & \hat{F}_T(Q)_R \hat{\times} \hat{F}_{S(Q)}(P/Q) & \\ {\scriptstyle \Delta_{P,S,Q}} \nearrow & & \searrow {\scriptstyle \Delta_{Q,T,R} \hat{\times} \mathrm{id}_{\hat{F}_{S(Q)}(P/Q)}} \\ \hat{F}_S(P)_R & & \hat{F}_U(R) \hat{\times} \hat{F}_{T(R)}(Q/R) \hat{\times} \hat{F}_{S(Q)}(P/Q) \\ {\scriptstyle \Delta_{P,S,R}} \searrow & & \nearrow {\scriptstyle \mathrm{id}_{\hat{F}_U(R)} \hat{\times} \Delta_{P/R,S(R),Q/R}} \\ & \hat{F}_U(R) \hat{\times} \hat{F}_{S(R)}(P/R)_{Q/R} & \end{array} \qquad 9.21.33;$$

similarly, for any $\hat{\psi} \in \hat{F}_T(Q)_R$ and any $\hat{\bar{\varphi}} \in \hat{F}_{S(R)}(P/R)_{Q/R}$, we already have

$$\begin{aligned}(\omega_{R,U}\hat{\times}\omega_{Q/R,T(R)})\big(\Delta_{Q,T,R}(\hat{\psi})\big) &= \omega_{Q,T}(\hat{\psi})\\ (\omega_{Q/R,T(R)} \hat{\times} \omega_{P/Q,S(Q)})\big(\Delta_{P/R,S(R),Q/R}(\hat{\bar{\varphi}})\big) &= \omega_{P/R,S(R)}(\hat{\bar{\varphi}})\end{aligned} \qquad 9.21.34;$$

now, as in equalities 9.21.32, it is easily checked that, for any $\hat{\varphi} \in \hat{F}_S(P)_R$, we still have

$$(\omega_{R,U} \hat{\times} \omega_{P/R,S(R)})\big(\Delta_{P,S,R}(\hat{\varphi})\big) = \omega_{P,S}(\hat{\varphi}) \qquad 9.21.35,$$

We are done.

Chapter 10

A gluing theorem for Dade P-algebras

10.1 Let P be a finite p-group and k an algebraically closed field of characteristic p. Mimicking the situation in Proposition 7.12 and Theorem 7.14, consider a P-stable nonempty set $\mathfrak{X}$ of subgroups of P such that any subgroup Q of P containing some $R \in \mathfrak{X}$ belongs to $\mathfrak{X}$, and assume that, for any $Q \in \mathfrak{X}$, we have an element $s(Q)$ in the Dade group $\mathcal{D}_k\big(\bar{N}_P(Q)\big)$ (cf. 1.20) in such a way that $s(Q^u)$ coincides with the image of $s(Q)$ in $\mathcal{D}_k\big(\bar{N}_P(Q^u)\big)$ by the isomorphism induced by $u \in P$, and that, for any normal subgroup R of Q belonging to $\mathfrak{X}$, we have

$$\mathrm{res}^{\bar{N}_P(Q)}_{\bar{\bar{N}}_{Q,R}}\big(s(Q)\big) = \widetilde{\mathfrak{Br}}^{\bar{N}_{Q,R}}_{\bar{Q}}\Big(\mathrm{res}^{\bar{N}_P(R)}_{\bar{N}_{Q,R}}\big(s(R)\big)\Big) \qquad 10.1.1,$$

where we denote by $\bar{N}_{Q,R}$ and $\bar{Q}$ the respective images of $N_P(Q) \cap N_P(R)$ and Q in $\bar{N}_P(R)$, and we set $\bar{\bar{N}}_{Q,R} = \bar{N}_{Q,R}/\bar{Q}$.

10.2 It is clear that, for any $s \in \mathcal{D}_k(P)$, the correspondence mapping $Q \in \mathfrak{X}$ on $\widetilde{\mathfrak{Br}}^{N_P(Q)}_{Q}\big(\mathrm{res}^P_{N_P(Q)}(s)\big)$ fulfills the conditions above (cf. equality 1.22.3), and we may ask for the correspondences above coming in this way from an element of $\mathcal{D}_k(P)$. If P is Abelian, simply all of them (cf. Proposition 3.6 in [45]) is the answer. In this chapter we prove that, whenever $p \neq 2$, in the *localized* group $\mathcal{D}_k(P)_p$ over the set $\{p^n\}_{n\in\mathbb{N}}$ of powers of p (cf. 8.10) the answer depends on the existence of the so-called *balanced functions* over the set of nontrivial elementary Abelian subgroups of P (cf. 10.10 below). Actually, for any $n \in 2\mathbb{Z}$, prime to p if $n \neq 0$, we may reformulate the analogous question in

$$\overline{\mathcal{D}}^n_k(P)_p = \mathcal{D}_k(P)_p/n{\cdot}\mathcal{D}_k(P)_p \qquad 10.2.1$$

and we obtain the analogous answer with the particularity that when $n = 2$ the corresponding *balanced functions* always exist. Although only interested in the cases where $n = 0$ or $n = 2$, we mainly formulate our statements in $\overline{\mathcal{D}}^n_k(P)_p$, keeping the same index for the corresponding maps. We need Carlson's result (cf. statement 8.21.2) that we restate below. Let us call *section* of P any pair (R,Q) of subgroups of P such that $R \triangleleft Q$, and we say that (R,Q) is *proper* if $|Q/R| \neq |P|$.

Proposition 10.3 *Assume that* $p \neq 2$ *and that* P *is not elementary Abelian. Then, in* $\overline{\mathcal{D}}^n_k(P)_p$ *we have*

$$\bigcap_{(R,Q)} \mathrm{Ker}\big(\widetilde{\mathfrak{Br}}^Q_R \circ \mathrm{res}^P_Q\big) = \{0\} \qquad 10.3.1,$$

where (R,Q) *runs over the set of proper sections of* P.

Proof: If the class of $s \in \mathcal{D}_k(P)_p$ in $\bar{\mathcal{D}}_k^n(P)_p$ belongs to this intersection, for any proper subgroup Q of P we get $\varpi_Q^P(s) = n{\cdot}t_Q$ for some $t_Q \in \mathcal{R}^\circ\mathcal{D}_k(Q)_p$; moreover, choosing $Q = P$ and $R = \Phi(P)$, it follows from Corollary 8.22 that $\varpi_P^P(s) = n{\cdot}t_P$ for some $t_P \in \mathcal{R}^\circ\mathcal{D}_k(P)_p$; further, it is quite clear that we may choose a P-stable family $\{t_Q\}_Q$ of such elements when Q runs over the set of subgroups of P; then, it follows from Proposition 8.11 that there exists $t \in \mathcal{D}_k(P)_p$ such that $s = n{\cdot}t$.

Remark 10.4 According to statements 8.15.4, 8.16.4 and 8.16.5, whenever P is elementary Abelian, equality 10.3.1 remains true provided $|P| \geq p^3$.

10.5 Note that it suffices to formulate our question when $\mathfrak{X}$ is the set of all the nontrivial subgroups of P, since otherwise we argue in $\bar{N}_P(R)$ for a subgroup R of P which is maximal outside $\mathfrak{X}$. Moreover, for any $Q \in \mathfrak{X}$, it is clear that we have an analogous correspondence, fulfilling the above conditions, for the set $\mathfrak{X}_Q$ of subgroups of Q in $\mathfrak{X}$. Thus, in the last step of a suitable induction argument, we may apply the following result.

Theorem 10.6 *Assume that $p \neq 2$ and that P is not elementary Abelian. For any proper section (R,Q) of P, let $\bar{s}(R,Q)$ be an element of $\bar{\mathcal{D}}_k^n(Q/R)_p$ in such a way that the following conditions hold*

10.6.1 *For any $u \in P$, $\bar{s}(R^u,Q^u)$ is the image of $\bar{s}(R,Q)$ by the isomorphism*

$$\bar{\mathcal{D}}_k^n\big(\bar{N}_P(Q/R)\big)_p \cong \bar{\mathcal{D}}_k^n\big(\bar{N}_P(Q^u/R^u)\big)_p$$

induced by the conjugation by u^{-1}.

10.6.2 *For any normal subgroup T of Q containing R, setting $\bar{Q} = Q/R$ and $\bar{T} = T/R$, we have*

$$\widetilde{\mathfrak{Br}}_{\bar{T}}^{\bar{Q}}\big(\bar{s}(R,Q)\big) = \bar{s}(T,Q) \quad \textit{and} \quad \mathrm{res}_{\bar{T}}^{\bar{Q}}\big(\bar{s}(R,Q)\big) = \bar{s}(R,T).$$

Then there is a unique element $\bar{s} \in \bar{\mathcal{D}}_k^n(P)_p$ such that

$$\widetilde{\mathfrak{Br}}_R^Q\big(\mathrm{res}_Q^P(\bar{s})\big) = \bar{s}(R,Q) \qquad 10.6.3$$

for any proper section (R,Q) of P.

Proof: For any proper section (R,Q) of P, note that either Q is contained in a maximal subgroup of P or R is normal in P, and therefore, in order to prove equality 10.6.3, it suffices to check that

$$\mathrm{res}_M^P(\bar{s}) = \bar{s}(1,M) \quad \text{and} \quad \widetilde{\mathfrak{Br}}_N^P(\bar{s}) = \bar{s}(N,P) \qquad 10.6.4$$

for any maximal subgroup M of P and, setting $Z = \Omega_1\big(Z(P)\big)$, for any nontrivial subgroup N of Z.

Thus, let us say that a section (R, Q) of P is *dominant* if

$$R \subset Z \subset Q \lhd P \qquad 10.6.5;$$

if (R, Q) and (R', Q') are two dominant sections of P, then we say that (R', Q') is *smaller* than (R, Q) and write

$$(R', Q') \leq (R, Q) \qquad 10.6.6$$

whenever $R \subset R' \subset Q' \subset Q$, which defines an order relation in the set of dominant sections. Moreover, setting

$$(R, Q) \cap (R', Q') = (R{\cdot}R', Q \cap Q') \qquad 10.6.7,$$

it is quite clear that $(R, Q) \cap (R', Q')$ is smaller than both (R, Q) and (R', Q'); conversely, if (U, T) is a dominant section of P such that

$$(U, T) \leq (R, Q) \quad \text{and} \quad (U, T) \leq (R', Q') \qquad 10.6.8$$

then we have

$$(U, T) \leq (R, Q) \cap (R', Q') \qquad 10.6.9.$$

Consider the Möbius function $\mu_{\mathfrak{D}}$ (cf. 1.32) over the set $\mathfrak{D}$ of dominant proper sections of P endowed with the *opposite* order; that is to say, $\mu_{\mathfrak{D}}(1, P) = 1$ and, for any dominant proper section (U, T), we have

$$\sum_{(R,Q)} \mu_{\mathfrak{D}}(R, Q) = 0 \qquad 10.6.10,$$

where (R, Q) runs over the set of dominant sections of P such that (U, T) is smaller than (R, Q). Then, for any pair of dominant proper sections (N, M) and (U, T) of P we claim that

$$\sum_{(R,Q)} \mu_{\mathfrak{D}}(R, Q) = -\delta_{(U,T)}^{(N,M)} \qquad 10.6.11,$$

when (R, Q) runs over the set of dominant proper sections such that

$$(N, M) \cap (R, Q) = (U, T) \qquad 10.6.12.$$

Indeed, we may assume that $(U, T) \leq (N, M)$ and then we have

$$\sum_{(U',T')} \sum_{(R',Q')} \mu_{\mathfrak{D}}(R', Q') = \sum_{(R'',Q'')} \mu_{\mathfrak{D}}(R'', Q'') = -1 \qquad 10.6.13,$$

where (U',T') runs over the set of dominant sections of P such that

$$(U,T) \leq (U',T') \leq (N,M) \qquad 10.6.14$$

and, for such a (U',T'), (R',Q') runs over the set of dominant proper sections of P such that

$$(N,M) \cap (R',Q') = (U',T') \qquad 10.6.15,$$

so that finally (R'',Q'') runs over the set of all the dominant proper sections of P; thus, arguing by induction on $|T/U|$, we may assume that

$$\sum_{(R',Q')} \mu_{\mathfrak{D}}(R',Q') = -\delta_{(U',T')}^{(N,M)} \qquad 10.6.16$$

for any $(U',T') \neq (U,T)$ fulfilling condition 10.6.14 and then it is clear that equality 10.6.13 forces equality 10.6.11.

Now, we claim that it suffices to choose

$$\bar{s} = \sum_{(R,Q)} \frac{|Q|}{|P|} \mu_{\mathfrak{D}}(R,Q)\cdot \mathrm{ten}_Q^P\Big(\mathrm{res}_{\pi_{(R,Q)}}\big(\bar{s}(R,Q)\big)\Big) \qquad 10.6.17,$$

where (R,Q) runs over the set of dominant proper sections of P and, for such a (R,Q), $\pi_{(R,Q)}\colon Q \to Q/R$ is the canonical map. Indeed, for any nontrivial subgroup N of Z we get

$$\widetilde{\mathfrak{Br}}_N^P(\bar{s}) = \sum_{(R,Q)} \frac{|Q|}{|P|} \mu_{\mathfrak{D}}(R,Q)\cdot(\widetilde{\mathfrak{Br}}_N^P \circ \mathrm{ten}_Q^P)\Big(\mathrm{res}_{\pi_{(R,Q)}}\big(\bar{s}(R,Q)\big)\Big) \qquad 10.6.18,$$

where (R,Q) runs over the set of dominant proper sections of P. But, according to equality 8.8.3 and to Corollary 8.22, we still get

$$\begin{aligned}
&(\widetilde{\mathfrak{Br}}_N^P \circ \mathrm{ten}_Q^P)\Big(\mathrm{res}_{\pi_{(R,Q)}}\big(\bar{s}(R,Q)\big)\Big) \\
&\quad = (\mathrm{ten}_{Q/N}^{P/N} \circ \widetilde{\mathfrak{Br}}_N^Q)\Big(\mathrm{res}_{\pi_{(R,Q)}}\big(\bar{s}(R,Q)\big)\Big) \qquad 10.6.19; \\
&\quad = (\mathrm{ten}_{Q/N}^{P/N} \circ \mathrm{res}_{\pi_{((N\cdot R)/N,Q/N)}})\Big(\widetilde{\mathfrak{Br}}_{(N\cdot R)/R}^{Q/R}\big(\bar{s}(R,Q)\big)\Big)
\end{aligned}$$

moreover, by condition 10.6.2, we have

$$\widetilde{\mathfrak{Br}}_{(N\cdot R)/R}^{Q/R}\big(\bar{s}(R,Q)\big) = \bar{s}(N\cdot R, Q) \qquad 10.6.20.$$

Consequently, denoting by $\bar X$ the quotient by N of any subgroup X of P containing N, from equality 10.6.11 we obtain

$$\begin{aligned}\widetilde{\mathfrak{Br}}{}_N^P(\bar s) &= \sum_{(R,Q)} \frac{|Q|}{|P|}\mu_{\mathfrak{D}}(R,Q)\cdot(\mathrm{ten}_{\bar Q}^{\bar P}\circ \mathrm{res}_{\pi_{(\overline{R\cdot N},\bar Q)}})\big(\bar s(R\cdot N,Q)\big)\\ &= \sum_{(U,T)}\Big(\sum_{(R',Q')}\frac{|Q'|}{|P|}\mu_{\mathfrak{D}}(R',Q')\Big)\cdot\mathrm{ten}_{\bar T}^{\bar P}\Big(\mathrm{res}_{\pi_{(\bar U,\bar T)}}\big(\bar s(U,T)\big)\Big) \qquad 10.6.21,\\ &= \bar s(N,P)\end{aligned}$$

where (R,Q) runs over the set of all the dominant proper sections of P, (U,T) runs over the set of those which are smaller than (N,P) and, for such a section (U,T), (R',Q') runs over the set of dominant proper sections of P fulfilling

$$(N,P)\cap(R',Q') = (U,T) \qquad 10.6.22$$

which implies that $Q' = T$.

It remains to prove that $\mathrm{res}_M^P(\bar s) = \bar s(1,M)$ for any maximal subgroup M of P. Firstly assume that M contains Z; it follows from the *Mackey formula* 8.8.1 that

$$\begin{aligned}\mathrm{res}_M^P(\bar s) &= \sum_{(R,Q)}\frac{|Q|}{|P|}\mu_{\mathfrak{D}}(R,Q)\cdot(\mathrm{res}_M^P\circ\mathrm{ten}_Q^P)\Big(\mathrm{res}_{\pi_{(R,Q)}}\big(\bar s(R,Q)\big)\Big)\\ &= \sum_{(R,Q)}\frac{|Q|}{|P|}\mu_{\mathfrak{D}}(R,Q)\cdot\sum_u \mathrm{ten}_{M\cap Q^u}^M\Big(\mathrm{res}_{\pi_{(R^u,M\cap Q^u)}}\big(\bar s(R^u,M\cap Q^u)\big)\Big) \qquad 10.6.23,\end{aligned}$$

where (R,Q) runs over the set of all the dominant proper sections of P and u over a set of representatives in M for $P\backslash M/Q$.

Hence, since $Z\subset M$, it follows from conditions 10.6.1 and 10.6.2, and from equality 10.6.11 that

$$\begin{aligned}&\mathrm{res}_M^P(\bar s)\\ &= \sum_{(R,Q)}\frac{|M\cap Q|}{|M|}\mu_{\mathfrak{D}}(R,Q)\cdot\mathrm{ten}_{M\cap Q}^M\Big(\mathrm{res}_{\pi_{(R,M\cap Q)}}\big(\bar s(R,M\cap Q)\big)\Big)\\ &= \sum_{(R,T)}\Big(\sum_{Q'}\frac{|M\cap Q'|}{|M|}\mu_{\mathfrak{D}}(R,Q')\Big)\cdot\mathrm{ten}_T^M\Big(\mathrm{res}_{\pi_{(R,T)}}\big(\bar s(R,T)\big)\Big) \qquad 10.6.24,\\ &= \bar s(1,M)\end{aligned}$$

where (R,Q) runs over the set of all the dominant proper sections of P, (R,T) runs over the set of those which are smaller than $(1,M)$ or, equivalently, fulfill

$T \subset M$, and for such a (R,T), Q' runs over the set of normal subgroups of P such that $M \cap Q' = T$ or, equivalently, fulfill

$$(1, M) \cap (R, Q') = (R, T) \qquad 10.6.25.$$

Now, assume that M does not contain Z, so that we have $P = Z{\cdot}M$. Let L be a subgroup of P such that $P = Z{\cdot}L$, so that L is not elementary Abelian and we have

$$\Phi(L) = L \cap \Phi(P) \quad \text{and} \quad \Omega_1\big(Z(L)\big) = L \cap Z \qquad 10.6.26;$$

arguing by induction on $|L \cap Z|$, we will prove that $\mathrm{res}^P_L(\bar{s}) = \bar{s}(1, L)$.

Let (R, Q) be a proper section of L. If $Q \neq L$ and K is a maximal subgroup of L containing Q, then either K contains $L \cap Z$, so that K is contained in a maximal subgroup H of P containing Z, and therefore, by condition 10.6.2, we get

$$\begin{aligned}\widetilde{\mathfrak{Br}}^Q_R\big(\mathrm{res}^P_Q(\bar{s})\big) &= \widetilde{\mathfrak{Br}}^Q_R\Big(\mathrm{res}^H_Q\big(\mathrm{res}^P_H(\bar{s})\big)\Big)\\ &= \widetilde{\mathfrak{Br}}^Q_R\Big(\mathrm{res}^H_Q\big(\bar{s}(1,H)\big)\Big) \qquad 10.6.27,\\ &= \widetilde{\mathfrak{Br}}^Q_R\big(\bar{s}(1,Q)\big) = \widetilde{\mathfrak{Br}}^Q_R\Big(\mathrm{res}^L_Q\big(\bar{s}(1,L)\big)\Big)\end{aligned}$$

or we have $L = (L\cap Z){\cdot}K$, so that $P = Z{\cdot}K$, and therefore, by the induction hypothesis, we get $\mathrm{res}^P_K(\bar{s}) = \bar{s}(1, K)$, so that we still get

$$\begin{aligned}\widetilde{\mathfrak{Br}}^Q_R\Big(\mathrm{res}^P_Q(\bar{s})\Big) &= \widetilde{\mathfrak{Br}}^Q_R\Big(\mathrm{res}^K_Q\big(\mathrm{res}^P_K(\bar{s})\big)\Big)\\ &= \widetilde{\mathfrak{Br}}^Q_R\Big(\mathrm{res}^K_Q\big(\bar{s}(1,K)\big)\Big) \qquad 10.6.28.\\ &= \widetilde{\mathfrak{Br}}^Q_R\big(\bar{s}(1,Q)\big) = \widetilde{\mathfrak{Br}}^Q_R\Big(\mathrm{res}^L_Q\big(\bar{s}(1,L)\big)\Big)\end{aligned}$$

Finally, if $Q = L$ then R is a nontrivial normal subgroup of P and, in particular, we have $N = R \cap Z \neq 1$, so that setting $\bar{Q} = Q/N$ and $\bar{R} = R/N$ we get (cf. 10.6.21 and equality 1.22.3)

$$\begin{aligned}\widetilde{\mathfrak{Br}}^Q_R\big(\mathrm{res}^P_Q(\bar{s})\big) &= \widetilde{\mathfrak{Br}}^{\bar{Q}}_{\bar{R}}\Big(\widetilde{\mathfrak{Br}}^Q_N\big(\mathrm{res}^P_Q(\bar{s})\big)\Big) = \widetilde{\mathfrak{Br}}^{\bar{Q}}_{\bar{R}}\big(\bar{s}(N,Q)\big)\\ &= \bar{s}(R,Q) = \widetilde{\mathfrak{Br}}^Q_R\big(\bar{s}(1,Q)\big).\end{aligned} \qquad 10.6.29.$$

In conclusion, the difference $\mathrm{res}^P_L(\bar{s}) - \bar{s}(1,L)$ belongs to the intersection in 10.3.1 for the p-group L; hence, it follows from Proposition 10.3 applied to L that $\mathrm{res}^P_L(\bar{s}) = \bar{s}(1, L)$. Thus, the element $\bar{s}$ in 10.6.17 fulfills statement 10.6.3.

Similarly, if $\bar{s}' \in \overline{\mathcal{D}}^n_k(P)_p$ is another element fulfilling statement 10.6.3, then the difference $\bar{s}' - \bar{s}$ belongs to the left-hand member of equality 10.3.1 for P and therefore we have $\bar{s}' = \bar{s}$. We are done.

10.7 Recall that when P is elementary Abelian, for any $s \in \mathcal{D}_k(P)_p$ we have (cf. equality 8.17.1)

$$s = \sum_Q z_Q \cdot j^P_Q \qquad 10.7.1,$$

where Q runs over the set of proper subgroups of P and, for such a Q, z_Q is an element of $\mathbb{Z}_p$, uniquely determined if $|P/Q| \geq p^2$; then, if $|P| \geq p^2$, we denote by

$$\varepsilon_P : \mathcal{D}_k(P)_p \longrightarrow \mathbb{Z}_p \qquad 10.7.2$$

the homomorphism mapping s on z_1, whereas if $|P| = p$ we denote by

$$\varepsilon_P : \mathcal{D}_k(P)_p \longrightarrow \overline{\mathbb{Z}}^2 = \mathbb{Z}_p/2\mathbb{Z}_p \qquad 10.7.3$$

the group homomorphism mapping s on the image $\bar{z}_1$ of z_1 in $\overline{\mathbb{Z}}^2$, which is uniquely determined. Accordingly, we denote by $\bar{\varepsilon}^n_P(s)$ the image of $\varepsilon_P(s)$ in $\overline{\mathbb{Z}}^n_p = \mathbb{Z}_p/n\mathbb{Z}_p$ or in $\overline{\mathbb{Z}}^2$, defining a map $\bar{\varepsilon}^n_P$ over $\overline{\mathcal{D}}^n_k(P)_p$.

Lemma 10.8 *If $p \neq 2$ and P is elementary Abelian, for any $s \in \mathcal{D}_k(P)_p$ we have*

$$s = \sum_T \varepsilon_{P/T}\big(\widetilde{\mathfrak{Br}}^P_T(s)\big) \cdot j^P_T \qquad 10.8.1$$

where T runs over the set of subgroups of P. Moreover, if Q and R are noncyclic subgroups of P, then we have

$$\begin{aligned} &\varepsilon_Q\big(\mathrm{res}^P_Q(s)\big) + \sum_T \varepsilon_{P/T}\big(\widetilde{\mathfrak{Br}}^P_T(s)\big) \\ &\qquad = \varepsilon_R\big(\mathrm{res}^P_R(s)\big) + \sum_U \varepsilon_{P/U}\big(\widetilde{\mathfrak{Br}}^P_U(s)\big) \end{aligned} \qquad 10.8.2,$$

where T and U respectively run over the set of subgroups of P fulfilling

$$T \cap R = 1 \neq T \cap Q \quad \textit{and} \quad U \cap Q = 1 \neq U \cap R \qquad 10.8.3.$$

If they are nontrivial then the analogous equality with the images in $\overline{\mathbb{Z}}^2$ of all the terms holds.

Proof: It follows from equality 10.7.1 that, for any nontrivial proper subgroup T of P, we have

$$\widetilde{\mathfrak{Br}}^{P}_{T}(s) = \sum_{Q} z_Q{\cdot}j^{P/T}_{Q/T} \qquad 10.8.4,$$

where Q runs over the set of proper subgroups of P containing T, and, according to definitions 10.7.2 and 10.7.3, $\varepsilon_{P/T}\big(\widetilde{\mathfrak{Br}}^{P}_{T}(s)\big)$ coincides with $\bar z_T$ or with z_T according to P/T is cyclic or not; then, equality 10.8.1 follows from equality 10.7.1.

Moreover, for any nontrivial subgroup Q of P, we have

$$\mathrm{res}^P_Q(j^P_T) = j^Q_{T\cap Q} \qquad 10.8.5$$

and therefore, if Q is not cyclic, we get

$$\varepsilon_Q\big(\mathrm{res}^P_Q(s)\big) = \sum_{T} z_T = \sum_{T\neq 1} \varepsilon_{P/T}\big(\widetilde{\mathfrak{Br}}^{P}_{T}(s)\big) + \varepsilon_P(s) \qquad 10.8.6,$$

whereas, in all the cases we still get

$$\bar\varepsilon^2_Q\big(\mathrm{res}^P_Q(s)\big) = \sum_{T} \bar z_T = \sum_{T\neq 1} \bar\varepsilon^2_{P/T}\big(\widetilde{\mathfrak{Br}}^{P}_{T}(s)\big) + \bar\varepsilon^2_P(s) \qquad 10.8.7,$$

where T runs over the set of subgroups of P such that $T\cap Q = 1$.

Now, if R is another nontrivial subgroup of P, the difference between the equalities 10.8.6 corresponding to Q and R yields equality 10.8.2 whenever they are not cyclic, whereas in all the cases the sum of the equalities 10.8.7 corresponding to Q and R yields the analogous equality in $\bar{\mathbb{Z}}^2$. We are done.

10.9 Let us come back to our starting setting and call $(P,\mathfrak{X},n)$-*gluing family* any family $\bar{\mathfrak{s}} = \{\bar s(Q)\}_{Q\in\mathfrak{X}}$ of elements $\bar s(Q)\in \bar{\mathcal{D}}^n_k\big(\bar N_P(Q)\big)_p$ fulfilling

$$\mathrm{res}^{\bar N_P(Q)}_{\bar{\bar N}_{Q,R}}\big(\bar s(Q)\big) = \widetilde{\mathfrak{Br}}^{\bar N_{Q,R}}_{\bar Q}\Big(\mathrm{res}^{\bar N_P(R)}_{\bar N_{Q,R}}\big(\bar s(R)\big)\Big) \qquad 10.9.1$$

for any $R\in\mathfrak{X}$ such that $R\triangleleft Q$, where we denote by $\bar N_{Q,R}$ and $\bar Q$ the respective images of $N_P(Q)\cap N_P(R)$ and Q in $\bar N_P(R)$, and where we set $\bar{\bar N}_{Q,R} = \bar N_{Q,R}/\bar Q$. As we show below, in order to determine a unique element $\bar s\in\bar{\mathcal{D}}^n_k(P)_p$ from the family $\bar{\mathfrak{s}}$ when $\mathfrak{X}$ is the set of all the nontrivial subgroups of P, we also need the function mapping any nontrivial elementary Abelian subgroup E of P on $\bar\varepsilon^n_E\big(\mathrm{res}^P_E(\bar s)\big)$ (cf. 10.7). We write $\widetilde{\mathfrak{Br}}^{P}_{Q}(\bar s)$ instead of $\widetilde{\mathfrak{Br}}^{N_P(Q)}_{Q}\big(\mathrm{res}^P_{N_P(Q)}(\bar s)\big)$ for short.

10.10 Let us be more explicit; denote by $\mathcal{A}(P)$ the set of nontrivial elementary Abelian subgroups of P and set

$$\varepsilon_{\bar{\mathfrak{s}}}(E,F) = \bar{\varepsilon}^n_{E/F}\Big(\mathrm{res}^{\bar{N}_P(F)}_{E/F}\big(\bar{s}(F)\big)\Big) \qquad 10.10.1$$

for any $E, F \in \mathcal{A}(P)$ such that $F \subset E$; having regard to Lemma 10.8, we call $(P,\bar{\mathfrak{s}})$*-balanced function* ε any P-stable map sending $E \in \mathcal{A}(P)$ to an element of $\overline{\mathbb{Z}}^2$ or of $\overline{\mathbb{Z}}_p^n$ according to whether E is cyclic or not, in such a way that, for any pair of elements E and F of $\mathcal{A}(P)$ centralizing each other, if they are noncyclic then in $\overline{\mathbb{Z}}_p^n$ we have

$$\varepsilon(E) + \sum_T \varepsilon_{\bar{\mathfrak{s}}}(E{\cdot}F, T) = \varepsilon(F) + \sum_U \varepsilon_{\bar{\mathfrak{s}}}(E{\cdot}F, U) \qquad 10.10.2$$

where T and U respectively run over the set of subgroups of $E{\cdot}F$ fulfilling

$$T \cap F = 1 \neq T \cap E \quad \text{and} \quad U \cap E = 1 \neq U \cap F \qquad 10.10.3,$$

whereas in all the cases it holds the analogous equality with the images of all the terms in $\overline{\mathbb{Z}}^2$.

Proposition 10.11 *Assume that* $p \neq 2$ *and that* $\mathfrak{X}$ *is the set of all the nontrivial subgroups of* P, *and let* $\bar{\mathfrak{s}} = \{\bar{s}(Q)\}_{Q\in\mathfrak{X}}$ *be a* $(P,\mathfrak{X},n)$*-gluing family. For any* $(P,\bar{\mathfrak{s}})$*-balanced function* ε *there is a unique element* $\bar{s} \in \overline{\mathcal{D}}^n_k(P)_p$ *fulfilling*

10.11.1 *For any nontrivial subgroup* Q *of* P, *we have* $\widetilde{\mathfrak{Br}}^P_Q(\bar{s}) = \bar{s}(Q)$.

10.11.2 *For any nontrivial elementary Abelian subgroup* E *of* P, *we have*

$$\bar{\varepsilon}^n_E\big(\mathrm{res}^P_E(\bar{s})\big) = \varepsilon(E)$$

Proof: If P is elementary Abelian, it follows from equality 10.8.1 and from conditions 10.11.1 and 10.11.2 above that the good choice of $\bar{s}$ should be

$$\bar{s} = \sum_T \bar{\varepsilon}^n_{P/T}\big(\bar{s}(T)\big){\cdot}\bar{\jmath}^P_T + \varepsilon(P){\cdot}\bar{\jmath}^P_1 \qquad 10.11.3,$$

where T runs over the set of nontrivial subgroups of P. Indeed, for any nontrivial subgroup Q of P, we have (cf. 10.9.1)

$$\begin{aligned}\widetilde{\mathfrak{Br}}^P_Q(\bar{s}) &= \sum_T \bar{\varepsilon}^n_{P/T}\big(\bar{s}(T)\big){\cdot}\bar{\jmath}^{P/Q}_{T/Q} \\ &= \sum_T \bar{\varepsilon}^n_{P/T}\Big(\widetilde{\mathfrak{Br}}^P_T\big(\bar{s}(Q)\big)\Big){\cdot}\bar{\jmath}^{P/Q}_{T/Q} = \bar{s}(Q)\end{aligned} \qquad 10.11.4,$$

where T runs over the set of subgroups of P containing Q.

Moreover, for any nontrivial subgroup E of P, we have

$$\mathrm{res}^P_E(\bar{s}) = \sum_T \bar{\varepsilon}^n_{P/T}\big(\bar{s}(T)\big)\cdot\bar{\jmath}^E_{T\cap E} + \varepsilon(P)\cdot\bar{\jmath}^E_1 \qquad 10.11.5,$$

where T runs over the set of nontrivial subgroups of P; in particular, if E is not cyclic, we get

$$\bar{\varepsilon}^n_E\big(\mathrm{res}^P_E(\bar{s})\big) = \sum_T \bar{\varepsilon}^n_{P/T}\big(s(T)\big) + \varepsilon(P) \qquad 10.11.6,$$

where T runs over the set of nontrivial subgroups of P fulfilling $T\cap E = 1$, whereas in all cases it holds the analogous equality with the images of all the terms in $\overline{\mathbb{Z}}^2$. Thus, we have

$$\bar{\varepsilon}^n_E\big(\mathrm{res}^P_E(\bar{s})\big) = \varepsilon(E) \qquad 10.11.7$$

since we are assuming that ε is $(P,\bar{\mathfrak{s}})$-balanced (cf. 10.10.2).

From now on, we assume that $\Phi(P)\neq 1$ and argue by induction on $|P|$. For any nontrivial proper subgroup Q of P, denoting by $\mathfrak{X}_Q$ the set of all the nontrivial subgroups of Q, it is easily checked that

$$\bar{\mathfrak{s}}_Q = \big\{\mathrm{res}^{\bar{N}_P(T)}_{\bar{N}_Q(T)}\big(\bar{s}(T)\big)\big\}_{T\in\mathfrak{X}_Q} \qquad 10.11.8$$

is a $(Q,\mathfrak{X}_Q,n)$-gluing family; hence, considering the $(Q,\bar{\mathfrak{s}}_Q)$-balanced function ε^Q determined by ε, it follows from our induction hypothesis that there is a unique element $\bar{s}_Q\in\bar{\mathcal{D}}^n_k(Q)_p$ fulfilling

$$\widetilde{\mathfrak{Br}}^Q_T(\bar{s}_Q) = \mathrm{res}^{\bar{N}_P(T)}_{\bar{N}_Q(T)}\big(\bar{s}(T)\big) \quad\text{and}\quad \bar{\varepsilon}^n_F\big(\mathrm{res}^Q_F(\bar{s}_Q)\big) = \varepsilon(F) \qquad 10.11.9$$

for any nontrivial subgroups T and F of Q, F being elementary Abelian.

Now, if Q and R are nontrivial proper subgroups of P such that $R\subset Q$, we claim that

$$\mathrm{res}^Q_R(\bar{s}_Q) = \bar{s}_R \qquad 10.11.10;$$

indeed, it is straightforward to prove that $\mathrm{res}^Q_R(\bar{s}_Q)$ and $\bar{s}_R$ determine the same $(R,\mathfrak{X}_R,n)$-gluing family; moreover, for any nontrivial elementary Abelian subgroup F of R, we have (cf. 10.11.9)

$$\bar{\varepsilon}^n_F\big(\mathrm{res}^Q_F(\bar{s}_Q)\big) = \varepsilon(F) = \bar{\varepsilon}^n_F\big(\mathrm{res}^R_F(\bar{s}_R)\big) \qquad 10.11.11;$$

hence, the claim follows from our induction hypothesis.

Then, for any proper section (R,Q) of P, we denote by $\bar{s}(R,Q)$ either $\widetilde{\mathfrak{Br}}_R^Q(\bar{s}_Q)$ if $Q \neq P$, or $\mathrm{res}_{\bar{Q}}^{\bar{N}_P(R)}\big(\bar{s}(R)\big)$ if $R \neq 1$; it follows from equalities 10.11.9 and 10.11.10 that the family $\{\bar{s}(R,Q)\}_{(R,Q)}$, where (R,Q) runs over the set of all the proper sections of P, is well-defined and fulfills conditions 10.6.1 and 10.6.2.

Consequently, by Theorem 10.6, there is a unique element $\bar{s} \in \bar{\mathcal{D}}_k^n(P)_p$ fulfilling

$$\widetilde{\mathfrak{Br}}_R^Q\big(\mathrm{res}_Q^P(\bar{s})\big) = \bar{s}(R,Q) \qquad 10.11.12$$

for any proper section (R,Q) of P; in particular, this element fulfills condition 10.11.1 and moreover, for any nontrivial elementary Abelian subgroup E of P, we have $\mathrm{res}_E^P(\bar{s}) = \bar{s}_E$, so that we still have

$$\bar{\varepsilon}_E^n\big(\mathrm{res}_Q^P(\bar{s})\big) = \bar{\varepsilon}_E^n(\bar{s}_E) = \varepsilon(E) \qquad 10.11.13.$$

Finally, the uniqueness of $\bar{s}$ follows from equality 10.11.3 whenever P is elementary Abelian, and from our induction argument together with Theorem 10.6 otherwise. We are done.

Corollary 10.12 *Assume that $p \neq 2$. Consider a $(P,\mathfrak{X},2)$-gluing family $\bar{\mathfrak{s}} = \{\bar{s}(Q)\}_{Q\in\mathfrak{X}}$ and an element $\bar{z}_R \in \bar{\mathbb{Z}}^2$ for any subgroup R of P outside $\mathfrak{X}$. Then, there exists a unique $\bar{s} \in \bar{\mathcal{D}}_k^2(P)_p$ such that*

10.12.1 *For any $Q \in \mathfrak{X}$, we have $\widetilde{\mathfrak{Br}}_Q^P(\bar{s}) = \bar{s}(Q)$.*

10.12.2 *For any subgroup R of P outside $\mathfrak{X}$, setting $\bar{Z}_R = \Omega_1\big(Z(\bar{N}_P(R))\big)$ we have*

$$\bar{\varepsilon}_{\bar{Z}_R}^2\Big(\mathrm{res}_{\bar{Z}_R}^{\bar{N}_P(R)}\big(\widetilde{\mathfrak{Br}}_R^P(\bar{s})\big)\Big) = \bar{z}_R\,.$$

Proof: We may assume that $\mathfrak{X}$ does not contain the trivial subgroup; let R be a subgroup of P which is maximal outside $\mathfrak{X}$, set $\bar{N} = \bar{N}_P(R)$, denote by $\mathfrak{Y}$ the set of all the nontrivial subgroups of $\bar{N}$ and consider the $(\bar{N},\mathfrak{Y},2)$-gluing family $\bar{\mathfrak{t}} = \{\bar{t}(\bar{Q})\}_{\bar{Q}\in\mathfrak{Y}}$ with

$$\bar{t}(\bar{Q}) = \mathrm{res}_{\bar{N}_{\bar{N}}(\bar{Q})}^{\bar{N}_P(Q)}\big(\bar{s}(Q)\big) \qquad 10.12.3$$

where Q is the converse image of $\bar{Q}$ in $N_P(R)$ and where we identify $\bar{N}_{\bar{N}}(\bar{Q})$ with its canonical image in $\bar{N}_P(Q)$.

Setting $\bar{Z} = \Omega_1\big(Z(\bar{N})\big)$, it easily follows from condition 10.10.2 over $\mathbb{Z}/2\mathbb{Z}$ that there is a unique $(\bar{N},\bar{\mathfrak{t}})$-balanced function ε such that $\varepsilon(\bar{Z}) = \bar{z}_R$; thus, according to Proposition 10.11, there is a unique element $\bar{s}(R) \in \mathcal{D}_k(\bar{N})$

such that, setting $\mathfrak{X}' = \mathfrak{X} \cup \{R^u\}_{u\in P}$, the extended family $\bar{\mathfrak{s}}' = \{\bar{s}(Q)\}_{Q\in\mathfrak{X}'}$ is a $(P,\mathfrak{X}',2)$-gluing family too and that we have

$$\bar{\varepsilon}^2_{z_R}\Big(\mathrm{res}^{\bar{N}}_{\bar{Z}}\big(\widetilde{\mathfrak{Br}}^P_R(\bar{s})\big)\Big) = \bar{z}_R \qquad 10.12.4.$$

Hence, arguing by induction on $|\mathfrak{X}|$, we get the existence of $\bar{s} \in \bar{\mathcal{D}}^2_k(P)_p$ fulfilling conditions 10.12.1 and 10.12.2; moreover, if $\bar{s}' \in \bar{\mathcal{D}}^2_k(P)_p$ fulfills the same conditions, it follows from the argument above that $\widetilde{\mathfrak{Br}}^P_R(\bar{s}') = \bar{s}(R)$; then, when $R = 1$ we get $\bar{s}' = \bar{s}$. We are done.

Corollary 10.13 *Assume that $p \neq 2$, that P is not cyclic and that $\mathfrak{X}$ is the set of all the nontrivial subgroups of P. Set $Z = \Omega_1\big(Z(P)\big)$ and let D be a normal elementary Abelian subgroup of P of order p^2. Consider a $(P,\mathfrak{X},0)$-gluing family $\mathfrak{s} = \{s(Q)\}_{Q\in\mathfrak{X}}$ such that $2{\cdot}s(Q) = 0$ for any nontrivial subgroup Q of P, and a P-stable function mapping any elementary Abelian subgroup E of P of order p^2 such that either $E = D$ or $\Omega_1\big(C_P(E)\big) = E$ on some element $z_E \in \mathbb{Z}_p$. Then, there exists a unique $s \in \mathcal{D}_k(P)_p$ such that*

10.13.1 *For any nontrivial subgroup Q of P, we have $\widetilde{\mathfrak{Br}}^P_Q(s) = s(Q)$.*

10.13.2 *We have $\varepsilon_D\big(\mathrm{res}^P_D(s)\big) = z_D$ and, for any elementary Abelian subgroup $E \neq D$ of P of order p^2 such that $\Omega_1\big(C_P(E)\big) = E$, denoting by a_E the element in $\{0,1\}$ fulfilling*

$$\bar{a}^2_E = \bar{\varepsilon}^2_Z\big(\mathrm{res}^P_Z(s)\big) + \sum_F \varepsilon_{\mathfrak{s}}(E,F)$$

where F runs over the set of proper subgroups of E not contained in Z, we have $\varepsilon_E\big(\mathrm{res}^P_E(s)\big) = a_E + 2z_E$.

Proof: First of all, note that our hypothesis on $\mathfrak{s}$ implies that $\varepsilon_{\mathfrak{s}}(E,F) = 0$ for any $E, F \in \mathcal{A}(P)$ such that $F \subset E$ and $|E/F| \geq p^2$ (cf. 8.16.5, 10.7 and 10.10.1). If $|Z| \geq p^2$ then condition 10.10.2 and this remark forces any $(P,\mathfrak{s})$-balanced function ε to fulfill the equality $\varepsilon(E) = \varepsilon(Z)$ for any noncyclic elementary Abelian subgroup E of P; thus, the value z_D and the analogous condition 10.10.2 over $\mathbb{Z}/2\mathbb{Z}$ determine a unique $(P,\mathfrak{s})$-balanced function ε and, according to the remark above, its existence is clear. Moreover, if E is an elementary Abelian subgroup of P of order p^2 such that $\Omega_1\big(C_P(E)\big) = E$, we have $E = Z = D$. Hence, in this case, the corollary follows from Proposition 10.11.

Assume that $|Z| = p$, so that $Z \subset D$; once again, condition 10.10.2 and our hypothesis on $\mathfrak{s}$ forces any $(P,\mathfrak{s})$-balanced function ε to fulfill the equality $\varepsilon(E) = \varepsilon(D)$ for any noncyclic elementary Abelian subgroup E of $C_P(D)$.

Moreover, for any elementary Abelian subgroup E of P of order p^2 such that $\Omega_1\big(C_P(E)\big) \neq E$, it is quite clear that the intersection $C_P(E) \cap C_P(D)$ contains an elementary Abelian subgroup F of order p^2 and that the same argument forces

$$\varepsilon(E) = \varepsilon(F) = \varepsilon(D) \qquad 10.13.3.$$

If $\Omega_1\big(C_P(E)\big) = E$ then E is a maximal elementary Abelian subgroup of P and the analogous condition 10.10.2 over $\overline{\mathbb{Z}}^2$ forces

$$\overline{\varepsilon(E)}^2 = \varepsilon(Z) + \sum_F \varepsilon_{\mathfrak{s}}(E,F) \qquad 10.13.4$$

where F runs over the set of proper subgroups of E not contained in Z. Consequently, in all the cases denoting by a_E the element in $\{0,1\}$ such that $\bar{a}_E^2$ coincides with right-hand member of equality 10.13.4, it is easily checked from the remark above that there exists indeed a unique $(P,\mathfrak{s})$-balanced function ε fulfilling $\varepsilon(D) = z_D$ and $\varepsilon(E) = a_E + 2{\cdot}z_E$ for any such a subgroup E. Then, the corollary follows from Proposition 10.11. We are done.

Remark 10.14 In [7] Serge Bouc and Jacques Thévenaz prove a more precise result. Note that, even fixing $z_D = 0 = z_E$ for all the subgroups E as above, $2{\cdot}s$ need not be zero since we have no control on the invariants a_E .

Chapter 11

The nilcentralized chain k^*-functor of a block

11.1 Let us come back to our setting in chapter 7. Let G be a finite group, b a block of G, (P,e) a maximal Brauer (b,G)-pair and $\mathcal{F}_{(b,G)}$ the associated Frobenius P-category, and let us denote by $(\mathcal{F}_{(b,G)})^{\mathrm{nc}}$ the full subcategory of $\mathcal{F}_{(b,G)}$ over the set of nilcentralized objects. In chapter 7 we have seen that, for any $\mathcal{F}_{(b,G)}$-nilcentralized subgroup Q of P, the group $\mathcal{F}_{(b,G)}(Q)$ can be *canonically* lifted to a k^*-group $\hat{\mathcal{F}}_{(b,G)}(Q)$. In this chapter, we discuss the possibility of lifting *whole* subcategories of $(\mathcal{F}_{(b,G)})^{\mathrm{nc}}$ to "k^*-categories"†. These results allow us to construct a suitable functor $\widehat{\mathfrak{aut}}_{(\mathcal{F}_{(b,G)})^{\mathrm{nc}}}$ from the *proper category of chains* of $(\mathcal{F}_{(b,G)})^{\mathrm{nc}}$ (cf. A2.8) to the category of *finite k^*-groups k^*-$\mathfrak{Gr}$*, which lifts the *automorphism functor* $\mathfrak{aut}_{(\mathcal{F}_{(b,G)})^{\mathrm{nc}}}$ introduced in Proposition A2.10 below. This is the main tool for the definition of the *Grothendieck group of the pair* $(\mathcal{F}_{(b,G)}, \widehat{\mathfrak{aut}}_{(\mathcal{F}_{(b,G)})^{\mathrm{nc}}})$ in chapter 14.

11.2 More explicitly, for any category $\mathfrak{C}$ and any Abelian group Z let us call *regular central Z-extension* of $\mathfrak{C}$ any category $\hat{\mathfrak{C}}$ over the same objects endowed with a *full* functor $\mathfrak{c}:\hat{\mathfrak{C}} \to \mathfrak{C}$ which is the identity over the objects, and, for any pair of $\mathfrak{C}$-objects A and B, with a *regular* action of Z over the *fibers* of the map

$$\hat{\mathfrak{C}}(A,B) \longrightarrow \mathfrak{C}(A,B) \qquad 11.2.1$$

induced by $\mathfrak{c}$, in such a way that these Z-actions are compatible with the composition of $\hat{\mathfrak{C}}$-morphisms. Note that, if $\mathfrak{C}'$ is a second category and $\mathfrak{f}:\mathfrak{C} \to \mathfrak{C}'$ an equivalence of categories, we easily can obtain a *regular central Z-extension* $\hat{\mathfrak{C}}'$ of $\mathfrak{C}'$ and a *Z-compatible equivalence of categories* $\hat{\mathfrak{f}}:\hat{\mathfrak{C}} \to \hat{\mathfrak{C}}'$. For short, we call *$k^*$-category* any *regular central k^*-extension* of a category.

11.3 Let H be a subgroup of G, β a point of H on kGb (cf. 1.12) and Q_δ a defect pointed group of H_β (cf. 1.13), and consider the divisible Q-category $\mathcal{F}_{(\beta,H)}$ (cf. 3.12); recall that we have defined a unique isomorphism class of faithful functors (cf. 3.12)

$$\mathfrak{i}^b_\beta : \mathcal{F}_{(\beta,H)} \longrightarrow \mathcal{F}_{(b,G)} \qquad 11.3.1.$$

† For the full subcategory over the selfcentralizing objects, the question on the existence of such an extension was raised by Markus Linckelmann at Durham in 2002.

Actually, in the second part of this chapter, it will be more handy to follow the point of view of Remark 3.10, replacing $\mathcal{F}_{(\beta,H)}$ and $\mathcal{F}_{(b,G)}$ by $\mathcal{F}^{\mathfrak{e}}_{(\beta,H)}$ and $\mathcal{F}^{\mathfrak{e}}_{(b,G)}$ respectively, so that we simply have an inclusion

$$\mathcal{F}^{\mathfrak{e}}_{(\beta,H)} \subset \mathcal{F}^{\mathfrak{e}}_{(b,G)} \qquad 11.3.2.$$

11.4 Assuming that Q is $\mathcal{F}_{(b,G)}$-nilcentralized, let $\mathfrak{X}$ be a nonempty set of $\mathcal{F}_{(b,G)}$-nilcentralized subgroups of Q containing any subgroup R of Q such that $\mathcal{F}_{(\beta,H)}(R,T) \neq \emptyset$ for some $T \in \mathfrak{X}$, and denote by $\mathcal{F}_{(\beta,H,\mathfrak{X})}$ the full subcategory of $\mathcal{F}_{(\beta,H)}$ over $\mathfrak{X}$. In this chapter, we discuss some sufficient conditions to guarantee the existence of a *regular central k^*-extension* $\hat{\mathcal{F}}_{(\beta,H,\mathfrak{X})}$ of $\mathcal{F}_{(\beta,H,\mathfrak{X})}$ such that, assuming that (P,e) contains the Brauer (b,G)-pair determined by Q_δ (cf. 1.15), for any $R \in \mathfrak{X}$ the inclusion $\mathcal{F}_{(\beta,H)}(R) \subset \mathcal{F}_{(b,G)}(R)$ can be lifted to a (noncanonical!) k^*-group homomorphism (cf. 7.4.3)

$$\hat{\mathcal{F}}_{(\beta,H,\mathfrak{X})}(R) \longrightarrow \hat{\mathcal{F}}_{(b,G)}(R) \qquad 11.4.1.$$

Remark 11.5 Note that, since k^* is *a p-divisible* group and the category $\mathcal{F}_{(b,G)}$ only involves a finite number of finite groups, there is $n \in \mathbb{N}$ such that, for any $\mathcal{F}_{(b,G)}$-nilcentralized subgroup R of P and any $m \in \mathbb{N}$, we have a k^*-group isomorphism

$$*^{p^{nm}} \hat{\mathcal{F}}_{(b,G)}(R) \cong \hat{\mathcal{F}}_{(b,G)}(R) \qquad 11.5.1,$$

where $*^{p^{nm}}\hat{\mathcal{F}}_{(b,G)}(R)$ denotes the *sum* as k^*-extensions of p^{nm} copies of the k^*-extension $\hat{\mathcal{F}}_{(b,G)}(R)$ (cf. 1.24); hence, it follows from Proposition 5.11 in [41] that, in our discussion, we may replace the H-algebra kGb by its *p^{nm}-th tensor power* $\bigotimes^{p^{nm}} kGb$.

11.6 In order to simplify our notation, we modify our choice of (P,e) in such a way that it contains the Brauer (b,G)-pair determined by Q_δ (cf. 1.15). Our sufficient conditions are expressed in terms of a suitable Dade Q-algebra (cf. 1.20) or, equivalently, in terms of a suitable element in the Dade group $\mathcal{D}_k(Q)$ (cf. 1.21), and it is handy to name each condition. We say that an element $s \in \mathcal{D}_k(Q)$ is *$\mathcal{F}_{(\beta,H,\mathfrak{X})}$-stable* if it fulfills

11.6.1 *For any $R \in \mathfrak{X}$ and any $\varphi \in \mathcal{F}_{(\beta,H)}(Q,R)$ we have* $\mathrm{res}_\varphi(s) = \mathrm{res}^Q_R(s)$.

Our second condition involves the family $\{s_R\}_R$ we have introduced in 7.7 of elements $s_R \in \mathcal{D}_k\big(\bar{N}_P(R)\big)$ where R runs over the set of $\mathcal{F}_{(b,G)}$-nilcentralized subgroups of P. For any $n \in \mathbb{N}$, we say that $s \in \mathcal{D}_k(Q)$ is a *$(\mathfrak{X},n)$-gluing element* if it fulfills

11.6.2 *For any $R \in \mathfrak{X}$ we have* $\widetilde{\mathfrak{Br}}_R^{N_Q(R)}\big(\mathrm{res}^Q_{N_Q(R)}(s)\big) = p^n\cdot\mathrm{res}^{\bar{N}_P(R)}_{\bar{N}_Q(R)}(s_R)$.

Of course, we are only interested in the integers $n \in \mathbb{N}$ fulfilling

$$*^{p^n} \hat{\mathcal{F}}_{(b,G)}(R) \cong \hat{\mathcal{F}}_{(b,G)}(R) \qquad 11.6.3,$$

called *(b,G)-admissible integers.*

Remark 11.7 In condition 11.6.1, we could avoid the choice of Q_δ by considering the obvious *contravariant* functor $\mathfrak{d}_k^\beta : \mathcal{F}^{\mathfrak{e}}_{(\beta,H)} \to \mathfrak{Ab}$ mapping any local pointed group R_ε on kGb contained in H_β on the Dade group $\mathcal{D}_k(R)$ and any $\mathcal{F}^{\mathfrak{e}}_{(\beta,H)}$-morphism on the corresponding *restriction homomorphism.* Then, moving from $\mathfrak{X}$ to the set $\mathfrak{X}^{\mathfrak{e}}$ of nilcentralized Brauer (β, H)-pairs $\mathcal{F}^{\mathfrak{e}}_{(\beta,H)}$-isomorphic to some element of $\mathfrak{X}$, condition 11.6.1 simply says that s belongs to the *inverse limit* of the restriction of $\mathfrak{d}_k^\beta$ to the full subcategory $\mathcal{F}^{\mathfrak{e}}_{(\beta,H,\mathfrak{X}^{\mathfrak{e}})}$ over $\mathfrak{X}^{\mathfrak{e}}$ (cf. 11.15 below).

Proposition 11.8 *With the notation above, fix $n \in \mathbb{N}$, let S and S' be Dade Q-algebras determining $(\mathfrak{X}, n)$-gluing elements in $\mathcal{D}_k(Q)$ and set*

$$\hat{B} = S^\circ \otimes_k \bigotimes{}^{p^n} (kGb)_\delta \quad \text{and} \quad S'' = S^\circ \otimes_k S' \qquad 11.8.1.$$

Any $R \in \mathfrak{X}$ has unique local points ε on $(kGb)_\delta$, $\varepsilon^{\hat{B}}$ on $\hat{B}$ and $\varepsilon^{S''}$ on S''. Then, denoting by $\delta^{\hat{B}}$ and $\delta^{S''}$ the respective local points of Q on $\hat{B}$ and S'', we have

$$\hat{B}_{\delta^{\hat{B}}}(R_{\varepsilon^{\hat{B}}}) \cong k \cong (S'')_{\delta^{S''}}(R) \qquad 11.8.2.$$

Proof: Since any $R \in \mathfrak{X}$ stabilizes bases on S and on kGb, it is well-known that (cf. Proposition 5.6 in [41])

$$\hat{B}(R) \cong S(R)^\circ \otimes_k \bigotimes{}^{p^n} (kGb)_\delta(R) \qquad 11.8.3.$$

Moreover, since R is $\mathcal{F}_{(b,G)}$-nilcentralized, the unique block g of $C_G(R)$ such that (P, e) contains (R, g) (cf. 1.15) is *nilpotent* (cf. Proposition 7.2) and therefore, up to isomorphism, there is a unique simple $kC_G(R)g$-module (cf. 1.13 and isomorphism 1.18.1) or, equivalently, a unique local point ε of R on kGb fulfilling $g\mathrm{Br}_R(\varepsilon) \neq \{0\}$ (cf. isomorphism 1.16.1).

In particular, we have $R_\varepsilon \subset Q_\delta$ (cf. 1.15) and ε is the unique local point of R on $(kGb)_\delta$, so that R has a unique local point $\varepsilon^{\hat{B}}$ on $\hat{B}$ (cf. Proposition 5.6 and Corollary 5.8 in [41]) and we still have

$$\hat{B}(R_{\varepsilon^{\hat{B}}}) \cong S(R)^\circ \otimes_k \bigotimes{}^{p^n} (kGb)_\delta(R_\varepsilon) \qquad 11.8.4.$$

Moreover, it follows from the definition of S_R and s_R in 7.7 that we have an $\bar{N}_Q(R)$-algebra embedding $S_R \to (kGb)_\delta(R_\varepsilon)$ (cf. embedding 1.12.2) and, by condition 11.6.2, the Dade $\bar{N}_Q(R)$-algebra $S(R)^\circ \otimes_k \bigotimes^{p^n} S_R$ determines the trivial element in the Dade group $\mathcal{D}_k\big(\bar{N}_Q(R)\big)$ (cf. 1.21), so that $\bar{N}_Q(R)$ fixes a primitive idempotent in this simple k-algebra. On the other hand, since S'' is a Dade Q-algebra (cf. 1.20), R has a unique local point $\varepsilon^{S''}$ on S'' (cf. 1.20) and a similar argument holds, proving that $\bar{N}_Q(R)$ fixes a primitive idempotent in $S''(R_{\varepsilon^{S''}}) = S''(R)$.

Consider either $D = \hat{B}_{\delta\hat{B}}$ or $D = (S'')_{\delta S''}$; in any case, in order to prove isomorphisms 11.8.2, it suffices to prove that the unity element is primitive in the k-algebra $D(R)$ for any $R \in \mathfrak{X}$; we argue by induction on $|Q:R|$ and the statement is clear when $R = Q$. Otherwise we may assume that, for any subgroup T of $N_Q(R)$ strictly containing R, the unity element is primitive in the k-algebra $D(T)$. But, we have proved above that $\bar{N}_Q(R)$ fixes a primitive idempotent $\bar{\ell}$ in $D(R)$ and then, setting $\bar{T} = T/R$, the primitivity of $\bar{\ell}$ forces $\mathrm{Br}_{\bar{T}}^{D(R)}(\bar{\ell}) \neq 0$ (cf. Theorem 7.2 in [51]); hence, we get

$$\mathrm{Br}_{\bar{T}}^{D(R)}(1 - \bar{\ell}) = 0 \qquad 11.8.5.$$

That is to say, the idempotent $1 - \bar{\ell}$ of $D(R)$ belongs to the intersection

$$\bigcap_{\bar{T}} \mathrm{Ker}(\mathrm{Br}_{\bar{T}}^{D(R)}) = D(R)_{\{1\}}^{\bar{N}_Q(R)} = \mathrm{Br}_R^D(D_R^Q) \qquad 11.8.6$$

where $\bar{T}$ runs over the set of nontrivial subgroups of $\bar{N}_Q(R)$ (cf. Lemmas 1.11 and 1.12 in [11]); but, the unity is primitive in D^Q and does not belong to $\mathrm{Ker}(\mathrm{Br}_Q^D) \supset D_R^Q$ since the involved point of Q is local, so that we have $D_R^Q \subset J(D^Q)$; hence, we still have $1 = \bar{\ell}$. We are done.

11.9 In the situation of this proposition, if moreover the element of $\mathcal{D}_k(Q)$ determined by S is $\mathcal{F}_{(\beta,H,\mathfrak{X})}$-stable, we claim that we can construct a *regular central k^*-extension* $\hat{\mathcal{F}}_{(\beta,H,\mathfrak{X})}$ of $\mathcal{F}_{(\beta,H,\mathfrak{X})}$; before proving it, let us explicitly describe this construction. With the notation above, set

$$B = \big(S^\circ \otimes_k \bigotimes{}^{p^n} kGb\big)_{\delta\hat{B}} \qquad 11.9.1$$

and, for any $R \in \mathfrak{X}$, denote by $\mathfrak{M}_R$ the kernel of the homomorphism $B^R \to k$ determined by the unique local point of R on B (cf. isomorphism 11.8.2), and choose an idempotent i_R in this point. For any $T \in \mathfrak{X}$ and any $\mathcal{F}_{(\beta,H)}$-morphism $\varphi: T \to R$, denote by $T_B^{\varphi}(R,T)$ the set of elements ai_T where $a \in B^*$ fulfills

$$\begin{gathered} i_T(i_R)^a = i_T = (i_R)^a i_T \\ ai_T v = \varphi(v) ai_T \text{ and } vi_T a^{-1} = i_T a^{-1}\varphi(v) \text{ for any } v \in T \end{gathered} \qquad 11.9.2,$$

and set $T' = \varphi(T)$; it is easily checked that the group $i_R + i_R\mathfrak{M}_{T'}i_R$ acts on the set $T_B^{\varphi}(R,T)$ by the multiplication on the left; then, what shall be the converse image $\hat{\mathcal{F}}^{\varphi}_{(\beta,H,\mathfrak{X})}(R,T)$ of φ in our claimed regular central k^*-extension $\hat{\mathcal{F}}_{(\beta,H,\mathfrak{X})}$ is defined by the quotient set

$$\hat{\mathcal{F}}^{\varphi}_{(\beta,H,\mathfrak{X})}(R,T) = (i_R + i_R\mathfrak{M}_{T'}i_R)\backslash T_B^{\varphi}(R,T) \qquad 11.9.3.$$

Theorem 11.10 *With the notation above, assume that there exists a Dade Q-algebra S determining a $\mathcal{F}_{(\beta,H,\mathfrak{X})}$-stable $(\mathfrak{X},n)$-gluing element s in $\mathcal{D}_k(Q)$ for some admissible integer n. Then, there exists a unique regular central k^*-extension $\hat{\mathcal{F}}_{(\beta,H,\mathfrak{X})}$ of $\mathcal{F}_{(\beta,H,\mathfrak{X})}$ such that, for any $R,T\in\mathfrak{X}$, the converse image of $\varphi\in\mathcal{F}_{(\beta,H,\mathfrak{X})}(R,T)$ in $\hat{\mathcal{F}}_{(\beta,H,\mathfrak{X})}(R,T)$ coincides with the quotient set*

$$(i_R+i_R\mathfrak{M}_{T'}i_R)\backslash T_B^{\varphi}(R,T) \qquad 11.10.1,$$

that the composition is induced by the product in $S^\circ\otimes_k\bigotimes^{p^n}kGb$ and that the inclusion $\mathcal{F}_{(\beta,H,\mathfrak{X})}(R)\subset\mathcal{F}_{(b,G)}(R)$ can be lifted to a k^*-group homomorphism

$$\hat{\mathcal{F}}_{(\beta,H,\mathfrak{X})}(R)\longrightarrow\hat{\mathcal{F}}_{(b,G)}(R) \qquad 11.10.2.$$

Moreover, the k^*-category $\hat{\mathcal{F}}_{(\beta,H,\mathfrak{X})}$ does not depend on the choice of S.

Proof: Denote by $\delta^{\hat{B}}$ the unique local point of Q on $\hat{B}=S^\circ\otimes_k\bigotimes^{p^n}(kGb)_\delta$ (cf. Proposition 11.8) and set

$$B=\big(S^\circ\otimes_k\bigotimes{}^{p^n}(kGb)_\delta\big)_{\delta^{\hat{B}}} \qquad 11.10.3.$$

Let R and T be elements of $\mathfrak{X}$ and respectively denote by ε, ε^S, μ and μ^S the local points of R and T on $(kGb)_\delta$ and S (cf. Proposition 11.8); we claim that the $\mathcal{F}_{(\beta,H,\mathfrak{X})}$-stability of s forces

$$\mathcal{F}_{(\beta,H)}(R,T)\subset F_S(R_{\varepsilon^S},T_{\mu^S}) \qquad 11.10.4;$$

indeed, for any $\varphi\in\mathcal{F}_{(\beta,H)}(R,T)$, the $\mathcal{F}_{(\beta,H,\mathfrak{X})}$-stability of s implies that the Dade T-algebras S_{μ^S} and $\mathrm{Res}_\varphi(S_{\varepsilon^S})$ are *similar* (cf. 1.21) or, equivalently, that there exists a T-algebra embedding $f_\varphi:S_{\mu^S}\to\mathrm{Res}_\varphi(S_{\varepsilon^S})$ (cf. 1.21) and, since S is a simple algebra, this amounts to saying that φ *is an S-fusion from T_{μ^S} to R_{ε^S}* (cf. 1.28 above and Corollary 2.4 in [37]).

On the other hand, it follows from Theorem 3.1 in [39] that (cf. 1.28)

$$\mathcal{F}_{(b,G)}(R,T)=F_{kGb}(R_\varepsilon,T_\mu) \qquad 11.10.5;$$

similarly, since (cf. Proposition 5.6 in [41])

$$\begin{aligned}\big(\textstyle\bigotimes^{p^n}kGb\big)(R)&\cong\textstyle\bigotimes^{p^n}(kGb)(R)\\ \big(\textstyle\bigotimes^{p^n}kGb\big)(T)&\cong\textstyle\bigotimes^{p^n}(kGb)(T)\end{aligned} \qquad 11.10.6,$$

ε and μ respectively determine local points $\otimes^{p^n}\varepsilon$ of R and $\otimes^{p^n}\mu$ of T on $\bigotimes^{p^n}kGb$ and then, the above result applied to the p^n-th direct product $\times^{p^n}G$ and to the diagonal p-subgroups determined by R and T yields

$$\mathcal{F}_{(b,G)}(R,T)=F_{\otimes^{p^n}kGb}(R_{\otimes^{p^n}\varepsilon},T_{\otimes^{p^n}\mu}) \qquad 11.10.7.$$

Moreover, denoting by ε^B, $\varepsilon^{\hat{B}}$, μ^B and $\mu^{\hat{B}}$ the respective local points of R and T on B and $\hat{B}$ (cf. Proposition 11.8), recall that (cf. Proposition 2.14 in [39])

$$F_B(R_{\varepsilon^B}, T_{\mu^B}) = F_{\hat{B}}(R_{\varepsilon^{\hat{B}}}, T_{\mu^{\hat{B}}}) \qquad 11.10.8$$

and that we actually identify $\varepsilon^{\hat{B}}$ with ε^B and $\mu^{\hat{B}}$ with μ^B. Consequently, it follows from inclusion 11.10.4, equality 11.10.7 and Theorem 5.3 in [41] that

$$\mathcal{F}_{(\beta,H)}(R,T) \subset F_B(R_{\varepsilon^B}, T_{\mu^B}) \qquad 11.10.9.$$

In 11.9 above, we have chosen $i_R \in \varepsilon^B$ and $i_T \in \mu^B$, and then, for any homomorphism $\varphi \in \mathcal{F}_{(\beta,H)}(R,T)$, we have considered the set $T_B^\varphi(R,T)$ of elements ai_T where $a \in B^*$ fulfills

$$\begin{gathered} i_T(i_R)^a = i_T = (i_R)^a i_T \\ ai_T v = \varphi(v) ai_T \text{ and } vi_T a^{-1} = i_T a^{-1}\varphi(v) \text{ for any } v \in T \end{gathered} \qquad 11.10.10;$$

but, by the very definition of the *B-fusions* (cf. 1.28), inclusion 11.10.9 implies that there is a T-interior algebra embedding $f_\varphi : B_{\mu^B} \to \mathrm{Res}_\varphi(B_{\varepsilon^B})$ such that the inclusion $B_{\mu^B} \subset B$ and the composition of f_φ with the inclusion $B_{\varepsilon^B} \subset B$ are B^*-conjugate, which amounts to saying that $T_B^\varphi(R,T)$ is not empty. Moreover, setting $T' = \varphi(T)$, it is again easily checked that the group $(i_R B^{T'} i_R)^*$ acts on this set by the multiplication on the left.

We claim that this action is transitive; indeed, if $a' \in B^*$ fulfills the above condition, for any $v \in T$ we have

$$a' i_T a^{-1} = \varphi(v) a' i_T a^{-1} \varphi(v)^{-1} \qquad 11.10.11,$$

so that $a' i_T a^{-1}$ and the idempotents $a i_T a^{-1}$ and $a' i_T a'^{-1}$ belong to $B^{T'}$; then, since we have

$$(a' i_T a'^{-1})(a' i_T a^{-1}) = a' i_T a^{-1} = (a' i_T a^{-1})(a i_T a^{-1}) \qquad 11.10.12$$

and the symmetric equalities on a and a' still hold, the left multiplication by $a' i_T a^{-1}$ determines a $kT' \otimes_k B^\circ$-module isomorphism

$$h : (a i_T a^{-1})B \cong (a' i_T a'^{-1})B \qquad 11.10.13.$$

Then, the $kT' \otimes_k B^\circ$-modules $(i_R - a i_T a^{-1})B$ and $(i_R - a' i_T a'^{-1})B$ are isomorphic too; choosing such an isomorphism, we get a $kT' \otimes_k B^\circ$-module automorphism of $i_R B$ extending h; but, such an automorphism is just the left multiplication by a suitable $c \in (i_R B^{T'} i_R)^*$ and, in particular, we get

$cai_{_T} = a'i_{_T}$, which proves the claim. Consequently, k^* acts transitively on the quotient set

$$\hat{\mathcal{F}}^{\varphi}_{(\beta,H,\mathfrak{X})}(R,T) = (i_R + i_R\mathfrak{M}_{T'}i_R)\backslash T^{\varphi}_B(R,T) \qquad 11.10.14$$

and it is easily checked that this action is actually regular.

If U is a third element of $\mathfrak{X}$ and $\psi : U \to T$ an $\mathcal{F}_{(\beta,H)}$-morphism, we claim that

$$T^{\varphi}_B(R,T)\cdot T^{\psi}_B(T,U) \subset T^{\varphi\circ\psi}_B(R,U) \qquad 11.10.15.$$

Indeed, if $a, c \in B^*$ fulfill the corresponding equalities 11.10.10, so that $ai_{_T}$ and $ci_{_U}$ respectively belong to $T^{\varphi}_B(R,T)$ and to $T^{\psi}_B(T,U)$, then we get

$$ai_{_T}ci_{_U} = ac(i_{_T})^c i_{_U} = aci_{_U} \qquad 11.10.16$$

and from these equalities we still get

$$\begin{aligned}
&i_{_U}(i_{_R})^{ac} = i_{_U}(i_{_T})^c(i_{_R})^{ac} = i_{_U}(i_{_T})^c = i_{_U}\\
&(i_{_R})^{ac}i_{_U} = (i_{_R})^{ac}(i_{_T})^c i_{_U} = i_{_U}\\
&aci_{_U}w = ai_{_T}\psi(w)ci_{_U} = (\varphi\circ\psi)(w)ai_{_T}ci_{_U} = (\varphi\circ\psi)(w)aci_{_U}\\
&wi_{_U}(ac)^{-1} = i_{_U}c^{-1}\psi(w)i_{_T}a^{-1} = i_{_U}(ca)^{-1}(\varphi\circ\psi)(w)
\end{aligned} \qquad 11.10.17.$$

Finally, we claim that the product in B induces a map

$$\hat{\mathcal{F}}^{\varphi}_{(\beta,H,\mathfrak{X})}(R,T) \times \hat{\mathcal{F}}^{\psi}_{(\beta,H,\mathfrak{X})}(T,U) \longrightarrow \hat{\mathcal{F}}^{\varphi\circ\psi}_{(\beta,H,\mathfrak{X})}(R,U) \qquad 11.10.18$$

compatible with the product in k^* ; indeed, first of all notice that, setting $U' = \psi(U)$ and $U'' = \varphi(U')$, and respectively denoting by μ'^B and η''^B the unique local points of T' and U'' on B, the inclusion $U'' \subset T'$ forces $U''_{\eta''^B} \subset T'_{\mu'^B}$, so that the image of μ'^B in $B(U''_{\eta''^B}) \cong k$ is equal to $\{1\}$ and therefore we have $\mathfrak{M}_{T'} \subset \mathfrak{M}_{U''}$.

Moreover, with the the notation above, we get

$$\begin{aligned}
&(i_{_R} + i_{_R}\mathfrak{M}_{T'}i_{_R})ai_{_T}(i_{_T} + i_{_T}\mathfrak{M}_{U'}i_{_T})ci_{_U}\\
&\quad = \big(i_{_R} + i_{_R}\mathfrak{M}_{T'}(i_{_T})^{a^{-1}} + i_{_R}\mathfrak{M}_{T'}(i_{_T}\mathfrak{M}_{U'}i_{_T})^{a^{-1}}\big)aci_{_U}
\end{aligned} \qquad 11.10.19;$$

then, one the one hand we have

$$i_{_R}\mathfrak{M}_{T'}(i_{_T})^{a^{-1}} \subset i_{_R}\mathfrak{M}_{U''}(i_{_T})^{a^{-1}} \subset i_{_R}\mathfrak{M}_{U''}i_{_R} \qquad 11.10.20;$$

on the other hand, since $U' \subset T$, it is easily checked from condition 11.10.10 that $(i_{_T}B^{U'}i_{_T})^{a^{-1}} = (i_{_T})^{a^{-1}}B^{U''}(i_{_T})^{a^{-1}}$ and therefore we get

$$(i_{_T}\mathfrak{M}_{U'}i_{_T})^{a^{-1}} = (i_{_T})^{a^{-1}}\mathfrak{M}_{U''}(i_{_T})^{a^{-1}} \qquad 11.10.21,$$

so that we also obtain $i_R \mathfrak{M}_{T'}(i_T \mathfrak{M}_{U'} i_T)^{a^{-1}} \subset i_R \mathfrak{M}_{U''} i_R$. We have proved the claim.

Now, it suffices to consider the disjoint union

$$\hat{\mathcal{F}}_{(\beta,H,\mathfrak{X})}(R,T) = \bigsqcup_{\varphi \in \mathcal{F}_{(\beta,H)}(R,T)} \hat{\mathcal{F}}^{\varphi}_{(\beta,H,\mathfrak{X})}(R,T) \qquad 11.10.22,$$

endowed with the *inverted* action of k^*, to get the announced regular central k^*-extension. Indeed, the maps 11.10.18 define the composition in $\hat{\mathcal{F}}_{(\beta,H,\mathfrak{X})}$, which is associative and compatible with the action of k^* since it is induced by the product in B. Moreover, if $T = R$ then it is clear that, for any $\varphi \in \mathcal{F}_{(\beta,H)}(R)$, we have

$$T^{\varphi}_B(R,R) \subset N_{(i_R B i_R)^*}(R{\cdot}i_R) \qquad 11.10.23,$$

so that, by its very definition, $\hat{F}_B(R_{\varepsilon^B})$ contains $\hat{\mathcal{F}}_{(\beta,H,\mathfrak{X})}(R)$ (cf. 1.29.2), and therefore, since $\hat{F}_S(R_{\varepsilon^S}) \cong k^* \times F_S(R_{\varepsilon^S})$ (cf. Theorem 9.21) and n is admissible, we get (cf. 11.10.9 and 1.24 above, and Proposition 5.11 in [41])

$$\begin{aligned}\hat{\mathcal{F}}_{(\beta,H,\mathfrak{X})}(R) \subset \hat{F}_B(R_{\varepsilon^B}) &\cong \hat{F}_S(R_{\varepsilon^S}) * \hat{F}_{\otimes^{p^n}(kGb)_\delta}(R_{\otimes^{p^n}\varepsilon}) \\ &\subset \hat{F}_{kGb}(R_\varepsilon)\end{aligned} \qquad 11.10.24.$$

Note that, up to a unique equivalence, this category $\hat{\mathcal{F}}_{(\beta,H,\mathfrak{X})}$ does not depend on the choice of $i_R \in \varepsilon^B$ for any $R \in \mathfrak{X}$.

Let S' be another Dade Q-algebra still determining a $\mathcal{F}_{(\beta,H,\mathfrak{X})}$-stable $(\mathfrak{X},n)$-gluing element s' in $\mathcal{D}_k(Q)$; *mutatis mutandis* denote by $\delta^{\hat{B}'}$ the unique local point of Q on $\hat{B}' = S'^{\circ} \otimes_k \bigotimes^{p^n}(kGb)_\delta$ (cf. Proposition 11.8), set

$$B' = \left(S'^{\circ} \otimes_k \bigotimes{}^{p^n}(kGb)_\delta\right)_{\delta^{\hat{B}'}} \qquad 11.10.25$$

and, for any $R \in \mathfrak{X}$, choose an idempotent i'_R in the unique local point $\varepsilon^{B'}$ of R on B' (cf. Proposition 11.8), denoting by $\mathfrak{M}'_R$ the kernel of the corresponding homomorphism. Again, for any $R, T \in \mathfrak{X}$ and any $\varphi \in \mathcal{F}_{(\beta,H)}(R,T)$, consider the corresponding quotient set

$$\hat{\mathcal{F}}'^{\varphi}_{(\beta,H,\mathfrak{X})}(R,T) = (i'_R + i'_R \mathfrak{M}'_{T'} i'_R)\backslash T^{\varphi}_{B'}(R,T) \qquad 11.10.26.$$

In order to find a relationship between the quotient sets $\hat{\mathcal{F}}'^{\varphi}(R,T)_{(\beta,H,\mathfrak{X})}$ and $\hat{\mathcal{F}}^{\varphi}(R,T)_{(\beta,H,\mathfrak{X})}$, set $S'' = S'^{\circ} \otimes_k S$ and consider the Q-algebra

$$B'' = S'^{\circ} \otimes_k S \otimes_k S^{\circ} \otimes_k \bigotimes{}^{p^n}(kGb)_\delta = S'' \otimes_k \hat{B} \qquad 11.10.27.$$

On the one hand, since we have a Q-algebra embedding $k \to S \otimes_k S^\circ$ (cf. isomorphism 1.21.3), we still have a Q-algebra embedding from B' to B'' and therefore, denoting by $\delta^{B''}$ the unique local point of Q on B'' (cf. Proposition 11.8), we can identify B' with $(B'')_{\delta^{B''}}$ and, in particular, the quotient set $\hat{\mathcal{F}}^{\prime\varphi}_{(\beta,H,\mathfrak{X})}(R,T)$ with the corresponding quotient set in B''.

On the other hand, choosing an idempotent j_Q in the unique local point $\delta^{S''}$ of Q on S'' (cf. 1.21), we have an injective Q-algebra homomorphism

$$t : B \longrightarrow \hat{B} = S^\circ \otimes_k \bigotimes{}^{p^n}(kGb)_\delta \longrightarrow B'' \qquad 11.10.28$$

determined by the canonical embedding $B = \hat{B}_{\delta^{\hat{B}}} \to \hat{B}$ (cf. 1.12) and by the injective homomorphism mapping $a \in \hat{B}$ on $j_Q \otimes a$; moreover, for any $R \in \mathfrak{X}$, t induces an injective k-algebra homomorphism

$$t(R) : B(R) \longrightarrow B''(R) \qquad 11.10.29$$

since, up to suitable identifications, we have the inclusion

$$t(B) \subset (S'')_{\delta^{S''}} \otimes_k B \qquad 11.10.30$$

and, according to Proposition 11.8, we get $(S'')_{\delta^{S''}}(R) \cong k$.

Actually, since Q has a unique local point on $(S'')_{\delta^{S''}} \otimes_k B$ (cf. Proposition 11.8 above and Proposition 5.6 in [41]) and the set $\hat{\mathcal{F}}^{\prime\varphi}_{(\beta,H,\mathfrak{X})}(R,T)$ does not depend on the choice of i'_R in $\varepsilon^{B'}$ for any $R \in \mathfrak{X}$, we may assume that we also have the inclusion

$$B' \subset (S'')_{\delta^{S''}} \otimes_k B = D \qquad 11.10.31$$

which allows us to replace B'' by D; note that $t(i_Q) = j_Q \otimes i_Q$ is the unity element of D and that then we still may assume that we have

$$i'_R t(i_R) = i'_R = t(i_R) i'_R \qquad 11.10.32$$

for any $R \in \mathfrak{X}$.

Furthermore, for any $a \in B^*$ fulfilling condition 11.10.10 with respect R, T and φ, it follows from this condition that $a i_T a^{-1}$ is a primitive idempotent in $B^{\varphi(T)}$ outside $\mathfrak{M}_{\varphi(T)}$, which is "contained" in i_R — namely, it appears in a *primitive pairwise orthogonal decomposition* of i_R in $B^{\varphi(T)}$; hence, it follows from equality 11.10.31 that $t(a) i'_T t(a^{-1})$ is a primitive idempotent in $D^{\varphi(T)}$ "contained" in $t(i_R)$, which has a nontrivial image in $D\big(\varphi(T)\big)$ by the injectivity of homomorphism 11.10.29.

But, denoting by ε^D the unique local point of R on D (cf. Proposition 11.8 above and Proposition 5.6 in [41]), it follows from Proposition 11.8 that

$$D(R_{\varepsilon^D}) \cong (S'')_{\delta S''}(R) \otimes_k B(R_{\varepsilon^B}) \cong k \qquad 11.10.33.$$

Thus, denoting by $\mathfrak{M}''_R$ the maximal ideal in D^R determined by ε^D, it follows from this isomorphism that there is an invertible element $c \in j_Q \otimes i_Q + \mathfrak{M}''_R$ such that the idempotent $t(a)i'_{T}t(a^{-1})$ is "contained" in $(i'_{R})^c$.

Moreover, since $t(i_Q) = j_Q \otimes i_Q$, $t(a)$ is invertible in D and therefore $d = ct(a)$ is an invertible element of D which fulfills

$$i'_{T}(i'_{R})^d = i'_{T} = (i'_{R})^d i'_T \qquad 11.10.34;$$

furthermore, for any $v \in T$, the multiplication of $t(ai_{T}v) = t(\varphi(v)ai_{T})$ by i'_{T} on the right and by c on the left yields

$$di'_{T}v = \varphi(v)di'_{T} \qquad 11.10.35$$

and *mutatis mutandis* we get $vi'_{T}d^{-1} = i'_{T}d^{-1}\varphi(v)$.

On the other hand, denoting by i'_{Q} the unity element of $B' \subset D$, so that $B' = i'_Q D i'_Q$, the multiplication by d on the left determines a D°-module isomorphism between $i'_{T}D$ and $(i'_{T})^{d^{-1}}D$ which are both direct summands of $i'_{Q}D$; thus, the D°-modules $(i'_{Q} - i'_{T})D$ and $(i'_{Q} - (i'_{T})^{d^{-1}})D$ are isomorphic too and, choosing such an isomorphism, we get a D°-module automorphism of $i'_{Q}D$ which can be extended to a D°-module automorphism of D; that is to say, we get an invertible element $d' \in D$ fulfilling

$$d'i'_{Q}d'^{-1} = i'_{Q} \quad , \quad d'i'_{T} = di'_{T} \quad \text{and} \quad i'_{T}d'^{-1} = i'_{T}d^{-1} \qquad 11.10.36.$$

In conclusion, di'_{T} belongs to $T^{\varphi}_{B'}(R,T)$ and its image in $\hat{\mathcal{F}}'^{\varphi}_{(\beta,H)}(R,T)$ does not depend on our choice of c since $i'_{R}\mathfrak{M}''_R i'_{R} = i'_{R}\mathfrak{M}'_R i'_{R}$; thus, we get a map from $T^{\varphi}_{B}(R,T)$ to $\hat{\mathcal{F}}'^{\varphi}_{(\beta,H)}(R,T)$ sending ai_{T} to the class of di'_{T}; finally, it is easily checked that this map factorizes throughout the quotient set $\hat{\mathcal{F}}^{\varphi}_{(\beta,H,\mathfrak{X})}(R,T)$, determining a new map

$$\hat{\mathcal{F}}^{\varphi}_{(\beta,H,\mathfrak{X})}(R,T) \longrightarrow \hat{\mathcal{F}}'^{\varphi}_{(\beta,H,\mathfrak{X})}(R,T) \qquad 11.10.37$$

which is compatible with the action of k^* and therefore it is bijective. Now, it is not difficult to check that these maps, when R and T run over $\mathfrak{X}$, are compatible with the products in B and D and therefore with the maps 11.10.18; consequently, they define a canonical isomorphism between $\hat{\mathcal{F}}_{(\beta,H,\mathfrak{X})}$ and the k^*-category constructed from S'. We are done.

11.11 With the notation above, let L be a subgroup of H, ξ a point of L on kGb such that $L_\xi \subset H_\beta$ and R_ε a defect pointed group of L_ξ; as in 3.12, it is easily checked that an element $y \in H$ such that $R_\varepsilon \subset (Q_\delta)^y$ determines a unique isomorphism class of faithful functors

$$\mathfrak{i}_\xi^\beta : \mathcal{F}_{(\xi,L)} \longrightarrow \mathcal{F}_{(\beta,H)} \qquad 11.11.1.$$

As above, assume that R is $\mathcal{F}_{(b,G)}$-nilcentralized and let $\mathfrak{Y}$ be a nonempty set of $\mathcal{F}_{(b,G)}$-nilcentralized subgroups of R, which contains any subgroup T of R such that $\mathcal{F}_{(\xi,L)}(T,U) \neq \emptyset$ for some $U \in \mathfrak{Y}$ and fulfills $\mathfrak{i}_\xi^\beta(\mathfrak{Y}) \subset \mathfrak{X}$; again, in order to simplify our notation, we modify our choices of Q_δ and (P,e) assuming that $R_\varepsilon \subset Q_\delta$, so that we have $\mathfrak{Y} \subset \mathfrak{X}$.

11.12 Whenever S is a Dade Q-algebra determining a $\mathcal{F}_{(\beta,H,\mathfrak{X})}$-stable $(\mathfrak{X},n)$-gluing element in $\mathcal{D}_k(Q)$ for some admissible integer n, it is quite clear that $\mathrm{Res}_R^Q(S)$ still determines a $\mathcal{F}_{(\xi,L,\mathfrak{Y})}$-stable $(\mathfrak{Y},n)$-gluing element in $\mathcal{D}_k(R)$, and the point is that we can exhibit a k^**-functor*

$$\hat{\mathfrak{i}}_{\xi,\mathfrak{Y}}^{\beta,\mathfrak{X}} : \hat{\mathcal{F}}_{(\xi,L,\mathfrak{Y})} \longrightarrow \hat{\mathcal{F}}_{(\beta,H,\mathfrak{X})} \qquad 11.12.1.$$

lifting the corresponding restriction of $\mathfrak{i}_\xi^\beta$; before proving it, let us explicitly describe this functor. Borrowing the notation in 11.9, for any $T,U \in \mathfrak{Y}$ and any $\mathcal{F}_{(\xi,L)}$-morphism $\varphi : U \to T$, we have defined a subset $T_B^\varphi(T,U)$ of $i_T B^* i_U$ where i_T and i_U respectively belong to the unique local points of T and U on B; moreover, denoting by $\varepsilon^{\hat{D}}$ the unique local point of R on the Q-algebra $\hat{D} = S^\circ \otimes_k \bigotimes^{p^n} kGb$ (cf. Proposition 11.8), we set

$$D = \big(S^\circ \otimes_k \bigotimes{}^{p^n} kGb\big)_{\varepsilon^{\hat{D}}} \qquad 11.12.2,$$

we choose the idempotents i_T and i_U in D, and we identify D with $i_R B i_R$; analogously, we have defined a subset $T_D^\varphi(T,U)$ of $i_T D^* i_U$ and, with these choices we have $i_T D i_U = i_T B i_U$. Actually, we have $T_D^\varphi(T,U) = T_B^\varphi(T,U)$ as we prove below, and thus, in this situation we simply get

$$\hat{\mathcal{F}}^\varphi_{(\xi,L,\mathfrak{Y})}(T,U) = \hat{\mathcal{F}}^\varphi_{(\beta,H,\mathfrak{X})}(T,U) \qquad 11.12.3.$$

Proposition 11.13 *With the notation and the hypothesis above, assume that there is a Dade Q-algebra S determining a $\mathcal{F}_{(\beta,H,\mathfrak{X})}$-stable $(\mathfrak{X},n)$-gluing element in $\mathcal{D}_k(Q)$ for some admissible integer n. Then, the Dade R-algebra $\mathrm{Res}_R^Q(S)$ determines a $\mathcal{F}_{(\xi,L,\mathfrak{Y})}$-stable $(\mathfrak{Y},n)$-gluing element in $\mathcal{D}_k(R)$ and the corresponding k^*-category $\hat{\mathcal{F}}_{(\xi,L,\mathfrak{Y})}$ is a k^*-subcategory of $\hat{\mathcal{F}}_{(\beta,H,\mathfrak{X})}$.*

Proof: The first statement is clear since conditions 11.6.1 and 11.6.2 for S and for the triple $(\beta, H, \mathfrak{X})$ imply that the same conditions for $\mathrm{Res}^Q_R(S)$ and for the triple $(\xi, L, \mathfrak{Y})$ hold.

Moreover, with the notation above, if $d \in D^*$ fulfills the corresponding condition 11.9.2 with respect T, U and φ, then di_U belongs to $T^\varphi_D(T,U)$ and $a = d + (i_Q - i_R)$ belongs to B^* and fulfills the same condition, so that the element $di_U = ai_U$ still belongs to $T^\varphi_B(T,U)$; consequently, we have $T^\varphi_D(T,U) \subset T^\varphi_B(T,U)$ and, since the group $(i_T D i_T)^* = (i_T B i_T)^*$ acts regularly by left multiplication on both sets, we get the announced equality.

In conclusion, we have the equality $\hat{\mathcal{F}}^\varphi_{(\xi,L,\mathfrak{Y})}(T,U) = \hat{\mathcal{F}}^\varphi_{(\beta,H,\mathfrak{X})}(T,U)$ for any $\varphi \in \mathcal{F}_{(\xi,L)}(T,U)$ and therefore $\hat{\mathcal{F}}_{(\xi,L,\mathfrak{Y})}(T,U)$ is just the converse image of $\mathcal{F}_{(\xi,L)}(T,U)$ in $\hat{\mathcal{F}}_{(\beta,H,\mathfrak{X})}(T,U)$. On the other hand, it is not difficult to see that this fact does not depend on the choice of S. We are done.

11.14 Furthermore, let M be a subgroup of L, ζ a point of M on kGb such that $M_\zeta \subset L_\xi$ and T_μ a defect pointed group of M_ζ (cf. 1.13); as above, we have unique isomorphism classes of faithful functors (cf. 3.12)

$$\mathfrak{i}^\xi_\zeta : \mathcal{F}_{(\zeta,M)} \longrightarrow \mathcal{F}_{(\xi,L)} \quad \text{and} \quad \mathfrak{i}^\beta_\zeta : \mathcal{F}_{(\zeta,M)} \longrightarrow \mathcal{F}_{(\beta,H)} \qquad 11.14.1$$

and it is clear that $\mathfrak{i}^\beta_\zeta \cong \mathfrak{i}^\beta_\xi \circ \mathfrak{i}^\xi_\zeta$. Once again, assume that T is $\mathcal{F}_{(b,G)}$-nilcentralized and let $\mathfrak{Z}$ be a nonempty set of $\mathcal{F}_{(b,G)}$-nilcentralized subgroups of T which contains any subgroup U of T such that $\mathcal{F}_{(\xi,L)}(U,V) \neq \emptyset$ for some $V \in \mathfrak{Z}$ and fulfills $\mathfrak{i}^\xi_\zeta(\mathfrak{Z}) \subset \mathfrak{Y}$. In this situation, if there is a Dade Q-algebra S determining a $\mathcal{F}_{(\beta,H,\mathfrak{X})}$-stable $(\mathfrak{X}, n)$-gluing element in $\mathcal{D}_k(Q)$ for some admissible integer n, then it follows from Proposition 11.13 applied three times that we have three faithful k^*-functors

$$\begin{gathered} \hat{\mathfrak{i}}^{\xi,\mathfrak{Y}}_{\zeta,\mathfrak{Z}} : \hat{\mathcal{F}}_{(\zeta,M,\mathfrak{Z})} \longrightarrow \hat{\mathcal{F}}_{(\xi,L,\mathfrak{Y})} \quad , \quad \hat{\mathfrak{i}}^{\beta,\mathfrak{X}}_{\xi,\mathfrak{Y}} : \hat{\mathcal{F}}_{(\xi,L,\mathfrak{Y})} \longrightarrow \hat{\mathcal{F}}_{(\beta,H,\mathfrak{X})} \\ \hat{\mathfrak{i}}^{\beta,\mathfrak{X}}_{\zeta,\mathfrak{Z}} : \hat{\mathcal{F}}_{(\xi,L,\mathfrak{Y})} \longrightarrow \hat{\mathcal{F}}_{(\beta,H,\mathfrak{X})} \end{gathered} \qquad 11.14.2$$

respectively lifting the corresponding restrictions of $\mathfrak{i}^\beta_\zeta$, $\mathfrak{i}^\beta_\xi$ and $\mathfrak{i}^\xi_\zeta$, and that we still have a natural isomorphism

$$\hat{\mathfrak{i}}^{\beta,\mathfrak{X}}_{\zeta,\mathfrak{Z}} \cong \hat{\mathfrak{i}}^{\beta,\mathfrak{X}}_{\xi,\mathfrak{Y}} \circ \hat{\mathfrak{i}}^{\xi,\mathfrak{Y}}_{\zeta,\mathfrak{Z}} \qquad 11.14.3.$$

11.15 Actually, following the point of view of Remark 3.10, we simply get inclusions of categories

$$\mathcal{F}^{\mathfrak{e}}_{(\zeta,M)} \subset \mathcal{F}^{\mathfrak{e}}_{(\xi,L)} \subset \mathcal{F}^{\mathfrak{e}}_{(\beta,H)} \qquad 11.15.1;$$

moreover, it is clear that the sets $\mathfrak{X}$, $\mathfrak{Y}$ and $\mathfrak{Z}$ respectively determine sets $\mathfrak{X}^{\mathfrak{e}}$, $\mathfrak{Y}^{\mathfrak{e}}$ and $\mathfrak{Z}^{\mathfrak{e}}$ of nilcentralized Brauer (b,G)-pairs, stable throughout $\mathcal{F}^{\mathfrak{e}}_{(\beta,H)}$-,

$\mathcal{F}^{\mathfrak{e}}_{(\xi,L)}$- and $\mathcal{F}^{\mathfrak{e}}_{(\zeta,M)}$-isomorphisms and then, denoting by $\mathcal{F}^{\mathfrak{e}}_{(\beta,H,\mathfrak{X}^{\mathfrak{e}})}$, $\mathcal{F}^{\mathfrak{e}}_{(\xi,L,\mathfrak{Y}^{\mathfrak{e}})}$ and $\mathcal{F}^{\mathfrak{e}}_{(\zeta,M,\mathfrak{Z}^{\mathfrak{e}})}$ the corresponding full subcategories over $\mathfrak{X}^{\mathfrak{e}}$, $\mathfrak{Y}^{\mathfrak{e}}$ and $\mathfrak{Z}^{\mathfrak{e}}$, it follows from 11.2 and Theorem 11.10 that we have k^*-categories $\hat{\mathcal{F}}^{\mathfrak{e}}_{(\beta,H,\mathfrak{X}^{\mathfrak{e}})}$, $\hat{\mathcal{F}}^{\mathfrak{e}}_{(\xi,L,\mathfrak{Y}^{\mathfrak{e}})}$ and $\hat{\mathcal{F}}^{\mathfrak{e}}_{(\zeta,M,\mathfrak{Z}^{\mathfrak{e}})}$ which are *regular k^*-extensions* of $\mathcal{F}^{\mathfrak{e}}_{(\beta,H,\mathfrak{X}^{\mathfrak{e}})}$, $\mathcal{F}^{\mathfrak{e}}_{(\xi,L,\mathfrak{Y}^{\mathfrak{e}})}$ and $\mathcal{F}^{\mathfrak{e}}_{(\zeta,M,\mathfrak{Z}^{\mathfrak{e}})}$ respectively. Hence, it follows from Proposition 11.13 applied three times that we have a commutative diagram of *faithful* k^*-functors

$$\begin{array}{ccc} \hat{\mathcal{F}}^{\mathfrak{e}}_{(\xi,L,\mathfrak{Y}^{\mathfrak{e}})} & \longrightarrow & \hat{\mathcal{F}}^{\mathfrak{e}}_{(\beta,H,\mathfrak{X}^{\mathfrak{e}})} \\ \nwarrow & & \nearrow \\ & \hat{\mathcal{F}}^{\mathfrak{e}}_{(\zeta,M,\mathfrak{Z}^{\mathfrak{e}})} & \end{array} \qquad 11.15.2$$

lifting inclusions 11.15.1.

11.16 The main situation we are interested in concerns the *normalizers* in G of the *totally normal sets* of *nilcentralized Brauer* (b,G)*-pairs* (cf. 7.4). We call *totally normal set* of nilcentralized Brauer (b,G)-pairs any nonempty set $\mathfrak{q}$ of nilcentralized Brauer (b,G)-pairs, which is totally ordered by the inclusion and fulfills $\mathfrak{q}(i) \lhd \mathfrak{q}(n)$ for any $1 \le i \le n$ where $n = |\mathfrak{q}|$ and $\mathfrak{q}(i)$ denote the i-th element of $\mathfrak{q}$ — that is to say, *totally normal sets* of nilcentralized Brauer (b,G)-pairs is nothing but a particular kind of *normal* $(\mathcal{F}_{(b,G)})^{\mathrm{nc}}$*-chains* (cf. 2.20); then, the *normalizer* $N_G(\mathfrak{q})$ of $\mathfrak{q}$ is the stabilizer of $\mathfrak{q}$ in G by G-conjugation. Moreover, if $\mathfrak{q}(n) = (Q,f)$ then, since

$$C_G(Q) \subset N_G(\mathfrak{q}) \subset N_G(Q,f) \qquad 11.16.1,$$

f is also a block of $N_G(\mathfrak{q})$ (cf. 1.10) and determines a point $\nu^{\mathfrak{q}}$ of $N_G(\mathfrak{q})$ on kGb (cf. 1.19).

11.17 Let $(P^{\mathfrak{q}})_{\gamma^{\mathfrak{q}}}$ be a defect pointed group of $N_G(\mathfrak{q})_{\nu^{\mathfrak{q}}}$ and denote by $\mathfrak{X}^{\mathfrak{q}}$ the set of subgroups of $P^{\mathfrak{q}}$ containing Q; recall that, since f is also a nilpotent block of $C_G(Q){\cdot}P^{\mathfrak{q}}$ by Proposition 6.5 in [34], it follows from the main theorem in [41] that there exists a Dade $P^{\mathfrak{q}}$-algebra $S^{\mathfrak{q}}$ such that we have a $P^{\mathfrak{q}}$-interior algebra isomorphism (cf. isomorphism 1.18.1)

$$\big(k(C_G(Q){\cdot}P^{\mathfrak{q}})\big)_{\gamma^{\mathfrak{q}}} \cong S^{\mathfrak{q}}P^{\mathfrak{q}} \qquad 11.17.1.$$

Proposition 11.18 *With the notation above, the Dade $P^{\mathfrak{q}}$-algebra $S^{\mathfrak{q}}$ determines a $\mathcal{F}_{(\nu^{\mathfrak{q}},N_G(\mathfrak{q}),\mathfrak{X}^{\mathfrak{q}})}$-stable $(\mathfrak{X}^{\mathfrak{q}},0)$-gluing element in $\mathcal{D}_k(P^{\mathfrak{q}})$.*

Proof: As in the notation above, we set $n = |\mathfrak{q}|$ and $\mathfrak{q}(n) = (Q,f)$; since $Q \subset P^{\mathfrak{q}}$, in particular we have $C_G(P^{\mathfrak{q}}) \subset C_G(Q)$ and we can identify $\gamma^{\mathfrak{q}}$ with a local point of $P^{\mathfrak{q}}$ on $kC_G(Q)f$ (cf. 1.19); then, it follows from Theorem 1.12 in [34] that, with the notation there, we have

$$\big(kN_G(\mathfrak{q})\big)_{\gamma^{\mathfrak{q}}} \cong S^{\mathfrak{q}} \otimes_k k_*\hat{L} \qquad 11.18.1$$

for a suitable k^*-group $\hat{L}$; in particular, for any $R \in \mathfrak{X}^{\mathfrak{q}}$, it follows from equality 3.13.1 above and from Theorem 3.1 in [39] that

$$\mathcal{F}_{(\nu^{\mathfrak{q}},N_G(\mathfrak{q}))}(P^{\mathfrak{q}},R) = F_{kN_G(\mathfrak{q})f}\big((P^{\mathfrak{q}})_{\gamma^{\mathfrak{q}}},R_\varepsilon\big) \qquad 11.18.2$$

where we denote by ε the unique local point of R on $\big(kN_G(\mathfrak{q})\big)_{\gamma^{\mathfrak{q}}}$ (cf. Proposition 11.8), and from Lemma 1.17 in [34] that

$$F_{kN_G(\mathfrak{q})f}\big((P^{\mathfrak{q}})_{\gamma^{\mathfrak{q}}},R_\varepsilon\big) \subset F_{S^{\mathfrak{q}}}\big((P^{\mathfrak{q}})_{\gamma^{S^{\mathfrak{q}}}},R_{\varepsilon^{S^{\mathfrak{q}}}}\big) \qquad 11.18.3$$

where we respectively denote by $\varepsilon^{S^{\mathfrak{q}}}$ and $\gamma^{S^{\mathfrak{q}}}$ the unique local points of R and P on $S^{\mathfrak{q}}$ (cf. 1.20).

Thus, by the very definition of $S^{\mathfrak{q}}$-fusions (cf. 1.28), for any homomorphism $\varphi \in \mathcal{F}_{(\nu^{\mathfrak{q}},N_G(\mathfrak{q}))}(P^{\mathfrak{q}},R)$ there is an R-interior algebra embedding

$$f_\varphi : (S^{\mathfrak{q}})_{\varepsilon^{S^{\mathfrak{q}}}} \longrightarrow \mathrm{Res}_\varphi(S^{\mathfrak{q}}) \qquad 11.18.4;$$

on the other hand, we have the structural embedding (1.12)

$$f_{\varepsilon^{S^{\mathfrak{q}}}} : (S^{\mathfrak{q}})_{\varepsilon^{S^{\mathfrak{q}}}} \longrightarrow \mathrm{Res}^P_R(S^{\mathfrak{q}}) \qquad 11.18.5;$$

consequently, $\mathrm{Res}_\varphi(S^{\mathfrak{q}})$ and $\mathrm{Res}^P_R(S^{\mathfrak{q}})$ determine the same element in $\mathcal{D}_k(R)$ (cf. 1.21), proving condition 11.6.1.

Note that, since Q is normal in $C_G(Q)\cdot P^{\mathfrak{q}}$, Q acts trivially on $S^{\mathfrak{q}}$ and, setting $\bar{P}^{\mathfrak{q}} = P^{\mathfrak{q}}/Q$, $S^{\mathfrak{q}}$ becomes a Dade $\bar{P}^{\mathfrak{q}}$-algebra; moreover, in order to prove condition 11.6.2, we may assume that (P,e) contains the Brauer (b,G)-pair determined by $(P^{\mathfrak{q}})_{\gamma^{\mathfrak{q}}}$; in this case, it is quite clear that the element of $\mathcal{D}_k(\bar{P}^{\mathfrak{q}})$ determined by $S^{\mathfrak{q}}$ coincides with $\mathrm{res}^{\bar{N}_P(Q)}_{\bar{P}^{\mathfrak{q}}}(s_Q)$ (cf. 7.7); then, condition 11.6.2 is an easy consequence of Theorem 7.14. We are done.

11.19 Consequently, for any totally normal set $\mathfrak{q}$ of nilcentralized Brauer (b,G)-pairs, it follows from Theorem 11.10 that we have the *regular* k^**-extension* $\hat{\mathcal{F}}_{(\nu^{\mathfrak{q}},N_G(\mathfrak{q}),\mathfrak{X}^{\mathfrak{q}})}$ of $\mathcal{F}_{(\nu^{\mathfrak{q}},N_G(\mathfrak{q}),\mathfrak{X}^{\mathfrak{q}})}$; similarly, denoting by $\mathfrak{B}^{\mathfrak{q}}$ the set of Brauer $(\nu^{\mathfrak{q}},N_G(\mathfrak{q}))$-pairs containing $\mathfrak{q}(n)$, we have the corresponding *regular* k^**-extension* $\hat{\mathcal{F}}^{\mathfrak{e}}_{(\nu^{\mathfrak{q}},N_G(\mathfrak{q}),\mathfrak{B}^{\mathfrak{q}})}$ of $\mathcal{F}^{\mathfrak{e}}_{(\nu^{\mathfrak{q}},N_G(\mathfrak{q}),\mathfrak{B}^{\mathfrak{q}})}$. Note that, if $\mathfrak{q}$ has a unique element (Q,f), clearly we have

$$\mathcal{F}^{\mathfrak{e}}_{(\nu^{\mathfrak{q}},N_G(\mathfrak{q}),\mathfrak{B}^{\mathfrak{q}})}(Q,f) = \mathcal{F}^{\mathfrak{e}}_{(b,G)}(Q,f) \qquad 11.19.1$$

and then Theorem 11.10 allows us to fix the choice of a k^*-group isomorphism

$$\hat{\theta}_{(Q,f)} : \hat{\mathcal{F}}^{\mathfrak{e}}_{(\nu^{\mathfrak{q}},N_G(\mathfrak{q}),\mathfrak{B}^{\mathfrak{q}})}(Q,f) \cong \hat{\mathcal{F}}^{\mathfrak{e}}_{(b,G)}(Q,f) \qquad 11.19.2,$$

which we may assume compatible with the G-conjugation when (Q,f) runs over the set of all nilcentralized Brauer (b,G)-pairs.

11.20 Moreover, for any totally normal set $\mathfrak{r}$ of nilcentralized Brauer (b,G)-pairs such that $\mathfrak{q}\subset\mathfrak{r}$ it is clear that we have $N_G(\mathfrak{r})\subset N_G(\mathfrak{q})$ and it is easily checked that we still have

$$N_G(\mathfrak{r})_{\nu^{\mathfrak{r}}}\subset N_G(\mathfrak{q})_{\nu^{\mathfrak{q}}}\quad\text{and}\quad \mathcal{F}^{\mathfrak{e}}_{(\nu^{\mathfrak{r}},N_G(\mathfrak{r}))}\subset \mathcal{F}^{\mathfrak{e}}_{(\nu^{\mathfrak{q}},N_G(\mathfrak{q}))} \qquad 11.20.1,$$

so that, since $\mathfrak{B}^{\mathfrak{r}}\subset\mathfrak{B}^{\mathfrak{q}}$, we have the k^*-functor

$$\hat{\mathcal{F}}^{\mathfrak{e}}_{(\nu^{\mathfrak{r}},N_G(\mathfrak{r}),\mathfrak{B}^{\mathfrak{r}})}\longrightarrow \hat{\mathcal{F}}^{\mathfrak{e}}_{(\nu^{\mathfrak{q}},N_G(\mathfrak{q}),\mathfrak{B}^{\mathfrak{q}})} \qquad 11.20.2$$

defined above (cf. 11.15), which lifts the inclusion. Finally, for any totally normal set $\mathfrak{t}$ of nilcentralized Brauer (b,G)-pairs such that $\mathfrak{r}\subset\mathfrak{t}$, in 11.15 above we already have exhibited a commutative diagram

$$\begin{array}{ccc}\hat{\mathcal{F}}^{\mathfrak{e}}_{(\nu^{\mathfrak{r}},N_G(\mathfrak{r}),\mathfrak{B}^{\mathfrak{r}})} & \longrightarrow & \hat{\mathcal{F}}^{\mathfrak{e}}_{(\nu^{\mathfrak{q}},N_G(\mathfrak{q}),\mathfrak{B}^{\mathfrak{q}})}\\ \nwarrow & & \nearrow\\ & \hat{\mathcal{F}}^{\mathfrak{e}}_{(\nu^{\mathfrak{t}},N_G(\mathfrak{t}),\mathfrak{B}^{\mathfrak{t}})} & \end{array} \qquad 11.20.3.$$

11.21 All these results allow us to exhibit a more explicit lifting in Theorem 7.16 above, namely a lifting fulfilling a suitable *transitive* condition. For this, we consider the particular case where $\mathfrak{q}$, $\mathfrak{r}$ and $\mathfrak{t}$ respectively have one, two and three elements. Explicitly, let (Q,f) and (R,g) be nilcentralized Brauer (b,G)-pairs such that

$$(R,g)\triangleleft(Q,f) \qquad 11.21.1$$

and set

$$\mathfrak{q}=\{(Q,f)\}\quad,\quad \mathfrak{r}=\{(R,g),(Q,f)\}\quad\text{and}\quad \mathfrak{u}=\{(R,g)\} \qquad 11.21.2;$$

for short, write $(\mathfrak{q})$, $(\mathfrak{r})$ and $(\mathfrak{u})$ instead of $(\nu^{\mathfrak{q}},N_G(\mathfrak{q}),\mathfrak{B}^{\mathfrak{q}})$, $(\nu^{\mathfrak{r}},N_G(\mathfrak{r}),\mathfrak{B}^{\mathfrak{r}})$ and $(\nu^{\mathfrak{u}},N_G(\mathfrak{u}),\mathfrak{B}^{\mathfrak{u}})$, and consider the k^*-functors defined in 11.20.2 above

$$\hat{\mathcal{F}}^{\mathfrak{e}}_{(\mathfrak{u})}\longleftarrow\hat{\mathcal{F}}^{\mathfrak{e}}_{(\mathfrak{r})}\longrightarrow\hat{\mathcal{F}}^{\mathfrak{e}}_{(\mathfrak{q})} \qquad 11.21.3;$$

then, since (Q,f) belongs to $\mathfrak{B}^{\mathfrak{q}}$, to $\mathfrak{B}^{\mathfrak{r}}$ and to $\mathfrak{B}^{\mathfrak{u}}$, we get injective k^*-group homomorphisms

$$\begin{aligned}\hat{\chi}^{\mathfrak{q}}_{\mathfrak{r}}&:\hat{\mathcal{F}}^{\mathfrak{e}}_{(\mathfrak{r})}(Q,f)\longrightarrow\hat{\mathcal{F}}^{\mathfrak{e}}_{(\mathfrak{q})}(Q,f)\\ \hat{\chi}^{\mathfrak{u}}_{\mathfrak{r}}&:\hat{\mathcal{F}}^{\mathfrak{e}}_{(\mathfrak{r})}(Q,f)\longrightarrow\hat{\mathcal{F}}^{\mathfrak{e}}_{(\mathfrak{u})}(Q,f)\end{aligned} \qquad 11.21.4.$$

11.22 Moreover, since (R,g) and (Q,f) belong to $\mathfrak{B}^{\mathfrak{u}}$, choosing a morphism $\hat{\imath}^{(Q,f)}_{(R,g)}\in\hat{\mathcal{F}}^{\mathfrak{e}}_{(\mathfrak{u})}\big((Q,f),(R,g)\big)$ lifting the inclusion of (R,g) in (Q,f), for any element $\hat{\sigma}^{\mathfrak{u}}$ in $\hat{\mathcal{F}}^{\mathfrak{e}}_{(\mathfrak{u})}(Q,f)$ there exists a unique $\hat{\tau}^{\mathfrak{u}} \in \hat{\mathcal{F}}^{\mathfrak{e}}_{(\mathfrak{u})}(R,g)$ such that, in the k^*-category $\hat{\mathcal{F}}^{\mathfrak{e}}_{(\mathfrak{u})}$, the following diagram is commutative

$$\begin{array}{ccc} (Q,f) & \xrightarrow{\hat{\sigma}^{\mathfrak{u}}} & (Q,f) \\ \hat{\iota}^{(Q,f)}_{(R,g)} \uparrow & & \uparrow \hat{\iota}^{(Q,f)}_{(R,g)} \\ (R,g) & \xrightarrow{\hat{\tau}^{\mathfrak{u}}} & (R,g) \end{array} \qquad 11.22.1;$$

thus, we get a k^*-group homomorphism

$$\hat{\mathfrak{u}}^{(R,g)}_{(Q,f)} : \hat{\mathcal{F}}^{\mathfrak{e}}_{(\mathfrak{u})}(Q,f) \longrightarrow \hat{\mathcal{F}}^{\mathfrak{e}}_{(\mathfrak{u})}(R,g) \qquad 11.22.2$$

such that, for any $\hat{\sigma}^{\mathfrak{u}} \in \hat{\mathcal{F}}^{\mathfrak{e}}_{(\mathfrak{u})}(Q,f)$, we have

$$\hat{\iota}^{(Q,f)}_{(R,g)} \circ \hat{\mathfrak{u}}^{(R,g)}_{(Q,f)}(\hat{\sigma}^{\mathfrak{u}}) = \hat{\sigma}^{\mathfrak{u}} \circ \hat{\iota}^{(Q,f)}_{(R,g)} \qquad 11.22.3,$$

which clearly does not depend on the choice of the lifting $\hat{\iota}^{(Q,f)}_{(R,g)}$. Denote by $\hat{\mathcal{F}}^{\mathfrak{e}}_{(b,G)}(Q,f)_R$ the stabilizer of R in $\hat{\mathcal{F}}^{\mathfrak{e}}_{(b,G)}(Q,f)$ and note that, with our choice in 11.19.2, we have

$$\mathrm{Im}\,(\hat{\theta}_{(Q,f)} \circ \hat{\chi}^{\mathfrak{q}}_{\mathfrak{r}}) = \hat{\mathcal{F}}^{\mathfrak{e}}_{(b,G)}(Q,f)_R \qquad 11.22.4.$$

Proposition 11.23 *With the notation above, there is a unique k^*-group homomorphism*

$$\hat{\alpha}^{(Q,f)}_{(R,g)} : \hat{\mathcal{F}}^{\mathfrak{e}}_{(b,G)}(Q,f)_R \longrightarrow \hat{\mathcal{F}}^{\mathfrak{e}}_{(b,G)}(R,g) \qquad 11.23.1.$$

lifting the restriction map and fulfilling

$$\hat{\alpha}^{(Q,f)}_{(R,g)} \circ \hat{\theta}_{(Q,f)} \circ \hat{\chi}^{\mathfrak{q}}_{\mathfrak{r}} = \hat{\theta}_{(R,g)} \circ \hat{\mathfrak{u}}^{(R,g)}_{(Q,f)} \circ \hat{\chi}^{\mathfrak{u}}_{\mathfrak{r}} \qquad 11.23.2.$$

Proof: Clearly, $\hat{\theta}_{(R,g)} \circ \hat{\mathfrak{u}}^{(R,g)}_{(Q,f)}$ is a k^*-group homomorphism from $\hat{\mathcal{F}}^{\mathfrak{e}}_{(\mathfrak{u})}(Q,f)$ to $\hat{\mathcal{F}}^{\mathfrak{e}}_{(b,G)}(R,g)$; moreover, since

$$\mathcal{F}^{\mathfrak{e}}_{(\nu^{\mathfrak{r}},N_G(\mathfrak{r}))}(Q,f) = \mathcal{F}^{\mathfrak{e}}_{(\nu^{\mathfrak{u}},N_G(\mathfrak{u}))}(Q,f) \qquad 11.23.3,$$

actually $\hat{\chi}^{\mathfrak{u}}_{\mathfrak{r}}$ is a k^*-group isomorphism; finally, according to equality 11.22.4 above, the composition $\hat{\theta}_{(Q,f)} \circ \hat{\chi}^{\mathfrak{q}}_{\mathfrak{r}}$ induces a k^*-group isomorphism between $\hat{\mathcal{F}}^{\mathfrak{e}}_{(\mathfrak{r})}(Q,f)$ and $\hat{\mathcal{F}}^{\mathfrak{e}}_{(b,G)}(Q,f)_R$. We are done.

Remark 11.24 The improvement with respect to Theorem 7.16 comes from the fact that here we *only* make independent choices over the objects.

11.25 Let (T, h) be another nilcentralized Brauer (b, G)-pair such that

$$(R, g) \triangleright (T, h) \triangleleft (Q, f) \qquad 11.25.1$$

and set

$$\begin{array}{lcl} \mathfrak{t} = \{(T,h),(R,g),(Q,f)\} & , & \mathfrak{v} = \{(T,h)\} \\ \mathfrak{p} = \{(T,h),(Q,f)\} & \text{and} & \mathfrak{s} = \{(T,h),(R,g)\} \end{array} \qquad 11.25.2;$$

for short, we write again $(\mathfrak{t})$, $(\mathfrak{v})$, $(\mathfrak{p})$ and $(\mathfrak{s})$ instead of $(\nu^{\mathfrak{t}}, N_G(\mathfrak{t}), \mathfrak{B}^{\mathfrak{t}})$, $(\nu^{\mathfrak{v}}, N_G(\mathfrak{v}), \mathfrak{B}^{\mathfrak{v}})$, $(\nu^{\mathfrak{p}}, N_G(\mathfrak{p}), \mathfrak{B}^{\mathfrak{p}})$ and $(\nu^{\mathfrak{s}}, N_G(\mathfrak{s}), \mathfrak{B}^{\mathfrak{s}})$, and consider the k^*-functors defined above together with the corresponding commutative diagrams (cf. 11.20.2 and 11.20.3)

$$\begin{array}{ccccccc} & & \hat{\mathcal{F}}^{\mathfrak{e}}_{(\mathfrak{s})} & \longrightarrow & \hat{\mathcal{F}}^{\mathfrak{e}}_{(\mathfrak{u})} & & \\ & \swarrow & & \nwarrow & & \nwarrow & \\ \hat{\mathcal{F}}^{\mathfrak{e}}_{(\mathfrak{v})} & & & & \hat{\mathcal{F}}^{\mathfrak{e}}_{(\mathfrak{t})} & \longrightarrow & \hat{\mathcal{F}}^{\mathfrak{e}}_{(\mathfrak{r})} \\ & \nwarrow & & \swarrow & & \swarrow & \\ & & \hat{\mathcal{F}}^{\mathfrak{e}}_{(\mathfrak{p})} & \longrightarrow & \hat{\mathcal{F}}^{\mathfrak{e}}_{(\mathfrak{q})} & & \end{array} \qquad 11.25.3.$$

11.26 Since (Q, f) belongs to $\mathfrak{B}^{\mathfrak{q}}$, $\mathfrak{B}^{\mathfrak{r}}$, $\mathfrak{B}^{\mathfrak{t}}$, $\mathfrak{B}^{\mathfrak{u}}$, $\mathfrak{B}^{\mathfrak{v}}$, $\mathfrak{B}^{\mathfrak{s}}$ and $\mathfrak{B}^{\mathfrak{p}}$, the diagram 11.25.3 above determines the following commutative $(k^*\text{-}\mathfrak{Gr})$-diagram

$$\begin{array}{ccccc} & \hat{\mathcal{F}}^{\mathfrak{e}}_{(\mathfrak{s})}(Q,f) & \xrightarrow{\hat{\chi}^{\mathfrak{u}}_{\mathfrak{s}}} & \hat{\mathcal{F}}^{\mathfrak{e}}_{(\mathfrak{u})}(Q,f) & \\ {\scriptstyle \hat{\chi}^{\mathfrak{v}}_{\mathfrak{s}}}\swarrow & & \nwarrow {\scriptstyle \hat{\chi}^{\mathfrak{s}}_{\mathfrak{t}}} & & \nwarrow {\scriptstyle \hat{\chi}^{\mathfrak{u}}_{\mathfrak{r}}} \\ \hat{\mathcal{F}}^{\mathfrak{e}}_{(\mathfrak{v})}(Q,f) & & \hat{\mathcal{F}}^{\mathfrak{e}}_{(\mathfrak{t})}(Q,f) & \xrightarrow{\hat{\chi}^{\mathfrak{r}}_{\mathfrak{t}}} & \hat{\mathcal{F}}^{\mathfrak{e}}_{(\mathfrak{r})}(Q,f) \\ {\scriptstyle \hat{\chi}^{\mathfrak{v}}_{\mathfrak{p}}}\nwarrow & & \swarrow {\scriptstyle \hat{\chi}^{\mathfrak{p}}_{\mathfrak{t}}} & & \swarrow {\scriptstyle \hat{\chi}^{\mathfrak{q}}_{\mathfrak{r}}} \\ & \hat{\mathcal{F}}^{\mathfrak{e}}_{(\mathfrak{p})}(Q,f) & \xrightarrow{\hat{\chi}^{\mathfrak{q}}_{\mathfrak{p}}} & \hat{\mathcal{F}}^{\mathfrak{e}}_{(\mathfrak{q})}(Q,f) & \end{array} \qquad 11.26.1;$$

similarly, since (R, g) belongs to $\mathfrak{B}^{\mathfrak{u}}$, $\mathfrak{B}^{\mathfrak{s}}$ and $\mathfrak{B}^{\mathfrak{v}}$, we get injective k^*-group homomorphisms

$$\begin{aligned} \hat{\rho}^{\mathfrak{u}}_{\mathfrak{s}} &: \hat{\mathcal{F}}^{\mathfrak{e}}_{(\mathfrak{s})}(R,g) \longrightarrow \hat{\mathcal{F}}^{\mathfrak{e}}_{(\mathfrak{u})}(R,g) \\ \hat{\rho}^{\mathfrak{v}}_{\mathfrak{s}} &: \hat{\mathcal{F}}^{\mathfrak{e}}_{(\mathfrak{s})}(R,g) \longrightarrow \hat{\mathcal{F}}^{\mathfrak{e}}_{(\mathfrak{v})}(R,g) \end{aligned} \qquad 11.26.2.$$

11.27 Moreover, since (T, h), (R, g) and (Q, f) belong to $\mathfrak{B}^{\mathfrak{v}}$, choosing

$$\begin{aligned} \hat{\imath}^{(R,g)}_{(T,h)} &\in \hat{\mathcal{F}}^{\mathfrak{e}}_{(\mathfrak{v})}\big((R,g),(T,h)\big) \\ \hat{\imath}^{(Q,f)}_{(T,h)} &\in \hat{\mathcal{F}}^{\mathfrak{e}}_{(\mathfrak{v})}\big((Q,f),(T,h)\big) \\ \hat{\imath}^{(Q,f)}_{(R,g)} &\in \hat{\mathcal{F}}^{\mathfrak{e}}_{(\mathfrak{v})}\big((Q,f),(R,g)\big) \end{aligned} \qquad 11.27.1$$

lifting the inclusions of (T,h) and (R,g) in (R,g) and (Q,f) respectively, we get k^*-group homomorphisms (cf. 11.22)

$$\begin{aligned} \hat{\mathfrak{v}}^{(T,h)}_{(R,g)} &: \hat{\mathcal{F}}^{\mathfrak{e}}_{(\mathfrak{v})}(R,g) \longrightarrow \hat{\mathcal{F}}^{\mathfrak{e}}_{(\mathfrak{v})}(T,h) \\ \hat{\mathfrak{v}}^{(T,h)}_{(Q,f)} &: \hat{\mathcal{F}}^{\mathfrak{e}}_{(\mathfrak{v})}(Q,f) \longrightarrow \hat{\mathcal{F}}^{\mathfrak{e}}_{(\mathfrak{v})}(T,h) \\ \hat{\mathfrak{v}}^{(R,g)}_{(Q,f)} &: \hat{\mathcal{F}}^{\mathfrak{e}}_{(\mathfrak{v})}(Q,f) \longrightarrow \hat{\mathcal{F}}^{\mathfrak{e}}_{(\mathfrak{v})}(R,g) \end{aligned} \qquad 11.27.2$$

such that, for any $\hat{\tau}^{\mathfrak{v}} \in \hat{\mathcal{F}}^{\mathfrak{e}}_{(\mathfrak{v})}(R,g)$ and any $\hat{\sigma}^{\mathfrak{v}} \in \hat{\mathcal{F}}^{\mathfrak{e}}_{(\mathfrak{v})}(Q,f)$, we have

$$\begin{aligned} \hat{\iota}^{(R,g)}_{(T,h)} \circ \hat{\mathfrak{v}}^{(T,h)}_{(R,g)}(\hat{\tau}^{\mathfrak{v}}) &= \hat{\tau}^{\mathfrak{v}} \circ \hat{\iota}^{(R,g)}_{(T,h)} \\ \hat{\iota}^{(Q,f)}_{(T,h)} \circ \hat{\mathfrak{v}}^{(T,h)}_{(Q,f)}(\hat{\sigma}^{\mathfrak{v}}) &= \hat{\sigma}^{\mathfrak{v}} \circ \hat{\iota}^{(Q,f)}_{(T,h)} \\ \hat{\iota}^{(Q,f)}_{(R,g)} \circ \hat{\mathfrak{v}}^{(R,g)}_{(Q,f)}(\hat{\sigma}^{\mathfrak{v}}) &= \hat{\sigma}^{\mathfrak{v}} \circ \hat{\iota}^{(Q,f)}_{(R,g)} \end{aligned} \qquad 11.27.3;$$

moreover, it is easily checked that $\hat{\mathfrak{v}}^{(T,h)}_{(Q,f)} = \hat{\mathfrak{v}}^{(T,h)}_{(R,g)} \circ \hat{\mathfrak{v}}^{(R,g)}_{(Q,f)}$.

11.28 Similarly, since (R,g) and (Q,f) belong to $\mathfrak{B}^{\mathfrak{s}}$, choosing a morphism $\hat{\iota}^{(Q,f)}_{(R,g)}$ in $\hat{\mathcal{F}}^{\mathfrak{e}}_{(\mathfrak{s})}\big((Q,f),(R,g)\big)$ lifting the inclusion of (R,g) in (Q,f), we get a k^*-group homomorphism (cf. 11.22)

$$\hat{\mathfrak{s}}^{(R,g)}_{(Q,f)} : \hat{\mathcal{F}}^{\mathfrak{e}}_{(\mathfrak{s})}(Q,f) \longrightarrow \hat{\mathcal{F}}^{\mathfrak{e}}_{(\mathfrak{s})}(R,g) \qquad 11.28.1$$

such that, for any $\hat{\sigma}^{\mathfrak{s}} \in \hat{\mathcal{F}}^{\mathfrak{e}}_{(\nu^{\mathfrak{s}},N_G(\mathfrak{s}),\mathfrak{B}^{\mathfrak{s}})}(Q,f)$, we have

$$\hat{\iota}^{(Q,f)}_{(R,g)} \circ \hat{\mathfrak{s}}^{(R,g)}_{(Q,f)}(\hat{\sigma}^{\mathfrak{s}}) = \hat{\sigma}^{\mathfrak{s}} \circ \hat{\iota}^{(Q,f)}_{(R,g)} \qquad 11.28.2.$$

On the other hand, denote by $\hat{\mathcal{F}}^{\mathfrak{e}}_{(b,G)}(R,g)_T$, $\hat{\mathcal{F}}^{\mathfrak{e}}_{(b,G)}(Q,f)_T$ and $\hat{\mathcal{F}}^{\mathfrak{e}}_{(b,G)}(Q,f)_{R,T}$ the respective stabilizers of T in $\hat{\mathcal{F}}^{\mathfrak{e}}_{(b,G)}(R,g)$, $\hat{\mathcal{F}}^{\mathfrak{e}}_{(b,G)}(Q,f)$ and $\hat{\mathcal{F}}^{\mathfrak{e}}_{(b,G)}(Q,f)_R$. Note that, by Proposition 11.23, we have the k^*-groups homomorphisms

$$\begin{aligned} \hat{\alpha}^{(R,g)}_{(T,h)} &: \hat{\mathcal{F}}^{\mathfrak{e}}_{(b,G)}(R,g)_T \longrightarrow \hat{\mathcal{F}}^{\mathfrak{e}}_{(b,G)}(T,h) \\ \hat{\alpha}^{(Q,f)}_{(T,h)} &: \hat{\mathcal{F}}^{\mathfrak{e}}_{(b,G)}(Q,f)_T \longrightarrow \hat{\mathcal{F}}^{\mathfrak{e}}_{(b,G)}(T,h) \end{aligned} \qquad 11.28.3.$$

Theorem 11.29 *With the notation above, we have*

$$\hat{\alpha}^{(Q,f)}_{(R,g)}\big(\hat{\mathcal{F}}^{\mathfrak{e}}_{(b,G)}(Q,f)_{R,T}\big) \subset \hat{\mathcal{F}}^{\mathfrak{e}}_{(b,G)}(R,g)_T \qquad 11.29.1$$

and $\hat{\alpha}^{(R,g)}_{(T,h)}\big(\hat{\alpha}^{(Q,f)}_{(R,g)}(\hat{\varphi})\big) = \hat{\alpha}^{(Q,f)}_{(T,h)}(\hat{\varphi})$ *for any* $\hat{\varphi} \in \hat{\mathcal{F}}^{\mathfrak{e}}_{(b,G)}(Q,f)_{R,T}$.

Proof: Since $\mathrm{Im}\,(\hat\theta_{(Q,f)} \circ \hat\chi^{\mathfrak q}_{\mathfrak r}) = \hat{\mathcal F}^{\mathfrak e}_{(\mathfrak b,G)}(Q,f)_R$ (cf. 11.22.4), it is clear that

$$(\hat\theta_{(Q,f)} \circ \hat\chi^{\mathfrak q}_{\mathfrak r})\big(\hat{\mathcal F}^{\mathfrak e}_{(\mathfrak r)}(Q,f)_T\big) = \hat{\mathcal F}^{\mathfrak e}_{(\mathfrak b,G)}(Q,f)_{R,T} \tag{11.29.2;}$$

moreover, since we have

$$\mathcal F^{\mathfrak e}_{(\nu^{\mathfrak r},N_G(\mathfrak r))}(Q,f)_T = \mathcal F^{\mathfrak e}_{(\nu^{\mathfrak t},N_G(\mathfrak t))}(Q,f) \tag{11.29.3,}$$

the k^*-group homomorphism $\hat\chi^{\mathfrak r}_{\mathfrak t}$ induces a k^*-group isomorphism

$$\hat{\mathcal F}^{\mathfrak e}_{(\mathfrak t)}(Q,f) \cong \hat{\mathcal F}^{\mathfrak e}_{(\mathfrak r)}(Q,f)_T \tag{11.29.4.}$$

Consequently, we have $\hat\varphi = (\hat\theta_{(Q,f)} \circ \hat\chi^{\mathfrak q}_{\mathfrak r} \circ \hat\chi^{\mathfrak r}_{\mathfrak t})(\hat\psi)$ for some $\hat\psi \in \hat{\mathcal F}^{\mathfrak e}_{(\mathfrak t)}(Q,f)$; then, according to Proposition 11.23, we get (cf. diagram 11.26.1)

$$\begin{aligned}\hat\alpha^{(R,g)}_{(T,h)}\big(\hat\alpha^{(Q,f)}_{(R,g)}(\hat\varphi)\big) &= \hat\alpha^{(R,g)}_{(T,h)}\big((\hat\alpha^{(Q,f)}_{(R,g)} \circ \hat\theta_{(Q,f)} \circ \hat\chi^{\mathfrak q}_{\mathfrak r} \circ \hat\chi^{\mathfrak r}_{\mathfrak t})(\hat\psi)\big)\\ &= \hat\alpha^{(R,g)}_{(T,h)}\big((\hat\theta_{(R,g)} \circ \hat{\mathfrak u}^{(R,g)}_{(Q,f)} \circ \hat\chi^{\mathfrak u}_{\mathfrak r} \circ \hat\chi^{\mathfrak r}_{\mathfrak t})(\hat\psi)\big)\\ &= \hat\alpha^{(R,g)}_{(T,h)}\big((\hat\theta_{(R,g)} \circ \hat{\mathfrak u}^{(R,g)}_{(Q,f)} \circ \hat\chi^{\mathfrak u}_{\mathfrak s} \circ \hat\chi^{\mathfrak s}_{\mathfrak t})(\hat\psi)\big)\end{aligned} \tag{11.29.5;}$$

but, since (Q,f) and (R,g) belong to both sets $\mathfrak B^{\mathfrak s}$ and $\mathfrak B^{\mathfrak u}$, the functor $\hat{\mathcal F}^{\mathfrak e}_{(\mathfrak s)} \to \hat{\mathcal F}^{\mathfrak e}_{(\mathfrak u)}$ defined above determines a map

$$\hat{\mathcal F}^{\mathfrak e}_{(\mathfrak s)}\big((Q,f),(R,g)\big) \longrightarrow \hat{\mathcal F}^{\mathfrak e}_{(\mathfrak u)}\big((Q,f),(R,g)\big) \tag{11.29.6}$$

preserving the morphisms which lift the inclusion $(R,g) \subset (Q,f)$; at this point, choosing in the right-hand term the image of the choice in the left-hand term, it is easily checked that the following diagram is commutative

$$\begin{array}{ccc}\hat{\mathcal F}^{\mathfrak e}_{(\mathfrak u)}(Q,f) & \xrightarrow{\hat{\mathfrak u}^{(R,g)}_{(Q,f)}} & \hat{\mathcal F}^{\mathfrak e}_{(\mathfrak u)}(R,g)\\ {\scriptstyle\hat\chi^{\mathfrak u}_{\mathfrak s}}\uparrow & & \uparrow{\scriptstyle\hat\rho^{\mathfrak u}_{\mathfrak s}}\\ \hat{\mathcal F}^{\mathfrak e}_{(\mathfrak s)}(Q,f) & \xrightarrow{\hat{\mathfrak s}^{(R,g)}_{(Q,f)}} & \hat{\mathcal F}^{\mathfrak e}_{(\mathfrak s)}(R,g)\end{array} \tag{11.29.7.}$$

Thus, from equalities 11.29.5, from the commutativity of this diagram and, once again, from Proposition 11.23 we get

$$\begin{aligned}\hat\alpha^{(R,g)}_{(T,h)}\big(\hat\alpha^{(Q,f)}_{(R,g)}(\hat\varphi)\big) &= \hat\alpha^{(R,g)}_{(T,h)}\big((\hat\theta_{(R,g)} \circ \hat\rho^{\mathfrak u}_{\mathfrak s} \circ \hat{\mathfrak s}^{(R,g)}_{(Q,f)} \circ \hat\chi^{\mathfrak s}_{\mathfrak t})(\hat\psi)\big)\\ &= (\hat\alpha^{(R,g)}_{(T,h)} \circ \hat\theta_{(R,g)} \circ \hat\rho^{\mathfrak u}_{\mathfrak s})\big((\hat{\mathfrak s}^{(R,g)}_{(Q,f)} \circ \hat\chi^{\mathfrak s}_{\mathfrak t})(\hat\psi)\big)\\ &= (\hat\theta_{(T,h)} \circ \hat{\mathfrak v}^{(T,h)}_{(R,g)} \circ \hat\rho^{\mathfrak v}_{\mathfrak s})\big((\hat{\mathfrak s}^{(R,g)}_{(Q,f)} \circ \hat\chi^{\mathfrak s}_{\mathfrak t})(\hat\psi)\big)\\ &= (\hat\theta_{(T,h)} \circ \hat{\mathfrak v}^{(T,h)}_{(R,g)} \circ \hat\rho^{\mathfrak v}_{\mathfrak s} \circ \hat{\mathfrak s}^{(R,g)}_{(Q,f)} \circ \hat\chi^{\mathfrak s}_{\mathfrak t})(\hat\psi)\end{aligned} \tag{11.29.8.}$$

Similarly, since (Q,f) and (R,g) belong to both sets $\mathfrak{B}^{\mathfrak{s}}$ and $\mathfrak{B}^{\mathfrak{v}}$, the functor $\hat{\mathcal{F}}^{\mathfrak{e}}_{(\mathfrak{s})} \to \hat{\mathcal{F}}^{\mathfrak{e}}_{(\mathfrak{v})}$ defined above determines a map

$$\hat{\mathcal{F}}^{\mathfrak{e}}_{(\mathfrak{s})}\big((Q,f),(R,g)\big) \longrightarrow \hat{\mathcal{F}}^{\mathfrak{e}}_{(\mathfrak{v})}\big((Q,f),(R,g)\big) \qquad 11.29.9,$$

and, once again, choosing our lifting of the inclusions from this map, it is easily checked that the following diagram is commutative

$$\begin{array}{ccc} \hat{\mathcal{F}}^{\mathfrak{e}}_{(\mathfrak{v})}(Q,f) & \xrightarrow{\hat{\mathfrak{v}}^{(R,g)}_{(Q,f)}} & \hat{\mathcal{F}}^{\mathfrak{e}}_{(\mathfrak{v})}(R,g) \\ \hat{\chi}^{\mathfrak{v}}_{\mathfrak{s}}\uparrow & & \uparrow\hat{\rho}^{\mathfrak{v}}_{\mathfrak{s}} \\ \hat{\mathcal{F}}^{\mathfrak{e}}_{(\mathfrak{s})}(Q,f) & \xrightarrow{\hat{\mathfrak{s}}^{(R,g)}_{(Q,f)}} & \hat{\mathcal{F}}^{\mathfrak{e}}_{(\mathfrak{s})}(R,g) \end{array} \qquad 11.29.10.$$

Finally, from equalities 11.29.8, from the commutativity of this diagram and from Proposition 11.23 again, we still get (cf. diagram 11.26.1)

$$\begin{aligned} \hat{\alpha}^{(R,g)}_{(T,h)}\big(\hat{\alpha}^{(Q,f)}_{(R,g)}(\hat{\varphi})\big) &= (\hat{\theta}_{(T,h)} \circ \hat{\mathfrak{v}}^{(T,h)}_{(R,g)} \circ \hat{\mathfrak{v}}^{(R,g)}_{(Q,f)} \circ \hat{\chi}^{\mathfrak{v}}_{\mathfrak{s}} \circ \hat{\chi}^{\mathfrak{s}}_{\mathfrak{t}})(\hat{\psi}) \\ &= (\hat{\theta}_{(T,h)} \circ \hat{\mathfrak{v}}^{(T,h)}_{(Q,f)} \circ \hat{\chi}^{\mathfrak{v}}_{\mathfrak{p}} \circ \hat{\chi}^{\mathfrak{p}}_{\mathfrak{t}})(\hat{\psi}) \\ &= (\hat{\alpha}^{(Q,f)}_{(T,h)} \circ \hat{\theta}_{(Q,f)} \circ \hat{\chi}^{\mathfrak{q}}_{\mathfrak{p}} \circ \hat{\chi}^{\mathfrak{p}}_{\mathfrak{t}})(\hat{\psi}) \\ &= \hat{\alpha}^{(Q,f)}_{(T,h)}(\hat{\varphi}) \end{aligned} \qquad 11.29.11.$$

We are done.

11.30 The *transitive* equality in Theorem 11.29 is the key for the construction of the announced functor (cf. 11.1) from the *proper category of* $(\mathcal{F}_{(b,G)})^{\rm nc}$*-chains* (cf. A2.8) to k^*-$\mathfrak{Gr}$. Recall that, for any $n \in \mathbb{N}$, a $(\mathcal{F}_{(b,G)})^{\rm nc}$-*chain* — more precisely called $(n,(\mathcal{F}_{(b,G)})^{\rm nc})$*-chain*, or n*-chain* for short — is a functor

$$\mathfrak{q} : \Delta_n \longrightarrow (\mathcal{F}_{(b,G)})^{\rm nc} \qquad 11.30.1$$

from the category Δ_n formed by the objects $0 \le i \le n$ and the morphisms $0 \le j \le i \le n$, with the obvious composition (cf. A2.2). Then, the *proper category of* $(\mathcal{F}_{(b,G)})^{\rm nc}$*-chains* (cf. A2.8) — denoted by $\mathfrak{ch}^*\big((\mathcal{F}_{(b,G)})^{\rm nc}\big)$ — is the category formed by all the pairs $(\mathfrak{q},\Delta_n)$, where $\mathfrak{q}$ is a $(\mathcal{F}_{(b,G)})^{\rm nc}$-chain, and by the morphisms $(\lambda,\delta) : (\mathfrak{r},\Delta_m) \to (\mathfrak{q},\Delta_n)$ — sometimes noted $(\lambda^{(\mathfrak{r})},\delta)$ to avoid confusion — where $\delta : \Delta_n \to \Delta_m$ is a functor, or equivalently an *order-preserving map*, and $\lambda : \mathfrak{r}\circ\delta \cong \mathfrak{q}$ a natural isomorphism, and where the composition with another morphism $(\mu,\varepsilon) : (\mathfrak{t},\Delta_\ell) \to (\mathfrak{r},\Delta_m)$ is defined by (cf. A2.6.3)

$$(\lambda,\delta)\circ(\mu,\varepsilon) = \big(\lambda\circ(\mu * \delta),\delta\circ\varepsilon\big) \qquad 11.30.2.$$

11.31 Often, we write $\mathfrak{q}$ instead of $(\mathfrak{q}, \Delta_n)$ and we denote by $\mathcal{F}_{(b,G)}(\mathfrak{q})$ the automorphism group of $(\mathfrak{q}, \Delta_n)$ in the category $\mathfrak{ch}^*\big((\mathcal{F}_{(b,G)})^{^{\rm nc}}\big)$ or, equivalently, the group of *natural automorphisms* of the functor $\mathfrak{q}$; note that $\mathcal{F}_{(b,G)}(\mathfrak{q})$ can be identified to a subgroup of $\mathcal{F}_{(b,G)}\big(\mathfrak{q}(n)\big)$ and we denote by $\hat{\mathcal{F}}_{(b,G)}(\mathfrak{q})$ the corresponding k^*-subgroup of $\hat{\mathcal{F}}_{(b,G)}\big(\mathfrak{q}(n)\big)$. On the other hand, recall that we have the *automorphism functor* (cf. Proposition A2.10)

$$\mathfrak{aut} = \mathfrak{aut}_{(\mathcal{F}_{(b,G)})^{\rm nc}} : \mathfrak{ch}^*\big((\mathcal{F}_{(b,G)})^{^{\rm nc}}\big) \longrightarrow \mathfrak{Gr} \qquad 11.31.1$$

mapping $\mathfrak{q}$ on $\mathcal{F}_{(b,G)}(\mathfrak{q})$.

Theorem 11.32 *With the notation above and our choice in* 11.19.2, *there is a unique functor*

$$\widehat{\mathfrak{aut}} = \widehat{\mathfrak{aut}}_{(\mathcal{F}_{(b,G)})^{\rm nc}} : \mathfrak{ch}^*\big((\mathcal{F}_{(b,G)})^{^{\rm nc}}\big) \longrightarrow k^*\text{-}\mathfrak{Gr} \qquad 11.32.1$$

which lifts the automorphism functor, maps any $(\mathcal{F}_{(b,G)})^{^{\rm nc}}$-chain $\mathfrak{q}$ on $\hat{\mathcal{F}}_{(b,G)}(\mathfrak{q})$ and, for any nilcentralized Brauer (b,G)-pairs (Q,f) and (R,g) contained in (P,e) such that $(R,g) \triangleleft (Q,f)$, sends to $\hat{\alpha}_{(R,g)}^{(Q,f)}$ the morphism determined by id_R from the 1-chain determined by the inclusion $R \subset Q$ to the 0-chain determined by R.

Proof: Thus, for any n-chain $\mathfrak{q}$ of $(\mathcal{F}_{(b,G)})^{^{\rm nc}}$, we set $\widehat{\mathfrak{aut}}(\mathfrak{q}) = \hat{\mathcal{F}}_{(b,G)}(\mathfrak{q})$. Let $(\lambda,\delta) : (\mathfrak{q}, \Delta_n) \to (\mathfrak{r}, \Delta_m)$ be a $\mathfrak{ch}^*\big((\mathcal{F}_{(b,G)})^{^{\rm nc}}\big)$-morphism, set $Q = \mathfrak{q}(n)$ and $R = \mathfrak{r}(m)$, and denote by $\varphi : R \to Q$ the composition of the inverse of the isomorphism $\lambda_m : \mathfrak{q}\big(\delta(m)\big) \cong \mathfrak{r}(m)$ with the $\mathfrak{q}$-image of the Δ_n-morphism from $\delta(m)$ to n — denoted by $\delta(m){\bullet}n$ (cf. A2.2).

If $R' = \varphi(R)$ is normal in Q then, denoting by f and g' the respective blocks of $C_G(Q)$ and $C_G(R')$ such that (P,e) contains (Q,f) and (R',g') (cf. 1.15), Q fixes g'; moreover, $\hat{\mathcal{F}}_{(b,G)}(\mathfrak{q})$ is a k^*-subgroup of $\hat{\mathcal{F}}_{(b,G)}(Q)_{R'}$ and, denoting by $\varphi_* : R \cong R'$ the isomorphism induced by φ, the restriction map sends $\mathcal{F}_{(b,G)}(\mathfrak{q}) \subset \mathcal{F}_{(b,G)}(Q)_{R'}$ to

$$\varphi_* \circ \mathcal{F}_{(b,G)}(\mathfrak{r}) \circ (\varphi_*)^{-1} \subset \mathcal{F}_{(b,G)}(R') \qquad 11.32.2.$$

In this case, denoting by

$$\hat{\mathcal{F}}_{(b,G)}(\varphi_*) : \hat{\mathcal{F}}_{(b,G)}(R) \cong \hat{\mathcal{F}}_{(b,G)}(R') \qquad 11.32.3$$

the k^*-group isomorphism determined by an element of G inducing φ_*, we define

$$\widehat{\mathfrak{aut}}(\lambda,\delta) : \hat{\mathcal{F}}_{(b,G)}(\mathfrak{q}) \longrightarrow \hat{\mathcal{F}}_{(b,G)}(\mathfrak{r}) \qquad 11.32.4$$

as the corresponding restriction of $\hat{\mathcal{F}}_{(b,G)}(\varphi_*)^{-1} \circ \hat{\alpha}_{(R',g')}^{(Q,f)}$ (cf. 11.23.1).

If $R' \neq Q$, consider the maximal $i \in \Delta_n$ such that $\mathfrak{q}(i\bullet n)$ maps $\mathfrak{q}(i)$ on a proper subgroup Q' of Q and then a $\mathcal{F}_{(b,G)}(\mathfrak{q})$-stable normal proper subgroup U of Q containing Q'; since Q' contains R', φ and $\mathfrak{q}(i \bullet n)$ respectively induce $\mathcal{F}_{(b,G)}$-morphisms $R \to U$ and $\mathfrak{q}(i) \to U$; these $\mathcal{F}_{(b,G)}$-morphisms and the inclusion $U \subset Q$ determine chains

$$\hat{\mathfrak{q}} : \Delta_{n+1} \longrightarrow \big(\mathcal{F}_{(b,G)}\big)^{\mathrm{nc}} \quad \text{and} \quad \hat{\mathfrak{r}} : \Delta_{m+1} \longrightarrow \big(\mathcal{F}_{(b,G)}\big)^{\mathrm{nc}} \qquad 11.32.5$$

fulfilling $\hat{\mathfrak{q}}\circ\delta^n_{i+1} = \mathfrak{q}$ and $\hat{\mathfrak{r}}\circ\delta^m_{m+1} = \mathfrak{r}$ (cf. A3.1) and respectively mapping $i+1$ and $m+1$ on U. In this situation, we have evident $\mathfrak{ch}^*\big(\big(\mathcal{F}_{(b,G)}\big)^{\mathrm{nc}}\big)$-morphisms building a commutative $\mathfrak{ch}^*\big(\big(\mathcal{F}_{(b,G)}\big)^{\mathrm{nc}}\big)$-diagram

$$\begin{array}{ccc} (\hat{\mathfrak{r}}, \Delta_{m+1}) & \xrightarrow{(\mathrm{id}_{\mathfrak{r}},\delta^m_{m+1})} & (\mathfrak{r}, \Delta_m) \\ {\scriptstyle (\hat{\lambda},\hat{\delta})}\uparrow & & \uparrow{\scriptstyle (\lambda,\delta)} \\ (\hat{\mathfrak{q}}, \Delta_{n+1}) & \xrightarrow{(\mathrm{id}_{\mathfrak{q}},\delta^n_{i+1})} & (\mathfrak{q}, \Delta_n) \end{array} \qquad 11.32.6$$

and, since U is normal in Q, the k^*-group homomorphism $\widehat{\mathfrak{aut}}(\hat{\lambda}, \hat{\delta})$ is already defined; moreover, arguing by induction on $|Q|/|R|$, we may assume that $\widehat{\mathfrak{aut}}(\mathrm{id}_{\mathfrak{r}}, \delta^m_{m+1})$ is defined too; further, since we have $\hat{\mathfrak{q}}(n+1) = \mathfrak{q}(n)$ and U is $\mathcal{F}_{(b,G)}(\mathfrak{q})$-stable, we get $\mathcal{F}_{(b,G)}(\hat{\mathfrak{q}}) = \mathcal{F}_{(b,G)}(\mathfrak{q})$ and therefore $\hat{\mathcal{F}}_{(b,G)}(\hat{\mathfrak{q}})$ and $\hat{\mathcal{F}}_{(b,G)}(\mathfrak{q})$ are the *same* k^*-subgroup of $\hat{\mathcal{F}}_{(b,G)}(Q)$, so that we have (cf. 11.32.4)

$$\widehat{\mathfrak{aut}}(\mathrm{id}_{\mathfrak{q}}, \delta^n_{i+1}) = \mathrm{id}_{\hat{\mathcal{F}}_{(b,G)}(\mathfrak{q})} \qquad 11.32.7.$$

Then, we define $\widehat{\mathfrak{aut}}(\lambda, \delta) = \widehat{\mathfrak{aut}}(\mathrm{id}_{\mathfrak{r}}, \delta^m_{m+1}) \circ \widehat{\mathfrak{aut}}(\hat{\lambda}, \hat{\delta})$.

We claim that $\widehat{\mathfrak{aut}}(\lambda, \delta)$ does not depend on the choice of U; indeed, for another choice U' of an $\mathcal{F}_{(b,G)}(\mathfrak{q})$-stable normal proper subgroup of Q containing Q', the intersection $U'' = U \cap U'$ is also a possible choice and therefore we may assume that $U' \subset U$; once again, we have chains

$$\hat{\mathfrak{q}}' : \Delta_{n+1} \longrightarrow \big(\mathcal{F}_{(b,G)}\big)^{\mathrm{nc}} \quad \text{and} \quad \hat{\mathfrak{r}}' : \Delta_{m+1} \longrightarrow \big(\mathcal{F}_{(b,G)}\big)^{\mathrm{nc}} \qquad 11.32.8$$

fulfilling $\hat{\mathfrak{q}}' \circ \delta^n_{i+1} = \mathfrak{q}$ and $\hat{\mathfrak{r}}' \circ \delta^m_{m+1} = \mathfrak{r}$ (cf. A3.1) and respectively mapping $i+1$ and $m+1$ on U', and we have evident $\mathfrak{ch}^*\big(\big(\mathcal{F}_{(b,G)}\big)^{\mathrm{nc}}\big)$-morphisms building a commutative $\mathfrak{ch}^*\big(\big(\mathcal{F}_{(b,G)}\big)^{\mathrm{nc}}\big)$-diagram

$$\begin{array}{ccc} (\hat{\mathfrak{r}}', \Delta_{m+1}) & \xrightarrow{(\mathrm{id}_{\mathfrak{r}}^{(\hat{\mathfrak{r}}')},\delta^m_{m+1})} & (\mathfrak{r}, \Delta_m) \\ {\scriptstyle (\hat{\lambda}',\hat{\delta})}\uparrow & & \uparrow{\scriptstyle (\lambda,\delta)} \\ (\hat{\mathfrak{q}}', \Delta_{n+1}) & \xrightarrow{(\mathrm{id}_{\mathfrak{q}}^{(\hat{\mathfrak{q}}')},\delta^n_{i+1})} & (\mathfrak{q}, \Delta_n) \end{array} \qquad 11.32.9.$$

But, the inclusion map $U' \to U$ still determines another pair of chains

$$\hat{\hat{\mathfrak{q}}} : \Delta_{n+2} \longrightarrow \big(\mathcal{F}_{(b,G)}\big)^{\mathrm{nc}} \quad \text{and} \quad \hat{\hat{\mathfrak{r}}} : \Delta_{m+2} \longrightarrow \big(\mathcal{F}_{(b,G)}\big)^{\mathrm{nc}} \qquad 11.32.10$$

fulfilling $\hat{\hat{\mathfrak{q}}} \circ \delta_{i+1}^{n+1} = \hat{\mathfrak{q}}$ and $\hat{\hat{\mathfrak{q}}} \circ \delta_{i+2}^{n+1} = \hat{\mathfrak{q}}'$, and analogously $\hat{\hat{\mathfrak{r}}} \circ \delta_{m+1}^{m+1} = \hat{\mathfrak{r}}$ and $\hat{\hat{\mathfrak{r}}} \circ \delta_{m+2}^{m+1} = \hat{\mathfrak{r}}'$; as above we have evident $\mathfrak{ch}^*\big((\mathcal{F}_{(b,G)})^{\mathrm{nc}}\big)$-morphisms building a commutative $\mathfrak{ch}^*\big((\mathcal{F}_{(b,G)})^{\mathrm{nc}}\big)$-diagram

$$\begin{array}{ccccc} (\hat{\mathfrak{r}}', \Delta_{m+1}) & \xleftarrow{(\mathrm{id}_{\hat{\mathfrak{r}}'}, \delta_{m+2}^{m+1})} & (\hat{\hat{\mathfrak{r}}}, \Delta_{m+2}) & \xrightarrow{(\mathrm{id}_{\hat{\mathfrak{r}}}, \delta_{m+1}^{m+1})} & (\hat{\mathfrak{r}}, \Delta_{m+1}) \\ {\scriptstyle (\hat{\lambda}', \hat{\delta})}\uparrow & & {\scriptstyle (\hat{\hat{\lambda}}, \hat{\hat{\delta}})}\uparrow & & \uparrow{\scriptstyle (\hat{\lambda}, \hat{\delta})} \\ (\hat{\mathfrak{q}}', \Delta_{n+1}) & \xleftarrow{(\mathrm{id}_{\hat{\mathfrak{q}}'}, \delta_{i+2}^{n+1})} & (\hat{\hat{\mathfrak{q}}}, \Delta_{n+2}) & \xrightarrow{(\mathrm{id}_{\hat{\mathfrak{q}}}, \delta_{i+1}^{n+1})} & (\hat{\mathfrak{q}}, \Delta_{n+1}) \end{array} \qquad 11.32.11;$$

moreover, as in the equality 11.32.7, we have

$$\widehat{\mathfrak{aut}}(\mathrm{id}_{\hat{\mathfrak{q}}'}, \delta_{i+2}^{n+1}) = \mathrm{id}_{\hat{\mathcal{F}}_{(b,G)}(\mathfrak{q})} = \widehat{\mathfrak{aut}}(\mathrm{id}_{\hat{\mathfrak{q}}}, \delta_{i+1}^{n+1}) \qquad 11.32.12.$$

On the other hand, since U and U' are normal in Q, we already have defined our functor to k^*-$\mathfrak{Gr}$ over all the morphisms in diagram 11.32.11, and it is easily checked from Theorem 11.29 that the corresponding (k^*-$\mathfrak{Gr}$)-diagram is commutative; consequently, we have

$$\begin{aligned} &\widehat{\mathfrak{aut}}(\mathrm{id}_{\mathfrak{r}}^{(\hat{\mathfrak{r}})}, \delta_{m+1}^{m}) \circ \widehat{\mathfrak{aut}}(\hat{\lambda}, \hat{\delta}) \\ &\qquad = \widehat{\mathfrak{aut}}(\mathrm{id}_{\mathfrak{r}}^{(\hat{\mathfrak{r}})}, \delta_{m+1}^{m}) \circ \widehat{\mathfrak{aut}}(\hat{\lambda}, \hat{\delta}) \circ \widehat{\mathfrak{aut}}(\mathrm{id}_{\hat{\mathfrak{q}}}, \delta_{i+1}^{n+1}) \\ &\qquad = \widehat{\mathfrak{aut}}(\mathrm{id}_{\mathfrak{r}}^{(\hat{\mathfrak{r}})}, \delta_{m+1}^{m}) \circ \widehat{\mathfrak{aut}}(\mathrm{id}_{\hat{\mathfrak{r}}}, \delta_{m+1}^{m+1}) \circ \widehat{\mathfrak{aut}}(\hat{\hat{\lambda}}, \hat{\hat{\delta}}) \\ &\widehat{\mathfrak{aut}}(\mathrm{id}_{\mathfrak{r}}^{(\hat{\mathfrak{r}}')}, \delta_{m+1}^{m}) \circ \widehat{\mathfrak{aut}}(\hat{\lambda}', \hat{\delta}) \\ &\qquad = \widehat{\mathfrak{aut}}(\mathrm{id}_{\mathfrak{r}}^{(\hat{\mathfrak{r}}')}, \delta_{m+1}^{m}) \circ \widehat{\mathfrak{aut}}(\hat{\lambda}', \hat{\delta}) \circ \widehat{\mathfrak{aut}}(\mathrm{id}_{\hat{\mathfrak{q}}}, \delta_{i+2}^{n+1}) \\ &\qquad = \widehat{\mathfrak{aut}}(\mathrm{id}_{\mathfrak{r}}^{(\hat{\mathfrak{r}}')}, \delta_{m+1}^{m}) \circ \widehat{\mathfrak{aut}}(\mathrm{id}_{\hat{\mathfrak{r}}'}, \delta_{m+2}^{m+1}) \circ \widehat{\mathfrak{aut}}(\hat{\hat{\lambda}}, \hat{\hat{\delta}}) \end{aligned} \qquad 11.32.13.$$

But, once again arguing by induction on $|Q|/|R|$, we may assume that the image in k^*-$\mathfrak{Gr}$ of the commutative $\mathfrak{ch}^*\big((\mathcal{F}_{(b,G)})^{\mathrm{nc}}\big)$-diagram

$$\begin{array}{ccc} (\hat{\mathfrak{r}}', \Delta_{m+1}) & \xrightarrow{(\mathrm{id}_{\mathfrak{r}}^{(\hat{\mathfrak{r}}')}, \delta_{m+1}^{m})} & (\mathfrak{r}, \Delta_{m}) \\ {\scriptstyle (\mathrm{id}_{\hat{\mathfrak{r}}'}, \delta_{m+2}^{m+1})}\uparrow & & \uparrow{\scriptstyle (\mathrm{id}_{\mathfrak{r}}^{(\hat{\mathfrak{r}})}, \delta_{m+1}^{m})} \\ (\hat{\hat{\mathfrak{r}}}, \Delta_{m+2}) & \xrightarrow{(\mathrm{id}_{\hat{\mathfrak{r}}}, \delta_{m+1}^{m+1})} & (\hat{\mathfrak{r}}, \Delta_{m+1}) \end{array} \qquad 11.32.14$$

is already defined and commutative; that is to say, we may assume that

$$\begin{aligned} &\widehat{\mathfrak{aut}}(\mathrm{id}_{\mathfrak{r}}^{(\hat{\mathfrak{r}})}, \delta_{m+1}^{m}) \circ \widehat{\mathfrak{aut}}(\mathrm{id}_{\hat{\mathfrak{r}}}, \delta_{m+1}^{m+1}) \\ &\qquad = \widehat{\mathfrak{aut}}(\mathrm{id}_{\mathfrak{r}}^{(\hat{\mathfrak{r}}')}, \delta_{m+1}^{m}) \circ \widehat{\mathfrak{aut}}(\mathrm{id}_{\hat{\mathfrak{r}}'}, \delta_{m+2}^{m+1}) \end{aligned} \qquad 11.32.15.$$

Consequently, it follows from equalities 11.32.13 that

$$\widehat{\mathfrak{aut}}(\mathrm{id}_{\mathfrak{r}}^{(\hat{\mathfrak{r}})}, \delta^m_{m+1}) \circ \widehat{\mathfrak{aut}}(\hat{\lambda}, \hat{\delta}) = \widehat{\mathfrak{aut}}(\mathrm{id}_{\mathfrak{r}}^{(\hat{\mathfrak{r}}')}, \delta^m_{m+1}) \circ \widehat{\mathfrak{aut}}(\hat{\lambda}', \hat{\delta}) \qquad 11.32.16$$

which proves that $\widehat{\mathfrak{aut}}(\lambda, \delta)$ does not depend on the choice of U.

Finally, let $(\mu, \varepsilon): (\mathfrak{r}, \Delta_m) \to (\mathfrak{t}, \Delta_\ell)$ be a second $\mathfrak{ch}^*\big((\mathcal{F}_{(b,G)})^{\mathrm{nc}}\big)$-morphism, set $T = \mathfrak{t}(\ell)$ and denote by $\psi : T \to R$ the composition of the inverse of the isomorphism $\mu_\ell : \mathfrak{r}\big(\varepsilon(\ell)\big) \cong \mathfrak{t}(\ell)$ with $\mathfrak{r}\big(\varepsilon(\ell)\bullet m\big)$; we will prove that

$$\widehat{\mathfrak{aut}}(\mu, \varepsilon) \circ \widehat{\mathfrak{aut}}(\lambda, \delta) = \widehat{\mathfrak{aut}}\big((\mu, \varepsilon) \circ (\lambda, \delta)\big) \qquad 11.32.17$$

arguing by induction on $|Q|/|T|$; set (cf. 11.30.2)

$$(\nu, \eta) = (\mu, \varepsilon) \circ (\lambda, \delta) = \big(\mu \circ (\lambda * \varepsilon), \varepsilon \circ \delta\big) \qquad 11.32.18.$$

If $|Q|/|T| = 1$ then $\widehat{\mathfrak{aut}}(\lambda, \delta)$ and $\widehat{\mathfrak{aut}}(\mu, \varepsilon)$ respectively coincide with the corresponding restrictions of the k^*-group isomorphisms $\hat{\mathcal{F}}_{(b,G)}(\varphi)$ and $\hat{\mathcal{F}}_{(b,G)}(\psi)$ induced by elements x and y of G respectively determining φ and ψ, whereas $\widehat{\mathfrak{aut}}(\nu, \eta)$ clearly coincides with the corresponding restriction of the k^*-group isomorphism $\hat{\mathcal{F}}_{(b,G)}(\varphi \circ \psi)$ induced by xy; thus, in this case we are done.

If $R' \neq Q$, we have defined

$$\widehat{\mathfrak{aut}}(\lambda, \delta) = \widehat{\mathfrak{aut}}(\mathrm{id}_{\mathfrak{r}}, \delta^m_{m+1}) \circ \widehat{\mathfrak{aut}}(\hat{\lambda}, \hat{\delta}) \qquad 11.32.19$$

from a suitable $\mathcal{F}_{(b,G)}(\mathfrak{q})$-stable normal proper subgroup U of Q; but, the *same* subgroup U of Q provides a good support in order to define $\widehat{\mathfrak{aut}}(\nu, \eta)$; that is to say, with the corresponding notation we still have

$$\widehat{\mathfrak{aut}}(\nu, \eta) = \widehat{\mathfrak{aut}}(\mathrm{id}_{\mathfrak{t}}, \delta^\ell_{\ell+1}) \circ \widehat{\mathfrak{aut}}(\hat{\nu}, \hat{\eta}) \qquad 11.32.20;$$

moreover, our induction hypothesis guarantees that

$$\widehat{\mathfrak{aut}}(\mu, \varepsilon) \circ \widehat{\mathfrak{aut}}(\mathrm{id}_{\mathfrak{r}}, \delta^m_{m+1}) = \widehat{\mathfrak{aut}}\big((\mu, \varepsilon) \circ (\mathrm{id}_{\mathfrak{r}}, \delta^m_{m+1})\big) \qquad 11.32.21$$

and therefore we get

$$\widehat{\mathfrak{aut}}(\mu, \varepsilon) \circ \widehat{\mathfrak{aut}}(\lambda, \delta) = \widehat{\mathfrak{aut}}\big((\mu, \varepsilon) \circ (\mathrm{id}_{\mathfrak{r}}, \delta^m_{m+1})\big) \circ \widehat{\mathfrak{aut}}(\hat{\lambda}, \hat{\delta}) \qquad 11.32.22.$$

Similarly, consider the corresponding chain $\hat{\mathfrak{t}} : \Delta_\ell \to (\mathcal{F}_{(b,G)})^{\mathrm{nc}}$ fulfilling $\hat{\mathfrak{t}} \circ \delta^\ell_{\ell+1} = \mathfrak{t}$ and mapping $\ell{+}1$ on U; as above, we have the corresponding $\mathcal{F}_{(b,G)}$-morphisms building a commutative $\mathfrak{ch}^*\big((\mathcal{F}_{(b,G)})^{\mathrm{nc}}\big)$-diagram

$$\begin{array}{ccc} (\hat{\mathfrak{t}}, \Delta_{\ell+1}) & \xrightarrow{(\mathrm{id}_{\mathfrak{t}}, \delta^\ell_{\ell+1})} & (\mathfrak{t}, \Delta_\ell) \\ {\scriptstyle (\hat{\mu}, \hat{\varepsilon})}\uparrow & & \uparrow{\scriptstyle (\mu, \varepsilon)} \\ (\hat{\mathfrak{r}}, \Delta_{m+1}) & \xrightarrow{(\mathrm{id}_{\mathfrak{r}}, \delta^m_{m+1})} & (\mathfrak{r}, \Delta_m) \end{array} \qquad 11.32.23$$

and our induction hypothesis still guarantees that

$$\widehat{\mathfrak{aut}}\big((\mu,\varepsilon)\circ(\mathrm{id}_{\mathfrak{r}},\delta^m_{m+1})\big)=\widehat{\mathfrak{aut}}(\mathrm{id}_{\mathfrak{t}},\delta^\ell_{\ell+1})\circ\widehat{\mathfrak{aut}}(\hat\mu,\hat\varepsilon) \qquad 11.32.24.$$

Moreover, from the commutative diagrams 11.32.6 and 11.32.23, it is easily checked that $(\hat\mu,\hat\varepsilon)\circ(\hat\lambda,\hat\delta)=(\hat\nu,\hat\eta)$; but, since U is normal in Q, $\widehat{\mathfrak{aut}}(\hat\lambda,\hat\delta)$ and $\widehat{\mathfrak{aut}}(\hat\nu,\hat\eta)$ are both defined from $\hat\alpha^{(Q,f)}_{(U,i)}$ (cf. 11.23.1), where (U,i) denotes the Brauer (b,G)-pair contained in (Q,f), and, since we assume that $\hat{\mathfrak{r}}(m+1)=U=\hat{\mathfrak{t}}(\ell+1)$, $\widehat{\mathfrak{aut}}(\hat\mu,\hat\varepsilon)$ is defined by the inclusion

$$\hat{\mathcal{F}}_{(b,G)}(\hat{\mathfrak{r}})\subset\hat{\mathcal{F}}_{(b,G)}(\hat{\mathfrak{t}})\subset\hat{\mathcal{F}}_{(b,G)}(U) \qquad 11.32.25;$$

consequently, we get

$$\widehat{\mathfrak{aut}}(\hat\mu,\hat\varepsilon)\circ\widehat{\mathfrak{aut}}(\hat\lambda,\hat\delta)=\widehat{\mathfrak{aut}}\big((\hat\mu,\hat\varepsilon)\circ(\hat\lambda,\hat\delta)\big)=\widehat{\mathfrak{aut}}(\hat\nu,\hat\eta) \qquad 11.32.26.$$

In this case, equality 11.32.17 follows from equalities 11.32.20, 11.32.22, 11.32.24 and 11.32.26.

If $R'=Q$ and $T'=\psi(T)\neq R$, in the argument above replacing $(\mathfrak{r},\Delta_m)$ by $(\mathfrak{t},\Delta_\ell)$ and (λ,δ) by (ν,η), we still may choose the same subgroup U of Q in the definition of $\widehat{\mathfrak{aut}}(\nu,\eta)$; again, we consider the chains

$$\hat{\mathfrak{q}}:\Delta_{n+1}\longrightarrow\big(\mathcal{F}_{(b,G)}\big)^{\mathrm{nc}}\quad\text{and}\quad\hat{\mathfrak{t}}:\Delta_{\ell+1}\longrightarrow\big(\mathcal{F}_{(b,G)}\big)^{\mathrm{nc}} \qquad 11.32.27$$

defined above and therefore equality 11.32.20 still holds; moreover, since we have $R'=Q$, $\widehat{\mathfrak{aut}}(\lambda,\delta)$ coincides with the restriction of the k^*-group isomorphism $\hat{\mathcal{F}}_{(b,G)}(\varphi)$ induced by a suitable element of G.

On the other hand, let $j\in\Delta_m$ be the maximal element such that the $\mathfrak{r}(j\bullet m)$ maps $\mathfrak{r}(j)$ on a proper subgroup R'' of R, so that $\varphi(R'')\subset Q'$; thus, R'' is an $\mathcal{F}_{(b,G)}(\mathfrak{r})$-stable subgroup of $\varphi^{-1}(U)\triangleleft R$ and therefore we can choose an $\mathcal{F}_{(b,G)}(\mathfrak{r})$-stable normal proper subgroup V of R containing R'' and contained in $\varphi^{-1}(U)$; as above, we have the corresponding chains

$$\tilde{\mathfrak{r}}:\Delta_{m+1}\longrightarrow\big(\mathcal{F}_{(b,G)}\big)^{\mathrm{nc}}\quad\text{and}\quad\tilde{\mathfrak{t}}:\Delta_{\ell+1}\longrightarrow\big(\mathcal{F}_{(b,G)}\big)^{\mathrm{nc}} \qquad 11.32.28$$

fulfilling $\tilde{\mathfrak{r}}\circ\delta^m_{j+1}=\mathfrak{r}$ and $\tilde{\mathfrak{t}}\circ\delta^\ell_{\ell+1}=\mathfrak{t}$ and respectively mapping $j+1$ and $\ell+1$ on V. Again, we have evident $\mathfrak{ch}^*\big((\mathcal{F}_{(b,G)})^{\mathrm{nc}}\big)$-morphisms building a commutative $\mathfrak{ch}^*\big((\mathcal{F}_{(b,G)})^{\mathrm{nc}}\big)$-diagram

$$\begin{array}{ccc}(\tilde{\mathfrak{t}},\Delta_{\ell+1}) & \xrightarrow{(\mathrm{id}^{(\tilde{\mathfrak{t}})}_{\mathfrak{t}},\delta^\ell_{\ell+1})} & (\mathfrak{t},\Delta_\ell)\\ {\scriptstyle(\tilde\mu,\tilde\varepsilon)}\big\uparrow & & \big\uparrow{\scriptstyle(\mu,\varepsilon)}\\ (\tilde{\mathfrak{r}},\Delta_{m+1}) & \xrightarrow{(\mathrm{id}^{(\tilde{\mathfrak{r}})}_{\mathfrak{r}},\delta^m_{j+1})} & (\mathfrak{r},\Delta_m)\end{array} \qquad 11.32.29$$

and, since $\widehat{\mathfrak{aut}}(\mathrm{id}^{(\tilde{\mathfrak{r}})}_{\mathfrak{r}},\delta^m_{j+1})=\mathrm{id}_{\mathcal{F}_{(b,G)}(\mathfrak{r})}$, we have defined (cf. 11.32.20)

$$\widehat{\mathfrak{aut}}(\mu,\varepsilon)=\widehat{\mathfrak{aut}}(\mathrm{id}^{(\tilde{\mathfrak{t}})}_{\mathfrak{t}},\delta^\ell_{\ell+1})\circ\widehat{\mathfrak{aut}}(\tilde\mu,\tilde\varepsilon) \qquad 11.32.30.$$

This time, the inverse of the isomorphism $\varphi^{-1}(U) \cong U$ determined by φ induces an $\mathcal{F}_{(b,G)}$-morphism $U \to R$ and we replace the chain $\hat{\mathfrak{r}}$ above by the corresponding chain

$$\hat{\mathfrak{r}}^\circ : \Delta_{m+1} \longrightarrow \big(\mathcal{F}_{(b,G)}\big)^{\mathrm{nc}} \qquad 11.32.31$$

fulfilling $\hat{\mathfrak{r}}^\circ \circ \delta^m_{j+1} = \mathfrak{r}$ and mapping $j+1$ on U; as above, we have evident $\mathfrak{ch}^*\big((\mathcal{F}_{(b,G)})^{\mathrm{nc}}\big)$-morphisms building a commutative $\mathfrak{ch}^*\big((\mathcal{F}_{(b,G)})^{\mathrm{nc}}\big)$-diagram

$$\begin{array}{ccc}
(\hat{\mathfrak{t}}, \Delta_{\ell+1}) & \xrightarrow{(\mathrm{id}_{\mathfrak{t}}^{(\hat{\mathfrak{t}})}, \delta^\ell_{\ell+1})} & (\mathfrak{t}, \Delta_\ell) \\
{\scriptstyle (\hat{\mu}^\circ, \hat{\varepsilon}^\circ)}\uparrow & & \uparrow{\scriptstyle (\mu,\varepsilon)} \\
(\hat{\mathfrak{r}}^\circ, \Delta_{m+1}) & \xrightarrow{(\mathrm{id}_{\mathfrak{r}}^{(\hat{\mathfrak{r}}^\circ)}, \delta^m_{j+1})} & (\mathfrak{r}, \Delta_m) \\
{\scriptstyle (\hat{\lambda}^\circ, \hat{\delta}^\circ)}\uparrow & & \uparrow{\scriptstyle (\lambda,\delta)} \\
(\hat{\mathfrak{q}}, \Delta_{n+1}) & \xrightarrow{(\mathrm{id}_{\mathfrak{q}}^{(\hat{\mathfrak{q}})}, \delta^n_{i+1})} & (\mathfrak{q}, \Delta_n)
\end{array} \qquad 11.32.32.$$

Once again, it is clear that

$$(\hat{\mu}^\circ, \hat{\varepsilon}^\circ) \circ (\hat{\lambda}^\circ, \hat{\delta}^\circ) = (\hat{\nu}, \hat{\eta}) \qquad 11.32.33;$$

moreover, since $\hat{\mathfrak{r}}^\circ(m+1) = R \overset{\varphi}{\cong} Q = \hat{\mathfrak{q}}(n+1)$, $\widehat{\mathfrak{aut}}(\hat{\lambda}^\circ, \hat{\delta}^\circ)$ coincides with the restriction of the k^*-group isomorphism $\hat{\mathcal{F}}_{(b,G)}(\varphi)$ and therefore we have

$$\widehat{\mathfrak{aut}}(\lambda, \delta) = \widehat{\mathfrak{aut}}(\mathrm{id}_{\mathfrak{r}}^{(\hat{\mathfrak{r}}^\circ)}, \delta^m_{j+1}) \circ \widehat{\mathfrak{aut}}(\hat{\lambda}^\circ, \hat{\delta}^\circ) \qquad 11.32.34$$

since $\widehat{\mathfrak{aut}}(\mathrm{id}_{\mathfrak{q}}^{(\hat{\mathfrak{q}})}, \delta^n_{i+1}) = \mathrm{id}_{\hat{\mathcal{F}}_{(b,G)}(\mathfrak{q})}$; similarly, since U is normal in Q, $\widehat{\mathfrak{aut}}(\hat{\nu}, \hat{\eta})$ and $\widehat{\mathfrak{aut}}(\hat{\mu}^\circ, \hat{\varepsilon}^\circ)$ are respectively defined from $\hat{\alpha}^{(Q,f)}_{(U,i)}$ and $\hat{\alpha}^{(Q,f)}_{(U,i)} \circ \hat{\mathcal{F}}_{(b,G)}(\varphi)^{-1}$ (cf. 11.23.1). Consequently, we get

$$\widehat{\mathfrak{aut}}(\hat{\nu}, \hat{\eta}) = \widehat{\mathfrak{aut}}(\hat{\mu}^\circ, \hat{\varepsilon}^\circ) \circ \widehat{\mathfrak{aut}}(\hat{\lambda}^\circ, \hat{\delta}^\circ) \qquad 11.32.35$$

and therefore, by its very definition (cf. 11.32.20), we still get

$$\widehat{\mathfrak{aut}}(\nu, \eta) = \widehat{\mathfrak{aut}}(\mathrm{id}_{\mathfrak{t}}^{(\hat{\mathfrak{t}})}, \delta^\ell_{\ell+1}) \circ \widehat{\mathfrak{aut}}(\hat{\mu}^\circ, \hat{\varepsilon}^\circ) \circ \widehat{\mathfrak{aut}}(\hat{\lambda}^\circ, \hat{\delta}^\circ) \qquad 11.32.36.$$

Finally, the $\hat{\mathcal{F}}_{(b,G)}$-morphism $V \to U$ induced by φ still determines another pair of chains

$$\hat{\hat{\mathfrak{r}}}: \Delta_{m+2} \longrightarrow \big(\mathcal{F}_{(b,G)}\big)^{\mathrm{nc}} \quad \text{and} \quad \hat{\hat{\mathfrak{t}}}: \Delta_{\ell+2} \longrightarrow \big(\mathcal{F}_{(b,G)}\big)^{\mathrm{nc}} \qquad 11.32.37$$

fulfilling $\hat{\hat{\mathfrak{r}}} \circ \delta^{m+1}_{j+2} = \tilde{\mathfrak{r}}$ and $\hat{\hat{\mathfrak{r}}} \circ \delta^{m+1}_{j+1} = \hat{\mathfrak{r}}^\circ$, and analogously $\hat{\hat{\mathfrak{t}}} \circ \delta^{\ell+1}_{\ell+2} = \tilde{\mathfrak{t}}$ and $\hat{\hat{\mathfrak{t}}} \circ \delta^{\ell+1}_{\ell+1} = \hat{\mathfrak{t}}$, and respectively mapping $(j+1) \bullet (j+2)$ and $(\ell+1) \bullet (\ell+2)$ on this homomorphism; as above, we have evident $\mathfrak{ch}^*\big((\mathcal{F}_{(b,G)})^{\mathrm{nc}}\big)$-morphisms building the commutative $\mathfrak{ch}^*\big((\mathcal{F}_{(b,G)})^{\mathrm{nc}}\big)$-diagram

$$\begin{array}{ccccccc}
 & & (\tilde{\mathfrak{t}},\Delta_{\ell+1}) & \xrightarrow{(\mathrm{id}_{\mathfrak{t}}^{(\tilde{\mathfrak{t}})},\delta^{\ell}_{\ell+1})} & & (\mathfrak{t},\Delta_{\ell}) & \\
 & {\scriptstyle(\tilde{\mu},\tilde{\varepsilon})}\nearrow & & & {\scriptstyle(\mu,\varepsilon)}\nearrow & & \\
(\tilde{\mathfrak{r}},\Delta_{m+1}) & \longrightarrow & \Big\uparrow{\scriptstyle(\mathrm{id}_{\tilde{\mathfrak{t}}},\delta^{\ell+1}_{\ell+2})} & & (\mathfrak{r},\Delta_m) & \Big\uparrow{\scriptstyle(\mathrm{id}_{\mathfrak{t}}^{(\hat{\mathfrak{t}})},\delta^{\ell}_{\ell+1})} & \\
{\scriptstyle(\mathrm{id}_{\tilde{\mathfrak{r}}},\delta^{m+1}_{j+2})}\Big\uparrow & & (\hat{\tilde{\mathfrak{t}}},\Delta_{\ell+2}) & \xrightarrow{(\mathrm{id}_{\hat{\mathfrak{t}}},\delta^{\ell+1}_{\ell+1})} & & (\hat{\mathfrak{t}},\Delta_{\ell+1}) & \\
 & {\scriptstyle(\hat{\tilde{\mu}},\hat{\tilde{\varepsilon}})}\nearrow & & & \uparrow & \nearrow{\scriptstyle(\hat{\mu}^{\circ},\hat{\varepsilon}^{\circ})} & \\
(\hat{\tilde{\mathfrak{r}}},\Delta_{m+2}) & & \xrightarrow{(\mathrm{id}_{\hat{\mathfrak{r}}^{\circ}},\delta^{m+1}_{j+1})} & & (\hat{\mathfrak{r}}^{\circ},\Delta_{m+1}) & &
\end{array} \qquad 11.32.38.$$

First of all, it follows from our induction hypothesis that the image in k^*-$\mathfrak{Gr}$ of this diagram is already defined and that it is also a commutative diagram; moreover, since $\mathcal{F}_{(b,G)}(\mathfrak{r})$ stabilizes V, we have

$$\hat{\mathcal{F}}_{(b,G)}(\hat{\tilde{\mathfrak{r}}}) = \hat{\mathcal{F}}_{(b,G)}(\hat{\mathfrak{r}}^{\circ}) \quad \text{and} \quad \widehat{\mathfrak{aut}}(\mathrm{id}_{\hat{\mathfrak{r}}^{\circ}},\delta^{m+1}_{j+1}) = \mathrm{id}_{\mathcal{F}_{(b,G)}(\hat{\mathfrak{r}}^{\circ})} \qquad 11.32.39;$$

consequently, we get (cf. definition 11.32.20 and diagram 11.32.38)

$$\begin{aligned}
&\widehat{\mathfrak{aut}}(\mathrm{id}_{\mathfrak{t}}^{(\hat{\mathfrak{t}})},\delta^{\ell}_{\ell+1}) \circ \widehat{\mathfrak{aut}}(\hat{\mu}^{\circ},\hat{\varepsilon}^{\circ}) \\
&\qquad = \widehat{\mathfrak{aut}}(\mathrm{id}_{\mathfrak{t}}^{(\hat{\mathfrak{t}})},\delta^{\ell}_{\ell+1}) \circ \widehat{\mathfrak{aut}}(\mathrm{id}_{\hat{\mathfrak{t}}},\delta^{\ell+1}_{\ell+1}) \circ \widehat{\mathfrak{aut}}(\hat{\tilde{\mu}},\hat{\tilde{\varepsilon}}) \\
&\qquad = \widehat{\mathfrak{aut}}(\mathrm{id}_{\mathfrak{t}}^{(\tilde{\mathfrak{t}})},\delta^{\ell}_{\ell+1}) \circ \widehat{\mathfrak{aut}}(\mathrm{id}_{\tilde{\mathfrak{t}}},\delta^{\ell+1}_{\ell+2}) \circ \widehat{\mathfrak{aut}}(\hat{\tilde{\mu}},\hat{\tilde{\varepsilon}}) \qquad 11.32.40 \\
&\qquad = \widehat{\mathfrak{aut}}(\mathrm{id}_{\mathfrak{t}}^{(\tilde{\mathfrak{t}})},\delta^{\ell}_{\ell+1}) \circ \widehat{\mathfrak{aut}}(\tilde{\mu},\tilde{\varepsilon}) \circ \widehat{\mathfrak{aut}}(\mathrm{id}_{\tilde{\mathfrak{r}}},\delta^{m+1}_{j+2}) \\
&\qquad = \widehat{\mathfrak{aut}}(\mu,\varepsilon) \circ \widehat{\mathfrak{aut}}(\mathrm{id}_{\tilde{\mathfrak{r}}},\delta^{m+1}_{j+2})
\end{aligned}$$

and therefore, since

$$\widehat{\mathfrak{aut}}(\mathrm{id}_{\mathfrak{r}}^{(\tilde{\mathfrak{r}})},\delta^{m}_{j+1}) = \mathrm{id}_{\mathcal{F}_{(b,G)}(\mathfrak{r})} \quad \text{and} \quad \widehat{\mathfrak{aut}}(\mathrm{id}_{\hat{\mathfrak{r}}''},\delta^{m+1}_{j+1}) = \mathrm{id}_{\mathcal{F}_{(b,G)}(\hat{\mathfrak{r}}'')} \qquad 11.32.41,$$

from equalities 11.32.34, 11.32.36 and 11.32.40 we finally obtain

$$\begin{aligned}
\widehat{\mathfrak{aut}}(\nu,\eta) &= \widehat{\mathfrak{aut}}(\mathrm{id}_{\mathfrak{t}}^{(\hat{\mathfrak{t}})},\delta^{\ell}_{\ell+1}) \circ \widehat{\mathfrak{aut}}(\hat{\mu}^{\circ},\hat{\varepsilon}^{\circ}) \circ \widehat{\mathfrak{aut}}(\hat{\lambda}^{\circ},\hat{\delta}^{\circ}) \\
&= \widehat{\mathfrak{aut}}(\mu,\varepsilon) \circ \widehat{\mathfrak{aut}}(\mathrm{id}_{\tilde{\mathfrak{r}}},\delta^{m+1}_{j+2}) \circ \widehat{\mathfrak{aut}}(\hat{\lambda}^{\circ},\hat{\delta}^{\circ}) \\
&= \widehat{\mathfrak{aut}}(\mu,\varepsilon) \circ \widehat{\mathfrak{aut}}(\mathrm{id}_{\mathfrak{r}}^{(\tilde{\mathfrak{r}})},\delta^{m}_{j+1}) \circ \widehat{\mathfrak{aut}}(\mathrm{id}_{\tilde{\mathfrak{r}}},\delta^{m+1}_{j+2}) \circ \widehat{\mathfrak{aut}}(\hat{\lambda}^{\circ},\hat{\delta}^{\circ}) \qquad 11.32.42. \\
&= \widehat{\mathfrak{aut}}(\mu,\varepsilon) \circ \widehat{\mathfrak{aut}}(\mathrm{id}_{\mathfrak{r}}^{(\hat{\mathfrak{r}}^{\circ})},\delta^{m}_{j+1}) \circ \widehat{\mathfrak{aut}}(\mathrm{id}_{\hat{\mathfrak{r}}^{\circ}},\delta^{m+1}_{j+1}) \circ \widehat{\mathfrak{aut}}(\hat{\lambda}^{\circ},\hat{\delta}^{\circ}) \\
&= \widehat{\mathfrak{aut}}(\mu,\varepsilon) \circ \widehat{\mathfrak{aut}}(\lambda,\delta)
\end{aligned}$$

We are done.

Remark 11.33 If we modify our choice of k^*-isomorphisms in 11.19.2, we clearly obtain an *isomorphic* functor.

Chapter 12

Quotients and normal subcategories in Frobenius P-categories

12.1 Let us come back to our abstract setting. Before going further into the blocks, we need to develop the analogy between our abstract setting and the groups by discussing *homomorphisms, quotients* and *normal* structures. Let P be a finite p-group and $\mathcal{F}$ a P-category; if P' is a second finite p-group and $\mathcal{F}'$ a P'-category, we say that a group homomorphism $\alpha: P \to P'$ is $(\mathcal{F},\mathcal{F}')$-*functorial* whenever, for any pair of subgroups Q and R of P and any $\varphi \in \mathcal{F}(Q,R)$, we have

$$\varphi\big(R \cap \mathrm{Ker}(\alpha)\big) \subset \mathrm{Ker}(\alpha) \qquad 12.1.1$$

and the group homomorphism $\varphi' : \alpha(R) \to \alpha(Q)$ determined by φ belongs to $\mathcal{F}'\big(\alpha(Q),\alpha(R)\big)$. In this case, α determines an evident functor

$$\mathfrak{f}_\alpha : \mathcal{F} \longrightarrow \mathcal{F}' \qquad 12.1.2$$

that we call *Frobenius functor*; clearly, the composition of Frobenius functors is a Frobenius functor. If $P' = P$ and $\mathcal{F}' = \mathcal{F}$ then id_P is obviously $(\mathcal{F},\mathcal{F})$-functorial and, if $\mathcal{F}$ is divisible, any $\sigma \in \mathcal{F}(P)$ is $(\mathcal{F},\mathcal{F})$-functorial and determines an evident *natural* isomorphism $\mathfrak{id}_\mathcal{F} \cong \mathfrak{f}_\sigma$.

12.2 We say that a subgroup U of P is $\mathcal{F}$-*stable*† if $\varphi(Q \cap U) \subset U$ for any subgroup Q of P and any element φ of $\mathcal{F}(P,Q)$; note that, in this case, U is normal in P, and that the $\mathcal{F}$-*stability* is a necessary condition to guarantee that U is the kernel of some $(\mathcal{F},\mathcal{F}')$-functorial homomorphism $\alpha: P \to P'$ where P' is a finite p-group and $\mathcal{F}'$ a P'-category; actually, the next result affirms that, in the Frobenius P-categories, it is also a sufficient condition. *In this chapter, from now on we assume that $\mathcal{F}$ is a Frobenius P-category.*

Proposition 12.3 *Let U be an $\mathcal{F}$-stable subgroup of P and set $\bar{P} = P/U$. There is a Frobenius $\bar{P}$-category $\bar{\mathcal{F}} = \mathcal{F}/U$ — called the U-quotient of $\mathcal{F}$ — such that, for any pair of subgroups $\bar{Q} = Q/U$ and $\bar{R} = R/U$ of $\bar{P}$, $\bar{\mathcal{F}}(\bar{Q},\bar{R})$ is the set of group homomorphisms $\bar{\varphi}: \bar{R} \to \bar{Q}$ induced by the homomorphism in $\mathcal{F}(Q,R)$. Moreover, the canonical homomorphism $P \to \bar{P}$ is $(\mathcal{F},\bar{\mathcal{F}})$-functorial.*

Proof: The above correspondence clearly defines a $\bar{P}$-category $\bar{\mathcal{F}}$; moreover, whenever $\bar{Q} = Q/U$, $\bar{R} = R/U$ and $\bar{T} = T/U$ are subgroups of $\bar{P}$, $\bar{\varphi}$ is an

† We borrow this term from [18, Ch. XII, section 9]; the term “strongly closed” which has been somewhat employed is inadequate: there is no “strong closure”!

element of $\bar{\mathcal{F}}(\bar{Q},\bar{R})$ and $\bar{\theta}:\bar{T}\to\bar{R}$ is a group homomorphism such that $\bar{\varphi}\circ\bar{\theta}$ belongs to $\bar{\mathcal{F}}(\bar{Q},\bar{T})$, then $\bar{\varphi}\circ\bar{\theta}$ can be lifted to some $\psi\in\mathcal{F}(Q,T)$ and in particular $\psi(T)$ is contained in $\varphi(R)$, so that there is a group homomorphism $\theta:R\to T$ fulfilling $\varphi\circ\theta=\psi$, which implies that θ belongs to $\mathcal{F}(R,T)$ since $\mathcal{F}$ is divisible, and therefore $\bar{\theta}$ belongs to $\bar{\mathcal{F}}(\bar{R},\bar{T})$; thus, $\bar{\mathcal{F}}$ is divisible too.

It is clear that $\bar{\mathcal{F}}$ fulfills condition 2.8.1. On the other hand, let Q be a subgroup of P containing U and φ an element of $\mathcal{F}(P,Q)$, set $\bar{Q}=Q/U$ and denote by $\bar{\varphi}:\bar{Q}\to\bar{P}$ the group homomorphism determined by φ; moreover, let $\bar{K}$ be a subgroup of $\mathrm{Aut}(\bar{Q})$ and denote by K the converse image of $\bar{K}$ in the stabilizer $\mathrm{Aut}(Q)_U$ of U in $\mathrm{Aut}(Q)$; although K need not map onto $\bar{K}$, it is clear that $N_{\bar{P}}^{\bar{K}}(\bar{Q})$ is the image of $N_P^K(Q)$. Set $Q'=\varphi(Q)$, $K'={}^{\varphi}K$, $\bar{Q}'=\bar{\varphi}(\bar{Q})$ and $\bar{K}'={}^{\bar{\varphi}}\bar{K}$, and assume that $\bar{Q}'$ is fully $\bar{K}'$-normalized in $\bar{\mathcal{F}}$.

Since $\mathcal{F}\big(P,Q'{\cdot}N_{\bar{P}}^{K'}(Q')\big)$ maps surjectively onto $\bar{\mathcal{F}}\big(\bar{P},\bar{Q}'{\cdot}N_{\bar{P}}^{\bar{K}'}(\bar{Q}')\big)$, it is not difficult to check that Q' is fully K'-normalized in $\mathcal{F}$, and therefore, since $\mathcal{F}$ is a Frobenius P-category, there are an $\mathcal{F}$-morphism

$$\zeta:Q{\cdot}N_{\bar{P}}^{K}(Q)\longrightarrow P \qquad 12.3.1$$

and an element $\chi\in K$ fulfilling $\zeta(u)=\varphi\big(\chi(u)\big)$ for any $u\in Q$; in particular, since $\chi(U)=U$, we get $\zeta(U)=U$, so that ζ determines a group homomorphism

$$\bar{\zeta}:\bar{Q}{\cdot}N_{\bar{P}}^{\bar{K}}(\bar{Q})\longrightarrow P \qquad 12.3.2$$

in $\bar{\mathcal{F}}\big(\bar{P},\bar{Q}{\cdot}N_{\bar{P}}^{\bar{K}}(\bar{Q})\big)$ fulfilling $\bar{\zeta}(\bar{u})=\bar{\varphi}\big(\bar{\chi}(\bar{u})\big)$ for any $\bar{u}\in\bar{Q}$.

It remains to prove that the canonical map $P\to\bar{P}$ is $(\mathcal{F},\bar{\mathcal{F}})$-functorial; since $\bar{\mathcal{F}}$ is divisible, it suffices to prove that, for any subgroup Q of P and any $\mathcal{F}$-morphism $\varphi:Q\to P$, setting $\bar{Q}=Q{\cdot}U/U$ the homomorphism $\bar{\varphi}:\bar{Q}\to\bar{P}$ induced by φ belongs to $\bar{\mathcal{F}}(\bar{P},\bar{Q})$, namely to the image of $\mathcal{F}(P,Q{\cdot}U)$ in the set of group homomorphisms from $\bar{Q}$ to $\bar{P}$. We argue by induction on $|P:Q|$ and on the length $\ell=\ell(\varphi)$ of φ (cf. 5.15), and it is clear that we may assume that $U\not\subset Q$; if $\ell=0$ then $\varphi=\sigma\circ\iota_Q^P$, where $\sigma\in\mathcal{F}(P)$ and $\iota_Q^P:Q\to P$ is the inclusion map (cf. 1.9), and it suffices to consider the group homomorphism $Q{\cdot}U\to P$ induced by σ.

Assume that $\ell\geq 1$, so that we have $\varphi=\iota_R^P\circ\sigma\circ\nu$, where R is an $\mathcal{F}$-essential subgroup of P fully normalized in $\mathcal{F}$, σ is a p'-element of $\mathcal{F}(R)$ and $\nu:Q\to R$ is an $\mathcal{F}$-morphism such that $\iota_R^P\circ\nu$ has length $\ell-1$ (cf. 5.15); thus, it follows from the induction hypothesis that there is an $\mathcal{F}$-morphism $\psi:Q{\cdot}U\to P$ inducing the same homomorphism as $\iota_R^P\circ\nu$ from $\bar{Q}$ to $\bar{P}$; in particular, we have $\bar{\psi}(\bar{Q})\subset\bar{R}$ and therefore we still have $\psi(Q{\cdot}U)\subset R{\cdot}U$; hence, since $\mathcal{F}$ is divisible, there is an $\mathcal{F}$-morphism $\eta:Q{\cdot}U\to R{\cdot}U$ such that $\bar{\eta}(\bar{u})=\bar{\psi}(\bar{u})=\bar{\nu}(\bar{u})$ for any $\bar{u}\in\bar{Q}$. If $U\subset R$ then $\bar{\sigma}$ belongs to $\bar{\mathcal{F}}(\bar{R})$ and therefore $\bar{\varphi}=\bar{\iota}_R\circ\bar{\sigma}\circ\bar{\eta}$ belongs to $\bar{\mathcal{F}}(\bar{P},\bar{Q})$.

Otherwise, denote by H the subgroup of elements of $\mathcal{F}(R)$ acting trivially on $\bar{R}$ and set $T = N_P^H(R)$; since $\mathcal{F}_P(R)$ is a Sylow p-subgroup of $\mathcal{F}(R)$, $\mathcal{F}_T(R)$ is a Sylow p-subgroup of H and, by the *Frattini argument*, we get

$$\mathcal{F}(R) = H{\cdot}N_{\mathcal{F}(R)}\big(\mathcal{F}_T(R)\big) \qquad 12.3.3;$$

but, since R is fully centralized in $\mathcal{F}$ (cf. Proposition 2.11), it follows from statement 2.10.1 that any element in $N_{\mathcal{F}(R)}\big(\mathcal{F}_T(R)\big)$ can be extended to T and actually determines an element of the stabilizer $\mathcal{F}(T)_R$ of R in $\mathcal{F}(T)$, which clearly stabilizes $U \cap T = N_U(R)$; consequently, $N_{\mathcal{F}(R)}\big(\mathcal{F}_T(R)\big)$ normalizes $\mathcal{F}_U(R)$ and therefore, setting

$$S = N_{R\cdot U}(R) = R{\cdot}(U \cap T) \qquad 12.3.4,$$

we still get

$$\mathcal{F}(R) = H{\cdot}N_{\mathcal{F}(R)}\big(\mathcal{F}_S(R)\big) \qquad 12.3.5;$$

again by statement 2.10.1, any element in $N_{\mathcal{F}(R)}\big(\mathcal{F}_S(R)\big)$ can be extended to S and actually determines an element of $\mathcal{F}(S)_R$. Thus, there are $\chi \in H$ and $\tau \in \mathcal{F}(P,S)$ such that $\iota_R^P \circ \sigma = \tau \circ \iota_R^S \circ \chi$; but, since $U \not\subset R$, R is properly contained in S; hence, $\bar{\tau}$ belongs to $\bar{\mathcal{F}}(\bar{P},\bar{S})$ and therefore $\bar{\varphi} = \bar{\tau} \circ \bar{\iota}_R^S \circ \bar{\eta}$ belongs to $\bar{\mathcal{F}}(\bar{P},\bar{Q})$. We are done.

Remark 12.4 Since U is clearly fully normalized in $\mathcal{F}$, $\mathcal{F}_P(U)$ is a Sylow p-subgroup of $\mathcal{F}(U)$ (cf. Proposition 2.11) and therefore $P/\big(U{\cdot}C_P(U)\big)$ is isomorphic to a Sylow p-subgroup of $\tilde{\mathcal{F}}(U)$ (cf. 1.3); thus, if U is $\mathcal{F}$-selfcentralizing (cf. 4.8), $\bar{P}$ is isomorphic to a Sylow p-subgroup of $\tilde{\mathcal{F}}(U)$ and it is easily checked that $\bar{\mathcal{F}}$ coincides with the *Frobenius category* of the group $\tilde{\mathcal{F}}(U)$ (cf. 1.8). More generally, even when U is $\mathcal{F}$-nilcentralized $\bar{\mathcal{F}}$ is the *Frobenius category* of a finite group; indeed, it follows from Theorem 18.6 below that there are a group L and two group homomorphisms

$$\tau_{_U} : P \longrightarrow L \quad \text{and} \quad \pi_{_U} : L \longrightarrow \mathcal{F}(U) \qquad 12.4.1$$

fulfilling (cf. 1.8, 2.14 and 13.2 below)

$$\begin{aligned} \mathrm{Ker}(\tau_{_U}) &= H_{C_{\mathcal{F}}(U)} \quad , \quad \mathrm{Ker}(\pi_{_U}) = \tau_{_U}\big(C_P(U)\big) \\ \mathrm{Im}(\pi_U) &= \mathcal{F}(U) \quad \text{and} \quad \pi_{_U} \circ \tau_{_U} = \kappa_{_U} \end{aligned} \qquad 12.4.2;$$

thus, if U is $\mathcal{F}$-nilcentralized (cf. 4.3), $\tau_{_U}$ is injective (cf. statement 13.2.2) and it is not difficult to check that $\bar{\mathcal{F}}$ coincides with the *Frobenius category* of the group $L/\tau_{_U}(U)$.

Corollary 12.5 *Let U be an $\mathcal{F}$-stable subgroup of P and set $\bar{P} = P/U$ and $\bar{\mathcal{F}} = \mathcal{F}/U$. The converse image in P of an $\bar{\mathcal{F}}$-stable subgroup of $\bar{P}$ is $\mathcal{F}$-stable. In particular, if V is an $\mathcal{F}$-stable subgroup of P then $U{\cdot}V$ is $\mathcal{F}$-stable too.*

Proof: If $\bar{V} = V/U$ is an $\bar{\mathcal{F}}$-stable subgroup of $\bar{P}$, setting $\bar{\bar{P}} = \bar{P}/\bar{V}$ it follows from Proposition 12.3 that we have a Frobenius $\bar{\bar{P}}$-category $\bar{\bar{\mathcal{F}}} = \bar{\mathcal{F}}/\bar{V}$; then, it is clear that the composition of the canonical functors

$$\mathcal{F} \longrightarrow \bar{\mathcal{F}} \longrightarrow \bar{\bar{\mathcal{F}}} \qquad 12.5.1$$

comes from the homomorphism $P \to \bar{\bar{P}} \cong P/V$ which is $(\mathcal{F}, \bar{\bar{\mathcal{F}}})$-functorial.

On the other hand, if V is an $\mathcal{F}$-stable subgroup of P, it is easily checked that $\bar{V} = V{\cdot}U/U$ is $\bar{\mathcal{F}}$-stable. We are done.

12.6 Let us discuss *normality*; it seems reasonable to call $\mathcal{F}$*-normal* any subgroup N of P fully normalized in $\mathcal{F}$ such that $N_{\mathcal{F}}(N) = \mathcal{F}$, and we consider them in chapter 19 when dealing with *solvability*. On the other hand, if P' is an $\mathcal{F}$-stable subgroup of P, we say that a divisible P'-subcategory $\mathcal{F}'$ of $\mathcal{F}$ is *normal* in $\mathcal{F}$ if $\mathcal{F}(P')$ stabilizes $\mathcal{F}'$ and, for any subgroup Q of P' and any $\varphi \in \mathcal{F}(P, Q)$, we have

$$\mathcal{F}'\big(\varphi(Q)\big) = {}^{\varphi}\big(\mathcal{F}'(Q)\big) \qquad 12.6.1;$$

in particular, $\mathcal{F}'(Q)$ is then a normal subgroup of $\mathcal{F}(Q)$. In this case,

12.6.2 *A subgroup Q of P' which is fully K-normalized in $\mathcal{F}$ for a subgroup K of* $\mathrm{Aut}(Q)$, *is fully K-normalized in $\mathcal{F}'$ too.*

Indeed, for any $\mathcal{F}$-morphism $\psi : Q{\cdot}N^K_{P'}(Q) \to P'$, setting $Q' = \psi(Q)$ and $K' = {}^{\psi}K$, the homomorphism $\varphi' : Q' \to P$ mapping $\psi(u)$ on u, for any $u \in Q$, composed with a suitable element $\chi \in K'$ can be extended to some $\mathcal{F}$-morphism $\zeta : Q'{\cdot}N^{K'}_{P'}(Q') \to P$ (cf. condition 2.8.2) and, since P' is $\mathcal{F}$-stable, we get

$$\zeta\big(Q'{\cdot}N^{K'}_{P'}(Q')\big) \subset \big(Q{\cdot}N^K_P(Q)\big) \cap P' = Q{\cdot}N^K_{P'}(Q) \qquad 12.6.3$$

which forces the equality.

12.7 In this chapter, we prove the existence of a normal Frobenius P-subcategory $\mathcal{F}^{\mathfrak{a}}$ of $\mathcal{F}$ — called the *adjoin subcategory* of $\mathcal{F}$† — which is the analogous of $\mathbf{O}^{p'}(G)$ in a finite group G (but $(\mathcal{F}_G)^{\mathfrak{a}}$ need not coincide with $\mathcal{F}_{\mathbf{O}^{p'}(G)}$!). We start by proving a general criterion on normality, which corresponds to the so-called *Frattini argument*.

Proposition 12.8 *Let P' be an $\mathcal{F}$-stable subgroup of P and $\mathcal{F}'$ a divisible P'-subcategory of $\mathcal{F}$. Then, $\mathcal{F}'$ is normal in $\mathcal{F}$ if and only if $\mathcal{F}(P')$ stabilizes $\mathcal{F}'$ and, for any subgroup Q of P' we have*

$$\mathcal{F}(P', Q) = \mathcal{F}(P') \circ \mathcal{F}'(P', Q) \qquad 12.8.1.$$

† The terminology comes from the Chevalley groups.

Proof: First of all, assume that $\mathcal{F}'$ is normal in $\mathcal{F}$, so that $\mathcal{F}(P')$ stabilizes $\mathcal{F}'$. We argue by induction on $|P'\colon Q|$ and may assume that $Q \neq P'$; let $\varphi\colon Q \to P'$ be an $\mathcal{F}$-morphism, set $Q' = \varphi(Q)$ and choose an $\mathcal{F}$-morphism $\psi\colon N_{P'}(Q') \to P$ such that $Q'' = \psi(Q')$ is fully $\mathcal{F}'(Q'')$-normalized in $\mathcal{F}$ (cf. Proposition 2.7). Hence, since $\mathcal{F}$ is a Frobenius P-category, setting $R = N_{P'}(Q)$ it is easily checked that there exists an $\mathcal{F}$-morphism $\zeta : R \to P'$ and $\chi' \in \mathcal{F}'(Q)$ such that $\psi\big(\varphi(u)\big) = \zeta\big(\chi'(u)\big)$ for any $u \in Q$; then, by our induction hypothesis, there are an $\mathcal{F}$-morphism $\zeta'\colon R \to P'$ and μ in $\mathcal{F}(P')$ such that $\zeta = \mu \circ \zeta'$, and therefore, denoting by $\varphi_*\colon Q \cong Q'$ the $\mathcal{F}$-isomorphism determined by φ, we get

$$\psi \circ \varphi_* = \iota_{P'}^{P} \circ \mu \circ (\zeta' \circ \iota_Q^R) \circ \chi' \qquad 12.8.2.$$

Similarly, by the induction hypothesis we still get $\psi = \iota_{P'}^{P} \circ \nu \circ \xi'$ for suitable $\nu \in \mathcal{F}(P')$ and $\xi' \in \mathcal{F}'(P', Q')$, and therefore we have

$$\nu \circ \xi' \circ \varphi_* = \mu \circ (\zeta' \circ \iota_Q^R) \circ \chi' \qquad 12.8.3;$$

that is to say, setting $\lambda = \nu^{-1} \circ \mu$ and denoting by

$$\delta : \mu^{-1}(Q'') \cong \nu^{-1}(Q'') \quad \text{and} \quad \varepsilon : \lambda^{-1}(Q') \cong Q' \qquad 12.8.4,$$

the $\mathcal{F}$-isomorphisms induced by λ, by

$$\xi'_* : Q' \cong \nu^{-1}(Q'') \qquad 12.8.5$$

the isomorphism determined by ξ', and by

$$\alpha' : Q \cong \mu^{-1}(Q'') \qquad 12.8.6$$

the isomorphism mapping $u \in Q$ on $\zeta'\big(\chi'(u)\big)$, we finally obtain $\xi'_* \circ \varphi_* = \delta \circ \alpha'$ and therefore we get

$$\begin{aligned} \varphi &= \iota_{Q'}^{P'} \circ \varphi_* = \iota_{Q'}^{P'} \circ \xi_*'^{-1} \circ \delta \circ \alpha' \\ &= \lambda \circ \iota_{\lambda^{-1}(Q')}^{P'} \circ (\varepsilon^{-1} \circ \xi_*'^{-1} \circ \delta) \circ \alpha' \end{aligned} \qquad 12.8.7,$$

where $\varepsilon^{-1} \circ \xi_*'^{-1} \circ \delta$ belongs to

$$\mathcal{F}'\big(\lambda^{-1}(Q'), \mu^{-1}(Q'')\big) = {}^{\lambda^{-1}}\mathcal{F}'\big(Q', \nu^{-1}(Q'')\big) \qquad 12.8.8$$

since $\mathcal{F}'$ is normal in $\mathcal{F}$.

Conversely, if $\mathcal{F}(P')$ stabilizes $\mathcal{F}'$ and equality 12.8.1 holds, then, for any subgroup Q of P' and any $\mathcal{F}$-morphism $\varphi\colon Q \to P$, we already know that $\varphi = \iota_{P'}^{P} \circ \psi$ for some $\mathcal{F}$-morphism $\psi\colon Q \to P'$ and therefore we have $\psi = \nu \circ \psi'$ for suitable $\nu \in \mathcal{F}(P')$ and $\psi' \in \mathcal{F}'(P', Q)$; thus, we get

$$\mathcal{F}'\big(\varphi(Q)\big) = \mathcal{F}'\Big(\nu\big(\psi'(Q)\big)\Big) = {}^{\nu}\Big(\mathcal{F}'\big(\psi'(Q)\big)\Big) = {}^{\varphi}\big(\mathcal{F}'(Q)\big) \qquad 12.8.9.$$

We are done.

12.9 Actually, the *Frattini argument* can be pushed further till we reach a *full functor*. Explicitly, let P' be an $\mathcal{F}$-stable subgroup of P and $\mathcal{F}'$ a normal Frobenius P'-subcategory of $\mathcal{F}$; consider the full subcategory $\mathcal{F}_{|P'}$ of $\mathcal{F}$ over the set of subgroups of P' — which is a divisible P'-category — and denote by $\mathcal{F}^*_{|P'}$ the subcategory of $\mathcal{F}_{|P'}$ formed by the same objects and by all the isomorphisms, and by $\mathcal{F}'^*$ the corresponding subcategory of $\mathcal{F}'$ (cf. 1.5); finally, denote by $\mathcal{F}^*_{|P'}/\mathcal{F}'^*$ the *quotient category* of $\mathcal{F}^*_{|P'}$ where, for any pair of $\mathcal{F}$-isomorphic subgroups Q and Q' of P', we have

$$(\mathcal{F}^*_{|P'}/\mathcal{F}'^*)(Q',Q) = \mathcal{F}(Q',Q)/\mathcal{F}'(Q) \qquad 12.9.1;$$

note that these quotient sets define indeed a category by the very definition of *normality*.

12.10 Always in order to iterate the *Frattini argument*, we have to consider the subgroups of P *fully highnormalized* in $\mathcal{F}$. We say that a subgroup Q of P is *fully highnormalized* in $\mathcal{F}$ whenever $N_0 = Q$ and $N_i = N_P(N_{i-1})$ for any $i \geq 1$ are fully normalized in $\mathcal{F}$; from Proposition 2.7 it is easily checked that, for any subgroup Q of P, there is $\varphi \in \mathcal{F}(P,Q)$ such that $\varphi(Q)$ is fully highnormalized in $\mathcal{F}$. Then, let us denote by $\mathcal{F}_{\mathcal{F}(P')|P'}$ the divisible P'-category formed by all the homomorphisms induced by the action of the elements in $\mathcal{F}(P')$ over the subgroups of P', and by $\mathcal{H}^*_{\mathcal{F}(P')|P'}$ the subcategory of $\mathcal{F}_{\mathcal{F}(P')|P'}$ formed by the subgroups of P' fully highnormalized in $\mathcal{F}$ and by all the $\mathcal{F}_{\mathcal{F}(P')|P'}$-isomorphisms between them. Similarly, we denote by $\mathcal{H}^*_{\mathcal{F}'(P')|P'}$ the corresponding subcategory and by

$$\mathcal{H}^*_{\mathcal{F}(P')|P'}/\mathcal{H}^*_{\mathcal{F}'(P')|P'} \qquad 12.10.1$$

the corresponding quotient. Note that, for any pair of subgroups Q and Q' of P' fully highnormalized in $\mathcal{F}$, we have the evident inclusion

$$\mathcal{H}^*_{\mathcal{F}(P')|P'}(Q',Q) \subset \mathcal{F}^*_{|P'}(Q',Q) \qquad 12.10.2.$$

Proposition 12.11 *Let P' be an $\mathcal{F}$-stable subgroup of P and $\mathcal{F}'$ a normal Frobenius P'-subcategory of $\mathcal{F}$. The inclusion $\mathcal{H}^*_{\mathcal{F}(P')|P'} \subset \mathcal{F}^*_{|P'}$ induces a full functor*

$$\mathcal{H}^*_{\mathcal{F}(P')|P'}/\mathcal{H}^*_{\mathcal{F}'(P')|P'} \longrightarrow \mathcal{F}^*_{|P'}/\mathcal{F}'^* \qquad 12.11.1$$

and the restricted functor over the $\mathcal{F}$-nilcentralized objects is faithful too.

Proof: Let Q and Q' be $\mathcal{F}$-isomorphic subgroups of P' fully highnormalized in $\mathcal{F}$ and $\theta : Q \cong Q'$ an $\mathcal{F}$-isomorphism; since Q' is fully normalized in $\mathcal{F}$ and $\mathcal{F}'(Q')$ is a normal subgroup of $\mathcal{F}(Q')$, it follows from Proposition 2.11

that Q' is also fully $\mathcal{F}'(Q')$-normalized in $\mathcal{F}$ and therefore, according to condition 2.8.2 and to equality 12.6.1, there are an $\mathcal{F}$-morphism

$$\zeta : N_P^{\mathcal{F}'(Q)}(Q) \longrightarrow P \qquad 12.11.2$$

and an element $\chi \in \mathcal{F}'(Q)$ fulfilling $\theta(u) = \zeta\big(\chi(u)\big)$ for any $u \in Q$; moreover, since Q is fully $\mathcal{F}'(Q)$-normalized in $\mathcal{F}$ too, we have

$$\zeta\big(N_P^{\mathcal{F}'(Q)}(Q)\big) = N_P^{\mathcal{F}'(Q')}(Q') \qquad 12.11.3;$$

finally, since P' is $\mathcal{F}$-stable and we have

$$P' \cap N_P^{\mathcal{F}'(Q)}(Q) = N_{P'}(Q) \quad \text{and} \quad P' \cap N_P^{\mathcal{F}'(Q')}(Q') = N_{P'}(Q') \qquad 12.11.4,$$

it is clear that ζ induces an $\mathcal{F}$-isomorphism

$$\xi : N_{P'}(Q) \cong N_{P'}(Q') \qquad 12.11.5.$$

That is to say, denoting by $\mathcal{F}\big(N_{P'}(Q'), N_{P'}(Q)\big)_{Q',Q}$ the set of $\mathcal{F}$-morphisms $\eta : N_{P'}(Q) \to N_{P'}(Q')$ such that $\eta(Q) = Q'$, we have proved that the map

$$\mathcal{F}\big(N_{P'}(Q'), N_{P'}(Q)\big)_{Q',Q}/\mathcal{F}'\big(N_{P'}(Q)\big)_{Q} \longrightarrow \mathcal{F}(Q',Q)/\mathcal{F}'(Q) \qquad 12.11.6$$

induced by the restriction is surjective; consequently, since $\mathcal{F}(P',P')_{Q',Q}$ maps $N_{P'}(Q)$ onto $N_{P'}(Q')$ and, by the very definition of $\mathcal{F}_{\mathcal{F}(P')|P'}$, we have

$$\mathcal{F}(P',P')_{Q',Q} = \mathcal{F}_{\mathcal{F}(P')|P'}(Q',Q) \qquad 12.11.7,$$

it suffices to argue by induction on $|P' : Q|$ to get the surjectivity of the map

$$\mathcal{F}_{\mathcal{F}(P')|P'}(Q',Q)/\mathcal{F}_{\mathcal{F}'(P')|P'}(Q) \longrightarrow \mathcal{F}(Q',Q)/\mathcal{F}'(Q) \qquad 12.11.8.$$

Moreover, if $\sigma \in \mathcal{F}(P')$ induces an isomorphism $\theta : Q \cong Q'$ belonging to $\mathcal{F}'(Q',Q)$, then, since Q and Q' are also fully highnormalized in $\mathcal{F}'$ by statement 12.6.2, and clearly we have

$$^{\theta}\mathcal{F}'_{N_{P'}(Q)}(Q) = \mathcal{F}'_{N_{P'}(Q')}(Q') \qquad 12.11.9,$$

it follows from statement 2.10.1 that there exists an $\mathcal{F}'$-isomorphism $\xi : N_{P'}(Q) \cong N_{P'}(Q')$ extending θ . But, the difference between ξ and the isomorphism induced by σ acts trivially on Q and therefore, according to 2.14

and to Proposition 2.16, it is an automorphism of $N_{P'}(Q)$ in the Frobenius $N_{P'}(Q)$-category

$$N_{P'}(Q)\cdot C_{\mathcal{F}}(Q) = N_{\mathcal{F}}^{\mathcal{F}_{P'}(Q)}(Q) \qquad 12.11.10$$

which coincides with $\mathcal{F}_{N_{P'}(Q)}$ whenever Q is $\mathcal{F}$-nilcentralized (cf. Remark 4.2); hence, this difference belongs to $\mathcal{F}'\big(N_{P'}(Q)\big)$, so that we may assume that ξ coincides with the isomorphism induced by σ. Consequently, once again arguing by induction on $|P':Q|$, we get $\sigma\in\mathcal{F}'(P')$. We are done.

Proposition 12.12 *Let H be a subgroup of* $\mathrm{Aut}(P)$ *containing $\mathcal{F}(P)$ and stabilizing $\mathcal{F}$. If $\bar H = H/\mathcal{F}(P)$ is a p'-group then there is a unique Frobenius P-category $\mathcal{F}^{\bar H}$ such that, for any subgroup Q of P, we have*

$$\mathcal{F}^{\bar H}(P,Q) = H\circ\mathcal{F}(P,Q) \qquad 12.12.1$$

and then $\mathcal{F}$ is normal in $\mathcal{F}^{\bar H}$.

Proof: For any pair of subgroups Q and R of P, we define $\mathcal{F}^{\bar H}(Q,R)$ as the set of group homomorphisms $\varphi: R\to Q$ such that

$$\iota_Q^P\circ\varphi\in H\circ\mathcal{F}(P,R) \qquad 12.12.2;$$

thus, if T is a subgroup of P and $\psi\in\mathcal{F}^{\bar H}(R,T)$, we have $\iota_R^P\circ\psi=\chi^{-1}\circ\eta$ for suitable $\chi\in H$ and $\eta\in\mathcal{F}(P,T)$, so that we still have

$$\iota_{\chi(R)}^P\circ(\chi_{_R}\circ\psi)=\eta \qquad 12.12.3$$

where $\chi_{_R}: R\cong\chi(R)$ denote the group isomorphism determined by χ, and therefore $\chi_{_R}\circ\psi$ belongs to $\mathcal{F}\big(\chi(R),T\big)$; hence, since

$$\begin{gathered}\chi\circ\iota_Q^P\circ(\varphi\circ\psi)=\chi\circ(\iota_Q^P\circ\varphi)\circ(\chi_{_R})^{-1}\circ(\chi_{_R}\circ\psi)\\ \chi\circ(\iota_Q^P\circ\varphi)\circ(\chi_{_R})^{-1}\in H\circ\mathcal{F}\big(P,\chi(R)\big)\end{gathered} \qquad 12.12.4,$$

$\iota_Q^P\circ(\varphi\circ\psi)$ belongs to $H\circ\mathcal{F}(P,T)$; that is to say, $\mathcal{F}^{\bar H}$ is a P-category and we claim that it is a Frobenius P-category too.

Indeed, if $\varphi\circ\theta\in\mathcal{F}^{\bar H}(Q,T)$ for some group homomorphism $\theta: T\to R$ then, for some $\zeta\in H$ and $\nu\in\mathcal{F}(P,T)$, we have

$$\iota_Q^P\circ\varphi\circ\theta=\zeta^{-1}\circ\nu \qquad 12.12.5;$$

moreover, since we have $\iota_Q^P\circ\varphi=\xi\circ\mu$ for suitable $\xi\in H$ and $\mu\in\mathcal{F}(P,R)$, denoting by $\rho: R\cong(\zeta\circ\xi)(R)$ the group isomorphism determined by $\zeta\circ\xi$, we get

$$(\zeta\circ\xi)\circ\mu\circ\rho^{-1}\circ(\rho\circ\theta)=\nu \qquad 12.12.6$$

and therefore, since $\mathcal{F}$ is divisible, $\rho\circ\theta$ belongs to $\mathcal{F}\big((\zeta\circ\xi)(R),T\big)$, so that the group homomorphism

$$\iota^P_{\rho(R)}\circ\rho\circ\theta=(\zeta\circ\xi)\circ\iota^P_R\circ\theta \qquad 12.12.7$$

belongs to $\mathcal{F}(P,T)$. Hence, $\mathcal{F}^{\bar{H}}$ is divisible too.

Since $\mathcal{F}^{\bar{H}}(P)=H$, $\mathcal{F}^{\bar{H}}$ fulfills condition 2.8.1. On the other hand, let Q be a subgroup of P, K a subgroup of $\mathrm{Aut}(Q)$ and φ an element of $\mathcal{F}^{\bar{H}}(P,Q)$ such that $\varphi(Q)$ is fully ${}^{\varphi}K$-normalized in $\mathcal{F}^{\bar{H}}$; thus, there are $\eta\in H$ and $\psi\in\mathcal{F}(P,Q)$ such that $\varphi=\eta\circ\psi$, and it is quite clear that $\psi(Q)$ is fully ${}^{\psi}K$-normalized in $\mathcal{F}^{\bar{H}}$, and *a fortiori* in $\mathcal{F}$ (since $\mathcal{F}^{\bar{H}}$ contains $\mathcal{F}$); consequently, since $\mathcal{F}$ is a Frobenius P-category, there are an $\mathcal{F}$-morphism

$$\zeta: Q{\cdot}N^K_P(Q)\longrightarrow P \qquad 12.12.8$$

and $\chi\in K$ such that $\zeta(u)=\psi\big(\chi(u)\big)$ for any $u\in Q$; hence, setting $\xi=\eta\circ\zeta$, we get $\xi(u)=\varphi\big(\chi(u)\big)$ for any $u\in Q$. It is quite clear that $\mathcal{F}$ is normal in $\mathcal{F}^{\bar{H}}$. We are done.

12.13 Recall that in chapter 5 we have introduced the $\mathcal{F}$*-essential* subgroups of P (cf. 5.7) and, for such a subgroup R, the normal subgroup $X_{\mathcal{F}}(R)$ of $\mathcal{F}(R)$ (cf. Corollary 5.13); in the analogy with the Chevalley groups we have evoked above, the corresponding quotients

$$X_{\tilde{\mathcal{F}}}(R)=X_{\mathcal{F}}(R)/\mathcal{F}_R(R) \qquad 12.13.1$$

play the role of *"Levi subgroups of rank* 1*"*.

Proposition 12.14 *If $\mathcal{F}'$ is a divisible P-category contained in $\mathcal{F}$ and, for any $\mathcal{F}$-essential subgroup R of P, $\mathcal{F}'(R)$ contains $X_{\mathcal{F}}(R)$, then, for any subgroup Q of P, we have*

$$\mathcal{F}(P,Q)=\mathcal{F}(P)\circ\mathcal{F}'(P,Q) \qquad 12.14.1.$$

In particular, for any subgroup K of $\mathrm{Aut}(Q)$, *Q is fully K-normalized in $\mathcal{F}'$ if and only if it is fully K-normalized in $\mathcal{F}$. Moreover, Q is $\mathcal{F}'$-selfcentralizing if and only if it is $\mathcal{F}$-selfcentralizing.*

Proof: Clearly we have

$$\mathcal{F}(P)\circ\mathcal{F}'(P,Q)\subset\mathcal{F}(P,Q) \qquad 12.14.2;$$

conversely, we will prove that any $\varphi\in\mathcal{F}(P,Q)$ belongs to $\mathcal{F}(P)\circ\mathcal{F}'(P,Q)$ arguing by induction on the length $\ell(\varphi)$ introduced in 5.15; since the inclusion $\iota^P_Q:Q\to P$ belongs to $\mathcal{F}'(P,Q)$, we may assume that $\ell(\varphi)\geq 1$ and therefore there are $\psi \in \mathcal{F}(P,Q)$, an $\mathcal{F}$-essential subgroup R of P fully normalized in $\mathcal{F}$, an element η of $\mathcal{F}(R,Q)$ and a p'-element τ of $X_{\mathcal{F}}(R)$ such that (cf. 5.15)

$$\ell(\psi) = \ell(\varphi) - 1 \quad , \quad \varphi = \iota_R^P \circ \tau \circ \eta \quad \text{and} \quad \psi = \iota_R^P \circ \eta \qquad 12.14.3;$$

by our induction hypothesis, we get $\psi = \sigma \circ \psi'$ for suitable $\sigma \in \mathcal{F}(P)$ and $\psi' \in \mathcal{F}'(P,Q)$ and thus, setting $R' = \sigma^{-1}(R)$ and denoting by $\theta \in \mathcal{F}(R',R)$ the isomorphism determined by σ^{-1}, we have $\psi' = \iota_{R'}^P \circ (\theta \circ \eta)$ which implies that $\theta \circ \eta$ belongs to $\mathcal{F}'(R',Q)$ since $\mathcal{F}'$ is divisible; but, denoting by τ' the image of τ in $X_{\mathcal{F}}(R')$ throughout θ, we get

$$\sigma^{-1} \circ \varphi = \iota_{R'}^P \circ \tau' \circ (\theta \circ \eta) \qquad 12.14.4$$

and therefore $\sigma^{-1} \circ \varphi$ belongs to $\mathcal{F}'(P,Q)$.

If Q is fully K-normalized in $\mathcal{F}$ and $\psi : Q{\cdot}N_P^K(Q) \to P$ is an $\mathcal{F}'$-morphism, we have

$$\psi(N_P^K(Q)) = N_P^{K_\psi}(\psi(Q)) \qquad 12.14.5$$

since ψ is an $\mathcal{F}$-morphism too. Conversely, if Q is fully K-normalized in $\mathcal{F}'$ and $\psi : Q{\cdot}N_P^K(Q) \to P$ is an $\mathcal{F}$-morphism, by the above argument we get $\psi = \sigma \circ \psi'$ for a suitable $\sigma \in \mathcal{F}(P)$ and some $\mathcal{F}'$-morphism

$$\psi' : Q{\cdot}N_P^K(Q) \longrightarrow P \qquad 12.14.6$$

and therefore we still get

$$\psi(N_P^K(Q)) = \sigma\Big(N_P^{\psi' K}(\psi'(Q))\Big) = N_P^{\psi K}(\psi(Q)) \qquad 12.14.7.$$

Moreover, if Q is fully centralized in both $\mathcal{F}$ and $\mathcal{F}'$, then Q is either $\mathcal{F}$- or $\mathcal{F}'$-selfcentralizing if and only if $C_P(Q) \subset Q$. We are done.

Corollary 12.15 *If $\mathcal{F}'$ is a Frobenius P-category contained in $\mathcal{F}$ and we have $X_{\mathcal{F}}(Q) \subset \mathcal{F}'(Q)$ for any $\mathcal{F}$-essential subgroup Q of P, then Q is $\mathcal{F}'$-essential too and we have*

$$X_{\mathcal{F}'}(Q) = X_{\mathcal{F}}(Q) \qquad 12.15.1.$$

Moreover, an $\mathcal{F}'$-essential subgroup R of P such that $\mathcal{F}'(R)$ is normal in $\mathcal{F}(R)$ is $\mathcal{F}$-essential.

Proof: If Q is $\mathcal{F}$-essential, it is $\mathcal{F}$-selfcentralizing (cf. Theorem 5.11), and we have (cf. equality 5.13.1)

$$\mathcal{F}'(Q) = X_{\mathcal{F}}(Q){\cdot}\mathcal{F}'(Q)_{\overline{\iota_Q^P}} \qquad 12.15.2;$$

thus, the stabilizer $\mathcal{F}'(Q)_{\overline{\iota_Q^P}}$ of $\overline{\iota_Q^P}$ is a proper subgroup of $\mathcal{F}'(Q)$ fulfilling the corresponding condition 5.11.2; moreover, Q is $\mathcal{F}'$-selfcentralizing by Proposition 12.14; hence, by Theorem 5.11 and Corollary 5.13, Q is $\mathcal{F}'$-essential too and we have $X_{\mathcal{F}'}(Q) = X_{\mathcal{F}}(Q)$.

Conversely, assume that Q is $\mathcal{F}'$-essential and that $\mathcal{F}'(Q) \triangleleft \mathcal{F}(Q)$; then, Q is $\mathcal{F}'$-selfcentralizing (cf. 5.11.1), so that it is $\mathcal{F}$-selfcentralizing by Proposition 12.14. Moreover, let M' be a proper subgroup of $\mathcal{F}'(Q)$ fulfilling condition 5.11.2; then, the *Frattini argument* proves that $N_{\mathcal{F}(Q)}(M')$ is a proper subgroup of $\mathcal{F}(Q)$ and it is easily checked that it fulfills the corresponding condition 5.11.2. We are done.

Theorem 12.16 *Let P' be an $\mathcal{F}$-stable subgroup of P and $\mathcal{F}'$ a divisible P'-subcategory of $\mathcal{F}$ such that $\mathcal{F}'_{P'}(P')$ is a Sylow p-subgroup of $\mathcal{F}'(P')$ and that, for any $\mathcal{F}'$-selfcentralizing subgroup Q of P', we have*

$$\mathcal{F}(P',Q) = \mathcal{F}(P') \circ \mathcal{F}'(P',Q) \qquad 12.16.1.$$

Then, $\mathcal{F}'$ is a Frobenius P'-category if and only if it fulfills the following two conditions

12.16.2 *For any element $\sigma \in \mathcal{F}(P')$ which, restricted to some $\mathcal{F}'$-selfcentralizing subgroup Q of P', determines an element of $\mathcal{F}'(P',Q)$, there is z in $C_P(Q)$ such that $\sigma \circ \kappa_{P'}(z)$ belongs to $\mathcal{F}'(P')$.*

12.16.3 *Any divisible P'-category $\mathcal{F}''$ fulfilling $\mathcal{F}''(P',Q) \supset \mathcal{F}'(P',Q)$ for any $\mathcal{F}'$-selfcentralizing subgroup Q of P' contains $\mathcal{F}'$.*

Proof: If $\mathcal{F}'$ is a Frobenius P-category and, for some $\mathcal{F}'$-selfcentralizing subgroup Q of P', the restriction $\rho: Q \to P'$ of $\sigma \in \mathcal{F}(P')$ to Q belongs to $\mathcal{F}'(P',Q)$, then, since clearly we have

$$^{\rho}\mathcal{F}'_{N_{P'}(Q)}(Q) \subset \mathcal{F}'_{P'}\big(\rho(Q)\big) \qquad 12.16.4,$$

it follows from statement 2.10.1 that there is an $\mathcal{F}'$-morphism

$$\zeta' : N_{P'}(Q) \longrightarrow P' \qquad 12.16.5$$

extending ρ. Thus, the restriction $\xi: N_{P'}(Q) \to P'$ of $\sigma \in \mathcal{F}(P')$ to $N_{P'}(Q)$ coincides with ζ' over Q; in particular, since Q is $\mathcal{F}'$-selfcentralizing, we get

$$\zeta'\big(N_{P'}(Q)\big) = \xi\big(N_{P'}(Q)\big) \qquad 12.16.6$$

and then, denoting by

$$\zeta'_* : N_{P'}(Q) \cong \zeta'\big(N_{P'}(Q)\big) \quad \text{and} \quad \xi_* : N_{P'}(Q) \cong \xi\big(N_{P'}(Q)\big) \qquad 12.16.7,$$

the respective isomorphisms determined by ζ' and ξ, $(\xi_*)^{-1} \circ \zeta'_*$ is an $\mathcal{F}$-automorphism of $N_{P'}(Q)$ acting trivially on Q.

But, since $C_{P'}(Q) = Z(Q)$, it follows from Theorem 3.2 in [28, Ch. 5] that, denoting by $\mathcal{F}\big(N_{P'}(Q)\big)_Q$ the stabilizer of Q in $\mathcal{F}\big(N_{P'}(Q)\big)$, the kernel of the restriction map

$$\mathcal{F}\big(N_{P'}(Q)\big)_Q \longrightarrow \mathcal{F}(Q) \qquad 12.16.8$$

is a p-group and thus it is contained in $\mathcal{F}_P\big(N_{P'}(Q)\big)$ (cf. Proposition 2.11). Consequently, there is $z \in C_P(Q)$ fulfilling $\zeta' = \xi \circ \kappa_{N_{P'}(Q)}(z)$ (cf. 1.8).

That is to say, the restriction of the $\mathcal{F}$-automorphism $\sigma \circ \kappa_{P'}(z)$ of P' to $N_{P'}(Q)$ belongs to $\mathcal{F}'\big(P', N_{P'}(Q)\big)$; at this point, arguing by induction on $|P' : Q|$, we get an element

$$z' \in C_P\big(N_{P'}(Q)\big) \subset C_P(Q) \qquad 12.16.9$$

such that $\sigma \circ \kappa_{P'}(zz')$ belongs to $\mathcal{F}'(P')$. On the other hand, if $\mathcal{F}'$ is a Frobenius P-category, condition 12.15.3 follows from Corollary 5.14, arguing by induction on the *length* of $\varphi' \in \mathcal{F}'(P', Q)$ (cf. 5.15).

From now on, we assume that $\mathcal{F}'$ fulfills these conditions; since $\mathcal{F}'_{P'}(P')$ is a Sylow p-subgroup of $\mathcal{F}'(P')$ and condition 12.16.3 implies condition 4.12.3, it suffices to prove that $\mathcal{F}'$ fulfills condition 4.12.2. Let Q be an $\mathcal{F}'$-selfcentralizing subgroup of P', R a subgroup of $N_{P'}(Q)$ which contains Q and $\varphi' : Q \to P'$ an $\mathcal{F}'$-morphism such that ${}^{\varphi'}\mathcal{F}_R(Q)$ is contained in $\mathcal{F}_{P'}\big(\varphi'(Q)\big)$. According to Proposition 2.7, there is an $\mathcal{F}$-morphism $\xi : R \to P$ such that $\bar{Q} = \xi(Q)$ is fully normalized in $\mathcal{F}$.

Set $\bar{R} = \xi(R)$ and denote by $\eta : \bar{Q} \cong Q$ the inverse of the isomorphism induced by ξ; since $\bar{Q}$ is also fully centralized in $\mathcal{F}$ (cf. Proposition 2.11) and φ' is an $\mathcal{F}$-morphism too, there is an $\mathcal{F}$-morphism $\bar{\psi} : \bar{R} \to P$ extending $\varphi' \circ \eta$ (cf. statement 2.10.1) and therefore $\psi = \bar{\psi} \circ \xi$ extends φ'. Then, since we have $\psi(R) \subset P'$, it follows from equality 12.16.1 above that we get $\psi = \iota_{P'}^P \circ \sigma \circ \psi'$ for a suitable $\sigma \in \mathcal{F}(P')$ and some $\mathcal{F}'$-morphism $\psi' : R \to P'$; hence, we still get $\sigma\big(\psi'(u)\big) = \varphi'(u)$ for any $u \in Q$, so that σ induces an $\mathcal{F}'$-isomorphism

$$Q' = \psi'(Q) \cong \varphi'(Q) \qquad 12.16.10.$$

That is to say, the restriction of σ to Q' belongs to $\mathcal{F}'(P', Q')$ and then, according to condition 12.16.2, there is $z \in C_P(Q')$ such that $\sigma \circ \kappa_{P'}(z)$ belongs to $\mathcal{F}'(P')$; finally, we obtain the $\mathcal{F}'$-morphism

$$\sigma \circ \kappa_{P'}(z) \circ \psi' : R \longrightarrow P' \qquad 12.16.11$$

which still extends φ'. We are done.

Corollary 12.17 *The set of Frobenius P-categories $\mathcal{F}'$ contained in $\mathcal{F}$ which fulfill $X_{\mathcal{F}}(Q) \subset \mathcal{F}'(Q)$ for any $\mathcal{F}$-essential subgroup Q of P has a smallest element $\mathcal{F}^{\mathfrak{a}}$. In particular, $\mathcal{F}^{\mathfrak{a}}$ is normal in $\mathcal{F}$, we have $(\mathcal{F}^{\mathfrak{a}})^{\mathfrak{a}} = \mathcal{F}^{\mathfrak{a}}$ and*

$$\mathbf{O}^{p'}\big(\mathcal{F}(R)\big) \subset \mathcal{F}^{\mathfrak{a}}(R) \qquad 12.17.1$$

for any subgroup R of P. Moreover, any divisible P-category contained in $\mathcal{F}$ and containing $\mathcal{F}^{\mathfrak{a}}$ is a Frobenius P-category.

Proof: In order to prove the first statement, it suffices to prove that if $\mathcal{F}'$ and $\mathcal{F}''$ are Frobenius P-categories which are contained in $\mathcal{F}$ and fulfill

$$X_{\mathcal{F}}(Q) \subset \mathcal{F}'(Q) \cap \mathcal{F}''(Q) \qquad 12.17.2$$

for any $\mathcal{F}$-essential subgroup Q of P, then the smallest element of the set of divisible P-categories $\mathcal{F}'''$ which are contained in $\mathcal{F}$ and, for any $\mathcal{F}$-selfcentralizing subgroup Q of P, fulfill

$$\mathcal{F}'''(P,Q) = \mathcal{F}'(P,Q) \cap \mathcal{F}''(P,Q) \qquad 12.17.3$$

is a Frobenius P-category.

Denote by $\mathcal{F}'''$ this smallest element; by Proposition 12.14, $\mathcal{F}'''$ fulfills equality 12.16.1 for any $\mathcal{F}'''$-selfcentralizing subgroup Q of P, and it is clear that $\mathcal{F}'''$ fulfills condition 12.16.3. Assume that $\sigma \in \mathcal{F}(P)$ restricted to Q belongs to $\mathcal{F}'''(P,Q)$; it follows again from Proposition 12.14 that $\mathcal{F}'$ and $\mathcal{F}''$ fulfill equality 12.16.1 and that Q is also $\mathcal{F}'$- and $\mathcal{F}''$-selfcentralizing. Then, it follows from Theorem 12.16 that σ belongs to the intersection

$$\mathcal{F}'(P) \cap \mathcal{F}''(P) = \mathcal{F}'''(P) \qquad 12.17.4;$$

that is to say, $\mathcal{F}'''$ also fulfills condition 12.16.2. Hence, the first statement follows from Theorem 12.16.

Now, since $\mathcal{F}^{\mathfrak{a}}$ is unique, it is clear that $\mathcal{F}(P)$ stabilizes $\mathcal{F}^{\mathfrak{a}}$; moreover, it follows from Proposition 12.14 that $\mathcal{F}^{\mathfrak{a}}$ fulfills condition 12.8.1 and therefore, according to Proposition 12.8, $\mathcal{F}^{\mathfrak{a}}$ is normal in $\mathcal{F}$. In particular, for any subgroup Q of P, $\mathcal{F}^{\mathfrak{a}}(Q)$ is a normal subgroup of $\mathcal{F}(Q)$; then, it follows from Corollary 12.15 that Q is $\mathcal{F}^{\mathfrak{a}}$-essential if and only if it is $\mathcal{F}$-essential and that, in this case, we have $X_{\mathcal{F}^{\mathfrak{a}}}(Q) = X_{\mathcal{F}}(Q)$; consequently, we get $(\mathcal{F}^{\mathfrak{a}})^{\mathfrak{a}} = \mathcal{F}^{\mathfrak{a}}$. On the other hand, it follows from Propositions 2.11 and 12.14 that $|\mathcal{F}^{\mathfrak{a}}(Q)|_p = |\mathcal{F}(Q)|_p$ and therefore, since $\mathcal{F}^{\mathfrak{a}}(Q) \triangleleft \mathcal{F}(Q)$, we still get the inclusion $\mathbf{O}^{p'}\big(\mathcal{F}(Q)\big) \subset \mathcal{F}^{\mathfrak{a}}(Q)$. The last statement follows from Propositions 12.12 and 12.14. We are done.

Corollary 12.18 *If P' is an $\mathcal{F}$-stable subgroup of P and $\mathcal{F}'$ is a normal Frobenius P'-subcategory of $\mathcal{F}$ then $\mathcal{F}'^{\mathfrak{a}}$ is contained in $\mathcal{F}^{\mathfrak{a}}$ and normal in $\mathcal{F}$.*

Proof: For any $\mathcal{F}'$-essential subgroup R of P', inclusion 12.17.1 implies that

$$X_{\mathcal{F}'}(R) \subset \mathbf{O}^{p'}\big(\mathcal{F}'(R)\big) \subset \mathbf{O}^{p'}\big(\mathcal{F}(R)\big) \subset \mathcal{F}^{\mathfrak{a}}(R) \qquad 12.18.1;$$

thus, in order to prove that $\mathcal{F}'^{\mathfrak{a}} \subset \mathcal{F}^{\mathfrak{a}}$, it suffices to prove that the minimal divisible P'-category $\mathcal{F}''$ fulfilling

$$\mathcal{F}''(P',Q) = \mathcal{F}'(P',Q) \cap \mathcal{F}^{\mathfrak{a}}(P',Q) \qquad 12.18.2,$$

for any $\mathcal{F}'$-selfcentralizing subgroup Q of P', is a Frobenius P'-category. Note that, according to Proposition 12.14 and to inclusion 12.18.1, $\mathcal{F}''$ fulfills equality 12.16.1 for any $\mathcal{F}'$-selfcentralizing subgroup of P', and that the minimality of $\mathcal{F}''$ implies condition 12.16.3 with respect to $\mathcal{F}'$.

Consequently, it suffices to prove that $\mathcal{F}''$ fulfills condition 12.16.2 too. Let σ' be an element of $\mathcal{F}'(P')$ which restricted to Q belongs to $\mathcal{F}''(P',Q)$ for some $\mathcal{F}''$-selfcentralizing subgroup Q of P'; setting $Q' = \sigma'(Q)$, consider an $\mathcal{F}^{\mathfrak{a}}$-morphism $\psi : N_{P'}(Q') \to P$ such that $Q'' = \psi(Q')$ is fully normalized in $\mathcal{F}^{\mathfrak{a}}$ (cf. Proposition 2.7); by Proposition 2.11, Q'' is fully centralized too and, since from the action of σ' we get

$$^{\sigma'}\mathcal{F}_{N_{P'}(Q)}(Q) = \mathcal{F}_{N_{P'}(Q')}(Q') \qquad 12.18.3,$$

it follows from statement 2.10.1 that there is an $\mathcal{F}^{\mathfrak{a}}$-morphism $\zeta : N_{P'}(Q) \to P$ fulfilling $\zeta(u) = \psi\big(\sigma'(u)\big)$ for any $u \in Q$.

Since $\mathcal{F}'$ is normal in $\mathcal{F}$, it follows from Proposition 12.8 that we still have $C_{P'}(Q'') = Z(Q'')$; hence, according to equality 12.18.3, we get

$$\zeta\big(N_{P'}(Q)\big) = \psi\big(N_{P'}(Q')\big) \qquad 12.18.4$$

and therefore there exists an $\mathcal{F}^{\mathfrak{a}}$-isomorphism

$$\xi : N_{P'}(Q) \cong N_{P'}(Q') \qquad 12.18.5$$

fulfilling $\xi(u) = \sigma'(u)$ for any $u \in Q$; thus, the difference between ξ and the isomorphism $N_{P'}(Q) \cong N_{P'}(Q')$ induced by σ' acts trivially on Q; but, since $C_{P'}(Q) = Z(Q)$, it follows from Theorem 3.4 in [28, Ch.5] that the kernel of the restriction map

$$\mathcal{F}\big(N_{P'}(Q)\big)_Q \longrightarrow \mathcal{F}(Q) \qquad 12.18.6$$

is a p-group and therefore it is contained in (cf. 12.17.1)

$$\mathbf{O}^{p'}\Big(\mathcal{F}\big(N_{P'}(Q)\big)\Big) \subset \mathcal{F}^{\mathfrak{a}}\big(N_{P'}(Q)\big) \qquad 12.18.7;$$

hence, the restriction of σ' to $N_{P'}(Q)$ belongs to $\mathcal{F}''\big(P', N_{P'}(Q)\big)$. Consequently, arguing by induction on $|P' : Q|$, we get $\sigma' \in \mathcal{F}''(P')$; thus, $\mathcal{F}''$ is a Frobenius P'-category by Theorem 12.16.

On the other hand, since $\mathcal{F}'$ is normal in $\mathcal{F}$, $\mathcal{F}(P')$ stabilizes $\mathcal{F}'$ and therefore it stabilizes $\mathcal{F}'^{\mathfrak{a}}$ too; moreover, since $\mathcal{F}'^{\mathfrak{a}}$ is normal in $\mathcal{F}'$, for any subgroup Q of P' we have (cf. Proposition 12.8)

$$\mathcal{F}(P',Q) = \mathcal{F}(P') \circ \mathcal{F}'(P',Q) = \mathcal{F}(P') \circ \mathcal{F}'^{\mathfrak{a}}(P',Q) \qquad 12.18.8;$$

now, it follows from Proposition 12.8 that $\mathcal{F}'^{\mathfrak{a}}$ is also normal in $\mathcal{F}$. We are done.

Corollary 12.19 *If U is an $\mathcal{F}$-stable subgroup of P then, setting $\bar{P} = P/U$ and $\bar{\mathcal{F}} = \mathcal{F}/U$, the image of $\mathcal{F}^{\mathfrak{a}}$ in $\bar{\mathcal{F}}$ contains $\bar{\mathcal{F}}^{\mathfrak{a}}$.*

Proof: On one hand, it is clear that U is also $\mathcal{F}^{\mathfrak{a}}$-stable, and the image of $\mathcal{F}^{\mathfrak{a}}$ in $\bar{\mathcal{F}}$ is just the U-quotient $\mathcal{F}^{\mathfrak{a}}/U$ of $\mathcal{F}^{\mathfrak{a}}$, which is a Frobenius $\bar{P}$-category by Proposition 12.3; on the other hand, for any subgroup Q of P containing U, $\mathbf{O}^{p'}\big(\mathcal{F}(Q)\big)$ maps onto $\mathbf{O}^{p'}\big(\bar{\mathcal{F}}(\bar{Q})\big)$, where $\bar{Q} = Q/U$, and therefore, according to inclusion 12.17.1, $(\mathcal{F}^{\mathfrak{a}}/U)(\bar{Q})$ contains $X_{\bar{\mathcal{F}}}(\bar{Q})$ whenever $\bar{Q}$ is $\bar{\mathcal{F}}$-essential. We are done.

12.20 We are ready to define the *simple Frobenius P-categories*; let us say that $\mathcal{F}$ is *simple* if $\mathcal{F} = \mathcal{F}^{\mathfrak{a}}$ and there is no nontrivial proper $\mathcal{F}$-stable subgroup in P; then, we have

12.20.1 *A finite group G such that $\mathcal{F}_G$ is simple and we have $\mathbf{O}_{p'}(G) = \{1\}$ is simple.*

Indeed, if P is a Sylow p-subgroup of G then, for any $N \triangleleft G$, $P \cap N$ is an $\mathcal{F}_G$-stable subgroup of P and therefore, if $\mathcal{F}_G$ is simple and $P \cap N \neq \{1\}$, N contains P which implies that N contains $\mathbf{O}^{p'}(G)$; thus, since we are assuming that $\mathcal{F}_G = (\mathcal{F}_G)^{\mathfrak{a}}$, N coincides with G.

Chapter 13

The hyperfocal subcategory of a Frobenius P-category

13.1 Let P be a finite p-group and $\mathcal{F}$ a Frobenius P-category. Denoting by $\mathfrak{i}_{\mathcal{F}}$ the *inclusion* functor from $\mathcal{F}$ to the category $\mathfrak{Gr}$ of groups, it is clear that we have a canonical surjective homomorphism from P to the *direct limit* of $\mathfrak{i}_{\mathcal{F}}$, and we call *$\mathcal{F}$-focal subgroup* of P the kernel $F_{\mathcal{F}}$ of this homomorphism

$$P \longrightarrow \varinjlim \mathfrak{i}_{\mathcal{F}} \qquad 13.1.1;$$

actually, it easily follows from Corollary 5.14 that $F_{\mathcal{F}}$ is generated by the union of the sets $\{u^{-1}\sigma(u)\}_{u\in Q}$ where Q runs over the set of subgroups of P and σ over $\mathcal{F}(Q)$. In the case we consider the Frobenius P-category $\mathcal{F}_G$ associated with a finite group G having P as a Sylow p-subgroup (cf. 1.8), it is well-known (for instance, see Theorems 3.1 and 4.1 in [28, Ch.7]) that $P/F_{\mathcal{F}_G}$ is canonically isomorphic to the maximal Abelian p-quotient of G .

13.2 The iteration of the above construction in the finite groups leads us to introduce the *$\mathcal{F}$-hyperfocal subgroup* as the subgroup $H_{\mathcal{F}}$ of P generated by the union of the sets $\{u^{-1}\sigma(u)\}_{u\in Q}$ where Q runs over the set of subgroups of P and σ over the set of *p'-elements* of $\mathcal{F}(Q)$. Indeed, it is clear that in a chain of normal subgroups of G starting on G and having successive Abelian p-quotients, all the terms contain all the p'-elements of G , and then, it is not difficult to prove that

$$H_{\mathcal{F}_G} = P \cap \mathbf{O}^p(G) \qquad 13.2.1.$$

In our abstract setting, it follows from Proposition 2.11 that

13.2.2 $H_{\mathcal{F}} = \{1\}$ *if and only if* $\mathcal{F} = \mathcal{F}_P$.

Moreover, it is clear that $\mathcal{F}(P)$ stabilizes $H_{\mathcal{F}}$ and that $\mathbf{O}^p\big(\mathcal{F}(P)\big)$ acts trivially on the quotient $P/H_{\mathcal{F}}$; more precisely, we have the following result.

Lemma 13.3 *For any subgroup Q of P and any $\varphi \in \mathcal{F}(P,Q)$, there is $w \in P$ such that, for any $u \in Q$, we have*

$$\varphi(u) \equiv u^w \mod H_{\mathcal{F}} \qquad 13.3.1.$$

In particular, we have $F_{\mathcal{F}} = H_{\mathcal{F}}{\cdot}[P,P]$.

Proof: We argue by induction on the length ℓ of φ (cf. 5.15); since $\mathbf{O}^p\big(\mathcal{F}(P)\big)$ acts trivially on the quotient $P/H_{\mathcal{F}}$, we may assume that $\ell \ge 1$ and therefore that we have $\varphi = \iota_R^P \circ \sigma \circ \nu$, where R is an $\mathcal{F}$-essential subgroup of P , σ is a p'-element of $\mathcal{F}(R)$ and ν is an element of $\mathcal{F}(R,Q)$ such that $\iota_R^P \circ \nu$ has a length equal to $\ell - 1$ (cf. 5.15); then, according to our induction hypothesis, there is $w \in P$ such that, for any $u \in Q$, we get

$$\varphi(u) \equiv \sigma\big(\nu(u)\big) \equiv \nu(u) \equiv u^w \mod H_{\mathcal{F}} \qquad 13.3.2.$$

13.4. Thus, *any normal subgroup P' of P containing the hyperfocal subgroup $H_{\mathcal{F}}$ is $\mathcal{F}$-stable* and, setting $\bar{P} = P/P'$, the corresponding P'-quotient $\bar{\mathcal{F}} = \mathcal{F}/P'$ clearly is the *Frobenius category* associated with the group $\bar{P}$ itself (cf. 1.8). In this chapter, we prove the existence of a Frobenius $H_{\mathcal{F}}$-subcategory $\mathcal{F}^{\mathfrak{h}}$ of $\mathcal{F}$ — called the *hyperfocal* subcategory of $\mathcal{F}$ — which, whenever $\mathcal{F}$ is the Frobenius category associated with a finite group G, is the Frobenius category associated with the group $\mathbf{O}^p(G)$. Firstly, we need the following result.

Lemma 13.5 *For any subgroup Q of P fully normalized in $\mathcal{F}$, $\mathcal{F}_{H_{\mathcal{F}}}(Q)$ contains a Sylow p-subgroup of $\mathbf{O}^p\big(\mathcal{F}(Q)\big)$.*

Proof: We already know that $\mathcal{F}_P(Q)$ is a Sylow p-subgroup of $\mathcal{F}(Q)$ (cf. Proposition 2.11) and that the intersection $\mathcal{F}_P(Q) \cap \mathbf{O}^p\big(\mathcal{F}(Q)\big)$ is generated by the union of the sets $\big[\mathcal{F}_R(Q), \sigma\big]$ where R runs over the set of subgroups of $N_P(Q)$ and σ over the set of p'-elements of $N_{\mathcal{F}(Q)}\big(\mathcal{F}_R(Q)\big)$ (cf. 13.2.1); but, since Q is fully centralized too (cf. Proposition 2.11), it follows from statement 2.10.1 that such a σ can be lifted to a p'-element τ of $\mathcal{F}(R)_Q$ and then $\big[\mathcal{F}_R(Q), \sigma\big]$ is the image of the set $\big\{v^{-1}\tau^{-1}(v)\big\}_{v\in R}$ which is contained in $H_{\mathcal{F}}$. We are done.

Theorem 13.6 *Let P' be a normal subgroup of P containing $H_{\mathcal{F}}$. Then, we have a normal Frobenius P'-subcategory $\mathcal{F}'$ of $\mathcal{F}$ such that*

$$\mathcal{F}'(Q') = \mathcal{F}_{P'}(Q') \circ \mathbf{O}^p\big(\mathcal{F}(Q')\big) \qquad 13.6.1$$

for any subgroup Q' of P' fully normalized in $\mathcal{F}$.

Proof:† For any subgroup Q of P', choose $\eta \in \mathcal{F}(P,Q)$ such that $Q' = \eta(Q)$ is fully normalized in $\mathcal{F}$ (cf. Proposition 2.7) and denote by $\mathcal{F}'(Q)$ the subgroup of $\mathcal{F}(Q)$ defined by

$$^{\eta}\mathcal{F}'(Q) = \mathcal{F}_{P'}(Q') \circ \mathbf{O}^p\big(\mathcal{F}(Q')\big) \qquad 13.6.2;$$

note that, since $\mathcal{F}_{H_{\mathcal{F}}}(Q')$ contains a Sylow p-subgroup of $\mathbf{O}^p\big(\mathcal{F}(Q')\big)$ by Lemma 13.5, we have

13.6.3 *$\mathcal{F}_{P'}(Q')$ is a Sylow p-subgroup of $\mathcal{F}'(Q')$*

and then we claim that $\mathcal{F}'(Q)$ does not depend on the choice of η.

Indeed, for another choice $\eta' \in \mathcal{F}(P,Q)$, setting $Q'' = \eta'(Q)$ and denoting by $\sigma : Q' \cong Q''$ the corresponding group isomorphism, we obviously have

$$^{\sigma}\mathbf{O}^p\big(\mathcal{F}(Q')\big) = \mathbf{O}^p\big(\mathcal{F}(Q'')\big) \qquad 13.6.4$$

† Our proof of this result in [52, Theorem 7.4] seems not correct.

and it follows from condition 2.8.2 that, for a suitable $\chi' \in \mathcal{F}(Q')$, there is an $\mathcal{F}$-morphism $\zeta' : N_P(Q') \to P$ extending $\sigma \circ \chi'$; thus, since Q' is fully normalized in $\mathcal{F}$ and P' is $\mathcal{F}$-stable, we get

$$\zeta'\big(N_{P'}(Q')\big) = N_{P'}(Q'') \qquad 13.6.5$$

and, since $\mathcal{F}'(Q')$ is normal in $\mathcal{F}(Q')$, we still get

$$^{\sigma}\big(\mathcal{F}_{P'}(Q') \circ \mathbf{O}^p\big(\mathcal{F}(Q')\big)\big) = \mathcal{F}_{P'}(Q'') \circ \mathbf{O}^p\big(\mathcal{F}(Q'')\big) \qquad 13.6.6.$$

Consequently, it is easily checked that, for any $\varphi \in \mathcal{F}(P,Q)$, we have

$$^{\varphi}\mathcal{F}'(Q) = \mathcal{F}'\big(\varphi(Q)\big) \qquad 13.6.7.$$

Moreover, if R is a subgroup of Q then we claim that:

13.6.8 *The action over R of the stabilizer $\mathcal{F}'(Q)_R$ of R in $\mathcal{F}'(Q)$ is contained in $\mathcal{F}'(R)$.*

Indeed, according to Proposition 2.7, we may assume that Q is fully normalized in $\mathcal{F}$ and then, according to Proposition 2.11 and to statement 13.6.3, $\mathcal{F}_P(Q)$ and $\mathcal{F}_{P'}(Q)$ are respective Sylow p-subgroups of $\mathcal{F}(Q)$ and $\mathcal{F}'(Q)$; thus, since $\mathcal{F}(Q) = \mathcal{F}_P(Q) \circ \mathcal{F}'(Q)$, up to $\mathcal{F}'(Q)$-conjugation we may assume that $\mathcal{F}_P(Q)_R$ is a Sylow p-subgroup of $\mathcal{F}(Q)_R$ and then, since

$$\mathcal{F}_P(Q)_R \cap \mathcal{F}'(Q) \subset \mathcal{F}_{P'}(Q) \qquad 13.6.9,$$

$\mathcal{F}_{P'}(Q)_R$ is a Sylow p-subgroup of $\mathcal{F}'(Q)_R$, so that we have

$$\mathcal{F}'(Q)_R = \mathcal{F}_{P'}(Q)_R \circ \mathbf{O}^p\big(\mathcal{F}'(Q)_R\big) \qquad 13.6.10;$$

but, the actions of $\mathcal{F}_{P'}(Q)_R$ and of $\mathbf{O}^p\big(\mathcal{F}'(Q)_R\big)$ over R are respectively contained in $\mathcal{F}_{P'}(R)$ and in $\mathbf{O}^p\big(\mathcal{F}(R)\big)$, so in $\mathcal{F}'(R)$.

For any pair Q and Q' of $\mathcal{F}$-isomorphic subgroups of P', consider the set $\mathcal{F}'(Q',Q)$ of $\varphi \in \mathcal{F}(Q',Q)$ fulfilling the following condition:

13.6.11 *There are two sequences $\{Q_i\}_{0\le i\le n}$ and $\{U_i\}_{1\le i\le n}$ of subgroups of P' such that*

$$Q_0 = Q \quad , \quad Q_n = Q' \quad \textit{and} \quad Q_{i-1}, Q_i \subset U_i \textit{ for } 1 \le i \le n \ ,$$

and a sequence $\{\varphi_i\}_{1\le i\le n}$ of $\mathcal{F}$-isomorphisms $\varphi_i : Q_{i-1} \cong Q_i$ induced by $\mathcal{F}'(U_i)$ such that their composition coincides with φ.

Note that φ^{-1} belongs to $\mathcal{F}'(Q,Q')$ for any $\varphi \in \mathcal{F}'(Q',Q)$, and that $\mathcal{F}'(Q)$ and $\mathcal{F}'(Q,Q)$ have been defined independently; moreover, it is clear that

$$\mathcal{F}'(Q) \subset \mathcal{F}'(Q,Q) \quad \text{and} \quad \mathcal{F}_{P'}(Q',Q) \subset \mathcal{F}'(Q',Q) \qquad 13.6.12.$$

Actually, by the tautological equality ${}^{\varphi}\sigma \circ \varphi = \varphi \circ \sigma$ for any $\sigma \in \mathcal{F}(Q)$, it follows from equality 13.6.7 and statement 13.6.8 that we always may assume that $Q_0 = U_1 = Q_1$ and that $Q_i \neq Q_{i+1}$ for any $1 \leq i < n$.

More generally, for any pair of subgroups Q and R of P' we set

$$\mathcal{F}'(Q,R) = \bigcup_{\varphi \in \mathcal{F}(Q,R)} \iota_{\varphi(R)}^{Q} \circ \mathcal{F}'\big(\varphi(R), R\big) \qquad 13.6.13$$

and we will prove that this correspondence $\mathcal{F}'$ is a divisible P'-category normal in $\mathcal{F}$ which fulfills the two conditions in Corollary 2.13, and that, for any subgroup Q of P', we have (cf. 13.6.2)

$$\mathcal{F}'(Q) = \mathcal{F}'(Q,Q) \qquad 13.6.14.$$

Actually, it is already clear that $\mathcal{F}'(P') = \mathcal{F}'(P',P')$ and therefore, by statement 13.6.3, that $\mathcal{F}'$ fulfills condition 2.8.1.

If Q, Q' and Q'' are three $\mathcal{F}$-isomorphic subgroups of P', φ an element of $\mathcal{F}'(Q',Q)$ and ψ an element of $\mathcal{F}'(Q'',Q')$, it is quite clear from condition 13.6.11 that $\psi \circ \varphi$ belongs to $\mathcal{F}'(Q'',Q)$. Moreover, if R and R' are subgroups of P' respectively containing Q and Q', and $\psi \in \mathcal{F}'(R',R)$ fulfills $\psi(Q) = Q'$, then it easily follows from the same condition that the $\mathcal{F}$-isomorphism $\varphi : Q \cong Q'$ induced by ψ belongs to $\mathcal{F}'(Q',Q)$.

Consequently, if R and T are subgroups of P', φ an element of $\mathcal{F}'(R,Q)$ and ψ an element of $\mathcal{F}'(T,R)$ then $\psi \circ \varphi$ belongs to $\mathcal{F}'(T,Q)$; indeed, setting $Q' = \varphi(Q)$ and $Q'' = \psi(Q')$, and denoting by $\varphi_* : Q \cong Q'$ and by $\psi_* : Q' \cong Q''$ the corresponding $\mathcal{F}$-isomorphisms, it follows from our definition that φ_* belongs to $\mathcal{F}'(Q',Q)$ and we already have proved that ψ_* and $\psi_* \circ \varphi_*$ respectively belong to $\mathcal{F}'(Q'',Q')$ and to $\mathcal{F}'(Q'',Q)$. This already proves that $\mathcal{F}'$ is a divisible P'-category.

Let Q and Q' be $\mathcal{F}'$-isomorphic subgroups of P' fully normalized and fully centralized in $\mathcal{F}'$; in order to prove that they fulfill conditions 2.13.1 and 2.13.2, we argue by induction on $|P' : Q|$ and, in particular, we may assume that all the subgroups of P' of smaller index fulfill condition 2.8.2 or, equivalently, condition 4.13.1 in Lemma 4.13.

Let $\varphi : Q \cong Q'$ be an $\mathcal{F}'$-morphism and choose two sequences $\{Q_i\}_{0 \leq i \leq n}$ and $\{U_i\}_{1 \leq i \leq n}$ of subgroups of P', and a sequence $\{\varphi_i\}_{1 \leq i \leq n}$ of $\mathcal{F}$-isomorphisms $\varphi_i : Q_{i-1} \cong Q_i$, as in condition 13.6.11; note that, in order to prove condition 2.13.1, in the remark above about our choice we may assume that $\varphi_1 = \{\mathrm{id}_{Q_0}\}$. According to Proposition 2.7, for any $1 \leq i \leq n$ we may choose an $\mathcal{F}'$-morphism

$$\eta_i : N_{P'}(Q_i) \longrightarrow P' \qquad 13.6.15$$

such that $Q'_i = \eta_i(Q_i)$ is fully $\mathcal{F}'(Q'_i)$-normalized in $\mathcal{F}'$, and actually may assume that $Q'_1 = Q$, $Q'_n = Q'$, $\eta_1 = \iota_{N_{P'}(Q)}^{P'}$ and $\eta_n = \iota_{N_{P'}(Q')}^{P'}$.

We claim that, for any $2 \leq i \leq n$, we can apply Lemma 4.13 to the subgroup Q'_{i-1} of P', to the subgroup $\mathcal{F}'(Q'_{i-1})$ of $\mathrm{Aut}(Q'_{i-1})$ and, denoting by η^*_{i-1} the inverse of the $\mathcal{F}'$-isomorphism induced by η_{i-1}

$$N_{P'}(Q_{i-1}) \cong \eta_{i-1}\big(N_{P'}(Q_{i-1})\big) \qquad 13.6.16,$$

to the $\mathcal{F}'$-isomorphism $\varphi'_i : Q'_{i-1} \cong Q'_i$ mapping $u \in Q'_{i-1}$ on $\eta_i\big(\varphi_i(\eta^*_{i-1}(u))\big)$. Indeed, according to condition 13.6.11, φ_i is induced by an element of $\mathcal{F}'(U_i)$ and therefore it is still induced by a suitable $\mathcal{F}'$-isomorphism

$$\psi_i : N_{U_i}(Q_{i-1}) \cong N_{U_i}(Q_i) \qquad 13.6.17;$$

thus, φ'_i can be extended to an $\mathcal{F}'$-isomorphism

$$\eta_{i-1}\big(N_{U_i}(Q_{i-1})\big) \cong \eta_i\big(N_{U_i}(Q_i)\big) \qquad 13.6.18$$

Note that, since we have chosen $Q_{i-1} \neq Q_i$, Q_i is properly contained in U_i and therefore in $N_{U_i}(Q_i)$.

Hence, it follows from this lemma that, for any $2 \leq i \leq n$, there are an $\mathcal{F}'$-morphism

$$\zeta_i : N_{P'}(Q'_{i-1}) \longrightarrow N_{P'}(Q'_i) \qquad 13.6.19$$

and an element $\chi_{i-1} \in \mathcal{F}'(Q'_{i-1})$ such that

$$\zeta_i\big(\chi_{i-1}(u)\big) = \eta_i\big(\varphi_i(\eta^*_{i-1}(u))\big) \qquad 13.6.20$$

for any $u \in Q'_{i-1}$; note that, since ζ_i determines a group isomorphism $\mathcal{F}'(Q'_{i-1}) \cong \mathcal{F}'(Q'_i)$ by equality 13.6.7, denoting by ${}^{\zeta_i}\chi_{i-1}$ the image of χ_{i-1} we still have

$$\zeta_i\big(\chi_{i-1}(u)\big) = {}^{\zeta_i}\chi_{i-1}\big(\zeta_i(u)\big) \qquad 13.6.21.$$

Consequently, the composition of the family $\{\zeta_i\}_{1 \leq i \leq n}$ defines an $\mathcal{F}'$-morphism

$$\zeta : N_{P'}(Q) \longrightarrow N_{P'}(Q') \qquad 13.6.22$$

and denoting by χ the composition of the family of the corresponding images in $\mathcal{F}'(Q)$ of all the elements $\chi_{i-1} \in \mathcal{F}'(Q'_{i-1})$ where $1 \leq i \leq n$, it is not difficult to check that

$$\zeta(u) = \varphi\big(\chi(u)\big) \qquad 13.6.23$$

for any $u \in Q$; in particular, we have $\zeta(Q) = Q'$ proving statement 2.13.1.

Moreover, allowing the possibility $\varphi_1 \neq \{\mathrm{id}\}_{Q_0}$ in the argument above, it still proves that, for any $\mathcal{F}'$-automorphism $\psi : Q \cong Q$ there are an $\mathcal{F}'$-automorphism

$$\xi : N_{P'}(Q) \cong N_{P'}(Q) \qquad 13.6.24$$

and $\theta \in \mathcal{F}'(Q)$ fulfilling $\xi\big(\theta(u)\big) = \psi(u)$ for any $u \in Q$; that is to say, $\mathcal{F}'(Q,Q)$ coincides with the set of compositions of all the elements in $\mathcal{F}'(Q)$ with all the elements of the image in $\mathcal{F}'(Q,Q)$ of the stabilizer of Q in $\mathcal{F}'\big(N_{P'}(Q), N_{P'}(Q)\big)$. But, since $\mathcal{F}'(P') = \mathcal{F}'(P',P')$, arguing by induction on $|P\!:\!Q|$ we may assume that we already have

$$\mathcal{F}'\big(N_{P'}(Q)\big) = \mathcal{F}'\big(N_{P'}(Q), N_{P'}(Q)\big) \qquad 13.6.25.$$

Then, it follows from statement 13.6.8 that this image is contained in $\mathcal{F}'(Q)$, so that we get $\mathcal{F}'(Q) = \mathcal{F}'(Q,Q)$. In particular, it follows from equality 13.6.7 that $\mathcal{F}'$ is normal in $\mathcal{F}$.

It remains to prove condition 2.13.2; let R be a subgroup of $N_{P'}(Q)$ containing $Q{\cdot}C_{P'}(Q)$; since the restriction induces a surjective group homomorphism (cf. statement 2.10.1)

$$\mathcal{F}(R)_Q \longrightarrow N_{\mathcal{F}(Q)}\big(\mathcal{F}_R(Q)\big) \qquad 13.6.26,$$

it maps $\mathbf{O}^p\big(\mathcal{F}(R)_Q\big)$ onto $\mathbf{O}^p\Big(N_{\mathcal{F}(Q)}\big(\mathcal{F}_R(Q)\big)\Big)$; but, we have

$$\mathbf{O}^p\big(\mathcal{F}(R)_Q\big) \subset \mathbf{O}^p\big(\mathcal{F}(R)\big) \subset \mathcal{F}'(R) \qquad 13.6.27$$

and therefore we still have $\mathbf{O}^p\big(\mathcal{F}(R)_Q\big) \subset \mathcal{F}'(R)_Q$; similarly, we get

$$\mathbf{O}^p\Big(N_{\mathcal{F}(Q)}\big(\mathcal{F}_R(Q)\big)\Big) \subset \mathbf{O}^p\big(\mathcal{F}(Q)\big) \subset \mathcal{F}'(Q) \qquad 13.6.28$$

and therefore we still get

$$\mathbf{O}^p\Big(N_{\mathcal{F}(Q)}\big(\mathcal{F}_R(Q)\big)\Big) \subset N_{\mathcal{F}'(Q)}\big(\mathcal{F}_R(Q)\big) \qquad 13.6.29$$

Hence, the image of the group homomorphism

$$\mathcal{F}'(R)_Q \longrightarrow N_{\mathcal{F}'(Q)}\big(\mathcal{F}_R(Q)\big) \qquad 13.6.30$$

contains $\mathbf{O}^p\Big(N_{\mathcal{F}'(Q)}\big(\mathcal{F}_R(Q)\big)\Big)$.

Consequently, in order to prove that homomorphism 13.6.30 is surjective, it suffices to prove that its image still contains a Sylow p-subgroup of $N_{\mathcal{F}'(Q)}\big(\mathcal{F}_R(Q)\big)$. But, according to statement 13.6.3, $\mathcal{F}_{P'}(Q)$ is a Sylow p-subgroup of $\mathcal{F}'(Q)$ and therefore it contains a Sylow p-subgroup of an $\mathcal{F}'(Q)$-conjugate of $N_{\mathcal{F}'(Q)}\big(\mathcal{F}_R(Q)\big)$; thus, since $\mathcal{F}$ is a Frobenius P-category and $\mathcal{F}'$ is normal in $\mathcal{F}$, we can choose an $\mathcal{F}$-morphism

$$\rho : N_{P'}(Q) \longrightarrow P \qquad 13.6.31$$

such that, setting $Q'' = \rho(Q)$ and $R'' = \rho(R)$, Q'' is fully normalized in $\mathcal{F}$ and in $\mathcal{F}'$ (cf. Proposition 2.7), R'' contains $\rho\big(C_{P'}(Q)\big) = C_{P'}(Q'')$

and $N_{\mathcal{F}_{P'}(Q'')}\big(\mathcal{F}_{R''}(Q'')\big)$ is a Sylow p-subgroup of $N_{\mathcal{F}'(Q'')}\big(\mathcal{F}_{R''}(Q'')\big)$ (cf. Proposition 2.11). On the other hand, since $C_{P'}(Q'') \subset R''$, $\mathcal{F}_{P'}(R'')_{Q''}$ and $N_{\mathcal{F}_{P'}(Q'')}\big(\mathcal{F}_{R''}(Q'')\big)$ are both images of $N_{P'}(R'') \cap N_{P'}(Q'')$ and therefore the subgroup $\mathcal{F}_{P'}(R'')_{Q''}$ of $\mathcal{F}'(R'')_{Q''}$ maps onto $N_{\mathcal{F}_{P'}(Q'')}\big(\mathcal{F}_{R''}(Q'')\big)$. We are done.

Remark 13.7† As a matter of fact, the argument above still proves the following more general fact.

Theorem *Let P' be an $\mathcal{F}$-stable subgroup of P and, for any subgroup Q of P', $\mathcal{F}'(Q)$ a (normal) subgroup of $\mathcal{F}(Q)$ in such a way that the following three conditions hold*

13.7.1 *$\mathcal{F}_{P'}(P')$ is a Sylow p-subgroup of $\mathcal{F}'(P')$.*

13.7.2 *For any subgroup Q of P' and any element $\varphi \in \mathcal{F}(P,Q)$, we have*

$$^{\varphi}\mathcal{F}'(Q) = \mathcal{F}'\big(\varphi(Q)\big).$$

13.7.3 *For any pair of subgroups Q and R of P' such that $C_{P'}(R) \subset Q$ and $R \lhd Q$, the restriction induces a surjective group homomorphism*

$$\mathcal{F}'(Q)_R \longrightarrow N_{\mathcal{F}'(R)}\big(\mathcal{F}_Q(R)\big)$$

where $\mathcal{F}'(Q)_R$ denotes the stabilizer of Q in $\mathcal{F}'(Q)$.
Then, there exists a normal Frobenius P'-subcategory $\mathcal{F}'$ of $\mathcal{F}$ such that $\mathcal{F}'(Q,Q) = \mathcal{F}'(Q)$ for any subgroup Q of P'.

In the corresponding argument, statement 13.6.8 is replaced by the *existence* of the group homomorphism in condition 13.7.3; moreover, the coincidence of condition 13.7.3 with condition 2.13.2 avoids the last part of the proof.

13.8 With the notation of Theorem 13.6, $\mathcal{F}'$ is uniquely determined by condition 13.6.1 and it follows from Corollary 5.14 that if P'' is a normal subgroup of P such that $H_{\mathcal{F}} \subset P'' \subset P'$ then the Frobenius P''-category fulfilling condition 13.6.1 is contained in $\mathcal{F}'$. In particular, whenever $P' = H_{\mathcal{F}}$ we set $\mathcal{F}' = \mathcal{F}^{\mathfrak{h}}$ and call *hyperfocal subcategory* of $\mathcal{F}$ this subcategory $\mathcal{F}^{\mathfrak{h}}$; note that

$$H_{\mathcal{F}^{\mathfrak{h}}} = H_{\mathcal{F}} \quad \text{and} \quad (\mathcal{F}^{\mathfrak{h}})^{\mathfrak{h}} = \mathcal{F}^{\mathfrak{h}} \qquad 13.8.1.$$

Proposition 13.9 *If P' is a subgroup of P and $\mathcal{F}'$ is a Frobenius P'-subcategory of $\mathcal{F}$ then we have*

$$H_{\mathcal{F}'} \subset H_{\mathcal{F}} \quad \textit{and} \quad \mathcal{F}'^{\mathfrak{h}} \subset \mathcal{F}^{\mathfrak{h}} \qquad 13.9.1.$$

Moreover, if P' is $\mathcal{F}$-stable and $\mathcal{F}'$ is normal in $\mathcal{F}$ then $H_{\mathcal{F}'}$ is also $\mathcal{F}$-stable and $\mathcal{F}'^{\mathfrak{h}}$ is still normal in $\mathcal{F}$.

---

† This remark has been motivated by Michael Aschbacher's preprint *"Normal subsystems of fusion systems"* [4].

Proof: The left inclusion is easily checked; moreover it follows from Proposition 2.7 that, for any subgroup Q of $H_{\mathcal{F}'}$, we can choose $\varphi \in \mathcal{F}'(P',Q)$ such that $Q' = \varphi(Q)$ is fully normalized in $\mathcal{F}'$; then, since Q' is also fully normalized in $\mathcal{F}'^{\mathfrak{h}}$ (cf. statement 12.6.2), we have (cf. equality 13.6.1)

$$\begin{aligned}\mathcal{F}'^{\mathfrak{h}}(Q') &= \mathcal{F}_{H_{\mathcal{F}'}}(Q')\mathbf{O}^p\big(\mathcal{F}'(Q')\big)\\ &\subset \mathcal{F}_{H_{\mathcal{F}}}(Q')\cdot\mathbf{O}^p\big(\mathcal{F}(Q')\big) \subset \mathcal{F}^{\mathfrak{h}}(Q')\end{aligned} \qquad 13.9.2$$

and therefore we get $\mathcal{F}'^{\mathfrak{h}}(Q) \subset \mathcal{F}^{\mathfrak{h}}(Q)$; now, the inclusion $\mathcal{F}'^{\mathfrak{h}} \subset \mathcal{F}^{\mathfrak{h}}$ follows from Corollary 5.14.

On the other hand, if P' is $\mathcal{F}$-stable and $\mathcal{F}'$ is normal in $\mathcal{F}$, it is clear that $\mathcal{F}(P')$ stabilizes $H_{\mathcal{F}'}$ and it also stabilizes $\mathcal{F}'^{\mathfrak{h}}$ by Theorem 13.6; moreover, it follows from Proposition 12.8 that for any subgroup Q of $H_{\mathcal{F}'}$ we have

$$\mathcal{F}(P',Q) = \mathcal{F}(P') \circ \mathcal{F}'(P',Q) \qquad 13.9.3$$

and therefore, since $\mathcal{F}(P')$ stabilizes $H_{\mathcal{F}'}$, it is easily checked that $H_{\mathcal{F}'}$ is also $\mathcal{F}$-stable and that we still have

$$\mathcal{F}(H_{\mathcal{F}'},Q) = \mathcal{F}(H_{\mathcal{F}'}) \circ \mathcal{F}'(H_{\mathcal{F}'},Q) \qquad 13.9.4;$$

finally, since $\mathcal{F}'^{\mathfrak{h}}$ is normal in $\mathcal{F}'$, the last statement follows from the corresponding equality obtained from equality 13.9.4 and Proposition 12.8. We are done.

Proposition 13.10 *If U is an $\mathcal{F}$-stable subgroup of P, setting $\bar{P} = P/U$ and $\bar{\mathcal{F}} = \mathcal{F}/U$, the natural homomorphism $P \to \bar{P}$ induces the group and the category isomorphisms*

$$H_{\bar{\mathcal{F}}} \cong H_{\mathcal{F}}/(H_{\mathcal{F}} \cap U) \quad \textit{and} \quad \bar{\mathcal{F}}^{\mathfrak{h}} \cong \mathcal{F}^{\mathfrak{h}}/(H_{\mathcal{F}} \cap U) \qquad 13.10.1.$$

Proof: It easily follows from Proposition 12.3 that the natural homomorphism $P \to \bar{P}$ maps $H_{\mathcal{F}}$ onto $H_{\bar{\mathcal{F}}}$, proving the first isomorphism. On the other hand, set $P' = U\cdot H_{\mathcal{F}}$ and denote by $\mathcal{F}'$ the Frobenius P'-category fulfilling equality 13.6.1, and by $\bar{\mathcal{F}}'$ the U-quotient of $\mathcal{F}'$; first of all, we will prove that $\bar{\mathcal{F}}' = \bar{\mathcal{F}}^{\mathfrak{h}}$.

According to Proposition 2.7, for any subgroup Q of P' containing U we can choose $\varphi \in \mathcal{F}(P,Q)$ such that $Q' = \varphi(Q)$ is fully normalized in $\mathcal{F}$, so that we have (cf. equality 13.6.1)

$$\mathcal{F}'(Q') = \mathcal{F}_{P'}(Q')\cdot\mathbf{O}^p\big(\mathcal{F}(Q')\big) \qquad 13.10.2.$$

But, setting $\bar{Q}' = Q'/U$, since $\mathcal{F}(Q')$ maps onto $\bar{\mathcal{F}}(\bar{Q}')$ by the very definition of $\bar{\mathcal{F}}$, $\mathbf{O}^p\big(\mathcal{F}(Q')\big)$ still maps onto $\mathbf{O}^p\big(\bar{\mathcal{F}}(\bar{Q}')\big) \subset \bar{\mathcal{F}}^{\mathfrak{h}}(\bar{Q}')$ and, since $U \subset Q'$,

$\mathcal{F}_P(Q')$ maps onto $\mathcal{F}_{\bar{P}}(\bar{Q}')$ too. Thus, $\mathcal{F}_{\bar{P}}(\bar{Q}')$ is a Sylow p-subgroup of $\bar{\mathcal{F}}(\bar{Q}')$ and therefore $\mathcal{F}_{H_{\bar{\mathcal{F}}}}(\bar{Q}')$ is a Sylow p-subgroup of $\bar{\mathcal{F}}^{\mathfrak{h}}(\bar{Q}')$; hence, since $\mathcal{F}_{P'}(Q')$ still maps onto $\mathcal{F}_{H_{\bar{\mathcal{F}}}}(\bar{Q}')$, we finally get the surjectivity of the canonical homomorphism $\mathcal{F}'(Q) \to \bar{\mathcal{F}}^{\mathfrak{h}}(\bar{Q})$, so that

$$\bar{\mathcal{F}}'(\bar{Q}) = \bar{\mathcal{F}}^{\mathfrak{h}}(\bar{Q}) \qquad 13.10.3,$$

and, since $\bar{\mathcal{F}}'$ and $\bar{\mathcal{F}}^{\mathfrak{h}}$ are both Frobenius $H_{\bar{\mathcal{F}}}$-categories, the equality $\bar{\mathcal{F}}' = \bar{\mathcal{F}}^{\mathfrak{h}}$ follows from Corollary 5.14.

Now, it suffices to prove that the inclusion $\mathcal{F}^{\mathfrak{h}} \subset \mathcal{F}'$ induces an isomorphism

$$\overline{\mathcal{F}^{\mathfrak{h}}} = \mathcal{F}^{\mathfrak{h}}/(H_{\mathcal{F}} \cap U) \cong \bar{\mathcal{F}}' \qquad 13.10.4;$$

but, if Q and R are subgroups of $H_{\mathcal{F}}$ containing the intersection $H_{\mathcal{F}} \cap U$ it is clear that

$$Q = (Q{\cdot}U) \cap H_{\mathcal{F}} \quad \text{and} \quad R = (R{\cdot}U) \cap H_{\mathcal{F}} \qquad 13.10.5$$

and therefore $\mathcal{F}(Q{\cdot}U, R{\cdot}U)$ maps R into Q ; hence, setting

$$\overline{Q} = Q/(H_{\mathcal{F}} \cap U) \quad \text{and} \quad \overline{R} = R/(H_{\mathcal{F}} \cap U) \qquad 13.10.6,$$

in Proposition 12.3 the sets $\overline{\mathcal{F}^{\mathfrak{h}}}(\overline{Q}, \overline{R})$ and $\bar{\mathcal{F}}'(\bar{Q}, \bar{R})$ are defined by the *same* quotient of $\mathcal{F}(Q{\cdot}U, R{\cdot}U)$. We are done.

13.11 A remarkable fact is that the hyperfocal subgroup $H_{\mathcal{F}}$ of P is *always* $\mathcal{F}$-nilcentralized; actually, we have the following slightly more general result.

Proposition 13.12 *For any subgroup Q of P fully centralized in $\mathcal{F}$, the subgroup $Q{\cdot}H_{C_{\mathcal{F}}(Q)}$ is $\mathcal{F}$-nilcentralized and fully centralized in $\mathcal{F}$. Moreover, for any subgroup R of $N_P(Q)$ containing $Q{\cdot}H_{C_{\mathcal{F}}(Q)}$ and any $\mathcal{F}$-morphism $\psi : R \to P$ such that $\psi(Q) = Q$, there is an $\mathcal{F}$-morphism*

$$\zeta : C_P(Q){\cdot}R \longrightarrow P \qquad 13.12.1$$

extending ψ. In particular, the restriction induces an exact sequence

$$C_P(Q{\cdot}H_{C_{\mathcal{F}}(Q)}) \longrightarrow \mathcal{F}\big(Q{\cdot}C_P(Q)\big)_Q \longrightarrow \mathcal{F}(Q{\cdot}H_{C_{\mathcal{F}}(Q)})_Q \longrightarrow 1 \qquad 13.12.2.$$

Proof: Actually, Q is also fully $\mathcal{F}_Q(Q)$-normalized in $\mathcal{F}$ (cf. 2.10) and thus, by Propositions 2.7 and 2.16, there is a morphism

$$\eta : Q{\cdot}H_{C_{\mathcal{F}}(Q)} \longrightarrow Q{\cdot}C_P(Q) \qquad 13.12.3$$

in the Frobenius $Q{\cdot}C_P(Q)$-category

$$Q{\cdot}C_{\mathcal{F}}(Q) = N_{\mathcal{F}}^{\mathcal{F}_Q(Q)}(Q) \qquad 13.12.4$$

such that $\eta(Q{\cdot}H_{C_{\mathcal{F}}(Q)})$ is fully centralized in this $Q{\cdot}C_P(Q)$-category; then, it follows from Lemma 2.17 that $\eta(Q{\cdot}H_{C_{\mathcal{F}}(Q)})$ is also fully centralized in $\mathcal{F}$ and, since η is an $N_{\mathcal{F}}^{\mathcal{F}_Q(Q)}(Q)$-morphism, it is clear that $\eta(Q{\cdot}H_{C_{\mathcal{F}}(Q)}) = Q{\cdot}H_{C_{\mathcal{F}}(Q)}$.

Then, if U is a subgroup of $C_P(Q{\cdot}H_{C_{\mathcal{F}}(Q)})$, any p'-subgroup K of $\mathcal{F}(U{\cdot}Q{\cdot}H_{C_{\mathcal{F}}(Q)})$ which normalizes U and centralizes $Q{\cdot}H_{C_{\mathcal{F}}(Q)}$, induces a p'-group of automorphisms of $U{\cdot}H_{C_{\mathcal{F}}(Q)}$ in the Frobenius $C_P(Q)$-category $C_{\mathcal{F}}(Q)$ and, in particular, from the very definition of $H_{C_{\mathcal{F}}(Q)}$, we have

$$[K, U{\cdot}H_{C_{\mathcal{F}}(Q)}] \subset H_{C_{\mathcal{F}}(Q)} \qquad 13.12.5;$$

hence, since K centralizes $H_{C_{\mathcal{F}}(Q)}$, K centralizes $U{\cdot}H_{C_{\mathcal{F}}(Q)}$ (cf. Theorem 3.2 in [28, Ch.5)]), so that we have $K = \{\mathrm{id}_{U\cdot Q\cdot H_{C_{\mathcal{F}}(Q)}}\}$. Consequently, the $C_{\mathcal{F}}(Q{\cdot}H_{C_{\mathcal{F}}(Q)})$-hyperfocal subgroup is trivial and the first statement follows from statement 13.2.2 and from the very definition of $\mathcal{F}$-nilcentralized subgroups in 4.3.

Moreover, if R is a subgroup of $N_P(Q)$ containing $Q{\cdot}H_{C_{\mathcal{F}}(Q)}$ and ψ an $\mathcal{F}$-morphism from R to P such that $\psi(Q) = Q$, it follows from the divisibility of $\mathcal{F}$ and from statement 2.10.1 that there is an $\mathcal{F}$-automorphism σ of $Q{\cdot}C_P(Q)$ fulfilling $\sigma(u) = \psi(u)$ for any $u \in Q$; thus, σ stabilizes $C_P(Q)$ and then the Frobenius $C_P(Q)$-category $C_{\mathcal{F}}(Q)$, so that it stabilizes $H_{C_{\mathcal{F}}(Q)}$ too.

Hence, denoting by

$$\bar{\sigma} : Q{\cdot}C_R(P) \longrightarrow Q{\cdot}\sigma(C_R(Q)) \qquad 13.12.6$$

the $\mathcal{F}$-isomorphism determined by σ, and by $\bar{\psi}$ the restriction to $Q{\cdot}C_R(Q)$ of ψ (cf. 2.3), the composition $\bar{\psi} \circ \bar{\sigma}^{-1}$ acts trivially on Q and therefore it defines a $Q{\cdot}C_{\mathcal{F}}(Q)$-morphism (cf. 13.12.4) from $Q{\cdot}\sigma(C_R(Q))$ to $Q{\cdot}C_P(Q)$. But, since $Q{\cdot}H_{C_{\mathcal{F}}(Q)}$ is $\mathcal{F}$-nilcentralized, the group $Q{\cdot}\sigma(C_R(Q))$ is also $\mathcal{F}$-nilcentralized (cf. Proposition 4.5) and therefore it is $Q{\cdot}C_{\mathcal{F}}(Q)$-nilcentralized too. Consequently, $\bar{\psi} \circ \bar{\sigma}^{-1}$ is determined by an element of $Q{\cdot}C_P(Q)$; that is to say, we may choose σ in such a way that we have $\sigma(u) = \psi(u)$ for any $u \in Q{\cdot}C_R(Q)$.

On the other hand, the product $Q{\cdot}C_P(Q)$ is $\mathcal{F}$-selfcentralizing and normal in $R{\cdot}C_P(Q)$; moreover, according to our choice, σ can be extended to a group homomorphism $R{\cdot}C_P(Q) \to P$ mapping $v{\cdot}z$ on $\psi(v){\cdot}\sigma(z)$ for any $v \in R$ and any $z \in C_P(Q)$; consequently, it follows from statement 2.10.1 that σ can be extended to an $\mathcal{F}$-morphism $\zeta : R{\cdot}C_P(Q) \to P$ and then, denoting by $\bar{\zeta}$ the restriction of ζ to R, the restrictions to $Q{\cdot}C_R(Q)$ of ψ and $\bar{\zeta}$ coincide each other; at this point, it follows from Proposition 4.6 that we can modify our choice of ζ in such a way that we have $\bar{\zeta} = \psi$. This proves the second statement above.

Furthermore, it is clear that any $\mathcal{F}$-automorphism of $Q{\cdot}C_P(Q)$ stabilizing Q still stabilizes $C_P(Q)$ and then the Frobenius $C_P(Q)$-category $C_{\mathcal{F}}(Q)$, so that it stabilizes $H_{C_{\mathcal{F}}(Q)}$ too; that is to say, the restriction induces a group homomorphism

$$\mathcal{F}\big(Q{\cdot}C_P(Q)\big)_Q \longrightarrow \mathcal{F}(Q{\cdot}H_{C_{\mathcal{F}}(Q)})_Q \qquad 13.12.7$$

which is surjective by the argument above applied to $R = Q{\cdot}H_{C_{\mathcal{F}}(Q)}$. Finally, if σ is a p'-element of the kernel of this homomorphism then, as above, it determines a p'-automorphism of $C_P(Q)$ in the Frobenius $C_P(Q)$-category $C_{\mathcal{F}}(Q)$ and then it follows from Lemma 13.3 that σ acts trivially on the quotient $C_P(Q)/H_{C_{\mathcal{F}}(Q)}$; since σ acts trivially on $H_{C_{\mathcal{F}}(Q)}$, we get $\sigma = \mathrm{id}_{Q{\cdot}C_P(Q)}$ (cf. Theorem 3.2 in [28, Ch. 5]) and therefore this kernel is a p-group. We are done.

13.13 If we restrict ourselves to the full subcategory $\mathcal{F}^{^{sc}}$ of $\mathcal{F}$ over the set of $\mathcal{F}$-selfcentralizing subgroups of P, the correspondence mapping any object Q in $\mathcal{F}^{^{sc}}$ on its center $Z(Q)$ is *functorial* since it is compatible with $\mathcal{F}$-isomorphisms and, for another object R in $\mathcal{F}^{^{sc}}$ contained in Q, we have

$$Z(Q) \subset Z(R) \qquad 13.13.1;$$

that is to say, we get a *contravariant* functor from $\mathcal{F}^{^{sc}}$ to the category of finitely generated Abelian groups $\mathfrak{Ab}$ noted

$$\mathfrak{z}_{\mathcal{F}^{sc}} : \mathcal{F}^{^{sc}} \longrightarrow \mathfrak{Ab} \qquad 13.13.2.$$

As a matter of fact, this *contravariant* functor can be extended to the whole Frobenius P-category $\mathcal{F}$, and in chapter 20 below we discuss the existence of a suitable *extension* of $\mathcal{F}$ by this extended *contravariant* functor.

Proposition 13.14 *There is a contravariant functor $\mathfrak{c}^{\mathfrak{h}}_{\mathcal{F}}$ from $\mathcal{F}$ to the exterior quotient $\widetilde{\mathfrak{Gr}}$ of the category of finite groups, unique up to natural isomorphisms, which maps any subgroup Q of P fully centralized in $\mathcal{F}$ on the quotient $C_P(Q)/H_{C_{\mathcal{F}}(Q)}$, and maps any $\mathcal{F}$-morphism $\varphi\colon R \to Q$, between subgroups of P fully centralized in $\mathcal{F}$, on the class of the group homomorphism*

$$\bar\zeta : C_P(Q)/H_{C_{\mathcal{F}}(Q)} \longrightarrow C_P(R)/H_{C_{\mathcal{F}}(R)} \qquad 13.14.1$$

induced by an $\mathcal{F}$-morphism

$$\zeta : \varphi(R){\cdot}C_P(Q) \to R{\cdot}C_P(R) \qquad 13.14.2$$

fulfilling $\zeta\big(\varphi(v)\big) = v$ for any $v \in R$. In particular, we have a contravariant functor

$$\mathfrak{c}^{\mathfrak{f}}_{\mathcal{F}} : \mathcal{F} \longrightarrow \mathfrak{Ab} \qquad 13.14.3$$

mapping Q on $C_P(Q)/F_{C_{\mathcal{F}}(Q)}$ and φ on the group homomorphism

$$C_P(Q)/F_{C_{\mathcal{F}}(Q)} \longrightarrow C_P(R)/F_{C_{\mathcal{F}}(R)} \qquad 13.14.4$$

determined by $\bar\zeta$.

Proof: Since $\mathcal{F}$ is equivalent to the full subcategory over the set of subgroups of P fully centralized in $\mathcal{F}$ (cf. Proposition 2.7), in order to define a *contravariant* functor $\mathcal{F} \to \widetilde{\mathfrak{Gr}}$ up to natural isomorphisms, it suffices to define it over this full subcategory.

With the notation above, the existence of ζ follows from statement 2.10.1 applied to the inverse of the isomorphism $R \cong \varphi(R)$ induced by φ; then, it is clear that $\zeta\big(C_P(Q)\big) \subset C_P(R)$ and the restriction $\xi : C_P(Q) \to C_P(R)$ of ζ is $\big(C_{\mathcal{F}}(Q), C_{\mathcal{F}}(R)\big)$-functorial since, for any subgroup U of $C_P(Q)$ and any $\mathcal{F}$-morphism $\theta : Q{\cdot}U \to Q{\cdot}C_P(Q)$ fulfilling $\theta(U) \subset C_P(Q)$ and $\theta(u) = u$ for any $u \in Q$, the group homomorphism $\eta : R{\cdot}\xi(U) \to R{\cdot}\xi\big(C_P(Q)\big)$ mapping $v\xi(u)$ on $\zeta\big(\varphi(v)\theta(u)\big)$, for any $v \in R$ and any $u \in U$, clearly induces a $C_{\mathcal{F}}(R)$-morphism from $\xi(U)$ to $\xi\big(C_P(Q)\big)$. Consequently, we have

$$\zeta(H_{C_{\mathcal{F}}(Q)}) \subset H_{C_{\mathcal{F}}(R)} \qquad 13.14.5$$

and therefore ζ induces the announced homomorphism 13.14.1.

Moreover, for another choice ζ' of ζ, it is quite clear that the isomorphism $\zeta\big(C_P(Q)\big) \cong \zeta'\big(C_P(Q)\big)$ induced by ζ and ζ' is a $C_{\mathcal{F}}(R)$-isomorphism and therefore, according to Lemma 13.3, there is $z \in C_P(R)$ such that, for any $u \in C_P(Q)$, we have

$$\zeta'(u) \equiv \zeta(u)^z \mod H_{C_{\mathcal{F}}(R)} \qquad 13.14.6,$$

so that the group homomorphisms induced by ζ and by ζ' determine the same $\widetilde{\mathfrak{Gr}}$-morphism

$$\mathfrak{c}^{\mathfrak{h}}_{\mathcal{F}}(\varphi) : C_P(Q)/H_{C_{\mathcal{F}}(Q)} \longrightarrow C_P(R)/H_{C_{\mathcal{F}}(R)} \qquad 13.14.7.$$

Finally, for a third subgroup T of P fully centralized in $\mathcal{F}$ and for any $\mathcal{F}$-morphism $\psi : T \to R$, it is easily checked that

$$\mathfrak{c}^{\mathfrak{h}}_{\mathcal{F}}(\varphi \circ \psi) = \mathfrak{c}^{\mathfrak{h}}_{\mathcal{F}}(\psi) \circ \mathfrak{c}^{\mathfrak{h}}_{\mathcal{F}}(\varphi) \qquad 13.14.8.$$

The last statement follows from the last statement in Lemma 13.3. We are done.

13.15 A last remark. Denoting by p-$\mathfrak{iGr}$ the subcategory of $\mathfrak{Gr}$ formed by the finite p-groups and by the injective group homomorphisms between them, there is a functor

$$\mathfrak{z}^\circ : p\text{-}\mathfrak{iGr} \longrightarrow \mathfrak{Ab} \qquad 13.15.1$$

mapping a finite p-group Q on $Z(Q)$, which restricted to $\mathcal{F}^{^{\mathrm{sc}}}$ induces an "almost" *Mackey complement* (cf. 6.24) of the *contravariant* functor (cf. 6.2)

$$\mathfrak{ac}(\tilde{\mathfrak{z}}_{\mathcal{F}^{\mathrm{sc}}}) : \mathfrak{ac}(\tilde{\mathcal{F}}^{^{\mathrm{sc}}}) \longrightarrow \mathfrak{Ab} \qquad 13.15.2$$

determined by $\mathfrak{z}_{\mathcal{F}^{\mathrm{sc}}}$, as we explain below.

13.16 Explicitly, if Q is a finite p-group, N a normal subgroup of Q and $U \subset Q$ a set of representatives for Q/N, then we have the group homomorphism (cf. 1.9)

$$\mathfrak{z}^\circ(\iota_N^Q) : Z(N) \longrightarrow Z(Q) \qquad 13.16.1$$

mapping $z \in Z(N)$ on

$$\mathrm{Tr}_N^Q(z) = \prod_{u \in U} uzu^{-1} \qquad 13.16.2$$

since all the terms of this product belong to $Z(R)$ and the product does not depend on the choice of U, so that it belongs to $Z(Q)$. For any proper subgroup R of Q, we argue by induction on $|Q:R|$ and, considering a proper normal subgroup N of Q containing R, we define

$$\mathfrak{z}^\circ(\iota_R^Q) = \mathfrak{z}^\circ(\iota_N^Q) \circ \mathfrak{z}^\circ(\iota_R^N) \qquad 13.16.3;$$

this composition does not depend on the choice of N since, for another proper normal subgroup N' of Q containing R, we clearly have

$$\mathrm{Tr}_{N'}^Q\big(\mathrm{Tr}_{N\cap N'}^{N'}(z)\big) = \mathrm{Tr}_{N\cap N'}^Q(z) = \mathrm{Tr}_N^Q\big(\mathrm{Tr}_{N\cap N'}^N(z)\big) \qquad 13.16.4$$

for any $z \in Z(N \cap N')$ and, by our induction hypothesis, we still have

$$\begin{aligned} \mathfrak{z}^\circ(\iota_R^N) &= \mathfrak{z}^\circ(\iota_{N\cap N'}^N) \circ \mathfrak{z}^\circ(\iota_R^{N\cap N'}) \\ \mathfrak{z}^\circ(\iota_R^{N'}) &= \mathfrak{z}^\circ(\iota_{N\cap N'}^{N'}) \circ \mathfrak{z}^\circ(\iota_R^{N\cap N'}) \end{aligned} \qquad 13.16.5.$$

13.17 Finally, for any isomorphism $\theta : Q \cong Q'$, we define

$$\mathfrak{z}^\circ(\theta) : Z(Q) \cong Z(Q') \qquad 13.17.1$$

as the isomorphism induced by θ. Since any injective group homomorphism $\varphi : R \to Q$ between two finite p-groups is the composition of an isomorphism $\varphi_* : R \cong \varphi(R)$ with the inclusion $\varphi(R) \subset Q$, it is easily checked that, setting

$$\mathfrak{z}^\circ(\varphi) = \mathfrak{z}^\circ(\iota_{\varphi(R)}^Q) \circ \mathfrak{z}^\circ(\varphi_*) \qquad 13.17.2,$$

we get the announced functor which maps any finite p-group Q on $Z(Q)$ and any injective group homomorphism $\varphi : R \to Q$ on $\mathfrak{z}^\circ(\varphi)$.

13.18 Consider the restriction of $\mathfrak{z}^\circ$ to $\mathcal{F}^{^{sc}}$

$$\mathfrak{z}^\circ_{\mathcal{F}^{sc}} : \mathcal{F}^{^{sc}} \longrightarrow \mathfrak{Ab} \qquad 13.18.1$$

and note that $\mathfrak{z}_{\mathcal{F}^{sc}}$ and $\mathfrak{z}^\circ_{\mathcal{F}^{sc}}$ obviously factorize throughout the *exterior quotient* $\tilde{\mathcal{F}}^{^{sc}}$ of $\mathcal{F}^{^{sc}}$ (cf. 1.3). Let us show that the corresponding *additive* functor (cf. A4.10)

$$\mathfrak{ac}(\tilde{\mathfrak{z}}^\circ_{\mathcal{F}^{sc}}) : \mathfrak{ac}(\tilde{\mathcal{F}}^{^{sc}}) \longrightarrow \mathfrak{Ab} \qquad 13.18.2$$

is an “almost” *Mackey complement* of $\mathfrak{ac}(\mathfrak{z}_{\mathcal{F}^{\mathrm{sc}}})$ (cf. 6.24) in the sense specified by the following lemma; recall that the category $\mathfrak{ac}(\tilde{\mathcal{F}}^{\mathrm{sc}})$ admits distributive *pull-backs* given by the *relative exterior intersection* (cf. Proposition 6.21).

Lemma 13.19 *With the notation above, for any $\tilde{\mathcal{F}}^{\mathrm{sc}}$-morphism $\tilde\varphi : R \to Q$ we have*

$$\tilde{\mathfrak{z}}^{\circ}_{\mathcal{F}^{\mathrm{sc}}}(\tilde\varphi) \circ \tilde{\mathfrak{z}}_{\mathcal{F}^{\mathrm{sc}}}(\tilde\varphi) = (\mathrm{id}_{Z(Q)})^{|Q|/|R|} \qquad 13.19.1.$$

Moreover, for another $\tilde{\mathcal{F}}^{\mathrm{sc}}$-morphism $\tilde\psi : T \to Q$ such that $\varphi(R) \cap \psi(T)$ is $\mathcal{F}$-selfcentralizing for any pair of representatives φ of $\tilde\varphi$ and ψ of $\tilde\psi$, we have the commutative diagram

$$\begin{array}{ccccc} & & Z(Q) & & \\ & {\scriptstyle \tilde{\mathfrak{z}}^{\circ}_{\mathcal{F}^{\mathrm{sc}}}(\tilde\varphi)}\nearrow & & \searrow{\scriptstyle \tilde{\mathfrak{z}}_{\mathcal{F}^{\mathrm{sc}}}(\tilde\psi)} & \\ Z(R) & & & & Z(T) \\ & {\scriptstyle \mathfrak{ac}(\tilde{\mathfrak{z}}_{\mathcal{F}^{\mathrm{sc}}})(\tilde\gamma)}\searrow & & \nearrow{\scriptstyle \mathfrak{ac}(\tilde{\mathfrak{z}}^{\circ}_{\mathcal{F}^{\mathrm{sc}}})(\tilde\delta)} & \\ & & \big(\mathfrak{ac}(\tilde{\mathfrak{z}}_{\mathcal{F}^{\mathrm{sc}}})\big)(R\tilde\cap_Q T) & & \end{array} \qquad 13.19.2$$

where $R\tilde\cap_Q T$ is the relative exterior intersection of R and T over Q and

$$R \xleftarrow{\ \tilde\gamma\ } R\tilde\cap_Q T \xrightarrow{\ \tilde\delta\ } T \qquad 13.19.3$$

are the structural $\mathfrak{ac}(\tilde{\mathcal{F}}^{\mathrm{sc}})$-morphisms.

Proof: Equality 13.19.1 is easily checked when we have $\varphi(R) \lhd Q$ for a representative φ of $\tilde\varphi$; in general, it suffices to argue by induction on $|Q|/|R|$.

For the second statement, we may assume that Q contains R and T, and that $\tilde\varphi$ and $\tilde\psi$ are induced by the inclusions. We argue by induction on $|Q : R|$ and may assume that $R \neq Q$; let N be a proper normal subgroup of Q containing R and, for any $w \in Q$, set $U_w = N \cap T$ and consider the group homomorphism $\kappa_w : U_w \to N$ determined by the conjugation by w; then, since $R \cap (U_w)^n = R \cap T^n$ is $\mathcal{F}$-selfcentralizing for any $n \in N$, it follows from the induction hypothesis that we have the commutative diagram

$$\begin{array}{ccccc} & & Z(N) & & \\ & {\scriptstyle \tilde{\mathfrak{z}}^{\circ}_{\mathcal{F}^{\mathrm{sc}}}(\tilde\iota_R^N)}\nearrow & & \searrow{\scriptstyle \tilde{\mathfrak{z}}_{\mathcal{F}^{\mathrm{sc}}}(\tilde\kappa_w)} & \\ Z(R) & & & & Z(U_w) \\ & {\scriptstyle \mathfrak{ac}(\tilde{\mathfrak{z}}_{\mathcal{F}^{\mathrm{sc}}})(\tilde\gamma_w)}\searrow & & \nearrow{\scriptstyle \mathfrak{ac}(\tilde{\mathfrak{z}}^{\circ}_{\mathcal{F}^{\mathrm{sc}}})(\tilde\varepsilon_w)} & \\ & & \big(\mathfrak{ac}(\tilde{\mathfrak{z}}_{\mathcal{F}^{\mathrm{sc}}})\big)(R\tilde\cap_N U_w) & & \end{array} \qquad 13.19.4$$

where $R\tilde\cap_N U_w$ is the *relative exterior intersection* of R and U_w over the morphisms $\tilde\iota_R^N$ and $\tilde\kappa_w$ (cf. 6.18), and

$$R \xleftarrow{\ \tilde\gamma_w\ } R\tilde\cap_N U_w \xrightarrow{\ \tilde\varepsilon_w\ } U_w \qquad 13.19.5$$

are the structural $\mathfrak{ac}(\tilde{\mathcal{F}}^{\mathrm{sc}})$-morphisms.

On the other hand, choosing a set of representatives $W \subset Q$ for $N\backslash Q/T$, it follows from Proposition 6.19 that

$$N \tilde{\cap}_Q T \cong \bigoplus_{w \in W} U_w \qquad 13.19.6$$

and that the structural $\mathfrak{ac}(\tilde{\mathcal{F}}^{^{sc}})$-morphisms $\tilde{\kappa}^N$ to N and $\tilde{\iota}^T$ to T are respectively defined by the families of $\tilde{\mathcal{F}}^{^{sc}}$-morphisms $\{\tilde{\kappa}_w\}_{w\in W}$ and $\{\tilde{\iota}^T_{U_w}\}_{w\in W}$; but, for any $z \in Z(N)$, we have

$$\mathrm{Tr}_N^Q(z) = \prod_u u^{-1}zu = \prod_{w\in W} \mathrm{Tr}^T_{U_w}(w^{-1}zw) \qquad 13.19.7$$

where $u \in Q$ runs over a set of representatives for $N\backslash Q$; hence, we still have the commutative diagram

$$\begin{array}{ccccc} & & Z(Q) & & \\ & {\scriptstyle \tilde{\mathfrak{z}}^{\circ}_{\mathcal{F}^{sc}}(\tilde{\iota}^Q_N)}\nearrow & & \searrow{\scriptstyle \tilde{\mathfrak{z}}_{\mathcal{F}^{sc}}(\tilde{\iota}^Q_T)} & \\ Z(N) & & & & Z(T) \\ & {\scriptstyle \mathfrak{ac}(\tilde{\mathfrak{z}}_{\mathcal{F}^{sc}})(\tilde{\kappa}^N)}\searrow & & \nearrow{\scriptstyle \mathfrak{ac}(\tilde{\mathfrak{z}}^{\circ}_{\mathcal{F}^{sc}})(\tilde{\iota}^T)} & \\ & & \big(\mathfrak{ac}(\tilde{\mathfrak{z}}_{\mathcal{F}^{sc}})\big)(N\tilde{\cap}_Q T) & & \end{array} \qquad 13.19.8.$$

Finally, on the one hand, the commutative diagrams 13.19.4 when w runs over W yield to the following commutative diagram

$$\begin{array}{ccccc} & & Z(N) & & \\ & {\scriptstyle \tilde{\mathfrak{z}}^{\circ}_{\mathcal{F}^{sc}}(\tilde{\iota}^N_R)}\nearrow & & \searrow & \\ Z(R) & & & & \prod_{w\in W} Z(U_w) \\ & \searrow & & \nearrow \; {\scriptstyle \prod_{w\in W} \mathfrak{ac}(\tilde{\mathfrak{z}}^{\circ}_{\mathcal{F}^{sc}})(\tilde{\varepsilon}_w)} & \\ & & \prod_{w\in W}\big(\mathfrak{ac}(\tilde{\mathfrak{z}}_{\mathcal{F}^{sc}})\big)(R\tilde{\cap}_N U_w) & & \end{array} \qquad 13.19.9;$$

on the other hand, since the *pull-backs* fulfill an obvious *transitive* condition, we have

$$R\tilde{\cap}_Q T \cong R\tilde{\cap}_N(N\tilde{\cap}_Q T) \cong \bigoplus_{w\in W} R\tilde{\cap}_N U_w \qquad 13.19.10$$

and the corresponding structural $\mathfrak{ac}(\tilde{\mathcal{F}}^{^{sc}})$-morphisms agree; consequently, the commutative diagram 13.19.9 becomes (cf. 13.19.6)

$$\begin{array}{ccccc} & & Z(N) & & \\ & {\scriptstyle \tilde{\mathfrak{z}}^{\circ}_{\mathcal{F}^{sc}}(\tilde{\iota}^N_R)}\nearrow & & \searrow{\scriptstyle \mathfrak{ac}(\tilde{\mathfrak{z}}_{\mathcal{F}^{sc}})(\tilde{\kappa}^N)} & \\ Z(R) & & & & \big(\mathfrak{ac}(\tilde{\mathfrak{z}}_{\mathcal{F}^{sc}})\big)(N\tilde{\cap}_Q T) \\ & {\scriptstyle \mathfrak{ac}(\tilde{\mathfrak{z}}_{\mathcal{F}^{sc}})(\tilde{\gamma})}\searrow & & \nearrow & \\ & & \big(\mathfrak{ac}(\tilde{\mathfrak{z}}_{\mathcal{F}^{sc}})\big)(R\tilde{\cap}_Q T) & & \end{array} \qquad 13.19.11.$$

Now, it is quite clear that the commutativity of diagram 13.19.2 follows from the commutativity of diagrams 13.19.9 and 13.19.11. We are done.

Chapter 14

The Grothendieck groups of a Frobenius P-category

14.1 Let P be a finite p-group and $\mathcal{F}$ a Frobenius P-category. As in chapter 11, let us denote by $\mathcal{F}^{^{\rm nc}}$ the full subcategory of $\mathcal{F}$ over the $\mathcal{F}$-nilcentralized subgroups of P (cf. 4.3) and let us consider the *proper category of $\mathcal{F}^{^{\rm nc}}$-chains* $\mathfrak{ch}^*(\mathcal{F}^{^{\rm nc}})$ (cf. A2.8); recall that we have the *automorphism functor* from $\mathfrak{ch}^*(\mathcal{F}^{^{\rm nc}})$ to the category of groups (cf. Proposition A2.10)

$$\mathfrak{aut}_{\mathcal{F}^{\rm nc}} : \mathfrak{ch}^*(\mathcal{F}^{^{\rm nc}}) \longrightarrow \mathfrak{Gr} \qquad 14.1.1$$

which maps any $\mathfrak{ch}^*(\mathcal{F}^{^{\rm nc}})$-object $(\mathfrak{q}, \Delta_n)$ on its $\mathfrak{ch}^*(\mathcal{F}^{^{\rm nc}})$-automorphism group — noted $\mathcal{F}(\mathfrak{q})$; explicitly, we identify $\mathcal{F}(\mathfrak{q})$ with the subgroup of $\mathcal{F}\big(\mathfrak{q}(n)\big)$ formed by the elements stabilizing the images $\mathfrak{q}(i)_n = \mathfrak{q}(i \bullet n)\big(\mathfrak{q}(i)\big)$ in $\mathfrak{q}(n)$ of all the groups $\mathfrak{q}(i)$ determined by the functor $\mathfrak{q} : \Delta_n \to \mathcal{F}^{^{\rm nc}}$ for $0 \le i \le n$.

14.2 In Theorem 11.32 above, we have proved that when $\mathcal{F}$ comes from a block b of a finite group G, the functor $\mathfrak{aut}_{(\mathcal{F}_{(b,G)})^{\rm nc}}$ can be lifted to a functor $\widehat{\mathfrak{aut}}_{(\mathcal{F}_{(b,G)})^{\rm nc}}$ from $\mathfrak{ch}^*(\mathcal{F}^{^{\rm nc}})$ to the category k^*-$\mathfrak{Gr}$ of *finite* k^*-groups (cf. 1.23); since our arguments mostly depend on the existence of such a lifting, in our abstract setting let us consider *a functor*

$$\widehat{\mathfrak{aut}}_{\mathcal{F}^{\rm nc}} : \mathfrak{ch}^*(\mathcal{F}^{^{\rm nc}}) \longrightarrow k^*\text{-}\mathfrak{Gr} \qquad 14.2.1$$

lifting the functor $\mathfrak{aut}_{\mathcal{F}^{\rm nc}}$ *to the category* k^*-$\mathfrak{Gr}$ and, for $(\mathfrak{q}, \Delta_n)$ as above, set

$$\hat{\mathcal{F}}(\mathfrak{q}) = \widehat{\mathfrak{aut}}_{\mathcal{F}^{\rm nc}}(\mathfrak{q}, \Delta_n) \qquad 14.2.2.$$

14.3 Let $\mathcal{O}$ be the *unramified complete discrete valuation ring* of characteristic zero admitting our field k (cf. 1.10) as residue class field, and denote by $\mathcal{O}$-$\mathfrak{mod}$ the category of finitely generated $\mathcal{O}$-modules. For any *finite* k^*-group $\hat{G}$, we denote by $\mathcal{G}_k(\hat{G})$ the *scalar extension* from $\mathbb{Z}$ to $\mathcal{O}$ of the so-called *Grothendieck group* (cf. Appendix in [54]) of the category of finite dimensional $k_*\hat{G}$-modules (cf. 1.25); it is well-known that we have a *contravariant* functor

$$\mathfrak{g}_k : k^*\text{-}\mathfrak{Gr} \longrightarrow \mathcal{O}\text{-}\mathfrak{mod} \qquad 14.3.1$$

mapping $\hat{G}$ on $\mathcal{G}_k(\hat{G})$ and any k^*-group homomorphism $\hat{\varphi} : \hat{G} \to \hat{G}'$ on the corresponding *restriction* map $\mathrm{res}_{\hat{\varphi}} : \mathcal{G}_k(\hat{G}') \to \mathcal{G}_k(\hat{G})$. Thus, we have the composed functor

$$\mathfrak{ch}^*(\mathcal{F}^{^{\rm nc}}) \xrightarrow{\widehat{\mathfrak{aut}}_{\mathcal{F}^{\rm nc}}} k^*\text{-}\mathfrak{Gr} \xrightarrow{\mathfrak{g}_k} \mathcal{O}\text{-}\mathfrak{mod} \qquad 14.3.2$$

and we define the *Grothendieck group of the pair* $(\mathcal{F}, \widehat{\mathfrak{aut}}_{\mathcal{F}^{\rm nc}})$ as the inverse limit

$$\mathcal{G}_k(\mathcal{F}, \widehat{\mathfrak{aut}}_{\mathcal{F}^{\rm nc}}) = \lim_{\longleftarrow} (\mathfrak{g}_k \circ \widehat{\mathfrak{aut}}_{\mathcal{F}^{\rm nc}}) \qquad 14.3.3.$$

14.4 It is a finitely generated free $\mathcal{O}$-module and the main purpose of this chapter is to determine its $\mathcal{O}$-rank (cf. Corollary 14.32 below); for this, we give alternative descriptions of this $\mathcal{O}$-module. First of all, recall that if R is a normal p-subgroup of $\hat{G}$ then the restriction *via* the canonical homomorphism $\hat{G} \to \hat{G}/R$ determines an $\mathcal{O}$-module isomorphism $\mathcal{G}_k(\hat{G}/R) \cong \mathcal{G}_k(\hat{G})$; in particular, for any $\mathfrak{ch}^*(\mathcal{F}^{^{\rm nc}})$-object $(\mathfrak{q}, \Delta_n)$, since the image $\mathfrak{q}(0)_n$ in $\mathfrak{q}(n)$ of $\mathfrak{q}(0)$ by $\mathfrak{q}(0 \bullet n)$ is contained, and therefore stabilizes, all the other images $\mathfrak{q}(i)_n$ for $1 \le i \le n$. we get

$$\mathcal{G}_k\Big(\hat{\mathcal{F}}(\mathfrak{q})/\mathcal{F}_{\mathfrak{q}(0)_n}\big(\mathfrak{q}(n)\big)\Big) \cong \mathcal{G}_k\big(\hat{\mathcal{F}}(\mathfrak{q})\big) \qquad 14.4.1.$$

14.5 This allows us to replace the subcategory $\mathcal{F}^{^{\rm nc}}$ by its *exterior quotient* $\tilde{\mathcal{F}}^{^{\rm nc}}$ (cf. 1.3); but, in chapter 6 we have seen that, rather than $\tilde{\mathcal{F}}^{^{\rm nc}}$, it is the full subcategory $\tilde{\mathcal{F}}^{^{\rm sc}}$ of $\tilde{\mathcal{F}}$ over the $\mathcal{F}$-selfcentralizing subgroups which has remarkable properties; luckily, the next result shows that the inverse limit does not change when we restrict our functor to $\mathcal{F}^{^{\rm sc}}$. Recall that, for any category $\mathfrak{C}$ and any *contravariant* functor $\mathfrak{m}\colon \mathfrak{ch}^*(\mathfrak{C}) \to \mathcal{O}\text{-}\mathfrak{mod}$, the inverse limit $\lim\limits_{\longleftarrow} \mathfrak{m}$ coincides with the 0-*cohomology group* of the *differential complex* given by the *differential maps* (cf. A3.11.2)

$$d_n : \prod_{\mathfrak{q}} \mathfrak{m}(\mathfrak{q}, \Delta_n) \longrightarrow \prod_{\mathfrak{r}} \mathfrak{m}(\mathfrak{r}, \Delta_{n+1}) \qquad 14.5.1,$$

where $\mathfrak{q}$ and $\mathfrak{r}$ respectively run over the sets of functors $\mathfrak{Fct}(\Delta_n, \mathfrak{C})$ and $\mathfrak{Fct}(\Delta_{n+1}, \mathfrak{C})$, sending any family $a = (a_{\mathfrak{q}})_{\mathfrak{q}}$ to the family $d_n(a) = \big(d_n(a)_{\mathfrak{r}}\big)_{\mathfrak{r}}$ defined by

$$d_n(a)_{\mathfrak{r}} = \sum_{i=0}^{n+1} (-1)^i \big(\mathfrak{m}(\mathrm{id}_{\mathfrak{r}\circ\delta_i^n}, \delta_i^n)\big)(a_{\mathfrak{r}\circ\delta_i^n}) \qquad 14.5.2.$$

Proposition 14.6 *Let $\mathfrak{X}$ be a nonempty set of $\mathcal{F}$-nilcentralized subgroups of P, denote by $\mathcal{F}^{^{\mathfrak{X}}}$ the full subcategory of $\mathcal{F}$ over $\mathfrak{X}$ and by $\mathfrak{i}^{\mathfrak{X}} : \mathcal{F}^{^{\mathfrak{X}}} \to \mathcal{F}^{^{\rm nc}}$ the inclusion functor, set $\widehat{\mathfrak{aut}}_{\mathcal{F}^{\mathfrak{X}}} = \widehat{\mathfrak{aut}}_{\mathcal{F}^{\rm nc}} \circ \mathfrak{ch}^*(\mathfrak{i}^{\mathfrak{X}})$ and assume that*

14.6.1 *For any $\mathcal{F}$-nilcentralized subgroup Q of P there are R in $\mathfrak{X}$ and ψ in $\mathcal{F}(R, Q)$ such that $\psi(Q)$ is fully centralized in $\mathcal{F}$, R contains $C_P\big(\psi(Q)\big)$ and we have*

$$\psi \circ \mathcal{F}(Q) \subset \mathcal{F}(R) \circ \psi \qquad .$$

Then, the functor $\mathfrak{ch}^*(\mathfrak{i}^{\mathfrak{X}})$ induces an $\mathcal{O}$-module isomorphism

$$\lim_{\longleftarrow} (\mathfrak{g}_k \circ \widehat{\mathfrak{aut}}_{\mathcal{F}^{\rm nc}}) \cong \lim_{\longleftarrow} \big(\mathfrak{g}_k \circ \widehat{\mathfrak{aut}}_{\mathcal{F}^{\mathfrak{X}}}\big) \qquad 14.6.2.$$

Proof: First of all, we prove that the homomorphism from $\varprojlim (\mathfrak{g}_k \circ \widehat{\mathfrak{aut}}_{\mathcal{F}^{\mathrm{nc}}})$ to $\varprojlim (\mathfrak{g}_k \circ \widehat{\mathfrak{aut}}_{\mathcal{F}^{\mathfrak{X}}})$ induced by $\mathfrak{ch}^*(\mathfrak{i}^{\mathfrak{X}})$ is injective; consider a family $X = (X_Q)_Q$, where Q runs over the set of all $\mathcal{F}$-nilcentralized subgroups of P and X_Q belongs to $\mathcal{G}_k\big(\hat{\mathcal{F}}(Q)\big)$, and assume that $d_0(X) = 0$ where d_0 denotes the corresponding differential map (cf. 14.5.2).

For such a Q, choose $R \in \mathfrak{X}$ and $\psi \in \mathcal{F}(R,Q)$ as in condition 14.6.1 above and consider the *chain* $\mathfrak{r}: \Delta_1 \to \mathcal{F}^{\mathrm{nc}}$ mapping 0 on Q, 1 on R and $0\bullet 1$ on ψ; then, we get (cf. 14.5.2)

$$0 = \mathrm{res}_{\widehat{\mathfrak{aut}}(\mathrm{id}_{\mathfrak{r}\circ\delta^0_0}, \delta^0_0)}(X_R) - \mathrm{res}_{\widehat{\mathfrak{aut}}(\mathrm{id}_{\mathfrak{r}\circ\delta^0_1}, \delta^0_1)}(X_Q) \qquad 14.6.3;$$

but, it follows from the inclusion $\psi \circ \mathcal{F}(Q) \subset \mathcal{F}(R) \circ \psi$ in condition 14.6.1 that the homomorphism $\mathcal{F}(R)_{\psi(Q)} \to \mathcal{F}(Q)$ induced by ψ is surjective and therefore the map

$$\mathrm{res}_{\widehat{\mathfrak{aut}}(\mathrm{id}_{\mathfrak{r}\circ\delta^0_1}, \delta^0_1)} : \mathcal{G}_k\big(\hat{\mathcal{F}}(Q)\big) \longrightarrow \mathcal{G}_k\big(\hat{\mathcal{F}}(\mathfrak{r})\big) \qquad 14.6.4$$

is injective; thus, if X belongs to the kernel of the homomorphism induced by $\mathfrak{ch}^*(\mathfrak{i}^{\mathfrak{X}})$, we get $X_Q = 0$.

In order to prove the surjectivity, assume that a family $Y = (Y_R)_{R\in\mathfrak{X}}$, where $Y_R \in \mathcal{G}_k\big(\hat{\mathcal{F}}(R)\big)$, belongs to the kernel of the corresponding differential map $d_0^{\mathfrak{X}}$; we have to extend Y to a family $X = (X_Q)_Q$ over the set of all $\mathcal{F}$-nilcentralized subgroups of P, fulfilling $d_0(X) = 0$. For any $\mathcal{F}$-nilcentralized subgroup Q of P, choosing $R \in \mathfrak{X}$ and $\psi \in \mathcal{F}(R,Q)$ as above, the kernel of the homomorphism $\mathcal{F}(R)_{\psi(Q)} \to \mathcal{F}(Q)$ coincides with $C_P\big(\psi(Q)\big)$ (cf. Corollary 4.7) and, in particular, homomorphism 14.6.4 is surjective too, so that with the same notation we can define

$$X_Q = \big(\mathrm{res}_{\widehat{\mathfrak{aut}}(\mathrm{id}_{\tilde{\mathfrak{r}}\circ\mathfrak{s}\delta^0_1}, \delta^0_1)}\big)^{-1}\big(\mathrm{res}_{\widehat{\mathfrak{aut}}(\mathrm{id}_{\tilde{\mathfrak{r}}\circ\mathfrak{s}\delta^0_0}, \delta^0_0)}(Y_R)\big) \qquad 14.6.5;$$

moreover, we assume that, for any subgroup of P which is $\mathcal{F}$-isomorphic to Q and any isomorphism between them, we choose the same $R \in \mathfrak{X}$ and the composed homomorphism; then, we claim that $d_0(X) = 0$.

Let $\mathfrak{q}: \Delta_1 \to \mathcal{F}^{\mathrm{sc}}$ be a *chain*, set $Q = \mathfrak{q}(0)$, $Q' = \mathfrak{q}(1)$ and $\varphi = \mathfrak{q}(0\bullet 1)$; arguing by induction on $|P\!:\!Q|$ and on $|Q'|/|Q|$, we will prove that we have $d_0(X)_{\mathfrak{q}} = 0$. Firstly note that, for another $\mathcal{F}$-nilcentralized subgroup Q'' of P and an $\mathcal{F}$-morphism $\varphi': Q' \to Q''$, considering the *chains*

$$\mathfrak{q}' : \Delta_1 \longrightarrow \mathcal{F}^{\mathrm{sc}} \quad , \quad \mathfrak{q}'' : \Delta_1 \longrightarrow \mathcal{F}^{\mathrm{sc}} \quad \text{and} \quad \mathfrak{q}^{Q''} : \Delta_2 \longrightarrow \mathcal{F}^{\mathrm{sc}} \qquad 14.6.6$$

respectively mapping 0 on Q', Q and Q, 1 on Q'', Q'' and Q', $0\bullet 1$ on φ', $\varphi'\circ\varphi$ and φ, 2 on Q'', and $1\bullet 2$ on φ', we have (cf. 14.5.2)

$$\begin{aligned}&\mathrm{res}_{\widehat{\mathfrak{aut}}(\mathrm{id}_{\mathfrak{q}^{Q''}\circ\delta_1^1},\delta_1^1)}\big(d_0(X)_{\mathfrak{q}''}\big)\\ &= \mathrm{res}_{\widehat{\mathfrak{aut}}(\mathrm{id}_{\mathfrak{q}^{Q''}\circ\delta_2^1},\delta_2^1)}\big(d_0(X)_{\mathfrak{q}}\big) + \mathrm{res}_{\widehat{\mathfrak{aut}}(\mathrm{id}_{\mathfrak{q}^{Q''}\circ\delta_0^1},\delta_0^1)}\big(d_0(X)_{\mathfrak{q}'}\big)\end{aligned} \qquad 14.6.7;$$

moreover, according to our choice above, we actually may assume that φ is not an isomorphism and then, according to our induction hypothesis, we already know that $d_0(X)_{\mathfrak{q}'} = 0$.

As above, let $R\in\mathfrak{X}$ and $\psi\in\mathcal{F}(R,Q)$ be our choice for Q in condition 14.6.1 and, setting $N = N_R\big(\psi(Q)\big)$, consider the *chains*

$$\mathfrak{n}:\Delta_1\longrightarrow\mathcal{F}^{^{\mathrm{sc}}}\quad,\quad \mathfrak{i}_N^R:\Delta_1\longrightarrow\mathcal{F}^{^{\mathrm{sc}}}\quad\text{and}\quad \mathfrak{r}:\Delta_1\longrightarrow\mathcal{F}^{^{\mathrm{sc}}} \qquad 14.6.8$$

where $\mathfrak{r}$ is defined as above, $\mathfrak{i}_N^R$ by the inclusion from R to N and $\mathfrak{n}$ by the restriction of ψ from Q to N; then, we have $d_0(X)_{\mathfrak{r}} = 0$ by equality 14.6.3, and it follows from the induction hypothesis that we still have $d_0(X)_{\mathfrak{i}_N^R} = 0$.

But, it follows from condition 14.6.1 above that the image in $\mathcal{F}(Q)$ of any element σ of $\mathcal{F}(N)_{\psi(Q)}$ can be lifted to an element ρ of $\mathcal{F}(R)$ which stabilizes $\psi(Q)$ and therefore it stabilizes N, so that the difference with σ of the restriction of ρ to N belongs to $\mathcal{F}_{C_P(\psi(Q))}(N)$ (cf. Proposition 4.6); consequently, since R contains $C_P\big(\psi(Q)\big)$, any element of $\mathcal{F}(N)_{\psi(Q)}$ can be lifted to an element of $\mathcal{F}(R)$. Thus, *mutatis mutandis,* considering the *chain* $\mathfrak{n}^R:\mathbf{\Delta}_2\longrightarrow\mathcal{F}^{^{\mathrm{sc}}}$ extending $\mathfrak{n}$, and mapping 2 on R and $1\bullet 2$ on the inclusion map $N\to R$, we have a k^*-group isomorphism

$$\widehat{\mathfrak{aut}}(\mathrm{id}_{\mathfrak{n}^R\circ\delta_2^1},\delta_2^1):\hat{\mathcal{F}}(\mathfrak{n}^R)\cong\hat{\mathcal{F}}(\mathfrak{n}) \qquad 14.6.9;$$

hence, since $\mathfrak{n}^R\circ\delta_1^1 = \mathfrak{r}$, it follows from equality 14.6.7 that $d_0(X)_{\mathfrak{n}} = 0$.

On the other hand, set $N' = N_{Q'}\big(\varphi(Q)\big)$ and consider the *chains*

$$\mathfrak{n}':\Delta_1\longrightarrow\mathcal{F}^{^{\mathrm{sc}}}\quad,\quad \mathfrak{i}_{N'}^{Q'}:\Delta_1\longrightarrow\mathcal{F}^{^{\mathrm{sc}}}\quad\text{and}\quad \mathfrak{q}:\Delta_1\longrightarrow\mathcal{F}^{^{\mathrm{sc}}} \qquad 14.6.10$$

where $\mathfrak{n}'$ is defined by the restriction of φ from Q to N' and $\mathfrak{i}_{N'}^{Q'}$ by the inclusion from N' to Q'; it follows from our induction hypothesis that we have $d_0(X)_{\mathfrak{i}_{N'}^{Q'}} = 0$, and it is clear that any element of $\mathcal{F}(Q')_{\varphi(Q)}$ stabilizes N'. As above, considering the *chain* $\mathfrak{n}'^{Q'}:\mathbf{\Delta}_2\to\mathcal{F}_{\mathrm{sc}}$ extending $\mathfrak{n}'$, and mapping 2 on Q' and $1\bullet 2$ on the inclusion map $N'\to Q'$, we still have a k^*-group isomorphism

$$\widehat{\mathfrak{aut}}(\mathrm{id}_{\mathfrak{n}'^{Q'}\circ\delta_1^1},\delta_1^1):\hat{\mathcal{F}}(\mathfrak{n}'^{Q'})\cong\hat{\mathcal{F}}(\mathfrak{q}) \qquad 14.6.11;$$

consequently, according to equality 14.6.7, in order to prove that $d_0(X)_{\mathfrak{q}} = 0$, it suffices to prove that $d_0(X)_{\mathfrak{n}'} = 0$.

That is to say, we may assume that $Q' = N'$ normalizes $\varphi(Q)$; in this case, it follows from Proposition 2.7 that there is an $\mathcal{F}$-morphism $\zeta' : Q' \to P$ such that $\zeta'\big(\varphi(Q)\big)$ is fully normalized in $\mathcal{F}$ and then, from condition 2.8.2, that there are an $\mathcal{F}$-morphism $\eta : N \to P$ and an element $\sigma \in \mathcal{F}(Q)$ fulfilling $\eta\Big(\psi\big(\sigma(u)\big)\Big) = \zeta'\big(\varphi(u)\big)$ for any $u \in Q$; actually, up to modifying our choice of ζ', we may assume that $\sigma = \mathrm{id}_Q$. Now, $\eta(N)$ and $\zeta'(Q')$ normalize $\zeta'\big(\varphi(Q)\big)$ and we consider the group $N'' = \langle \eta(N), \zeta'(Q')\rangle$ and the *chains*

$$\mathfrak{e} : \Delta_1 \longrightarrow \mathcal{F}^{^{\mathrm{sc}}} \quad , \quad \mathfrak{n}'' : \Delta_1 \longrightarrow \mathcal{F}^{^{\mathrm{sc}}} \quad \text{and} \quad \mathfrak{n}^{N''} : \Delta_2 \longrightarrow \mathcal{F}^{^{\mathrm{sc}}} \qquad 14.6.12,$$

where $\mathfrak{e}$ and $\mathfrak{n}''$ are respectively defined by the homomorphisms $N \to N''$ and $Q \to N''$ determined by η and $\zeta' \circ \varphi$, and $\mathfrak{n}^{N''}$ extends $\mathfrak{n}$ mapping 2 on N'' and $1 \bullet 2$ on the homomorphism $N \to N''$ determined by η.

We already know that $d_0(X)_{\mathfrak{n}} = 0$ and it follows from our induction hypothesis that $d_0(X)_{\mathfrak{e}} = 0$; but, as above, it follows from condition 14.6.1 that any element of $\mathcal{F}(Q)$ can be lifted *via* ψ to an element of $\mathcal{F}(R)$ which stabilizes N. In particular, for any element σ in $\mathcal{F}(N'')_{\zeta'(\varphi(Q))} = \mathcal{F}(\mathfrak{n}'')$, we have an element ρ in $\mathcal{F}(N)$ such that $\eta \circ \rho$ coincides with $\sigma \circ \eta$ over $\psi(Q)$ and therefore it follows from Proposition 4.6 that it suffices to modify our choice of ρ by composing it with the conjugation by a suitable element of $C_P\big(\psi(Q)\big) \subset N$ to get $\eta \circ \rho = \sigma \circ \eta$. In conclusion, the k^*-group homomorphism

$$\widehat{\mathfrak{aut}}(\mathrm{id}_{\mathfrak{n}^{N''} \circ \delta_1^1}, \delta_1^1) : \hat{\mathcal{F}}(\mathfrak{n}^{N''}) \longrightarrow \hat{\mathcal{F}}(\mathfrak{n}'') \qquad 14.6.13$$

is an isomorphism; consequently, by equality 14.6.7, we get $d_0(\xi)_{\mathfrak{n}''} = 0$.

Similarly, consider the *chains*

$$\mathfrak{z}' : \Delta_1 \longrightarrow \mathcal{F}^{^{\mathrm{sc}}} \quad , \quad \mathfrak{n}'' : \Delta_1 \longrightarrow \mathcal{F}^{^{\mathrm{sc}}} \quad \text{and} \quad \mathfrak{q}^{N''} : \Delta_2 \longrightarrow \mathcal{F}^{^{\mathrm{sc}}} \qquad 14.6.14$$

where $\mathfrak{z}'$ is defined by the homomorphism $Q' \to N''$ determined by ζ', and $\mathfrak{q}^{N''}$ extends $\mathfrak{q} = \mathfrak{n}'$ mapping 2 on N'' and $1 \bullet 2$ on the homomorphism $Q' \to N''$ determined by ζ'; it follows from our induction hypothesis that $d_0(X)_{\mathfrak{z}'} = 0$ and we already know that $d_0(X)_{\mathfrak{n}''} = 0$.

Since N'' contains

$$\eta\Big(C_P\big(\psi(Q)\big)\Big) = C_P\Big(\zeta'\big(\varphi(Q)\big)\Big) \qquad 14.6.15,$$

it follows from statement 2.10.1 that the automorphism of $\zeta'\big(\varphi(Q)\big)$ determined by any $\sigma' \in \mathcal{F}(Q')_{\varphi(Q)}$ *via* ζ' can be extended to an *automorphism* σ'' of N'' and then, arguing as above, it follows again from Proposition 4.6

that we may choose σ'' fulfilling $\sigma''\big(\zeta'(u')\big) = \zeta'\big(\sigma'(u')\big)$ for any $u' \in Q'$. In conclusion, the k^*-group homomorphism

$$\widehat{\mathfrak{aut}}(\mathrm{id}_{\mathfrak{q}^{N''}\circ\delta_2^1}, \delta_2^1) : \hat{\mathcal{F}}(\mathfrak{q}^{N''}) \longrightarrow \hat{\mathcal{F}}(\mathfrak{q}) \qquad 14.6.16$$

is an isomorphism; consequently, again by equality 14.6.7, we get $d_0(\xi)_{\mathfrak{q}} = 0$. We are done.

Corollary 14.7 *The inclusion $\mathcal{F}^{^{\mathrm{sc}}} \subset \mathcal{F}^{^{\mathrm{nc}}}$ induces an $\mathcal{O}$-module isomorphism*

$$\varprojlim\, (\mathfrak{g}_k \circ \widehat{\mathfrak{aut}}_{\mathcal{F}^{\mathrm{sc}}}) \cong \mathcal{G}_k(\mathcal{F}, \widehat{\mathfrak{aut}}_{\mathcal{F}^{\mathrm{nc}}}) \qquad 14.7.1.$$

Proof: If Q is a $\mathcal{F}$-nilcentralized subgroup of P and ψ an element of $\mathcal{F}(P,Q)$ such that $Q' = \psi(Q)$ is fully normalized in $\mathcal{F}$ (cf. Proposition 2.7), we already know that $R = Q'{\cdot}C_P(Q')$ is $\mathcal{F}$-selfcentralizing (cf. 4.8) and it follows from statement 2.10.1 that any $\sigma \in \mathcal{F}(Q')$ can be extended to some $\rho \in \mathcal{F}(P,R)$; since $C_P(Q') \subset R$, we have $\rho(R) = R$, so that ρ determines $\tau \in \mathcal{F}(R)$ (cf. 2.3) fulfilling $\tau \circ \psi = \psi \circ \sigma$. That is to say, the set of $\mathcal{F}$-selfcentralizing subgroups of P fulfills condition 14.6.1 and therefore the corollary follows from Proposition 14.6.

14.8 We are ready to move from $\mathcal{F}^{^{\mathrm{nc}}}$ to the *exterior quotient* $\tilde{\mathcal{F}}^{^{\mathrm{sc}}}$ (cf. 1.3); as in 14.1.1, we have the corresponding *automorphism functor* (cf. Proposition A2.10)

$$\mathfrak{aut}_{\tilde{\mathcal{F}}^{\mathrm{sc}}} : \mathfrak{ch}^*(\tilde{\mathcal{F}}^{^{\mathrm{sc}}}) \longrightarrow \mathfrak{Gr} \qquad 14.8.1$$

mapping any $\mathfrak{ch}^*(\tilde{\mathcal{F}}^{^{\mathrm{sc}}})$-object $(\tilde{\mathfrak{q}}, \Delta_n)$ on its $\mathfrak{ch}^*(\tilde{\mathcal{F}}^{^{\mathrm{sc}}})$-automorphism group — still noted $\tilde{\mathcal{F}}(\tilde{\mathfrak{q}})$; this time, it follows from Corollary 4.9 that $\tilde{\mathcal{F}}(\tilde{\mathfrak{q}})$ can be identified with the subgroup of $\tilde{\mathcal{F}}\big(\tilde{\mathfrak{q}}(0)\big)$ formed by the $\tilde{\mathcal{F}}$-isomorphisms $\tilde{\sigma}_0 : \tilde{\mathfrak{q}}(0) \cong \tilde{\mathfrak{q}}(0)$ which, for any $i \in \Delta_n$, fulfill

$$\tilde{\mathfrak{q}}(0\bullet i) \circ \tilde{\sigma}_0 \in \tilde{\mathcal{F}}\big(\tilde{\mathfrak{q}}(i)\big) \circ \tilde{\mathfrak{q}}(0\bullet i) \qquad 14.8.2.$$

Indeed, for any element $(\tilde{\sigma}, \mathrm{id}_{\Delta_n})$ of $\tilde{\mathcal{F}}(\tilde{\mathfrak{q}})$, since $\tilde{\sigma} : \tilde{\mathfrak{q}} \cong \tilde{\mathfrak{q}}$ is a natural isomorphism, for any $0 \le i \le n$ we have an element $\tilde{\sigma}_i \in \tilde{\mathcal{F}}\big(\tilde{\mathfrak{q}}(i)\big)$ in such a way that

$$\tilde{\mathfrak{q}}(0\bullet i) \circ \tilde{\sigma}_0 = \tilde{\sigma}_i \circ \tilde{\mathfrak{q}}(0\bullet i) \qquad 14.8.3;$$

but, according to Corollary 4.9, $\tilde{\sigma}_i$ is then determined by $\tilde{\sigma}_0$ for any $1 \le i \le n$.

14.9 Actually, denoting by $\mathfrak{e}^{\mathrm{sc}} : \mathcal{F}^{^{\mathrm{sc}}} \to \tilde{\mathcal{F}}^{^{\mathrm{sc}}}$ the canonical functor, it is quite clear that $\tilde{\mathfrak{q}} = \mathfrak{e}^{\mathrm{sc}} \circ \mathfrak{q}$ for some *chain* $\mathfrak{q} : \Delta_n \to \mathcal{F}^{^{\mathrm{sc}}}$, and that a suitable representative $\sigma_n \in \mathcal{F}\big(\mathfrak{q}(n)\big)$ of $\tilde{\sigma}_n$ determines a natural isomorphism $\sigma : \mathfrak{q} \cong \mathfrak{q}$

lifting $\tilde{\sigma}$; that is to say, since $C_{\mathfrak{q}(0)_n}\big(\mathfrak{q}(n)\big) = Z\big(\mathfrak{q}(n)\big)$ (cf. 14.1), we get an evident exact sequence

$$0 \longrightarrow Z\big(\mathfrak{q}(n)\big) \longrightarrow \mathfrak{q}(0) \longrightarrow \mathcal{F}(\mathfrak{q}) \longrightarrow \tilde{\mathcal{F}}(\tilde{\mathfrak{q}}) \longrightarrow 0 \qquad 14.9.1.$$

Moreover, if $(\tilde{\mathfrak{r}}, \Delta_m)$ is another $\mathfrak{ch}^*(\tilde{\mathcal{F}}^{^{\rm sc}})$-object and $(\tilde{\nu}, \delta) : (\tilde{\mathfrak{r}}, \Delta_m) \to (\tilde{\mathfrak{q}}, \Delta_n)$ a $\mathfrak{ch}^*(\tilde{\mathcal{F}}^{^{\rm sc}})$-morphism, for any *chain* $\mathfrak{r} : \Delta_m \to \mathcal{F}^{^{\rm sc}}$ such that $\tilde{\mathfrak{r}} = \mathfrak{e}^{\rm sc} \circ \mathfrak{r}$, it is not difficult to check that $(\tilde{\nu}, \delta)$ can be lifted to a $\mathfrak{ch}^*(\mathcal{F}^{^{\rm sc}})$-morphism $(\nu, \delta) : (\mathfrak{r}, \Delta_m) \to (\mathfrak{q}, \Delta_n)$ and that we get a commutative diagram

$$\begin{array}{ccccccccc} 0 \longrightarrow & Z\big(\mathfrak{r}(n)\big) & \longrightarrow & \mathfrak{r}(0) & \longrightarrow & \mathcal{F}(\mathfrak{r}) & \longrightarrow & \tilde{\mathcal{F}}(\tilde{\mathfrak{r}}) & \longrightarrow 0 \\ & \downarrow & & \downarrow & & \downarrow \alpha & & \downarrow \bar{\alpha} & \\ 0 \longrightarrow & Z\big(\mathfrak{q}(n)\big) & \longrightarrow & \mathfrak{q}(0) & \longrightarrow & \mathcal{F}(\mathfrak{q}) & \longrightarrow & \tilde{\mathcal{F}}(\tilde{\mathfrak{q}}) & \longrightarrow 0 \end{array} \qquad 14.9.2.$$

where $\bar{\alpha} = \mathfrak{aut}_{\tilde{\mathcal{F}}^{\rm sc}}(\tilde{\nu}, \delta)$ and $\alpha = \mathfrak{aut}_{\mathcal{F}^{\rm sc}}(\nu, \delta)$.

14.10 In other words, respectively denoting by

$$\mathfrak{v}_{\mathcal{F}^{\rm sc}} : \mathfrak{ch}^*(\mathcal{F}^{^{\rm sc}}) \longrightarrow \mathcal{F}^{^{\rm sc}} \subset \mathfrak{Gr} \quad \text{and} \quad \mathfrak{z}_{\mathfrak{ch}^*(\mathcal{F}^{\rm sc})} : \mathfrak{ch}^*(\mathcal{F}^{^{\rm sc}}) \longrightarrow \mathfrak{Gr} \qquad 14.10.1$$

the *evaluation functor* mapping $(\mathfrak{q}, \Delta_n)$ on $\mathfrak{q}(0)$ (cf. A3.7.3) and the evident functor mapping $(\mathfrak{q}, \Delta_n)$ on $Z\big(\mathfrak{q}(n)\big)$ (cf. 13.13), we have obtained the exact sequence of functors

$$1 \longrightarrow \mathfrak{z}_{\mathfrak{ch}^*(\mathcal{F}^{\rm sc})} \longrightarrow \mathfrak{v}_{\mathcal{F}^{\rm sc}} \longrightarrow \mathfrak{aut}_{\mathcal{F}^{\rm sc}} \longrightarrow \mathfrak{aut}_{\tilde{\mathcal{F}}^{\rm sc}} \circ \mathfrak{ch}^*(\mathfrak{e}^{\rm sc}) \longrightarrow 1 \qquad 14.10.2.$$

In particular, considering the restricted functor (cf. 14.2.1)

$$\widehat{\mathfrak{aut}}_{\mathcal{F}^{\rm sc}} : \mathfrak{ch}^*(\mathcal{F}^{^{\rm sc}}) \longrightarrow k^*\text{-}\mathfrak{Gr} \qquad 14.10.3,$$

the *converse image* over $\mathfrak{v}_{\mathcal{F}^{\rm sc}}$ of the natural map $\widehat{\mathfrak{aut}}_{\mathcal{F}^{\rm sc}} \to \mathfrak{aut}_{\mathcal{F}^{\rm sc}}$ is *split* since $\mathfrak{v}_{\mathcal{F}^{\rm sc}}$ maps any $\mathfrak{ch}^*(\mathcal{F}^{^{\rm sc}})$-object on a p-group, and therefore the natural map $\mathfrak{v}_{\mathcal{F}^{\rm sc}} \to \mathfrak{aut}_{\mathcal{F}^{\rm sc}}$ can be lifted to $\widehat{\mathfrak{aut}}_{\mathcal{F}^{\rm sc}}$, defining a functor

$$\widehat{\mathfrak{aut}}_{\tilde{\mathcal{F}}^{\rm sc}} : \mathfrak{ch}^*(\tilde{\mathcal{F}}^{^{\rm sc}}) \longrightarrow k^*\text{-}\mathfrak{Gr} \qquad 14.10.4$$

which lifts the *automorphism functor* (cf. 14.8.1); as above, for any $\mathfrak{ch}^*(\tilde{\mathcal{F}}^{^{\rm sc}})$-object $(\tilde{\mathfrak{q}}, \Delta_n)$, we set

$$\widehat{\mathfrak{aut}}_{\tilde{\mathcal{F}}^{\rm sc}}(\tilde{\mathfrak{q}}, \Delta_n) = \hat{\tilde{\mathcal{F}}}(\tilde{\mathfrak{q}}) \qquad 14.10.5.$$

14.11 As announced, we now can replace $\widehat{\mathfrak{aut}}_{\mathcal{F}^{\rm sc}}$ by $\widehat{\mathfrak{aut}}_{\tilde{\mathcal{F}}^{\rm sc}}$; indeed, for any $\mathfrak{ch}^*(\mathcal{F}^{^{\rm sc}})$-object $(\mathfrak{q}, \Delta_n)$, from the exact sequence 14.9.1 and this *splitting* we get a surjective k^*-group homomorphism $\hat{\mathcal{F}}(\mathfrak{q}) \to \hat{\tilde{\mathcal{F}}}(\tilde{\mathfrak{q}})$ and therefore the restriction induces an injective $\mathcal{O}$-module homomorphism

$$\mathcal{G}_k\big(\hat{\tilde{\mathcal{F}}}(\tilde{\mathfrak{q}})\big) \longrightarrow \mathcal{G}_k\big(\hat{\mathcal{F}}(\mathfrak{q})\big) \qquad 14.11.1$$

which is actually an isomorphism since the kernel of that k^*-group homomorphism is a p-group; that is to say, we get a natural isomorphism

$$\mathfrak{g}_k \circ \widehat{\mathfrak{aut}}_{\tilde{\mathcal{F}}^{\mathrm{sc}}} \circ \mathfrak{ch}^*(\mathfrak{e}^{\mathrm{sc}}) \cong \mathfrak{g}_k \circ \widehat{\mathfrak{aut}}_{\mathcal{F}^{\mathrm{sc}}} \qquad 14.11.2$$

and therefore, since the functor $\mathfrak{ch}^*(\mathfrak{e}^{\mathrm{sc}})$ is *full* and *essentially surjective*, it follows from Corollary 14.7 that

$$\mathcal{G}_k(\mathcal{F}, \widehat{\mathfrak{aut}}_{\mathcal{F}^{\mathrm{nc}}}) \cong \lim_{\longleftarrow} (\mathfrak{g}_k \circ \widehat{\mathfrak{aut}}_{\tilde{\mathcal{F}}^{\mathrm{sc}}}) \qquad 14.11.3.$$

14.12 In order to determine the $\mathcal{O}$-rank of $\mathcal{G}_k(\mathcal{F}, \widehat{\mathfrak{aut}}_{\mathcal{F}^{\mathrm{nc}}})$, the other ingredient we need is the *character decomposition* of the functor $\mathfrak{g}_k$ (cf. 14.3). First of all, we introduce some notation; for any $h \in \mathbb{N} - p\mathbb{N}$, fix a primitive h-th root of unity ξ_h in k and set $U_h = \langle \xi_h \rangle$ and $\hat{U}_h = k^* \times U_h$; note that the kernel of the canonical group homomorphism

$$\mathrm{Aut}_{k^*}(\hat{U}_h) \longrightarrow \mathrm{Aut}(U_h) \qquad 14.12.1$$

can be identified with $\mathrm{Hom}(U_h, k^*)$ and that ξ_h determines a group isomorphism $\mathrm{Hom}(U_h, k^*) \cong U_h$. Recall that we have a canonical group isomorphism (cf. Proposition 6 in [55, Ch. II])

$$\mathcal{O}^* \cong k^* \times \big(1 + J(\mathcal{O})\big) \qquad 14.12.2$$

and, as in 1.25, the inclusion $\mathcal{O}^* \subset \mathcal{O}$ and this isomorphism determine an evident $\mathcal{O}$-algebra homomorphism $\mathcal{O}k^* \to \mathcal{O}$.

14.13 As in 1.25, for any k^*-group $\hat{G}$ with finite k^*-quotient G (cf 1.23), denoting by $\mathcal{O}\hat{G}$ the *group algebra* of the group $\hat{G}$, we set

$$\mathcal{O}_*\hat{G} = \mathcal{O} \otimes_{\mathcal{O}k^*} \mathcal{O}\hat{G} \qquad 14.13.1$$

which is an $\mathcal{O}$-algebra of $\mathcal{O}$-rank $|G|$. Now, for any injective k^*-group homomorphism $\hat{\eta}: \hat{U}_h \to \hat{G}$, the *modular characters* introduced by Brauer (cf. 18 in [54]) determine an $\mathcal{O}$-module homomorphism $\mathcal{G}_k(\hat{G}) \to \mathcal{O}$ mapping the class of $k \otimes_{\mathcal{O}} M$ in $\mathcal{G}_k(\hat{G})$ on the linear trace $\mathrm{tr}_M\big(\hat{\eta}(1, \xi_h)\big)$ for any $\mathcal{O}_*\hat{G}$-module M ; moreover, if we have $\hat{\sigma}(1, \xi_h) = (\lambda, \xi_h)$ for some element $\hat{\sigma}$ in the kernel of homomorphism 14.12.1 then we still have

$$\mathrm{tr}_M\big((\hat{\eta} \circ \hat{\sigma})(1, \xi_h)\big) = \lambda{\cdot}\mathrm{tr}_M\big(\hat{\eta}(1, \xi_h)\big) \qquad 14.13.2;$$

in particular, denoting by $\mathrm{Mon}_{k^*}(\hat{U}_h, \hat{G})$ the set of injective k^*-homomorphisms from $\hat{U}_h$ to $\hat{G}$, U_h acts on this set *via* the identification above with the kernel of homomorphism 14.12.1, and acts on $\mathcal{O}$ *via* isomorphism 14.12.2 together with the inclusion $U_h \subset k^*$.

14.14 Thus, denoting by $\mathcal{F}ct_{U_h}\big(\mathrm{Mon}_{k^*}(\hat U_h,\hat G),\mathcal{O}\big)$ the set of $\mathcal{O}$-valued functions which preserve the corresponding U_h-actions, we have obtained an $\mathcal{O}$-module homomorphism

$$\mathcal{G}_k(\hat G) \longrightarrow \mathcal{F}ct_{U_h}\big(\mathrm{Mon}_{k^*}(\hat U_h,\hat G),\mathcal{O}\big) \qquad 14.14.1;$$

moreover, G acts by conjugation on $\mathrm{Mon}_{k^*}(\hat U_h,\hat G)$ centralizing the action of U_h , so that it acts on $\mathcal{F}ct_{U_h}\big(\mathrm{Mon}_{k^*}(\hat U_h,\hat G),\mathcal{O}\big)$; actually, since $\hat G = k^*{\cdot}G'$ for a suitable finite subgroup G' of $\hat G$ and then we have $\mathcal{G}_k(\hat G)\cong \mathcal{G}_k(kG'e')$ where we set $Z' = k^*\cap G'$ and $e' = \frac{1}{|Z'|}\cdot\sum_{z'\in Z'} z'$ (cf Proposition 5.15 in [42]), it is not difficult to prove that homomorphisms 14.14.1 when h runs over $\mathbb{N}-p\mathbb{N}$ determine an $\mathcal{O}$-module isomorphism

$$\mathcal{G}_k(\hat G) \cong \prod_{h\in\mathbb{N}-p\mathbb{N}} \mathcal{F}ct_{U_h}\big(\mathrm{Mon}_{k^*}(\hat U_h,\hat G),\mathcal{O}\big)^G \qquad 14.14.2.$$

Note that any function $\chi\in\mathcal{F}ct_{U_h}\big(\mathrm{Mon}_{k^*}(\hat U_h,\hat G),\mathcal{O}\big)^G$ *vanishes* on the k^*-group monomorphisms $\hat\eta:\hat U_h\to\hat G$ mapping $(1,\xi_h)$ on an element $\hat x\in\hat G$ such that the image of $C_{\hat G}(\hat x)$ in G is a proper subgroup of $C_G(x)$ where $x = k^*{\cdot}\hat x$.

14.15 Pushing it further, denote by $\mathrm{Mon}(U_h,G)$ the set of injective group homomorphisms from U_h to G and by

$$\varpi_{h,\hat G} : \mathrm{Mon}_{k^*}(\hat U_h,\hat G) \longrightarrow \mathrm{Mon}(U_h,G) \qquad 14.15.1$$

the canonical map, so that U_h acts *regularly* on the *fibers*; thus, we still have the obvious decomposition

$$\mathcal{F}ct_{U_h}\big(\mathrm{Mon}_{k^*}(\hat U_h,\hat G),\mathcal{O}\big) \cong \prod_{\eta\in\mathrm{Mon}(U_h,G)} \mathcal{F}ct_{U_h}\big((\varpi_{h,\hat G})^{-1}(\eta),\mathcal{O}\big) \qquad 14.15.2$$

and therefore we get

$$\mathcal{G}_k(\hat G) \cong \Big(\prod_{h\in\mathbb{N}-p\mathbb{N}}\ \prod_{\eta\in\mathrm{Mon}(U_h,G)} \mathcal{F}ct_{U_h}\big((\varpi_{h,\hat G})^{-1}(\eta),\mathcal{O}\big)\Big)^G \qquad 14.15.3.$$

Then, for any $\eta\in\mathrm{Mon}(U_h,G)$, the term $\mathcal{F}ct_{U_h}\big((\varpi_{h,\hat G})^{-1}(\eta),\mathcal{O}\big)$ is a *free $\mathcal{O}$-module of rank* 1, but note that the quotient of $C_G\big(\eta(U_h)\big)$ by the image of $C_{\hat G}\big(\hat\eta(\hat U_h)\big)$, where $\hat\eta\in(\varpi_{h,\hat G})^{-1}(\eta)$, acts *faithfully* on it.

14.16 As a matter of fact, isomorphisms 14.14.2 and 14.15.3 are *natural*. In order to prove this *naturality*, we have to develop a suitable functorial framework; respectively denote by k^*-$\mathfrak{iGr}$ and by $\mathfrak{iGr}$ the subcategories of k^*-$\mathfrak{Gr}$ and $\mathfrak{Gr}$ formed by the same objects and by the homomorphisms which are injective *over the corresponding set of p'-elements* — called

p'-injective homomorphisms — and by $\mathfrak{qt}: k^*\text{-}\mathfrak{iGr} \to \mathfrak{iGr}$ the obvious *k^*-quotient* functor (cf. 1.23). As in 6.3 above, denote by ${}^{U_h}\aleph$ the category of finite sets endowed with a U_h-action and by

$$\mathfrak{res}_1^{U_h} : {}^{U_h}\aleph \longrightarrow \aleph \tag{14.16.1}$$

the corresponding *forgetful* functor; then, we have two functors

$$\hat{\mathfrak{u}}_h : k^*\text{-}\mathfrak{iGr} \longrightarrow {}^{U_h}\aleph \quad \text{and} \quad \mathfrak{u}_h : \mathfrak{iGr} \longrightarrow \aleph \tag{14.16.2}$$

which map any *finite* k^*-group $\hat{G}$ (cf. 1.23) on the U_h-set $\mathrm{Mon}_{k^*}(\hat{U}_h, \hat{G})$ and a finite group G on $\mathrm{Mon}(U_h, G)$, and still have a natural map sending $\hat{G}$ to $\varpi_{h,\hat{G}}$ (cf. 14.15.1)

$$\varpi_h : \mathfrak{res}_1^{U_h} \circ \hat{\mathfrak{u}}_h \longrightarrow \mathfrak{u}_h \circ \mathfrak{qt} \tag{14.16.3.}$$

14.17 Moreover, for any h' dividing h, choosing the identification between $\mathrm{Hom}(U_{h'}, k^*)$ and $U_{h'}$ determined by $(\xi_h)^{h/h'}$, the inclusion $U_{h'} \subset U_h$ induces a functor and two natural maps

$$\begin{gathered} \mathfrak{res}_{U_{h'}}^{U_h} : {}^{U_h}\aleph \longrightarrow {}^{U_{h'}}\aleph \\ \hat{\rho}_{h',h} : \mathfrak{res}_{U_{h'}}^{U_h} \circ \hat{\mathfrak{u}}_h \longrightarrow \hat{\mathfrak{u}}_{h'} \quad \text{and} \quad \rho_{h',h} : \mathfrak{u}_h \longrightarrow \mathfrak{u}_{h'} \end{gathered} \tag{14.17.1,}$$

and it is easily checked that (cf. A1.5.1)

$$\varpi_{h'} \circ (\hat{\rho}_{h',h} * \mathfrak{res}_{U_{h'}}^{U_h}) = (\rho_{h',h} * \mathfrak{qt}) \circ \varpi_h \tag{14.17.2.}$$

Furthermore, identifying the category of sets $\aleph$ with the full subcategory of the *category of small categories* $\mathfrak{CC}$ (cf. A1.6) over the small categories which have no other morphisms than the corresponding *identity* morphisms, we can consider the functor

$$\mathfrak{u}_h \circ \mathfrak{qt} : k^*\text{-}\mathfrak{iGr} \longrightarrow \aleph \subset \mathfrak{CC} \tag{14.17.3}$$

as a *representation* of $k^*\text{-}\mathfrak{iGr}$ (cf. A2.2) and thus we can apply to it the construction of the corresponding *semidirect product* (cf. A2.7)

$${}^{\mathfrak{u}_h}(k^*\text{-}\mathfrak{iGr}) = (\mathfrak{u}_h \circ \mathfrak{qt}) \rtimes (k^*\text{-}\mathfrak{iGr}) \tag{14.17.4}$$

where the objects are the pairs $(\eta, \hat{G})$ formed by a k^*-group $\hat{G}$ with finite k^*-quotient G, and by an injective group homomorphism $\eta: U_h \to G$, and the morphisms from $(\eta, \hat{G})$ to another object $(\eta', \hat{G}')$ are the p'-injective homomorphisms $\hat{\varphi}: \hat{G} \to \hat{G}'$ fulfilling $\eta' = \mathfrak{qt}(\hat{\varphi}) \circ \eta$. Moreover, for any h'

dividing h, the natural map $\rho_{h',h}$ above determines a functor (cf. Proposition A2.17)

$$(\rho_{h',h} * \mathfrak{qt}) \rtimes \mathfrak{id}_{k^*\text{-}\mathfrak{iGr}} : {}^{\mathfrak{u}_h}(k^*\text{-}\mathfrak{iGr}) \longrightarrow {}^{\mathfrak{u}_{h'}}(k^*\text{-}\mathfrak{iGr}) \qquad 14.17.5$$

mapping $(\eta, \hat{G})$ on $(\eta', \hat{G})$ where $\eta' : U_{h'} \to \hat{G}$ is the restriction of η.

Proposition 14.18 *With the notation above, we have a functor*

$$\hat{\mathfrak{w}}_h : {}^{\mathfrak{u}_h}(k^*\text{-}\mathfrak{iGr}) \longrightarrow {}^{U_h}\aleph \qquad 14.18.1$$

mapping any ${}^{\mathfrak{u}_h}(k^*\text{-}\mathfrak{iGr})$*-object* $(\eta, \hat{G})$ *on the regular* U_h*-set* $(\varpi_{h,\hat{G}})^{-1}(\eta)$ *and any* ${}^{\mathfrak{u}_h}(k^*\text{-}\mathfrak{iGr})$*-morphism* $\hat{\varphi} : (\eta, \hat{G}) \to (\eta', \hat{G}')$ *on the bijective map*

$$(\varpi_{h,\hat{G}})^{-1}(\eta) \cong (\varpi_{h,\hat{G}'})^{-1}(\eta') \qquad 14.18.2$$

determined by $\hat{\mathfrak{u}}_h(\hat{\varphi})$. *Moreover,* $\hat{\mathfrak{u}}_h$ *is the direct image of* $\hat{\mathfrak{w}}_h$ *via the structural functor*

$$\mathfrak{p}_h : {}^{\mathfrak{u}_h}(k^*\text{-}\mathfrak{iGr}) \longrightarrow k^*\text{-}\mathfrak{iGr} \qquad 14.18.3.$$

Proof: As we mention above, the ${}^{\mathfrak{u}_h}(k^*\text{-}\mathfrak{iGr})$-objects are the pairs formed by a k^*-$\mathfrak{iGr}$-object $\hat{G}$ and by an object of the "category" $\mathrm{Mon}(U_h, G)$ where $G = \mathfrak{qt}(\hat{G})$, namely an injective group homomorphism $\eta : U_h \to G$; then, in the map (cf. 14.15.1)

$$\varpi_{h,\hat{G}} : \mathrm{Mon}_{k^*}(\hat{U}_h, \hat{G}) \longrightarrow \mathrm{Mon}(U_h, G) \qquad 14.18.4,$$

U_h acts on the left end stabilizing and acting regularly on the fibers, so that $(\varpi_{h,\hat{G}})^{-1}(\eta)$ is indeed a regular U_h*-set*. Analogously, the ${}^{\mathfrak{u}_h}(k^*\text{-}\mathfrak{iGr})$-morphisms between two objects $(\eta, \hat{G})$ and $(\eta', \hat{G}')$ are the pairs formed by a k^*-$\mathfrak{iGr}$-morphism $\hat{\varphi} : \hat{G} \to \hat{G}'$ and by a "$\mathrm{Mon}(U_h, G')$-morphism" from the image of η by the functor $(\mathfrak{u}_h \circ \mathfrak{qt})(\hat{\varphi})$ to η', which actually forces the equality $\mathfrak{qt}(\hat{\varphi}) \circ \eta = \eta'$. Then, it is clear that the map

$$\hat{\mathfrak{u}}(\hat{\varphi}) : \mathrm{Mon}_{k^*}(\hat{U}_h, \hat{G}) \longrightarrow \mathrm{Mon}_{k^*}(\hat{U}_h, \hat{G}') \qquad 14.18.5$$

sends $(\varpi_{h,\hat{G}})^{-1}(\eta)$ bijectively onto $(\varpi_{h,\hat{G}'})^{-1}(\eta')$, determining a U_h*-set map*. The proofs of the functoriality and of the last statement are straightforward.

Remark 14.19 Note that, for any $\xi \in U_h$, the inner k^*-group automorphism of $\hat{G}$ determined by an element lifting $\eta(\xi)$ to $\hat{G}$ acts trivially on $(\varpi_{h,\hat{G}})^{-1}(\eta)$.

Proposition 14.20 *With the notation above, for any h' dividing h we have a natural map*

$$\hat{\tau}_{h',h} : \mathfrak{res}^{U_h}_{U_{h'}} \circ \hat{\mathfrak{w}}_h \longrightarrow \hat{\mathfrak{w}}_{h'} \circ \big((\rho_{h',h} * \mathfrak{qt}) \rtimes \mathfrak{id}_{k^*\text{-}\mathfrak{i}\mathfrak{Gr}}\big) \qquad 14.20.1$$

sending any ${}^{\mathfrak{u}_h}(k^*\text{-}\mathfrak{i}\mathfrak{Gr})$-object $(\eta, \hat{G})$ to the $U_{h'}$-morphism

$$\mathfrak{res}^{U_h}_{U_{h'}}\big((\varpi_{h,\hat{G}})^{-1}(\eta)\big) \longrightarrow (\varpi_{h',\hat{G}})^{-1}\big(\mathrm{Res}^{U_h}_{U_{h'}}(\eta)\big) \qquad 14.20.2$$

which maps any $\hat{\eta} \in (\varpi_{h,\hat{G}})^{-1}(\eta) \subset \mathrm{Mon}_{k^*}(\hat{U}_h, \hat{G})$ on its restriction to $\hat{U}_{h'}$.

Proof: For any k^*-group homomorphism $\hat{\varphi}: \hat{G} \to \hat{G}'$, setting $\varphi = \mathfrak{qt}(\hat{\varphi})$ the functor $\mathfrak{res}^{U_h}_{U_{h'}} \circ \hat{\mathfrak{w}}_h$ maps the ${}^{\mathfrak{u}_h}(k^*\text{-}\mathfrak{i}\mathfrak{Gr})$-morphism $(\eta, \hat{G}) \to (\varphi \circ \eta, \hat{G}')$ on the $U_{h'}$-set map

$$\mathfrak{res}^{U_h}_{U_{h'}}\big((\varpi_{h,\hat{G}})^{-1}(\eta)\big) \longrightarrow \mathfrak{res}^{U_h}_{U_{h'}}\big((\varpi_{h,\hat{G}'})^{-1}(\varphi \circ \eta)\big) \qquad 14.20.3$$

sending $\hat{\eta} \in (\varpi_{h,\hat{G}})^{-1}(\eta)$ to $\hat{\varphi} \circ \hat{\eta}$ whereas, setting $\eta' = \mathrm{Res}^{U_h}_{U_{h'}}(\eta)$, the functor $\hat{\mathfrak{w}}_{h'} \circ \big((\rho_{h',h} * \mathfrak{qt}) \rtimes \mathfrak{id}_{k^*\text{-}\mathfrak{i}\mathfrak{Gr}}\big)$ maps this morphism on the analogous $U_{h'}$-set map

$$(\varpi_{h',\hat{G}})^{-1}(\eta') \longrightarrow (\varpi_{h',\hat{G}'})^{-1}(\varphi \circ \eta') \qquad 14.20.4;$$

thus, the corresponding diagram is indeed commutative since the restriction to $U_{h'}$ is compatible with the composition with $\hat{\varphi}$ on the left.

14.21 We are ready to discuss the *naturality* of isomorphisms 14.14.2 and 14.15.3. First of all, consider the evident *contravariant* functor

$$\mathcal{F}ct_{U_h} : {}^{U_h}\aleph \longrightarrow \mathcal{O}\text{-}\mathfrak{mod} \qquad 14.21.1$$

mapping any finite U_h-set X on the $\mathcal{O}$-module $\mathcal{F}ct_{U_h}(X, \mathcal{O})$ of the $\mathcal{O}$-valued functions over X preserving the U_h-actions; note that if $\xi \cdot x = x$ for some $x \in X$ and some $\xi \in U_h - \{1\}$ then we have $f(x) = 0$ for any $f \in \mathcal{F}ct_{U_h}(X, \mathcal{O})$. On the other hand, note that if we have a *contravariant* functor

$$\mathfrak{m} : k^*\text{-}\mathfrak{i}\mathfrak{Gr} \longrightarrow \mathcal{O}\text{-}\mathfrak{mod} \qquad 14.21.2,$$

for any k^*-group $\hat{G}$ with finite k^*-quotient G (cf. 1.23) $\mathfrak{m}(\hat{G})$ has an obvious $\mathcal{O}G$-module structure, so that it makes sense to consider

$$\mathbb{H}^0\big(G, \mathfrak{m}(\hat{G})\big) = \mathfrak{m}(\hat{G})^G \qquad 14.21.3;$$

further, if $\hat{\varphi}: \hat{G} \to \hat{G}'$ is a p'-injective k^*-group homomorphism (cf. 14.16), it is easily checked that $\mathfrak{m}(\hat{\varphi})$ maps $\mathfrak{m}(\hat{G}')^{G'}$ on an $\mathcal{O}$-submodule of $\mathfrak{m}(\hat{G})^G$;

that is to say, we get a new *contravariant* functor from $k^*\text{-}\mathfrak{iGr}$ to $\mathcal{O}\text{-}\mathfrak{mod}$ — denoted by $\mathfrak{h}^0(\mathfrak{m})$ — mapping $\hat{G}$ on $\mathfrak{m}(\hat{G})^G$ and $\hat{\varphi}$ on the map from $\mathfrak{m}(\hat{G}')^{G'}$ to $\mathfrak{m}(\hat{G})^G$ induced by $\mathfrak{m}(\hat{\varphi})$.

14.22 With all this notation, it is now clear that isomorphism 14.14.2 actually defines a natural isomorphism

$$\mathfrak{g}_k \cong \prod_{h\in\mathbb{N}-p\mathbb{N}} \mathfrak{h}^0(\mathcal{F}ct_{U_h} \circ \hat{\mathfrak{u}}_h) \qquad 14.22.1.$$

Consequently, isomorphism 14.11.3 becomes

$$\mathcal{G}_k(\mathcal{F}, \widehat{\mathfrak{aut}}_{\mathcal{F}^{\mathrm{nc}}}) \cong \varprojlim \Big(\prod_{h\in\mathbb{N}-p\mathbb{N}} \mathfrak{h}^0(\mathcal{F}ct_{U_h} \circ \hat{\mathfrak{u}}_h) \circ \widehat{\mathfrak{aut}}_{\tilde{\mathcal{F}}^{\mathrm{sc}}} \Big) \qquad 14.22.2;$$

that is to say, we have

$$\mathcal{G}_k(\mathcal{F}, \widehat{\mathfrak{aut}}_{\mathcal{F}^{\mathrm{nc}}}) \cong \prod_{h\in\mathbb{N}-p\mathbb{N}} \mathcal{G}_k(\mathcal{F}, \widehat{\mathfrak{aut}}_{\mathcal{F}^{\mathrm{nc}}})_h \qquad 14.22.3$$

where, for any $h \in \mathbb{N} - p\mathbb{N}$, we set

$$\mathcal{G}_k(\mathcal{F}, \widehat{\mathfrak{aut}}_{\mathcal{F}^{\mathrm{nc}}})_h = \varprojlim \big(\mathfrak{h}^0(\mathcal{F}ct_{U_h} \circ \hat{\mathfrak{u}}_h) \circ \widehat{\mathfrak{aut}}_{\tilde{\mathcal{F}}^{\mathrm{sc}}} \big) \qquad 14.22.4$$

which, as it is easily checked, yields

$$\mathcal{G}_k(\mathcal{F}, \widehat{\mathfrak{aut}}_{\mathcal{F}^{\mathrm{nc}}})_h \cong \varprojlim (\mathcal{F}ct_{U_h} \circ \hat{\mathfrak{u}}_h \circ \widehat{\mathfrak{aut}}_{\tilde{\mathcal{F}}^{\mathrm{sc}}}) \qquad 14.22.5.$$

14.23 Recall that, according to Proposition 14.18, the functor

$$\hat{\mathfrak{u}}_h : k^*\text{-}\mathfrak{iGr} \longrightarrow {}^{U_h}\aleph \qquad 14.23.1$$

is the *direct image* of the functor (cf. 1.6)

$$\hat{\mathfrak{w}}_h : {}^{\mathfrak{u}_h}(k^*\text{-}\mathfrak{iGr}) \longrightarrow {}^{U_h}\aleph \qquad 14.23.2$$

throughout the structural functor (cf. A2.7.1)

$$\mathfrak{p}_h : {}^{\mathfrak{u}_h}(k^*\text{-}\mathfrak{iGr}) \longrightarrow k^*\text{-}\mathfrak{iGr} \qquad 14.23.3.$$

Moreover, since $\mathfrak{u}_h \circ \mathfrak{qt} \circ \widehat{\mathfrak{aut}}_{\tilde{\mathcal{F}}^{\mathrm{sc}}} = \mathfrak{u}_h \circ \mathfrak{aut}_{\tilde{\mathcal{F}}^{\mathrm{sc}}}$, similarly considering the semi-direct product

$${}^{\mathfrak{u}_h}\mathfrak{ch}^*(\tilde{\mathcal{F}}^{\mathrm{sc}}) = (\mathfrak{u}_h \circ \mathfrak{aut}_{\tilde{\mathcal{F}}^{\mathrm{sc}}}) \rtimes \mathfrak{ch}^*(\tilde{\mathcal{F}}^{\mathrm{sc}}) \qquad 14.23.4,$$

we have the commutative diagram of functors

$$\begin{array}{ccc} {}^{\mathfrak{u}_h}(k^*\text{-}\mathfrak{iGr}) & \xrightarrow{\mathfrak{p}_h} & k^*\text{-}\mathfrak{iGr} \\ {\scriptstyle \mathrm{id}_{\mathfrak{u}_h\circ\mathfrak{aut}_{\tilde{\mathcal{F}}^{\mathrm{sc}}}}\rtimes\widehat{\mathfrak{aut}}_{\tilde{\mathcal{F}}^{\mathrm{sc}}}}\uparrow & & \uparrow{\scriptstyle \widehat{\mathfrak{aut}}_{\tilde{\mathcal{F}}^{\mathrm{sc}}}} \\ {}^{\mathfrak{u}_h}\mathfrak{ch}^*(\tilde{\mathcal{F}}^{\mathrm{sc}}) & \longrightarrow & \mathfrak{ch}^*(\tilde{\mathcal{F}}^{\mathrm{sc}}) \end{array} \qquad 14.23.5.$$

14.24 Then, it is not difficult to check that the composition $\hat{\mathfrak{u}}_h \circ \widehat{\mathfrak{aut}}_{\tilde{\mathcal{F}}^{\mathrm{sc}}}$ is also the *direct image* of the composition (cf. 1.6)

$$\hat{\mathfrak{w}}_h^{\mathfrak{aut}} = \hat{\mathfrak{w}}_h \circ (\mathrm{id}_{\mathfrak{u}_h\circ\mathfrak{aut}_{\tilde{\mathcal{F}}^{\mathrm{sc}}}} \rtimes \widehat{\mathfrak{aut}}_{\tilde{\mathcal{F}}^{\mathrm{sc}}}) \qquad 14.24.1$$

throughout the bottom structural functor in diagram 14.23.5; further, the *direct image* is clearly compatible with the functor $\mathcal{F}ct_{U_h} : {}^{U_h}\aleph \longrightarrow \mathcal{O}\text{-}\mathfrak{mod}$. Consequently, we finally get

$$\begin{aligned} \mathcal{G}_k(\mathcal{F}, \widehat{\mathfrak{aut}}_{\mathcal{F}^{\mathrm{nc}}})_h &\cong \lim_{\longleftarrow} (\mathcal{F}ct_{U_h} \circ \hat{\mathfrak{u}}_h \circ \widehat{\mathfrak{aut}}_{\tilde{\mathcal{F}}^{\mathrm{sc}}}) \\ &\cong \lim_{\longleftarrow} (\mathcal{F}ct_{U_h} \circ \hat{\mathfrak{w}}_h^{\mathfrak{aut}}) \end{aligned} \qquad 14.24.2.$$

14.25 The point is that the semidirect product ${}^{\mathfrak{u}_h}\mathfrak{ch}^*(\tilde{\mathcal{F}}^{\mathrm{sc}})$ admits another description in terms of the subcategory ${}^{h}(\tilde{\mathcal{F}}^{\mathrm{sc}}) \subset {}^{U_h}(\tilde{\mathcal{F}}^{\mathrm{sc}})$ of *faithful U_h-objects* of $\tilde{\mathcal{F}}^{\mathrm{sc}}$ (cf. 6.3). Precisely, we denote by ${}^{h}(\tilde{\mathcal{F}}^{\mathrm{sc}})$ the full subcategory of ${}^{U_h}(\tilde{\mathcal{F}}^{\mathrm{sc}})$ over the objects Q^η formed by an $\mathcal{F}$-selfcentralizing subgroup Q of P and by an *injective* group homomorphism $\eta : U_h \to \tilde{\mathcal{F}}(Q)$; we set for short

$$\mathfrak{v}_h = \mathfrak{v}_{{}^h(\tilde{\mathcal{F}}^{\mathrm{sc}})} : \mathfrak{ch}^*\big({}^h(\tilde{\mathcal{F}}^{\mathrm{sc}})\big) \longrightarrow {}^h(\tilde{\mathcal{F}}^{\mathrm{sc}}) \qquad 14.25.1$$

for the evaluation functor of this subcategory. Note that we have a *faithful forgetful* functor ${}^{h}(\tilde{\mathcal{F}}^{\mathrm{sc}}) \to \tilde{\mathcal{F}}^{\mathrm{sc}}$; moreover, for any h' dividing h, it is clear that the inclusion $U_{h'} \subset U_h$ induces a faithful functor

$$\mathfrak{r}_{h',h} : {}^{h}(\tilde{\mathcal{F}}^{\mathrm{sc}}) \longrightarrow {}^{h'}(\tilde{\mathcal{F}}^{\mathrm{sc}}) \qquad 14.25.2.$$

14.26 On the other hand, since the category of chains $\mathfrak{ch}(\tilde{\mathcal{F}}^{\mathrm{sc}})$ is already a semidirect product (cf. A2.8), the ${}^{\mathfrak{u}_h}\mathfrak{ch}^*(\tilde{\mathcal{F}}^{\mathrm{sc}})$-objects can be identified with the triples $(\eta, \tilde{\mathfrak{q}}, \Delta_n)$ formed by a functor $\tilde{\mathfrak{q}} : \Delta_n \to \tilde{\mathcal{F}}^{\mathrm{sc}}$ and by an injective group homomorphism $\eta : U_h \to \tilde{\mathcal{F}}(\tilde{\mathfrak{q}})$. Moreover, for any $i \in \Delta_n$, denote by $\iota_i^{\tilde{\mathfrak{q}}} : \tilde{\mathcal{F}}(\tilde{\mathfrak{q}}) \to \tilde{\mathcal{F}}\big(\tilde{\mathfrak{q}}(i)\big)$ the structural map (cf. 14.8), which actually coincides with the image by the functor $\mathfrak{aut}_{\tilde{\mathcal{F}}_{\mathrm{sc}}}$ of the obvious $\mathfrak{ch}^*(\tilde{\mathcal{F}}^{\mathrm{sc}})$-morphism from $(\mathfrak{q}, \Delta_n)$ to $(\mathfrak{q}(i), \Delta_0)$.

Proposition 14.27 *For any $h \in \mathbb{N} - p\mathbb{N}$, we have an equivalence of categories*

$$\mathfrak{j}_h : {}^{\mathfrak{u}_h}\mathfrak{ch}^*(\tilde{\mathcal{F}}^{\rm sc}) \cong \mathfrak{ch}^*\big({}^h(\tilde{\mathcal{F}}^{\rm sc})\big) \qquad 14.27.1$$

which maps any ${}^{\mathfrak{u}_h}\mathfrak{ch}^*(\tilde{\mathcal{F}}^{\rm sc})$-object $(\eta, \tilde{\mathfrak{q}}, \Delta_n)$ on the chain $\tilde{\mathfrak{q}}^\eta : \Delta_n \to {}^h(\tilde{\mathcal{F}}^{\rm sc})$ sending any $i \in \Delta_n$ to the ${}^h(\tilde{\mathcal{F}}^{\rm sc})$-object $\tilde{\mathfrak{q}}(i)^{\iota_i^{\tilde{\mathfrak{q}}} \circ \eta}$ and any Δ_n-morphism $j \bullet i$ to $\tilde{\mathfrak{q}}(j \bullet i)$. Moreover, for any h' dividing h, we have the commutative diagram

$$\begin{array}{ccc} {}^{\mathfrak{u}_{h'}}\mathfrak{ch}^*(\tilde{\mathcal{F}}^{\rm sc}) & \overset{\mathfrak{j}_{h'}}{\cong} & \mathfrak{ch}^*\big({}^{h'}(\tilde{\mathcal{F}}^{\rm sc})\big) \\ {\scriptstyle (\rho_{h',h} * \mathfrak{aut}_{\tilde{\mathcal{F}}^{\rm sc}}) \rtimes \mathfrak{id}_{\mathfrak{ch}^*(\tilde{\mathcal{F}}^{\rm sc})}} \uparrow & & \uparrow {\scriptstyle \mathfrak{ch}^*(\mathfrak{r}_{h',h})} \\ {}^{\mathfrak{u}_h}\mathfrak{ch}^*(\tilde{\mathcal{F}}^{\rm sc}) & \overset{\mathfrak{j}_h}{\cong} & \mathfrak{ch}^*\big({}^h(\tilde{\mathcal{F}}^{\rm sc})\big) \end{array} \qquad 14.27.2.$$

Proof: It is easily checked that $\tilde{\mathfrak{q}}(j \bullet i)$ is indeed a morphism between the ${}^h(\tilde{\mathcal{F}}^{\rm sc})$-objects $\tilde{\mathfrak{q}}(j)^{\iota_j^{\tilde{\mathfrak{q}}} \circ \eta}$ and $\tilde{\mathfrak{q}}(i)^{\iota_i^{\tilde{\mathfrak{q}}} \circ \eta}$. Moreover, a ${}^{\mathfrak{u}_h}\mathfrak{ch}^*(\tilde{\mathcal{F}}^{\rm sc})$-morphism to $(\eta, \tilde{\mathfrak{q}}, \Delta_n)$ from a ${}^{\mathfrak{u}_h}\mathfrak{ch}^*(\tilde{\mathcal{F}}^{\rm sc})$-object $(\theta, \tilde{\mathfrak{r}}, \Delta_m)$ is defined by a $\mathfrak{ch}^*(\tilde{\mathcal{F}}^{\rm sc})$-morphism

$$(\tilde{\mu}, \delta) : (\tilde{\mathfrak{r}}, \Delta_m) \longrightarrow (\tilde{\mathfrak{q}}, \Delta_n) \qquad 14.27.3$$

fulfilling $\mathfrak{aut}_{\tilde{\mathcal{F}}^{\rm sc}}(\tilde{\mu}, \delta) \circ \theta = \eta$ (cf. condition A2.6.2), and therefore $(\tilde{\mu}, \delta)$ is also a morphism from $(\tilde{\mathfrak{r}}^\theta, \Delta_m)$ to $(\tilde{\mathfrak{q}}^\eta, \Delta_n)$. Thus, we have a functor

$$\mathfrak{j}_h : {}^{\mathfrak{u}_h}\mathfrak{ch}^*(\tilde{\mathcal{F}}^{\rm sc}) \longrightarrow \mathfrak{ch}^*\big({}^h(\tilde{\mathcal{F}}^{\rm sc})\big) \qquad 14.27.4.$$

On the other hand, any ${}^h(\tilde{\mathcal{F}}^{\rm sc})$-chain $\hat{\tilde{\mathfrak{q}}} : \Delta_n \to {}^h(\tilde{\mathcal{F}}^{\rm sc})$ clearly determines a functor $\tilde{\mathfrak{q}} : \Delta_n \to \tilde{\mathcal{F}}^{\rm sc}$; consequently, by the very definition of ${}^h(\tilde{\mathcal{F}}^{\rm sc})$, we have $\hat{\tilde{\mathfrak{q}}}(i) = \tilde{\mathfrak{q}}(i)^{\eta_i}$ where $\eta_i : U_h \to \tilde{\mathcal{F}}\big(\tilde{\mathfrak{q}}(i)\big)$ is an injective group homomorphism for any $i \in \Delta_n$, and, since $\hat{\tilde{\mathfrak{q}}}(j \bullet i)$ is a morphism from $\tilde{\mathfrak{q}}(j)^{\eta_j}$ to $\tilde{\mathfrak{q}}(i)^{\eta_i}$ for any $0 \le j \le i \le n$, $\eta_0(U_h)$ is actually contained in $\tilde{\mathcal{F}}(\tilde{\mathfrak{q}})$, so that there is a unique $\eta : U_h \to \tilde{\mathcal{F}}(\tilde{\mathfrak{q}})$ fulfilling $\iota_i^{\tilde{\mathfrak{q}}} \circ \eta = \eta_i$ for any $i \in \Delta_n$.

Similarly, if $\hat{\tilde{\mathfrak{r}}} : \Delta_m \to {}^h(\tilde{\mathcal{F}}^{\rm sc})$ is an ${}^h(\tilde{\mathcal{F}}^{\rm sc})$-chain and, for any $j \in \Delta_m$, we have $\hat{\tilde{\mathfrak{r}}}(j) = \tilde{\mathfrak{r}}(j)^{\iota_j^{\tilde{\mathfrak{r}}} \circ \theta}$ for an injective group homomorphism $\theta : U_h \to \tilde{\mathcal{F}}(\tilde{\mathfrak{r}})$, then any $\mathfrak{ch}^*\big({}^h(\tilde{\mathcal{F}}^{\rm sc})\big)$-morphism $(\hat{\tilde{\mu}}, \delta) : (\hat{\tilde{\mathfrak{r}}}, \Delta_m) \to (\hat{\tilde{\mathfrak{q}}}, \Delta_n)$ induces a $\mathfrak{ch}^*(\tilde{\mathcal{F}}^{\rm sc})$-morphism $(\tilde{\mu}, \delta) : (\tilde{\mathfrak{r}}, \Delta_m) \to (\tilde{\mathfrak{q}}, \Delta_n)$ such that

$$\mathfrak{aut}_{\tilde{\mathcal{F}}^{\rm sc}}(\tilde{\mu}_i, \mathrm{id}_{\Delta_0}) \circ \iota_{\delta(i)}^{\tilde{\mathfrak{r}}} \circ \theta = \eta_i = \iota_i^{\tilde{\mathfrak{q}}} \circ \eta \qquad 14.27.5$$

for any $i \in \Delta_n$ and therefore, since (cf. 14.25)

$$\mathfrak{aut}_{\tilde{\mathcal{F}}^{\rm sc}}(\tilde{\mu}_i, \mathrm{id}_{\Delta_0}) \circ \iota_{\delta(i)}^{\tilde{\mathfrak{r}}} = \iota_i^{\tilde{\mathfrak{q}}} \circ \mathfrak{aut}_{\tilde{\mathcal{F}}^{\rm sc}}(\tilde{\mu}, \delta) \qquad 14.27.6,$$

we get $\mathfrak{aut}_{\tilde{\mathcal{F}}^{\mathrm{sc}}}(\tilde{\mu},\delta)\circ\theta=\eta$ by the uniqueness of η. Consequently, the functor $\mathfrak{j}_h$ is an equivalence of categories. The commutativity of the diagram is easily checked. We are done.

Proposition 14.28 *For any $h\in\mathbb{N}-p\mathbb{N}$, we have a factorization of $\hat{\mathfrak{w}}_h^{\mathrm{sc}}$*

$$\begin{array}{ccc} {}^{\mathfrak{u}_h}\mathfrak{ch}^*(\tilde{\mathcal{F}}^{\mathrm{sc}}) & \xrightarrow{\hat{\mathfrak{w}}_h^{\mathfrak{aut}}} & {}^{U_h}\aleph \\ \mathfrak{j}_h\wr\| & & \uparrow\mathfrak{s}_h \\ \mathfrak{ch}^*({}^h(\tilde{\mathcal{F}}^{\mathrm{sc}})) & \xrightarrow{\mathfrak{v}_h} & {}^h(\tilde{\mathcal{F}}^{\mathrm{sc}}) \end{array} \qquad 14.28.1$$

throughout a functor $\mathfrak{s}_h : {}^h(\tilde{\mathcal{F}}^{\mathrm{sc}})\to{}^{U_h}\aleph$ mapping any ${}^h(\tilde{\mathcal{F}}^{\mathrm{sc}})$-object Q^ρ on the U_h-set $(\varpi_{h,\hat{\tilde{\mathcal{F}}}(Q)})^{-1}(\rho)$ and any ${}^h(\tilde{\mathcal{F}}^{\mathrm{sc}})$-morphism $\tilde{\varphi}: R^\sigma\to Q^\rho$ on a bijection

$$(\varpi_{h,\hat{\tilde{\mathcal{F}}}(R)})^{-1}(\sigma)\cong(\varpi_{h,\hat{\tilde{\mathcal{F}}}(Q)})^{-1}(\rho) \qquad 14.28.2.$$

In particular, we have

$$\mathcal{G}_k(\mathcal{F},\widehat{\mathfrak{aut}}_{\mathcal{F}^{\mathrm{nc}}})_h\cong\mathbb{H}^0\big({}^h(\tilde{\mathcal{F}}^{\mathrm{sc}}),\mathcal{F}ct_{U_h}\circ\mathfrak{s}_h\big) \qquad 14.28.3.$$

Moreover, for any h' dividing h we have a natural map

$$\theta_{h',h}:\mathfrak{res}^{U_h}_{U_{h'}}\circ\mathfrak{s}_h\longrightarrow\mathfrak{s}_{h'}\circ\mathfrak{r}_{h',h} \qquad 14.28.4$$

sending Q^ρ to the $U_{h'}$-morphism

$$\mathfrak{res}^{U_h}_{U_{h'}}\big((\varpi_{h,\hat{\tilde{\mathcal{F}}}(Q)})^{-1}(\rho)\big)\longrightarrow(\varpi_{h',\hat{\tilde{\mathcal{F}}}(Q)})^{-1}\big(\mathrm{Res}^{U_h}_{U_{h'}}(\rho)\big) \qquad 14.28.5.$$

induced by the restriction throughout the inclusion $U_{h'}\subset U_h$.

Proof: Let us denote by R and Q the obvious 0-chains and by $\tilde{\varphi}$ the 1-chain mapping 0 on R, 1 on Q and the Δ_1-morphism $0\bullet 1$ on $\tilde{\varphi}$; thus, we have evident $\mathfrak{ch}^*(\tilde{\mathcal{F}}^{\mathrm{sc}})$-morphisms

$$(R,\Delta_0)\longleftarrow(\tilde{\varphi},\Delta_1)\longrightarrow(Q,\Delta_0) \qquad 14.28.6$$

and then the natural map $\widehat{\mathfrak{aut}}_{\tilde{\mathcal{F}}^{\mathrm{sc}}}\to\mathfrak{aut}_{\tilde{\mathcal{F}}^{\mathrm{sc}}}$ (cf. 14.10) sends them to the commutative diagram

$$\begin{array}{ccccc} \tilde{\mathcal{F}}(R) & \longleftarrow & \tilde{\mathcal{F}}(\tilde{\varphi}) & \longrightarrow & \tilde{\mathcal{F}}(Q) \\ \uparrow & & \uparrow & & \uparrow \\ \hat{\tilde{\mathcal{F}}}(R) & \longleftarrow & \hat{\tilde{\mathcal{F}}}(\tilde{\varphi}) & \longrightarrow & \hat{\tilde{\mathcal{F}}}(Q) \end{array} \qquad 14.28.7.$$

Since we have $\rho(\xi)\circ\tilde{\varphi}=\tilde{\varphi}\circ\sigma(\xi)$ for any $\xi\in U_h$, and we have identified $\tilde{\mathcal{F}}(\tilde{\varphi})$ with its image in $\tilde{\mathcal{F}}(R)$ (cf. 14.8), actually we get $\sigma(U_h)\subset\tilde{\mathcal{F}}(\tilde{\varphi})$

and the right-top group homomorphism maps $\sigma(\xi)$ on $\rho(\xi)$; hence, the right-bottom k^*-group homomorphism induces a U_h-set isomorphism

$$\mathfrak{s}_h(\tilde\varphi) : (\varpi_{h,\hat{\tilde{\mathcal{F}}}(R)})^{-1}(\sigma) \cong (\varpi_{h,\hat{\tilde{\mathcal{F}}}(Q)})^{-1}(\rho) \qquad 14.28.8.$$

Now, we claim that the correspondence mapping the ${}^h(\tilde{\mathcal{F}}^{^{sc}})$-morphism $\tilde\varphi$ on the U_h-set isomorphism $\mathfrak{s}_h(\tilde\varphi)$ defines a functor; indeed, for a third U_h-object T^τ of $\tilde{\mathcal{F}}^{^{sc}}$ and an ${}^h(\tilde{\mathcal{F}}^{^{sc}})$-morphism $\tilde\psi : T^\tau \to R^\sigma$, we have the evident commutative $\mathfrak{ch}^*(\tilde{\mathcal{F}}^{^{sc}})$-diagram

$$\begin{array}{ccccccc}
(Q,\Delta_0) & = & (Q,\Delta_0) & & & & \\
 & & \uparrow & & & & \\
 & & (\tilde\varphi,\Delta_1) & \rightarrow & (R,\Delta_0) & & \\
\uparrow & & \uparrow & & \uparrow & & \\
 & & (\tilde{\mathfrak{c}},\Delta_2) & \rightarrow & (\tilde\psi,\Delta_1) & \rightarrow & (T,\Delta_0) \\
 & \swarrow & & & & & \| \\
(\tilde\varphi\circ\psi,\Delta_1) & & & & \rightarrow & & (T,\Delta_0)
\end{array} \qquad 14.28.9$$

where the $\tilde{\mathcal{F}}^{^{sc}}$-chain $\tilde{\mathfrak{c}} : \Delta_2 \to \tilde{\mathcal{F}}^{^{sc}}$ maps 0 on T , 1 on R , 2 on Q , $0\bullet 1$ on $\tilde\psi$ and $1\bullet 2$ on $\tilde\varphi$; once again, the functor $\mathfrak{aut}_{\tilde{\mathcal{F}}^{sc}}$ maps it on the commutative diagram of groups

$$\begin{array}{ccccccc}
\tilde{\mathcal{F}}^{^{sc}}(Q) & = & \tilde{\mathcal{F}}^{^{sc}}(Q) & & & & \\
 & & \uparrow & & & & \\
 & & \tilde{\mathcal{F}}^{^{sc}}(\tilde\varphi) & \rightarrow & \tilde{\mathcal{F}}^{^{sc}}(R) & & \\
\uparrow & & \uparrow & & \uparrow & & \\
 & & \tilde{\mathcal{F}}^{^{sc}}(\tilde{\mathfrak{c}}) & \rightarrow & \tilde{\mathcal{F}}^{^{sc}}(\tilde\psi) & \rightarrow & \tilde{\mathcal{F}}^{^{sc}}(T) \\
 & \swarrow & & & & & \| \\
\tilde{\mathcal{F}}^{^{sc}}(\tilde\varphi\circ\psi) & & & & \rightarrow & & \tilde{\mathcal{F}}^{^{sc}}(T)
\end{array} \qquad 14.28.10.$$

But, in this diagram it is quite clear that the middle square is a *pull-back*; thus, since $\sigma(u)$ and $\tau(u)$ respectively belong to $\tilde{\mathcal{F}}^{^{sc}}(\tilde\varphi)$ and to $\tilde{\mathcal{F}}^{^{sc}}(\tilde\psi)$ for any $u \in U_h$, and then $\tau(u)$ is mapped on $\sigma(u)$, actually $\tau(u)$ belongs to $\tilde{\mathcal{F}}^{^{sc}}(\tilde{\mathfrak{c}})$. At this point, it is easily checked that $\mathfrak{s}_h(\tilde\varphi\circ\tilde\psi) = \mathfrak{s}_h(\tilde\varphi)\circ\mathfrak{s}_h(\tilde\psi)$.

Moreover, it follows from Proposition 14.18 that the functor $\mathfrak{w}_h^{\mathfrak{aut}}$ maps $(\eta,\tilde{\mathfrak{q}},\Delta_n)$ on the U_h-set $(\varpi_{h,\hat{\tilde{\mathcal{F}}}(\tilde{\mathfrak{q}})})^{-1}(\eta)$; but, the image by $\widehat{\mathfrak{aut}}_{\tilde{\mathcal{F}}_{sc}}$ of the obvious $\mathfrak{ch}^*(\tilde{\mathcal{F}}^{^{sc}})$-morphism $(\tilde{\mathfrak{q}},\Delta_n) \to (\tilde{\mathfrak{q}}(0),\Delta_0)$ determines a lifting of $\iota_0^{\tilde{\mathfrak{q}}}$

$$\hat\iota_0^{\tilde{\mathfrak{q}}} : \hat{\tilde{\mathcal{F}}}(\tilde{\mathfrak{q}}) \longrightarrow \hat{\tilde{\mathcal{F}}}(\tilde{\mathfrak{q}}(0)) \qquad 14.28.11;$$

then, since $\hat\iota_0^{\tilde{\mathfrak{q}}}$ is injective (cf. 14.8), it is quite clear that

$$\hat\iota_0^{\tilde{\mathfrak{q}}} \circ (\varpi_{h,\hat{\tilde{\mathcal{F}}}(\tilde{\mathfrak{q}})})^{-1}(\eta) = (\varpi_{h,\hat{\tilde{\mathcal{F}}}(\tilde{\mathfrak{q}}(0))})^{-1}(\iota_0^{\tilde{\mathfrak{q}}}\circ\eta) \qquad 14.28.12,$$

where the left member denotes the set of compositions of $\hat\iota_0^{\tilde{\mathfrak{q}}}$ with all the elements of $(\varpi_{h,\hat{\tilde{\mathcal{F}}}(\tilde{\mathfrak{q}})})^{-1}(\eta)$. On the other hand, by the very definition of $\mathfrak{s}_h$, we actually have

$$(\varpi_{h,\hat{\tilde{\mathcal{F}}}(\tilde{\mathfrak{q}}(0))})^{-1}(\iota_0^{\tilde{\mathfrak{q}}}\circ\eta) = (\mathfrak{s}_h\circ\mathfrak{v}_h\circ\mathfrak{j}_h)(\eta,\tilde{\mathfrak{q}},\Delta_n) \qquad 14.28.13.$$

From these equalities it is easily checked that we have a natural isomorphism

$$\hat{\mathfrak{w}}_h^{\mathfrak{aut}} \cong \mathfrak{s}_h\circ\mathfrak{v}_h\circ\mathfrak{j}_h \qquad 14.28.14.$$

Finally, the $\mathcal{O}$-module isomorphism 14.28.3 follows from the $\mathcal{O}$-module isomorphism 14.24.2, from Proposition 14.27 and from this natural isomorphism (cf. A3.9). The proof of the last statement is straightforward.

Remark 14.29 By Remark 14.19, for any $\xi\in U_h$ the functor $\mathfrak{s}_h$ maps the ${}^h(\tilde{\mathcal{F}}^{\mathrm{sc}})$-automorphism $\rho(\xi)$ of Q^ρ on the identity map of $(\varpi_{h,\hat{\tilde{\mathcal{F}}}(Q)})^{-1}(\rho)$; in particular, $\mathfrak{s}_h$ factorizes *via* the *exterior quotient* ${}^{\tilde{h}}(\tilde{\mathcal{F}}^{\mathrm{sc}})$ of ${}^h(\tilde{\mathcal{F}}^{\mathrm{sc}})$ introduced in 6.3. Moreover, since we have (cf. A3.9.1)

$$\mathbb{H}^0\big({}^h(\tilde{\mathcal{F}}^{\mathrm{sc}}),\mathcal{F}ct_{U_h}\circ\mathfrak{s}_h\big) = \varprojlim\,(\mathcal{F}ct_{U_h}\circ\mathfrak{s}_h) \qquad 14.29.1,$$

we still have

$$\mathcal{G}_k(\mathcal{F},\widehat{\mathfrak{aut}}_{\mathcal{F}^{\mathrm{nc}}})_h \cong \mathcal{F}ct_{U_h}(\varinjlim\mathfrak{s}_h,\mathcal{O}) \qquad 14.29.2$$

for any $h\in\mathbb{N}-p\mathbb{N}$. In particular, we still get

$$k\otimes_{\mathcal{O}}\mathcal{G}_k(\mathcal{F},\widehat{\mathfrak{aut}}_{\mathcal{F}^{\mathrm{nc}}}) \cong \varprojlim\,\big((k\otimes_{\mathcal{O}}\mathfrak{g}_k)\circ\widehat{\mathfrak{aut}}_{\mathcal{F}^{\mathrm{nc}}}\big) \qquad 14.29.3.$$

Theorem 14.30 *For any $h\in\mathbb{N}-p\mathbb{N}$ and any $n\geq 1$ we have*

$$\mathbb{H}^n\big({}^h(\tilde{\mathcal{F}}^{\mathrm{sc}}),\mathcal{F}ct_{U_h}\circ\mathfrak{s}_h\big) = \{0\} \qquad 14.30.1.$$

Proof: In order to apply Theorem 6.26, let us consider the full subcategory ${}^h\mathfrak{F}$ of ${}^{U_h}\mathfrak{ac}(\tilde{\mathcal{F}}^{\mathrm{sc}})$ — the category of U_h-objects of $\mathfrak{ac}(\tilde{\mathcal{F}}^{\mathrm{sc}})$ (cf. 6.3) — over the set of *faithful indecomposable* U_h-objects, namely over the *indecomposable U_h-objects Q^ρ* of $\mathfrak{ac}(\tilde{\mathcal{F}}^{\mathrm{sc}})$ (cf. 6.23) such that the group homomorphism $\rho: U_h\to\tilde{\mathcal{F}}(Q)$ is injective. Note that the indecomposable U_h-object $\big(\bigoplus_{u\in U_h}P\big)^\pi$ defined in 6.25 from the *regular* action of U_h on itself, is faithful.

Moreover, if $Q^\rho = \big(\bigoplus_{i\in I} Q_i\big)^\rho$ and $R^\sigma = \big(\bigoplus_{j\in J} R_j\big)^\sigma$ are faithful indecomposable ${}^{U_h}\mathfrak{ac}(\tilde{\mathcal{F}}^{^{\rm sc}})$-objects then, according to 6.11 and 6.13, the exterior intersection of $Q = \bigoplus_{i\in I} Q_i$ and $R = \bigoplus_{j\in J} R_j$ in $\mathfrak{ac}(\tilde{\mathcal{F}}^{^{\rm sc}})$ yields

$$Q \,\tilde{\cap}\, R = \bigoplus_{(i,j)\in I\times J}\ \bigoplus_{(\tilde\alpha,T,\tilde\beta)\in\check{\mathfrak{T}}_{Q_i,R_j}} T \tag{14.30.2}$$

and, for any $\xi \in U_h$, $\rho(\xi)$ and $\sigma(\xi)$ induce an automorphism of this intersection. In particular, they induce a permutation of the disjoint union $\bigsqcup_{(i,j)\in I\times J}\check{\mathfrak{T}}_{Q_i,R_j}$, and if $(\tilde\alpha,T,\tilde\beta)\in\check{\mathfrak{T}}_{Q_i,R_j}$ is a fixed element then $\rho(\xi)$ and $\sigma(\xi)$ respectively fix i and j, and induce $\tilde{\mathcal{F}}$-automorphisms $\rho_i(\xi)$ of Q_i, $\sigma_j(\xi)$ of R_j and $\tau(\xi)$ of T fulfilling

$$\rho_i(\xi)\circ\tilde\alpha = \tilde\alpha\circ\tau(\xi) \quad\text{and}\quad \sigma_j(\xi)\circ\tilde\beta = \tilde\beta\circ\tau(\xi) \tag{14.30.3.}$$

In particular, if $\tau(\xi)$ is trivial then it follows from Corollary 4.9 that $\rho_i(\xi)$ and $\sigma_j(\xi)$ are trivial too and therefore we get $\xi = 1$; thus, the exterior intersection $Q^\rho \,\tilde\cap\, R^\sigma$ in the category ${}^{U_h}\mathfrak{ac}(\tilde{\mathcal{F}}^{^{\rm sc}})$ is a *direct sum* of *faithful indecomposable U_h-objects*. Consequently, since an indecomposable direct summand of a *pull-back* is also a direct summand of some exterior intersection, *the subcategory $\mathfrak{ac}({}^h\mathfrak{F})$ of ${}^{U_h}\mathfrak{ac}(\tilde{\mathcal{F}}^{^{\rm sc}})$ is closed by direct products and pull-backs.*

On the other hand, it is clear that ${}^h(\tilde{\mathcal{F}}^{^{\rm sc}})$ is a full subcategory of ${}^h\mathfrak{F}$ and we claim that the *contravariant* functor (cf. Proposition 14.28)

$$\mathfrak{n}_h = \mathcal{F}ct_{U_h}\circ\mathfrak{s}_h : {}^h(\tilde{\mathcal{F}}^{^{\rm sc}}) \longrightarrow \mathcal{O}\text{-}\mathfrak{mod} \tag{14.30.4}$$

can be extended to a *contravariant* functor $\mathfrak{m}_h$ from ${}^h\mathfrak{F}$ to $\mathcal{O}$-$\mathfrak{mod}$. Indeed, for any faithful indecomposable ${}^{U_h}\mathfrak{ac}(\tilde{\mathcal{F}}^{^{\rm sc}})$-object $Q^\rho = \big(\bigoplus_{i\in I} Q_i\big)^\rho$ choose $i\in I$; it is clear that ρ induces a group homomorphism $\rho_i : U_{h_i} \to \tilde{\mathcal{F}}(Q_i)$ where U_{h_i} denotes the stabilizer of i in U_h, and then we define

$$\mathfrak{m}_h(Q^\rho) = \mathcal{F}ct_{U_{h_i}}\Big(\big(\omega_{h_i,\hat{\tilde{\mathcal{F}}}(Q_i)}\big)^{-1}(\rho_i),\mathcal{O}\Big) \tag{14.30.5.}$$

For any faithful indecomposable ${}^{U_h}\mathfrak{ac}(\tilde{\mathcal{F}}^{^{\rm sc}})$-object $R^\sigma = \big(\bigoplus_{j\in J} R_j\big)^\sigma$ and any ${}^h\mathfrak{F}$-morphism $\tilde\varphi : R^\sigma \to Q^\rho$, $\tilde\varphi$ induces a necessarily surjective U_h-set map $f : J \to I$ and, for the chosen $j\in J$, an $\tilde{\mathcal{F}}^{^{\rm sc}}$-morphism $\tilde\varphi_j : R_j \to Q_{f(j)}$; in particular, it is clear that $U_{h_j} \subset U_{h_{f(j)}}$ or, equivalently, that h_j divides $h_{f(j)}$ and therefore, denoting by ρ_j the restriction to U_{h_j} of the group homomorphism

$$\rho_{f(j)} : U_{h_{f(j)}} \longrightarrow \tilde{\mathcal{F}}(Q_{f(j)}) \tag{14.30.6,}$$

$\tilde\varphi_j$ becomes an ${}^{h_j}(\tilde{\mathcal{F}}^{\mathrm{sc}})$-morphism from $(R_j)^{\sigma_j}$ to $(Q_{f(j)})^{\rho_j}$, so that from Proposition 14.28 we get a U_{h_j}-set isomorphism

$$\mathfrak{s}_{h_j}(\tilde\varphi_j) : (\varpi_{h_j,\hat{\tilde{\mathcal{F}}}(R_j)})^{-1}(\sigma_j) \cong (\varpi_{h_j,\hat{\tilde{\mathcal{F}}}(Q_{f(j)})})^{-1}(\rho_j) \qquad 14.30.7.$$

On the other hand, it is clear that the inclusion $U_{h_j} \subset U_{h_{f(j)}}$ determines the commutative diagram (cf. 14.17.2)

$$\begin{array}{ccc} \mathrm{Mon}_{k^*}(\hat U_{h_j}, \hat{\tilde{\mathcal{F}}}(Q_{f(j)})) & \xrightarrow{\varpi_{h_j,\hat{\tilde{\mathcal{F}}}(Q_{f(j)})}} & \mathrm{Mon}(U_{h_j}, \tilde{\mathcal{F}}(Q_{f(j)})) \\ \uparrow & & \uparrow \\ \mathrm{Mon}_{k^*}(\hat U_{h_{f(j)}}, \hat{\tilde{\mathcal{F}}}(Q_{f(j)})) & \xrightarrow{\varpi_{h_{f(j)},\hat{\tilde{\mathcal{F}}}(Q_{f(j)})}} & \mathrm{Mon}(U_{h_{f(j)}}, \tilde{\mathcal{F}}(Q_{f(j)})) \end{array} \qquad 14.30.8;$$

then, by applying $\mathcal{F}ct_{U_{h_j}}$ and the inclusion $\mathcal{F}ct_{U_{h_{f(j)}}} \subset \mathcal{F}ct_{U_{h_j}} \circ \mathfrak{res}_{U_{h_j}}^{U_{h_{f(j)}}}$ to this diagram and to the bottom map respectively, we immediately get an $\mathcal{O}$-module isomorphism

$$\begin{aligned} \mathrm{res}_{f,j}^{Q^\rho} : \mathcal{F}ct_{U_{h_j}}\big((\varpi_{h_j,\hat{\tilde{\mathcal{F}}}(Q_{f(j)})})^{-1}(\rho_j), \mathcal{O}\big) \\ \cong \mathcal{F}ct_{U_{h_{f(j)}}}\big((\varpi_{h_{f(j)},\hat{\tilde{\mathcal{F}}}(Q_{f(j)})})^{-1}(\rho_{f(j)}), \mathcal{O}\big) \end{aligned} \qquad 14.30.9$$

since both members are free $\mathcal{O}$-modules of rank 1 (cf. 14.15).

Moreover, there is $\xi \in U_h$ such that $\rho(\xi)$ maps $f(j)$ on the chosen element $i \in I$ and therefore we have $h_{f(j)} = h_i$ and ξ induces an ${}^{h_i}(\tilde{\mathcal{F}}^{\mathrm{sc}})$-morphism

$$\rho(\xi)_i^{f(j)} : (Q_{f(j)})^{\rho_{f(j)}} \longrightarrow (Q_i)^{\rho_i} \qquad 14.30.10,$$

so that it follows again from Proposition 14.28 that we get a U_{h_i}-set isomorphism

$$\mathfrak{s}_{h_i}\big(\rho(\xi)_i^{f(j)}\big) : (\varpi_{h_i,\hat{\tilde{\mathcal{F}}}(Q_{f(j)})})^{-1}(\rho_{f(j)}) \cong (\varpi_{h_i,\hat{\tilde{\mathcal{F}}}(Q_i)})^{-1}(\rho_i) \qquad 14.30.11$$

which clearly does not depend on the choice of ξ. Finally, we consider the composition

$$\mathfrak{m}_h(\tilde\varphi) = \mathfrak{n}_{h_j}(\tilde\varphi_j) \circ (\mathrm{res}_{f,j}^{Q^\rho})^{-1} \circ \mathfrak{n}_{h_i}\big(\rho(\xi)_i^{f(j)}\big) \qquad 14.30.12$$

which defines an $\mathcal{O}$-module isomorphism $\mathfrak{m}_h(Q^\rho) \cong \mathfrak{m}_h(R^\sigma)$; we claim that this correspondence $\mathfrak{m}_h$ is the announced *contravariant* functor.

Indeed, it is clear that it extends $\mathfrak{n}_h$ and, for any faithful indecomposable ${}^{U_h}\mathfrak{ac}(\tilde{\mathcal{F}}^{\mathrm{sc}})$-object $T^\tau = \big(\bigoplus_{\ell\in L} T_\ell\big)^\tau$ and any ${}^h\mathfrak{F}$-morphism $\tilde{\psi}: T^\tau \to R^\sigma$, *mutatis mutandis* we have a surjective U_h-set map $g: L \to J$, a chosen element ℓ in L and an ${}^{h_\ell}(\tilde{\mathcal{F}}^{\mathrm{sc}})$-morphism $\tilde{\psi}_\ell: (T_\ell)^{\tau_\ell} \to (R_{g(\ell)})^{\sigma_\ell}$, together with a U_{h_ℓ}-set isomorphism

$$\mathfrak{s}_{h_\ell}(\tilde{\psi}_\ell): (\varpi_{h_\ell,\hat{\tilde{\mathcal{F}}}(T_\ell)})^{-1}(\tau_\ell) \cong (\varpi_{h_\ell,\hat{\tilde{\mathcal{F}}}(R_{g(\ell)})})^{-1}(\sigma_\ell) \qquad 14.30.13.$$

Analogously, we have an $\mathcal{O}$-module isomorphism

$$\begin{aligned}\mathrm{res}^{R^\sigma}_{g,\ell}: \mathcal{F}ct_{U_{h_\ell}}\big((\varpi_{h_\ell,\hat{\tilde{\mathcal{F}}}(R_{g(\ell)})})^{-1}(\sigma_\ell),\mathcal{O}\big)&\\ \cong \mathcal{F}ct_{U_{h_{g(\ell)}}}\big((\varpi_{h_{g(\ell)},\hat{\tilde{\mathcal{F}}}(R_{g(\ell)})})^{-1}(\sigma_{g(\ell)}),\mathcal{O}\big)&\end{aligned} \qquad 14.30.14\,.$$

Moreover, choosing $\zeta \in U_h$ such that $\sigma(\zeta)$ maps $g(\ell)$ on j, as above we have $h_{g(\ell)} = h_j$ and ζ induces an ${}^{h_j}(\tilde{\mathcal{F}}^{\mathrm{sc}})$-morphism

$$\sigma(\zeta)^{g(\ell)}_j : (R_{g(\ell)})^{\sigma_{g(\ell)}} \longrightarrow (R_j)^{\sigma_j} \qquad 14.30.15,$$

and therefore we get a U_{h_j}-set isomorphism (cf. Proposition 14.28)

$$\mathfrak{s}_{h_j}\big(\sigma(\zeta)^{g(\ell)}_j\big): (\varpi_{h_{g(\ell)},\hat{\tilde{\mathcal{F}}}(R_{g(\ell)})})^{-1}(\sigma_{g(\ell)}) \cong (\varpi_{h_j,\hat{\tilde{\mathcal{F}}}(R_j)})^{-1}(\sigma_j) \qquad 14.30.16.$$

Finally, we also get

$$\mathfrak{m}_h(\tilde{\psi}) = \mathfrak{n}_{h_\ell}(\tilde{\psi}_\ell) \circ (\mathrm{res}^{R^\sigma}_{g,\ell})^{-1} \circ \mathfrak{n}_{h_j}\big(\sigma(\zeta)^{g(\ell)}_j\big) \qquad 14.30.17.$$

In order to compute $\mathfrak{m}_h(\tilde{\psi}) \circ \mathfrak{m}_h(\tilde{\varphi})$ note that, since f is a U_h-set map, $\rho(\zeta)$ maps $f\big(g(\ell)\big)$ on $f(j)$ and we have $h_{f(g(\ell))} = h_{f(j)} = h_i$; thus, ζ induces an ${}^{h_{f(j)}}(\tilde{\mathcal{F}}^{\mathrm{sc}})$-morphism

$$\rho(\zeta)^{f(g(\ell))}_{f(j)} : (Q_{f(g(\ell))})^{\rho_{f(g(\ell))}} \longrightarrow (Q_{f(j)})^{\rho_{f(j)}} \qquad 14.30.18$$

and, as above, denoting by $\rho_{g(\ell)}$ the restriction to U_{h_j} of the group homomorphism $\rho_{f(g(\ell))}$, the ${}^h\mathfrak{F}$-morphism $\tilde{\varphi}: R^\sigma \to Q^\rho$ forces the following commutative diagram (cf. 14.30.15 and 14.30.18)

$$\begin{array}{ccc} (Q_{f(g(\ell))})^{\rho_{g(\ell)}} & \longrightarrow & (Q_{f(j)})^{\rho_j} \\ {\scriptstyle \tilde{\varphi}_{g(\ell)}}\uparrow & & \uparrow{\scriptstyle \tilde{\varphi}_j} \\ (R_{g(\ell)})^{\sigma_{g(\ell)}} & \longrightarrow & (R_j)^{\sigma_j} \end{array} \qquad 14.30.19;$$

hence, since $\mathfrak{n}_{h_j}$ is a *contravariant* functor, we get

$$\begin{aligned}\mathfrak{n}_{h_j}\big(\sigma(\zeta)_j^{g(\ell)}\big)\circ\mathfrak{n}_{h_j}(\tilde\varphi_j) &= \mathfrak{n}_{h_j}\big(\tilde\varphi_j\circ\sigma(\zeta)_j^{g(\ell)}\big)\\ &= \mathfrak{n}_{h_j}\big(\rho(\zeta)_{f(j)}^{f(g(\ell))}\circ\tilde\varphi_{g(\ell)}\big)\\ &= \mathfrak{n}_{h_j}(\tilde\varphi_{g(\ell)})\circ\mathfrak{n}_{h_j}\big(\rho(\zeta)_{f(j)}^{f(g(\ell))}\big)\end{aligned} \qquad 14.30.20.$$

On the other hand, the natural map θ_{h_ℓ,h_j} in Proposition 14.28 applied to the ${}^{h_j}(\tilde{\mathcal F}^{\mathrm{sc}})$-morphism $\tilde\varphi_{g(\ell)}:(R_{g(\ell)})^{\sigma_{g(\ell)}}\to(Q_{f(g(\ell))})^{\rho_{g(\ell)}}$ yields the following commutative diagram

$$\begin{array}{ccc}(\varpi_{h_j,\hat{\tilde{\mathcal F}}(R_{g(\ell)})})^{-1}(\sigma_{g(\ell)}) & \overset{\mathfrak{s}_{h_j}(\tilde\varphi_{g(\ell)})}{\cong} & (\varpi_{h_j,\hat{\tilde{\mathcal F}}(Q_{f(g(\ell))})})^{-1}(\rho_{g(\ell)})\\ \downarrow & & \downarrow\\ (\varpi_{h_\ell,\hat{\tilde{\mathcal F}}(R_{g(\ell)})})^{-1}(\sigma_\ell) & \overset{\mathfrak{s}_{h_\ell}(\tilde\varphi_{g(\ell)})}{\cong} & (\varpi_{h_\ell,\hat{\tilde{\mathcal F}}(Q_{f(g(\ell))})})^{-1}(\rho_\ell)\end{array} \qquad 14.30.21$$

where, as above, ρ_ℓ is the restriction of the group homomorphism $\rho_{f(g(\ell))}$ to U_{h_ℓ}; then, as in 14.30.9 above, considering the $\mathcal O$-module isomorphism determined by the inclusion $U_{h_\ell}\subset U_{h_j}$

$$\begin{aligned}\mathrm{res}^{Q^\rho}_{g,\ell}:\mathcal{F}ct_{U_{h_\ell}}\big((\varpi_{h_\ell,\hat{\tilde{\mathcal F}}(Q_{f(g(\ell))})})^{-1}(\rho_\ell),\mathcal O\big)\\ \cong\mathcal{F}ct_{U_{h_j}}\big((\varpi_{h_j,\hat{\tilde{\mathcal F}}(Q_{f(g(\ell))})})^{-1}(\rho_{g(\ell)}),\mathcal O\big)\end{aligned} \qquad 14.30.22,$$

we still get

$$\mathfrak{n}_{h_j}(\tilde\varphi_{g(\ell)})\circ\mathrm{res}^{Q^\rho}_{g,\ell}=\mathrm{res}^{R^\sigma}_{g,\ell}\circ\mathfrak{n}_{h_\ell}(\tilde\varphi_{g(\ell)}) \qquad 14.30.23.$$

Moreover, note that the composition $\mathrm{res}^{Q^\rho}_{f,j}\circ\mathrm{res}^{Q^\rho}_{g,\ell}$ coincides with the $\mathcal O$-module isomorphism determined by the inclusion $U_{h_\ell}\subset U_{h_i}$

$$\begin{aligned}\mathrm{res}^{Q^\rho}_{g\circ f,\ell}:\mathcal{F}ct_{U_{h_\ell}}\big((\varpi_{h_\ell,\hat{\tilde{\mathcal F}}(Q_{f(g(\ell))})})^{-1}(\rho_\ell),\mathcal O\big)\\ \cong\mathcal{F}ct_{U_{h_i}}\big((\varpi_{h_i,\hat{\tilde{\mathcal F}}(Q_{f(g(\ell))})})^{-1}(\rho_{f(g(\ell))}),\mathcal O\big)\end{aligned} \qquad 14.30.24.$$

Finally, applying the natural inclusion $\mathcal{F}ct_{U_{h_i}}\subset\mathcal{F}ct_{U_{h_j}}\circ\mathfrak{res}^{U_{h_i}}_{U_{h_j}}$ to the U_{h_i}-set isomorphism $\mathfrak{s}_{h_i}\big(\rho(\zeta)_{f(j)}^{f(g(\ell))}\big)$ (cf. 14.30.18), we easily get

$$\mathfrak{n}_{h_j}\big(\rho(\zeta)_{f(j)}^{f(g(\ell))}\big)\circ\mathrm{res}^{Q^\rho}_{f,j}=\mathrm{res}^{Q^\rho}_{f,j}\circ\mathfrak{n}_{h_i}\big(\rho(\zeta)_{f(j)}^{f(g(\ell))}\big) \qquad 14.30.25.$$

In conclusion, from equalities 14.30.20, 14.30.23 and 14.30.25 we get

$$\begin{aligned}&(\mathrm{res}^{R^\sigma}_{g,\ell})^{-1}\circ\mathfrak{n}_{h_j}\big(\sigma(\zeta)^{g(\ell)}_j\big)\circ\mathfrak{n}_{h_j}(\tilde\varphi_j)\circ(\mathrm{res}^{Q^\rho}_{f,j})^{-1}\\ &=(\mathrm{res}^{R^\sigma}_{g,\ell})^{-1}\circ\mathfrak{n}_{h_j}(\tilde\varphi_{g(\ell)})\circ\mathfrak{n}_{h_j}\big(\rho(\zeta)^{f(g(\ell))}_{f(j)}\big)\circ(\mathrm{res}^{Q^\rho}_{f,j})^{-1}\\ &=\mathfrak{n}_{h_\ell}(\tilde\varphi_{g(\ell)})\circ(\mathrm{res}^{Q^\rho}_{g,\ell})^{-1}\circ(\mathrm{res}^{Q^\rho}_{f,g(\ell)})^{-1}\circ\mathfrak{n}_{h_j}\big(\rho(\zeta)^{f(g(\ell))}_{f(j)}\big)\\ &=\mathfrak{n}_{h_\ell}(\tilde\varphi_{g(\ell)})\circ(\mathrm{res}^{Q^\rho}_{f\circ g,\ell})^{-1}\circ\mathfrak{n}_{h_j}\big(\rho(\zeta)^{f(g(\ell))}_{f(j)}\big)\end{aligned}\qquad 14.30.26$$

and therefore, from equalities 14.30.12 and 14.30.17 we obtain

$$\begin{aligned}&\mathfrak{m}_h(\tilde\psi)\circ\mathfrak{m}_h(\tilde\varphi)\\ &\qquad=\mathfrak{n}_{h_\ell}(\tilde\varphi_{g(\ell)}\circ\tilde\psi_\ell)\circ(\mathrm{res}^{Q^\rho}_{f\circ g,\ell})^{-1}\circ\mathfrak{n}_{h_i}\big(\rho(\zeta\xi)^{f(g(\ell))}_\ell\big)\\ &\qquad=\mathfrak{m}_h(\tilde\varphi\circ\tilde\psi)\end{aligned}\qquad 14.30.27$$

which proves our claim. Moreover, it follows from the very definition of $\mathfrak{m}_h$ that it maps any ${}^h\mathfrak{F}$-morphism on an $\mathcal{O}$-module isomorphism; hence, it follows from Theorem 6.26 that, for any $n\geq 1$, we have

$$\mathbb{H}^n({}^h\mathfrak{F},\mathfrak{m}_h)=\{0\}\qquad 14.30.28.$$

If $h=0$ then ${}^0\mathfrak{F}={}^0(\tilde{\mathcal{F}}^{\mathrm{sc}})=\tilde{\mathcal{F}}^{\mathrm{sc}}$ and we are done. Otherwise, we consider the full subcategory ${}^h\mathfrak{E}$ of ${}^h\mathfrak{F}$ over the faithful indecomposable ${}^{U_h}\mathfrak{ac}(\tilde{\mathcal{F}}^{\mathrm{sc}})$-objects $Q^\rho=\big(\bigoplus_{i\in I}Q_i\big)^\rho$ such that $|I|>1$, namely over all the ${}^h\mathfrak{F}$-objects which are *not* ${}^h(\tilde{\mathcal{F}}^{\mathrm{sc}})$-objects; note that there is no ${}^h\mathfrak{F}$-morphism from a ${}^h(\tilde{\mathcal{F}}^{\mathrm{sc}})$-object to an ${}^h\mathfrak{E}$-object. Denoting by $\mathfrak{l}_h:{}^h\mathfrak{E}\to\mathcal{O}\text{-}\mathfrak{mod}$ the restriction of $\mathfrak{m}_h$ to ${}^h\mathfrak{E}$, it is quite clear that Theorem 6.26 applies again to ${}^h\mathfrak{E}$ and $\mathfrak{l}_h$, so that for any $n\geq 1$ we get

$$\mathbb{H}^n({}^h\mathfrak{E},\mathfrak{l}_h)=\{0\}\qquad 14.30.29.$$

Recall that, denoting by $\mathbf{\Delta}$ the *simplicial* 2-*category* (cf. A1.7), the cohomology groups $\mathbb{H}^n({}^h\mathfrak{F},\mathfrak{m}_h)$, $\mathbb{H}^n({}^h\mathfrak{E},\mathfrak{l}_h)$ and $\mathbb{H}^n\big({}^h(\tilde{\mathcal{F}}^{\mathrm{sc}}),\mathfrak{n}_h\big)$ are nothing but the homology groups of the respective obvious functors $\mathfrak{c}_{\mathfrak{m}_h}$, $\mathfrak{c}_{\mathfrak{l}_h}$ and $\mathfrak{c}_{\mathfrak{n}_h}$ from $\mathbf{\Delta}$ to $\mathcal{O}\text{-}\mathfrak{mod}$, mapping Δ_n on (cf. A3.8)

$$\begin{aligned}\mathbb{C}^n({}^h\mathfrak{F},\mathfrak{m}_h)&=\prod_{\tilde{\mathfrak{q}}\in\mathfrak{Fct}(\Delta_n,{}^h\mathfrak{F})}\mathfrak{m}_h\big(\tilde{\mathfrak{q}}(0)\big)\\ \mathbb{C}^n({}^h\mathfrak{E},\mathfrak{l}_h)&=\prod_{\tilde{\mathfrak{q}}\in\mathfrak{Fct}(\Delta_n,{}^h\mathfrak{E})}\mathfrak{m}_h\big(\tilde{\mathfrak{q}}(0)\big)\\ \mathbb{C}^n\big({}^h(\tilde{\mathcal{F}}^{\mathrm{sc}}),\mathfrak{n}_h\big)&=\prod_{\tilde{\mathfrak{q}}\in\mathfrak{Fct}(\Delta_n,{}^h(\tilde{\mathcal{F}}^{\mathrm{sc}}))}\mathfrak{m}_h\big(\tilde{\mathfrak{q}}(0)\big)\end{aligned}\qquad 14.30.30;$$

then, the inclusion ${}^{h}(\tilde{\mathcal{F}}^{sc}) \subset {}^{h}\mathfrak{F}$ clearly determines a surjective natural map

$$\mu_h : \mathfrak{c}_{\mathfrak{m}_h} \longrightarrow \mathfrak{c}_{\mathfrak{n}_h} \qquad 14.30.31,$$

so that we obtain a fourth functor

$$\mathfrak{Ker}(\mu_h) : \mathbf{\Delta} \longrightarrow \mathcal{O}\text{-}\mathfrak{mod} \qquad 14.30.32.$$

Moreover, since there is no ${}^{h}\mathfrak{F}$-morphism from a ${}^{h}(\tilde{\mathcal{F}}^{sc})$-object to any ${}^{h}\mathfrak{E}$-object, we have an $\mathcal{O}$-module isomorphism

$$\big(\mathfrak{Ker}(\mu_h)\big)(\Delta_n) \cong \prod_{\tilde{\mathfrak{q}}} \mathfrak{m}_h\big(\tilde{\mathfrak{q}}(0)\big) \qquad 14.30.33,$$

where $\tilde{\mathfrak{q}}$ runs over the set $\mathcal{E}^h(\Delta_n)$ of functors from Δ_n to ${}^{h}\mathfrak{F}$ such that $\tilde{\mathfrak{q}}(0)$ is a ${}^{h}\mathfrak{E}$-object and therefore the inclusion ${}^{h}\mathfrak{E} \subset {}^{h}\mathfrak{F}$ also determines a surjective natural map

$$\lambda_h : \mathfrak{Ker}(\mu_h) \longrightarrow \mathfrak{c}_{\mathfrak{l}_h} \qquad 14.30.34.$$

But on the one hand, we already know that $\mathbb{H}_n(\mathfrak{c}_{\mathfrak{m}_h}) = \{0\} = \mathbb{H}_n(\mathfrak{c}_{\mathfrak{l}_h})$ for any $n \geq 1$ (cf. equalities 14.30.28 and 14.30.29) and on the other hand, setting

$$\mathbb{H}_{-n}(\mathfrak{c}_{\mathfrak{n}_h}) = \mathbb{H}_{-n}(\mathfrak{c}_{\mathfrak{m}_h}) = \mathbb{H}_{-n}\big(\mathfrak{Ker}(\mu_h)\big) = \{0\} \qquad 14.30.35$$

for any $n > 0$, there is a 1-*graded connecting homomorphism* (cf. A3.3.4)

$$\delta : \bigoplus_{n\in\mathbb{Z}} \mathbb{H}_n(\mathfrak{c}_{\mathfrak{n}_h}) \longrightarrow \bigoplus_{n\in\mathbb{Z}} \mathbb{H}_n\big(\mathfrak{Ker}(\mu_h)\big) \qquad 14.30.36$$

such that we have the following *exact triangle* (cf. A3.3.5)

$$\begin{array}{ccc} \bigoplus_{n\in\mathbb{Z}} \mathbb{H}_n(\mathfrak{c}_{\mathfrak{n}_h}) & \stackrel{\delta}{\longrightarrow} & \bigoplus_{n\in\mathbb{Z}} \mathbb{H}_n\big(\mathfrak{Ker}(\mu_h)\big) \\ {\scriptstyle \oplus_{n\in\mathbb{Z}}\mathbb{H}_n(\mu_h)} \nwarrow & & \swarrow \\ & \bigoplus_{n\in\mathbb{Z}} \mathbb{H}_n(\mathfrak{c}_{\mathfrak{m}_h}) & \end{array} \qquad 14.30.37.$$

Consequently, in order to prove that $\mathbb{H}_n(\mathfrak{c}_{\mathfrak{n}_h}) = \{0\}$ for any $n \geq 1$, it suffices to show that $\mathbb{H}_n(\lambda_h)$ is injective. Actually, since the functor $\mathfrak{s}_h$ factorizes throughout the exterior quotient ${}^{\tilde{h}}(\tilde{\mathcal{F}}^{sc})$ of ${}^{h}(\tilde{\mathcal{F}}^{sc})$ (cf. Remark 14.29), it is easily checked that the *contravariant* functors $\mathfrak{n}_h$, $\mathfrak{m}_h$ and $\mathfrak{l}_h$ respectively determine *contravariant* functors

$$\begin{gathered} \tilde{\mathfrak{n}}_h : {}^{\tilde{h}}(\tilde{\mathcal{F}}^{sc}) \longrightarrow \mathcal{O}\text{-}\mathfrak{mod} \\ \tilde{\mathfrak{m}}_h : {}^{\tilde{h}}\mathfrak{F} \longrightarrow \mathcal{O}\text{-}\mathfrak{mod} \quad \text{and} \quad \tilde{\mathfrak{l}}_h : {}^{\tilde{h}}\mathfrak{E} \longrightarrow \mathcal{O}\text{-}\mathfrak{mod} \end{gathered} \qquad 14.30.38,$$

where ${}^{\tilde h}\mathfrak{F}$ and ${}^{\tilde h}\mathfrak{E}$ denote the corresponding *exterior quotients* (cf. 6.3). Coherently, we get the corresponding functors $\mathfrak{c}_{\tilde{\mathfrak{n}}_h}$, $\mathfrak{c}_{\tilde{\mathfrak{m}}_h}$ and $\mathfrak{c}_{\tilde{\mathfrak{l}}_h}$ (cf. 14.30.30), and the corresponding natural maps (cf. 14.30.31 and 14.30.34)

$$\tilde\mu_h : \mathfrak{c}_{\tilde{\mathfrak{m}}_h} \longrightarrow \mathfrak{c}_{\tilde{\mathfrak{n}}_h} \quad \text{and} \quad \tilde\lambda_h : \mathfrak{Ker}(\tilde\mu_h) \longrightarrow \mathfrak{c}_{\tilde{\mathfrak{l}}_h} \qquad 14.30.39.$$

Then, it follows from equality 6.25.1 that, for any $n \geq 1$, it suffices to prove that $\mathbb{H}_n(\tilde\lambda_h)$ is injective. First of all note that, up to isomorphisms, any ${}^{\tilde h}\mathfrak{F}$-object has the *canonical* form $Q^\rho = \big(\bigoplus_{\bar\xi\in U_h/U} Q_{\bar\xi}\big)^\rho$ for a suitable subgroup U of U_h , and, denoting by $\tilde{\mathfrak{j}}_h : {}^{\tilde h}\mathfrak{E} \to {}^{\tilde h}\mathfrak{F}$ the *inclusion* functor, we claim that we have a functor and a natural map

$$\tilde{\mathfrak{e}}_h : {}^{\tilde h}\mathfrak{F} \longrightarrow {}^{\tilde h}\mathfrak{E} \quad \text{and} \quad \tilde\varepsilon_h : \tilde{\mathfrak{j}}_h \circ \tilde{\mathfrak{e}}_h \longrightarrow \mathfrak{id}_{{}^{\tilde h}\mathfrak{F}} \qquad 14.30.40$$

which respectively map Q^ρ on the ${}^{U_h}\mathfrak{ac}(\tilde{\mathcal{F}}^{^{\mathrm{sc}}})$-object $\hat Q^{\hat\rho}$ formed by

$$\hat Q = \bigoplus\nolimits_{\xi\in U_h} Q_{\bar\xi} \qquad 14.30.41$$

and by the k^*-group homomorphism $\hat\rho : U_h \to \mathfrak{F}(\hat Q)$ defined by the regular action of U_h on itself, together with the $\tilde{\mathcal{F}}^{^{\mathrm{sc}}}$-isomorphisms from the ξ-summand to the ξ'-summand induced by $\rho(\xi'\xi^{-1})$, and on the ${}^{\tilde h}\mathfrak{F}$-morphism

$$(\tilde\varepsilon_h)_{Q^\rho} : \hat Q^{\hat\rho} \longrightarrow Q^\rho \qquad 14.30.42$$

defined by the canonical map $U_h \to U_h/U$ and by the identity automorphism of $Q_{\bar\xi}$ for any $\xi \in U_h$.

Indeed, if $R^\sigma = \big(\bigoplus_{\tilde\xi\in U_h/V} R_{\tilde\xi}\big)^\sigma$ is another ${}^{\tilde h}\mathfrak{F}$-object, an ${}^{\tilde h}\mathfrak{F}$-morphism from R^σ to Q^ρ forces $V \subset U$ and admits a *canonical* representative formed by the canonical map $U_h/V \to U_h/U$ and by an $\tilde{\mathcal{F}}^{^{\mathrm{sc}}}$-morphism $\tilde\varphi_{\tilde\xi} : R_{\tilde\xi} \to Q_{\bar\xi}$ for any $\tilde\xi \in U_h/V$; then, the functor $\tilde{\mathfrak{e}}_h$ above maps this ${}^{\tilde h}\mathfrak{F}$-morphism on the ${}^{\tilde h}\mathfrak{E}$-morphism

$$\hat R^{\hat\sigma} = \big(\bigoplus\nolimits_{\xi\in U_h} R_{\tilde\xi}\big)^{\hat\sigma} \longrightarrow \hat Q^{\hat\rho} = \big(\bigoplus\nolimits_{\xi\in U_h} Q_{\bar\xi}\big)^{\hat\rho} \qquad 14.30.43$$

admitting a representative formed by the identity map of U_h and by the $\tilde{\mathcal{F}}^{^{\mathrm{sc}}}$-morphism $\tilde\varphi_{\tilde\xi} : R_{\tilde\xi} \to Q_{\bar\xi}$ for any $\xi \in U_h$; it is quite clear that this correspondence preserves the composition of ${}^{\tilde h}\mathfrak{F}$-morphisms and is compatible with the ${}^{\tilde h}\mathfrak{F}$-morphisms 14.30.42.

Secondly, we consider the *contravariant* functor $\tilde{\mathfrak{m}}_h \circ \tilde{\mathfrak{j}}_h \circ \tilde{\mathfrak{e}}_h = \tilde{\mathfrak{l}}_h \circ \tilde{\mathfrak{e}}_h$ which, as above, determines a functor (cf. 14.30.30)

$$\mathfrak{c}_{\tilde{\mathfrak{l}}_h\circ\tilde{\mathfrak{e}}_h} : \boldsymbol{\Delta} \longrightarrow \mathcal{O}\text{-}\mathfrak{mod} \qquad 14.30.44$$

mapping Δ_n on

$$\mathbb{C}^n(^{\tilde h}\mathfrak{F}, \tilde{\mathfrak{l}}_h \circ \tilde{\mathfrak{e}}_h) = \prod_{\tilde{\mathfrak{q}}\in\mathfrak{Fct}(\Delta_n,{}^{\tilde h}\mathfrak{F})} \tilde{\mathfrak{m}}_h\Big(\hat{\mathfrak{e}}_h\big(\tilde{\mathfrak{q}}(0)\big)\Big) \qquad 14.30.45,$$

and then we get a natural map $\kappa_{\tilde{\mathfrak{e}}_h} : \mathfrak{c}_{\tilde{\mathfrak{m}}_h} \to \mathfrak{c}_{\tilde{\mathfrak{l}}_h\circ\tilde{\mathfrak{e}}_h}$ sending Δ_n to the $\mathcal{O}$-module homomorphism

$$(\kappa_{\tilde{\mathfrak{e}}_h})_n : \prod_{\tilde{\mathfrak{q}}\in\mathfrak{Fct}(\Delta_n,{}^{\tilde h}\mathfrak{F})} \tilde{\mathfrak{m}}_h\big(\tilde{\mathfrak{q}}(0)\big) \longrightarrow \prod_{\tilde{\mathfrak{q}}\in\mathfrak{Fct}(\Delta_n,{}^{\tilde h}\mathfrak{F})} \tilde{\mathfrak{m}}_h\Big(\hat{\mathfrak{e}}_h\big(\tilde{\mathfrak{q}}(0)\big)\Big) \qquad 14.30.46$$

mapping $m = (m_{\tilde{\mathfrak{q}}})_{\tilde{\mathfrak{q}}\in\mathfrak{Fct}(\Delta_n,{}^{\tilde h}\mathfrak{F})}$ on $(m_{\tilde{\mathfrak{j}}_h\circ\tilde{\mathfrak{e}}_h\circ\tilde{\mathfrak{q}}})_{\tilde{\mathfrak{q}}\in\mathfrak{Fct}(\Delta_n,{}^{\tilde h}\mathfrak{F})}$. Note that if m belongs to $\mathrm{Ker}\big((\tilde\lambda_h)_{\Delta_n}\big)$ (cf. 14.30.39) then, for any $\tilde{\mathfrak{q}} \in \mathfrak{Fct}(\Delta_n, {}^{\tilde h}\mathfrak{E})$, we have $m_{\tilde{\mathfrak{q}}} = 0$ and therefore we get $(\kappa_{\tilde{\mathfrak{e}}_h})_n(m) = 0$. Moreover, since $\tilde{\mathfrak{m}}_h$ maps any ${}^{\tilde h}\mathfrak{F}$-morphism on an $\mathcal{O}$-module isomorphism (cf. 14.30.12), the natural map (cf. 14.30.40)

$$\tilde{\mathfrak{m}}_h * \tilde\varepsilon_h : \tilde{\mathfrak{m}}_h \longrightarrow \tilde{\mathfrak{l}}_h \circ \tilde{\mathfrak{e}}_h \qquad 14.30.47$$

determines a natural isomorphism

$$\kappa_{\tilde\varepsilon_h} : \mathfrak{c}_{\tilde{\mathfrak{m}}_h} \cong \mathfrak{c}_{\tilde{\mathfrak{l}}_h\circ\tilde{\mathfrak{e}}_h} \qquad 14.30.48$$

which sends Δ_n to the $\mathcal{O}$-module isomorphism

$$(\kappa_{\tilde\varepsilon_h})_n : \prod_{\tilde{\mathfrak{q}}\in\mathfrak{Fct}(\Delta_n,{}^{\tilde h}\mathfrak{F})} \tilde{\mathfrak{m}}_h\big(\tilde{\mathfrak{q}}(0)\big) \cong \prod_{\tilde{\mathfrak{q}}\in\mathfrak{Fct}(\Delta_n,{}^{\tilde h}\mathfrak{F})} \tilde{\mathfrak{m}}_h\Big(\hat{\mathfrak{e}}_h\big(\tilde{\mathfrak{q}}(0)\big)\Big) \qquad 14.30.49$$

mapping $m = (m_{\tilde{\mathfrak{q}}})_{\tilde{\mathfrak{q}}\in\mathfrak{Fct}(\Delta_n,{}^{\tilde h}\mathfrak{F})}$ on $\bigg(\Big(\tilde{\mathfrak{m}}_h\big((\tilde\varepsilon_h)_{\tilde{\mathfrak{q}}(0)}\big)\Big)(m_{\tilde{\mathfrak{q}}})\bigg)_{\tilde{\mathfrak{q}}\in\mathfrak{Fct}(\Delta_n,{}^{\tilde h}\mathfrak{F})}$.

At this point, for any $n \geq 1$, following the notation introduced in Lemma A4.2 below we consider the $\mathcal{O}$-module homomorphism

$$h_{n-1} : \mathfrak{c}_{\tilde{\mathfrak{m}}_h}(\Delta_n) \longrightarrow \mathfrak{c}_{\tilde{\mathfrak{l}}_h\circ\tilde{\mathfrak{e}}_h}(\Delta_{n-1}) \qquad 14.30.50$$

mapping $m = (m_{\tilde{\mathfrak{q}}})_{\tilde{\mathfrak{q}}\in\mathfrak{Fct}(\Delta_n,{}^{\tilde h}\mathfrak{F})} \in \mathfrak{c}_{\tilde{\mathfrak{m}}_h}(\Delta_n)$ on

$$h_{n-1}(m) = \Big(\sum_{i=0}^{n-1}(-1)^i\, m_{\mathfrak{h}_i^{n-1}(\tilde\varepsilon_h*\tilde{\mathfrak{r}})}\Big)_{\tilde{\mathfrak{r}}\in\mathfrak{Fct}(\Delta_n,{}^{\tilde h}\mathfrak{F})} \qquad 14.30.51$$

Then, respectively denoting by d_n and $\hat d_n$ the differential maps for the functors $\mathfrak{c}_{\tilde{\mathfrak{m}}_h}$ and $\mathfrak{c}_{\tilde{\mathfrak{l}}_h\circ\tilde{\mathfrak{e}}_h}$ (cf. A3.2), we claim that

$$(\kappa_{\tilde\varepsilon_h})_n(m) - (\kappa_{\tilde{\mathfrak{e}}_h})_n(m) = (\hat d_{n-1} \circ h_{n-1} + h_n \circ d_n)(m) \qquad 14.30.52;$$

indeed, setting $\theta = \mathfrak{m}_h\Big(\tilde{\mathfrak{e}}_h\big(\tilde{\mathfrak{q}}(0\bullet 1)\big)\Big)$, for any $\tilde{\mathfrak{q}} \in \mathfrak{Fct}(\Delta_n, {}^{\tilde h}\mathfrak{F})$ we have

$$\begin{aligned}\hat{d}_{n-1}\big(h_{n-1}(m)\big)_{\tilde{\mathfrak{q}}} &\\ &= \sum_{i=0}^{n-1}(-1)^i \hat{d}_{n-1}\big((m_{\mathfrak{h}_i^{n-1}(\tilde\varepsilon_h * \tilde{\mathfrak{r}})})_{\tilde{\mathfrak{r}}\in\mathfrak{Fct}(\Delta_{n-1},{}^{\tilde h}\mathfrak{F})}\big)_{\tilde{\mathfrak{q}}} \\ &= \sum_{i=0}^{n-1}(-1)^i\big(\theta(m_{\mathfrak{h}_i^{n-1}(\tilde\varepsilon_h*(\tilde{\mathfrak{q}}\circ\delta_0^{n-1}))}) \\ &\qquad\qquad + \sum_{j=1}^{n}(-1)^j m_{\mathfrak{h}_i^{n-1}(\tilde\varepsilon_h*(\tilde{\mathfrak{q}}\circ\delta_j^{n-1}))}\big)\end{aligned} \qquad 14.30.53;$$

analogously, we still have

$$\begin{aligned}h_n\big(d_n(m)\big)_{\tilde{\mathfrak{q}}} &= h_n\Big(\big(\theta(m_{\tilde{\mathfrak{r}}\circ\delta_0^n})\big)_{\mathfrak{r}\in\mathfrak{Fct}(\Delta_{n+1},{}^{\tilde h}\mathfrak{F})}\Big)_{\tilde{\mathfrak{q}}} \\ &\qquad + \sum_{j=1}^{n+1}(-1)^j h_n\big((m_{\mathfrak{r}\circ\delta_j^n})_{\mathfrak{r}\in\mathfrak{Fct}(\Delta_{n+1},{}^{\tilde h}\mathfrak{F})}\big)_{\tilde{\mathfrak{q}}} \qquad 14.30.54. \\ &= \sum_{i=0}^{n}(-1)^i\big(\theta(m_{\mathfrak{h}_i^n(\tilde\varepsilon_h*\tilde{\mathfrak{q}})\circ\delta_0^n}) + \sum_{j=1}^{n+1}(-1)^j\, m_{\mathfrak{h}_i^n(\tilde\varepsilon_h*\tilde{\mathfrak{q}})\circ\delta_j^n}\big)\end{aligned}$$

But from Lemma A4.2 we know that

$$\begin{aligned}\mathfrak{h}_{i+1}^n(\tilde\varepsilon_h * \tilde{\mathfrak{q}})\circ\delta_{i+1}^n &= \mathfrak{h}_i^n(\tilde\varepsilon_h * \tilde{\mathfrak{q}})\circ\delta_{i+1}^n \\ \mathfrak{h}_i^{n-1}\big(\tilde\varepsilon_h * (\tilde{\mathfrak{q}}\circ\delta_j^{n-1})\big) &= \begin{cases}\mathfrak{h}_{i+1}^n(\tilde\varepsilon_h * \tilde{\mathfrak{q}})\circ\delta_j^n & \text{if } j\le i \\ \mathfrak{h}_i^n(\tilde\varepsilon_h * \tilde{\mathfrak{q}})\circ\delta_{j+1}^n & \text{if } i<j\,.\end{cases}\end{aligned} \qquad 14.30.55.$$

Consequently, in equality 14.30.54 the terms where $j = i$ and $j = i+1$ cancel with each other for any $1 \le j \le n$; moreover, the term (i,j) in equality 14.30.53 cancels either with the term $(i+1,j)$ if $1 \le j \le i \le n-1$, or with the term $(i,j+1)$ if $0 \le i < j \le n$ in equality 14.30.54. Finally, the term $(i,0)$ in equality 14.30.53 cancels with the term $(i+1,0)$ in equality 14.30.54 for any $0 \le i \le n-1$, whereas the terms $(0,0)$ and $(n,n+1)$ respectively coincide with $(\kappa_{\tilde\varepsilon_h})_n(m)$ and $-(\kappa_{\tilde{\mathfrak{e}}_h})_n(m)$, which proves the claim.

In conclusion, if $m \in \mathrm{Ker}(d_n)$ lifts an element of $\mathbb{H}_n(\tilde\lambda_n)$ then we may assume that $m \in \mathrm{Ker}\big((\lambda_h)_{\Delta_n}\big)$, so that we get $(\kappa_{\tilde{\mathfrak{e}}_h})_n(m) = 0$, and therefore it follows from isomorphism 14.30.48 and from equality 14.30.52 that

$$\begin{aligned}(\kappa_{\tilde\varepsilon_h})_n(m) &= \hat{d}_{n-1}\big(h_{n-1}(m)\big) \\ &= \hat{d}_{n-1}\Big(\big((\kappa_{\tilde\varepsilon_h})_{n-1}\circ\big((\kappa_{\tilde\varepsilon_h})_{n-1}\big)^{-1}\circ h_{n-1}\big)(m)\Big) \qquad 14.30.56 \\ &= \big((\kappa_{\tilde\varepsilon_h})_n\circ d_{n-1}\big)\Big(\big(\big((\kappa_{\tilde\varepsilon_h})_{n-1}\big)^{-1}\circ h_{n-1}\big)(m)\Big)\end{aligned}$$

which proves that m belongs to $\mathrm{Im}(d_{n-1})$. We are done.

14.31 Let us say that a $\mathfrak{ch}^*(\mathcal{F}^{^{\mathrm{sc}}})$-object $(\mathfrak{q},\Delta_n)$ is *regular* if $\mathfrak{q}(i-1,i)$ is *not* an isomorphism for any $1\le i\le n$ (cf. A5.2); note that there is a canonical bijection between a set of representatives for the isomorphism classes of regular $\mathfrak{ch}^*(\mathcal{F}^{^{\mathrm{sc}}})$-objects and a set of representatives for the $\mathcal{F}$-isomorphism classes of nonempty sets of $\mathcal{F}$-selfcentralizing subgroups of P, which are totally ordered by the inclusion.

Corollary 14.32 *We have*

$$\mathrm{rank}_{\mathcal{O}}\big(\mathcal{G}_k(\mathcal{F},\widehat{\mathfrak{aut}}_{\mathcal{F}^{\mathrm{nc}}})\big)=\sum_{(\mathfrak{q},\Delta_n)}(-1)^n\,\mathrm{rank}_{\mathcal{O}}\Big(\mathcal{G}_k\big(\hat{\mathcal{F}}(\mathfrak{q})\big)\Big) \qquad 14.32.1$$

where $(\mathfrak{q},\Delta_n)$ *runs over a set of representatives for the set of isomorphism classes of regular* $\mathfrak{ch}^*(\mathcal{F}^{^{\mathrm{sc}}})$*-objects.*

Proof: Denote by $\mathcal{K}$ the field of quotients of $\mathcal{O}$ and by ${}^{\mathcal{K}}\mathfrak{n}_h$ the extension of $\mathfrak{n}_h=\mathcal{F}ct_{U_h}\circ\mathfrak{s}_h$ from $\mathcal{O}$ to $\mathcal{K}$ for any $h\in\mathbb{N}-p\mathbb{N}$. It is more or less well-known that for any $n\in\mathbb{N}$ we have (cf. Propositions A4.13 and A5.7)

$$\mathbb{H}^n\big({}^h(\tilde{\mathcal{F}}^{^{\mathrm{sc}}}),{}^{\mathcal{K}}\mathfrak{n}_h\big)\cong\mathbb{H}^n_*\big({}^h(\tilde{\mathcal{F}}^{^{\mathrm{sc}}}),{}^{\mathcal{K}}\mathfrak{n}_h\big)\cong\mathbb{H}^n_{\mathrm{r}}\big({}^h(\tilde{\mathcal{F}}^{^{\mathrm{sc}}}),{}^{\mathcal{K}}\mathfrak{n}_h\big) \qquad 14.32.2.$$

In particular, it follows from Theorem 14.30 that for any $n\ge 1$ we get

$$\mathbb{H}^n_{\mathrm{r}}\big({}^h(\tilde{\mathcal{F}}^{^{\mathrm{sc}}}),{}^{\mathcal{K}}\mathfrak{n}_h\big)=\{0\} \qquad 14.32.3$$

which amounts to saying that we have an infinite exact sequence

$$0\longrightarrow\mathbb{H}^0_{\mathrm{r}}\big({}^h(\tilde{\mathcal{F}}^{^{\mathrm{sc}}}),{}^{\mathcal{K}}\mathfrak{n}_h\big)\longrightarrow\ldots\longrightarrow C^n\to C^{n+1}\longrightarrow\ldots \qquad 14.32.4$$

where $C^n=\mathbb{C}^n_{\mathrm{r}}\big({}^h(\tilde{\mathcal{F}}^{^{\mathrm{sc}}}),{}^{\mathcal{K}}\mathfrak{n}_h\big)$ is the set of elements

$$(\chi_{\tilde{\mathfrak{q}}})_{\tilde{\mathfrak{q}}\in\mathfrak{Fct}_{\mathrm{r}}(\Delta_n,{}^h(\tilde{\mathcal{F}}^{\mathrm{sc}}))}\in\prod_{\tilde{\mathfrak{q}}\in\mathfrak{Fct}_{\mathrm{r}}(\Delta_n,{}^h(\tilde{\mathcal{F}}^{\mathrm{sc}}))}{}^{\mathcal{K}}\mathfrak{n}_h\big(\tilde{\mathfrak{q}}(0)\big) \qquad 14.32.5$$

such that, for any natural isomorphism $\tilde{\nu}:\tilde{\mathfrak{q}}\cong\tilde{\mathfrak{q}}'$ between *regular* ${}^h(\tilde{\mathcal{F}}^{^{\mathrm{sc}}})$*-valued* n*-chains* $\tilde{\mathfrak{q}}$ and $\tilde{\mathfrak{q}}'$, ${}^{\mathcal{K}}\mathfrak{n}_h(\tilde{\nu}_0)$ maps $\chi_{\tilde{\mathfrak{q}}'}$ on $\chi_{\tilde{\mathfrak{q}}}$. That is to say, we actually have

$$\mathbb{C}^n_{\mathrm{r}}\big({}^h(\tilde{\mathcal{F}}^{^{\mathrm{sc}}}),{}^{\mathcal{K}}\mathfrak{n}_h\big)\cong\prod_{\tilde{\mathfrak{q}}}\mathcal{F}ct_{U_h}\Big(\mathfrak{s}_k\big(\tilde{\mathfrak{q}}(0)\big),\mathcal{K}\Big)^{\tilde{\mathcal{F}}(\tilde{\mathfrak{q}})} \qquad 14.32.6$$

where $\tilde{\mathfrak{q}}$ runs over a set of representatives for the set of isomorphism classes in $\mathfrak{Fct}_{\mathrm{r}}\big(\Delta_n,{}^h(\tilde{\mathcal{F}}^{^{\mathrm{sc}}})\big)$ (cf. A5.3).

On the other hand, it is clear that for n big enough there are no *regular* ${}^h(\tilde{\mathcal{F}}^{\mathrm{sc}})$*-valued* n*-chains* and therefore, in the exact sequence above, only finitely many terms are not zero; thus, we still get

$$\mathrm{rank}_{\mathcal{O}}\Big(\mathbb{H}^0\big({}^h(\tilde{\mathcal{F}}^{\mathrm{sc}}),\mathfrak{n}_h\big)\Big) = \sum_{(\tilde{\mathfrak{q}},\Delta_n)} (-1)^n \dim_{\mathcal{K}}\Big(\mathcal{F}ct_{U_h}\Big(\mathfrak{s}_k\big(\tilde{\mathfrak{q}}(0)\big),\mathcal{K}\Big)^{\tilde{\mathcal{F}}(\tilde{\mathfrak{q}})}\Big) \qquad 14.32.7$$

where $(\tilde{\mathfrak{q}},\Delta_n)$ runs over a set of representatives for the isomorphism classes of *regular* $\mathfrak{ch}^*\big({}^h(\tilde{\mathcal{F}}^{\mathrm{sc}})\big)$-objects (cf. A5.3).

Consequently, it follows from Propositions 14.27 and 14.28 that, denoting by $\tilde{\mathcal{F}}(\tilde{\mathfrak{q}})_\eta$ the stabilizer of η in $\tilde{\mathcal{F}}(\tilde{\mathfrak{q}})$ for any ${}^{\mathfrak{u}_h}\mathfrak{ch}^*(\tilde{\mathcal{F}}^{\mathrm{sc}})$-object $(\eta,\tilde{\mathfrak{q}},\Delta_n)$ (cf. 14.23.4), since we have

$$(\varpi_{h,\hat{\tilde{\mathcal{F}}}(\tilde{\mathfrak{q}}(0))})^{-1}(\iota_0^{\tilde{\mathfrak{q}}}\circ\eta) \cong (\varpi_{h,\hat{\tilde{\mathcal{F}}}(\tilde{\mathfrak{q}})})^{-1}(\eta) \qquad 14.32.8,$$

we actually have

$$\mathrm{rank}_{\mathcal{O}}\big(\mathcal{G}_k(\mathcal{F},\widehat{\mathfrak{aut}}_{\mathcal{F}^{\mathrm{nc}}})\big) = \sum_{(\eta,\tilde{\mathfrak{q}},\Delta_n)} (-1)^n \sum_h \dim_{\mathcal{K}}\Big(\mathcal{F}ct_{U_h}\big((\varpi_{h,\hat{\tilde{\mathcal{F}}}(\tilde{\mathfrak{q}})})^{-1}(\eta),\mathcal{K}\big)^{\tilde{\mathcal{F}}(\tilde{\mathfrak{q}})_\eta}\Big) \qquad 14.32.9$$

where h runs over $\mathbb{N}-p\mathbb{N}$ and $(\eta,\tilde{\mathfrak{q}},\Delta_n)$ runs over a set of representatives for the isomorphism classes of ${}^{\mathfrak{u}_h}\mathfrak{ch}^*(\tilde{\mathcal{F}}^{\mathrm{sc}})$-objects such that $(\tilde{\mathfrak{q}},\Delta_n)$ is a regular $\mathfrak{ch}^*(\tilde{\mathcal{F}}^{\mathrm{sc}})$-object.

On the other hand, for any regular $\mathfrak{ch}^*(\tilde{\mathcal{F}}^{\mathrm{sc}})$-object $(\tilde{\mathfrak{q}},\Delta_n)$, it follows from isomorphism 14.15.3 that

$$\begin{aligned}&\bigoplus_{h\in\mathbb{N}-p\mathbb{N}}\bigoplus_{\eta}\mathcal{F}ct_{U_h}\big((\varpi_{h,\hat{\tilde{\mathcal{F}}}(\tilde{\mathfrak{q}})})^{-1}(\eta),\mathcal{O}\big)^{\tilde{\mathcal{F}}(\tilde{\mathfrak{q}})_\eta}\\ &\cong\Big(\bigoplus_{h\in\mathbb{N}-p\mathbb{N}}\ \bigoplus_{\rho\in\mathrm{Mon}(U_h,\tilde{\mathcal{F}}(\tilde{\mathfrak{q}}))}\mathcal{F}ct_{U_h}\big((\varpi_{h,\hat{\tilde{\mathcal{F}}}(\tilde{\mathfrak{q}})})^{-1}(\rho),\mathcal{O}\big)\Big)^{\tilde{\mathcal{F}}(\tilde{\mathfrak{q}})} \qquad 14.32.10\\ &\cong\mathcal{G}_k\big(\hat{\tilde{\mathcal{F}}}(\tilde{\mathfrak{q}})\big)\cong\mathcal{G}_k\big(\hat{\mathcal{F}}(\mathfrak{q})\big)\end{aligned}$$

where η runs over a set of representatives for the orbits of $\tilde{\mathcal{F}}(\tilde{\mathfrak{q}})$ on the set $\mathrm{Mon}\big(U_h,\tilde{\mathcal{F}}(\tilde{\mathfrak{q}})\big)$ and $\mathfrak{q}:\Delta_n\to\mathcal{F}^{\mathrm{sc}}$ is a functor lifting $\tilde{\mathfrak{q}}$. We are done.

Chapter 15

Reduction results for Grothendieck groups

15.1 Let k be an algebraically closed field of characteristic p and $\mathcal{O}$ the *unramified complete discrete valuation ring* of characteristic zero with residue field k ; denote by $\mathcal{O}$-$\mathfrak{mod}$ the category of finitely generated $\mathcal{O}$-modules. Let $\hat{G}$ be a *finite* k^*-group — we already have seen the interest of replacing finite groups by k^*-groups with a finite k^*-quotient G (cf. 1.23) — and b be a primitive idempotent of $Z(k_*\hat{G})$ — called a *block* of $\hat{G}$ (cf. 1.25). Recall that, according to Proposition 5.15 in [42], b is also a block of a finite subgroup G' of $\hat{G}$ fulfilling $kG'b = k_*\hat{G}b$ and therefore all the terminology developed for G' can be easily translated to $\hat{G}$. Thus, let (P, e) be a maximal Brauer $(b, \hat{G})$-pair (cf. 1.16 and 1.28) and $\mathcal{F} = \mathcal{F}_{(b,\hat{G})}$ the corresponding Frobenius P-category (cf. Theorem 3.7); recall that we denote by $\mathcal{F}^{^{\rm nc}}$ the full subcategory of $\mathcal{F}$ over the set of $\mathcal{F}$-nilcentralized subgroups of P (cf. 11.1).

15.2 As in chapter 14, let us denote by $\mathcal{G}_k(\hat{G}, b)$ the scalar extension from $\mathbb{Z}$ to $\mathcal{O}$ of the *Grothendieck group* of the category of $k_*\hat{G}b$-modules of finite dimension. Applying again the remark above, it follows from Theorem 11.32 that we have a functor

$$\widehat{\mathfrak{aut}}_{\mathcal{F}^{\rm nc}} : \mathfrak{ch}^*(\mathcal{F}^{^{\rm nc}}) \longrightarrow k^*\text{-}\mathfrak{Gr} \tag{15.2.1}$$

lifting $\mathfrak{aut}_{\mathcal{F}^{\rm nc}}$ (cf. Proposition A2.10) and, in this situation, in chapter 14 we have defined the *Grothendieck group* $\mathcal{G}_k(\mathcal{F}, \widehat{\mathfrak{aut}}_{\mathcal{F}^{\rm nc}})$ of this pair $(\mathcal{F}, \widehat{\mathfrak{aut}}_{\mathcal{F}^{\rm nc}})$. As we explain in the Introduction (cf. I31), Corollary 14.32 shows that the $\mathcal{O}$-rank of $\mathcal{G}_k(\mathcal{F}, \widehat{\mathfrak{aut}}_{\mathcal{F}^{\rm nc}})$ coincides with the $\mathcal{O}$-rank that Jon Alperin conjectures for $\mathcal{G}_k(\hat{G}, b)$ and therefore Alperin's Conjecture is equivalent to the existence of an $\mathcal{O}$-module isomorphism $\mathcal{G}_k(\hat{G}, b) \cong \mathcal{G}_k(\mathcal{F}, \widehat{\mathfrak{aut}}_{\mathcal{F}^{\rm nc}})$.

15.3 Let us say that a *block* $(c, \hat{H})$ (cf. 1.10) is a *normal sub-block* of $(b, \hat{G})$ when $\hat{H}$ is a normal k^*-subgroup of $\hat{G}$ and we have $cb \neq 0$ — in 15.7 below we show that we actually may assume that $\hat{G}$ fixes c . In this chapter, we prove that — independently of Alperin's Conjecture — both *Grothendieck groups* $\mathcal{G}_k(\hat{G}, b)$ and $\mathcal{G}_k(\mathcal{F}, \widehat{\mathfrak{aut}}_{\mathcal{F}^{\rm nc}})$ have an analogous relationship with the *Grothendieck groups* corresponding to suitable normal sub-blocks of $(b, \hat{G})$. The point is that we prove enough such reduction results to show — in the next chapter — that a suitable reformulation strengthening the above conjecture can be verified just testing the blocks "around" the simple groups.

15.4 As a matter of fact, some of the reduction steps really force us to employ *finite k^*-groups* instead of ordinary finite groups. First of all, it is convenient to clarify the definition of the functor $\widehat{\mathfrak{aut}}_{\mathcal{F}^{\mathrm{nc}}}$ for the finite k^*-group $\hat G$; let (R,g) be a nilcentralized Brauer $(b,\hat G)$-pair (cf. 7.4) contained in (P,e) (cf. 1.15); as in 7.4 above, the quotient $k_*C_{\hat G}(R)g/J\big(k_*C_{\hat G}(R)g\big)$ is a simple k-algebra — in particular, g determines a unique local point ε of R on $k_*\hat G b$ (cf. 1.19) — and therefore the action of the normalizer $N_{\hat G}(R,g)$ on this k-algebra determines a *k^*-group* $\hat N_G(R,g)$ together with a k-algebra homomorphism (cf. 1.25)

$$k_*\hat N_G(R,g) \longrightarrow k_*C_{\hat G}(R)g/J\big(k_*C_{\hat G}(R)g\big) \qquad 15.4.1;$$

moreover, since we have an obvious $N_G(R,g)$-stable k-algebra homomorphism

$$k_*C_{\hat G}(R) \longrightarrow k_*C_{\hat G}(R)g/J\big(k_*C_{\hat G}(R)g\big) \qquad 15.4.2,$$

we still have an $N_G(R,g)$-stable k^*-group homomorphism $C_{\hat G}(R) \to \hat N_G(R,g)$ lifting the inclusion $C_G(R) \subset N_G(R,g)$.

15.5 Consequently, we have an evident group homomorphism (cf. 1.23)

$$\Delta_{(R,g)} : C_G(R) \longrightarrow N_{\hat G}(R,g) * \hat N_G(R,g)^\circ \qquad 15.5.1$$

having a normal image and then, since (cf. 1.26)

$$\mathcal{F}(R) \cong E_G(R,g) = N_G(R,g)/C_G(R) \qquad 15.5.2$$

where we identify the Brauer $(b,\hat G)$-pair (R,g) with the unique local pointed group R_ε , we set

$$\hat{\mathcal{F}}(R) \cong \hat E_G(R,g) = \big(N_{\hat G}(R,g) * \hat N_G(R,g)^\circ\big)/\Delta_{(R,g)}\big(C_G(R)\big) \qquad 15.5.3.$$

Similarly, if $\hat F$ is a k^*-group containing $\hat G$ as a normal k^*-subgroup, denoting by $C_{\hat F}(R,g)$ the stabilizer of g in $C_{\hat F}(R)$, we define

$$C_{\hat G}^{\hat F}(R,g) = \big(C_{\hat F}(R,g) * \hat C_F(R,g)^\circ\big)/\Delta_{(R,g)}\big(C_G(R)\big) \qquad 15.5.4.$$

15.6 Then, for any $\mathcal{F}^{^{\mathrm{nc}}}$*-chain* $\mathfrak{q} : \Delta_n \to \mathcal{F}^{^{\mathrm{nc}}}$ (cf. 2.18), identifying $\mathcal{F}(\mathfrak{q})$ to its image in $\mathcal{F}\big(\mathfrak{q}(n)\big)$, as in 11.31 above we denote by $\hat{\mathcal{F}}(\mathfrak{q})$ the corresponding k^*-subgroup of $\hat{\mathcal{F}}\big(\mathfrak{q}(n)\big)$; as mentioned above, it follows from Theorem 11.32 that this correspondence can be extended to a functor $\widehat{\mathfrak{aut}}_{\mathcal{F}^{\mathrm{nc}}}$ lifting $\mathfrak{aut}_{\mathcal{F}^{\mathrm{nc}}}$. Thus, as in 14.3, we have the composed functor

$$\mathfrak{ch}^*(\mathcal{F}^{^{\mathrm{nc}}}) \xrightarrow{\widehat{\mathfrak{aut}}_{\mathcal{F}^{\mathrm{nc}}}} k^*\text{-}\mathfrak{Gr} \xrightarrow{\mathfrak{g}_k} \mathcal{O}\text{-}\mathfrak{mod} \qquad 15.6.1$$

and we have defined the *Grothendieck group of the pair* $(\mathcal{F}, \widehat{\mathfrak{aut}}_{\mathcal{F}^{\rm nc}})$ as the inverse limit

$$\mathcal{G}_k(\mathcal{F}, \widehat{\mathfrak{aut}}_{\mathcal{F}^{\rm nc}}) = \lim_{\longleftarrow} (\mathfrak{g}_k \circ \widehat{\mathfrak{aut}}_{\mathcal{F}^{\rm nc}}) \qquad 15.6.2.$$

15.7 In this chapter, our *standard setting* is the pair formed by the block $(b, \hat{G})$ and a normal sub-block $(c, \hat{H})$; so, $\hat{H}$ is a normal k^*-subgroup of $\hat{G}$ and we assume that $cb \neq 0$. In particular, we have $b\mathrm{Tr}_{\hat{G}_c}^{\hat{G}}(c) = b$ where $\hat{G}_c$ denotes the stabilizer of c in $\hat{G}$; since we know that $e\mathrm{Br}_P(b) \neq 0$ (cf. 1.16), we may assume that P stabilizes c; then, considering the G-stable semisimple k-subalgebra $\sum_x k{\cdot}bc^x$ of $k_*\hat{G}$, where $x \in G$ runs over a set of representatives for $\hat{G}/\hat{G}_c$, it follows from Proposition 2.5 in [53] that (bc, G_c) is a block (cf. 1.10) and, from Proposition 2.2 and Corollary 2.3 again in [53], that P remains a defect p-subgroup of this block and we have

$$\mathcal{F}_{(bc,\hat{G}_c)} = \mathcal{F}_{(b,\hat{G})} \quad \text{and} \quad \mathcal{G}_k(bc, \hat{G}_c) \cong \mathcal{G}_k(b, \hat{G}) \qquad 15.7.1.$$

Thus, for our purposes, in our standard setting we may assume that $\hat{G}$ fixes c and we have $bc = b$.

15.8 Furthermore, we choose a maximal Brauer $(c, \hat{H})$-pair (Q, f); note that $\alpha = \{c\}$ is also a point of $\hat{G}$ on $k_*\hat{H}$ and that, denoting by e^H a block of $C_{\hat{H}}(P)$ such that $e^He \neq 0$, (P, e^H) and (Q, f) are *Brauer* (α, G)*-pairs* (cf. 1.14). Thus, in our standard setting, we may assume that (P, e^H) and (Q, f) are both contained in a maximal Brauer (α, G)-pair (O, o) (cf. 1.15); moreover, setting $Q' = \hat{H} \cap O$, since there is a block f' of $C_{\hat{H}}(Q')$ such that (cf. 1.15)

$$(Q, f) \subset (Q', f') \subset (O, o) \qquad 15.8.1,$$

we actually have

$$Q = \hat{H} \cap O \qquad 15.8.2.$$

Then, we consider the Frobenius P- and Q-categories

$$\mathcal{F} = \mathcal{F}_{(b,\hat{G})} \quad \text{and} \quad \mathcal{H} = \mathcal{F}_{(c,\hat{H})} \qquad 15.8.3;$$

although, in general, there is no easy relationship between $\mathcal{F}$ and $\mathcal{H}$, in the situations we consider below we obtain some connection between $\mathcal{F}^{\rm nc}$ or $\mathcal{F}^{\rm sc}$ and $\mathcal{H}^{\rm nc}$ or $\mathcal{H}^{\rm sc}$, which allows us to relate the corresponding Grothendieck groups $\mathcal{G}_k(\mathcal{F}, \widehat{\mathfrak{aut}}_{\mathcal{F}^{\rm sc}})$ and $\mathcal{G}_k(\mathcal{H}, \widehat{\mathfrak{aut}}_{\mathcal{H}^{\rm sc}})$.

Proposition 15.9 *In our standard setting, we have*

$$Q = \hat{H} \cap P \quad \textit{and} \quad \mathrm{Br}_P(f) \neq 0 \qquad 15.9.1.$$

Proof: Setting $R = Q \cap P$ and denoting by g the block of $C_{\hat{G}}(R)$ such that (P,e) contains (R,g) (cf. 1.15), a maximal Brauer $\big(g, R{\cdot}C_{\hat{G}}(R)\big)$-pair (T,h) is a Brauer $(b,\hat{G})$-pair too and therefore there is $x \in \hat{G}$ such that (cf. 1.15)

$$(R,g)^x \subset (T,h)^x \subset (P,e) \qquad 15.9.2\,;$$

thus, we have $R^x \subset \hat{H} \cap P$ which forces $R^x = R$ and then $g^x = g$ (cf. 1.15). That is to say, $\bar{C}_P(R)$ is a *defect group* of the image $\bar{g}$ of g in $\bar{C}_{\hat{G}}(R)$ (cf. 1.9.2), so that $\bar{g}$ belongs to $k_*\bar{C}_{\hat{G}}(R)^{\bar{C}_{\hat{G}}(R)}_{\bar{C}_P(R)}$ (cf. Theorem 1.14 in [11]), and therefore, since we have $\bar{C}_{\hat{H}}(R) \cap \bar{C}_P(R) = \{1\}$, the restriction of any simple $k_*\bar{C}_{\hat{G}}(R)\bar{g}$-module to $k_*\bar{C}_{\hat{H}}(R)$ is both semisimple (cf. Theorem 4.1 in [28, Ch. 3]) and projective.

Consequently, since the block g^H of $C_{\hat{H}}(R)$ such that (P,e^H) contains (R,g^H) (cf. 1.15) is P-stable and we have $\mathrm{Br}_P(g^H)e \neq 0$ (cf. Theorem 1.8 in [11]), we still have $g^H g \neq 0$ and therefore a simple $k_*\bar{C}_{\hat{H}}(R)\bar{g}^H$-module is projective, so that $\bar{g}^H$ has defect zero (cf. 1.17) and the Brauer $(c,\hat{H})$-pair (R,g^H) is selfcentralizing (cf. 7.4); but, since

$$(R,g^H) \subset (P,e^H) \subset (O,o) \quad \text{and} \quad (Q,f) \subset (O,o) \qquad 15.9.3,$$

we have $(R,g^H) \subset (Q,f)$ (cf. 1.15) and therefore R is $\mathcal{H}$-selfcentralizing (cf. Corollary 7.3).

Moreover, since we have

$$(P,e^H) \subset (O,o) \supset (Q,f) \qquad 15.9.4,$$

$N_Q(P)$ fixes e^H (cf. 1.15) and the quotient $\bar{N}_Q(P) = N_Q(P)/R$ (cf. 1.9.2) acts on $k_*\bar{C}_{\hat{G}}(P)\bar{e}^H$; but, since the block $\bar{e}^H$ has defect zero (cf. 1.17), it follows from Proposition 3.7 in [53] that we have a $\bar{N}_Q(P)$-algebra isomorphism (cf. definition 15.5.4)

$$k_*\bar{C}_{\hat{G}}(P)\bar{e}^H \cong k_*\bar{C}_{\hat{H}}(P)\bar{e}^H \otimes_k k_*C^{\hat{G}}_{\hat{H}}(P,e^H) \qquad 15.9.5$$

and, in particular, we still have a $\bar{N}_Q(P)$-algebra isomorphism

$$Z\big(k_*\bar{C}_{\hat{G}}(P)\bar{e}^H\big) \cong Z\big(k_*C^{\hat{G}}_{\hat{H}}(P,e^H)\big) \qquad 15.9.6;$$

on the other hand, since the *p-group* $\bar{N}_Q(P)$ acts trivially on G/H, it acts trivially on $C^{\hat{G}}_{\hat{H}}(P,e^H)$. Hence, $\bar{N}_Q(P)$ acts trivially on $Z\big(k_*\bar{C}_{\hat{G}}(P)\bar{e}^H\big)$ and therefore it fixes e; consequently, since we have $C_Q(P) \subset C_Q(R) = Z(R)$, this *p-group* maps into the *p'-group* $N_{\hat{G}}(P,e)/P{\cdot}C_{\hat{G}}(P)$ (cf. 1.16) which forces $\bar{N}_Q(P) = \{1\}$ and therefore, since P acts on Q/R and contains R, this implies that $Q = R$; in particular, we get $(Q,f) \subset (P,e^H)$. We are done.

Proposition 15.10 *In our standard setting, assume that $C_{\hat{G}}(Q,f)/C_{\hat{H}}(Q)$ is a p-group. Then, we have*

$$b = c \quad and \quad C_{\hat{G}}(Q,f) = C_{\hat{H}}(Q){\cdot}C_P(Q) \qquad 15.10.1,$$

P/Q is a Sylow p-subgroup of $\hat{G}/\hat{H}$ and, for any selfcentralizing Brauer $(c,\hat{H})$-pair (R,g), the quotient $C_{\hat{G}}(R,g)/C_{\hat{H}}(R)$ is a p-group and, setting $g^G = \mathrm{Tr}_{C_{\hat{G}}(R,g)}^{C_{\hat{G}}(R)}(g)$, (R,g^G) is a nilcentralized Brauer $(b,\hat{G})$-pair. In particular, $\mathcal{H}^{^{\mathrm{sc}}}$ and $\mathcal{H}$ are respective subcategories of $\mathcal{F}^{^{\mathrm{nc}}}$ and $\mathcal{F}$.

Proof: We may assume that $(R,g) \subset (Q,f)$; since the correspondence sending (R,g) to (R,g^G) preserves the inclusion of Brauer pairs, the inclusion $\mathcal{H}^{^{\mathrm{sc}}} \subset \mathcal{F}^{^{\mathrm{nc}}}$ will follow from the previous statement and then the inclusion $\mathcal{H} \subset \mathcal{F}$ will follow from Corollary 5.14. Moreover, in order to prove that (R,g^G) is a nilcentralized Brauer $(b,\hat{G})$-pair, it suffices to prove that $C_{\hat{G}}(R,g)/C_{\hat{H}}(R)$ is a p-group; indeed, since the block $\big(g, C_{\hat{H}}(R)\big)$ is nilpotent (cf. 1.10 and 7.4), it follows from Proposition 6.5 in [34] that g remains a nilpotent block of $C_{\hat{G}}(R,g)$ and then, by Proposition 3.5 in [53], g^G is a nilpotent block of $C_{\hat{G}}(R)$.

Denote by R_ε the local pointed group on $k_*\hat{H}c$ determined by (R,g) (cf. 15.4), by λ the point of $R{\cdot}C_{\hat{H}}(R)$ on $k_*\hat{H}$ such that $R_\varepsilon \subset R{\cdot}C_{\hat{H}}(R)_\lambda$ (cf. 1.19) and by T_ν a defect pointed group of $R{\cdot}C_{\hat{H}}(R)_\lambda$ (cf. 1.15); we argue by induction on $|Q:R|$ and, since $C_{\hat{G}}(Q,f)/C_{\hat{H}}(Q)$ is a p-group, may assume that $R \neq Q$. By the *Frattini argument*, we get

$$R{\cdot}C_{\hat{G}}(R_\varepsilon) = R{\cdot}C_{\hat{H}}(R){\cdot}\big(N_{\hat{G}}(T_\nu) \cap C_{\hat{G}}(R_\varepsilon)\big) \qquad 15.10.2;$$

but, by Thompson's Lemma (cf. Theorem 3.4 in [28, Ch. 5]), any k^*-subgroup $\hat{K}$ of $N_{\hat{G}}(T_\nu) \cap C_{\hat{G}}(R_\varepsilon)$ with k^*-quotient of order prime to p, centralizes T and therefore, by our induction hypothesis, it is contained in $\hat{H}$.

In particular, since f remains a block of $C_{\hat{G}}(Q,f)$ and any idempotent of $Z\big(k_*N_{\hat{G}}(Q,f)\big)$ belongs to $Z\big(k_*C_{\hat{G}}(Q,f)\big)$ (cf. 1.19), f remains a block of $N_{\hat{G}}(Q,f)$ too and therefore it follows again from Proposition 3.5 in [53] that $\mathrm{Tr}_{N_{\hat{G}}(Q,f)}^{N_{\hat{G}}(Q)}(f)$ is a block of $N_{\hat{G}}(Q)$, so that we get (cf. 1.15)

$$\mathrm{Br}_Q(c) = \mathrm{Tr}_{N_{\hat{H}}(Q,f)}^{N_{\hat{H}}(Q)}(f) = \mathrm{Tr}_{N_{\hat{G}}(Q,f)}^{N_{\hat{G}}(Q)}(f) \qquad 15.10.3$$

which forces $\mathrm{Br}_Q(b) = \mathrm{Br}_Q(c)$. Similarly, for any block b' of $\hat{G}$ such that $b'c = b'$ we still get $\mathrm{Br}_Q(b') = \mathrm{Br}_Q(c)$ which forces $b' = b$, so that $b = c$.

Then, since $(k_*\hat{H}c)(O)$ (cf. 15.8) maps into $(k_*\hat{G}b)(O)$ (cf. 1.19), the Brauer $(\alpha,\hat{G})$-pair (P,e^H) in 15.8 above is maximal and therefore it follows from Proposition 5.3 and Corollary 6.3 in [34] that P/Q is a Sylow p-subgroup of $\hat{G}/\hat{H}$; moreover, P normalizes (Q,f) (cf. 1.15) and therefore P/Q contains the p-subgroup $C_{\hat{G}}(Q,f)/C_{\hat{H}}(Q)$ of $\hat{G}/\hat{H}$, proving the right-hand equality in 15.10.1.

15.11 Thus, if $C_{\hat{G}}(Q,f)/C_{\hat{H}}(Q)$ is a p-group and in our standard setting we consider a k^*-subgroup $\hat{K}$ of $\hat{G}$ containing $\hat{H}$, then $b=c$ is still a block of $\hat{K}$ and, setting $\mathcal{K}=\mathcal{F}_{(b,\hat{K})}$, it is possible to connect the functors $\widehat{\mathfrak{aut}}_{\mathcal{F}^{\mathrm{nc}}}$ and $\widehat{\mathfrak{aut}}_{\mathcal{K}^{\mathrm{nc}}}$, suitably restricted to the $\mathcal{F}$- and $\mathcal{K}$-*chains* whose terms contain $\mathcal{H}$-selfcentralizing subgroups of Q; actually, we only need this connection when $\hat{K}/\hat{H}$ is a p'-group, which is easier to we discuss than the general case. If $\hat{K}/\hat{H}$ is a p'-group then $\mathcal{K}$ is a Frobenius Q-category and a subgroup of Q is $\mathcal{K}$-selfcentralizing if and only if it is $\mathcal{H}$-selfcentralizing; more explicitly, if (R,g) is a selfcentralizing Brauer $(c,\hat{H})$-pair then, according to Proposition 15.10, (R,g^K) is a selfcentralizing Brauer $(c,\hat{K})$-pair and, by the *Frattini argument*, we have

$$N_{\hat{K}}(R,g^K)=C_{\hat{K}}(R)\cdot N_{\hat{K}}(R,g) \qquad 15.11.1;$$

hence, we get $\mathcal{K}(R)\subset\mathcal{F}(R)$ and then it follows from Corollary 5.14 and Proposition 15.10 that $\mathcal{K}^{^{\mathrm{sc}}}$ is a subcategory of $\mathcal{F}^{^{\mathrm{nc}}}$; in particular, denoting by $\mathfrak{j}:\mathcal{K}^{^{\mathrm{sc}}}\to\mathcal{F}^{^{\mathrm{nc}}}$ the inclusion functor, we have a functor and a natural map (cf. A2.8 and Proposition A2.10)

$$\begin{aligned}\mathfrak{ch}^*(\mathfrak{j}):\mathfrak{ch}^*(\mathcal{K}^{^{\mathrm{sc}}})&\longrightarrow\mathfrak{ch}^*(\mathcal{F}^{^{\mathrm{nc}}})\\ \nu:\mathfrak{aut}_{\mathcal{K}^{\mathrm{sc}}}&\longrightarrow\mathfrak{aut}_{\mathcal{F}^{\mathrm{nc}}}\circ\mathfrak{ch}^*(\mathfrak{j})\end{aligned} \qquad 15.11.2.$$

Proposition 15.12 *In our standard setting, assume that $C_{\hat{G}}(Q,f)/C_{\hat{H}}(Q)$ is a p-group and let $\hat{K}$ be a k^*-subgroup of $\hat{G}$ containing $\hat{H}$ such that $\hat{K}/\hat{H}$ is a p'-group. Then, with the notation above, ν can be lifted to a natural map*

$$\hat{\nu}:\widehat{\mathfrak{aut}}_{\mathcal{K}^{\mathrm{sc}}}\longrightarrow\widehat{\mathfrak{aut}}_{\mathcal{F}^{\mathrm{nc}}}\circ\mathfrak{ch}^*(\mathfrak{j}) \qquad 15.12.1.$$

Proof: Let (R,g) be a selfcentralizing Brauer $(c,\hat{H})$-pair; it follows from Proposition 15.10 that the relative traces

$$g^K=\mathrm{Tr}_{C_{\hat{K}}(R,g)}^{C_{\hat{K}}(R)}(g)\quad\text{and}\quad g^G=\mathrm{Tr}_{C_{\hat{G}}(R,g)}^{C_{\hat{G}}(R)}(g) \qquad 15.12.2$$

are respective nilpotent blocks of $C_{\hat{K}}(R)$ and $C_{\hat{G}}(R)$; actually, since R is also $\mathcal{K}$-selfcentralizing, g^K is a block of $C_{\hat{K}}(R)$ of *defect zero* (cf. Corollary 7.3);

hence, we get canonical embeddings (cf. 1.16.1 and 7.7.1 above and Proposition 3.5 in [53])

$$\begin{array}{ccc} & & k_*C_{\hat{K}}(R)g^K/J\big(k_*C_{\hat{K}}(R)g^K\big) \\ & \nearrow & \\ k_*\bar{C}_{\hat{H}}(R)\bar{g} & & \\ & \searrow & \\ & & k_*C_{\hat{G}}(R)g^G/J\big(k_*C_{\hat{G}}(R)g^G\big) \end{array} \qquad 15.12.3;$$

consequently, the actions of $N_{\hat{K}}(R,g)$ on these three simple k-algebras determine the *same* k^*-group $\hat{N}_{\hat{K}}(R,g)$ (cf. 1.25) and therefore from the equality 15.11.1 successively applied to $N_{\hat{K}}(R,g^K)$ and to $N_{\hat{G}}(R,g^G)$ we get a canonical k^*-group homomorphism

$$\hat{\nu}_R : \hat{\mathcal{K}}(R) \longrightarrow \hat{\mathcal{F}}(R) \qquad 15.12.4$$

which is clearly compatible with the $\mathcal{K}$-isomorphisms.

Moreover, for another selfcentralizing Brauer $(c,\hat{H})$-pair (T,h) contained in (R,g), it is not difficult to check that in 11.22.2 the k^*-group homomorphisms $\hat{\mathfrak{u}}_{(R,g^K)}^{(T,h^K)}$ and $\hat{\mathfrak{u}}_{(R,g^K)}^{(T,h^K)}$, respectively defined from $\hat{K}$ and $\hat{G}$, agree and therefore, for coherent choices in 11.19.2, the uniqueness of the k^*-group homomorphisms $\hat{\alpha}_{(R,g^K)}^{(T,h^K)}$ and $\hat{\alpha}_{(R,g^G)}^{(T,h^G)}$ obtained from Proposition 11.23 forces the commutative diagram

$$\begin{array}{ccc} \hat{\mathcal{F}}(R)_T & \xrightarrow{\hat{\alpha}_{(R,g^G)}^{(T,h^G)}} & \hat{\mathcal{F}}(T) \\ \hat{\nu}_R \uparrow & & \uparrow \hat{\nu}_T \\ \hat{\mathcal{K}}(R)_T & \xrightarrow{\hat{\alpha}_{(R,g^K)}^{(T,h^K)}} & \hat{\mathcal{K}}(T) \end{array} \qquad 15.12.5.$$

Pushing it further, for any $\mathcal{K}^{^{\rm sc}}$-chain $\mathfrak{q}:\Delta_n \to \mathcal{K}^{^{\rm sc}}$, we have the $\mathcal{F}^{^{\rm nc}}$-chain $\mathfrak{j}\circ\mathfrak{q}:\Delta_n \to \mathcal{F}^{^{\rm nc}}$ and a k^*-group homomorphism

$$\hat{\nu}_{\mathfrak{q}} : \hat{\mathcal{K}}(\mathfrak{q}) \longrightarrow \hat{\mathcal{F}}(\mathfrak{j}\circ\mathfrak{q})$$

determined by $\hat{\nu}_{\mathfrak{q}(n)}$ (cf. 15.6); consequently, since from the composition $\widehat{\mathfrak{aut}}_{\mathcal{F}^{\rm nc}} \circ \mathfrak{ch}^*(\mathfrak{j})$ we can construct a functor

$$\mathfrak{ch}^*(\mathcal{K}^{^{\rm sc}}) \longrightarrow k^*\text{-}\mathfrak{Gr} \qquad 15.12.6$$

lifting $\mathfrak{aut}_{\mathcal{K}^{\rm sc}}$ and fulfilling all the conditions in Theorem 11.32, suitably restated over the set of $\mathcal{K}$-selfcentralizing subgroups of Q, it follows from the uniqueness part of this theorem that we get the announced natural map

$$\hat{\nu} : \widehat{\mathfrak{aut}}_{\mathcal{K}^{\rm sc}} \longrightarrow \widehat{\mathfrak{aut}}_{\mathcal{F}^{\rm nc}} \circ \mathfrak{ch}^*(\mathfrak{j}) \qquad 15.12.7.$$

Proposition 15.13 *In our standard setting, set $\hat{K} = \hat{H}\cdot C_{\hat{G}}(Q,f)$ and assume that $C = \hat{G}/\hat{K}$ is a p'-group and that $C_{\hat{G}}(Q,f)/C_{\hat{H}}(Q)$ is a p-group. Denote by $\hat{H}^C$ the converse image in $\hat{G}$ of a complement in $\hat{G}/\hat{H}$ of the Sylow p-subgroup, and by $\mathcal{H}^C$ the Frobenius Q-category determined by the canonical isomorphism*

$$C \cong N_{\hat{G}}(Q,f)/N_{\hat{K}}(Q,f) \cong \mathcal{F}(Q)/\mathcal{H}(Q) \qquad 15.13.1.$$

Then, $b = c$ is a block of $\hat{H}^C$, we have $\mathcal{F}_{(b,\hat{H}^C)} = \mathcal{H}^C$, $(\mathcal{H}^C)^{\text{sc}}$ is a full subcategory of $\mathcal{F}^{\text{nc}}$ and, denoting by $\mathfrak{j}: (\mathcal{H}^C)^{\text{sc}} \to \mathcal{F}^{\text{nc}}$ the inclusion functor, we have a natural isomorphism

$$\hat{\nu} : \widehat{\mathfrak{aut}}_{(\mathcal{H}^C)^{\text{sc}}} \cong \widehat{\mathfrak{aut}}_{\mathcal{F}^{\text{nc}}} \circ \mathfrak{ch}^*(\mathfrak{j}) \qquad 15.13.2.$$

Moreover, the pair $(\mathfrak{j},\hat{\nu})$ and the restriction induce $\mathcal{O}$-module isomorphisms

$$\mathcal{G}_k(\mathcal{F},\widehat{\mathfrak{aut}}_{\mathcal{F}^{\text{nc}}}) \cong \mathcal{G}_k(\mathcal{H}^C,\widehat{\mathfrak{aut}}_{(\mathcal{H}^C)^{\text{nc}}}) \quad \text{and} \quad \mathcal{G}_k(\hat{G},b) \cong \mathcal{G}_k(\hat{H}^C,b) \quad 15.13.3.$$

Proof: It follows from Proposition 15.10 that $b = c$, that $\hat{K} = \hat{H}\cdot C_P(Q)$ and that $\mathcal{H}^{\text{sc}}$ is a subcategory of $\mathcal{F}^{\text{nc}}$; moreover, from the obvious isomorphism 15.13.1, we may identify $\mathcal{F}(Q)/\mathcal{H}(Q)$ with the p'-group C and then it follows from Proposition 12.12 that we have indeed a Frobenius Q-category $\mathcal{H}^C$ fulfilling $\mathcal{H}^C(Q) = \mathcal{F}(Q)$; thus, since a subgroup of Q is $\mathcal{H}^C$-selfcentralizing if and only if it is $\mathcal{H}$-selfcentralizing, it follows from Proposition 12.8 that $(\mathcal{H}^C)^{\text{sc}}$ is also a subcategory of $\mathcal{F}^{\text{nc}}$.

Denoting by δ the local point of Q on $k_*\hat{H}$ determined by f (cf. 1.17), recall that the group $\widetilde{\text{Aut}}_Q\big((k_*\hat{H})_\delta\big)$ of *outer automorphisms* of the *Q-interior source algebra* $(k_*\hat{H})_\delta$ (cf. 1.13 and 1.27) of the block $(b,\hat{H})$ is a p'-group (cf. Proposition 14.9 in [42]). On the other hand, for any element z of

$$C_{\hat{G}}(Q_\delta) = C_{\hat{G}}(Q,f) = C_{\hat{H}}(Q)\cdot C_P(Q) \qquad 15.13.4,$$

if $(k_*\hat{H})_\delta = j(k_*\hat{H})j$ where $j \in \delta$ (cf. 1.12), we have $j^z = j^a$ for a suitable invertible element $a \in (k_*\hat{H})^Q$ and then za^{-1} determines an element of $\text{Aut}_Q\big((k_*\hat{H})_\delta\big)$; it is clear that its image in $\widetilde{\text{Aut}}_Q\big((k_*\hat{H})_\delta\big)$ only depends on z and that this correspondence defines a group homomorphism from $C_{\hat{G}}(Q_\delta)/C_{\hat{H}}(Q)$ to $\widetilde{\text{Aut}}_Q\big((k_*\hat{H})_\delta)\big)$; hence, this group homomorphism is the trivial one and we may assume that za^{-1} centralizes $(k_*\hat{H})_\delta$.

In particular, since $\hat{K} = \hat{H}\cdot C_P(Q)$, $\hat{K}$ stabilizes all the isomorphism classes of simple $k_*\hat{H}b$-modules (cf. 1.13) and, since $\hat{K}/\hat{H}$ is a p-group, any simple $k_*\hat{H}b$-module can be extended to $\hat{K}$ in a unique way (cf. 1.25); moreover, this bijection between the respective sets $\text{Irr}_k(\hat{H},b)$ and $\text{Irr}_k(\hat{K},b)$ of isomorphism classes of simple $k_*\hat{H}$- and $k_*\hat{K}$-modules (cf. 1.25) is clearly compatible with the action of

$$C = \hat{G}/\hat{K} \cong \hat{H}^C/\hat{H} \qquad 15.13.5.$$

Since C is a p'-group, the right-hand isomorphism in 15.13.3 follows easily from the so-called *Clifford Theory* (cf. §17 in [31, Ch. V]) applied to the normal k^*-subgroups $\hat{K}$ of $\hat{G}$ and $\hat{H}$ of $\hat{H}^C$; indeed, we get a *canonical* bijection, preserving the dimensions, between the set of elements of $\mathrm{Irr}_k(\hat{H}^C, b)$ such that their restriction to $\hat{H}$ is a multiple of $\chi \in \mathrm{Irr}_k(\hat{H}, b)$, and the set of elements of $\mathrm{Irr}_k(\hat{G}, b)$ such that their restriction to $\hat{H}$ is a multiple of the element $\hat{\chi}$ of $\mathrm{Irr}_k(\hat{K}, b)$ extending χ.

On the other hand, it follows from Proposition 15.10, applied to the pair $(\hat{H}^C, \hat{H})$, that $c = b$ is also a block of $\hat{H}^C$ and that $\mathcal{H}$ is a normal subcategory of $\mathcal{F}_{(b,\hat{H}^C)}$; then, since $\mathcal{H}^C(Q) = \mathcal{F}(Q)$, it follows from Proposition 12.8 that we have $\mathcal{H}^C = \mathcal{F}_{(b,\hat{H}^C)}$; moreover, since $(\mathcal{H}^C)^{^{\mathrm{sc}}}$ is a subcategory of $\mathcal{F}^{^{\mathrm{nc}}}$, for any pair of $\mathcal{H}^C$-selfcentralizing subgroups R and T of Q, we have

$$\mathcal{H}^C(R, T) \subset \mathcal{F}(R, T) \qquad 15.13.6;$$

but, since $\hat{G} = \hat{H}^C \cdot C_P(Q)$, respectively denoting by g and h the blocks of $C_{\hat{H}}(R)$ and $C_{\hat{H}}(T)$ such that (Q, f) contains (R, g) and (T, h), for any $x \in \hat{G}$ fulfilling $(T, h) \subset (R, g)^x$ we have $x = y \cdot z$ for suitable $y \in \hat{H}^C$ and $z \in C_P(Q)$, and therefore we still have $(T, h) \subset (R, g)^y$, both elements inducing the same group homomorphism since z centralizes (T, h) (cf. 1.15); hence, we have the equality

$$\mathcal{H}^C(R, T) = \mathcal{F}(R, T) \qquad 15.13.7.$$

In particular, for any $\mathcal{H}^C$-selfcentralizing subgroup R of Q, we have $\mathcal{H}^C(R) = \mathcal{F}(R)$ and therefore the natural map 15.12.1 here becomes a natural isomorphism

$$\hat{\nu} : \widehat{\mathfrak{aut}}_{(\mathcal{H}^C)^{\mathrm{sc}}} \cong \widehat{\mathfrak{aut}}_{\mathcal{F}^{\mathrm{nc}}} \circ \mathfrak{ch}^*(\mathfrak{j}) \qquad 15.13.8.$$

Hence, since $\mathcal{H}$ fulfills condition 15.14.2 in Lemma 15.14 below and $\mathfrak{j}$ induces an equivalence between $\mathcal{H}^C$ and the full subcategory $\mathcal{F}^{^{\mathfrak{S}}}$ of $\mathcal{F}^{^{\mathrm{nc}}}$ over the set $\mathfrak{S}$ of $\mathcal{H}$-selfcentralizing subgroups of Q, from this lemma and from the inverse limit of the natural isomorphism $\mathfrak{g}_k * \hat{\nu}$ we get the following $\mathcal{O}$-module isomorphisms

$$\mathcal{G}_k(\mathcal{F}, \widehat{\mathfrak{aut}}_{\mathcal{F}^{\mathrm{nc}}}) \cong \lim_{\longleftarrow} \left(\mathfrak{g}_k \circ \widehat{\mathfrak{aut}}_{\mathcal{F}^{\mathrm{nc}}} \circ \mathfrak{ch}^*(\mathfrak{j})\right) \cong \mathcal{G}_k(\mathcal{H}^C, \widehat{\mathfrak{aut}}_{(\mathcal{H}^C)^{\mathrm{nc}}}) \qquad 15.13.9.$$

We are done.

Lemma 15.14 *Let P be a finite p-group, $\mathcal{F}$ a Frobenius P-category, S an $\mathcal{F}$-stable subgroup of P and*

$$\widehat{\mathfrak{aut}}_{\mathcal{F}^{\mathrm{nc}}} : \mathfrak{ch}^*(\mathcal{F}^{\mathrm{nc}}) \longrightarrow k^*\text{-}\mathfrak{Gr} \qquad 15.14.1$$

a functor lifting $\mathfrak{aut}_{\mathcal{F}^{\mathrm{nc}}}$. Assume that the following condition holds

15.14.2 *We have a normal Frobenius S-subcategory $\mathcal{H}$ of $\mathcal{F}$ such that any $\mathcal{H}$-selfcentralizing subgroup of S is $\mathcal{F}$-nilcentralized.*

Then, denoting by $\mathcal{F}^{\mathfrak{S}}$ the full subcategory of $\mathcal{F}^{\mathrm{nc}}$ over the set $\mathfrak{S}$ of $\mathcal{H}$-selfcentralizing subgroups of S, and by $\widehat{\mathfrak{aut}}_{\mathcal{F}^{\mathfrak{S}}} : \mathfrak{ch}^*(\mathcal{F}^{\mathfrak{S}}) \to k^*\text{-}\mathfrak{Gr}$ the corresponding restricted functor, the restriction induces an $\mathcal{O}$-module isomorphism

$$\mathcal{G}_k(\mathcal{F}, \widehat{\mathfrak{aut}}_{\mathcal{F}^{\mathrm{nc}}}) \cong \lim_{\longleftarrow} (\mathfrak{g}_k \circ \widehat{\mathfrak{aut}}_{\mathcal{F}^{\mathfrak{S}}}) \qquad 15.14.3.$$

Proof: Let $\mathfrak{X}$ be the set of $\mathcal{F}$-nilcentralized subgroups R of P such that $R \cap S$ is $\mathcal{H}$-selfcentralizing (cf. 4.8); then, for any $\mathcal{F}$-nilcentralized subgroup Q of P (cf. 4.3), choose $\psi \in \mathcal{F}(P, Q)$ such that $T = \psi(Q \cap S)$ is fully normalized in $\mathcal{F}$ (cf. Proposition 2.7); moreover, considering the Frobenius $N_P(T)$-category $\mathcal{N} = N_{\mathcal{F}}(T)$ (cf. Proposition 2.16), up to replacing $U = \psi(Q)$ by an $\mathcal{N}$-isomorphic subgroup, we may assume that U is fully normalized in $\mathcal{N}$ (cf. Proposition 2.7).

Since T is also fully normalized in $\mathcal{H}$, $T{\cdot}C_S(T)$ is $\mathcal{H}$-selfcentralizing (cf. Proposition 2.11); in particular, $R = U{\cdot}C_S(T)$ belongs to $\mathfrak{X}$ and, denoting by $\varphi : Q \to R$ the homomorphism determined by ψ, we claim that

$$\varphi \circ \mathcal{F}(Q) \subset \mathcal{F}(R) \circ \varphi \qquad 15.14.4,$$

so that $\mathfrak{X}$ fulfills condition 14.6.1. Indeed, consider the respective Frobenius $N_{N_P(T)}(U)$- and R-categories $N_{\mathcal{N}}(U)$ and $\mathcal{C} = U{\cdot}C_{\mathcal{H}}(T) = N_{\mathcal{H}}^U(T)$ (cf. Proposition 2.16); it is easily checked that $\mathcal{C}$ is normal in $N_{\mathcal{N}}(U)$ (cf. 12.5); thus, it follows from Proposition 12.8 that we have

$$\big(N_{\mathcal{N}}(U)\big)(R, U) = \big(N_{\mathcal{N}}(U)\big)(R) \circ \mathcal{C}(R, U) \qquad 15.14.5.$$

On the other hand, for any element $\sigma \in \mathcal{F}(Q)$ we have $\tau = {}^{\varphi}\sigma \in \mathcal{F}(U)$ fulfilling $\tau\big(\varphi(u)\big) = \varphi\big(\sigma(u)\big)$ for any $u \in Q$ and, since $T = U \cap S$, the element of $\mathcal{F}(R, U)$ determined by τ and the inclusion $U \subset R$ belongs to the left-hand member of equality 15.14.5. Hence, there are $\rho \in \big(N_{\mathcal{N}}(U)\big)(R)$ and $\theta \in \mathcal{C}(R, U)$ fulfilling $\tau\big(\varphi(u)\big) = (\rho \circ \theta)\big(\varphi(u)\big)$ for any $u \in Q$ and, since $\tau(U) = U$, we have $\theta(U) = U$; but, since $\mathcal{H}$ is normal in $\mathcal{F}$, for any p'-subgroup K of $\mathcal{C}(U)$ we have $[K, U] \subset T$ and therefore $\big[K, [K, U]\big] = \{1\}$ which forces $K = \{\mathrm{id}_U\}$ (cf. Theorem in [28, Ch. 5]); consequently, θ is induced by some $v \in R$ and finally, denoting by $\kappa_R(v)$ the conjugation by v on R, we get

$$\varphi\big(\sigma(u)\big) = \big(\rho \circ \kappa_R(v)\big)\big(\varphi(u)\big) \qquad 15.14.6$$

for any $u \in Q$, which proves inclusion 15.14.4.

In conclusion, it follows from Proposition 14.6 that we have an $\mathcal{O}$-module isomorphism

$$\mathcal{G}_k(\mathcal{F}, \widehat{\mathfrak{aut}}_{\mathcal{F}^{\mathrm{nc}}}) \cong \varprojlim (\mathfrak{g}_k \circ \widehat{\mathfrak{aut}}_{\mathcal{F}^{\mathfrak{X}}}) \qquad 15.14.7,$$

where $\widehat{\mathfrak{aut}}_{\mathcal{F}^{\mathfrak{X}}} : \mathfrak{ch}^*(\mathcal{F}^{\mathfrak{X}}) \to k^*\text{-}\mathfrak{Gr}$ denotes the corresponding restricted functor. Moreover, according to our definition, the inclusion functor $\mathfrak{i}_S : \mathcal{F}^{\mathfrak{S}} \to \mathcal{F}^{\mathfrak{X}}$ admits a section $\mathfrak{s}_S : \mathcal{F}^{\mathfrak{X}} \to \mathcal{F}^{\mathfrak{S}}$ mapping any $Q \in \mathfrak{X}$ on $Q \cap S$ and any $\mathcal{F}^{\mathfrak{X}}$-morphism on the corresponding restriction; thus, the $\mathcal{O}$-module homomorphism induced by $\mathfrak{i}_S$ (cf. A3.9 and A3.11)

$$\varprojlim (\mathfrak{g}_k \circ \widehat{\mathfrak{aut}}_{\mathcal{F}^{\mathfrak{X}}}) \longrightarrow \varprojlim (\mathfrak{g}_k \circ \widehat{\mathfrak{aut}}_{\mathcal{F}^{\mathfrak{S}}}) \qquad 15.14.8$$

is surjective.

Actually, we claim that it is also injective; indeed, if $X = (X_Q)_{Q \in \mathfrak{X}}$ belongs to $\varprojlim (\mathfrak{g}_k \circ \widehat{\mathfrak{aut}}_{\mathcal{F}^{\mathfrak{X}}})$, namely if $X_Q \in \mathcal{G}_k\big(\hat{\mathcal{F}}(Q)\big)$ for any $Q \in \mathfrak{X}$ and we have (cf. A3.6 and A3.11)

$$d_0^{(p^\circ_{\mathfrak{ch}(\mathcal{F}^{\mathfrak{X}})})_*(\mathfrak{g}_k \circ \widehat{\mathfrak{aut}}_{\mathcal{F}^{\mathfrak{X}}})}(X) = 0 \qquad 15.14.9,$$

it suffices to consider the $\mathcal{F}^{\mathfrak{X}}$-chain $\mathfrak{q} : \Delta_1 \to \mathcal{F}^{\mathfrak{X}}$ mapping 0 on $Q \cap S$, 1 on Q and $0 \bullet 1$ on the inclusion map $Q \cap S \to Q$, to get

$$0 = \mathrm{res}_{\widehat{\mathfrak{aut}}_{\mathcal{F}^{\mathfrak{X}}}(\mathrm{id}_{\mathfrak{q}\circ\delta_0^0}, \delta_0^0)}(X_Q) - \mathrm{res}_{\widehat{\mathfrak{aut}}_{\mathcal{F}^{\mathfrak{X}}}(\mathrm{id}_{\mathfrak{q}\circ\delta_1^0}, \delta_1^0)}(X_{Q\cap S}) \qquad 15.14.10;$$

but, since S is $\mathcal{F}$-stable, $\hat{\mathcal{F}}(Q)$ stabilizes $Q \cap S$ and therefore the restriction map $\mathrm{res}_{\widehat{\mathfrak{aut}}_{\mathcal{F}^{\mathfrak{X}}}(\mathrm{id}_{\mathfrak{q}\circ\delta_0^0}, \delta_0^0)}$ is an isomorphism; hence, if X also belongs to the kernel of the $\mathcal{O}$-module homomorphism 15.14.8 then we obtain $X_Q = 0$. We are done.

15.15 In order to discuss statements analogous to Proposition 15.13 when $C_{\hat{G}}(Q, f)/C_{\hat{H}}(Q)$ is a p'-group, we need the following auxiliary result on the relationship between the local pointed groups on $k_*\hat{G}b$ and on $k_*\hat{H}c$ (cf. 1.28), which essentially depends on *Fong's reduction* for interior algebras (cf. §3 in [53]); as in 15.5 above, if R_ε is a local pointed group on $k_*\hat{H}$, the stabilizer $C_G(R_\varepsilon)$ of ε in $C_G(R)$ acts on the simple algebra $(k_*\hat{H})(R_\varepsilon)$ (cf. 1.12), determining a k^*-group $\hat{C}_G(R_\varepsilon)$ (cf. 1.25), and then we set (cf. 15.5.4)

$$C_{\hat{H}}^{\hat{G}}(R_\varepsilon) = \big(C_{\hat{G}}(R_\varepsilon) * \hat{C}_G(R_\varepsilon)^\circ\big)/\Delta_{R_\varepsilon}\big(C_H(R)\big) \qquad 15.15.1$$

where, as above, Δ_{R_ε} is the group homomorphism mapping $x \in C_H(R)$ on the product of the two images of $\hat{x}$ and $\hat{x}^{-1}$ in $C_{\hat{G}}(R_\varepsilon)\hat{\times}\hat{C}_G(R_\varepsilon)^\circ$ where $\hat{x}$ is a lifting to $C_{\hat{H}}(R_\varepsilon) = \hat{C}_H(R_\varepsilon)$ of x.

Lemma 15.16 *In our standard setting, assume that $\hat{G}/\hat{H}$ is a p'-group. For any local pointed group R_ε on $k_*\hat{H}$, ε splits into a set $\{(\varepsilon,\varphi)\}_{\varphi\in\mathcal{P}(k_*C_{\hat{H}}^{\hat{G}}(R_\varepsilon))}$ of local points of R on $k_*\hat{G}$ and, for any $\varphi \in \mathcal{P}\big(k_*C_{\hat{H}}^{\hat{G}}(R_\varepsilon)\big)$, we have a $C_{\hat{G}}(R)$-interior k-algebra isomorphism*

$$(k_*\hat{G})(R_{(\varepsilon,\varphi)}) \cong \mathrm{Ind}_{C_{\hat{G}}(R_\varepsilon)}^{C_{\hat{G}}(R)}\Big(\big(k_*C_{\hat{H}}^{\hat{G}}(R_\varepsilon)\big)(\varphi) \otimes_k (k_*\hat{H})(R_\varepsilon)\Big) \qquad 15.16.1.$$

Moreover, the Brauer $\hat{H}$-pair determined by R_ε is selfcentralizing if and only if the Brauer $\hat{G}$-pair determined by $R_{(\varepsilon,\varphi)}$ is selfcentralizing, and then we have

$$\mathbf{O}_p\big(E_H(R_\varepsilon)\big) = \mathbf{O}_p\big(E_G(R_{(\varepsilon,\varphi)})\big) \qquad 15.16.2.$$

Proof: The inclusion $\hat{H} \subset \hat{G}$ induces an injective k-algebra homomorphism

$$k_*C_{\hat{H}}(R) \longrightarrow k_*C_{\hat{G}}(R) \qquad 15.16.3$$

and, since $C_{\hat{G}}(R)$ acts on the semisimple k-algebra $k_*C_{\hat{H}}(R)/J\big(k_*C_{\hat{H}}(R)\big)$, setting $J = J\big(k_*C_{\hat{H}}(R)\big)$ it is easily checked from Proposition 3.2 in [53] that

$$k_*C_{\hat{G}}(R)/k_*C_{\hat{G}}(R)\cdot J \cong \prod_{(\varepsilon,\varphi)} \mathrm{Ind}_{C_{\hat{G}}(R_\varepsilon)}^{C_{\hat{G}}(R)}\big(S_\varphi \otimes_k (k_*\hat{H})(R_\varepsilon)\big) \qquad 15.16.4,$$

where ε runs over a set of representatives for the set of orbits of $C_{\hat{G}}(R)$ on the set of local points of R on $k_*\hat{H}$ (cf. 1.18.1) and for such an ε, since $C_{\hat{G}}(R_\varepsilon)/C_{\hat{H}}(R)$ is a p'-group, φ runs over the set of points of $k_*C_{\hat{H}}^{\hat{G}}(R_\varepsilon)$ (cf. 1.12) and S_φ denotes the corresponding $C_{\hat{H}}^{\hat{G}}(R_\varepsilon)$-interior simple algebra (cf. 1.27 and 15.15.1). Moreover, since the right-hand member of isomorphism 15.16.4 is semisimple, we get the first statement and isomorphism 15.16.1.

On the other hand, if the Brauer $\hat{H}$-pair determined by R_ε is selfcentralizing then the corresponding simple $k_*\bar{C}_{\hat{H}}(R)$-module is also projective (cf. 1.17 and 4.7) and therefore, since $C_{\hat{G}}(R)/C_{\hat{H}}(R)$ is a p'-group, according to isomorphism 15.16.1, the simple $k_*\bar{C}_{\hat{G}}(R)$-module which corresponds to the local pointed group $R_{(\varepsilon,\varphi)}$ is projective too, so that the Brauer $\hat{G}$-pair determined by $R_{(\varepsilon,\varphi)}$ is selfcentralizing too (cf. 1.17 and 4.7).

Conversely, the projectivity of the simple $k_*\bar{C}_{\hat{G}}(R)$-module corresponding to the local pointed group $R_{(\varepsilon,\varphi)}$ on $k_*\hat{G}$ forces the projectivity of the simple $k_*\bar{C}_{\hat{H}}(R)$-module corresponding to the local pointed group R_ε on $k_*\hat{H}$. Finally, since $\mathbf{O}^{p'}\big(N_H(R_\varepsilon)\big)$ acts trivially on $k_*C_{\hat{H}}^{\hat{G}}(R_\varepsilon)$, it fixes any point φ and therefore $\mathbf{O}^{p'}\big(E_{\hat{H}}(R_\varepsilon)\big)$ is a normal subgroup of $E_{\hat{G}}(R_{(\varepsilon,\varphi)})$ of index prime to p; hence, $\mathbf{O}^{p'}\big(E_{\hat{H}}(R_\varepsilon)\big)$ contains $\mathbf{O}_p\big(E_{\hat{G}}(R_{(\varepsilon,\varphi)})\big)$ and therefore we get

$$\mathbf{O}_p\big(E_{\hat{H}}(R_\varepsilon)\big) = \mathbf{O}_p\big(E_{\hat{G}}(R_{(\varepsilon,\varphi)})\big) \qquad 15.16.5.$$

We are done.

15.17 Now, in our standard setting, assume that $\hat{G}/\hat{H}$ is a p'-group; moreover, set $\hat{K} = \hat{H}{\cdot}C_{\hat{G}}(Q,f)$ and $C = \hat{G}/\hat{K}$, and assume that

15.17.1 *The canonical k^*-group homomorphism*

$$C_{\hat{H}}(Q) \longrightarrow \big(k_*\bar{C}_{\hat{H}}(Q)\bar{f}\big)^*$$

can be extended to $N_{\hat{K}}(Q,f)$.

In this situation, setting $D = C_G(Q,f)/C_H(Q)$ it is clear that

$$C_{\hat{H}}^{\hat{G}}(Q,f) \cong k^* \times D \qquad 15.17.2,$$

and that D acts trivially on $\hat{E}_H(Q,f)$. In particular, the group

$$N_{\hat{K}}(Q,f) = N_{\hat{H}}(Q,f){\cdot}C_{\hat{G}}(Q,f) \qquad 15.17.3$$

acts trivially on $\mathcal{P}\big(k_*C_{\hat{H}}^{\hat{G}}(Q,f)\big)$ (cf. 1.12) and, since $\hat{G} = \hat{H}{\cdot}N_{\hat{G}}(Q,f)$ by the *Frattini argument* (cf. 1.15), the group $C \cong N_{\hat{G}}(Q,f)/N_{\hat{K}}(Q,f)$ acts on the set $\mathcal{P}\big(k_*C_{\hat{H}}^{\hat{G}}(Q,f)\big)$.

15.18 On the other hand, denote by δ the local point of Q on $k_*\hat{H}c$ determined by f (cf. 1.17) and choose $j \in \delta$ such that $j(k_*\hat{H})j = (k_*\hat{H}c)_\delta$ (cf. 1.12); setting $B = (k_*\hat{H}c)_\delta$, recall that the exact sequence

$$1 \longrightarrow Q{\cdot}\big(j + J(B^Q)\big) \longrightarrow N_{B^*}(Q{\cdot}j) \longrightarrow \hat{E}_H(Q_\delta)^\circ \longrightarrow 1 \qquad 15.18.1$$

is *split*; more precisely, it follows from Theorem 12.8 in [51] that there is a unique $(B^Q)^*$-conjugacy class of sections†

$$\hat{\sigma}_\delta : \hat{E}_H(Q_\delta)^\circ \longrightarrow N_{B^*}(Q{\cdot}j) \qquad 15.18.2$$

† Note that our notation in [51] and here do not agree; precisely, definition 12.6.2 in [51] corresponds to definition 1.29.2 above.

of the canonical k^*-group homomorphism $N_{B^*}(Q{\cdot}j) \to \hat{E}_H(Q_\delta)^\circ$ (cf. 1.29). Similarly, since we are assuming $P = Q$, it follows from Lemma 15.16 that the block e of $C_{\hat{G}}(Q)$ (cf. 15.1) determines a *point* φ of $\mathcal{P}\big(k_*C_{\hat{H}}^{\hat{G}}(Q,f)\big)$ such that $Q_{(\delta,\varphi)}$ is a defect pointed group of $\hat{G}_{\{b\}}$ (cf. 1.13) and then, choosing $i \in (\delta,\varphi)$ such that $i(k_*\hat{G}b)i = (k_*\hat{G}b)_{(\delta,\varphi)}$ and setting $A = (k_*\hat{G}b)_{(\delta,\varphi)}$, there is a unique $(A^Q)^*$-conjugacy class of sections

$$\hat{\sigma}_{(\delta,\varphi)} : \hat{E}_G(Q_{(\delta,\varphi)})^\circ \longrightarrow N_{A^*}(Q{\cdot}i) \qquad 15.18.3$$

of the canonical k^*-group homomorphism $N_{A^*}(Q{\cdot}i) \to \hat{E}_G(Q_{(\delta,\varphi)})^\circ$ (cf. 1.29 above and Theorem 12.8 in [51]).

Proposition 15.19 *In our standard setting, assume that $\hat{G}/\hat{H}$ is a p'-group and that condition* 15.17.1 *holds. With the notation above, B is isomorphic to an $\hat{E}_G(Q_{(\delta,\varphi)})^\circ$-stable Q-interior subalgebra of A and, up to suitable identifications, $\hat{\sigma}_{(\delta,\varphi)}$ coincides with $\hat{\sigma}_\delta$ over $\hat{E}_H(Q_\delta)^\circ$ and we have*

$$A = \bigoplus_{\hat{x}} B{\cdot}\hat{\sigma}_{(\delta,\varphi)}(\hat{x}) \qquad 15.19.1$$

where $\hat{x} \in \hat{E}_G(Q_{(\delta,\varphi)})^\circ$ runs over a set of representatives for the quotient $\hat{E}_G(Q_{(\delta,\varphi)})^\circ/\hat{E}_H(Q_\delta)^\circ$. In particular, denoting by $C_{(\delta,\varphi)}$ the stabilizer of (δ,φ) in C, we have $\mathcal{H}^{C_{(\delta,\varphi)}} = \mathcal{F}$.

Proof: For any $\hat{x} \in N_{\hat{G}}(Q_\delta)$ there is an invertible element $a_{\hat{x}} \in B^Q$ such that we have $\hat{x}{\cdot}j{\cdot}\hat{x}^{-1} = a_{\hat{x}}j(a_{\hat{x}})^{-1}$, so that $(a_{\hat{x}})^{-1}{\cdot}\hat{x}$ centralizes j; but, since we have $\hat{G} = \hat{H}{\cdot}N_{\hat{G}}(Q,f)$ by the *Frattini argument*, we get

$$k_*\hat{G} = \bigoplus_{\hat{x}\in\mathcal{X}} k_*\hat{H}\hat{x} \qquad 15.19.2$$

where $\mathcal{X}$ is a set of representatives for $N_{\hat{G}}(Q_\delta)/N_{\hat{H}}(Q_\delta)$ in $N_{\hat{G}}(Q_\delta)$; consequently, we still get

$$j(k_*\hat{G})j = \bigoplus_{\hat{x}\in\mathcal{X}} j(k_*\hat{H})j(a_{\hat{x}})^{-1}{\cdot}\hat{x} \qquad 15.19.3;$$

actually, setting $d_{\hat{x}} = j(a_{\hat{x}})^{-1}{\cdot}\hat{x}$ which is an element of $j(k_*\hat{G})j$, it follows from 15.18 that we can modify our choice of $a_{\hat{x}}$ in such a way that $d_{\hat{x}}$ normalizes $\hat{\sigma}_\delta\big(\hat{E}_{\hat{H}}(Q_\delta)^\circ\big)$.

Moreover, since $C_{\hat{H}}^{\hat{G}}(Q,f)$ (cf. 15.5.4) acts trivially on $\hat{E}_{\hat{H}}(Q_\delta)$, it follows from Proposition 14.9 in [42] that $d_{\hat{z}}$ has an inner action on B for any $\hat{z}$

in $C_{\hat{G}}(Q_\delta)$; that is to say, we can modify our choice of $a_{\hat{z}}$ in such a way that $d_{\hat{z}}$ centralizes B. In this situation, denoting by $\mathcal{Z}$ a set of representatives $\hat{z} \in C_{\hat{G}}(Q_\delta)$ for $D = C_{\hat{G}}(Q_\delta)/C_{\hat{H}}(Q)$, it is easily checked that

$$C_{j(k_*\hat{G})j}(B) = \bigoplus_{\hat{z}\in\mathcal{Z}} Z(B)d_{\hat{z}} \qquad 15.19.4;$$

actually, since D is a p'-group, for a suitable k^*-group $\hat{D}$ of k^*-quotient D we get an isomorphism

$$C_{j(k_*\hat{G})j}(B) \cong Z(B) \otimes_k k_*\hat{D} \qquad 15.19.5.$$

Consequently, the point $\varphi \in \mathcal{P}\big(k_* C_{\hat{H}}^{\hat{G}}(Q,f)\big)$ (cf. 1.12) determined by e (cf. 15.1) determines a direct factor $Z(B) \otimes_k (k_*\hat{D})(\varphi)$ of $C_{j(k_*\hat{G})j}(B)$ (cf. 15.17.2) and, for any $\hat{x} \in N_{\hat{G}}(Q_{(\delta,\varphi)})$, we may assume that $d_{\hat{x}}$ normalizes this direct factor; then, since we may assume that $d_{\hat{x}}$ induces a p'-action on $(k_*\hat{D})(\varphi)$, it is not difficult to prove that $d_{\hat{x}}$ centralizes a primitive idempotent in $Z(B)\otimes_k (k_*\hat{D})(\varphi)$; since all the primitive idempotents in this direct factor are pairwise conjugate, we can modify our choice in such a way that, for any $\hat{x} \in N_{\hat{G}}(Q_{(\delta,\varphi)})$, $d_{\hat{x}}$ centralizes the *same* primitive idempotent i in $Z(B) \otimes_k (k_*\hat{D})(\varphi)$.

Finally, it is clear that the multiplication by i induces a Q-interior algebra isomorphism $B \cong Bi$ mapping j on i and, since we still have

$$iC_{j(k_*\hat{G})j}(B)i = Z(B)i \quad \text{and} \quad d_{\hat{x}}d_{\hat{x}'}i \in d_{\hat{x}\hat{x}'}iB^{Q\cdot\hat{E}_H(Q_\delta)} \qquad 15.19.6$$

for any $\hat{x}, \hat{x}' \in N_{\hat{G}}(Q_{(\delta,\varphi)})$, it is easily checked that

$$A = i(k_*\hat{G})i = \bigoplus_{\hat{x}} Bd_{\hat{x}}i \qquad 15.19.7$$

where $\hat{x} \in N_{\hat{G}}(Q_{(\delta,\varphi)})$ runs over a set of representatives for $C_{(\delta,\varphi)}$.

Now, it is quite clear that in equality 15.19.7 we can replace $d_{\hat{x}}i$ by $\hat{\sigma}_{(\delta,\varphi)}(\tilde{\hat{x}})$ where $\tilde{\hat{x}}$ denotes the image of $\hat{x}$ in $\hat{E}_G(Q_{(\delta,\varphi)})^\circ$ (cf. 15.18.3), that $\hat{E}_G(Q_{(\delta,\varphi)})^\circ$ stabilizes Bi and contains the k^*-group $\hat{E}_H(Q_\delta)^\circ$ (cf. Proposition 6.21 in [42]), and that we have

$$\begin{gathered} \hat{E}_G(Q_{(\delta,\varphi)})^\circ/\hat{E}_H(Q_\delta)^\circ \cong C_{(\delta,\varphi)} \\ Bi \cap \hat{\sigma}_{(\delta,\varphi)}\big(\hat{E}_G(Q_{(\delta,\varphi)})^\circ\big) = \hat{\sigma}_{(\delta,\varphi)}\big(\hat{E}_H(Q_\delta)^\circ\big) \end{gathered} \qquad 15.19.8$$

which proves equality 15.19.1 and allows us to assume that $\hat{\sigma}_\delta$ coincides with the restriction of $\hat{\sigma}_{(\delta,\varphi)}$ to $\hat{E}_H(Q_\delta)$. Since the Frobenius Q-category $\mathcal{H}$ is normal in $\mathcal{F}$, the last statement follows from Propositions 12.8 and 12.12. We are done.

Corollary 15.20 *In our standard setting, assume that $\hat{G}/\hat{H}$ is a p'-group and that condition* 15.17.1 *holds. With the notation above, let $\hat{G}'$ be a k^*-subgroup of $\hat{G}$ containing $\hat{H}$, b' a block of $\hat{G}'$ such that $f\mathrm{Br}_Q(b') \neq 0$, and f' a block of $C_{\hat{G}'}(Q)$ fulfilling $ff' \neq 0$ and $f'\mathrm{Br}_Q(b') = f'$. Setting $\hat{K}' = \hat{K} \cap \hat{G}'$ and $C' = \hat{G}'/\hat{K}'$, and denoting by $\varphi' \in \mathcal{P}\big(k_* C_{\hat{H}}^{\hat{G}'}(Q,f)\big)$ the point determined by f', assume that $C'_{(\delta,\varphi')} = C_{(\delta,\varphi)}$. Then $Q_{(\delta,\varphi')}$ is a defect pointed group of $\hat{G}'_{\{b'\}}$, the Frobenius Q-category $\mathcal{F}' = \mathcal{F}_{(b',\hat{G}')}$ coincides with $\mathcal{F}$, we have a natural isomorphism $\widehat{\mathfrak{aut}}_{\mathcal{F}'^{\mathrm{nc}}} \cong \widehat{\mathfrak{aut}}_{\mathcal{F}^{\mathrm{nc}}}$ inducing an $\mathcal{O}$-module isomorphism*

$$\mathcal{G}_k(\mathcal{F}, \widehat{\mathfrak{aut}}_{\mathcal{F}^{\mathrm{nc}}}) \cong \mathcal{G}_k(\mathcal{F}', \widehat{\mathfrak{aut}}_{\mathcal{F}'^{\mathrm{nc}}}) \qquad 15.20.1,$$

and the restrictions to the respective source algebras induce an $\mathcal{O}$-module isomorphism

$$\mathcal{G}_k(\hat{G}, b) \cong \mathcal{G}_k(\hat{G}', b') \qquad 15.20.2.$$

Proof: It follows from Proposition 15.19 that the block algebras $k_*\hat{G}'b'$ and $k_*\hat{G}b$ have isomorphic *source algebras* and that $\mathcal{F}' = \mathcal{H}^{C_{(\delta,\varphi)}} = \mathcal{F}$; since $\widehat{\mathfrak{aut}}_{\mathcal{F}'^{\mathrm{nc}}}$ and $\widehat{\mathfrak{aut}}_{\mathcal{F}^{\mathrm{nc}}}$, $\mathcal{G}_k(\mathcal{F}, \widehat{\mathfrak{aut}}_{\mathcal{F}^{\mathrm{nc}}})$ and $\mathcal{G}_k(\mathcal{F}', \widehat{\mathfrak{aut}}_{\mathcal{F}'^{\mathrm{nc}}})$, and also $\mathcal{G}_k(\hat{G}, b)$ and $\mathcal{G}_k(\hat{G}', b')$ are completely determined from the respective source algebras (cf. Theorem 3.1 in [39] and Proposition 6.21 in [42]), we are done.

15.21 Always in our standard setting, let us denote by $\hat{H}^C$ the converse image in $\hat{G}$ of a cyclic p'-subgroup C of $\hat{G}/\hat{H}$, and by H^C its k^*-quotient; recall that $\mathcal{G}_k(C)$ and $\mathcal{G}_k(H^C)$ have ring structures and that the restriction determines a ring homomorphism

$$\mathcal{G}_k(C) \longrightarrow \mathcal{G}_k(H^C) \qquad 15.21.1,$$

so that $\mathcal{G}_k(\hat{H}^C)$ becomes a $\mathcal{G}_k(C)$-module (cf. 5.13 in [42]); since clearly we have $C^* = \mathrm{Hom}(C, k^*) \subset \mathcal{G}_k(C)$, the group C^* acts on $\mathcal{G}_k(\hat{H}^C)$ and it is elementary to check that the restriction determines an $\mathcal{O}$-module isomorphism

$$\mathcal{G}_k(\hat{H}^C)^{C^*} \cong \mathcal{G}_k(\hat{H})^C \qquad 15.21.2.$$

15.22 For our purposes, it is very useful to consider the intersection — denoted by $\mathcal{R}_{\hat{H}}\mathcal{G}_k(\hat{H}^C)$ and called *C-residual Grothendieck group of $\hat{H}^C$* — of the kernels of all the $\mathcal{O}$-module homomorphisms determined by the restriction

$$\mathcal{G}_k(\hat{H}^C) \longrightarrow \mathcal{G}_k(\hat{H}^D) \qquad 15.22.1$$

where D runs over the set of proper subgroups of C. Note that if $C_{\hat{G}}(Q_\delta) \subset \hat{H}$ then it follows from Corollary 15.13 that $(c, \hat{H}^D)$ is a block for any subgroup D of C (cf. 1.10) and therefore the *C-residual Grothendieck group $\mathcal{R}_{\hat{H}}\mathcal{G}_k(\hat{H}^C, c)$ of $(c, \hat{H}^C)$* makes sense.

15.23 Actually, the intersection — denoted by $\mathcal{R}\mathcal{G}_k(C)$ — of the kernels of all the $\mathcal{O}$-module homomorphisms $\mathcal{G}_k(C) \to \mathcal{G}_k(D)$ determined by the restriction, where D runs over the set of proper subgroups of C, still makes sense and it is isomorphic to the following residue class ring of $\mathcal{G}_k(C)$ (cf. Proposition 3.4 in [42])

$$\mathcal{R}\mathcal{G}_k(C) \cong \mathcal{G}_k(C)/\sum_D \mathrm{Ind}_D^C(\mathcal{G}_k(D)) \qquad 15.23.1,$$

where D runs again over the set of proper subgroups of C. Now, it is quite clear that $\mathcal{R}_{\hat{H}}\mathcal{G}_k(\hat{H}^C)$ becomes a $\mathcal{R}\mathcal{G}_k(C)$-module and then it is easily checked that

$$\mathcal{R}_{\hat{H}}\mathcal{G}_k(\hat{H}^C) = \bigoplus_X \mathcal{R}\mathcal{G}_k(C)\cdot X \cong \mathcal{R}\mathcal{G}_k(C) \otimes_{\mathcal{G}_k(C)} \mathcal{G}_k(\hat{H}^C) \qquad 15.23.2,$$

where X runs over a set of representatives for the set of C^*-orbits on the set of isomorphism classes of the simple $k_*\hat{H}^C$-modules which remain simple restricted to $\hat{H}$. Moreover, it is clear that C acts on $\mathcal{G}_k(\hat{H}^D)$ for any subgroup D of C, and it follows from isomorphism 15.21.2 that the restriction induces an $\mathcal{O}$-module isomorphism

$$\mathcal{R}\mathcal{G}_k(D) \otimes_{\mathcal{G}_k(C)} \mathcal{G}_k(\hat{H}^C) \cong \mathcal{R}_{\hat{H}}\mathcal{G}_k(\hat{H}^D)^C \qquad 15.23.3;$$

hence, since $\mathcal{G}_k(C) \cong \bigoplus_D \mathcal{R}\mathcal{G}_k(D)$ (cf. Proposition 3.4 in [42]), we get

$$\mathcal{G}_k(\hat{H}^C) \cong \bigoplus_D \mathcal{R}_{\hat{H}}\mathcal{G}_k(\hat{H}^D)^C \qquad 15.23.4$$

where D runs over the set of subgroups of C.

15.24 More generally, let us denote by $\mathcal{C}_{\hat{G}/\hat{H}}$ the category where the objects are the cyclic p'-subgroups C of $\hat{G}/\hat{H}$ and the morphisms from C to another cyclic p'-subgroup C' of $\hat{G}/\hat{H}$ are the elements $\bar{x} \in \hat{G}/\hat{H}$ fulfilling $C \subset C'^{\bar{x}}$, with the composition determined by the product in this quotient. Clearly we have a *contravariant* functor

$$\mathcal{G}_k(\hat{H}^\bullet) : \mathcal{C}_{\hat{G}/\hat{H}} \longrightarrow \mathcal{O}\text{-}\mathfrak{mod} \qquad 15.24.1$$

mapping C on $\mathcal{G}_k(\hat{H}^C)$ and the $\mathcal{C}_{\hat{G}/\hat{H}}$-morphism $\bar{x}: C \to C'$ on the $\mathcal{O}$-module homomorphism $\mathcal{G}_k(\hat{H}^{C'}) \to \mathcal{G}_k(\hat{H}^C)$ determined by the restriction *via* the conjugation by an element $x \in \hat{G}$ lifting $\bar{x}$; moreover, the restriction induces an $\mathcal{O}$-module homomorphism

$$\mathcal{G}_k(\hat{G}) \longrightarrow \varprojlim \mathcal{G}_k(\hat{H}^\bullet) \qquad 15.24.2.$$

Note that the systematic consideration of isomorphisms 15.23.4 proves that

$$\varprojlim \mathcal{G}_k(\hat{H}^\bullet) \cong \Big(\bigoplus_C \mathcal{R}_{\hat{H}}\mathcal{G}_k(\hat{H}^C)\Big)^{\hat{G}/\hat{H}} \qquad 15.24.3,$$

where C runs over the set of cyclic p'-subgroups of $\hat{G}/\hat{H}$.

Proposition 15.25 *With the notation above, the restriction induces an $\mathcal{O}$-module isomorphism*

$$\mathcal{G}_k(\hat{G}) \cong \lim_{\longleftarrow} \mathcal{G}_k(\hat{H}^\bullet) \qquad 15.25.1.$$

Proof: Borrowing our notation in 14.12, 14.13 and 14.14, we have a canonical $\mathcal{O}$-module isomorphism (cf. isomorphism 14.14.2)

$$\mathcal{G}_k(\hat{G}) \cong \prod_{h\in\mathbb{N}-p\mathbb{N}} \mathcal{F}ct_{U_h}\big(\mathrm{Mon}_{k^*}(\hat{U}_h, \hat{G}), \mathcal{O}\big)^G \qquad 15.25.2;$$

in particular, for any $h \in \mathbb{N} - p\mathbb{N}$, denoting by

$$\mathcal{F}ct_{U_h}\big(\mathrm{Mon}_{k^*}(\hat{U}_h, \hat{H}^\bullet), \mathcal{O}\big)^{\hat{H}^\bullet} : \mathcal{C}_{\hat{G}/\hat{H}} \longrightarrow \mathcal{O}\text{-}\mathfrak{mod} \qquad 15.25.3$$

the *contravariant* functor mapping any cyclic p'-subgroup C of $\hat{G}/\hat{H}$ on the $\mathcal{O}$-module $\mathcal{F}ct_{U_h}\big(\mathrm{Mon}_{k^*}(\hat{U}_h, \hat{H}^C), \mathcal{O}\big)^{\hat{H}^C}$ and any $\mathcal{C}_{\hat{G}/\hat{H}}$-morphism on the obvious $\mathcal{O}$-module homomorphism, we have a natural isomorphism

$$\mathcal{G}_k(\hat{H}^\bullet) \cong \prod_{h\in\mathbb{N}-p\mathbb{N}} \mathcal{F}ct_{U_h}\big(\mathrm{Mon}_{k^*}(\hat{U}_h, \hat{H}^\bullet), \mathcal{O}\big)^{\hat{H}^\bullet} \qquad 15.25.4.$$

On the other hand, for any $h \in \mathbb{N} - p\mathbb{N}$ it is clear that (cf. 1.9)

$$\mathrm{Mon}_{k^*}(\hat{U}_h, \hat{G}) = \bigcup_C \iota_{\hat{H}^C}^{\hat{G}} \circ \mathrm{Mon}_{k^*}(\hat{U}_h, \hat{H}^C) \qquad 15.25.5,$$

where C runs over the set of cyclic p'-subgroups of $\hat{G}/\hat{H}$, and therefore it is easily checked that

$$\mathcal{F}ct_{U_h}\big(\mathrm{Mon}_{k^*}(\hat{U}_h, \hat{G}), \mathcal{O}\big)^G \cong \lim_{\longleftarrow}\Big(\mathcal{F}ct_{U_h}\big(\mathrm{Mon}_{k^*}(\hat{U}_h, \hat{H}^\bullet), \mathcal{O}\big)^{\hat{H}^\bullet}\Big) \qquad 15.25.6.$$

We are done.

Corollary 15.26 *In our standard setting, assume that $C_{\hat{G}}(Q_\delta) \subset \hat{H}$. Then, we have $b = c$ and the restriction induces an $\mathcal{O}$-module isomorphism*

$$\mathcal{G}_k(\hat{G}, b) \cong \Big(\bigoplus_C \mathcal{R}_{\hat{H}} \mathcal{G}_k(\hat{H}^C, c)\Big)^{\hat{G}/\hat{H}} \qquad 15.26.1$$

where C runs over the set of cyclic p'-subgroups of $\hat{G}/\hat{H}$.

Proof: From isomorphisms 15.24.3 and 15.25.1 we get an $\mathcal{O}$-module isomorphism

$$\mathcal{G}_k(\hat{G}) \cong \Big(\bigoplus_C \mathcal{R}_{\hat{H}}\mathcal{G}_k(\hat{H}^C)\Big)^{\hat{G}/\hat{H}} \qquad 15.26.2$$

and it is easily checked that this isomorphism is compatible with the "multiplication" by $b = c$ (cf. Proposition 15.10).

15.27 In order to obtain an analogous isomorphism for $\mathcal{G}_k(\mathcal{F}, \widehat{\mathfrak{aut}}_{\mathcal{F}^{\mathrm{nc}}})$, in our standard setting assume that $C_{\hat{G}}(Q,f) \subset \hat{H}$, so that $b = c$ (cf. Proposition 15.10). As in 15.13.1 above, by the *Frattini argument* and this inclusion we get

$$\hat{G}/\hat{H} \cong N_{\hat{G}}(Q,f)/N_{\hat{H}}(Q,f) \cong \mathcal{F}(Q)/\mathcal{H}(Q) \qquad 15.27.1.$$

Let us identify both ends with each other; then, it follows from Proposition 12.12 that, for any cyclic p'-subgroup C of $\hat{G}/\hat{H}$, we have a Frobenius Q-category $\mathcal{H}^C$ containing $\mathcal{H}$ as a normal Q-subcategory and from Proposition 15.10 that c is also a block of $\hat{H}^C$ and that then we have (cf. 3.2 and Proposition 12.8)

$$\mathcal{H}^C = \mathcal{F}_{(c,\hat{H}^C)} \qquad 15.27.2.$$

15.28 Now, as in 15.11 above, we have an *inclusion* functor

$$\mathfrak{j}_C : (\mathcal{H}^C)^{\mathrm{sc}} \longrightarrow \mathcal{F}^{\mathrm{nc}} \qquad 15.28.1$$

and then a functor and a natural map

$$\begin{aligned} \mathfrak{ch}^*(\mathfrak{j}_C) &: \mathfrak{ch}^*\big((\mathcal{H}^C)^{\mathrm{sc}}\big) \longrightarrow \mathfrak{ch}^*(\mathcal{F}^{\mathrm{nc}}) \\ \nu_C &: \mathfrak{aut}_{(\mathcal{H}^C)^{\mathrm{sc}}} \longrightarrow \mathfrak{aut}_{\mathcal{F}^{\mathrm{nc}}} \circ \mathfrak{ch}^*(\mathfrak{j}_C) \end{aligned} \qquad 15.28.2;$$

moreover, it follows from Proposition 15.12 that this natural map can be lifted to a new natural map

$$\hat{\nu}_C : \widehat{\mathfrak{aut}}_{(\mathcal{H}^C)^{\mathrm{sc}}} \longrightarrow \widehat{\mathfrak{aut}}_{\mathcal{F}^{\mathrm{nc}}} \circ \mathfrak{ch}^*(\mathfrak{j}_C) \qquad 15.28.3;$$

as usual, for any $(\mathcal{H}^C)^{\mathrm{sc}}$-chain $\mathfrak{r} : \Delta_n \to (\mathcal{H}^C)^{\mathrm{sc}}$, we write $\hat{\mathcal{H}}^C(\mathfrak{r})$ instead of $\widehat{\mathfrak{aut}}_{(\mathcal{H}^C)^{\mathrm{sc}}}(\mathfrak{r}, \Delta_n)$. Finally, it follows from Corollary 14.7 that the natural map $\mathfrak{g}_k * \hat{\nu}_C$ determines an $\mathcal{O}$-module homomorphism

$$\mathrm{res}^{\mathcal{F}}_{\mathcal{H}^C} : \mathcal{G}_k(\mathcal{F}, \widehat{\mathfrak{aut}}_{\mathcal{F}^{\mathrm{nc}}}) \longrightarrow \mathcal{G}_k(\mathcal{H}^C, \widehat{\mathfrak{aut}}_{(\mathcal{H}^C)^{\mathrm{nc}}}) \qquad 15.28.4.$$

15.29 Similarly, if C' is a cyclic p'-subgroup of $\hat{G}/\hat{H}$ containing C then $\mathcal{H}^C$ is a subcategory of $\mathcal{H}^{C'}$ and, as above, we successively get an *inclusion* functor and a natural map

$$\begin{aligned} \mathfrak{j}_C^{C'} &: (\mathcal{H}^C)^{\mathrm{sc}} \longrightarrow (\mathcal{H}^{C'})^{\mathrm{sc}} \\ \hat{\nu}_C^{C'} &: \widehat{\mathfrak{aut}}_{(\mathcal{H}^C)^{\mathrm{sc}}} \longrightarrow \widehat{\mathfrak{aut}}_{(\mathcal{H}^{C'})^{\mathrm{sc}}} \circ \mathfrak{ch}^*(\mathfrak{j}_C^{C'}) \end{aligned} \qquad 15.29.1;$$

it is clear that $\mathfrak{j}_{C'} \circ \mathfrak{j}_C^{C'} = \mathfrak{j}_C$ and, once again, it follows from the uniqueness part of Theorem 11.32, suitably restated over the set of $\mathcal{H}$-selfcentralizing subgroups of Q, that we get a natural isomorphism

$$\big(\hat{\nu}_{C'} * \mathfrak{ch}^*(\mathfrak{j}_C^{C'})\big) \circ \hat{\nu}_C^{C'} \cong \hat{\nu}_C \qquad 15.29.2.$$

Then, it follows again from Corollary 14.7 that the natural map $\mathfrak{g}_k * \hat{\nu}_C^{C'}$ determines an $\mathcal{O}$-module homomorphism

$$\mathrm{res}_{\mathcal{H}^C}^{\mathcal{H}^{C'}} : \mathcal{G}_k(\mathcal{H}^{C'}, \widehat{\mathfrak{aut}}_{(\mathcal{H}^{C'})^{\mathrm{nc}}}) \longrightarrow \mathcal{G}_k(\mathcal{H}^{C}, \widehat{\mathfrak{aut}}_{(\mathcal{H}^{C})^{\mathrm{nc}}}) \qquad 15.29.3;$$

finally, it follows from isomorphism 15.29.2 that we have

$$\mathrm{res}_{\mathcal{H}^C}^{\mathcal{H}^{C'}} \circ \mathrm{res}_{\mathcal{H}^{C'}}^{\mathcal{F}} = \mathrm{res}_{\mathcal{H}^C}^{\mathcal{F}} \qquad 15.29.4.$$

15.30 Moreover, we claim that the quotient $\mathcal{F}(Q)/\mathcal{H}(Q)$ acts, up to natural isomorphisms, on the family of categories $\{\mathcal{H}^C\}_C$, where C runs over the set of cyclic p'-subgroups of $\mathcal{F}(Q)/\mathcal{H}(Q)$; explicitly, for any $\sigma \in \mathcal{F}(Q)$, it is clear that the group automorphism $\sigma : Q \cong Q$ is $(\mathcal{H}^{C^{\bar{\sigma}}}, \mathcal{H}^C)$-*functorial* (cf. 12.1), where $\bar{\sigma}$ denotes the image of σ in $\mathcal{F}(Q)/\mathcal{H}(Q)$, so that it determines a *Frobenius functor* (cf. 12.1.2)

$$\mathfrak{f}_\sigma : \mathcal{H}^{C^{\bar{\sigma}}} \longrightarrow \mathcal{H}^C \qquad 15.30.1,$$

which is naturally isomorphic to the *identity functor* whenever σ belongs to $\mathcal{H}(Q)$ (cf. 12.1). As above, $\mathfrak{f}_\sigma$ induces a functor $\mathfrak{j}_\sigma$ from $(\mathcal{H}^{C^{\bar{\sigma}}})^{\mathrm{sc}}$ to $(\mathcal{H}^C)^{\mathrm{sc}}$ and σ still determines a natural isomorphism

$$\hat{\nu}_\sigma : \widehat{\mathfrak{aut}}_{(\mathcal{H}^{C^{\bar{\sigma}}})^{\mathrm{sc}}} \cong \widehat{\mathfrak{aut}}_{(\mathcal{H}^C)^{\mathrm{sc}}} \circ \mathfrak{ch}^*(\mathfrak{j}_\sigma) \qquad 15.30.2,$$

so that we get an $\mathcal{O}$-module isomorphism

$$\begin{aligned} \mathcal{G}_k(\mathcal{H}^C, \widehat{\mathfrak{aut}}_{(\mathcal{H}^C)^{\mathrm{sc}}}) &\cong \lim_{\longleftarrow} (\mathfrak{g}_k \circ \widehat{\mathfrak{aut}}_{(\mathcal{H}^C)^{\mathrm{sc}}}) \\ &\cong \lim_{\longleftarrow} \big(\mathfrak{g}_k \circ \widehat{\mathfrak{aut}}_{(\mathcal{H}^C)^{\mathrm{sc}}} \circ \mathfrak{ch}^*(\mathfrak{j}_\sigma)\big) \qquad 15.30.3 \\ &\cong \lim_{\longleftarrow} (\mathfrak{g}_k \circ \widehat{\mathfrak{aut}}_{(\mathcal{H}^{C^{\bar{\sigma}}})^{\mathrm{sc}}}) \cong \mathcal{G}_k(\mathcal{H}^{C^{\bar{\sigma}}}, \widehat{\mathfrak{aut}}_{(\mathcal{H}^{C^{\bar{\sigma}}})^{\mathrm{sc}}}) \end{aligned}$$

which is the identity whenever $\sigma \in \mathcal{H}(Q)$ (cf. Proposition A4.5).

15.31 That is to say, as in 15.24 above, we get a *contravariant* functor

$$\mathcal{G}_k(\mathcal{H}^\bullet, \widehat{\mathfrak{aut}}_{(\mathcal{H}^\bullet)^{\mathrm{sc}}}) : \mathcal{C}_{\hat{G}/\hat{H}} \longrightarrow \mathcal{O}\text{-}\mathfrak{mod} \qquad 15.31.1$$

which maps any cyclic p'-subgroup of $\hat{G}/\hat{H}$ on $\mathcal{G}_k(\mathcal{H}^C, \widehat{\mathfrak{aut}}_{(\mathcal{H}^C)^{\mathrm{sc}}})$ and any $\mathcal{C}_{\hat{G}/\hat{H}}$-morphism $\bar{x} : C \to C'$ on the $\mathcal{O}$-module homomorphism

$$\mathrm{res}_{(C,\bar{x},C')} : \mathcal{G}_k(\mathcal{H}^{C'}, \widehat{\mathfrak{aut}}_{(\mathcal{H}^{C'})^{\mathrm{sc}}}) \longrightarrow \mathcal{G}_k(\mathcal{H}^{C}, \widehat{\mathfrak{aut}}_{(\mathcal{H}^{C})^{\mathrm{sc}}}) \qquad 15.31.2$$

obtained as the composition of the $\mathcal{O}$-module isomorphism above (cf. 15.30.3)

$$\mathcal{G}_k(\mathcal{H}^{C'}, \widehat{\mathfrak{aut}}_{(\mathcal{H}^{C'})^{\mathrm{sc}}}) \cong \mathcal{G}_k(\mathcal{H}^{C'^{\bar{x}}}, \widehat{\mathfrak{aut}}_{(\mathcal{H}^{C'^{\bar{x}}})^{\mathrm{sc}}}) \qquad 15.31.3$$

with the $\mathcal{O}$-module homomorphism $\mathrm{res}^{\mathcal{H}^{C'^{\bar{x}}}}_{\mathcal{H}^C}$ (cf. 15.29.3); the proof of the functoriality of this correspondence is straightforward. Moreover, since we clearly have a natural isomorphism $\mathfrak{j}_{C'} \circ \mathfrak{j}_{\bar{x}} \cong \mathfrak{j}_{C'^{\bar{x}}}$, it follows again from Proposition A4.5 that

$$\mathrm{res}_{(C,\bar{x},C')} \circ \mathrm{res}^{\mathcal{F}}_{\mathcal{H}^{C'}} = \mathrm{res}^{\mathcal{F}}_{\mathcal{H}^C} \qquad 15.31.4,$$

so that we finally obtain an $\mathcal{O}$-module homomorphism

$$\mathcal{G}_k(\mathcal{F}, \widehat{\mathfrak{aut}}_{\mathcal{F}^{\mathrm{nc}}}) \longrightarrow \varprojlim \mathcal{G}_k(\mathcal{H}^\bullet, \widehat{\mathfrak{aut}}_{(\mathcal{H}^\bullet)^{\mathrm{sc}}}) \qquad 15.31.5.$$

Theorem 15.32 *In our standard setting, assume that $C_{\hat{G}}(Q,f) \subset \hat{H}$. Then, the $\mathcal{O}$-module homomorphisms $\mathrm{res}^{\mathcal{F}}_{\mathcal{H}^C}$, where C runs over the set of cyclic p'-subgroups of $\hat{G}/\hat{H}$, induce an $\mathcal{O}$-module isomorphism*

$$\mathcal{G}_k(\mathcal{F}, \widehat{\mathfrak{aut}}_{\mathcal{F}^{\mathrm{nc}}}) \cong \varprojlim \mathcal{G}_k(\mathcal{H}^\bullet, \widehat{\mathfrak{aut}}_{(\mathcal{H}^\bullet)^{\mathrm{nc}}}) \qquad 15.32.1.$$

Proof: Denoting by $\mathfrak{S}$ the set of $\mathcal{H}$-selfcentralizing subgroups of Q and by $\mathcal{F}^{\mathfrak{S}}$ the full subcategory of $\mathcal{F}$ over $\mathfrak{S}$, it follows from Lemma 15.14 that

$$\mathcal{G}_k(\mathcal{F}, \widehat{\mathfrak{aut}}_{\mathcal{F}^{\mathrm{nc}}}) \cong \varprojlim (\mathfrak{g}_k \circ \widehat{\mathfrak{aut}}_{\mathcal{F}^{\mathfrak{S}}}) \qquad 15.32.2;$$

thus, denoting by $\mathfrak{j}: \mathcal{F}^{\mathfrak{S}} \to \mathcal{F}^{\mathrm{nc}}$ the *inclusion functor* and setting

$$\widehat{\mathfrak{aut}}_{\mathcal{F}^{\mathfrak{S}}} = \widehat{\mathfrak{aut}}_{\mathcal{F}^{\mathrm{nc}}} \circ \mathfrak{ch}^*(\mathfrak{j}) \qquad 15.32.3,$$

it suffices to prove that the $\mathcal{O}$-module homomorphisms $\mathrm{res}^{\mathcal{F}}_{\mathcal{H}^C}$, where C runs over the set of cyclic p'-subgroups of $\hat{G}/\hat{H}$, induce an injective $\mathcal{O}$-module homomorphism

$$\varprojlim (\mathfrak{g}_k \circ \widehat{\mathfrak{aut}}_{\mathcal{F}^{\mathfrak{S}}}) \longrightarrow \prod_{C \in \mathrm{Obj}(\mathcal{C}_{\hat{G}/\hat{H}})} \varprojlim (\mathfrak{g}_k \circ \widehat{\mathfrak{aut}}_{(\mathcal{H}^C)^{\mathrm{sc}}}) \qquad 15.32.4$$

such that the image coincides with the corresponding inverse limit.

Let $Z = (Z_R)_{R \in \mathfrak{S}}$ be an element of the kernel of homomorphism 15.32.4; that is to say, for any $R \in \mathfrak{S}$, Z_R is an element of $\mathcal{G}_k\big(\hat{\mathcal{F}}(R)\big)$ which has restriction zero in $\mathcal{G}_k\big(\hat{\mathcal{H}}^C(R)\big)$, for any cyclic p'-subgroup C of $\hat{G}/\hat{H} \cong \mathcal{F}(Q)/\mathcal{H}(Q)$; we claim that $Z_R = 0$. Indeed, since Z is *stable* by $\mathcal{F}$-isomorphisms, we may assume that R is *fully highnormalized* in $\mathcal{H}$ (cf. 12.10) and then, since R is $\mathcal{F}$-nilcentralized (cf. Proposition 15.10), it follows from Proposition 12.11

that the restriction induces a group isomorphism

$$\mathcal{F}(Q)_R/\mathcal{H}(Q)_R \cong \mathcal{F}(R)/\mathcal{H}(R) \qquad 15.32.5,$$

where $\mathcal{F}(Q)_R$ and $\mathcal{H}(Q)_R$ denote the corresponding stabilizers of R; hence, for any k^*-subgroup $\hat{D}$ of $\hat{\mathcal{F}}(R)$ with a cyclic k^*-quotient D of order prime to p, we can find a cyclic p'-subgroup C of $\mathcal{F}(Q)/\mathcal{H}(Q)$ such that $\hat{D}$ is contained in $\hat{\mathcal{H}}^C(R)$ and therefore Z_R still has restriction zero in $\mathcal{G}_k(\hat{D})$. Now, the claim follows from isomorphism 14.14.2.

From equalities 15.31.4, we already know that the image of homomorphism 15.32.4 is contained in the corresponding inverse limit. Conversely, let $Y = (Y_C)_{C \in \mathrm{Obj}(\mathcal{C}_{\hat{G}/\hat{H}})}$ be an element of this limit; that is to say, for any cyclic p'-subgroup C of $\mathcal{F}(Q)/\mathcal{H}(Q) \cong \hat{G}/\hat{H}$, we have $Y_C = (Y_{C,R})_{R \in \mathfrak{S}}$ where, for any $R \in \mathfrak{S}$, $Y_{C,R}$ is an element of $\mathcal{G}_k(\hat{\mathcal{H}}^C(R))$ and, for another $R' \in \mathfrak{S}$ and any $\psi \in \mathcal{H}^C(R',R)$, in $\mathcal{G}_k(\hat{\mathcal{H}}^C(\mathfrak{r}))$ we have

$$\mathrm{res}_{\widehat{\mathfrak{aut}}_{(\mathcal{H}^C)^{\mathrm{sc}}}(\mathrm{id}_{\mathfrak{r}\circ\delta_1^0},\delta_1^0)}(Y_{C,R}) = \mathrm{res}_{\widehat{\mathfrak{aut}}_{(\mathcal{H}^C)^{\mathrm{sc}}}(\mathrm{id}_{\mathfrak{r}\circ\delta_0^0},\delta_0^0)}(Y_{C,R'}) \qquad 15.32.6,$$

where $\mathfrak{r}\colon \Delta_1 \to (\mathcal{H}^C)^{\mathrm{sc}}$ is the obvious $(\mathcal{H}^C)^{\mathrm{sc}}$-chain defined by ψ.

Moreover, for any $\sigma \in \mathcal{F}(Q)$, we have $\mathrm{res}_{(C^{\bar{\sigma}},\bar{\sigma},C)}(Y_C) = Y_{C^{\bar{\sigma}}}$ (cf. 15.31.2) or, equivalently, for any $R \in \mathfrak{S}$ we still have

$$\mathrm{res}_{(\hat{\nu}_\sigma)_R}(Y_{C,R}) = Y_{C^{\bar{\sigma}},\sigma^{-1}(R)} \qquad 15.32.7,$$

where $\bar{\sigma}$ denotes the image of σ in $\mathcal{F}(Q)/\mathcal{H}(Q)$ and

$$(\hat{\nu}_\sigma)_R : \hat{\mathcal{H}}^{C^{\bar{\sigma}}}\big(\sigma^{-1}(R)\big) \cong \hat{\mathcal{H}}^C(R) \qquad 15.32.8$$

is the k^*-group isomorphism determined by natural isomorphism $\hat{\nu}_\sigma$ above; finally, for any subgroup D of C and any $R \in \mathfrak{S}$, we have

$$\mathrm{res}^{\hat{\mathcal{H}}^C(Q)}_{\hat{\mathcal{H}}^D(Q)}(Y_{C,Q}) = Y_{D,Q} \qquad 15.32.9.$$

Thus, fixing $R \in \mathfrak{S}$ fully highnormalized in $\mathcal{H}$, for any cyclic p'-subgroup C of $\mathcal{F}(Q)_R/\mathcal{H}(Q)_R$ we have the element $Y_{C,R}$ of $\mathcal{G}_k(\hat{\mathcal{H}}^C(R))$ and, moreover, for any $\sigma \in \mathcal{F}(R)$ and any subgroup D of C, we already know that

$$\mathrm{res}_{(\hat{\nu}_\sigma)_R}(Y_{C,R}) = Y_{C^{\bar{\sigma}},R} \quad \text{and} \quad \mathrm{res}^{\hat{\mathcal{H}}^C(R)}_{\hat{\mathcal{H}}^D(R)}(Y_{C,R}) = Y_{D,R} \qquad 15.32.10;$$

consequently, it follows from Proposition 15.25 and isomorphism 15.32.5 that there exists $X_R \in \mathcal{G}_k(\hat{\mathcal{F}}(R))$ fulfilling

$$\mathrm{res}^{\hat{\mathcal{F}}(R)}_{\hat{\mathcal{H}}^C(R)}(X_R) = Y_{C,R} \qquad 15.32.11$$

for any cyclic p'-subgroup C of $\mathcal{F}(Q)_R/\mathcal{H}(Q)_R$.

More generally, for any cyclic p'-subgroup C of $\mathcal{F}(Q)/\mathcal{H}(Q)$ and any k^*-subgroup $\hat{D}$ of $\hat{\mathcal{H}}^C(R) \subset \hat{\mathcal{F}}(R)$ with a cyclic k^*-quotient D of order prime to p, denoting by $\bar{D}$ the image of $\hat{D}$ in $\mathcal{F}(Q)_R/\mathcal{H}(Q)_R$ *via* the composed group homomorphism

$$\hat{\mathcal{H}}^C(R)/\hat{\mathcal{H}}(R) \subset \mathcal{F}(R)/\mathcal{H}(R) \cong \mathcal{F}(Q)_R/\mathcal{H}(Q)_R \qquad 15.32.12,$$

we have $\hat{D} \subset \hat{\mathcal{H}}^{\bar{D}}(R) \subset \hat{\mathcal{H}}^C(R)$ and therefore we get (cf. 15.32.9)

$$\begin{aligned} \mathrm{res}_{\hat{D}}^{\hat{\mathcal{H}}^C(R)}\big(\mathrm{res}_{\hat{\mathcal{H}}^C(R)}^{\hat{\mathcal{F}}(R)}(X_R)\big) &= \mathrm{res}_{\hat{D}}^{\hat{\mathcal{H}}^{\bar{D}}(R)}\big(\mathrm{res}_{\hat{\mathcal{H}}^{\bar{D}}(R)}^{\hat{\mathcal{F}}(R)}(X_R)\big) \\ &= \mathrm{res}_{\hat{D}}^{\hat{\mathcal{H}}^{\bar{D}}(R)}(Y_{\bar{D},R}) = \mathrm{res}_{\hat{D}}^{\hat{\mathcal{H}}^C(R)}(Y_{C,R}) \end{aligned} \qquad 15.32.13.$$

Thus, equality 15.32.11 holds for any cyclic p'-subgroup C of $\mathcal{F}(Q)/\mathcal{H}(Q)$.

Now, if $R' \in \mathfrak{S}$ is another element also fully highnormalized in $\mathcal{H}$ and $\theta : R \cong R'$ is an $\mathcal{F}$-isomorphism, we claim that

$$\mathrm{res}_{\widehat{\mathfrak{aut}}_{\mathcal{F}^{\mathfrak{S}}}(\theta,\mathrm{id}_{\Delta_0})}(X_{R'}) = X_R \qquad 15.32.14;$$

note that, according to Proposition 12.8, we may assume either that θ is induced by an element σ of $\mathcal{F}(Q)$ or that it is an $\mathcal{H}$-isomorphism. Let C be a cyclic p'-subgroup of $\mathcal{F}(R)/\mathcal{H}(R)$; in the first case, it follows from equality 15.32.7 that

$$\begin{aligned} &\mathrm{res}_{\hat{\mathcal{H}}^C(R)}^{\hat{\mathcal{F}}(R)}\big(\mathrm{res}_{\widehat{\mathfrak{aut}}_{\mathcal{F}^{\mathfrak{S}}}(\theta,\mathrm{id}_{\Delta_0})}(X_{R'})\big) \\ &\qquad = (\mathrm{res}_{(\hat{\nu}_\sigma)_R})^{-1}\big(\mathrm{res}_{\hat{\mathcal{H}}^{C^{\bar{\sigma}}}(R')}^{\hat{\mathcal{F}}(R')}(X_{R'})\big) \\ &\qquad = (\mathrm{res}_{(\hat{\nu}_\sigma)_R})^{-1}(Y_{C^{\bar{\sigma}},R'}) = Y_{C,R} = \mathrm{res}_{\hat{\mathcal{H}}^C(R)}^{\hat{\mathcal{F}}(R)}(X_R) \end{aligned} \qquad 15.32.15.$$

In the second case, denoting by $\mathfrak{r} : \Delta_1 \to (\mathcal{H}^C)^{\mathrm{sc}}$ the obvious $(\mathcal{H}^C)^{\mathrm{sc}}$-chain defined by the $\mathcal{H}$-isomorphism θ, and identifying $\hat{\mathcal{H}}^C(\mathfrak{r})$ with $\hat{\mathcal{H}}^C(R)$, it follows from equality 15.32.6 that

$$\begin{aligned} &\mathrm{res}_{\hat{\mathcal{H}}^C(R)}^{\hat{\mathcal{F}}(R)}\big(\mathrm{res}_{\widehat{\mathfrak{aut}}_{\mathcal{F}^{\mathfrak{S}}}(\theta,\mathrm{id}_{\Delta_0})}(X_{R'})\big) \\ &\qquad = \mathrm{res}_{\widehat{\mathfrak{aut}}_{(\mathcal{H}^C)^{\mathrm{sc}}}(\mathrm{id}_{\mathfrak{q}\circ\delta_0^0},\delta_0^0)}\big(\mathrm{res}_{\hat{\mathcal{H}}^C(R')}^{\hat{\mathcal{F}}(R')}(X_{R'})\big) \\ &\qquad = \mathrm{res}_{\widehat{\mathfrak{aut}}_{(\mathcal{H}^C)^{\mathrm{sc}}}(\mathrm{id}_{\mathfrak{q}\circ\delta_0^0},\delta_0^0)}(Y_{C,R'}) \\ &\qquad = \mathrm{res}_{\widehat{\mathfrak{aut}}_{(\mathcal{H}^C)^{\mathrm{sc}}}(\mathrm{id}_{\mathfrak{q}\circ\delta_1^0},\delta_1^0)}(Y_{C,R}) = \mathrm{res}_{\hat{\mathcal{H}}^C(R)}^{\hat{\mathcal{F}}(R)}(X_R) \end{aligned} \qquad 15.32.16.$$

By the transitivity of the equalities, the claim is proved.

Consequently, we can define a unique element $X = (X_R)_{R\in\mathfrak{S}}$, *stable* by $\mathcal{F}$-isomorphisms, by setting

$$X_R = \mathrm{res}_{\widehat{\mathfrak{aut}}_{\mathcal{F}^{\mathfrak{S}}}(\theta,\mathrm{id}_{\Delta_0})}(X_{R'}) \qquad 15.32.17$$

for a choice of $R' \in \mathfrak{S}$ fully highnormalized in $\mathcal{H}$ admitting an $\mathcal{F}$-isomorphism $\theta : R \cong R'$; actually, the arguments above also prove that, for any cyclic p'-subgroup C of $\mathcal{F}(Q)/\mathcal{H}(Q)$, we still get

$$\mathrm{res}^{\hat{\mathcal{F}}(R)}_{\hat{\mathcal{H}}^C(R)}(X_R) = Y_{C,R} \qquad 15.32.18.$$

Finally, we claim that the element X belongs to $\lim\limits_{\longleftarrow}(\mathfrak{g}_k \circ \widehat{\mathfrak{aut}}_{\mathcal{F}^{\mathfrak{S}}})$. Indeed, for an $\mathcal{F}^{\mathfrak{S}}$-chain $\mathfrak{r} : \Delta_1 \to \mathcal{F}^{\mathfrak{S}}$, set $T = \mathfrak{r}(0)$, $R = \mathfrak{r}(1)$ and $\psi = \mathfrak{r}(0\bullet 1)$, and choose $R' \in \mathfrak{S}$ fully highnormalized in $\mathcal{H}$ admitting an $\mathcal{H}$-isomorphism $\theta : R' \cong R$ (cf. 12.10); according to Proposition 12.8, we have

$$\iota^Q_{R'} \circ \theta^{-1} \circ \psi = \tau \circ \psi' \qquad 15.32.19$$

for suitable $\tau \in \mathcal{F}(Q)$ and $\psi' \in \mathcal{H}(Q,T)$; set $R'' = \tau^{-1}(R')$ and denote by $\eta : R'' \cong R'$ the isomorphism induced by τ, by $\psi'' : T \to R''$ the homomorphism determined by ψ', and by $\mathfrak{r}'' : \Delta_1 \to \mathcal{F}^{\mathfrak{S}}$ the *chain* defined by ψ''. Then, it suffices to prove the equalities

$$\begin{gathered}\mathrm{res}_{\widehat{\mathfrak{aut}}_{\mathcal{F}^{\mathfrak{S}}}(\mathrm{id}_{\mathfrak{r}''\circ\delta^0_1},\delta^0_1)}(X_T) = \mathrm{res}_{\widehat{\mathfrak{aut}}_{\mathcal{F}^{\mathfrak{S}}}(\mathrm{id}_{\mathfrak{r}''\circ\delta^0_0},\delta^0_0)}(X_{R''}) \\ X_{R''} = \mathrm{res}_{\widehat{\mathfrak{aut}}_{\mathcal{F}^{\mathfrak{S}}}(\theta\circ\eta,\mathrm{id}_{\Delta_0})}(X_R)\end{gathered} \qquad 15.32.20.$$

The bottom equality follows from equality 15.32.14; moreover, since we have (cf. Proposition 12.11)

$$\hat{\mathcal{F}}(\mathfrak{r}'') = \hat{\mathcal{F}}(R'')_{\psi''(T)} \quad \text{and} \quad \mathcal{F}(Q)_{R''}/\mathcal{H}(Q)_{R''} \cong \mathcal{F}(R'')/\mathcal{H}(R'') \qquad 15.32.21,$$

it suffices to prove the restriction of the top equality to $\hat{\mathcal{H}}^C(\mathfrak{r}'')$ for any cyclic p'-subgroup C of $\mathcal{F}(Q)/\mathcal{H}(Q)$; but, according to equalities 15.32.6 and 15.32.18, we get

$$\begin{aligned}&\mathrm{res}^{\hat{\mathcal{F}}(\mathfrak{r}'')}_{\hat{\mathcal{H}}^C(\mathfrak{r}'')}\big(\mathrm{res}_{\widehat{\mathfrak{aut}}_{\mathcal{F}^{\mathfrak{S}}}(\mathrm{id}_{\mathfrak{r}''\circ\delta^0_1},\delta^0_1)}(X_T)\big) \\ &= \mathrm{res}_{\widehat{\mathfrak{aut}}_{(\mathcal{H}^C)^{\mathrm{sc}}}(\mathrm{id}_{\mathfrak{r}''\circ\delta^0_1},\delta^0_1)}(Y_{C,T}) = \mathrm{res}_{\widehat{\mathfrak{aut}}_{(\mathcal{H}^C)^{\mathrm{sc}}}(\mathrm{id}_{\mathfrak{r}''\circ\delta^0_0},\delta^0_0)}(Y_{C,R''}) \\ &= \mathrm{res}^{\hat{\mathcal{F}}(\mathfrak{r}'')}_{\hat{\mathcal{H}}^C(\mathfrak{r}'')}\big(\mathrm{res}_{\widehat{\mathfrak{aut}}_{\mathcal{F}^{\mathfrak{S}}}(\mathrm{id}_{\mathfrak{r}''\circ\delta^0_0},\delta^0_0)}(X_{R''})\big)\end{aligned} \qquad 15.32.22.$$

We have proved the claim. At this point, equality 15.32.18 shows that X maps on Y. We are done.

15.33 As above, in our standard setting assume that $C_{\hat{G}}(Q,f) \subset \hat{H}$. Let C be a cyclic p'-subgroup of $\hat{G}/\hat{H} \cong \mathcal{F}(Q)/\mathcal{H}(Q)$; for any $(\mathcal{H}^C)^{\mathrm{sc}}$-chain $\mathfrak{r}:\Delta_n \to (\mathcal{H}^C)^{\mathrm{sc}}$, since $\mathfrak{r}(n)$ is also $\mathcal{H}$-selfcentralizing (cf. Lemma 15.16) and we have identified $\mathcal{H}^C(\mathfrak{r})$ with a subgroup of $\mathcal{H}^C\big(\mathfrak{r}(n)\big)$, it makes sense to consider $\mathcal{H}(\mathfrak{r}) \subset \mathcal{H}\big(\mathfrak{r}(n)\big)$ which is a normal subgroup of $\mathcal{H}^C(\mathfrak{r})$, and then we set $C_\mathfrak{r} = \mathcal{H}^C(\mathfrak{r})/\mathcal{H}(\mathfrak{r})$ which can be identified with a subgroup of C. Thus, since $\mathcal{G}_k\big(\hat{\mathcal{H}}^C(\mathfrak{r})\big)$ has a canonical $\mathcal{G}_k(C_\mathfrak{r})$-module structure (cf. 15.21), $\mathcal{G}_k\big(\hat{\mathcal{H}}^C(\mathfrak{r})\big)$ becomes a $\mathcal{G}_k(C)$-module *via* the restriction map $\mathcal{G}_k(C) \to \mathcal{G}_k(C_\mathfrak{r})$; furthermore, it is easily checked that these $\mathcal{G}_k(C)$-module structures are compatible with the $\mathfrak{ch}^*\big((\mathcal{H}^C)^{\mathrm{sc}}\big)$-morphisms; finally, from its very definition (cf. 14.3.3), $\mathcal{G}_k(\mathcal{H}^C, \widehat{\mathfrak{aut}}_{(\mathcal{H}^C)^{\mathrm{nc}}})$ becomes a $\mathcal{G}_k(C)$-module. Moreover, according to 15.29 and 15.30, the group C acts on $\mathcal{G}_k(\mathcal{H}, \widehat{\mathfrak{aut}}_{\mathcal{H}^{\mathrm{nc}}})$ and we have a *restriction* $\mathcal{O}$-module homomorphism

$$\mathcal{G}_k(\mathcal{H}^C, \widehat{\mathfrak{aut}}_{(\mathcal{H}^C)^{\mathrm{nc}}}) \longrightarrow \mathcal{G}_k(\mathcal{H}, \widehat{\mathfrak{aut}}_{\mathcal{H}^{\mathrm{nc}}})^C \qquad 15.33.1$$

since C acts trivially on $\mathcal{G}_k(\mathcal{H}^C, \widehat{\mathfrak{aut}}_{(\mathcal{H}^C)^{\mathrm{nc}}})$; then, we claim we have an isomorphism analogous to isomorphism 15.21.2.

Proposition 15.34 *In our standard setting, assume that $C_{\hat{G}}(Q,f) \subset \hat{H}$. For any cyclic p'-subgroup C of $\hat{G}/\hat{H}$, the restriction induces an $\mathcal{O}$-module isomorphism*

$$\mathcal{G}_k(\mathcal{H}^C, \widehat{\mathfrak{aut}}_{(\mathcal{H}^C)^{\mathrm{nc}}})^{C^*} \cong \mathcal{G}_k(\mathcal{H}, \widehat{\mathfrak{aut}}_{\mathcal{H}^{\mathrm{nc}}})^C \qquad 15.34.1.$$

Proof: It is clear that any $(\mathcal{H}^C)^{\mathrm{sc}}$-chain $\mathfrak{r}:\Delta_n \to (\mathcal{H}^C)^{\mathrm{sc}}$ is $\mathcal{H}^C$-isomorphic to a $(\mathcal{H}^C)^{\mathrm{sc}}$-chain $\mathfrak{r}':\Delta_n \to (\mathcal{H}^C)^{\mathrm{sc}}$ such that $\mathfrak{r}'(i-1) \subset \mathfrak{r}'(i)$ and $\mathfrak{r}'(i-1\bullet i)$ is the inclusion map for any $1 \le i \le n$; thus, since any subgroup of Q is $\mathcal{H}^C$-selfcentralizing if and only if it is $\mathcal{H}$-selfcentralizing (cf. Lemma 15.16), the set $\mathfrak{C}$ of these $(\mathcal{H}^C)^{\mathrm{sc}}$-chains coincides with the corresponding set of $\mathcal{H}^{\mathrm{sc}}$-chains and we have injective $\mathcal{O}$-module homomorphisms

$$\begin{aligned} \mathcal{G}_k(\mathcal{H}^C, \widehat{\mathfrak{aut}}_{(\mathcal{H}^C)^{\mathrm{nc}}}) &\longrightarrow \prod_{\mathfrak{r}\in\mathfrak{C}} \mathcal{G}_k\big(\hat{\mathcal{H}}^C(\mathfrak{r})\big) \\ \mathcal{G}_k(\mathcal{H}, \widehat{\mathfrak{aut}}_{\mathcal{H}^{\mathrm{nc}}}) &\longrightarrow \prod_{\mathfrak{r}\in\mathfrak{C}} \mathcal{G}_k\big(\hat{\mathcal{H}}(\mathfrak{r})\big) \end{aligned} \qquad 15.34.2$$

clearly compatible with the corresponding restriction homomorphisms.

Moreover, it is clear that the actions of $N_{\hat{H}^C}(Q_\delta)$ on Q, $\hat{H}^C$ and $\hat{H}$ determine an action of this group on both $\mathcal{O}$-module homomorphisms, which is trivial on $\mathcal{G}_k(\mathcal{H}^C, \widehat{\mathfrak{aut}}_{(\mathcal{H}^C)^{\mathrm{nc}}})$ whereas it coincides with the action of C on

$\mathcal{G}_k(\mathcal{H},\widehat{\mathfrak{aut}}_{\mathcal{H}^{\rm nc}})$; hence, choosing a set of representatives $\mathfrak{R}$ for the set of orbits of $N_{\hat{H}^C}(Q_\delta)$ over $\mathfrak{C}$, we have a commutative diagram

$$\begin{array}{ccc} \prod_{\mathfrak{r}\in\mathfrak{R}}\mathcal{G}_k\big(\hat{\mathcal{H}}^C(\mathfrak{r})\big) & \longrightarrow & \prod_{\mathfrak{r}\in\mathfrak{R}}\mathcal{G}_k\big(\hat{\mathcal{H}}(\mathfrak{r})\big)^{C_\mathfrak{r}} \\ \uparrow & & \uparrow \\ \mathcal{G}_k(\mathcal{H}^C,\widehat{\mathfrak{aut}}_{(\mathcal{H}^C)^{\rm nc}}) & \longrightarrow & \mathcal{G}_k(\mathcal{H},\widehat{\mathfrak{aut}}_{\mathcal{H}^{\rm nc}})^C \end{array} \qquad 15.34.3$$

where the vertical arrows are still injective.

On the other hand, the multiplicative subgroup $C^*\subset\mathcal{G}_k(C)$ (cf. 15.21) acts on $\mathcal{G}_k(\mathcal{H}^C,\widehat{\mathfrak{aut}}_{(\mathcal{H}^C)^{\rm nc}})$ and, for any $\mathfrak{r}\in\mathfrak{R}$, it acts on $\mathcal{G}_k\big(\hat{\mathcal{H}}^C(\mathfrak{r})\big)$ *via* its image in $(C_\mathfrak{r})^*$ (cf. 15.33); consequently, we still have a commutative diagram (cf. 15.21.2)

$$\begin{array}{ccc} \prod_{\mathfrak{r}\in\mathfrak{R}}\mathcal{G}_k\big(\hat{\mathcal{H}}^C(\mathfrak{r})\big)^{(C_\mathfrak{r})^*} & \cong & \prod_{\mathfrak{r}\in\mathfrak{R}}\mathcal{G}_k\big(\hat{\mathcal{H}}(\mathfrak{r})\big)^{C_\mathfrak{r}} \\ \uparrow & & \uparrow \\ \mathcal{G}_k(\mathcal{H}^C,\widehat{\mathfrak{aut}}_{(\mathcal{H}^C)^{\rm nc}})^{C^*} & \longrightarrow & \mathcal{G}_k(\mathcal{H},\widehat{\mathfrak{aut}}_{\mathcal{H}^{\rm nc}})^C \end{array} \qquad 15.34.4$$

Finally, the obvious commutative diagrams

$$\begin{array}{ccc} \mathcal{G}_k\big(\hat{\mathcal{H}}^C(\mathfrak{r})\big)^{(C_\mathfrak{r})^*} & \cong & \mathcal{G}_k\big(\hat{\mathcal{H}}(\mathfrak{r})\big)^{C_\mathfrak{r}} \\ {\scriptstyle \mathrm{res}_{\widehat{\mathfrak{aut}}_{(\mathcal{H}^C)^{\rm sc}}(\nu,\delta)}}\downarrow & & \downarrow{\scriptstyle \mathrm{res}_{\widehat{\mathfrak{aut}}_{\mathcal{H}^{\rm sc}}(\nu,\delta)}} \\ \mathcal{G}_k\big(\hat{\mathcal{H}}^C(\mathfrak{t})\big)^{(C_\mathfrak{t})^*} & \cong & \mathcal{G}_k\big(\hat{\mathcal{H}}(\mathfrak{t})\big)^{C_\mathfrak{t}} \end{array} \qquad 15.34.5,$$

where $\mathfrak{r}$ and $\mathfrak{t}$ run over $\mathfrak{R}$ and $(\nu,\delta)\colon(\mathfrak{t},\Delta_m)\to(\mathfrak{r},\Delta_n)$ over the set of $\mathfrak{ch}^*(\mathcal{H}_{\rm sc})$-morphisms from $(\mathfrak{t},\Delta_m)$ to $(\mathfrak{r},\Delta_n)$, proves the surjectivity of the bottom homomorphism in diagram 15.34.4. We are done.

15.35 At this point, we also consider the *C-residual Grothendieck group of the pair* $(\mathcal{H}^C,\widehat{\mathfrak{aut}}_{(\mathcal{H}^C)^{\rm nc}})$ — denoted by $\mathcal{R}_\mathcal{H}\mathcal{G}_k(\mathcal{H}^C,\widehat{\mathfrak{aut}}_{(\mathcal{H}^C)^{\rm nc}})$ — as the intersection of kernels of all the restriction $\mathcal{O}$-module homomorphisms

$$\mathcal{G}_k(\mathcal{H}^C,\widehat{\mathfrak{aut}}_{(\mathcal{H}^C)^{\rm nc}})\longrightarrow\mathcal{G}_k(\mathcal{H}^D,\widehat{\mathfrak{aut}}_{(\mathcal{H}^D)^{\rm nc}}) \qquad 15.35.1,$$

where D runs over the set of proper subgroups of C; since they actually are $\mathcal{G}_k(C)$-module homomorphisms *via* the restriction maps $\mathcal{G}_k(C)\to\mathcal{G}_k(D)$, the intersection $\mathcal{R}_\mathcal{H}\mathcal{G}_k(\mathcal{H}^C,\widehat{\mathfrak{aut}}_{(\mathcal{H}^C)^{\rm nc}})$ is a $\mathcal{G}_k(C)$-submodule.

15.36 As the full *Grothendieck group* $\mathcal{G}_k(\mathcal{H}^C,\widehat{\mathfrak{aut}}_{(\mathcal{H}^C)^{\rm nc}})$, we also can describe $\mathcal{R}_\mathcal{H}\mathcal{G}_k(\mathcal{H}^C,\widehat{\mathfrak{aut}}_{(\mathcal{H}^C)^{\rm nc}})$ as an inverse limit. In order to show this, for any $(\mathcal{H}^C)^{\rm sc}$-chain $\mathfrak{r}\colon\Delta_n\to(\mathcal{H}^C)^{\rm sc}$ such that $C_\mathfrak{r}=C$, consider the $\mathcal{O}$-module $\mathcal{R}_{\hat{\mathcal{H}}(\mathfrak{r})}\mathcal{G}_k\big(\hat{\mathcal{H}}^C(\mathfrak{r})\big)$; moreover, for any subgroup D of C and any

$\mathfrak{ch}^*\big((\mathcal{H}^D)^{^{\mathrm{sc}}}\big)$-morphism $(\mu,\delta)\colon(\mathfrak{t},\Delta_m)\to(\mathfrak{r},\Delta_n)$ such that $C_{\mathfrak{t}}=C_{\mathfrak{r}}=C$, note that we still have the restriction homomorphism

$$\mathrm{res}_{\widehat{\mathfrak{aut}}_{(\mathcal{H}^D)^{\mathrm{sc}}}(\mu,\delta)}:\mathcal{G}_k\big(\hat{\mathcal{H}}^D(\mathfrak{r})\big)\longrightarrow\mathcal{G}_k\big(\hat{\mathcal{H}}^D(\mathfrak{t})\big)\qquad 15.36.1;$$

then, it is quite clear that

$$\mathrm{res}_{\widehat{\mathfrak{aut}}_{(\mathcal{H}^C)^{\mathrm{sc}}}(\mu,\delta)}\Big(\mathcal{R}_{\hat{\mathcal{H}}(\mathfrak{r})}\mathcal{G}_k\big(\hat{\mathcal{H}}^C(\mathfrak{r})\big)\Big)\subset\mathcal{R}_{\hat{\mathcal{H}}(\mathfrak{t})}\mathcal{G}_k\big(\hat{\mathcal{H}}^C(\mathfrak{t})\big)\qquad 15.36.2.$$

That is to say, denoting by $\mathfrak{ch}^*_C\big((\mathcal{H}^C)^{^{\mathrm{sc}}}\big)$ the full subcategory of $\mathfrak{ch}^*\big((\mathcal{H}^C)^{^{\mathrm{sc}}}\big)$ over the set of $(\mathcal{H}^C)^{^{\mathrm{sc}}}$-chains $\mathfrak{r}\colon\Delta_n\to(\mathcal{H}^C)^{^{\mathrm{sc}}}$ such that $C_{\mathfrak{r}}=C$, we get a *contravariant* functor

$$\mathcal{R}_{\hat{\mathcal{H}}(\bullet)}\mathcal{G}_k\big(\hat{\mathcal{H}}^C(\bullet)\big):\mathfrak{ch}^*_C\big((\mathcal{H}^C)^{^{\mathrm{sc}}}\big)\longrightarrow\mathcal{O}\text{-}\mathfrak{mod}\qquad 15.36.3.$$

On the other hand, for any subgroup D of C, from the very definition of $\mathcal{G}_k(\mathcal{H}^D,\widehat{\mathfrak{aut}}_{(\mathcal{H}^D)^{\mathrm{nc}}})$ (cf. 14.3.3) we get structural $\mathcal{O}$-module homomorphisms

$$\mathcal{G}_k(\mathcal{H}^D,\widehat{\mathfrak{aut}}_{(\mathcal{H}^D)^{\mathrm{nc}}})\longrightarrow\mathcal{G}_k\big(\hat{\mathcal{H}}^D(\mathfrak{r})\big)\qquad 15.36.4$$

compatible with homomorphisms 15.36.1 and finally, after routine verifications, we obtain an $\mathcal{O}$-module homomorphism

$$\mathcal{R}_{\mathcal{H}}\mathcal{G}_k(\mathcal{H}^C,\widehat{\mathfrak{aut}}_{(\mathcal{H}^C)^{\mathrm{nc}}})\longrightarrow\varprojlim\mathcal{R}_{\hat{\mathcal{H}}(\bullet)}\mathcal{G}_k\big(\hat{\mathcal{H}}^C(\bullet)\big)\qquad 15.36.5.$$

Proposition 15.37 *In our standard setting, assume that $C_{\hat{G}}(Q,f)\subset\hat{H}$. For any cyclic p'-subgroup C of $\hat{G}/\hat{H}$, the structural homomorphisms induce $\mathcal{O}$-module isomorphisms*

$$\begin{aligned}\mathcal{R}_{\mathcal{H}}\mathcal{G}_k(\mathcal{H}^C,\widehat{\mathfrak{aut}}_{(\mathcal{H}^C)^{\mathrm{nc}}})&\cong\varprojlim\mathcal{R}_{\hat{\mathcal{H}}(\bullet)}\mathcal{G}_k\big(\hat{\mathcal{H}}^C(\bullet)\big)\\&\cong\mathcal{R}\mathcal{G}_k(C)\otimes_{\mathcal{G}_k(C)}\mathcal{G}_k(\mathcal{H}^C,\widehat{\mathfrak{aut}}_{(\mathcal{H}^C)^{\mathrm{nc}}})\end{aligned}\qquad 15.37.1.$$

Proof: Actually, for any $(\mathcal{H}^C)^{^{\mathrm{sc}}}$-chain $\mathfrak{r}\colon\Delta_n\to(\mathcal{H}^C)^{^{\mathrm{sc}}}$ we have

$$\mathcal{R}\mathcal{G}_k(C)\otimes_{\mathcal{G}_k(C)}\mathcal{G}_k\big(\hat{\mathcal{H}}^C(\mathfrak{r})\big)=\begin{cases}\mathcal{R}_{\hat{\mathcal{H}}(\mathfrak{r})}\mathcal{G}_k\big(\hat{\mathcal{H}}^C(\mathfrak{r})\big) & \text{if } C_{\mathfrak{r}}=C\\ \{0\} & \text{otherwise}\end{cases}\qquad 15.37.2$$

since, by definition, the restriction $\mathcal{G}_k\big(\hat{\mathcal{H}}^C(\mathfrak{r})\big)\to\mathcal{G}_k\big(\hat{\mathcal{H}}^{C_{\mathfrak{r}}}(\mathfrak{r})\big)$ is an $\mathcal{O}$-module isomorphism. Moreover, since $\mathcal{R}\mathcal{G}_k(C)$ is a direct factor of $\mathcal{G}_k(C)$, the functor $\mathcal{R}\mathcal{G}_k(C)\otimes_{\mathcal{G}_k(C)}\mathcal{G}_k\big(\hat{\mathcal{H}}^C(\bullet)\big)$ is a direct summand of $\mathfrak{g}_k\circ\widehat{\mathfrak{aut}}_{(\mathcal{H}^C)^{\mathrm{sc}}}$. Consequently, it is quite clear that we have

$$\begin{aligned}&\varprojlim\mathcal{R}\mathcal{G}_k(C)\otimes_{\mathcal{G}_k(C)}\mathcal{G}_k\big(\hat{\mathcal{H}}^C(\bullet)\big)\\&\qquad\cong\mathcal{R}\mathcal{G}_k(C)\otimes_{\mathcal{G}_k(C)}\mathcal{G}_k(\mathcal{H}^C,\widehat{\mathfrak{aut}}_{(\mathcal{H}^C)^{\mathrm{nc}}})\end{aligned}\qquad 15.37.3.$$

But, on the other hand, we clearly have

$$\mathcal{R}\mathcal{G}_k(C) \otimes_{\mathcal{G}_k(C)} \mathcal{G}_k(\mathcal{H}^C, \widehat{\mathfrak{aut}}_{(\mathcal{H}^C)^{\mathrm{nc}}}) \subset \mathcal{R}_{\mathcal{H}}\mathcal{G}_k(\mathcal{H}^C, \widehat{\mathfrak{aut}}_{(\mathcal{H}^C)^{\mathrm{nc}}}) \qquad 15.37.4.$$

Then, the existence of homomorphism 15.36.5 and of equalities 15.37.2 completes the proof.

Proposition 15.38 *In our standard setting, assume that $C_{\hat{G}}(Q,f) \subset \hat{H}$. Any cyclic p'-subgroup C of $\hat{G}/\hat{H}$ acts on $\mathcal{G}_k(\mathcal{H}^D, \widehat{\mathfrak{aut}}_{(\mathcal{H}^D)^{\mathrm{nc}}})$ for any subgroup D of C and, denoting by $\mathfrak{S}_C$ the set of them, the restriction induces an $\mathcal{O}$-module isomorphism*

$$\mathcal{G}_k(\mathcal{H}^C, \widehat{\mathfrak{aut}}_{(\mathcal{H}^C)^{\mathrm{nc}}}) \cong \bigoplus_{D\in\mathfrak{S}_C} \mathcal{R}_{\mathcal{H}}\mathcal{G}_k(\mathcal{H}^D, \widehat{\mathfrak{aut}}_{(\mathcal{H}^D)^{\mathrm{nc}}})^C \qquad 15.38.1.$$

Proof: By the *Frattini argument*, we have $H^C = H{\cdot}N_{H^C}(Q,f)$ and the conjugation by an element $x \in N_{H^C}(Q,f)$ defines an $(\mathcal{H}^D, \mathcal{H}^D)$-functorial automorphism of Q and therefore a Frobenius functor $\mathfrak{f}_x : \mathcal{H}^D \to \mathcal{H}^D$ (cf. 12.1); as in 15.30 above, it follows from the uniqueness part of Theorem 11.32 that x determines a natural isomorphism

$$\hat{\nu}_x : \widehat{\mathfrak{aut}}_{(\mathcal{H}^D)^{\mathrm{sc}}} \cong \widehat{\mathfrak{aut}}_{(\mathcal{H}^D)^{\mathrm{sc}}} \circ \mathfrak{ch}^*(\mathfrak{f}_x) \qquad 15.38.2.$$

Consequently, we get an $\mathcal{O}$-module automorphism

$$\alpha_x : \mathcal{G}_k(\mathcal{H}^D, \widehat{\mathfrak{aut}}_{(\mathcal{H}^D)^{\mathrm{nc}}}) \cong \mathcal{G}_k(\mathcal{H}^D, \widehat{\mathfrak{aut}}_{(\mathcal{H}^D)^{\mathrm{nc}}}) \qquad 15.38.3$$

which coincides with the identity map whenever x belongs to H since, in this case, $\mathfrak{f}_x$ is naturally isomorphic to the identity functor (cf. 12.1 and Proposition A4.5). Since $\alpha_{x'} \circ \alpha_x = \alpha_{x'x}$ for any $x, x' \in N_{H^C}(Q_\delta)$, we obtain an action of $C \cong N_{H^C}(Q_\delta)/N_H(Q_\delta)$ on $\mathcal{G}_k(\mathcal{H}^D, \widehat{\mathfrak{aut}}_{(\mathcal{H}^D)^{\mathrm{nc}}})$.

Moreover, as in 15.23 above, in order to prove isomorphism 15.38.1 it suffices to prove the $\mathcal{O}$-module isomorphism

$$\mathcal{R}\mathcal{G}_k(D) \otimes_{\mathcal{G}_k(C)} \mathcal{G}_k(\mathcal{H}^C, \widehat{\mathfrak{aut}}_{(\mathcal{H}^C)^{\mathrm{nc}}}) \cong \mathcal{R}_{\mathcal{H}}\mathcal{G}_k(\mathcal{H}^D, \widehat{\mathfrak{aut}}_{(\mathcal{H}^D)^{\mathrm{nc}}})^C \qquad 15.38.4$$

for any D in $\mathfrak{S}_C$. But, it follows from Proposition 15.34 that

$$\mathcal{G}_k(\mathcal{H}^C, \widehat{\mathfrak{aut}}_{(\mathcal{H}^C)^{\mathrm{nc}}})^{(C/D)^*} \cong \mathcal{G}_k(\mathcal{H}^D, \widehat{\mathfrak{aut}}_{(\mathcal{H}^D)^{\mathrm{nc}}})^{C/D} \qquad 15.38.5$$

and, since $(C/D)^*$ is a cyclic p'-group and its group $\mathcal{O}$-algebra is isomorphic to $\mathcal{G}_k(C/D)$, it is quite clear that

$$\mathcal{G}_k(\mathcal{H}^C, \widehat{\mathfrak{aut}}_{(\mathcal{H}^C)^{\mathrm{nc}}})^{(C/D)^*} \cong \mathcal{G}_k(D) \otimes_{\mathcal{G}_k(C)} \mathcal{G}_k(\mathcal{H}^C, \widehat{\mathfrak{aut}}_{(\mathcal{H}^C)^{\mathrm{nc}}}) \qquad 15.38.6.$$

On the other hand, since C/D is a p'-group, it follows from Proposition 15.37 that

$$\begin{aligned}\mathcal{R}_{\mathcal{H}}\mathcal{G}_k(\mathcal{H}^D, \widehat{\mathfrak{aut}}_{(\mathcal{H}^D)^{\mathrm{nc}}})^C \\ \cong \mathcal{R}\mathcal{G}_k(D) \otimes_{\mathcal{G}_k(D)} \mathcal{G}_k(\mathcal{H}^D, \widehat{\mathfrak{aut}}_{(\mathcal{H}^D)^{\mathrm{nc}}})^{C/D}\end{aligned} \qquad 15.38.7.$$

Now, isomorphism 15.38.4 follows from isomorphism 15.38.7 and from the isomorphisms obtained applying the functor $\mathcal{R}\mathcal{G}_k(D)\otimes_{\mathcal{G}_k(D)}$ to isomorphisms 15.38.5 and 15.38.6.

Corollary 15.39 *In our standard setting, assume that* $C_{\hat{G}}(Q, f) \subset \hat{H}$. *Then, the restriction induces an* $\mathcal{O}$*-module isomorphism*

$$\mathcal{G}_k(\mathcal{F}, \widehat{\mathfrak{aut}}_{\mathcal{F}^{\mathrm{nc}}}) \cong \Big(\bigoplus_C \mathcal{R}_{\mathcal{H}}\mathcal{G}_k(\mathcal{H}^C, \widehat{\mathfrak{aut}}_{(\mathcal{H}^C)^{\mathrm{nc}}})\Big)^{\mathcal{F}(Q)/\mathcal{H}(Q)} \qquad 15.39.1$$

where C runs over the set of cyclic p'-subgroups of $\mathcal{F}(Q)/\mathcal{H}(Q)$.

Proof: From Proposition 15.38, it is not difficult to compute the *inverse limit* in the right-hand member of isomorphism 15.32 since all the homomorphisms we have to consider are direct sums of either zero homomorphisms or isomorphisms, and then we obtain the right-hand member of isomorphism 15.39.1 above.

15.40 Now, let us "double" our standard setting; that is to say, besides $(b, \hat{G})$ and $(c, \hat{H})$, (P, e) and (Q, f), and $\mathcal{F}$ and $\mathcal{H}$, consider $(b', \hat{G}')$ and $(c', \hat{H}')$, (P', e') and (Q', f'), and $\mathcal{F}'$ and $\mathcal{H}'$ fulfilling the analogous hypothesis. Let C be a cyclic p'-group; we are only interested in the case where

$$C_{\hat{G}}(Q, f) \subset \hat{H} \quad , \quad C_{\hat{G}'}(Q', f') \subset \hat{H}' \quad \text{and} \quad \hat{G}/\hat{H} \cong C \cong \hat{G}'/\hat{H}' \qquad 15.40.1;$$

thus, we already know that $b = c$, $P = Q$, $b' = c'$ and $P' = Q'$ (cf. Proposition 15.10)

15.41 More precisely, we are interested in the *pull-back*

$$\hat{G}'' = \hat{G} \hat{\times}_C \hat{G}' \qquad 15.41.1$$

which contains the *direct sum* of k^*-groups $\hat{H}'' = \hat{H} \hat{\times} \hat{H}'$ (cf. 1.24) as a normal k^*-subgroup; moreover, it is clear that $c'' = c \otimes c'$ is a block of $k_*\hat{H}'' \cong k_*\hat{H} \otimes_k k_*\hat{H}'$ (cf. 1.24 and 1.25), that setting

$$Q'' = Q \times Q' \quad \text{and} \quad f'' = f \otimes f' \qquad 15.41.2$$

(Q'', f'') is a maximal Brauer $(c'', \hat{H}'')$-pair (cf. 1.16), and that we have

$$C_{\hat{G}''}(Q'', f'') \subset \hat{H}'' \qquad 15.41.3.$$

In particular, $(c'', \hat{G}'')$ is also a block (cf. Proposition 15.13), and we consider the Frobenius Q''-categories

$$\mathcal{F}'' = \mathcal{F}_{(c'',\hat{G}'')} \quad \text{and} \quad \mathcal{H}'' = \mathcal{H}_{(c'',\hat{H}'')} \qquad 15.41.4.$$

15.42 First of all, since $k_*(\hat{G} \hat{\times} \hat{G}') \cong k_*\hat{G} \otimes_k k_*\hat{G}'$ (cf. 1.24 and 1.25), we get

$$\mathcal{G}_k(\hat{G} \hat{\times} \hat{G}') \cong \mathcal{G}_k(\hat{G}) \otimes_{\mathcal{O}} \mathcal{G}_k(\hat{G}') \qquad 15.42.1,$$

so that we still get

$$\mathcal{G}_k(\hat{G} \hat{\times} \hat{G}', c'') \cong \mathcal{G}_k(\hat{G}, c) \otimes_{\mathcal{O}} \mathcal{G}_k(\hat{G}', c') \qquad 15.42.2.$$

Thus, since $\hat{G}'' \subset \hat{G} \hat{\times} \hat{G}'$, we have a *restriction* $\mathcal{O}$-module homomorphism

$$\mathrm{res}^{\hat{G} \hat{\times} \hat{G}'}_{\hat{G}''} : \mathcal{G}_k(\hat{G}, c) \otimes_{\mathcal{O}} \mathcal{G}_k(\hat{G}', c') \longrightarrow \mathcal{G}_k(\hat{G}'', c'') \qquad 15.42.3.$$

15.43 On the other hand, it is easily checked that the projection maps

$$Q \longleftarrow Q'' \longrightarrow Q' \qquad 15.43.1$$

are $(\mathcal{F}'', \mathcal{F})$- and $(\mathcal{F}'', \mathcal{F}')$-functorial respectively, so that we have *Frobenius functors*

$$\mathcal{F} \longleftarrow \mathcal{F}'' \longrightarrow \mathcal{F}' \qquad 15.43.2,$$

and therefore we get a *faithful* functor $\mathfrak{i}: \mathcal{F}'' \to \mathcal{F} \times \mathcal{F}'$ (cf. 1.5) which clearly sends $\mathcal{F}''^{\mathrm{sc}}$ to $\mathcal{F}^{\mathrm{sc}} \times \mathcal{F}'^{\mathrm{sc}}$. Then, consider the obvious functor

$$\widehat{\mathfrak{aut}}_{\mathcal{F}^{\mathrm{sc}}} \hat{\times} \widehat{\mathfrak{aut}}_{\mathcal{F}'^{\mathrm{sc}}} : \mathfrak{ch}^*(\mathcal{F}^{\mathrm{sc}} \times \mathcal{F}'^{\mathrm{sc}}) \longrightarrow k^*\text{-}\mathfrak{Gr} \qquad 15.43.3$$

mapping any $\mathcal{F}^{\mathrm{sc}} \times \mathcal{F}'^{\mathrm{sc}}$-chain $\mathfrak{r} \times \mathfrak{r}' : \Delta_n \to \mathcal{F}^{\mathrm{sc}} \times \mathcal{F}'^{\mathrm{sc}}$ on $\hat{\mathcal{F}}(\mathfrak{r}) \hat{\times} \hat{\mathcal{F}}'(\mathfrak{r}')$, where $\mathfrak{r}$ and $\mathfrak{r}'$ are respective $\mathcal{F}^{\mathrm{sc}}$- and $\mathcal{F}'^{\mathrm{sc}}$-chains; since for any $\mathcal{F}''^{\mathrm{sc}}$-chain $\mathfrak{r}''$ which $\mathfrak{i}$ maps on $\mathfrak{r} \times r'$ we have an evident injective k^*-group homomorphism

$$\hat{\mathcal{F}}''(\mathfrak{r}'') \longrightarrow \hat{\mathcal{F}}(\mathfrak{r}) \hat{\times} \hat{\mathcal{F}}'(\mathfrak{r}') \qquad 15.43.4,$$

it follows from Proposition 15.12 that we have an injective natural map

$$\hat{\nu}^{\mathcal{F} \times \mathcal{F}'}_{\mathcal{F}''} : \widehat{\mathfrak{aut}}_{\mathcal{F}''^{\mathrm{sc}}} \longrightarrow (\widehat{\mathfrak{aut}}_{\mathcal{F}^{\mathrm{sc}}} \hat{\times} \widehat{\mathfrak{aut}}_{\mathcal{F}'^{\mathrm{sc}}}) \circ \mathfrak{ch}^*(\mathfrak{i}) \qquad 15.43.5.$$

15.44 Moreover, it is clear that the structural functors from $\mathcal{F}^{^{\mathrm{sc}}} \times \mathcal{F}'^{^{\mathrm{sc}}}$ to $\mathcal{F}^{^{\mathrm{sc}}}$ and to $\mathcal{F}'^{^{\mathrm{sc}}}$ determine a functor

$$\mathfrak{ch}^*(\mathcal{F}^{^{\mathrm{sc}}} \times \mathcal{F}'^{^{\mathrm{sc}}}) \longrightarrow \mathfrak{ch}^*(\mathcal{F}^{^{\mathrm{sc}}}) \times \mathfrak{ch}^*(\mathcal{F}'^{^{\mathrm{sc}}}) \qquad 15.44.1$$

and that the functor $\widehat{\mathfrak{aut}}_{\mathcal{F}^{\mathrm{sc}}} \hat{\times} \widehat{\mathfrak{aut}}_{\mathcal{F}'^{\mathrm{sc}}}$ above can be extended to the direct product $\mathfrak{ch}^*(\mathcal{F}^{^{\mathrm{sc}}}) \times \mathfrak{ch}^*(\mathcal{F}'^{^{\mathrm{sc}}})$ (cf. 1.5) mapping an object $(\mathfrak{r}, \Delta_n) \times (\mathfrak{r}', \Delta_{n'})$ on the k^*-group $\hat{\mathcal{F}}(\mathfrak{r}) \hat{\times} \hat{\mathcal{F}}'(\mathfrak{r}')$; thus, since we already know that (cf. 15.42.1)

$$\mathcal{G}_k\big(\hat{\mathcal{F}}(\mathfrak{r}) \hat{\times} \hat{\mathcal{F}}'(\mathfrak{r}')\big) \cong \mathcal{G}_k\big(\hat{\mathcal{F}}(\mathfrak{r})\big) \otimes_{\mathcal{O}} \mathcal{G}_k\big(\hat{\mathcal{F}}'(\mathfrak{r}')\big) \qquad 15.44.2,$$

it is quite clear that we get an $\mathcal{O}$-module homomorphism

$$\begin{array}{c} \lim\limits_{\longleftarrow}\big(\mathfrak{g}_k \circ (\widehat{\mathfrak{aut}}_{\mathcal{F}^{\mathrm{sc}}} \hat{\times} \widehat{\mathfrak{aut}}_{\mathcal{F}'^{\mathrm{sc}}})\big) \\ \uparrow \\ \mathcal{G}_k(\mathcal{F}, \widehat{\mathfrak{aut}}_{\mathcal{F}^{\mathrm{nc}}}) \otimes_{\mathcal{O}} \mathcal{G}_k(\mathcal{F}', \widehat{\mathfrak{aut}}_{\mathcal{F}'^{\mathrm{nc}}}) \end{array} \qquad 15.44.3.$$

15.45 But, we have the canonical $\mathcal{O}$-module homomorphism (cf. A3.9)

$$\begin{array}{c} \lim\limits_{\longleftarrow}\big(\mathfrak{g}_k \circ (\widehat{\mathfrak{aut}}_{\mathcal{F}^{\mathrm{sc}}} \hat{\times} \widehat{\mathfrak{aut}}_{\mathcal{F}'^{\mathrm{sc}}}) \circ \mathfrak{ch}^*(\mathfrak{i})\big) \\ \uparrow \\ \lim\limits_{\longleftarrow}\big(\mathfrak{g}_k \circ (\widehat{\mathfrak{aut}}_{\mathcal{F}^{\mathrm{sc}}} \hat{\times} \widehat{\mathfrak{aut}}_{\mathcal{F}'^{\mathrm{sc}}})\big) \end{array} \qquad 15.45.1$$

and the natural map $\hat{\nu}_{\mathcal{F}''}^{\mathcal{F}\times\mathcal{F}'}$ above determines an $\mathcal{O}$-module homomorphism

$$\lim_{\longleftarrow}\big(\mathfrak{g}_k \circ (\widehat{\mathfrak{aut}}_{\mathcal{F}^{\mathrm{sc}}} \hat{\times} \widehat{\mathfrak{aut}}_{\mathcal{F}'^{\mathrm{sc}}}) \circ \mathfrak{ch}^*(\mathfrak{i})\big) \longrightarrow \mathcal{G}_k(\mathcal{F}'', \widehat{\mathfrak{aut}}_{\mathcal{F}''^{\mathrm{nc}}}) \qquad 15.45.2.$$

Consequently, by composition, we get the $\mathcal{O}$-module homomorphism

$$\mathcal{G}_k(\mathcal{F}, \widehat{\mathfrak{aut}}_{\mathcal{F}^{\mathrm{nc}}}) \otimes_{\mathcal{O}} \mathcal{G}_k(\mathcal{F}', \widehat{\mathfrak{aut}}_{\mathcal{F}'^{\mathrm{nc}}}) \longrightarrow \mathcal{G}_k(\mathcal{F}'', \widehat{\mathfrak{aut}}_{\mathcal{F}''^{\mathrm{nc}}}) \qquad 15.45.3$$

that we denote by $\mathrm{res}_{\mathcal{F}''}^{\mathcal{F}\times\mathcal{F}'}$. Moreover, denoting by $\mathfrak{S}_{\mathcal{H}}$ and $\mathfrak{S}_{\mathcal{H}'}$ the respective sets of $\mathcal{H}$- and $\mathcal{H}'$-selfcentralizing subgroups of Q and Q' , it follows from Lemma 15.14 that the structural maps determine injective $\mathcal{O}$-module homomorphisms

$$\begin{array}{l} \mathcal{G}_k(\mathcal{F}, \widehat{\mathfrak{aut}}_{\mathcal{F}^{\mathrm{nc}}}) \longrightarrow \prod\limits_{R\in\mathfrak{S}_{\mathcal{H}}} \mathcal{G}_k\big(\hat{\mathcal{F}}(R)\big) \\ \mathcal{G}_k(\mathcal{F}', \widehat{\mathfrak{aut}}_{\mathcal{F}'^{\mathrm{nc}}}) \longrightarrow \prod\limits_{R'\in\mathfrak{S}_{\mathcal{H}'}} \mathcal{G}_k\big(\hat{\mathcal{F}}'(R')\big) \\ \mathcal{G}_k(\mathcal{F}'', \widehat{\mathfrak{aut}}_{\mathcal{F}''^{\mathrm{nc}}}) \longrightarrow \prod\limits_{(R,R')\in\mathfrak{S}_{\mathcal{H}}\times\mathfrak{S}_{\mathcal{H}'}} \mathcal{G}_k\big(\hat{\mathcal{F}}''(R\times R')\big) \end{array} \qquad 15.45.4.$$

Theorem 15.46 *With the notation and the hypothesis above, assume that C is trivial. Then $\mathrm{res}^{\mathcal{F}\times\mathcal{F}'}_{\mathcal{F}''}$ induces an $\mathcal{O}$-module isomorphism*

$$\mathcal{G}_k(\mathcal{F},\widehat{\mathfrak{aut}}_{\mathcal{F}^{\mathrm{nc}}})\otimes_{\mathcal{O}}\mathcal{G}_k(\mathcal{F}',\widehat{\mathfrak{aut}}_{(\mathcal{F}')^{\mathrm{nc}}})\cong\mathcal{G}_k(\mathcal{F}'',\widehat{\mathfrak{aut}}_{(\mathcal{F}'')^{\mathrm{nc}}}) \qquad 15.46.1.$$

Proof: If $C=\{1\}$ then, for any $R\in\mathfrak{S}_{\mathcal{H}}$ and any $R'\in\mathfrak{S}_{\mathcal{H}'}$, we have

$$\hat{\mathcal{F}}''(R\times R')\cong\hat{\mathcal{F}}(R)\,\hat{\times}\,\hat{\mathcal{F}}'(R') \qquad 15.46.2;$$

thus, according to isomorphism 15.42.1 and to the injectivity in 15.45.4, it is quite clear that we get the following commutative diagram with injective vertical arrows

$$\begin{array}{ccc}\prod\limits_{(R,R')}\mathcal{G}_k(\hat{\mathcal{F}}(R))\otimes_{\mathcal{O}}\mathcal{G}_k(\mathcal{F}'(R')) & \cong & \prod\limits_{(R,R')}\mathcal{G}_k(\hat{\mathcal{F}}''(R\times R'))\\ \uparrow & & \uparrow\\ \mathcal{G}_k(\mathcal{F},\widehat{\mathfrak{aut}}_{\mathcal{F}^{\mathrm{nc}}})\otimes_{\mathcal{O}}\mathcal{G}_k(\mathcal{F}',\widehat{\mathfrak{aut}}_{\mathcal{F}'^{\mathrm{nc}}}) & \longrightarrow & \mathcal{G}_k(\mathcal{F}'',\widehat{\mathfrak{aut}}_{\mathcal{F}''^{\mathrm{nc}}})\end{array} \qquad 15.46.3,$$

where (R,R') runs over $(\mathfrak{S}_{\mathcal{H}})^C\times(\mathfrak{S}_{\mathcal{H}'})^C$.

Moreover, according to Remark 14.29, the three k-vector spaces

$$\begin{gathered}k\otimes_{\mathcal{O}}\mathcal{G}_k(\mathcal{F},\widehat{\mathfrak{aut}}_{\mathcal{F}^{\mathrm{nc}}})\\ k\otimes_{\mathcal{O}}\mathcal{G}_k(\mathcal{F}',\widehat{\mathfrak{aut}}_{\mathcal{F}'^{\mathrm{nc}}})\quad\text{and}\quad k\otimes_{\mathcal{O}}\mathcal{G}_k(\mathcal{F}'',\widehat{\mathfrak{aut}}_{\mathcal{F}''^{\mathrm{nc}}})\end{gathered} \qquad 15.46.4$$

still coincide with the corresponding *inverse limits* and therefore, applying the functor $k\otimes_{\mathcal{O}}\bullet={}^{k\otimes}\bullet$ to diagram 15.46.3, the vertical arrows are injective too; hence, the bottom arrow yields an injective k-linear homomorphism

$${}^{k\otimes}\mathcal{G}_k(\mathcal{F},\widehat{\mathfrak{aut}}_{\mathcal{F}^{\mathrm{nc}}})\otimes_k{}^{k\otimes}\mathcal{G}_k(\mathcal{F}',\widehat{\mathfrak{aut}}_{\mathcal{F}'^{\mathrm{nc}}})\longrightarrow{}^{k\otimes}\mathcal{G}_k(\mathcal{F}'',\widehat{\mathfrak{aut}}_{\mathcal{F}''^{\mathrm{nc}}}) \qquad 15.46.5.$$

Consequently, in order to prove isomorphism 15.46.1, it suffices to prove that the $\mathcal{O}$-ranks of both sides coincide. But, according to Corollary 14.32, we have

$$\begin{aligned}\mathrm{rank}_{\mathcal{O}}\big(\mathcal{G}_k(\mathcal{F},\widehat{\mathfrak{aut}}_{\mathcal{F}^{\mathrm{nc}}})\big)&=\sum_{(\mathfrak{r},\Delta_n)}(-1)^n\,\mathrm{rank}_{\mathcal{O}}\Big(\mathcal{G}_k\big(\hat{\mathcal{F}}(\mathfrak{r})\big)\Big)\\ \mathrm{rank}_{\mathcal{O}}\big(\mathcal{G}_k(\mathcal{F}',\widehat{\mathfrak{aut}}_{\mathcal{F}'^{\mathrm{nc}}})\big)&=\sum_{(\mathfrak{r}',\Delta_{n'})}(-1)^{n'}\,\mathrm{rank}_{\mathcal{O}}\Big(\mathcal{G}_k\big(\hat{\mathcal{F}}'(\mathfrak{r}')\big)\Big)\end{aligned} \qquad 15.46.6$$

where $(\mathfrak{r},\Delta_n)$ and $(\mathfrak{r}',\Delta_{n'})$ respectively run over sets of representatives for the isomorphism classes of *regular* $\mathfrak{ch}^*(\mathcal{F}^{^{\mathrm{sc}}})$- and $\mathfrak{ch}^*(\mathcal{F}'^{^{\mathrm{sc}}})$-objects (cf. A5.2);

actually, we may assume that all the morphisms $\mathfrak{r}(j \bullet i)$ and $\mathfrak{r}'(j' \bullet i')$ are inclusion maps. Thus, we still have (cf. 15.42.1)

$$\begin{aligned}
&\mathrm{rank}_{\mathcal{O}}\big(\mathcal{G}_k(\mathcal{F}, \widehat{\mathfrak{aut}}_{\mathcal{F}^{\mathrm{nc}}}) \otimes_{\mathcal{O}} \mathcal{G}_k(\mathcal{F}', \widehat{\mathfrak{aut}}_{\mathcal{F}'^{\mathrm{nc}}})\big) \\
&\quad = \sum_{(\mathfrak{r},\Delta_n)} \sum_{(\mathfrak{r}',\Delta_{n'})} (-1)^{n+n'} \mathrm{rank}_{\mathcal{O}}\Big(\mathcal{G}_k\big(\hat{\mathcal{F}}(\mathfrak{r})\big) \otimes_{\mathcal{O}} \mathcal{G}_k\big(\hat{\mathcal{F}}'(\mathfrak{r}')\big)\Big) \\
&\quad = \sum_{(\mathfrak{r},\Delta_n)} \sum_{(\mathfrak{r}',\Delta_{n'})} (-1)^{n+n'} \mathrm{rank}_{\mathcal{O}}\Big(\mathcal{G}_k\big(\hat{\mathcal{F}}(\mathfrak{r}) \hat{\times} \hat{\mathcal{F}}'(\mathfrak{r}')\big)\Big)
\end{aligned} \qquad 15.46.7$$

where $(\mathfrak{r}, \Delta_n)$ and $(\mathfrak{r}', \Delta_{n'})$ run over the same sets of representatives.

On the other hand, from any regular $\mathfrak{ch}^*(\mathcal{F}^{^{\mathrm{sc}}})$- and $\mathfrak{ch}^*(\mathcal{F}'^{^{\mathrm{sc}}})$-objects $(\mathfrak{r}, \Delta_n)$ and $(\mathfrak{r}', \Delta_{n'})$, we have the regular $\mathfrak{ch}^*(\mathcal{F}''^{^{\mathrm{sc}}})$-chain $\mathfrak{r}*\mathfrak{r}' : \Delta_{n+n'} \to \mathcal{F}''^{^{\mathrm{sc}}}$ mapping any $i \in \Delta_n$ on $\mathfrak{r}(i) \times \mathfrak{r}'(0)$, any $n + i' \in \Delta_{n+n'}$ on $\mathfrak{r}(n) \times \mathfrak{r}'(i')$, any Δ_n-morphism $j\bullet i$ on $\mathfrak{r}(j\bullet i) \times \mathrm{id}_{\mathfrak{r}'(0)}$ and any $\Delta_{n+n'}$-morphism $(n+j')\bullet (n+i')$ on $\mathrm{id}_{\mathfrak{r}(n)} \times \mathfrak{r}'(j' \bullet i')$; then, it is easily checked that we have

$$\hat{\mathcal{F}}(\mathfrak{r}) \hat{\times} \hat{\mathcal{F}}'(\mathfrak{r}') \cong \hat{\mathcal{F}}''(\mathfrak{r} * \mathfrak{r}') \qquad 15.46.8$$

and therefore, in order to prove isomorphism 15.46.1 it suffices to prove that we still have

$$\begin{aligned}
&\mathrm{rank}_{\mathcal{O}}\big(\mathcal{G}_k(\mathcal{F}'', \widehat{\mathfrak{aut}}_{\mathcal{F}''^{\mathrm{nc}}})\big) \\
&\quad = \sum_{(\mathfrak{r},\Delta_n)} \sum_{(\mathfrak{r}',\Delta_{n'})} (-1)^{n+n'} \mathrm{rank}_{\mathcal{O}}\Big(\mathcal{G}_k\big(\hat{\mathcal{F}}''(\mathfrak{r} * \mathfrak{r}')\big)\Big)
\end{aligned} \qquad 15.46.9$$

where $(\mathfrak{r}, \Delta_n)$ and $(\mathfrak{r}', \Delta_{n'})$ run over the same sets of representatives.

But, it is clear that any regular $\mathfrak{ch}^*(\mathcal{F}''^{^{\mathrm{sc}}})$-chain $\mathfrak{r}'' : \Delta_{n''} \to \mathcal{F}''^{^{\mathrm{sc}}}$ such that all the morphisms $\mathfrak{r}''(j'' \bullet i'')$ are inclusion maps, composed with both functors 15.43.2 above determine regular $\mathfrak{ch}^*(\mathcal{F}^{^{\mathrm{sc}}})$- and $\mathfrak{ch}^*(\mathcal{F}'^{^{\mathrm{sc}}})$-chains

$$\mathfrak{r} : \Delta_n \longrightarrow \mathcal{F}^{^{\mathrm{sc}}} \quad \text{and} \quad \mathfrak{r}' : \Delta_{n'} \longrightarrow \mathcal{F}'^{^{\mathrm{sc}}} \qquad 15.46.10$$

where once again all the morphisms $\mathfrak{r}(j \bullet i)$ and $\mathfrak{r}'(j' \bullet i')$ are inclusion maps. Consequently, it suffices to prove that, for any such regular $\mathfrak{ch}^*(\mathcal{F}^{^{\mathrm{sc}}})$- and $\mathfrak{ch}^*(\mathcal{F}'^{^{\mathrm{sc}}})$-chains $\mathfrak{r}$ and $\mathfrak{r}'$, we have

$$\begin{aligned}
&\sum_{(\mathfrak{r}'',\Delta_{n''})} (-1)^{n''} \mathrm{rank}_{\mathcal{O}}\Big(\mathcal{G}_k\big(\hat{\mathcal{F}}''(\mathfrak{r}'')\big)\Big) \\
&\qquad\qquad = (-1)^{n+n'} \mathrm{rank}_{\mathcal{O}}\Big(\mathcal{G}_k\big(\hat{\mathcal{F}}''(\mathfrak{r} * \mathfrak{r}')\big)\Big)
\end{aligned} \qquad 15.46.11$$

where $(\mathfrak{r}'', \Delta_{n''})$ runs over a set of representatives $\mathfrak{R}''$ for the isomorphism classes of $\mathfrak{ch}^*(\mathcal{F}''^{^{\mathrm{sc}}})$-objects such that $\mathfrak{r}''$ is a regular $\mathfrak{ch}^*(\mathcal{F}''^{^{\mathrm{sc}}})$-chain, where all the morphisms $\mathfrak{r}''(j'' \bullet i'')$ are inclusion maps, which maps on the pair $(\mathfrak{r}, \mathfrak{r}')$.

First of all, let us prove that

$$\sum_{(\mathfrak{r}'',\Delta_{n''})} (-1)^{n''} \mathrm{rank}_{\mathcal{O}}\Big(\mathcal{G}_k\big(\hat{\mathcal{F}}''(\mathfrak{r}'')\big)\Big) = 0 \qquad 15.46.12$$

when $(\mathfrak{r}'', \Delta_{n''}) \in \mathfrak{R}''$ runs over the set of elements such that $\mathfrak{r}''(i'')$ is *not* a direct product for some $i'' \in \Delta_{n''}$; it suffices to exhibit an involutive permutation τ of this set such that if $\tau(\mathfrak{r}'', \Delta_{n''}) = (\mathfrak{t}'', \Delta_{m''})$ then we have

$$\hat{\mathcal{F}}''(\mathfrak{t}'') \cong \hat{\mathcal{F}}''(\mathfrak{r}'') \quad \text{and} \quad (-1)^{m''} = -(-1)^{n''} \qquad 15.46.13.$$

Consider the biggest $i'' \in \Delta_{n''}$ such that $\mathfrak{r}''(i'')$ it is *not* a direct product and denote by R and R' the respective images of $\mathfrak{r}''(i'')$ in Q and Q' by the projection maps 15.43.1; if $i'' \neq n''$ and $\mathfrak{r}''(i''+1) = R \times R'$ then we set

$$m'' = n'' - 1 \quad \text{and} \quad \mathfrak{t}'' = \mathfrak{r}'' \circ \delta^{n''-1}_{i''+1} \qquad 15.46.14;$$

otherwise, if either $i'' = n''$ or $i'' \neq n''$ and $\mathfrak{r}''(i''+1) \neq R \times R'$ then we set $m'' = n'' + 1$ and denote by $\mathfrak{t}'' : \Delta_{n''} \longrightarrow \mathcal{F}''^{\mathrm{sc}}$ the regular $\mathfrak{ch}^*(\mathcal{F}''^{\mathrm{sc}})$-chain, with inclusion maps, fulfilling

$$\mathfrak{t}'' \circ \delta^{n''}_{i''+1} = \mathfrak{r}'' \quad \text{and} \quad \mathfrak{t}''(i''+1) = R \times R' \qquad 15.46.15.$$

Since $i'' \in \Delta_{m''}$ is also the biggest element such that $\mathfrak{t}''(i'')$ it is *not* a direct product, it is easily checked that $\tau(\mathfrak{t}'', \Delta_{m''}) = (\mathfrak{r}'', \Delta_{n''})$; moreover, it is quite clear that we have $\hat{\mathcal{F}}''(\mathfrak{t}'') \cong \hat{\mathcal{F}}''(\mathfrak{r}'')$.

Now, in order to prove equality 15.46.11, it suffices to exhibit again an involutive permutation — still denoted by τ — fulfilling condition 15.46.13, over the set of elements $(\mathfrak{r}'', \Delta_{n''}) \in \mathfrak{R}''$ such that $\mathfrak{r}''(i'')$ is a direct product for any $i'' \in \Delta_{n''}$, whereas $\mathfrak{r}''$ does not coincide with $\mathfrak{r} * \mathfrak{r}'$. Consider the smallest $i'' \in \Delta_{n''}$ such that we have

$$\mathfrak{r}''(i'') = \mathfrak{r}(n) \times \mathfrak{r}'(i') \qquad 15.46.16$$

for some $i' \in \Delta_{n'}$, which actually forces $i' = n' - n'' + i''$; after, consider the smallest $j'' \leq i''$ fulfilling $\mathfrak{r}''(j'') = \mathfrak{r}(i) \times \mathfrak{r}'(i')$ for some $i \in \Delta_n$, which similarly forces $i = n - i'' + j''$.

Note that if $j'' = 0$ then $i = 0 = i'$ and $\mathfrak{r}'' = \mathfrak{r} * \mathfrak{r}'$, so that we get $j'' \neq 0$; moreover, $\mathfrak{r}''(j''-1)$ is equal either to $\mathfrak{r}(i) \times \mathfrak{r}'(i'-1)$ or to $\mathfrak{r}(i-1) \times \mathfrak{r}'(i'-1)$; in the first case, we set

$$m'' = n'' - 1 \quad \text{and} \quad \mathfrak{t}'' = \mathfrak{r}'' \circ \delta^{n''-1}_{j''} \qquad 15.46.17;$$

in the second case, we set $m'' = n''+1$ and denote by $\mathfrak{t}'' : \Delta_{n''} \longrightarrow \mathcal{F}''^{\mathrm{sc}}$ the regular $\mathfrak{ch}^*(\mathcal{F}''^{\mathrm{sc}})$-chain, with inclusion maps, fulfilling

$$\mathfrak{t}'' \circ \delta^{n''}_{j''} = \mathfrak{r}'' \quad \text{and} \quad \mathfrak{t}''(j'') = \mathfrak{r}(i) \times \mathfrak{r}'(i'-1) \qquad 15.46.18.$$

Then, it is easily checked that $i''+m''-n'' \in \Delta_{m''}$ is the smallest element such that we have

$$\mathfrak{t}''(i''+m''-n'') = \mathfrak{r}(n) \times \mathfrak{r}'(i') \qquad 15.46.19$$

for some $i' \in \Delta_{n'}$ which coincides with the corresponding element above, whereas the same j'' — which actually fulfills $j'' \le i''+m''-n''$ — is the smallest element such that $\mathfrak{t}''(j'') = \mathfrak{r}(j) \times \mathfrak{r}'(i')$ for some $j \in \Delta_n$ and this time we get

$$j = n - i'' - m'' + n'' + j'' = i - m'' + n'' \qquad 15.46.20.$$

Thus, we still get

$$\mathfrak{t}''(j''-1) = \begin{cases} \mathfrak{r}(i)\times\mathfrak{r}'(i'-1) = \mathfrak{r}(j-1)\times\mathfrak{r}'(i'-1) & \text{if } m''=n''-1 \\ \\ \mathfrak{r}(i-1)\times\mathfrak{r}'(i'-1) = \mathfrak{r}(j)\times\mathfrak{r}'(i'-1) & \text{if } m''=n''+1 \end{cases} \qquad 15.46.21$$

and then it is easily checked that $\tau(\mathfrak{t}'', \Delta_{m''}) = (\mathfrak{r}'', \Delta_{n''})$. We are done.

Corollary 15.47 *With the notation and the hypothesis above, the $\mathcal{O}$-module homomorphisms* $\mathrm{res}_{\hat G''}^{\hat G \hat\times \hat G'}$ *and* $\mathrm{res}_{\mathcal{F}''}^{\mathcal{F}\times\mathcal{F}'}$ *induce $\mathcal{O}$-module isomorphisms*

$$\begin{aligned} \mathcal{R}_{\hat H}\mathcal{G}_k(\hat G, c) \otimes_{\mathcal{R}\mathcal{G}_k(C)} \mathcal{R}_{\hat H'}\mathcal{G}_k(\hat G', c') \\ \cong \mathcal{R}_{\hat H''}\mathcal{G}_k(\hat G'', c'') \\ \mathcal{R}_{\mathcal{H}}\mathcal{G}_k(\mathcal{F}, \widehat{\mathfrak{aut}}_{\mathcal{F}^{\mathrm{nc}}}) \otimes_{\mathcal{R}\mathcal{G}_k(C)} \mathcal{R}_{\mathcal{H}'}\mathcal{G}_k(\mathcal{F}', \widehat{\mathfrak{aut}}_{(\mathcal{F}')^{\mathrm{nc}}}) \\ \cong \mathcal{R}_{\mathcal{H}''}\mathcal{G}_k(\mathcal{F}'', \widehat{\mathfrak{aut}}_{(\mathcal{F}'')^{\mathrm{nc}}}) \end{aligned} \qquad 15.47.1.$$

Proof: Setting $\nabla(C) = (C\times C)/\Delta(C)$ where $\Delta(C)$ is the *diagonal* subgroup, it is clear that $\nabla(C)$ is canonically isomorphic to the quotient $(\hat G \hat\times \hat G')/\hat G''$, so that we have $\hat G \hat\times \hat G' = \hat G''^{\nabla(C)}$ according to the notation above; hence, it follows from isomorphism 15.21.2 that

$$\mathcal{G}_k(\hat G \hat\times \hat G')^{\nabla(C)^*} \cong \mathcal{G}_k(\hat G'')^{\nabla(C)} \qquad 15.47.2;$$

moreover, up to suitable identifications, $\nabla(C)^*$ coincides with $\Delta(C^*)$ and therefore, since they are p'-groups and we have $\mathcal{O}C^* \cong \mathcal{G}_k(C)$, we get the $\mathcal{O}$-module isomorphism

$$\mathcal{G}_k(\hat G) \otimes_{\mathcal{G}_k(C)} \mathcal{G}_k(\hat G') \cong \mathcal{G}_k(\hat G \hat\times \hat G')^{\nabla(C)^*} \qquad 15.47.3.$$

On the other hand, for any simple $k_*\hat G$- and $k_*\hat G'$-modules M and M' which remain simple restricted to $\hat H$ and to $\hat H'$ respectively, the composed homomorphism from $\mathcal{G}_k(\hat G)\otimes_{\mathcal{G}_k(C)}\mathcal{G}_k(\hat G')$ to $\mathcal{G}_k(\hat G'')$ maps the tensor product

$X\otimes X'$ of their respective isomorphism classes X and X', on the isomorphism class of the restriction of $M\otimes_k M'$ to $k_*\hat G''$, which remains simple restricted to $\hat H''$. Moreover, in this form we obtain *all* the simple $k_*\hat G''$-modules which remain simple restricted to $\hat H''$ and this correspondence is compatible with the blocks. At this point, since isomorphisms 15.47.2 and 15.47.3 are actually $\mathcal{G}_k(C)$-module isomorphisms, the top isomorphism in 15.47.1 follows from the equality in 15.23.2.

In order to prove the bottom isomorphism, let us denote by $\mathcal{F}\otimes\mathcal{F}'$ the Frobenius Q''-category associated with the block $(c'',\hat G\,\hat\times\,\hat G')$; it is clear that the Frobenius Q''-category $\mathcal{F}''$ is a normal subcategory of $\mathcal{F}\otimes\mathcal{F}'$, and in fact we have $\mathcal{F}\otimes\mathcal{F}' = \mathcal{F}''^{\nabla(C)}$ (cf. Proposition 12.12) since $\nabla(C)$ is canonically isomorphic to $(\mathcal{F}\otimes\mathcal{F}')(Q'')/\mathcal{F}''(Q'')$; moreover, it follows from Theorem 15.46 that we have the canonical isomorphism

$$\mathcal{G}_k(\mathcal{F},\widehat{\mathfrak{aut}}_{\mathcal{F}^{\mathrm{nc}}})\otimes_{\mathcal{O}}\mathcal{G}_k(\mathcal{F}',\widehat{\mathfrak{aut}}_{(\mathcal{F}')^{\mathrm{nc}}})\cong\mathcal{G}_k(\mathcal{F}\otimes\mathcal{F}',\widehat{\mathfrak{aut}}_{(\mathcal{F}\otimes\mathcal{F}')^{\mathrm{nc}}}) \qquad 15.47.4.$$

Consequently, replacing isomorphism 15.21.2 by isomorphism 15.34.1, we get as in 15.47.2 and 15.47.3 above the $\mathcal{O}$-module isomorphism

$$\mathcal{G}_k(\mathcal{F},\widehat{\mathfrak{aut}}_{\mathcal{F}^{\mathrm{nc}}})\otimes_{\mathcal{G}_k(C)}\mathcal{G}_k(\mathcal{F}',\widehat{\mathfrak{aut}}_{(\mathcal{F}')^{\mathrm{nc}}})\cong\mathcal{G}_k(\mathcal{F}'',\widehat{\mathfrak{aut}}_{\mathcal{F}''^{\mathrm{nc}}})^{\nabla(C)} \qquad 15.47.5.$$

On the other hand, it follows from Proposition 15.37 that, as in 15.45.4 above, we have an injective $\mathcal{O}$-module homomorphism

$$\mathcal{R}_{\mathcal{H}''}\mathcal{G}_k(\mathcal{F}'',\widehat{\mathfrak{aut}}_{\mathcal{F}''^{\mathrm{nc}}})\longrightarrow\prod_{(R,R')}\mathcal{R}_{\hat{\mathcal{H}}''(R\times R')}\mathcal{G}_k\big(\hat{\mathcal{F}}''(R\times R')\big) \qquad 15.47.6$$

where R and R' respectively run over the subsets $\mathfrak{S}^C_{\mathcal{H}}\subset\mathfrak{S}_{\mathcal{H}}$ and $\mathfrak{S}^C_{\mathcal{H}'}\subset\mathfrak{S}_{\mathcal{H}'}$ of elements fulfilling $C_R = C = C_{R'}$ (cf. 15.36); but, according to isomorphisms 15.47.2 and 15.47.3 above, $\nabla(C)$ acts trivially on each right-hand factor $\mathcal{R}_{\hat{\mathcal{H}}''(R\times R')}\mathcal{G}_k\big(\hat{\mathcal{F}}''(R\times R')\big)$; hence, $\nabla(C)$ still acts trivially on the left-hand term $\mathcal{R}_{\mathcal{H}''}\mathcal{G}_k(\mathcal{F}'',\widehat{\mathfrak{aut}}_{\mathcal{F}''^{\mathrm{nc}}})$ and therefore it is quite clear that

$$\mathcal{R}_{\mathcal{H}''}\mathcal{G}_k(\mathcal{F}'',\widehat{\mathfrak{aut}}_{\mathcal{F}''^{\mathrm{nc}}})\cong\mathcal{R}\mathcal{G}_k(C)\otimes_{\mathcal{G}_k(C)}\mathcal{G}_k(\mathcal{F}'',\widehat{\mathfrak{aut}}_{\mathcal{F}''^{\mathrm{nc}}})^{\nabla(C)} \qquad 15.47.7;$$

thus, according to Proposition 15.37, applying the functor $\mathcal{R}\mathcal{G}_k(C)\otimes_{\mathcal{G}_k(C)}\bullet$ to isomorphism 15.47.5, we are done.

15.48 Besides our standard setting, we need reduction results on the *quotients of k^*_groups by central p-subgroups*. More generally, let Z be a normal Abelian p-subgroup of $\hat G$ such that the centralizer $\hat H = C_{\hat G}(Z)$ contains a Sylow p-subgroup of $\hat G$, so that the quotient $\hat G/\hat H$ is a p'-group; it is clear that $\hat{\bar G} = \hat G/Z$ remains a k^*-group with k^*-quotient $\bar G = G/Z$ and, for any p-subgroup R of $\hat{G}$, we denote by $\bar{R}$ its image in $\hat{\bar{G}}$; recall that we set $\bar{C}_{\hat{G}}(Z) = C_{\hat{G}}(Z)/Z$ (cf. 1.9.2) and coherently we also set $\hat{\bar{H}} = \hat{H}/Z$. Then, we claim that the canonical H-algebra homomorphism $k_*\hat{H} \to k_*\hat{\bar{H}}$ is a *strict semicovering* (cf. 1.11); indeed, for any p-subgroup R of $\hat{H}$, any p'-subgroup of $C_{\hat{\bar{H}}}(\bar{R})$ can be lifted to a p'-subgroup of $C_{\hat{H}}(R)$ (cf. Theorem 3.2 in [28, Ch. 5]) and therefore the kernel and the cokernel of the canonical k^*-group homomorphism (cf. 1.19.1)

$$C_{\hat{H}}(R) \longrightarrow C_{\hat{\bar{H}}}(\bar{R}) \qquad 15.48.1$$

are both p-groups. Hence, if follows from Example 3.9 in [34] that the corresponding k^*-group algebra homomorphism is a *strict semicovering* of k-algebras; that is to say, for any p-subgroup R of $\hat{H}$, the canonical k-algebra homomorphism

$$(k_*\hat{H})(R) \longrightarrow (k_*\hat{\bar{H}})(\bar{R}) \qquad 15.48.2$$

is a *strict semicovering*; now, the claim follows from Theorem 3.16 in [34].

Lemma 15.49 *In our standard setting, let Z be a normal Abelian p-subgroup of $\hat{G}$ and assume that $\hat{H} = C_{\hat{G}}(Z)$ and that $\hat{H}$ contains a Sylow p-subgroup of $\hat{G}$. For any local pointed group R_ε on $k_*\hat{G}$ such that $Z \subset R$, $\varepsilon \cap k_*\hat{H}$ is a local point of R on $k_*\hat{H}$ and the image of $\varepsilon \cap k_*\hat{H}$ in $k_*\hat{\bar{H}}$ is contained in a local point $\bar{\varepsilon}$ of $\bar{R}$ over $k_*\hat{\bar{H}}$ which splits into a set $\{(\bar{\varepsilon}, \psi)\}_{\psi \in \mathcal{P}(k_* C^{\hat{\bar{G}}}_{\hat{\bar{H}}}(\bar{R}_{\bar{\varepsilon}}))}$ of local points of $\bar{R}$ on $k_*\hat{\bar{G}}$. Moreover, the Brauer $\hat{\bar{G}}$-pair determined by $\bar{R}_{(\bar{\varepsilon},\psi)}$ is selfcentralizing if and only if the Brauer $\hat{G}$-pair determined by R_ε is so and fulfills $\mathbf{O}_p(\tilde{E}_G(R_\varepsilon)) \cong \mathbf{O}_p(\tilde{E}_{\bar{H}}(\bar{R}_{\bar{\varepsilon}}))$. In this case, we have*

$$\mathcal{G}_k(\hat{E}_G(R_\varepsilon)) \cong \Big(\bigoplus_{\psi \in \mathcal{P}\left(k_* C^{\hat{\bar{G}}}_{\hat{\bar{H}}}(\bar{R}_{\bar{\varepsilon}})\right)} \mathcal{G}_k(\hat{E}_{\bar{G}}(\bar{R}_{(\bar{\varepsilon},\psi)}))\Big)^{N_G(R_\varepsilon)} \qquad 15.49.1.$$

Proof: If R_ε is a local pointed group over $k_*\hat{G}$ such that $Z \subset R$, we have $C_{\hat{G}}(R) \subset \hat{H}$, so that we still have (cf. 15.5.4)

$$C^{\hat{G}}_{\hat{H}}(R_\varepsilon) \cong k^* \qquad 15.49.2,$$

and therefore, since $\hat{G}/\hat{H}$ is a p'-group, it follows from 1.18 and from isomorphism 15.16.1 that $\varepsilon \cap k_*\hat{H}$ is a local point of R over $k_*\hat{H}$; moreover, since the H-algebra homomorphism $k_*\hat{H} \to k_*\hat{\bar{H}}$ is a *strict semicovering* (cf. 15.48), the image of $\varepsilon \cap k_*\hat{H}$ in $k_*\hat{\bar{H}}$ is contained in a local point $\bar{\varepsilon}$ of $\bar{R}$ over $k_*\hat{\bar{H}}$ (cf. Proposition 3.15 in [34]); then, it follows from Lemma 15.16 again that $\bar{R}_{\bar{\varepsilon}}$ splits into a set $\{\bar{R}_{(\bar{\varepsilon},\psi)}\}_{\psi \in \mathcal{P}(k_* C^{\hat{\bar{G}}}_{\hat{\bar{H}}}(\bar{R}_{\bar{\varepsilon}}))}$ of local pointed groups over $k_*\hat{\bar{G}}$.

Moreover, since $C_{\hat{G}}(R) = C_{\hat{H}}(R)$, the Brauer $\hat{G}$-pair determined by R_ε is selfcentralizing if and only if the Brauer $\hat{H}$-pair determined by $R_{\varepsilon\cap k_*\hat{H}}$ is so (cf. 7.4) and then, according to Lemma 15.16, we have

$$\mathbf{O}_p\big(E_G(R_\varepsilon)\big) = \mathbf{O}_p\big(E_H(R_{\varepsilon\cap k_*\hat{H}})\big) \qquad 15.49.3.$$

On the other hand, it follows from Corollary 2.13 in [50] that the Brauer $\hat{H}$-pair determined by $R_{\varepsilon\cap k_*\hat{H}}$ is selfcentralizing and fulfills (cf. 1.3)

$$\tilde{E}_H(R_{\varepsilon\cap k_*\hat{H}}) \cong \tilde{E}_{\bar{H}}(\bar{R}_{\bar{\varepsilon}}) \qquad 15.49.4$$

if and only if the Brauer $\hat{\bar{H}}$-pair determined by $\bar{R}_{\bar{\varepsilon}}$ is selfcentralizing. But, in any case, it follows from Theorem 2.9 in [50] and Theorem 3.1 in [39] that the group homomorphism

$$E_H(R_{\varepsilon\cap k_*\hat{H}}) \longrightarrow E_{\bar{H}}(\bar{R}_{\bar{\varepsilon}}) \qquad 15.49.5$$

is surjective and moreover, since $Z \subset Z(H)$, its kernel is a p-group (cf. Theorem 3.2 in [28, Ch. 5]); hence, isomorphism 15.49.4 is equivalent to (cf. 1.3)

$$\mathbf{O}_p\big(\tilde{E}_H(R_{\varepsilon\cap k_*\hat{H}})\big) \cong \mathbf{O}_p\big(\tilde{E}_{\bar{H}}(\bar{R}_{\bar{\varepsilon}})\big) \qquad 15.49.6.$$

Finally, it follows from Lemma 15.16 that the Brauer $\hat{\bar{H}}$-pair determined by $\bar{R}_{\bar{\varepsilon}}$ is selfcentralizing if and only if the Brauer $\hat{\bar{G}}$-pair determined by $\bar{R}_{(\bar{\varepsilon},\psi)}$ is so, and then we have

$$\mathbf{O}_p\big(E_{\bar{H}}(\bar{R}_{\bar{\varepsilon}})\big) = \mathbf{O}_p\big(E_{\bar{G}}(\bar{R}_{(\bar{\varepsilon},\psi)})\big) \qquad 15.49.7.$$

Consequently, the Brauer $\hat{\bar{G}}$-pair determined by $\bar{R}_{(\bar{\varepsilon},\psi)}$ is selfcentralizing if and only if the Brauer $\hat{G}$-pair determined by R_ε is so and fulfills (cf. 1.3)

$$\mathbf{O}_p\big(\tilde{E}_G(R_\varepsilon)\big) \cong \mathbf{O}_p\big(\tilde{E}_{\bar{H}}(\bar{R}_{\bar{\varepsilon}})\big) \qquad 15.49.8.$$

In this case, denoting by $b(\varepsilon)$ the block of $C_{\hat{G}}(R)$ determined by ε, and by $\bar{b}(\varepsilon)$ the image of $b(\varepsilon)$ in $k_*\bar{C}_{\hat{\bar{G}}}(\bar{R})$ (cf. 1.10), it is quite clear that

$$\bar{b}(\varepsilon) = \sum_{\psi\in\mathcal{P}(k_*C^{\hat{\bar{G}}}_{\hat{\bar{H}}}(\bar{R}_{\bar{\varepsilon}}))} \bar{b}(\bar{\varepsilon},\psi) \qquad 15.49.9,$$

since any local point $(\bar{\varepsilon},\psi)$ of $\bar{R}$ on $k_*\hat{\bar{G}}$ involved in $\bar{R}_{\bar{\varepsilon}}$ determines and is determined by the block of defect zero $\big(\bar{b}(\bar{\varepsilon},\psi), \bar{C}_{\hat{\bar{G}}}(\bar{R})\big)$ (cf. 1.17 and 7.4).

Then, on the one hand, since the block $\big(\bar b(\varepsilon), \bar C_{\hat{\bar G}}(R)\big)$ has defect zero (cf. 7.4) so that the k-algebra $k_*\bar C_{\hat{\bar G}}(R)\bar b(\varepsilon)$ is simple (cf. 1.17), we have (cf. Theorem 3.7 in [53] and definition 15.5.3 above)

$$k_*\bar N_{\hat G}(R_\varepsilon)\bar b(\varepsilon) \cong k_*\bar C_{\hat G}(R)\bar b(\varepsilon) \otimes_k k_*\hat E_G(R_\varepsilon) \qquad 15.49.10$$

and therefore we still have

$$\mathcal{G}_k\big(N_{\hat G}(R_\varepsilon), b(\varepsilon)\big) \cong \mathcal{G}_k\big(\hat E_G(R_\varepsilon)\big) \qquad 15.49.11.$$

On the other hand, since $N_G(R_\varepsilon)$ stabilizes the decomposition 15.49.9, it follows from Proposition 3.2 and Theorem 3.7 in [53] that (cf. definition 15.5.3 above)

$$\begin{aligned} &k_*\bar N_{\hat G}(R_\varepsilon)\bar b(\varepsilon) \\ &\quad\cong \prod_\psi \mathrm{Ind}_{\bar N_{\hat G}(\bar R_{(\bar\varepsilon,\psi)})}^{\bar N_{\hat G}(R_\varepsilon)}\big(k_*\bar C_{\hat{\bar G}}(\bar R)\bar b(\bar\varepsilon,\psi) \otimes_k k_*\hat E_{\bar G}(\bar R_{(\bar\varepsilon,\psi)})\big) \end{aligned} \qquad 15.49.12,$$

where $\psi \in \mathcal{P}\big(k_*C_{\hat{\bar H}}^{\hat{\bar G}}(\bar R_{\bar\varepsilon})\big)$ runs over a set of representatives for the set of orbits of $N_G(R_\varepsilon)$, which this time implies that

$$\mathcal{G}_k\big(N_{\hat G}(R_\varepsilon), b(\varepsilon)\big) \cong \Big(\bigoplus_{\psi\in\mathcal{P}\big(k_*C_{\hat{\bar H}}^{\hat{\bar G}}(\bar R_{\bar\varepsilon})\big)} \mathcal{G}_k\big(\hat E_{\bar G}(\bar R_{(\bar\varepsilon,\psi)})\big)\Big)^{N_G(R_\varepsilon)} \qquad 15.49.13.$$

We are done.

Proposition 15.50 *In our standard setting, let Z be a normal Abelian p-subgroup of $\hat G$ and assume that $\hat H = C_{\hat G}(Z)$ and that $\hat H$ contains a Sylow p-subgroup of $\hat G$. With the notation above, denote by P_γ a defect pointed group of b and by $\bar\gamma$ the local point of $\bar P = P/Z$ on $k_*\hat{\bar H}$ containing the image of $\gamma \cap k_*\hat H$. Set $\mathcal{P}_{\bar\gamma} = \mathcal{P}\big(k_*C_{\hat{\bar H}}^{\hat{\bar G}}(\bar P_{\bar\gamma})\big)$ and denote by $\bar{\mathcal{P}}_{\bar\gamma}$ the set of $N_{\hat G}(P_\gamma)$-orbits on $\mathcal{P}_{\bar\gamma}$ and by $\bar\varphi$ the $N_{\hat G}(P_\gamma)$-orbit of $\varphi \in \mathcal{P}_{\bar\gamma}$. Then, there is a bijection from $\bar{\mathcal{P}}_{\bar\gamma}$ to the primitive decomposition of the image $\bar b$ of b in $Z(k_*\hat{\bar G})$ such that, for any $\varphi \in \mathcal{P}_{\bar\gamma}$, $\bar P_{(\bar\gamma,\varphi)}$ is a defect pointed group of the block $\bar b_{\bar\varphi}$ of $\hat{\bar G}$ corresponding to $\bar\varphi \in \bar{\mathcal{P}}_{\bar\gamma}$. Moreover, setting $\bar{\mathcal{F}}_{\bar\varphi} = \mathcal{F}_{(\bar b_{\bar\varphi},\hat{\bar G})}$ for any $\bar\varphi \in \bar{\mathcal{P}}_{\bar\gamma}$, the restriction induces $\mathcal{O}$-module isomorphisms*

$$\begin{aligned} \mathcal{G}_k(\hat G, b) &\cong \bigoplus_{\bar\varphi\in\bar{\mathcal{P}}_{\bar\gamma}} \mathcal{G}_k(\hat{\bar G}, \bar b_{\bar\varphi}) \\ \mathcal{G}_k(\mathcal{F}, \widehat{\mathfrak{aut}}_{\mathcal{F}^{\mathrm{nc}}}) &\cong \bigoplus_{\bar\varphi\in\bar{\mathcal{P}}_{\bar\gamma}} \mathcal{G}_k(\bar{\mathcal{F}}_{\bar\varphi}, \widehat{\mathfrak{aut}}_{(\bar{\mathcal{F}}_{\bar\varphi})^{\mathrm{nc}}}) \end{aligned} \qquad 15.50.1.$$

Proof: First of all note that, since any point of Z on $k_*\hat{G}$ is local (cf. 1.19 above and Proposition 5.15 in [42]), P contains Z. It is clear that, for any p-subgroup R of $\hat{G}$, we have $\mathrm{Br}_{\bar{R}}(\bar{b}) \neq 0$ only if $\mathrm{Br}_R(b) \neq 0$, and therefore $\bar{R}$ has a local point on $k_*\hat{\bar{G}}\bar{b}$ only if R has a local point on $k_*\hat{G}b$ (cf. 1.13); thus, it follows from Lemma 15.49 that any maximal local pointed group on $k_*\hat{\bar{G}}\bar{b}$ is G-conjugate to $\bar{P}_{(\bar{\gamma},\varphi)}$ for some $\varphi \in \mathcal{P}\big(k_*C_{\hat{\bar{H}}}^{\hat{\bar{G}}}(\bar{P}_{\bar{\gamma}})\big)$; then, the first statement follows from the Brauer First Main Theorem (cf. 1.16) and the top isomorphism in 15.50.1 is an immediate consequence of this statement.

Let R_ε and T_η be local pointed groups over $k_*\hat{G}$ contained in P_γ such that R and T are $\mathcal{F}$-selfcentralizing, so that R and T contain Z; we know that any $\alpha \in \mathcal{F}(R,T)$ is induced by an element $x \in \hat{G}$ fulfilling $T_\eta \subset (R_\varepsilon)^x$ (cf. 3.2) and therefore, denoting by $\bar{R}$, $\bar{T}$ and $\bar{x}$ the respective images of R, T and x in $\hat{\bar{G}}$, we have

$$C_{\hat{\bar{G}}}(\bar{R})^{\bar{x}} \subset C_{\hat{\bar{G}}}(\bar{T}) \quad \text{and} \quad C_{\hat{\bar{H}}}(\bar{R})^{\bar{x}} \subset C_{\hat{\bar{H}}}(\bar{T}) \qquad 15.50.2,$$

so that $\bar{x}^{-1}$ determines a group homomorphism

$$\mathfrak{c}_x : C_{\bar{G}}(\bar{R})/C_{\bar{H}}(\bar{R}) \longrightarrow C_{\bar{G}}(\bar{T})/C_{\bar{H}}(\bar{T}) \qquad 15.50.3$$

which only depends on α since $C_{\hat{G}}(T) \subset \hat{H}$. That is to say, we get a *contravariant* functor

$$\mathfrak{c} : \mathcal{F}^{^{\mathrm{sc}}} \longrightarrow \mathfrak{Gr} \qquad 15.50.4$$

mapping R on the quotient $C_{\bar{G}}(\bar{R})/C_{\bar{H}}(\bar{R})$ and α on $\mathfrak{c}_x$.

Moreover, in 15.48 we have proved that the quotient $\bar{C}_{\hat{\bar{H}}}(\bar{R})/\bar{C}_{\hat{H}}(R)$ is a p-group; hence, since the image $\bar{b}(\varepsilon)$ in $k_*\bar{C}_{\hat{H}}(R)$ of the block $b(\varepsilon)$ determined by ε has defect zero (cf. 7.4), it follows from Proposition 6.5 in [34] that the image $\bar{b}(\bar{\varepsilon})$ in $k_*\bar{C}_{\hat{\bar{H}}}(\bar{R})$ of the block $b(\bar{\varepsilon})$ of $C_{\hat{\bar{H}}}(\bar{R})$ determined by the local point $\bar{\varepsilon}$ of $\bar{R}$ on $k_*\hat{\bar{H}}$ containing the image of $\varepsilon \cap k_*\hat{H}$ is *nilpotent* and then, it follows from Proposition 3.5 in [53] that we have an *embedding* (cf. 1.11)

$$k_*\bar{C}_{\hat{H}}(R)\bar{b}(\varepsilon) \longrightarrow k_*\bar{C}_{\hat{\bar{H}}}(\bar{R})\bar{b}(\bar{\varepsilon})/J\big(k_*\bar{C}_{\hat{\bar{H}}}(\bar{R})\bar{b}(\bar{\varepsilon})\big) \qquad 15.50.5.$$

Consider the k^*-group $C_{\hat{\bar{H}}}^{\hat{\bar{G}}}(\bar{R}_{\bar{\varepsilon}})$ which lifts the quotient $C_{\bar{G}}(\bar{R})/C_{\bar{H}}(\bar{R})$ (cf. 15.5.4); we claim that the functor $\mathfrak{c}$ above can be lifted to a functor

$$\hat{\mathfrak{c}} : \mathcal{F}^{^{\mathrm{sc}}} \longrightarrow k^*\text{-}\mathfrak{Gr} \qquad 15.50.6$$

mapping R on $C_{\hat{\bar{H}}}^{\hat{\bar{G}}}(\bar{R}_{\bar{\varepsilon}})$. Indeed, since $Z \subset R$, we have $C_{\hat{G}}(R) = C_{\hat{H}}(R)$ and therefore we still have (cf. isomorphism 1.16.1)

$$(k_*\hat{G})(R_\varepsilon) \cong k_*\bar{C}_{\hat{G}}(R)\bar{b}(\varepsilon) = k_*\bar{C}_{\hat{H}}(R)\bar{b}(\varepsilon) \qquad 15.50.7;$$

consequently, from embedding 15.50.5 and this isomorphism, it follows that the k^*-group $\hat{C}_{\hat{\bar{G}}}(\bar{R}_{\bar{\varepsilon}})$ obtained from the action of $C_{\hat{\bar{G}}}(\bar{R}_{\bar{\varepsilon}})$ on the simple algebra $k_*\bar{C}_{\hat{\bar{H}}}(\bar{R})\bar{b}(\bar{\varepsilon})/J\big(k_*\bar{C}_{\hat{\bar{H}}}(\bar{R})\bar{b}(\bar{\varepsilon})\big)$ (cf. 15.5.4) is canonically isomorphic to a k^*-subgroup of $\hat{N}_{\hat{G}}(R_\varepsilon)/Z$ (cf. 1.26 above and Proposition 2.14 in [39]) and therefore $C^{\hat{\bar{G}}}_{\hat{\bar{H}}}(\bar{R}_{\bar{\varepsilon}})$ is (canonically isomorphic to) a k^*-subgroup of the quotient of $\hat{\mathcal{F}}(R)$ by $\bar{C}_{\bar{H}}(\bar{R})/\bar{C}_H(R)$ (cf. 15.5).

Now, if R_ε and T_η are local pointed groups over $k_*\hat{G}$ contained in P_γ such that R and T are $\mathcal{F}$-selfcentralizing, and $\alpha : T \to R$ is an $\mathcal{F}$-morphism, still denoting by α the $\mathcal{F}^{^{\rm sc}}$-*chain* $\Delta_1 \to \mathcal{F}^{^{\rm sc}}$ mapping 0 on T, 1 on R and $0{\bullet}1$ on α (cf. A2.2), the functor $\widehat{\mathfrak{aut}}_{\mathcal{F}^{\rm nc}}$ supplied by Theorem 11.32, applied to the obvious $\mathfrak{ch}(\mathcal{F}^{^{\rm nc}})$-morphisms from the 1-chain α to the 0-chains determined by R and T produces two k^*-group homomorphisms

$$\hat{\mathcal{F}}(T) \longleftarrow \hat{\mathcal{F}}(\alpha) \longrightarrow \hat{\mathcal{F}}(R) \qquad 15.50.8;$$

but, it is quite clear that the converse image of $C^{\hat{\bar{G}}}_{\hat{\bar{H}}}(\bar{R}_{\bar{\varepsilon}})$ in $\hat{\mathcal{F}}(R)$ is contained in the image of $\hat{\mathcal{F}}(\alpha)$; then, it is not difficult to prove that the left-hand k^*-group homomorphism in 15.50.8 induces a k^*-group homomorphism

$$\hat{\mathfrak{c}}(\alpha) : C^{\hat{\bar{G}}}_{\hat{\bar{H}}}(\bar{R}_{\bar{\varepsilon}}) \longrightarrow C^{\hat{\bar{G}}}_{\hat{\bar{H}}}(\bar{T}_{\bar{\eta}}) \qquad 15.50.9$$

and that this correspondence defines the announced functor $\hat{\mathfrak{c}}$.

We claim that, if the Brauer $(\bar{b}, \hat{\bar{H}})$-pair determined by $\bar{T}_{\bar{\eta}}$ is selfcentralizing too, $\hat{\mathfrak{c}}(\alpha)$ is actually an isomorphism; indeed, we may assume that $T \subset R$ and that α is the inclusion map; moreover, arguing by induction on $|R : T|$, we also may assume that T is normal in R and then, that T is *fully normalized* in $\mathcal{F}$ (cf. Proposition 2.7); finally, arguing by induction on $|P : T|$, we still may assume that we have $R = N_P(T)$; thus, denoting by ν the *unique* point of $N_{\hat{H}}(R_\varepsilon)$ on $k_*\hat{G}$ determined by η (cf. 1.19), R_ε is a defect pointed group of $N_{\hat{H}}(T_\eta)_\nu$ (cf. Proposition 3.5).

Consequently, respectively denoting by $\hat{C}_R$ and $\hat{C}_T$ the converse images of $C_{\hat{\bar{G}}}(\bar{R})$ and $C_{\hat{\bar{G}}}(\bar{T})$ in $\hat{G}$, and by λ the *unique* point of $R{\cdot}(\hat{H} \cap \hat{C}_T)$ on $k_*\hat{G}$ determined by η (cf. 1.19), R_ε is also a defect pointed group of $R{\cdot}(\hat{H} \cap \hat{C}_T)_\lambda$ and therefore, by the *Frattini argument*, we have

$$R{\cdot}\hat{C}_T = R{\cdot}(\hat{H} \cap \hat{C}_T){\cdot}N_{\hat{C}_T}(R_\varepsilon) \qquad 15.50.10.$$

But, since we assume that the Brauer $(\bar{b}, \hat{\bar{H}})$-pair determined by $\bar{T}_{\bar{\eta}}$ is selfcentralizing, we have $C_{\bar{P}}(\bar{R}) = Z(\bar{R}) \subset \bar{T}$ (cf. Corollary 7.3); thus, it follows from Theorem 3.4 in [28, Ch. 5] that any p'-subgroup of $N_{\hat{C}_R}(T_\tau)$ centralizes $\bar{T}$, so that it is contained in $\hat{C}_T$; hence, since $\hat{G}/\hat{H}$ is a p'-group, we get $\hat{C}_R = (\hat{H} \cap \hat{C}_R)\cdot\hat{C}_T$ and therefore we still get

$$\begin{aligned} C_{\bar{G}}(\bar{R})/C_{\bar{H}}(\bar{R}) &\cong \hat{C}_R/(\hat{H} \cap \hat{C}_R) \cong \hat{C}_T/(\hat{H} \cap \hat{C}_T) \\ &\cong C_{\bar{G}}(\bar{T})/C_{\bar{H}}(\bar{T}) \end{aligned} \qquad 15.50.11,$$

so that $\mathfrak{c}(\alpha)$ and $\hat{\mathfrak{c}}(\alpha)$ are isomorphisms, which proves our claim.

On the other hand, for any local pointed group R_ε on $k_*\hat{G}$ which determines a selfcentralizing Brauer $(b, \hat{G})$-pair and fulfills $\mathbf{O}_p\big(\tilde{E}_G(R_\varepsilon)\big) = \{1\}$, we have $\mathbf{O}_p\big(\tilde{E}_H(R_{\varepsilon\cap k_*\hat{H}})\big) = \{1\}$ (cf. Lemma 15.16) and always denoting by $\bar{\varepsilon}$ the local point of $\bar{R}$ on $k_*\hat{\bar{H}}$ containing the image of $\varepsilon \cap k_*\hat{H}$, we still have $\mathbf{O}_p\big(\tilde{E}_{\bar{H}}(\bar{R}_{\bar{\varepsilon}})\big) = \{1\}$ since the kernel of the group homomorphism

$$\tilde{E}_H(R_{\varepsilon\cap(k_*\hat{H})}) \longrightarrow \tilde{E}_{\bar{H}}(\bar{R}_{\bar{\varepsilon}}) \qquad 15.50.12$$

is a p-group (cf. Theorem 3.2 in [28, Ch. 5]); then, it follows from Lemma 15.49 that $\bar{R}_{\bar{\varepsilon}}$ determines a selfcentralizing Brauer $(\bar{b}, \hat{\bar{H}})$-pair. Conversely, it follows from the same lemma that in this way we obtain all the subgroups $\bar{R}$ of $\bar{P}$ which, for some $\bar{\varphi} \in \bar{\mathcal{P}}_{\bar{\gamma}}$, are $\bar{\mathcal{F}}_{\bar{\varphi}}$-selfcentralizing and fulfill $\mathbf{O}_p\big(\tilde{\bar{\mathcal{F}}}_{\bar{\varphi}}(\bar{R})\big) = \{1\}$ and that, in this case, this holds for any $\bar{\varphi} \in \bar{\mathcal{P}}_{\bar{\gamma}}$.

Finally, denoting by $\bar{\mathfrak{X}}$ the set of such subgroups of $\bar{P}$, by $\mathfrak{X}$ the set of their converse images in P, by $\mathcal{F}^{\mathfrak{X}}$ the full subcategory of $\mathcal{F}$ over $\mathfrak{X}$, by $\bar{\mathcal{F}}$ and $\bar{\mathcal{F}}^{\bar{\mathfrak{X}}}$ the respective quotients of $\mathcal{F}$ and $\mathcal{F}^{\mathfrak{X}}$ by Z (cf. Proposition 12.3), and by $(\bar{\mathcal{F}}_{\bar{\varphi}})^{\bar{\mathfrak{X}}}$ the full subcategory of $\bar{\mathcal{F}}_{\bar{\varphi}}$ over $\bar{\mathfrak{X}}$ for any $\bar{\varphi} \in \bar{\mathcal{P}}_{\bar{\gamma}}$, on the one hand it follows from Lemma 15.51 below and from Proposition 14.6 that we have

$$\begin{aligned} \mathcal{G}_k(\mathcal{F}, \widehat{\mathfrak{aut}}_{\mathcal{F}^{\mathrm{nc}}}) &\cong \lim_{\longleftarrow}(\mathfrak{g}_k \circ \widehat{\mathfrak{aut}}_{\bar{\mathcal{F}}^{\bar{\mathfrak{X}}}}) \\ \mathcal{G}_k(\bar{\mathcal{F}}_{\bar{\varphi}}, \widehat{\mathfrak{aut}}_{(\bar{\mathcal{F}}_{\bar{\varphi}})^{\mathrm{nc}}}) &\cong \lim_{\longleftarrow}(\mathfrak{g}_k \circ \widehat{\mathfrak{aut}}_{(\bar{\mathcal{F}}_{\bar{\varphi}})^{\bar{\mathfrak{X}}}}) \text{ for any } \bar{\varphi} \in \bar{\mathcal{P}}_{\bar{\gamma}} \end{aligned} \qquad 15.50.13.$$

On the other hand, it is clear that the restriction to $\mathcal{F}^{\mathfrak{X}}$ of the functor $\hat{\mathfrak{c}}$ above factorizes throughout $\bar{\mathcal{F}}^{\bar{\mathfrak{X}}}$ giving a functor

$$\hat{\bar{\mathfrak{c}}}^{\bar{\mathfrak{X}}} : \bar{\mathcal{F}}^{\bar{\mathfrak{X}}} \longrightarrow k^*\text{-}\mathfrak{Gr} \qquad 15.50.14$$

and we already know that it maps any $\bar{\mathcal{F}}^{\bar{\mathfrak{X}}}$-morphism on a k^*-group isomorphism; in particular, it induces a functor from $\bar{\mathcal{F}}^{\bar{\mathfrak{X}}}$ to the category of finite sets (cf. 1.1)

$$\hat{\bar{\mathfrak{c}}}^{\bar{\mathfrak{X}},\mathcal{P}} : \bar{\mathcal{F}}^{\bar{\mathfrak{X}}} \longrightarrow \aleph \qquad 15.50.15$$

mapping $\bar R \in \bar{\mathfrak X}$ on $\mathcal P\big(k_* C_{\hat{\bar H}}^{\hat{\bar G}}(\bar R_{\bar\varepsilon})\big)$ where $\bar\varepsilon$ is the unique local point of $\bar R$ on $k_*\hat{\bar H}$ such that $\bar R_{\bar\varepsilon} \subset \bar P_{\bar\gamma}$ (cf. 1.17 and Corollary 7.3).

As usual, we can identify $\aleph$ with the 2-subcategory of $\mathfrak{CC}$ (cf. A1.6) formed by finite categories with no other morphisms than the identity ones, and then $\hat{\bar{\mathfrak c}}^{\bar{\mathfrak X},\mathcal P}$ becomes a *representation* of $\bar{\mathcal F}^{\bar{\mathfrak X}}$ (cf. A2.2) which allows us to consider the *semidirect product* $\hat{\bar{\mathfrak c}}^{\bar{\mathfrak X},\mathcal P} \rtimes \bar{\mathcal F}^{\bar{\mathfrak X}}$ together with the structural functor (cf. A2.7.1)

$$\mathfrak p_{\bar{\mathcal F}^{\bar{\mathfrak X}}} : \hat{\bar{\mathfrak c}}^{\bar{\mathfrak X},\mathcal P} \rtimes \bar{\mathcal F}^{\bar{\mathfrak X}} \longrightarrow \bar{\mathcal F}^{\bar{\mathfrak X}} \qquad 15.50.16.$$

Explicitly, a $\hat{\mathfrak c}^{\bar{\mathfrak X},\mathcal P} \rtimes \mathcal F^{\bar{\mathfrak X}}$-object is a pair $(\psi, \bar R)$ formed by an element $\bar R$ of $\bar{\mathfrak X}$ and an element ψ of $\mathcal P\big(k_* C_{\hat{\bar H}}^{\hat{\bar G}}(\bar R_{\bar\varepsilon})\big)$; but, according to Lemma 15.49, the pair $(\psi, \bar R)$ determines and is determined by the local pointed group $\bar R_{(\bar\varepsilon,\psi)}$ over $k_*\hat{\bar G}$; analogously, it is easily checked that any $\hat{\mathfrak c}^{\bar{\mathfrak X},\mathcal P} \rtimes \mathcal F^{\bar{\mathfrak X}}$-morphism determines and is determined by the inclusion and the $\hat{\bar G}$-conjugation between the corresponding local pointed groups, so that it determines a block of $\hat{\bar G}$.

Consequently, the category $\hat{\bar{\mathfrak c}}^{\bar{\mathfrak X},\mathcal P} \rtimes \bar{\mathcal F}^{\bar{\mathfrak X}}$ is the *disjoint union* of the categories $(\bar{\mathcal F}_{\bar\varphi})^{\bar{\mathfrak X}}$ — the $\hat{\bar{\mathfrak c}}^{\bar{\mathfrak X},\mathcal P} \rtimes \bar{\mathcal F}^{\bar{\mathfrak X}}$-objects and the $\hat{\bar{\mathfrak c}}^{\bar{\mathfrak X},\mathcal P} \rtimes \bar{\mathcal F}^{\bar{\mathfrak X}}$-morphisms are the disjoint union of the $(\bar{\mathcal F}_{\bar\varphi})^{\bar{\mathfrak X}}$-objects and the $(\bar{\mathcal F}_{\bar\varphi})^{\bar{\mathfrak X}}$-morphisms respectively — when $\bar\varphi$ runs over $\bar{\mathcal P}_{\bar\gamma}$, and therefore we still have

$$\mathfrak{ch}^*\big(\hat{\bar{\mathfrak c}}^{\bar{\mathfrak X},\mathcal P} \rtimes \bar{\mathcal F}^{\bar{\mathfrak X}}\big) = \bigsqcup_{\bar\varphi\in\bar{\mathcal P}_{\bar\gamma}} \mathfrak{ch}^*\big((\bar{\mathcal F}_{\bar\varphi})^{\bar{\mathfrak X}}\big) \qquad 15.50.17.$$

In particular, we have an evident *union* functor

$$\widehat{\mathfrak{aut}}_{\hat{\mathfrak c}^{\bar{\mathfrak X},\mathcal P}\rtimes\bar{\mathcal F}^{\bar{\mathfrak X}}} = \bigsqcup_{\bar\varphi\in\bar{\mathcal P}_{\bar\gamma}} \widehat{\mathfrak{aut}}_{(\bar{\mathcal F}_{\bar\varphi})^{\bar{\mathfrak X}}} : \mathfrak{ch}^*\big(\hat{\bar{\mathfrak c}}^{\bar{\mathfrak X},\mathcal P} \rtimes \bar{\mathcal F}^{\bar{\mathfrak X}}\big) \longrightarrow k^*\text{-}\mathfrak{Gr} \qquad 15.50.18$$

mapping any $\hat{\bar{\mathfrak c}}^{\bar{\mathfrak X},\mathcal P} \rtimes \bar{\mathcal F}^{\bar{\mathfrak X}}$-chain $\bar{\mathfrak q}^{\mathcal P} : \Delta_n \to \hat{\bar{\mathfrak c}}^{\bar{\mathfrak X},\mathcal P} \rtimes \bar{\mathcal F}^{\bar{\mathfrak X}}$ on $\hat{\bar{\mathcal F}}_{\bar\varphi}(\mathfrak p_{\bar{\mathcal F}^{\bar{\mathfrak X}}} \circ \bar{\mathfrak q}^{\mathcal P})$, where $b_{\bar\varphi}$ is the block of $\hat{\bar G}$ determined by $\bar{\mathfrak q}^{\mathcal P}$.

Moreover, it is easily checked that the *direct image* (cf. 1.6) of the *contravariant* functor $\mathfrak g \circ \widehat{\mathfrak{aut}}_{\hat{\mathfrak c}^{\bar{\mathfrak X},\mathcal P}\rtimes\bar{\mathcal F}^{\bar{\mathfrak X}}}$ *via* the structural functor $\mathfrak{ch}^*(\mathfrak p_{\bar{\mathcal F}^{\bar{\mathfrak X}}})$ maps any $\mathfrak{ch}^*(\bar{\mathcal F}^{\bar{\mathfrak X}})$-object $(\bar{\mathfrak q}, \Delta_n)$ such that $\bar{\mathfrak q}(i-1) \subset \bar{\mathfrak q}(i)$ and $\bar{\mathfrak q}(i-1 \bullet i)$ is the inclusion map for any $1 \le i \le n$, on the direct sum

$$\Big(\big(\mathfrak{ch}^*(\mathfrak p_{\bar{\mathcal F}^{\bar{\mathfrak X}}})\big)_*\big(\mathfrak g_k \circ \widehat{\mathfrak{aut}}_{\hat{\mathfrak c}^{\bar{\mathfrak X},\mathcal P}\rtimes\bar{\mathcal F}^{\bar{\mathfrak X}}}\big)\Big)(\bar{\mathfrak q}, \Delta_n) = \bigoplus_{\psi\in\hat{\mathfrak c}^{\bar{\mathfrak X},\mathcal P}(\bar{\mathfrak q}(n))} \mathcal G_k\big(\hat{\bar{\mathcal F}}_{\bar\varphi_\psi}(\bar{\mathfrak q})\big) \qquad 15.50.19$$

where, denoting by R the converse image of $\bar{\mathfrak{q}}(n)$ in P — which is $\mathcal{F}$-selfcentralizing — and by ε the unique local point of R on $k_*\hat{G}$ such that $R_\varepsilon \subset P_\gamma$ (cf. 1.17 and Corollary 7.3), $\bar{\varphi}_\psi$ is the element in $\bar{\mathcal{P}}_{\bar{\gamma}}$ fulfilling $\bar{R}_{(\bar{\varepsilon},\psi)} \subset \bar{P}_{(\bar{\gamma},\varphi_\psi)}$ for some $\varphi_\psi \in \bar{\varphi}_\psi$ and any $\psi \in \hat{\mathfrak{c}}^{\bar{\mathfrak{x}},\mathcal{P}}\big(\bar{\mathfrak{q}}(n)\big) = \mathcal{P}\big(k_* C^{\hat{\bar{G}}}_{\hat{\bar{H}}}(\bar{R}_{\bar{\varepsilon}})\big)$, and we identify $\bar{\mathfrak{q}}$ with the corresponding $\hat{\bar{\mathcal{F}}}_{\bar{\varphi}_\psi}$-chain.

At this point, with our notation in 14.21, from Lemma 15.49 it is not difficult to check that we have a natural isomorphism

$$\mathfrak{g}_k \circ \widehat{\mathfrak{aut}}_{\bar{\mathcal{F}}^{\bar{\mathfrak{x}}}} \cong \mathfrak{h}^0\Big(\big(\mathfrak{ch}^*(\mathfrak{p}_{\bar{\mathcal{F}}^{\bar{\mathfrak{x}}}})\big)_*(\mathfrak{g}_k \circ \widehat{\mathfrak{aut}}_{\hat{\mathfrak{c}}^{\bar{\mathfrak{x}},\mathcal{P}\rtimes\bar{\mathcal{F}}^{\bar{\mathfrak{x}}}}})\Big) \qquad 15.50.20;$$

then, as in 14.22 above, it is quite clear that

$$\begin{aligned}\lim_{\longleftarrow} \mathfrak{h}^0\Big(\big(\mathfrak{ch}^*(\mathfrak{p}_{\bar{\mathcal{F}}^{\bar{\mathfrak{x}}}})\big)_*(\mathfrak{g}_k \circ \widehat{\mathfrak{aut}}_{\hat{\mathfrak{c}}^{\bar{\mathfrak{x}},\mathcal{P}\rtimes\bar{\mathcal{F}}^{\bar{\mathfrak{x}}}}})\Big) & \\ \cong \lim_{\longleftarrow} \big(\mathfrak{ch}^*(\mathfrak{p}_{\bar{\mathcal{F}}^{\bar{\mathfrak{x}}}})\big)_*(\mathfrak{g}_k \circ \widehat{\mathfrak{aut}}_{\hat{\mathfrak{c}}^{\bar{\mathfrak{x}},\mathcal{P}\rtimes\bar{\mathcal{F}}^{\bar{\mathfrak{x}}}}}) & \qquad 15.50.21. \\ \cong \lim_{\longleftarrow} (\mathfrak{g}_k \circ \widehat{\mathfrak{aut}}_{\hat{\mathfrak{c}}^{\bar{\mathfrak{x}},\mathcal{P}\rtimes\bar{\mathcal{F}}^{\bar{\mathfrak{x}}}}}) &\end{aligned}$$

Now, since clearly we have

$$\lim_{\longleftarrow} (\mathfrak{g}_k \circ \widehat{\mathfrak{aut}}_{\hat{\mathfrak{c}}^{\bar{\mathfrak{x}},\mathcal{P}\rtimes\bar{\mathcal{F}}^{\bar{\mathfrak{x}}}}}) \cong \bigoplus_{\bar{\varphi}\in\bar{\mathcal{P}}_{\bar{\gamma}}} \lim_{\longleftarrow} (\mathfrak{g}_k \circ \widehat{\mathfrak{aut}}_{(\bar{\mathcal{F}}_{\bar{\varphi}})^{\bar{\mathfrak{x}}}}) \qquad 15.50.22,$$

the bottom isomorphism in 15.50.1 follows from the isomorphisms 15.50.13, 15.50.20 and 15.50.21. We are done.

Lemma 15.51 *Let P be a finite p-group and $\mathcal{F}$ a Frobenius P-category. For any $\mathcal{F}$-nilcentralized subgroup Q of P there are an $\mathcal{F}$-selfcentralizing subgroup R fulfilling $\mathbf{O}_p\big(\tilde{\mathcal{F}}(R)\big) = \{1\}$ and an $\mathcal{F}$-morphism $\psi\colon Q \to R$ such that $\psi(Q)$ is fully centralized in $\mathcal{F}$, R contains $C_P\big(\psi(Q)\big)$ and we have*

$$\psi \circ \mathcal{F}(Q) \subset \mathcal{F}(R) \circ \psi \qquad 15.51.1.$$

Proof: We argue by induction on $|P\colon Q|$; let $\psi\colon Q \to P$ be an $\mathcal{F}$-morphism such that $Q' = \psi(Q)$ is fully normalized in $\mathcal{F}$ (cf. Proposition 2.7); we already know that $R = Q'{\cdot}C_P(Q')$ is $\mathcal{F}$-selfcentralizing (cf. Proposition 2.11) and it follows from statement 2.10.1 that any $\sigma \in \mathcal{F}(Q')$ can be extended to some $\mathcal{F}$--morphism $\rho\colon R \to P$ and then we have $\rho(R) = R$, so that ρ determines an element $\tau \in \mathcal{F}(R)$ (cf. 2.3) fulfilling $\tau \circ \psi = \psi \circ \sigma$.

If Q is not an $\mathcal{F}$-selfcentralizing subgroup of P, it follows from the induction hypothesis that there are an $\mathcal{F}$-selfcentralizing subgroup T fulfilling $\mathbf{O}_p\big(\tilde{\mathcal{F}}(T)\big) = \{1\}$ and an $\mathcal{F}$-morphism $\eta\colon R \to T$ such that T contains $C_P\big(\eta(R)\big)$ and we have

$$\eta \circ \mathcal{F}(R) \subset \mathcal{F}(T) \circ \eta \qquad 15.51.2;$$

then, it is clear that $(\eta \circ \psi)(Q) = \eta(Q')$ is fully centralized in $\mathcal{F}$, we have $C_P\big(\eta(Q')\big) \subset \eta(R)$ since Q' is fully centralized too (cf. Proposition 2.11) and we get

$$\eta \circ \psi \circ \mathcal{F}(Q) \subset \eta \circ \mathcal{F}(R) \circ \psi \in \mathcal{F}(T) \circ \eta \circ \psi \qquad 15.51.3.$$

Thus, we may assume that Q is $\mathcal{F}$-selfcentralizing but $\mathbf{O}_p\big(\tilde{\mathcal{F}}(Q)\big) \neq \{1\}$; then, $\mathcal{F}_P(Q')$ is a Sylow p-subgroup of $\mathcal{F}(Q')$ (cf. Proposition 2.11) and, denoting by U the converse image of $\mathbf{O}_p\big(\tilde{\mathcal{F}}(Q')\big)$ in P, it follows again from statement 2.10.1 that any $\sigma \in \mathcal{F}(Q')$ can be extended to some $\mathcal{F}$-morphism $\zeta : U \to P$ which this time fulfills $\zeta(U) = U$, so that ζ determines $\xi \in \mathcal{F}(U)$ (cf. 2.3) fulfilling $\xi \circ \psi = \psi \circ \sigma$; since $Q \neq U$, once again it suffices to apply the induction hypothesis.

Chapter 16

The local-global question: reduction to the simple groups

16.1 In this chapter we keep all the notation introduced in chapter 15. In particular, for a *finite* k^*-group $\hat{G}$ (cf. 1.23) and a primitive idempotent b of $Z(k_*\hat{G})$ (cf. 1.25), we consider the *Grothendieck groups* $\mathcal{G}_k(\hat{G}, b)$ and $\mathcal{G}_k(\mathcal{F}, \widehat{\mathfrak{aut}}_{\mathcal{F}^{\mathrm{nc}}})$ — actually, they are $\mathcal{O}$-modules — respectively coming from the category of $k_*\hat{G}b$-modules of finite dimension and, choosing a maximal Brauer $(b, \hat{G})$-pair (P, e) and denoting by $\mathcal{F} = \mathcal{F}_{(b,\hat{G})}$ the corresponding Frobenius P-category, from the functor (cf. Theorem 11.32)

$$\widehat{\mathfrak{aut}}_{\mathcal{F}^{\mathrm{nc}}} : \mathfrak{ch}^*(\mathcal{F}^{\mathrm{nc}}) \longrightarrow k^*\text{-}\mathfrak{Gr} \qquad 16.1.1$$

lifting $\mathfrak{aut}_{\mathcal{F}^{\mathrm{nc}}}$ (cf. Proposition A2.10).

16.2 It is clear that the group of k^*-group automorphisms $\mathrm{Aut}_{k^*}(\hat{G})$ of $\hat{G}$ acts on the group algebra $k_*\hat{G}$ (cf. 1.25); then, since for any element σ of the stabilizer $\mathrm{Aut}_{k^*}(\hat{G})_b$ of b and any $k_*\hat{G}b$-module M we have the $k_*\hat{G}b$-module $\mathrm{Res}_\sigma(M)$, $\mathrm{Aut}_{k^*}(\hat{G})_b$ acts on the category of $k_*\hat{G}b$-modules of finite dimension and therefore, since the action of an inner automorphism determines a functor isomorphic to the *identity*, the corresponding group of *exterior automorphisms* (cf. 1.3)

$$\widetilde{\mathrm{Aut}}_{k^*}(\hat{G})_b = \mathrm{Out}_{k^*}(\hat{G})_b \qquad 16.2.1$$

still acts on the Grothendieck group $\mathcal{G}_k(\hat{G}, b)$, so that $\mathcal{G}_k(\hat{G}, b)$ becomes an $\mathcal{O}\mathrm{Out}_{k^*}(\hat{G})_b$-module.

16.3 On the other hand, it is clear that the stabilizer $\mathrm{Aut}_{k^*}(\hat{G})_{(P,e)}$ of (P, e) acts on the Frobenius P-category $\mathcal{F}$ throughout *Frobenius functors* (cf. 12.1) and, once again, the *Frobenius functors* from $\mathcal{F}$ to $\mathcal{F}$ induced by the elements of $N_{\hat{G}}(P, e)$ are isomorphic to the *identity* (cf. 12.1). Moreover, it is clear that any Frobenius functor $\mathfrak{f}: \mathcal{F} \to \mathcal{F}$ stabilizes the full subcategory $\mathcal{F}^{\mathrm{nc}}$ and therefore it induces a functor

$$\mathfrak{ch}^*(\mathfrak{f}^{\mathrm{nc}}) : \mathfrak{ch}^*(\mathcal{F}^{\mathrm{nc}}) \longrightarrow \mathfrak{ch}^*(\mathcal{F}^{\mathrm{nc}}) \qquad 16.3.1;$$

then, it follows from the uniqueness part of Theorems 11.10 and 11.32 above that we have a natural isomorphism

$$\hat{\nu}_{\mathfrak{f}} : \widehat{\mathfrak{aut}}_{\mathcal{F}^{\mathrm{nc}}} \cong \widehat{\mathfrak{aut}}_{\mathcal{F}^{\mathrm{nc}}} \circ \mathfrak{ch}^*(\mathfrak{f}^{\mathrm{nc}}) \qquad 16.3.2.$$

16.4 Hence, for any $\sigma \in \mathrm{Aut}_{k^*}(\hat G)_{(P,e)}$, we get the $\mathcal{O}$-module isomorphisms (cf. A3.9)

$$\begin{array}{c}\mathbb{H}^0_{\mathfrak{o}}(\mathcal{F}^{^{\rm nc}}, \mathfrak{g}_k \circ \widehat{\mathfrak{aut}}_{\mathcal{F}^{\rm nc}}) \\ \wr\| \quad \mathbb{H}^0_{\mathfrak{o}}(\mathfrak{f}^{^{\rm nc}}_\sigma, \mathfrak{g}_k \circ \widehat{\mathfrak{aut}}_{\mathcal{F}^{\rm nc}}) \\ \mathbb{H}^0_{\mathfrak{o}}\big(\mathcal{F}^{^{\rm nc}}, \mathfrak{g}_k \circ \widehat{\mathfrak{aut}}_{\mathcal{F}^{\rm nc}} \circ \mathfrak{ch}^*(\mathfrak{f}^{^{\rm nc}}_\sigma)\big) \\ \wr\| \quad \mathbb{H}^0_{\mathfrak{o}}(\mathcal{F}^{^{\rm nc}}, \nu_{\mathfrak{f}_\sigma}) \\ \mathbb{H}^0_{\mathfrak{o}}(\mathcal{F}^{^{\rm nc}}, \mathfrak{g}_k \circ \widehat{\mathfrak{aut}}_{\mathcal{F}^{\rm nc}})\end{array} \qquad 16.4.1$$

and it is easily checked that this correspondence defines an action of the stabilizer $\mathrm{Aut}_{k^*}(\hat G)_{(P,e)}$ on the $\mathcal{O}$-module (cf. A3.9)

$$\mathbb{H}^0_{\mathfrak{o}}(\mathcal{F}^{^{\rm nc}}, \mathfrak{g}_k \circ \widehat{\mathfrak{aut}}_{\mathcal{F}^{\rm nc}}) \cong \varprojlim (\mathfrak{g}_k \circ \widehat{\mathfrak{aut}}_{\mathcal{F}^{\rm nc}}) = \mathcal{G}_k(\mathcal{F}, \widehat{\mathfrak{aut}}_{\mathcal{F}^{\rm nc}}) \qquad 16.4.2.$$

Moreover, if σ is induced by an element of $N_{\hat G}(P,e)$ then we have $\mathfrak{f}_\sigma \cong \mathfrak{id}_{\mathcal{F}}$ (cf. 12.1) and thus the corresponding automorphism of $\mathbb{H}^0_{\mathfrak{o}}(\mathcal{F}^{^{\rm nc}}, \mathfrak{g}_k \circ \widehat{\mathfrak{aut}}_{\mathcal{F}^{\rm nc}})$ is the *identity*. On the other hand, $\hat G$ acts transitively on the set of maximal Brauer $(b, \hat G)$-pairs (cf. 1.15) and therefore we get an exact sequence

$$N_{\hat G}(P,e) \longrightarrow \mathrm{Aut}_{k^*}(\hat G)_{(P,e)} \longrightarrow \mathrm{Out}_{k^*}(\hat G)_b \longrightarrow 1 \qquad 16.4.3.$$

Consequently, $\mathcal{G}_k(\mathcal{F}, \widehat{\mathfrak{aut}}_{\mathcal{F}^{\rm nc}})$ becomes an $\mathcal{O}\mathrm{Out}_{k^*}(\hat G)_b$-module too.

16.5 Since Alperin's Conjecture is equivalent to the existence of an $\mathcal{O}$-module isomorphism $\mathcal{G}_k(\hat G, b) \cong \mathcal{G}_k(\mathcal{F}, \widehat{\mathfrak{aut}}_{\mathcal{F}^{\rm nc}})$ (cf I32), in this new context it seems reasonable to go farther and try to discuss the statement:

(Q) *There is an $\mathcal{O}\mathrm{Out}_{k^*}(\hat G)_b$-module isomorphism*

$$\mathcal{G}_k(\hat G, b) \cong \mathcal{G}_k(\mathcal{F}, \widehat{\mathfrak{aut}}_{\mathcal{F}^{\rm nc}}) \qquad 16.5.1.$$

Although stronger than Alperin's Conjecture, in this chapter we show that this statement has the advantage of admitting the possibility of an *inductive proof*; that is to say, whether it is true, its proof can be reduced to the verification of some conditions in the case where the k^*-quotient G of $\hat G$ (cf. 1.23) contains a normal noncommutative simple subgroup S such that $C_G(S) = \{1\}$ and p divides $|S|$ without dividing $|G\!:\!S|$ (cf. Theorem 16.45 below).

Proposition 16.6 *With the notation above, assume that there is a nontrivial characteristic k^*-subgroup $\hat N$ of $\hat G$ and a non-$\hat G$-stable block c of $\hat N$ such that $bc \neq 0$, and denote by $\hat H$ the stabilizer of c in $\hat G$. The product bc is a block of $\hat H$ and if* (Q) *holds for* $(bc, \hat H)$ *then it holds for* $(b, \hat G)$.

Proof: It is clear that $\mathrm{Tr}_{\hat{H}}^{\hat{G}}(bc) = b\mathrm{Tr}_{\hat{H}}^{\hat{G}}(c) = b$ and then it follows from Proposition 3.5 in [53] that bc is a block of $\hat{H}$, that bc belongs to a point β of $\hat{H}$ on $k\hat{G}b$ and that we have $\hat{G}$- and $\hat{H}$-interior algebra isomorphism (cf. 1.30)

$$k_*\hat{G}b \cong \mathrm{Ind}_{\hat{H}}^{\hat{G}}(k_*\hat{H}bc) \quad \text{and} \quad (k_*\hat{G}b)_\beta \cong k_*\hat{H}bc \qquad 16.6.1;$$

moreover, it follows from the same proposition that we may assume that a defect pointed group P_γ of $\hat{H}_\beta$ (cf. 1.15) determines the chosen maximal Brauer $(b, \hat{G})$-pair (P, e); then, always from the same proposition, P is also a defect group of the block bc of $\hat{H}$ and the Frobenius P-category $\mathcal{H} = \mathcal{F}_{(bc,\hat{H})}$ associated with this block coincides with $\mathcal{F}$.

On the other hand, since we assume that a k^*-group automorphism of $\hat{G}$ stabilizes $\hat{N}$, if it stabilizes b then it still stabilizes the $\hat{G}$-orbit of c and it is clear that the stabilizer $\mathrm{Aut}_{k^*}(\hat{G})_c$ of c acts on $\hat{H}$; hence, we get a canonical group homomorphism

$$\mathrm{Out}_{k^*}(\hat{G})_b \longrightarrow \mathrm{Out}_{k^*}(\hat{H})_{bc} \qquad 16.6.2;$$

more generally, we get as above a commutative diagram of exact sequences

$$\begin{array}{ccccccc} N_{\hat{H}}(P_\gamma) & \longrightarrow & \mathrm{Aut}_{k^*}(\hat{H})_\gamma & \longrightarrow & \mathrm{Out}_{k^*}(\hat{H})_{bc} & \longrightarrow 1 \\ \| & & \uparrow & & \uparrow & \\ N_{\hat{G}}(P_\gamma) & \longrightarrow & \mathrm{Aut}_{k^*}(\hat{G})_\gamma & \longrightarrow & \mathrm{Out}_{k^*}(\hat{G})_b & \longrightarrow 1 \end{array} \qquad 16.6.3.$$

Pushing it further, from the *Frattini argument* it is not difficult to get group homomorphisms from $\mathrm{Out}_{k^*}(\hat{G})_b$ and $\mathrm{Out}_{k^*}(\hat{H})_{bc}$ to the groups of *exterior automorphisms* of the *source P-interior algebras* of the blocks b of $\hat{G}$ and bc of $\hat{H}$ respectively. But, it follows from the isomorphism $(k_*\hat{G}b)_\beta \cong k_*\hat{H}bc$ that the *source P-interior algebras* of the blocks b of $\hat{G}$ and bc of $\hat{H}$ are isomorphic throughout an isomorphism compatible with homomorphism 16.6.2. Then, since $\widehat{\mathfrak{aut}}_{\mathcal{F}^{\mathrm{nc}}}$ and $\widehat{\mathfrak{aut}}_{\mathcal{H}^{\mathrm{nc}}}$, $\mathcal{G}_k(\mathcal{F}, \widehat{\mathfrak{aut}}_{\mathcal{F}^{\mathrm{nc}}})$ and $\mathcal{G}_k(\mathcal{H}, \widehat{\mathfrak{aut}}_{\mathcal{H}^{\mathrm{nc}}})$, and $\mathcal{G}_k(\hat{G}, b)$ and $\mathcal{G}_k(\hat{H}, bc)$ are completely determined from the respective source algebras (cf. Theorem 3.1 in [39] and Proposition 6.21 in [42]), we are done.

Proposition 16.7 *With the notation above, assume that any block involved in b of any characteristic k^*-subgroup of $\hat{G}$ is G-stable, and that there is a nontrivial characteristic k^*-subgroup $\hat{N}$ of $\hat{G}$ with a block c of defect zero such that $bc = b$. Then there are a k^*-group $\grave{G}$ with k^*-quotient G and a k^*-group homomorphism $\grave{G} \to (k_*\hat{N}c)^*$ extending the canonical map $\hat{N} \to k_*\hat{N}c$ such that, setting $\hat{\hat{G}} = \hat{G} * \grave{G}^\circ/\Delta(N)$, we have a $\hat{G}$-interior algebra isomorphism*

$$k_*\hat{G}b \cong k_*\hat{N}c \otimes_k k_*\hat{\hat{G}}\bar{b} \qquad 16.7.1$$

for a suitable block $\bar{b}$ of $\hat{\hat{G}}$ and if (Q) *holds for $(\bar{b}, \hat{\hat{G}})$, it holds for $(b, \hat{G})$.*

Proof: Since $k_*\hat{N}c$ is a simple algebra (cf. 1.17), the action of G can be lifted to a k^*-group homomorphism $\rho: \hat{}\!G \to (k_*\hat{N}c)^*$ extending the canonical map $\hat{N} \to k_*\hat{N}c$ where $\hat{}\!G$ is a suitable k^*-group containing $\hat{N}$, with the same k^*-quotient G (cf. 1.25); then, the canonical decomposition $\hat{N} * \hat{N}^\circ \cong k^* \times N$ (cf. 1.24) determines a normal subgroup $\Delta(N)$ of $\hat{G} * \hat{}\!G^\circ$ (cf. 1.24) isomorphic to N and, setting $\hat{\hat{G}} = \hat{G} * \hat{}\!G^\circ/\Delta(N)$, the tensor product $k_*\hat{N}c \otimes_k k_*\hat{\hat{G}}$ has an evident $\hat{G}$-interior algebra structure (cf. 1.27) and it follows from Theorem 3.7 in [53] that the obvious k^*-group isomorphism $\hat{G} \cong \hat{}\!G * \big(\hat{G} * \hat{}\!G^\circ\big)$ induces a $\hat{G}$-interior algebra isomorphism

$$k_*\hat{G}b \cong k_*\hat{N}c \otimes_k k_*\hat{\hat{G}}\bar{b} \qquad 16.7.2$$

for a suitable block $\bar{b}$ of $\hat{\hat{G}}$, so that the categories of $k_*\hat{G}b$- and $k_*\hat{\hat{G}}\bar{b}$-modules are equivalent. Moreover, it follows from the same theorem that P is isomorphic to its image $\bar{P}$ in $\hat{\hat{G}}$, that γ determines a local point $\bar{\gamma}$ of $\bar{P}$ on $k_*\hat{\hat{G}}\bar{b}$, that $\bar{P}_{\bar{\gamma}}$ is a defect pointed group of $\bar{b}$ and that this isomorphism $\alpha: P \cong \bar{P}$ induces a *Frobenius equivalence*

$$\mathfrak{f}_\alpha : \mathcal{F} \cong \mathcal{F}_{(\bar{b},\hat{\hat{G}})} \qquad 16.7.3.$$

As in the argument above, since we assume that a k^*-group automorphism of $\hat{G}$ stabilizes $\hat{N}$, if it stabilizes b then it fixes c and also induces a k^*-group automorphism of $\hat{}\!G$, so that it induces a k^*-group automorphism of $\hat{\hat{G}}$ stabilizing $\bar{b}$; hence, we get a canonical group homomorphism

$$\omega : \mathrm{Out}_{k^*}(\hat{G})_b \longrightarrow \mathrm{Out}_{k^*}(\hat{\hat{G}})_{\bar{b}} \qquad 16.7.4.$$

Pushing it further, from the *Frattini argument* it is not difficult to get group homomorphisms from $\mathrm{Out}_{k^*}(\hat{G})_b$ and $\mathrm{Out}_{k^*}(\hat{\hat{G}})_{\bar{b}}$ to the groups of *exterior automorphisms* of the *source P- and $\bar{P}$-interior algebras* $(k_*\hat{G}b)_\gamma$ and $(k_*\hat{\hat{G}}\bar{b})_{\bar{\gamma}}$ of the blocks b of $\hat{G}$ and $\bar{b}$ of $\hat{\hat{G}}$ respectively (cf. 1.13).

But, it follows from isomorphism 16.7.2 that we have a canonical P-interior algebra embedding

$$(k_*\hat{G}b)_\gamma \longrightarrow k_*\hat{N}c \otimes_k \mathrm{Res}_\alpha(k_*\hat{\hat{G}}\bar{b})_{\bar{\gamma}} \qquad 16.7.5,$$

and, since we have a P-interior algebra embedding (cf. 1.21.3)

$$k \longrightarrow k_*\hat{N}c \otimes_k (k_*\hat{N}c)^\circ \qquad 16.7.6,$$

it is easily checked that we still have a canonical P-interior algebra embedding

$$\mathrm{Res}_\alpha(k_*\hat{\hat{G}}\bar{b})_{\bar{\gamma}} \longrightarrow (k_*\hat{N}c)^\circ \otimes_k (k_*\hat{G}b)_\gamma \qquad 16.7.7.$$

Then, since the restriction induces Morita equivalences from $(k_*\hat{G}b)_\gamma$ and $(k_*\hat{\bar{G}}\bar{b})_{\bar{\gamma}}$ to $k_*\hat{G}b$ and $k_*\hat{\bar{G}}\bar{b}$ respectively (cf. 1.13), from the embeddings above we get an $\mathcal{O}\mathrm{Out}_{k^*}(\hat{G})_b$-module isomorphism

$$\mathcal{G}_k(\hat{G}, b) \cong \mathrm{Res}_\omega\big(\mathcal{G}_k(\hat{\bar{G}}, \bar{b})\big) \qquad 16.7.8.$$

Moreover, it follows from the uniqueness part of Theorems 11.10 and 11.32, and from the embeddings above that we have a natural isomorphism

$$\widehat{\mathfrak{aut}}_{\mathcal{F}^{\mathrm{nc}}} \cong \widehat{\mathfrak{aut}}_{(\mathcal{F}_{(\bar{b},\hat{\bar{G}})})^{\mathrm{nc}}} \circ \mathfrak{f}_\alpha \qquad 16.7.9$$

and therefore it is clear that we get an $\mathcal{O}\mathrm{Out}_{k^*}(\hat{G})_b$-module isomorphism

$$\mathcal{G}_k(\mathcal{F}, \widehat{\mathfrak{aut}}_{\mathcal{F}^{\mathrm{nc}}}) \cong \mathrm{Res}_\omega\big(\mathcal{G}_k(\mathcal{F}_{(\bar{b},\hat{\bar{G}})}, \widehat{\mathfrak{aut}}_{(\mathcal{F}_{(\bar{b},\hat{\bar{G}})})^{\mathrm{nc}}})\big) \qquad 16.7.10\,.$$

Now, the $\mathcal{O}\mathrm{Out}_{k^*}(\hat{G})_b$-module isomorphisms 16.7.8 and 16.7.10 prove the last statement. We are done.

Proposition 16.8 *With the notation above, assume that any block involved in b of any nontrivial characteristic k^*-subgroup of $\hat{G}$ is G-stable and has positive defect. Set $Z = Z\big(\mathbf{O}_p(\hat{G})\big)$ and $\hat{H} = C_{\hat{G}}(Z)$ and, for any cyclic p'-subgroup C of $\hat{G}/\hat{H}$, denote by $\hat{H}^C$ the converse image of C in $\hat{G}$. Then, b is a block of $\hat{H}^C$ and if* (Q) *holds for $(b, \hat{H}^C)$ for any cyclic p'-subgroup C of $\hat{G}/\hat{H}$, it holds for $(b, \hat{G})$.*

Proof: Set $U = \mathbf{O}_p(\hat{G})$; since b is contained in $kC_{\hat{G}}(U)$ (cf. 1.19 above and Proposition 5.15 in [42]), it follows from our hypothesis that b is also a block of $\hat{H}$ and of $C_{\hat{G}}(U)$; in particular, (U,b) is a Brauer $(b,\hat{G})$-pair contained in (P,e) (cf. 1.15) and therefore, setting $Q = \hat{H} \cap P$, there is a block f of $C_{\hat{G}}(Q) = C_{\hat{H}}(Q)$ such that (cf. 1.15)

$$(U,b) \subset (Q,f) \subset (P,e) \qquad 16.8.1;$$

actually, (U,b) and (Q,f) are also Brauer $(b,\hat{H})$-pairs and (Q,f) is a maximal one (cf. 1.15 and 1.19); moreover, according to Proposition 15.10, b is a block of $\hat{H}^C$ for any cyclic p'-subgroup C of $\hat{G}/\hat{H}$.

Denote by $\mathcal{H} = \mathcal{F}_{(b,\hat{H})}$ the Frobenius Q-category associated with $(b,\hat{H})$, which is clearly a subcategory of $\mathcal{F}$. Since $C_{\hat{G}}(Q) \subset \hat{H}$, it follows again from Proposition 15.10 that any $\mathcal{H}$-selfcentralizing subgroup of Q is $\mathcal{F}$-nilcentralized, so that $\mathcal{H}^{^{\mathrm{sc}}}$ is also a subcategory of $\mathcal{F}^{^{\mathrm{nc}}}$, and by the *Frattini argument* we get (cf. 1.15)

$$\mathcal{F}(Q)/\mathcal{H}(Q) \cong N_{\hat{G}}(Q,f)/N_{\hat{H}}(Q,f) \cong \hat{G}/\hat{H} \qquad 16.8.2.$$

Moreover, for any cyclic p'-subgroup C of $\mathcal{F}(Q)/\mathcal{H}(Q)$, it follows from Proposition 12.12 that there is a unique Frobenius Q-category $\mathcal{H}^C$, containing and normalizing $\mathcal{H}$, such that $\mathcal{H}^C(Q)$ is the converse image of C in $\mathcal{F}(Q)$; moreover, identifying $\mathcal{F}(Q)/\mathcal{H}(Q)$ with $\hat{G}/\hat{H}$, it follows from Proposition 12.8 that $\mathcal{H}^C$ coincides with the Frobenius Q-category associated with the block b of $\hat{H}^C$.

That is to say, we are in the situation discussed in 15.24 and therefore, considering the action of the stabilizer $\mathrm{Aut}_{k^*}(\hat{G})_{(P,e)}$ on the set of cyclic p'-subgroups of $\mathcal{F}(Q)/\mathcal{H}(Q) \cong \hat{G}/\hat{H}$, it follows from Corollaries 15.26 and 15.39 that we get $\mathcal{O}\mathrm{Out}_{k^*}(\hat{G})_b$-module isomorphisms

$$\begin{aligned} \mathcal{G}_k(\hat{G}, b) &\cong \Big(\bigoplus_C \mathcal{R}_{\hat{H}}\mathcal{G}_k(\hat{H}^C, b)\Big)^{\hat{G}/\hat{H}} \\ \mathcal{G}_k(\mathcal{F}, \widehat{\mathfrak{aut}}_{\mathcal{F}^{\mathrm{nc}}}) &\cong \Big(\bigoplus_C \mathcal{R}_{\mathcal{H}}\mathcal{G}_k(\mathcal{H}^C, \widehat{\mathfrak{aut}}_{(\mathcal{H}^C)^{\mathrm{nc}}})\Big)^{\mathcal{F}(Q)/\mathcal{H}(Q)} \end{aligned} \qquad 16.8.3$$

where C runs over the set of cyclic p'-subgroups of $\mathcal{F}(Q)/\mathcal{H}(Q) \cong \hat{G}/\hat{H}$.

Now, let us assume that we have an $\mathcal{O}\mathrm{Out}_{k^*}(\hat{H}^C)_b$-module isomorphism

$$\mathcal{G}_k(\hat{H}^C, b) \cong \mathcal{G}_k(\mathcal{H}^C, \widehat{\mathfrak{aut}}_{(\mathcal{H}^C)^{\mathrm{nc}}}) \qquad 16.8.4$$

for any cyclic p'-subgroup C of $\mathcal{F}(Q)/\mathcal{H}(Q) \cong \hat{G}/\hat{H}$; since any group homomorphism from C to k^* determines a homomorphism to k^* from the k^*-quotient H^C of $\hat{H}^C$, and therefore it determines a k^*-group automorphism of $\hat{H}^C$ in the center of $\mathrm{Aut}_{k^*}(\hat{H}^C)$, we have an $\mathcal{O}$-algebra homomorphism

$$\mathcal{G}_k(C) \cong \mathcal{O}\mathrm{Hom}(C, k^*) \longrightarrow Z\big(\mathcal{O}\mathrm{Out}_{k^*}(\hat{H}^C)_b\big) \qquad 16.8.5$$

compatible with the $\mathcal{G}_k(C)$-module structures of both members in isomorphism 16.8.4 (cf. 15.21 and 15.33).

Consequently, for any cyclic p'-subgroup C of $\mathcal{F}(Q)/\mathcal{H}(Q) \cong \hat{G}/\hat{H}$, we still have an $\mathcal{O}\mathrm{Out}_{k^*}(\hat{H}^C)_b$-module isomorphism (cf. isomorphisms 15.23.2 and 15.37.1)

$$\begin{array}{ccc} \mathcal{R}\mathcal{G}_k(C) \otimes_{\mathcal{G}_k(C)} \mathcal{G}_k(\hat{H}^C, b) & \cong & \mathcal{R}\mathcal{G}_k(C) \otimes_{\mathcal{G}_k(C)} \mathcal{G}_k(\mathcal{H}^C, \widehat{\mathfrak{aut}}_{(\mathcal{H}^C)^{\mathrm{nc}}}) \\ \wr\| & & \wr\| \\ \mathcal{R}_{\hat{H}}\mathcal{G}_k(\hat{H}^C, b) & & \mathcal{R}_{\mathcal{H}}\mathcal{G}_k(\mathcal{H}^C, \widehat{\mathfrak{aut}}_{(\mathcal{H}^C)^{\mathrm{nc}}}) \end{array} \qquad 16.8.6.$$

At this point, it is easy to construct an $\mathcal{O}\mathrm{Out}_{k^*}(\hat{G})_b$-module isomorphism

$$\Big(\bigoplus_C \mathcal{R}_{\hat{H}}\mathcal{G}_k(\hat{H}^C, b)\Big)^{\hat{G}/\hat{H}} \cong \Big(\bigoplus_C \mathcal{R}_{\mathcal{H}}\mathcal{G}_k(\mathcal{H}^C, \widehat{\mathfrak{aut}}_{(\mathcal{H}^C)^{\mathrm{nc}}})\Big)^{\mathcal{F}(Q)/\mathcal{H}(Q)} \qquad 16.8.7$$

where C runs over the set of cyclic p'-subgroups of $\mathcal{F}(Q)/\mathcal{H}(Q) \cong \hat{G}/\hat{H}$. We are done.

Proposition 16.9 *With the same notation as above, set $Z = Z(\mathbf{O}_p(\hat{G}))$ and $\hat{\bar{G}} = \hat{G}/Z$, and denote by $\bar{b}$ the image of b in $Z(k_*\hat{\bar{G}})$. Assume that any block involved in b of any nontrivial characteristic k^*-subgroup of $\hat{G}$ is G-stable and has positive defect, and that the quotient $\hat{G}/C_{\hat{G}}(Z)$ is a cyclic p'-group. Then, if (Q) holds for $(\bar{c}, \hat{\bar{G}})$ for any block $\bar{c}$ of $\hat{\bar{G}}$ such that $\bar{c}\bar{b} = \bar{c}$, it holds for $(b, \hat{G})$.*

Proof: Set $\hat{H} = C_{\hat{G}}(Z)$ and $\hat{\bar{H}} = \hat{H}/Z$, and denote by P_γ the defect pointed group of b determined by (P, e), and by $\bar{\gamma}$ the local point of $\bar{P} = P/Z$ on $k_*\hat{\bar{H}}$ containing the image of $\gamma \cap k_*\hat{H}$ (cf. Lemma 15.49). It follows from Proposition 15.50 that, setting $\mathcal{P}_{\bar{\gamma}} = \mathcal{P}\big(k_* C^{\hat{\bar{G}}}_{\hat{\bar{H}}}(\bar{P}_{\bar{\gamma}})\big)$ and denoting by $\bar{\mathcal{P}}_{\bar{\gamma}}$ the set of $N_{\hat{G}}(P_\gamma)$-orbits in $\mathcal{P}_{\bar{\gamma}}$ and by $\bar{\varphi}$ the $N_{\hat{G}}(P_\gamma)$-orbit of $\varphi \in \mathcal{P}_{\bar{\gamma}}$, we have

$$\bar{b} = \sum_{\bar{\varphi} \in \bar{\mathcal{P}}_{\bar{\gamma}}} \bar{b}_{\bar{\varphi}} \qquad 16.9.1$$

where, for any $\varphi \in \mathcal{P}_{\bar{\gamma}}$, $\bar{b}_{\bar{\varphi}}$ is a block of $\hat{\bar{G}}$; moreover, according to the same proposition, $\bar{P}_{(\bar{\gamma},\varphi)}$ is a defect pointed group of $\bar{b}_{\bar{\varphi}}$ and, setting $\bar{\mathcal{F}}_{\bar{\varphi}} = \mathcal{F}_{(\bar{b}_{\bar{\varphi}}, \hat{\bar{G}})}$, we have the following $\mathcal{O}$-module isomorphisms

$$\begin{aligned} \mathcal{G}_k(\hat{G}, b) &\cong \bigoplus_{\bar{\varphi} \in \bar{\mathcal{P}}_{\bar{\gamma}}} \mathcal{G}_k(\hat{\bar{G}}, \bar{b}_{\bar{\varphi}}) \\ \mathcal{G}_k(\mathcal{F}, \widehat{\mathfrak{aut}}_{\mathcal{F}^{\mathrm{nc}}}) &\cong \bigoplus_{\bar{\varphi} \in \bar{\mathcal{P}}_{\bar{\gamma}}} \mathcal{G}_k(\bar{\mathcal{F}}_{\bar{\varphi}}, \widehat{\mathfrak{aut}}_{(\bar{\mathcal{F}}_{\bar{\varphi}})^{\mathrm{nc}}}) \end{aligned} \qquad 16.9.2.$$

More precisely, it is clear that $\mathrm{Aut}_{k^*}(\hat{G})$ acts on the k^*-group $\hat{\bar{G}}$ inducing an $\mathcal{O}$-algebra homomorphism

$$\mathcal{O}\mathrm{Out}_{k^*}(\hat{G}) \longrightarrow \mathcal{O}\mathrm{Out}_{k^*}(\hat{\bar{G}}) \qquad 16.9.3$$

and, since the stabilizer $\mathrm{Aut}_{k^*}(\hat{G})_{P_\gamma}$ of P_γ in $\mathrm{Aut}_{k^*}(\hat{G})$ clearly acts on the k^*-group $C^{\hat{\bar{G}}}_{\hat{\bar{H}}}(\bar{P}_{\bar{\gamma}})$ (cf. definition 15.5.4), by the exact sequence 16.4.3 we get an action of $\mathrm{Out}_{k^*}(\hat{G})_b$ on $\bar{\mathcal{P}}_{\bar{\gamma}}$; then, for any $\bar{\varphi} \in \bar{\mathcal{P}}$, denoting by $\mathrm{Out}_{k^*}(\hat{G})_{b,\bar{\varphi}}$ the stabilizer of $\bar{\varphi}$ in $\mathrm{Out}_{k^*}(\hat{G})_b$, on the one hand we get a new $\mathcal{O}$-algebra homomorphism

$$\mathcal{O}\mathrm{Out}_{k^*}(\hat{G})_{b,\bar{\varphi}} \longrightarrow \mathcal{O}\mathrm{Out}_{k^*}(\hat{\bar{G}})_{\bar{b}_{\bar{\varphi}}} \qquad 16.9.4;$$

on the other hand, it is quite clear that $\mathrm{Out}_{k^*}(\hat{G})_b$ acts on the two families of $\mathcal{O}$-modules

$$\{\mathcal{G}_k(\hat{\bar{G}}, \bar{b}_{\bar{\varphi}})\}_{\bar{\varphi} \in \bar{\mathcal{P}}_{\bar{\gamma}}} \quad \text{and} \quad \{\mathcal{G}_k(\bar{\mathcal{F}}_{\bar{\varphi}}, \widehat{\mathfrak{aut}}_{(\bar{\mathcal{F}}_{\bar{\varphi}})^{\mathrm{nc}}})\}_{\bar{\varphi} \in \bar{\mathcal{P}}_{\bar{\gamma}}} \qquad 16.9.5$$

and it is easily checked that these actions are compatible with the isomorphisms 16.9.2; hence, we obtain $\mathcal{O}\mathrm{Out}_{k^*}(\hat{G})_b$-module isomorphisms

$$\begin{aligned} \mathcal{G}_k(\hat{G}, b) &\cong \bigoplus_{\bar\varphi} \mathrm{Ind}_{\mathrm{Out}_{k^*}(\hat{G})_{b,\bar\varphi}}^{\mathrm{Out}_{k^*}(\hat{G})_b} \big(\mathcal{G}_k(\hat{\bar{G}}, \bar{b}_{\bar\varphi})\big) \\ \mathcal{G}_k(\mathcal{F}, \widehat{\mathfrak{aut}}_{\mathcal{F}^{\mathrm{nc}}}) &\cong \bigoplus_{\bar\varphi} \mathrm{Ind}_{\mathrm{Out}_{k^*}(\hat{G})_{b,\bar\varphi}}^{\mathrm{Out}_{k^*}(\hat{G})_b} \big(\mathcal{G}_k(\bar{\mathcal{F}}_{\bar\varphi}, \widehat{\mathfrak{aut}}_{(\bar{\mathcal{F}}_{\bar\varphi})^{\mathrm{nc}}})\big) \end{aligned} \qquad 16.9.6$$

where $\bar\varphi$ runs over a set of representatives for the set of $\mathrm{Out}_{k^*}(\hat{G})_b$-orbits on $\bar{\mathcal{P}}_{\bar\gamma}$.

Consequently, if we assume that we have an $\mathcal{O}\mathrm{Out}_{k^*}(\hat{\bar{G}})_{b_{\bar\varphi}}$-module isomorphism

$$\mathcal{G}_k(\hat{\bar{G}}, \bar{b}_{\bar\varphi}) \cong \mathcal{G}_k(\bar{\mathcal{F}}_{\bar\varphi}, \widehat{\mathfrak{aut}}_{(\bar{\mathcal{F}}_{\bar\varphi})^{\mathrm{nc}}}) \qquad 16.9.7$$

for any $\bar\varphi \in \bar{\mathcal{P}}_{\bar\gamma}$, it suffices to consider the restrictions throughout homomorphism 16.9.4 of a set of representatives for the $\mathrm{Out}_{k^*}(\hat{G})_b$-orbits on $\bar{\mathcal{P}}_{\bar\gamma}$ to obtain an $\mathcal{O}\mathrm{Out}_{k^*}(\hat{G})_b$-module isomorphism $\mathcal{G}_k(\hat{G}, b) \cong \mathcal{G}_k(\mathcal{F}, \widehat{\mathfrak{aut}}_{\mathcal{F}^{\mathrm{nc}}})$. We are done.

16.10 At this point, arguing by induction on $|G|$, in order to prove that a block b of a nontrivial finite k^*-group $\hat{G}$ fulfills (Q) we may assume that any block involved in b of any nontrivial characteristic k^*-subgroup of $\hat{G}$ is G-stable and has positive defect, which forces $\mathbf{O}_{p'}(\hat{G}) = k^*$, and that we also have (cf. 1.31)

$$\mathbf{O}_p(\hat{G}) = \{1\} \qquad 16.10.1.$$

Then, it is well-known that the product H of all the minimal nontrivial normal subgroups of the k^*-quotient G of $\hat{G}$ is a *characteristic* subgroup of G isomorphic to a direct product

$$H \cong \prod_{i\in I} H_i \qquad 16.10.2$$

of a finite family of noncommutative simple groups H_i of order divisible by p (cf. Theorem 1.5 in [28]).

16.11 Denoting by $\hat{H}$ and $\hat{H}_i$ the respective converse images of H and H_i in $\hat{G}$, it is quite clear that

$$\hat{H} = \widehat{\prod}_{i\in I} \hat{H}_i \quad \text{and} \quad C_{\hat{G}}(\hat{H}) = k^* \qquad 16.11.1;$$

moreover, since this decomposition is unique, the action of $\mathrm{Aut}_{k^*}(\hat{G})$ on $\hat{H}$ induces an $\mathrm{Aut}_{k^*}(\hat{G})$-action on I and, denoting by $\hat{W}$ the kernel of the action of $\hat{G}$ on I, we have $\hat{H} \subset \hat{W}$ and get an injective group homomorphism

$$\hat{W}/\hat{H} \longrightarrow \prod_{i\in I} \mathrm{Out}_{k^*}(\hat{H}_i) \qquad 16.11.2;$$

thus, admitting the announced *Classification of the Finite Simple Groups*, the quotient $\hat{W}/\hat{H}$ is *solvable*.

16.12 Let c be the block of $\hat{H}$ such that $cb = b$, recall that (P, e) is a maximal Brauer $(b, \hat{G})$-pair and set $Q = P \cap \hat{H}$; as in our standard setting in 15.7, it follows from Proposition 15.9 that Q is a defect group of c and that there is a block f of $C_{\hat{H}}(Q)$ such that we have $e\mathrm{Br}_P(f) \neq 0$ and (Q, f) is a maximal Brauer $(c, \hat{H})$-pair; then, we consider the Frobenius P- and Q-categories

$$\mathcal{F} = \mathcal{F}_{(b,\hat{G})} \quad \text{and} \quad \mathcal{H} = \mathcal{F}_{(c,\hat{H})} \qquad 16.12.1.$$

Since clearly $c = \otimes_{i\in I} c_i$ where c_i is a block of $\hat{H}_i$, we have $Q = \prod_{i\in I} Q_i$ where Q_i is a defect group of c_i, and $f = \otimes_{i\in I} f_i$ where f_i is a block of $C_{\hat{H}_i}(Q_i)$ and (Q_i, f_i) is a maximal Brauer $(c_i, \hat{H}_i)$-pair.

16.13 Moreover, since we are assuming that any block involved in b of any nontrivial characteristic k^*-subgroup of $\hat{G}$ has positive defect, for any $i \in I$ the defect group Q_i is *nontrivial*; in particular, since any $\mathcal{H}$-selfcentralizing subgroup T of Q contains $Z(Q) = \prod_{i\in I} Z(Q_i)$, $C_{\hat{G}}(T)$ centralizes $Z(Q_i) \neq \{1\}$ for any $i \in I$ and therefore we get

$$C_{\hat{G}}(T) \subset \hat{W} \qquad 16.13.1.$$

Thus, $\hat{N} = \hat{H}{\cdot}C_{\hat{G}}(Q, f)$ is a normal subgroup of $\hat{G}$ contained in $\hat{W}$ and therefore the quotient $\hat{N}/\hat{H}$ is solvable (cf. 16.11).

16.14 As usual, we define $\mathbf{O}_{p',p}(\hat{N}/\hat{H})$ by (cf. 1.31)

$$\mathbf{O}_{p',p}(\hat{N}/\hat{H})/\mathbf{O}_{p'}(\hat{N}/\hat{H}) = \mathbf{O}_p\big((\hat{N}/\hat{H})/\mathbf{O}_{p'}(\hat{N}/\hat{H})\big) \qquad 16.14.1.$$

Consider the converse image $\hat{K}$ of $\mathbf{O}_{p',p}(\hat{N}/\hat{H})$ in $\hat{G}$ and the block d of $\hat{K}$ fulfilling $db = b$, and set $R = \hat{K} \cap P$; it follows from Proposition 15.9 that R is a defect group of d and then, from Proposition 5.3 and Corollary 6.3 in [34], that R/Q is a Sylow p-subgroup of $\mathbf{O}_{p',p}(\hat{N}/\hat{H})$. But, according to Proposition 15.10, (Q, f) remains a Brauer $(d, \hat{K})$-pair since $f^K = f$, and, according to Proposition 15.9, we may assume that there exists a block g of $C_{\hat{K}}(R)$ such that $(Q, f) \subset (R, g)$ as Brauer $(d, \hat{K})$-pairs (cf. 1.15) and therefore we still have $(R, g^G) \subset (P, e)$ (cf. Proposition 15.10). Then, it is clear that $C_{\hat{G}}(R, g)$

is contained in $C_{\hat{G}}(Q,f)$ and therefore, since $C_{\hat{G}}(R)$ acts trivially on R/Q and the quotient $\hat{N}/\hat{H}$ is solvable, we get (cf. Theorem 3.2 in [28, Ch. 6])

$$C_{\hat{G}}(R,g) \subset \hat{K} \qquad 16.14.2.$$

At this point, it follows from Proposition 15.10 that $d = b$ and it is clear that (R,g) is also a maximal Brauer $(b,\hat{K})$-pair; denote by $\mathcal{K} = \mathcal{F}_{(b,\hat{K})}$ the Frobenius R-category associated with the block $(b,\hat{K})$, which is clearly a subcategory of $\mathcal{F}$. From now on in this chapter, all this will be our *standard setting*.

Proposition 16.15 *In our standard setting, for any cyclic p'-subgroup C of $\hat{G}/\hat{K}$, denote by $\hat{K}^C$ the converse image of C in $\hat{G}$. Then, b is a block of $\hat{K}^C$ and if* (Q) *holds for $(b,\hat{K}^C)$ for any cyclic p'-subgroup C of $\hat{G}/\hat{K}$, it holds for $(b,\hat{G})$.*

Proof: We already know that b is a block of $\hat{K}$ (cf. 16.14) and therefore, since $C_{\hat{G}}(R,g) \subset \hat{K}$, it follows from Proposition 15.10 that b is also a block of $\hat{K}^C$ for any cyclic p'-subgroup C of $\hat{G}/\hat{K}$ and that any $\mathcal{K}$-selfcentralizing subgroup of R is $\mathcal{F}$-nilcentralized, so that $\mathcal{K}^{^{\rm sc}}$ is a subcategory of $\mathcal{F}^{^{\rm nc}}$, and together with the *Frattini argument* we get (cf. 1.15)

$$\mathcal{F}(R)/\mathcal{K}(R) \cong N_{\hat{G}}(R,g)/N_{\hat{K}}(R,g) \cong \hat{G}/\hat{K} \qquad 16.15.1$$

since (R,g) is a maximal Brauer $(b,\hat{K})$-pair.

Now, for any cyclic p'-subgroup C of $\mathcal{F}(R)/\mathcal{K}(R)$, it follows from Proposition 12.12 that there is a unique Frobenius R-category $\mathcal{K}^C$, containing and normalizing $\mathcal{K}$, such that $\mathcal{K}^C(R)$ is the converse image of C in $\mathcal{F}(R)$; moreover, identifying $\mathcal{F}(R)/\mathcal{K}(R)$ with $\hat{G}/\hat{K}$, it follows from Proposition 12.8 that $\mathcal{K}^C$ is the Frobenius R-category associated with the block b of $\hat{K}^C$.

That is to say, we are in the situation discussed in 15.24 above and therefore, considering the action of the stabilizer $\mathrm{Aut}_{k^*}(\hat{G})_{(P,e)}$ on the set of cyclic p'-subgroups of $\mathcal{F}(R)/\mathcal{K}(R) \cong \hat{G}/\hat{K}$, it follows from Corollaries 15.26 and 15.39 that we get $\mathcal{O}\mathrm{Out}_{k^*}(\hat{G})_b$-module isomorphisms

$$\begin{aligned} \mathcal{G}_k(\hat{G},b) &\cong \Big(\bigoplus_C \mathcal{R}_{\hat{K}}\mathcal{G}_k(\hat{K}^C,b)\Big)^{\hat{G}/\hat{K}} \\ \mathcal{G}_k(\mathcal{F},\widehat{\mathfrak{aut}}_{\mathcal{F}^{\rm nc}}) &\cong \Big(\bigoplus_C \mathcal{R}_{\mathcal{K}}\mathcal{G}_k(\mathcal{K}^C,\widehat{\mathfrak{aut}}_{(\mathcal{K}^C)^{\rm nc}})\Big)^{\mathcal{F}(R)/\mathcal{K}(R)} \end{aligned} \qquad 16.15.2$$

where C runs over the set of cyclic p'-subgroups of $\mathcal{F}(R)/\mathcal{K}(R) \cong \hat{G}/\hat{K}$.

Now, let us assume that we have an $\mathcal{O}\mathrm{Out}_{k^*}(\hat{K}^C)_b$-module isomorphism

$$\mathcal{G}_k(\hat{K}^C, b) \cong \mathcal{G}_k(\mathcal{K}^C, \widehat{\mathfrak{aut}}_{(\mathcal{K}^C)^{\mathrm{nc}}}) \qquad 16.15.3$$

for any cyclic p'-subgroup C of $\mathcal{F}(R)/\mathcal{K}(R) \cong \hat{G}/\hat{K}$; once again, since any group homomorphism from C to k^* determines a homomorphism to k^* from the k^*-quotient K^C of $\hat{K}^C$, and therefore it determines a k^*-group automorphism of $\hat{K}^C$ in the center of $\mathrm{Aut}_{k^*}(\hat{K}^C)$, we get an $\mathcal{O}$-algebra homomorphism

$$\mathcal{G}_k(C) \cong \mathcal{O}\mathrm{Hom}(C, k^*) \longrightarrow Z\big(\mathcal{O}\mathrm{Out}_{k^*}(\hat{K}^C)_b\big) \qquad 16.15.4$$

compatible with the $\mathcal{G}_k(C)$-module structures of both members in isomorphism 16.15.3 (cf. 15.21 and 15.33).

Consequently, for any cyclic p'-subgroup C of $\mathcal{F}(R)/\mathcal{K}(R) \cong \hat{G}/\hat{K}$, we still have an $\mathcal{O}\mathrm{Out}_{k^*}(\hat{K}^C)_b$-module isomorphism (cf. isomorphisms 15.23.2 and 15.37.1)

$$\begin{array}{ccc}\mathcal{R}\mathcal{G}_k(C) \otimes_{\mathcal{G}_k(C)} \mathcal{G}_k(\hat{K}^C, b) & \cong & \mathcal{R}\mathcal{G}_k(C) \otimes_{\mathcal{G}_k(C)} \mathcal{G}_k(\mathcal{K}^C, \widehat{\mathfrak{aut}}_{(\mathcal{K}^C)^{\mathrm{nc}}}) \\ \wr\| & & \wr\| \\ \mathcal{R}_{\hat{K}}\mathcal{G}_k(\hat{K}^C, b) & & \mathcal{R}_{\mathcal{K}}\mathcal{G}_k(\mathcal{K}^C, \widehat{\mathfrak{aut}}_{(\mathcal{K}^C)^{\mathrm{nc}}})\end{array} \qquad 16.15.5.$$

At this point, it is not difficult to construct an $\mathcal{O}\mathrm{Out}_{k^*}(\hat{G})_b$-module isomorphism

$$\Big(\bigoplus_C \mathcal{R}_{\hat{K}}\mathcal{G}_k(\hat{K}^C, b)\Big)^{\hat{G}/\hat{K}} \cong \Big(\bigoplus_C \mathcal{R}_{\mathcal{K}}\mathcal{G}_k(\mathcal{K}^C, \widehat{\mathfrak{aut}}_{(\mathcal{K}^C)^{\mathrm{nc}}})\Big)^{\mathcal{F}(R)/\mathcal{K}(R)} \qquad 16.15.6$$

where C runs over the set of cyclic p'-subgroups of $\mathcal{F}(R)/\mathcal{K}(R) \cong \hat{G}/\hat{K}$. We are done.

16.16 In our standard setting, it follows from this proposition that now it suffices to discuss the case where $\hat{G}/\hat{K}$ is a cyclic p'-group, so that in particular we have $P = R$; let us denote by $\hat{L}$ the converse image of $\mathbf{O}_{p'}(\hat{K}/\hat{H})$ in $\hat{K}$; since we have

$$\hat{K}/\hat{L} \cong P/Q \qquad 16.16.1$$

and $b = \sum_x \lambda_x \cdot \hat{x}$ where x runs over the set of p'-elements of the k^*-quotient K of $\hat{K}$ and $\hat{x}$ is a lifting of x to $\hat{K}$ (cf. 16.2 in [54]), b is also a block of $\hat{L}$ and we denote by $\mathcal{L} = \mathcal{F}_{(b,\hat{L})}$ the corresponding Frobenius Q-category. Moreover, $\hat{G}$ acts on $\bar{P} = P/Q$ and $\hat{K}$ acts trivially on the *Frattini space* $V = \bar{P}/\Phi(\bar{P})$; thus, considering the semidirect product $\hat{G}^V = V \rtimes \hat{G}$, it is clear that b remains a block of the subgroups

$$\hat{K}^V = V \times \hat{K} \quad \text{and} \quad \hat{L}^V = V \times \hat{L} \qquad 16.16.2,$$

that the p-subgroups $P^{^V} = V \times P$ and $Q^{^V} = V \times Q$ are respective defect groups of b in $\hat K^{^V}$ and $\hat L^{^V}$, and that e remains a block of $C_{\hat K^V}(P^{^V})$. Then, since we still have (cf. 16.14.2)

$$C_{\hat G^V}(P^{^V}, e) \subset \hat K^{^V} \qquad 16.16.3,$$

it follows from Proposition 15.10 that b is also a block of $\hat G^{^V}$ and it is clear that the restriction throughout the canonical homomorphism $\hat G^{^V} \to \hat G$ determines an $\mathcal{O}\mathrm{Out}(\hat G)_b$-module isomorphism

$$\mathcal{G}_k(\hat G, b) \cong \mathcal{G}_k(\hat G^{^V}, b) \qquad 16.16.4.$$

16.17 Analogously, respectively denote by

$$\mathcal{F}^{^V} = \mathcal{F}_{(b,\hat G^V)} \quad , \quad \mathcal{K}^{^V} = \mathcal{F}_{(b,\hat K^V)} \quad \text{and} \quad \mathcal{L}^{^V} = \mathcal{F}_{(b,\hat L^V)} \qquad 16.17.1$$

the Frobenius $P^{^V}$- and $Q^{^V}$-categories of the blocks $(b, \hat G^{^V})$, $(b, \hat K^{^V})$ and $(b, \hat L^{^V})$; since $\hat G^{^V}/\hat K^{^V}$ is a p'-group, a subgroup of $P^{^V}$ is $\mathcal{F}^{^V}$-selfcentralizing if and only if it is $\mathcal{K}^{^V}$-selfcentralizing; moreover, it is clear that it is $\mathcal{K}^{^V}$-selfcentralizing if and only if it has the form $T^{^V} = V \times T$ where T is a $\mathcal{K}$-selfcentralizing or, equivalently, an $\mathcal{F}$-selfcentralizing subgroup of P. In this case, denoting by h the block of $C_{\hat K}(T)$ such that $(T,h) \subset (P,e)$ (cf. 16.14), h is also a block of

$$C_{\hat K^V}(T^{^V}) = V \times C_{\hat K}(T) \qquad 16.17.2$$

and, since $\hat G/\hat K$ is a p'-group, it follows from Proposition 15.10 that we have

$$C_{\hat G^V}(T^{^V}, h) = V \times C_{\hat K}(T) = V \times C_{\hat G}(T,h) \qquad 16.17.3$$

and that the relative traces

$$h^{G^V} = \mathrm{Tr}_{C_{\hat G^V}(T^V,h)}^{C_{\hat G^V}(T^V)}(h) \quad \text{and} \quad h^G = \mathrm{Tr}_{C_{\hat G}(T,h)}^{C_{\hat G}(T)}(h) \qquad 16.17.4$$

are blocks of $C_{\hat G^V}(T^{^V})$ and $C_{\hat G}(T)$ respectively.

16.18 Consequently, we get

$$\begin{aligned} \mathcal{F}^{^V}(T^{^V}) &\cong N_{\hat G^V}(T^{^V}, h^{G^V})/C_{\hat G^V}(T^{^V}) \\ &\cong N_{\hat G^V}(T^{^V}, h)/C_{\hat G^V}(T^{^V}, h) \\ &\cong N_{\hat G}(T,h)/C_{\hat G}(T,h) \cong \mathcal{F}(T) \end{aligned} \qquad 16.18.1;$$

moreover, since we have (cf. Proposition 3.5 in [53])

$$k_*\bar{C}_{\hat{G}^V}(T^V)\bar{h}^{G^V} \cong \mathrm{Ind}_{C_{\hat{G}^V}(T^V,h)}^{C_{\hat{G}^V}(T^V)}\big(k_*\bar{C}_{\hat{K}}(T)\bar{h}\big)$$
$$k_*\bar{C}_{\hat{G}}(T)\bar{h}^G \cong \mathrm{Ind}_{C_{\hat{G}}(T,h)}^{C_{\hat{G}}(T)}\big(k_*\bar{C}_{\hat{K}}(T)\bar{h}\big) \qquad 16.18.2,$$

it is easily checked that we still get a canonical k^*-group isomorphism (cf. Proposition 6.18 in [42])

$$\hat{\mathcal{F}}^V(T^V) \cong \hat{\mathcal{F}}(T) \qquad 16.18.3;$$

then, denoting by $\mathfrak{p}:\mathcal{F}^V \to \mathcal{F}$ the obvious *Frobenius functor*, it is not difficult to check that we get a natural isomorphism

$$\widehat{\mathfrak{aut}}_{(\mathcal{F}^V)^{\mathrm{sc}}} \cong \widehat{\mathfrak{aut}}_{\mathcal{F}^{\mathrm{sc}}} \circ \mathfrak{ch}^*(\mathfrak{p}^{\mathrm{sc}}) \qquad 16.18.4.$$

In conclusion, according to Corollary 14.7, we have also an $\mathcal{O}\mathrm{Out}_{k^*}(\hat{G})_b$-module isomorphism

$$\mathcal{G}_k(\mathcal{F},\widehat{\mathfrak{aut}}_{\mathcal{F}^{\mathrm{nc}}}) \cong \mathcal{G}_k(\mathcal{F}^V,\widehat{\mathfrak{aut}}_{(\mathcal{F}^V)^{\mathrm{nc}}}) \qquad 16.18.5.$$

Proposition 16.19 *In our standard setting, assume that* $C = \hat{G}/\hat{K}$ *is a cyclic* p'*-group and denote by* $\hat{L}$ *the converse image of* $\mathbf{O}_{p'}(\hat{K}/\hat{H})$ *in* $\hat{K}$ *and by* $\hat{L}^C$ *the converse image in* $\hat{G}$ *of a complement of the Sylow* p*-subgroup of* $\hat{G}/\hat{L}$. *Then,* b *is a block of* $\hat{L}^C$ *and if* (Q) *holds for* $(b,\hat{L}^C)$, *it holds for* $(b,\hat{G})$.

Proof: Since, from the very definition of $\hat{K}$, we have $\hat{K} = \hat{H}\cdot C_{\hat{K}}(Q,f)$ (cf. 16.14), we still have $\hat{K}^V = \hat{L}^V\cdot C_{\hat{K}^V}(Q^V,f)$ (cf. 16.17.2) and the quotient $\hat{G}^V/(\hat{L}^V\cdot C_{\hat{G}^V}(Q^V,f))$ is also a cyclic p'-group. Moreover, since $C_{G^V}(Q^V,f)$ acts trivially on $V=\bar{P}/\Phi(\bar{P})$, a p'-subgroup of $C_{G^V}(Q^V,f)$ acts trivially on (cf. Theorem 1.4 in [28, Ch. 5])

$$\bar{P} = P/Q \cong \hat{K}/\hat{L} \cong \mathbf{O}_{p',p}(\hat{G}/\hat{H})/\mathbf{O}_{p'}(\hat{G}/\hat{H}) \qquad 16.19.1$$

and therefore it is contained in $C_{L^V}(Q^V,f)$ (cf. Theorem 3.2 in [28, Ch. 6]), so that the quotient $C_{G^V}(Q^V,f)/C_{L^V}(Q^V,f)$ is a p-group; thus, we have $C_{\hat{G}^V}(Q^V,f) = C_{\hat{K}^V}(Q^V,f)$ and therefore we still have

$$\begin{aligned} C &\cong \hat{G}^V/(\hat{L}^V\cdot C_{\hat{G}^V}(Q^V,f)) \\ &\cong N_{\hat{G}^V}(Q^V,f)/(N_{\hat{L}^V}(Q^V,f)\cdot C_{\hat{G}^V}(Q^V,f)) \qquad 16.19.2. \\ &\cong \mathcal{F}^V(Q^V)/\mathcal{L}^V(Q^V) \end{aligned}$$

Consequently, denoting by $(\mathcal{L}^{^V})^C$ the Frobenius $Q^{^V}$-category determined by this isomorphism (cf. Proposition 12.12), it follows from Proposition 15.13 that we have $\mathcal{O}$-isomorphisms

$$\begin{gathered}\mathcal{G}_k(\mathcal{F}^{^V},\widehat{\mathfrak{aut}}_{(\mathcal{F}^V)^{\rm nc}}) \cong \mathcal{G}_k\big((\mathcal{L}^{^V})^C,\widehat{\mathfrak{aut}}_{((\mathcal{L}^V)^C)^{\rm nc}}\big)\\ \mathcal{G}_k(\hat G^{^V},b) \cong \mathcal{G}_k\big((\hat L^{^V})^C,b\big)\end{gathered} \qquad 16.19.3.$$

On the other hand, since $\hat G$ acts transitively on the set of choices for $\hat L^C$, it follows from the *Frattini argument* that we have canonical group homomorphisms

$$\begin{gathered}\mathrm{Out}_{k^*}(\hat G)_b \longrightarrow \mathrm{Out}_{k^*}(\hat L^C)_b\\ \mathrm{Out}_{k^*}(\hat G^{^V})_b \longrightarrow \mathrm{Out}_{k^*}\big((\hat L^{^V})^C\big)_b\end{gathered} \qquad 16.19.4.$$

Moreover, since it follows from Proposition 15.13 that the bottom isomorphism in 16.19.3 is induced by the restriction, this isomorphism is actually an $\mathcal{O}\mathrm{Out}_{k^*}(\hat G^{^V})_b$-module isomorphism; but, since $\mathrm{Aut}_{k^*}(\hat G)$ stabilizes $\hat H$, we have a canonical group homomorphism from $\mathrm{Out}_{k^*}(\hat G)_b$ to $\mathrm{Out}_{k^*}(\hat G^{^V})_b$. Thus, we get $\mathcal{O}\mathrm{Out}_{k^*}(\hat G)_b$-module isomorphisms (cf. 16.16.4)

$$\mathcal{G}_k(\hat G,b) \cong \mathcal{G}_k(\hat G^{^V},b) \cong \mathcal{G}_k\big((\hat L^{^V})^C,b\big) \cong \mathcal{G}_k(\hat L^C,b) \qquad 16.19.5.$$

Similarly, we claim that we have $\mathcal{O}\mathrm{Out}_{k^*}(\hat G)_b$-module isomorphisms

$$\begin{aligned}\mathcal{G}_k(\mathcal{F},\widehat{\mathfrak{aut}}_{\mathcal{F}^{\rm nc}}) &\cong \mathcal{G}_k(\mathcal{F}^{^V},\widehat{\mathfrak{aut}}_{(\mathcal{F}^V)^{\rm nc}})\\ &\cong \mathcal{G}_k\big((\mathcal{L}^{^V})^C,\widehat{\mathfrak{aut}}_{((\mathcal{L}^V)^C)^{\rm nc}}\big) \cong \mathcal{G}_k\big(\mathcal{L}^C,\widehat{\mathfrak{aut}}_{(\mathcal{L}^C)^{\rm nc}}\big)\end{aligned} \qquad 16.19.6;$$

indeed, it follows from 16.3 and 16.4 that the groups

$$\begin{matrix}\mathrm{Out}_{k^*}(\hat G)_b & , & \mathrm{Out}_{k^*}(\hat G^{^V})_b\\ \mathrm{Out}_{k^*}\big((\hat L^{^V})^C\big)_b & \text{and} & \mathrm{Out}_{k^*}(\hat L^C)_b\end{matrix} \qquad 16.19.7$$

respectively act on the Grothendieck groups

$$\begin{matrix}\mathcal{G}_k(\mathcal{F},\widehat{\mathfrak{aut}}_{\mathcal{F}^{\rm nc}}) & , & \mathcal{G}_k(\mathcal{F}^{^V},\widehat{\mathfrak{aut}}_{(\mathcal{F}^V)^{\rm nc}})\\ \mathcal{G}_k\big((\mathcal{L}^{^V})^C,\widehat{\mathfrak{aut}}_{((\mathcal{L}^V)^C)^{\rm nc}}\big) & \text{and} & \mathcal{G}_k\big(\mathcal{L}^C,\widehat{\mathfrak{aut}}_{(\mathcal{L}^C)^{\rm nc}}\big)\end{matrix} \qquad 16.19.8;$$

thus, according to all the group homomorphisms above, $\mathrm{Out}_{k^*}(\hat G)_b$ acts on all of these Grothendieck groups.

Then, in 16.19.6 the first $\mathcal{O}\mathrm{Out}_{k^*}(\hat{G})_b$-module isomorphism is just isomorphism 16.18.5. The second $\mathcal{O}\mathrm{Out}_{k^*}(\hat{G})_b$-module isomorphism follows from the naturality of isomorphism 15.13.2 in Proposition 15.13. Finally, note that the arguments in 16.17 and 16.18 still apply to $\hat{L}^C$ and $\hat{L}$ in the place of $\hat{G}$ and $\hat{K}$, and therefore the last $\mathcal{O}\mathrm{Out}_{k^*}(\hat{G})_b$-module isomorphism follows again from isomorphism 16.18.5.

At this point, condition (Q) for $(b, \hat{L}^C)$ guarantees the existence of a $\mathcal{O}\mathrm{Out}_{k^*}(\hat{L}^C)_b$-module isomorphism

$$\mathcal{G}_k(\hat{L}^C, b) \cong \mathcal{G}_k\big(\mathcal{L}^C, \widehat{\mathfrak{aut}}_{(\mathcal{L}^C)^{\mathrm{nc}}}\big) \qquad 16.19.9$$

and then the $\mathcal{O}\mathrm{Out}_{k^*}(\hat{G})_b$-module isomorphisms 16.19.5 and 16.19.6 imply condition (Q) for $(b, \hat{G})$. We are done.

16.20 In our standard setting, it follows from this proposition that it suffices to discuss the case where $\hat{G}/\hat{H}$ is a p'-group and $\hat{G}/\hat{K}$ is cyclic; in this case, we actually have $\hat{K} = \hat{H}{\cdot}C_{\hat{G}}(Q, f)$ and, since $\hat{W}$ contains $\hat{K}$ (cf. 16.11), $\hat{G}/\hat{W}$ is cyclic too. For any $i \in I$, we respectively denote by G^i and K^i the respective images of W and $\hat{K}$ in $\mathrm{Aut}_{k^*}(\hat{H}_i)$; note that, since

$$K^i = H_i{\cdot}C_{K^i}(Q_i, f_i) \qquad 16.20.1$$

and G^i/K^i is cyclic, Lemmas 16.21 and 16.27 below successively determine a k^*-group $\hat{K}^i$ containing $\hat{H}_i$ of k^*-quotient K^i and a k^*-group $\hat{G}^i$ containing $\hat{K}$ of k^*-quotient G^i for any $i \in I$. Then, we set

$$\hat{K}^\star = \widehat{\prod}_{i\in I} \hat{K}^i \quad \text{and} \quad \hat{W}^\star = \widehat{\prod}_{i\in I} \hat{G}^i \qquad 16.20.2.$$

Lemma 16.21 *Let $\hat{H}$ be a finite k^*-group such that $Z(\hat{H}) = k^*$, c a block of $\hat{H}$, (Q, f) a maximal Brauer $(c, \hat{H})$-pair and K a subgroup of $\mathrm{Aut}_{k^*}(\hat{H})_c$ such that $H \subset K$ and that $K = H{\cdot}C_K(Q, f)$. Then there exists an essentially unique central k^*-extension $\hat{K}$ of K containing $\hat{H}$, lifting the inclusion $H \subset K$ and inducing a k^*-group isomorphism*

$$C_{\hat{K}}(Q, f) \longrightarrow \big(k_*\bar{C}_{\hat{H}}(Q)\bar{f}\big)^* \qquad 16.21.1$$

which extends the canonical k^*-group homomorphism from $C_{\hat{H}}(Q)$ and lifts the action of $C_{\hat{K}}(Q, f)$ over the simple algebra $k_*\bar{C}_{\hat{H}}(Q)\bar{f}$.

Proof: The action of $C_K(Q, f)$ on the simple k-algebra $k_*\bar{C}_{\hat{H}}(Q)\bar{f}$ (cf. 1.16) defines a k^*-group $\hat{C}_K(Q, f)$ containing $C_{\hat{H}}(Q, f)$ as a normal k^*-subgroup (cf. 1.25) and then it suffices to consider

$$\hat{K} = \big(\hat{H} \rtimes \hat{C}_K(Q, f)\big)/\Delta^*\big(\hat{C}_H(Q, f)\big) \qquad 16.21.2$$

where $\Delta^*\big(\hat{C}_H(Q,f)\big)$ denotes the "inverse diagonal" subgroup (cf. definition 16.22.3 below); the uniqueness is clear.

16.22 At this point, similarly as in 15.17 above, assume that

16.22.1 *For any $i \in I$, the canonical k^*-group homomorphism*

$$C_{\hat{H}_i}(Q_i) \longrightarrow \big(k_* C_{\hat{H}_i}(Q_i) f_i\big)^*$$

can be extended to an $\mathrm{Aut}_{k^*}(\hat{K}^i)_{(Q_i,f_i)}$-stable k^*-group homomorphism from $N_{\hat{K}^i}(Q_i, f_i)$ where $\mathrm{Aut}_{k^*}(\hat{K}^i)_{(Q_i,f_i)}$ is the stabilizer of (Q_i, f_i) in $\mathrm{Aut}_{k^*}(\hat{K}^i)$.

Then it follows from Lemma 16.21 that $\hat{K}^\star$ *essentially contains* $\hat{K}$ and that the canonical k^*-group homomorphism

$$C_{\hat{H}}(Q) \longrightarrow \big(k_* C_{\hat{H}}(Q) f\big)^* \cong \Big(\bigotimes_{i\in I} k_* C_{\hat{H}_i}(Q_i) f_i\Big)^* \qquad 16.22.2$$

can be extended to $N_{\hat{K}}(Q,f)$; analogously, it follows from Lemma 16.27 below that $\hat{W}^\star$ *essentially contains* $\hat{W}$. Moreover, it is clear that we may assume that $\hat{G}$ acts on $\hat{W}^\star$; then, in the semidirect product $\hat{W}^\star \rtimes \hat{G}$, the *inverse diagonal* subgroup

$$\Delta^*(\hat{W}) = \{(\hat{x}, \hat{x}^{-1})\}_{\hat{x}\in\hat{W}} \qquad 16.22.3$$

is a normal subgroup; finally, we consider the k^*-group

$$\hat{G}^\star = (\hat{W}^\star \rtimes \hat{G})/\Delta^*(\hat{W}) \qquad 16.22.4$$

which, up to suitable identifications, contains $\hat{G}$ and $\hat{W}^\star$.

Proposition 16.23 *In our standard setting, assume that $\hat{G}/\hat{H}$ is a p'-group, that $C = \hat{G}/\hat{K}$ is cyclic and that condition* 16.22.1 *holds. Denote by δ the local point of Q on $k_*\hat{H}$ determined by f, by φ the point of $k_* C_{\hat{H}}^{\hat{G}}(Q,f)$ such that (δ,φ) is the local point of $Q = P$ on $k_*\hat{G}b$ determined by e, and by $\hat{G}^\varphi$ the converse image in $\hat{G}$ of the stabilizer $C_{(\delta,\varphi)}$ of (δ,φ) in C. Then, b is a block of $\hat{G}^\varphi$ and if* (Q) *holds for $(b,\hat{G}^\varphi)$, it holds for $(b,\hat{G})$.*

Proof: According to Lemma 15.16, the pair (δ,φ) has been identified indeed with a local point of $Q = P$ on $k_*\hat{G}b$, so that e determines φ; moreover, according to 15.17, C acts on the set of points of $k_* C_{\hat{H}}^{\hat{G}}(Q,f)$ and therefore it makes sense to consider the converse image $\hat{G}^\varphi$ of $C_{(\delta,\varphi)}$ in $\hat{G}$.

Since $C_{\hat{G}}(Q,f) \subset \hat{K} \subset \hat{G}^\varphi$, b is also a block of $\hat{G}^\varphi$ (cf. Proposition 15.10); moreover, since by the *Frattini argument* the stabilizer $\mathrm{Aut}(\hat{G})_{(P,e)}$ of (P,e)

in $\mathrm{Aut}(\hat{G})_b$ *covers* $\mathrm{Out}_{k^*}(\hat{G})_b$ (cf. 1.15), we get a canonical group homomorphism

$$\mathrm{Out}_{k^*}(\hat{G})_b \longrightarrow \mathrm{Out}_{k^*}(\hat{G}^\varphi)_b \qquad 16.23.1.$$

Now, setting $\mathcal{F}^\varphi = \mathcal{F}_{(b,\hat{G}^\varphi)}$, it is easily checked from Corollary 15.20 that we have $\mathcal{O}\mathrm{Out}(\hat{G})_b$-module isomorphisms

$$\mathcal{G}_k(\mathcal{F}, \widehat{\mathfrak{aut}}_{\mathcal{F}^{\mathrm{nc}}}) \cong \mathcal{G}_k(\mathcal{F}^\varphi, \widehat{\mathfrak{aut}}_{(\mathcal{F}^\varphi)^{\mathrm{nc}}}) \quad \text{and} \quad \mathcal{G}_k(\hat{G}, b) \cong \mathcal{G}_k(\hat{G}^\varphi, b) \qquad 16.23.2.$$

Thus, if there is an $\mathcal{O}\mathrm{Out}_{k^*}(\hat{G}^\varphi)_b$-module isomorphism

$$\mathcal{G}_k(\mathcal{F}^\varphi, \widehat{\mathfrak{aut}}_{(\mathcal{F}^\varphi)^{\mathrm{nc}}}) \cong \mathcal{G}_k(\hat{G}^\varphi, b) \qquad 16.23.3$$

then we get an $\mathcal{O}\mathrm{Out}_{k^*}(\hat{G})_b$-module isomorphism $\mathcal{G}_k(\mathcal{F}, \widehat{\mathfrak{aut}}_{\mathcal{F}^{\mathrm{nc}}}) \cong \mathcal{G}_k(\hat{G}, b)$. We are done.

16.24 With the notation and the hypothesis in the proposition above, it follows from this proposition that we are reduced to discuss the case where C fixes (δ, φ); but, for any $i \in I$, choosing an $\mathrm{Aut}_{k^*}(\hat{K}^i)$-stable k^*-group homomorphism

$$\hat{\rho}_i : N_{\hat{K}^i}(Q_i, f_i) \longrightarrow \big(k_* C_{\hat{H}_i}(Q_i) f_i\big)^* \qquad 16.24.1$$

which extends the canonical k^*-group homomorphism

$$C_{\hat{H}_i}(Q_i) \longrightarrow \big(k_* C_{\hat{H}_i}(Q_i) f_i\big)^* \qquad 16.24.2,$$

we a get k^*-group isomorphism (cf. 15.17)

$$C_{\hat{H}_i}^{\hat{K}^i}(Q_i, f_i) \cong k^* \times D_i \qquad 16.24.3;$$

moreover, clearly we can choose a family $\{\rho_i\}_{i\in I}$ which is $\mathrm{Aut}_{k^*}(\hat{G})$-stable. In this situation, the family of trivial representations of D_i for any $i \in I$ determines a $\hat{G}^\star/\hat{K}^\star$-stable point $\varphi^\star$ of the k-algebra (cf. definition 16.22.4)

$$k_* C_{\hat{H}}^{\hat{G}^\star}(Q, f) \cong \bigotimes_{i\in I} k_* C_{\hat{H}_i}^{\hat{K}^i}(Q_i, f_i) \cong \bigotimes_{i\in I} k D_i \qquad 16.24.4$$

and the local point $(\delta, \varphi^\star)$ of Q on $k_*\hat{G}^\star$ (cf. Lemma 15.16) determines a block $b^\star$ of $\hat{G}^\star$ with defect pointed group $Q_{(\delta,\varphi^\star)}$; we set $\mathcal{F}^\star = \mathcal{F}_{(b^\star,\hat{G}^\star)}$.

Proposition 16.25 *In our standard setting, assume that* $\hat{G}/\hat{H}$ *is a* p'*-group, that* $C = \hat{G}/\hat{K}$ *is cyclic and that condition* 16.22.1 *holds. Moreover, with the notation above, assume that* C *fixes* (δ, φ). *If* (Q) *holds for* $(b^\star, \hat{G}^\star)$ *then it holds for* $(b, \hat{G})$.

Proof: Since $\mathrm{Aut}(\hat{G})$ stabilizes both $\hat{H}$ and the family $\{\hat{H}_i\}_{i\in I}$, $\mathrm{Aut}(\hat{G})_b$ still stabilizes $\hat{K}$ and therefore it acts on the family $\{K^i\}_{i\in I}$; hence, according to Lemmas 16.21 and 16.27 below, $\mathrm{Aut}_{k^*}(\hat{G})_b$ still acts on the family $\{\hat{G}^i\}_{i\in I}$ (cf. 16.20), and therefore it acts on $\hat{G}^\star$ (cf. definition 16.22.4); moreover, from our hypothesis it is easily checked that $\mathrm{Aut}_{k^*}(\hat{G})_b$ stabilizes $b^\star$. Consequently, we get a group homomorphism

$$\mathrm{Out}_{k^*}(\hat{G})_b \longrightarrow \mathrm{Out}_{k^*}(\hat{G}^\star)_{b^\star} \qquad 16.25.1.$$

and, as above, it follows from Corollary 15.20 that we have $\mathcal{O}\mathrm{Out}_{k^*}(\hat{G})_b$-module isomorphisms

$$\mathcal{G}_k(\mathcal{F}, \widehat{\mathfrak{aut}}_{\mathcal{F}^{\mathrm{nc}}}) \cong \mathcal{G}_k(\mathcal{F}^\star, \widehat{\mathfrak{aut}}_{(\mathcal{F}^\star)^{\mathrm{nc}}}) \quad \text{and} \quad \mathcal{G}_k(\hat{G}, b) \cong \mathcal{G}_k(\hat{G}^\star, b^\star) \qquad 16.25.2.$$

Hence, if there is an $\mathcal{O}\mathrm{Out}_{k^*}(\hat{G}^*)_{b^*}$-module isomorphism

$$\mathcal{G}_k(\mathcal{F}^*, \widehat{\mathfrak{aut}}_{(\mathcal{F}^*)^{\mathrm{nc}}}) \cong \mathcal{G}_k(\hat{G}^*, b) \qquad 16.25.3$$

then we get an $\mathcal{O}\mathrm{Out}_{k^*}(\hat{G})_b$-module isomorphism $\mathcal{G}_k(\mathcal{F}, \widehat{\mathfrak{aut}}_{\mathcal{F}^{\mathrm{nc}}}) \cong \mathcal{G}_k(\hat{G}, b)$. We are done.

16.26 From now on, in our standard setting let us assume that $\hat{G}/\hat{H}$ is a p'-group, that $C = \hat{G}/\hat{K}$ is cyclic, that condition 16.22.1 holds, and that $\hat{G}^*$ coincides with $\hat{G}$ (cf. 16.22.4), so that we have (cf. 16.20)

$$W = \prod_{i\in I} G^i \quad , \quad K = \prod_{i\in I} K^i \quad \text{and} \quad b = \bigotimes_{i\in I} b_i \qquad 16.26.1,$$

where, for any $i \in I$, b_i is a block of $\hat{K}^i$ and $\hat{G}^i$; we set

$$\mathcal{W} = \mathcal{F}_{(b,\hat{W})} \quad , \quad \mathcal{F}^i = \mathcal{F}_{(b_i,\hat{G}^i)} \quad \text{and} \quad \mathcal{K}^i = \mathcal{F}_{(b_i,\hat{K})} \qquad 16.26.2.$$

It is clear that $\bar{C} = G/W$ acts on I; if I decomposes on a disjoint union of two nonempty $\bar{C}$-stable subsets I' and I'' then, setting

$$\hat{W}' = \hat{\prod_{i'\in I'}} \hat{G}^{i'} \quad \text{and} \quad \hat{W}'' = \hat{\prod_{i''\in I''}} \hat{G}^{i''} \qquad 16.26.3,$$

the following lemma implies that there exist essentially unique k^*-groups $\hat{G}'$ and $\hat{G}''$, respectively containing and normalizing $\hat{W}'$ and $\hat{W}''$, such that

$$\hat{G}'/\hat{W}' \cong \bar{C} \cong \hat{G}''/\hat{W}'' \quad \text{and} \quad \hat{G}' \hat{\times}_{\bar{C}} \hat{G}'' \cong \hat{G} \qquad 16.26.4.$$

Moreover, setting $b' = \otimes_{i' \in I'} b_{i'}$ and $b'' = \otimes_{i'' \in I''} b_{i''}$, it follows from Proposition 15.10 that b' and b'' are respective blocks of $\hat{G}'$ and $\hat{G}''$; we set

$$\begin{aligned} \mathcal{F}' &= \mathcal{F}_{(b',\hat{G}')} \quad \text{and} \quad \mathcal{F}'' = \mathcal{F}_{(b'',\hat{G}'')} \\ \mathcal{W}' &= \mathcal{F}_{(b',\hat{W}')} \quad \text{and} \quad \mathcal{W}'' = \mathcal{F}_{(b'',\hat{W}'')} \end{aligned} \qquad 16.26.5.$$

Note that, for any subgroup $\bar{D}$ of $\bar{C}$, we have an analogous situation with the respective converse images $\hat{W}^{\bar{D}}$, $\hat{W}'^{\bar{D}}$ and $\hat{W}''^{\bar{D}}$ of $\bar{D}$ in $\hat{G}$, $\hat{G}'$ and $\hat{G}''$.

Lemma 16.27 *Let n be a positive integer, $\hat{H}$ a central $(k^*)^n$-extension of a finite group H such that $Z(\hat{H}) = (k^*)^n$, and K a subgroup of $\mathrm{Aut}_{(k^*)^n}(\hat{H})$ such that, up to a suitable identification, we have $H \subset K$ and K/H is cyclic. There exists an essentially unique central $(k^*)^n$-extension $\hat{K}$ of K containing $\hat{H}$ and lifting the inclusion map $H \to K$. In particular, any automorphism $\hat{\sigma}$ of $\hat{H}$ stabilizing K can be lifted to an automorphism of $\hat{K}$.*

Proof: We choose a cyclic subgroup C of K such that $K = H{\cdot}C$ and set $D = C \cap H$; since the converse image $\hat{D}$ of D in $\hat{H}$ is split, choosing a splitting, we can define an "inverse diagonal" subgroup $\Delta^*(D)$ (cf. definition 16.22.3) which is actually contained in the center of the semidirect product $\hat{H} \rtimes C$; then, it suffices to consider

$$\hat{K} = (\hat{H} \rtimes C)/\Delta^*(D) \qquad 16.27.1.$$

Moreover, if $\hat{\hat{K}}$ is a central $(k^*)^n$-extension of K containing $\hat{H}$ and lifting the inclusion map $H \to K$, then $\hat{\hat{K}} * \hat{K}^\circ$ contains $\hat{H} * \hat{H}^\circ$ which is *canonically* isomorphic to $(k^*)^n \times H$ and therefore, up to suitable identifications, the quotient $(\hat{\hat{K}} * \hat{K}^\circ)/H$ is a central $(k^*)^n$-extension of the cyclic group K/H, which is split; hence, $\hat{\hat{K}} * \hat{K}^\circ$ is also split and, more precisely, there is an isomorphism $\hat{\hat{K}} \cong \hat{K}$ inducing the identity on $\hat{H}$.

In particular, since $\hat{\sigma}$ induces an automorphism σ of K, it induces a group isomorphism $\hat{H} \rtimes C \cong \hat{H} \rtimes \sigma(C)$ mapping $\Delta^*(D)$ onto $\Delta^*\big(\sigma(D)\big)$ and therefore it determines an isomorphism

$$\hat{K} \cong \big(\hat{H} \rtimes \sigma(C)\big)/\Delta^*\big(\sigma(D)\big) \qquad 16.27.2;$$

but, the right member of this isomorphism is also a central $(k^*)^n$-extension of $\hat{H}$ lifting the inclusion map $H \to K$ and therefore it admits an isomorphism to $\hat{K}$ inducing the identity on $\hat{H}$. We are done.

Proposition 16.28 *With the notation and the hypothesis in* 16.26 *above, assume that the group $\mathrm{Aut}_{k^*}(\hat{G})_b$ stabilizes I' and I''. If* (Q) *holds for $(b', \hat{W}'^{\bar{D}})$ and $(b'', \hat{W}''^{\bar{D}})$ for any subgroup $\bar{D}$ of $\bar{C}$, then it holds for $(b, \hat{G})$.*

Proof: According to our hypothesis, we have canonical group homomorphisms

$$\mathrm{Out}_{k^*}(\hat{G}')_{b'} \longleftarrow \mathrm{Out}_{k^*}(\hat{G})_b \longrightarrow \mathrm{Out}_{k^*}(\hat{G}'')_{b''} \qquad 16.28.1;$$

then, since the homomorphisms from $\bar{C}$ to k^* induce k^*-group automorphisms of $\hat{G}$, $\hat{G}'$ and $\hat{G}''$, which are contained in the centers of $\mathrm{Aut}_{k^*}(\hat{G})$, $\mathrm{Aut}_{k^*}(\hat{G}')$ and $\mathrm{Aut}_{k^*}(\hat{G}'')$ respectively, it follows from Corollary 15.47 that we have $\mathcal{O}\mathrm{Out}_{k^*}(\hat{G})_b$-module isomorphisms

$$\begin{aligned}\mathcal{R}_{\hat{W}'}\mathcal{G}_k(\hat{G}',b') \otimes_{\mathcal{R}\mathcal{G}_k(\bar{C})} \mathcal{R}_{\hat{W}''}\mathcal{G}_k(\hat{G}'',b'') &\\ \cong \mathcal{R}_{\hat{W}}\mathcal{G}_k(\hat{G},b)&\\ \mathcal{R}_{\mathcal{W}'}\mathcal{G}_k(\mathcal{F}',\widehat{\mathfrak{aut}}_{\mathcal{F}'^{\mathrm{nc}}}) \otimes_{\mathcal{R}\mathcal{G}_k(\bar{C})} \mathcal{R}_{\mathcal{W}''}\mathcal{G}_k(\mathcal{F}'',\widehat{\mathfrak{aut}}_{\mathcal{F}''^{\mathrm{nc}}}) &\\ \cong \mathcal{R}_{\mathcal{W}}\mathcal{G}_k(\mathcal{F},\widehat{\mathfrak{aut}}_{\mathcal{F}^{\mathrm{nc}}})&\end{aligned} \qquad 16.28.2.$$

On the other hand, if we have $\mathcal{O}\mathrm{Out}_{k^*}(\hat{G}')_{b'}$- and $\mathcal{O}\mathrm{Out}_{k^*}(\hat{G}'')_{b''}$-module isomorphisms

$$\mathcal{G}_k(\mathcal{F}',\widehat{\mathfrak{aut}}_{\mathcal{F}'^{\mathrm{nc}}}) \cong \mathcal{G}_k(\hat{G}',b') \quad \text{and} \quad \mathcal{G}_k(\mathcal{F}'',\widehat{\mathfrak{aut}}_{\mathcal{F}''^{\mathrm{nc}}}) \cong \mathcal{G}_k(\hat{G}'',b'') \quad 16.28.3,$$

since the restriction induces compatible $\mathcal{G}_k(\bar{C})$-module structures on all the members of these isomorphisms (cf. 15.21 and 15.33), it follows from isomorphisms 15.23.2 and 15.37.1 that we still have $\mathcal{R}\mathcal{G}_k(\bar{C})\mathrm{Out}_{k^*}(\hat{G})_b$-module isomorphisms

$$\begin{aligned}\mathcal{R}_{\mathcal{W}'}\mathcal{G}_k(\mathcal{F}',\widehat{\mathfrak{aut}}_{\mathcal{F}'^{\mathrm{nc}}}) &\cong \mathcal{R}_{\hat{W}'}\mathcal{G}_k(\hat{G}',b')\\ \mathcal{R}_{\mathcal{W}''}\mathcal{G}_k(\mathcal{F}'',\widehat{\mathfrak{aut}}_{\mathcal{F}''^{\mathrm{nc}}}) &\cong \mathcal{R}_{\hat{W}''}\mathcal{G}_k(\hat{G}'',b'')\end{aligned} \qquad 16.28.4.$$

Then, from isomorphisms 16.28.2 we get an $\mathcal{O}\mathrm{Out}_{k^*}(\hat{G})_b$-module isomorphism

$$\mathcal{R}_{\mathcal{W}}\mathcal{G}_k(\mathcal{F},\widehat{\mathfrak{aut}}_{\mathcal{F}^{\mathrm{nc}}}) \cong \mathcal{R}_{\hat{W}}\mathcal{G}_k(\hat{G},b) \qquad 16.28.5.$$

Consequently, according to our hypothesis, for any subgroup $\bar{D}$ of $\bar{C}$ we have an $\mathcal{O}\mathrm{Out}_{k^*}(\hat{W}^{\bar{D}})_b$-module isomorphism

$$\mathcal{R}_{\mathcal{W}}\mathcal{G}_k(\mathcal{W}^{\bar{D}},\widehat{\mathfrak{aut}}_{(\mathcal{W}^{\bar{D}})^{\mathrm{nc}}}) \cong \mathcal{R}_{\hat{W}}\mathcal{G}_k(\hat{W}^{\bar{D}},b) \qquad 16.28.6;$$

but, since $\mathrm{Aut}_{k^*}(\hat{G})$ stabilizes $\hat{W}$, we have evident group homomorphisms

$$\bar{C} \longrightarrow \mathrm{Out}_{k^*}(\hat{W}^{\bar{D}})_b \longleftarrow \mathrm{Aut}_{k^*}(\hat{G})_b \qquad 16.28.7$$

and it is clear that the image of $\mathrm{Aut}_{k^*}(\hat{G})_b$ contains and normalizes the image of $\bar{C}$; hence, we still have an $\mathcal{O}\mathrm{Out}_{k^*}(\hat{G})_b$-module isomorphism

$$\mathcal{R}_{\mathcal{W}}\mathcal{G}_k(\mathcal{W}^{\bar{D}},\widehat{\mathfrak{aut}}_{(\mathcal{W}^{\bar{D}})^{\mathrm{nc}}})^{\bar{C}} \cong \mathcal{R}_{\hat{W}}\mathcal{G}_k(\hat{W}^{\bar{D}},b)^{\bar{C}} \qquad 16.28.8.$$

Then, it follows from isomorphisms 15.23.4 and 15.38.1 that the direct sum of isomorphisms 16.28.8 when $\bar D$ runs over the set of subgroups of $\bar C$ supplies an $\mathcal{O}\mathrm{Out}_{k^*}(\hat G)_b$-module isomorphism $\mathcal{G}_k(\mathcal{F},\widehat{\mathfrak{aut}}_{\mathcal{F}^{\mathrm{nc}}})\cong\mathcal{G}_k(\hat G,b)$. We are done.

16.29 Now, with the notation and the hypothesis in 16.26 above, we may assume that the group $\mathrm{Aut}_{k^*}(\hat G)_b$ acts transitively on I; in particular, it acts transitively on the set of $\bar C$-orbits of I and, for any $\bar C$-orbit O we consider the k^*-groups and the block

$$\hat W^O=\prod_{i\in O}G^i\quad,\quad\hat K^O=\prod_{i\in O}K^i\quad\text{and}\quad b^O=\bigotimes_{i\in O}b_i\qquad 16.29.1;$$

it follows from Lemma 16.27 that there exists an essentially unique k^*-group $\hat G^O$ containing $\hat W^O$ and fulfilling $\bar C\cong\hat G^O/\hat W^O$; then, it follows from Proposition 15.10 that b^O is also a block of $\hat G^O$ and we set

$$\mathcal{F}^O=\mathcal{F}_{(b_O,\hat G^O)}\quad,\quad\mathcal{W}^O=\mathcal{F}_{(b_O,\hat W^O)}\quad\text{and}\quad\mathcal{K}^O=\mathcal{F}_{(b_O,\hat K^O)}\qquad 16.29.2.$$

Moreover, $\hat G$ is isomorphic to the *direct sum over* $\bar C$ of the family of k^*-groups $\hat G^O$ when O runs over the set of $\bar C$-orbits of I.

Proposition 16.30 *With the notation and the hypothesis in* 16.26 *above, assume that the group* $\mathrm{Aut}_{k^*}(\hat G)_b$ *acts transitively on* I *and let* O *be a* $\bar C$*-orbit of* I. *If* (Q) *holds for* $\big(b^O,(\hat W^O)^{\bar D}\big)$ *for any subgroup* $\bar D$ *of* $\bar C$, *then it holds for* $(b,\hat G)$.

Proof: It is clear that the action of $\mathrm{Aut}_{k^*}(\hat G)_b$ on I induces an action of $\mathrm{Out}_{k^*}(\hat G)_b$ on the set $\bar I$ of $\bar C$-orbits of I; moreover, denoting by $\mathrm{Out}_{k^*}(\hat G)_{b,O}$ the stabilizer of O in $\mathrm{Out}_{k^*}(\hat G)_b$, it is quite clear that the restriction induces a group homomorphism

$$\mathrm{Out}_{k^*}(\hat G)_{b,O}\longrightarrow\mathrm{Out}_{k^*}(\hat G^O)_{b^O}\qquad 16.30.1.$$

On the other hand, it is quite clear that $\mathrm{Out}_{k^*}(\hat G)_b$ acts on the two families of $\mathcal{O}$-modules

$$\{\mathcal{G}_k(\hat G^{O'},b^{O'})\}_{O'\in\bar I}\quad\text{and}\quad\{\mathcal{R}_{\mathcal{W}^{O'}}\mathcal{G}_k(\mathcal{F}^{O'},\widehat{\mathfrak{aut}}_{(\mathcal{F}^{O'})^{\mathrm{nc}}})\}_{O'\in\bar I}\qquad 16.30.2$$

and then it is not difficult to check from the canonical isomorphisms defined in Corollary 15.47 that we have $\mathcal{O}\mathrm{Out}_{k^*}(\hat G)_b$-module isomorphisms

$$\begin{aligned}&{}^{\mathcal{RG}_k(\bar C)}\mathbf{Ten}^{\mathrm{Out}_{k^*}(\hat G)_b}_{\mathrm{Out}_{k^*}(\hat G)_{b,O}}\big(\mathcal{R}_{\hat W^O}\mathcal{G}_k(\hat G^O,b^O)\big)\\ &\qquad\cong\mathcal{R}_{\hat W}\mathcal{G}_k(\hat G,b)\\ &{}^{\mathcal{RG}_k(\bar C)}\mathbf{Ten}^{\mathrm{Out}_{k^*}(\hat G)_b}_{\mathrm{Out}_{k^*}(\hat G)_{b,O}}\big(\mathcal{R}_{\mathcal{W}^O}\mathcal{G}_k(\mathcal{F}^O,\widehat{\mathfrak{aut}}_{(\mathcal{F}^O)^{\mathrm{nc}}})\big)\\ &\qquad\cong\mathcal{R}_{\mathcal{W}}\mathcal{G}_k(\mathcal{F},\widehat{\mathfrak{aut}}_{\mathcal{F}^{\mathrm{nc}}})\end{aligned}\qquad 16.30.3$$

where ${}^{\mathcal{R}\mathcal{G}_k(\bar C)}\mathbf{Ten}$ denotes the *tensor induction of $\mathcal{R}\mathcal{G}_k(\bar C)$-modules* (cf. 8.2 above and A2.2 in [48]).

But, assume that we have an $\mathcal{O}\mathrm{Out}_{k^*}(\hat G^O)_{b^O}$-module isomorphism

$$\mathcal{G}_k(\mathcal{F}^O, \widehat{\mathfrak{aut}}_{(\mathcal{F}^O)^{\mathrm{nc}}}) \cong \mathcal{G}_k(\hat G^O, b^O) \qquad 16.30.4;$$

then, since the restriction induces compatible $\mathcal{G}_k(\bar C)$-module structures on both members of this isomorphism (cf. 15.21 and 15.33), it follows from isomorphisms 15.23.2 and 15.37.1 that we still have an $\mathcal{R}\mathcal{G}_k(\bar C)\mathrm{Out}_{k^*}(\hat G)_{b,\mathcal{O}}$-module isomorphism

$$\mathcal{R}_{\mathcal{W}^O}\mathcal{G}_k(\mathcal{F}^O, \widehat{\mathfrak{aut}}_{(\mathcal{F}^O)^{\mathrm{nc}}}) \cong \mathcal{R}_{\hat W^O}\mathcal{G}_k(\hat G^O, b^O) \qquad 16.30.5.$$

Thus, from isomorphisms 16.30.3 we get an $\mathcal{O}\mathrm{Out}_{k^*}(\hat G)_b$-module isomorphism

$$\mathcal{R}_{\mathcal{W}}\mathcal{G}_k(\mathcal{F}, \widehat{\mathfrak{aut}}_{\mathcal{F}^{\mathrm{nc}}}) \cong \mathcal{R}_{\hat W}\mathcal{G}_k(\hat G, b) \qquad 16.30.6.$$

Consequently, according to our hypothesis, for any subgroup $\bar D$ of $\bar C$ we have an $\mathcal{O}\mathrm{Out}_{k^*}\big((\hat W)^{\bar D}\big)_b$-module isomorphism

$$\mathcal{R}_{\mathcal{W}}\mathcal{G}_k\big((\mathcal{W})^{\bar D}, \widehat{\mathfrak{aut}}_{((\mathcal{W})^{\bar D})^{\mathrm{nc}}}\big) \cong \mathcal{R}_{\hat W}\mathcal{G}_k\big((\hat W)^{\bar D}, b\big) \qquad 16.30.7;$$

but, since $\mathrm{Aut}_{k^*}(\hat G)$ stabilizes $\hat W$, we have evident group homomorphisms

$$\bar C \longrightarrow \mathrm{Out}_{k^*}(\hat W^{\bar D})_b \longleftarrow \mathrm{Aut}_{k^*}(\hat G)_b \qquad 16.30.8$$

and it is clear that the image of $\mathrm{Aut}_{k^*}(\hat G)_b$ contains and normalizes the image of $\bar C$; hence, we still have an $\mathcal{O}\mathrm{Out}_{k^*}(\hat G)_b$-module isomorphism

$$\mathcal{R}_{\mathcal{W}}\mathcal{G}_k\big((\mathcal{W})^{\bar D}, \widehat{\mathfrak{aut}}_{((\mathcal{W})^{\bar D})^{\mathrm{nc}}}\big)^{\bar C} \cong \mathcal{R}_{\hat W}\mathcal{G}_k\big((\hat W)^{\bar D}, b\big)^{\bar C} \qquad 16.30.9.$$

Then, it follows from isomorphisms 15.23.4 and 15.38.1 that the direct sum of isomorphisms 16.30.9 when $\bar D$ runs over the set of subgroups of $\bar C$ supplies an $\mathcal{O}\mathrm{Out}_{k^*}(\hat G)_b$-module isomorphism $\mathcal{G}_k(\mathcal{F}, \widehat{\mathfrak{aut}}_{\mathcal{F}^{\mathrm{nc}}}) \cong \mathcal{G}_k(\hat G, b)$. We are done.

16.31 With the notation and the hypothesis in 16.26 above, in the last step of our reduction, we may assume that the group $\bar C = \hat G/\hat W$ acts regularly on I; in this case, since $W = \prod_{i\in I} G^i$ and $\bar C$ is a cyclic p'-group, fixing an element $i \in I$ it is elementary to check that (cf. 1.32)

$$G \cong G^i \wr \bar C \quad \text{and} \quad \hat G \cong \hat W \rtimes \bar C \qquad 16.31.1$$

where we may choose the right isomorphism in such a way that $\bar{C}$ stabilizes (Q, f). Actually, many arguments we develop in this last situation could have been discussed in the more general setting of chapter 15, as the interested reader may check.

16.32 In this situation, we have to consider the *direct product* of groups

$$\hat{\hat{W}} = \prod_{i \in I} \hat{G}^i \qquad 16.32.1;$$

since G/W is cyclic, it follows from Lemma 16.27 that there exists an essentially unique central $(k^*)^I$-extension $\hat{\hat{G}}$ of G containing $\hat{\hat{W}}$ and, fixing an element $i \in I$ as above, we have again (cf. 1.32)

$$\hat{\hat{G}} \cong \hat{G}^i \wr \bar{C} \qquad 16.32.2.$$

Moreover, denoting by $\nabla_{k^*} : (k^*)^I \to k^*$ the group homomorphism induced by the product and considering the *group algebras* of $(k^*)^I$ and $\hat{\hat{G}}$ over k and the k-algebra homomorphism $k(k^*)^I \to k$ determined by ∇_{k^*}, it is quite clear that we have a k^*-group and a k-algebra isomorphisms

$$\hat{\hat{G}}/\mathrm{Ker}(\nabla_{k^*}) \cong \hat{G} \quad \text{and} \quad k \otimes_{k(k^*)^I} k\hat{\hat{G}} \cong k_*\hat{G} \qquad 16.32.3.$$

16.33 In particular, since $\hat{\hat{G}}$ acts transitively on the family $\{\hat{G}^i\}_{i\in I}$ and, for any $i \in I$, b_i is a block of $\hat{G}^i$ (cf. 16.26.1), by the *Frattini argument* we get canonical group homomorphisms

$$\mathrm{Out}_{k^*}(\hat{G})_b \longrightarrow \mathrm{Out}_{k^*}(\hat{G}^i)_{b_i} \qquad 16.33.1.$$

Moreover, fixing an element $i \in I$, it follows from equality 16.32.1 that any $k_*\hat{G}_i$-module M_i can be considered as a $k\hat{\hat{W}}$-module, and the key point is that, according to isomorphisms 16.32.3, *the tensor induction* $\mathrm{Ten}_{\hat{\hat{W}}}^{\hat{\hat{G}}}(M_i)$ *becomes a* $k_*\hat{G}$*-module* (cf. 8.2).

Proposition 16.34 *With the notation and hypothesis above, let* i *be an element of* I. *The correspondence mapping any* $k_*\hat{G}_i$*-module* M_i *on the* $k_*\hat{G}$*-module* $\mathrm{Ten}_{\hat{\hat{W}}}^{\hat{\hat{G}}}(M_i)$ *induces an* $\mathcal{O}\mathrm{Out}_{k^*}(\hat{G})_b$*-module isomorphism*

$$\mathcal{R}\mathcal{G}_k(\bar{C}) \otimes_{\mathcal{O}} \mathcal{G}_k(\hat{G}^i, b_i) \cong \mathcal{R}_{\hat{\hat{W}}}\mathcal{G}_k(\hat{G}, b) \qquad 16.34.1.$$

Proof: Recall that the *tensor induction* of M_i from $\hat{\hat{W}}$ to $\hat{\hat{G}}$ is the $k\hat{\hat{G}}$-module (cf. 8.2 above and A2.2 in [48])

$$\mathrm{Ten}_{\hat{\hat{W}}}^{\hat{\hat{G}}}(M_i) = \bigotimes_{X\in\hat{\hat{G}}/\hat{\hat{W}}} (kX \otimes_{k\hat{\hat{W}}} M_i) \qquad 16.34.2,$$

where kX denotes the k-vector space over the (right) $\hat{\hat{W}}$-class $X \subset \hat{\hat{G}}$, endowed with the (right) $k\hat{\hat{W}}$-module structure determined by the multiplication on the right; actually, in our situation the image of $\bar{C}$ in $\hat{\hat{G}}$ by isomorphism 16.32.2 supplies a multiplicative closed set of representatives. It is clear that in $\mathrm{Ten}_{\hat{\hat{W}}}^{\hat{\hat{G}}}(M_i)$ the multiplication by $(\lambda_i)_{i\in I} \in (k^*)^I$ coincides with the multiplication by $\prod_{i\in I} \lambda_i \in k^*$, so that, according to isomorphisms 16.32.3, $\mathrm{Ten}_{\hat{\hat{W}}}^{\hat{\hat{G}}}(M_i)$ becomes a $k_*\hat{G}$-module.

Moreover, setting $c = |\bar{C}| = |I|$ and denoting by χ_i the *modular character* of M_i (cf. 18 in [54]), it is well-known and easily checked that, for any p'-element $\hat{\hat{x}} \in \hat{\hat{G}}$ such that $\hat{\hat{G}} = \hat{\hat{W}}\cdot\langle\hat{\hat{x}}\rangle$, we have

$$\big(\mathrm{Ten}_{\hat{\hat{W}}}^{\hat{\hat{G}}}(\chi_i)\big)(\hat{\hat{x}}) = \chi_i\big(\overline{(\hat{\hat{x}})^c}^{\,i}\big) \qquad 16.34.3$$

where $\overline{(\hat{\hat{x}})^c}^{\,i}$ denotes the *projection* of $(\hat{\hat{x}})^c \in \hat{\hat{W}}$ in $\hat{G}^i$; in particular, this equality shows that the tensor induction $\mathrm{Ten}_{\hat{\hat{W}}}^{\hat{\hat{G}}}$ induces an $\mathcal{O}$-module homomorphism from $\mathcal{G}_k(\hat{G}_i)$ to the $\mathcal{O}$-module of *modular characters* of $\hat{G}$ restricted to the set of p'-elements $\hat{x} \in \hat{G}$ such that $\hat{G} = \hat{W}\cdot\langle\hat{x}\rangle$; but, it is easily checked that this restriction is equivalent to the projection (cf. 15.21.4)

$$\mathcal{G}_k(\hat{G}) \longrightarrow \mathcal{R}_{\hat{W}}\mathcal{G}_k(\hat{G}) \qquad 16.34.4;$$

that is to say, we finally get an $\mathcal{O}$-module homomorphism

$$\mathcal{G}_k(\hat{G}^i) \longrightarrow \mathcal{R}_{\hat{W}}\mathcal{G}_k(\hat{G}) \qquad 16.34.5$$

and, since $\mathcal{R}_{\hat{W}}\mathcal{G}_k(\hat{G})$ has an $\mathcal{R}\mathcal{G}_k(\bar{C})$-module structure (cf. 15.23), we still get an $\mathcal{R}\mathcal{G}_k(\bar{C})$-module homomorphism

$$\mathcal{R}\mathcal{G}_k(\bar{C}) \otimes_{\mathcal{O}} \mathcal{G}_k(\hat{G}^i) \longrightarrow \mathcal{R}_{\hat{W}}\mathcal{G}_k(\hat{G}) \qquad 16.34.6$$

which we claim is bijective.

Indeed, if M_i is simple then, since we have $\hat{W} = \widehat{\prod}_{i\in I}\hat{G}^i$ and therefore $k_*\hat{W} \cong \otimes_{i\in I} k_*\hat{G}^i$, it is clear that the restriction of $\mathrm{Ten}_{\hat{\hat{W}}}^{\hat{\hat{G}}}(M_i)$ to $k_*\hat{W}$ is simple too; moreover, since the group $\mathrm{Hom}(\bar{C}, k^*)$ determines k^*-group automorphisms of $\hat{G}$, from 16.32 above it is not difficult to check that this correspondence induces a bijection between the set of isomorphism classes of simple $k_*\hat{G}^i$-modules and the set of $\mathrm{Hom}(\bar{C}, k^*)$-orbits of isomorphism classes of the simple $k_*\hat{G}$-modules which remain simple restricted to $\hat{W}$. Then, the bijectivity of homomorphism 16.34.6 follows from the equality in 15.23.2; moreover, it is easily checked that M_i is associated with the block b_i if and only if the restriction of $\mathrm{Ten}_{\hat{\hat{W}}}^{\hat{\hat{G}}}(M_i)$ to $k_*\hat{W}$ is associated with the block b; hence, isomorphism 16.34.6 induces the $\mathcal{O}$-module isomorphism 16.34.1.

On the other hand, since any k^*-automorphism $\hat{\sigma}$ of $\hat{G}$ stabilizes the family $\{\hat{H}_j\}_{j\in I}$, if $\hat{\sigma}$ stabilizes b then it also stabilizes both families $\{\hat{K}^j\}_{j\in I}$ and $\{\hat{G}^j\}_{j\in I}$, and therefore, according to Lemma 16.27, $\hat{\sigma}$ can be lifted to an automorphism $\hat{\hat{\sigma}}$ of $\hat{\hat{G}}$; moreover, since $\hat{G}$ acts transitively on I, up to a modification of $\hat{\sigma}$ by an inner automorphism of $\hat{G}$, we may assume that $\hat{\sigma}$ fixes i and then $\hat{\hat{\sigma}}$ determines a k^*-automorphism $\hat{\sigma}_i$ of $\hat{G}^i$; in this case, assuming that M_i is associated with the block b_i, it is quite clear that

$$\mathrm{Ten}_{\hat{\hat{W}}}^{\hat{\hat{G}}}\big(\mathrm{Res}_{\hat{\sigma}_i}(M_i)\big) \cong \mathrm{Res}_{\hat{\hat{\sigma}}}\big(\mathrm{Ten}_{\hat{\hat{W}}}^{\hat{\hat{G}}}(M_i)\big) \qquad 16.34.7$$

and therefore, since homomorphism 16.33.1 maps the class of $\hat{\sigma}_i$ on the class of $\hat{\sigma}$, homomorphism 16.34.6 is actually an $\mathcal{O}\mathrm{Out}_{k^*}(\hat{G})_b$-module homomorphism. We are done.

16.35 Now, always fixing an element i in I, we need an analogous result connecting $\mathcal{R}_{\mathcal{W}}\mathcal{G}_k(\mathcal{F}, \widehat{\mathfrak{aut}}_{\mathcal{F}^{\mathrm{nc}}})$ with $\mathcal{G}_k(\mathcal{F}^i, \widehat{\mathfrak{aut}}_{(\mathcal{F}^i)^{\mathrm{nc}}})$. Let us borrow our notation in 15.33 with $\hat{W}$ in the place of $\hat{H}$. First of all recall that, since $\mathcal{F} = \mathcal{W}^{\bar{C}}$ (cf. Proposition 12.12), denoting by $\mathfrak{ch}^*_{\bar{C}}\big((\mathcal{W}^{\bar{C}})^{\mathrm{sc}}\big)$ the full subcategory of $\mathfrak{ch}^*\big((\mathcal{W}^{\bar{C}})^{\mathrm{sc}}\big)$ over the set of $(\mathcal{W}^{\bar{C}})^{\mathrm{sc}}$-chains $\mathfrak{r}: \Delta_n \to (\mathcal{W}^{\bar{C}})^{\mathrm{sc}}$ such that $\bar{C}_{\mathfrak{r}} = \bar{C}$ and considering the *contravariant* functor (cf. 15.36.3)

$$\mathcal{R}_{\hat{\mathcal{W}}(\bullet)}\mathcal{G}_k\big(\hat{\mathcal{W}}^{\bar{C}}(\bullet)\big) : \mathfrak{ch}^*_{\bar{C}}\big((\mathcal{W}^{\bar{C}})^{\mathrm{sc}}\big) \longrightarrow \mathcal{O}\text{-}\mathfrak{mod} \qquad 16.35.1,$$

it follows from Proposition 15.37 that we have

$$\mathcal{R}_{\mathcal{W}}\mathcal{G}_k(\mathcal{F}, \widehat{\mathfrak{aut}}_{\mathcal{F}^{\mathrm{nc}}}) \cong \varprojlim \mathcal{R}_{\hat{\mathcal{W}}(\bullet)}\mathcal{G}_k\big(\hat{\mathcal{W}}^{\bar{C}}(\bullet)\big) \qquad 16.35.2.$$

On the other hand, isomorphism 16.32.2 determines a functor $\mathfrak{tr}_i$ from $\mathcal{F}^i$ to $\mathcal{W}^{\bar{C}}$ mapping any subgroup R_i of Q_i on the unique $\bar{C}$-stable subgroup R

of Q fulfilling $R\cap Q_i = R_i$, where we identify Q_i with the obvious subgroup of Q, and mapping any $\mathcal{F}^i$-morphism $\psi_i : T_i \to R_i$ on the unique $\bar{C}$-stable $\mathcal{W}^{\bar{C}}$-morphism $\psi : T \to R$ such that $T\cap Q_i = T_i$, that $R\cap Q_i = R_i$ and that the restriction of ψ to T_i coincides with ψ_i; then, it is quite clear that the restriction $\mathfrak{tr}_i^{\mathrm{sc}}$ of $\mathfrak{tr}_i$ to $(\mathcal{F}^i)^{\mathrm{sc}}$ induces a new functor

$$\mathfrak{ch}^*(\mathfrak{tr}_i^{\mathrm{sc}}) : \mathfrak{ch}^*\big((\mathcal{F}^i)^{\mathrm{sc}}\big) \longrightarrow \mathfrak{ch}^*_{\bar{C}}\big((\mathcal{W}^{\bar{C}})^{\mathrm{sc}}\big) \qquad 16.35.3.$$

16.36 The key point, proved below, is that there is a natural map

$$\tau_i : \mathfrak{g}_k \circ \widehat{\mathfrak{aut}}_{(\mathcal{F}^i)^{\mathrm{sc}}} \longrightarrow \mathcal{R}_{\hat{\mathcal{W}}(\bullet)}\mathcal{G}_k\big(\hat{\mathcal{W}}^{\bar{C}}(\bullet)\big) \circ \mathfrak{ch}^*(\mathfrak{tr}_i^{\mathrm{sc}}) \qquad 16.36.1$$

inducing an $\mathcal{R}\mathcal{G}_k(\bar{C})$-module isomorphism

$$\begin{aligned}\mathcal{R}\mathcal{G}_k(\bar{C}) \otimes_{\mathcal{O}} \mathcal{G}_k(\mathcal{F}^i, \widehat{\mathfrak{aut}}_{(\mathcal{F}^i)^{\mathrm{nc}}})& \\ \cong \varprojlim \big(\mathcal{R}_{\hat{\mathcal{W}}(\bullet)}\mathcal{G}_k\big(\hat{\mathcal{W}}^{\bar{C}}(\bullet)\big) \circ \mathfrak{ch}^*(\mathfrak{tr}_i^{\mathrm{sc}})\big)&\end{aligned} \qquad 16.36.2,$$

and that the functor $\mathfrak{ch}^*(\mathfrak{tr}_i^{\mathrm{sc}})$ induces an $\mathcal{O}$-module isomorphism

$$\varprojlim \mathcal{R}_{\hat{\mathcal{W}}(\bullet)}\mathcal{G}_k\big(\hat{\mathcal{W}}^{\bar{C}}(\bullet)\big) \cong \varprojlim \big(\mathcal{R}_{\hat{\mathcal{W}}(\bullet)}\mathcal{G}_k\big(\hat{\mathcal{W}}^{\bar{C}}(\bullet)\big) \circ \mathfrak{ch}^*(\mathfrak{tr}_i^{\mathrm{sc}})\big) \qquad 16.36.3.$$

Then, from isomorphism 16.35.2 we get

$$\mathcal{R}\mathcal{G}_k(\bar{C}) \otimes_{\mathcal{O}} \mathcal{G}_k(\mathcal{F}^i, \widehat{\mathfrak{aut}}_{(\mathcal{F}^i)^{\mathrm{nc}}}) \cong \mathcal{R}_{\mathcal{W}}\mathcal{G}_k(\mathcal{F}, \widehat{\mathfrak{aut}}_{\mathcal{F}^{\mathrm{nc}}}) \qquad 16.36.4.$$

Firstly, we prove the second isomorphism.

16.37 Recall that, by the very definition of the *inverse limit*, up to suitable identifications we have

$$\varprojlim \mathcal{R}_{\hat{\mathcal{W}}(\bullet)}\mathcal{G}_k\big(\hat{\mathcal{W}}^{\bar{C}}(\bullet)\big) \subset \prod_{\mathfrak{r}} \mathcal{R}_{\hat{\mathcal{W}}(\mathfrak{r})}\mathcal{G}_k\big(\hat{\mathcal{F}}(\mathfrak{r})\big) \qquad 16.37.1$$

where $\mathfrak{r} : \Delta_n \to \mathcal{F}^{\mathrm{sc}}$ runs over the set of $\mathcal{F}^{\mathrm{sc}}$-*chains* such that $\bar{C}_{\mathfrak{r}} = \bar{C}$ (cf. 15.33), whereas the right-hand member in 16.36.3 is contained in the subproduct where, identifying $\bar{C}$ with its image in $\hat{G}$ by isomorphism 16.32.2, $\mathfrak{r} : \Delta_n \to \mathcal{F}^{\mathrm{sc}}$ runs over the set of $\bar{C}$-*stable* $\mathcal{F}^{\mathrm{sc}}$-*chains*; moreover, it is clear that the *projection map* from the whole direct product to this subproduct induces an $\mathcal{O}$-module homomorphism

$$\varprojlim \mathcal{R}_{\hat{\mathcal{W}}(\bullet)}\mathcal{G}_k\big(\hat{\mathcal{W}}^{\bar{C}}(\bullet)\big) \longrightarrow \varprojlim \big(\mathcal{R}_{\hat{\mathcal{W}}(\bullet)}\mathcal{G}_k\big(\hat{\mathcal{W}}^{\bar{C}}(\bullet)\big) \circ \mathfrak{ch}^*(\mathfrak{tr}_i^{\mathrm{sc}})\big) \qquad 16.37.2.$$

16.38 On the other hand, since $\mathrm{Out}_{k^*}(\hat{G})_b$ stabilizes the normal k^*-subgroup $\hat{W}$, it follows from 16.3 and 16.4 that this group acts on the above direct product stabilizing $\lim\limits_{\longleftarrow} \mathcal{R}_{\hat{\mathcal{W}}(\bullet)}\mathcal{G}_k(\hat{\mathcal{W}}^{\bar{C}}(\bullet))$. Actually, since $\bar{C}$ acts regularly on I, we still can define an action of $\mathrm{Out}_{k^*}(\hat{G})_b$ on the above subproduct stabilizing $\lim\limits_{\longleftarrow}\big(\mathcal{R}_{\hat{\mathcal{W}}(\bullet)}\mathcal{G}_k(\hat{\mathcal{W}}^{\bar{C}}(\bullet)) \circ \mathfrak{ch}^*(\mathfrak{tr}_i^{\mathrm{sc}})\big)$ and it is easily checked that homomorphism 16.37.2 becomes an $\mathcal{O}\mathrm{Out}_{k^*}(\hat{G})_b$-module homomorphism.

Proposition 16.39 *With the notation and the hypothesis above, the functor* $\mathfrak{ch}^*(\mathfrak{tr}_i^{\mathrm{sc}})$ *induces an* $\mathcal{O}\mathrm{Out}_{k^*}(\hat{G})_b$*-module isomorphism*

$$\lim_{\longleftarrow} \mathcal{R}_{\hat{\mathcal{W}}(\bullet)}\mathcal{G}_k(\hat{\mathcal{W}}^{\bar{C}}(\bullet)) \cong \lim_{\longleftarrow}\big(\mathcal{R}_{\hat{\mathcal{W}}(\bullet)}\mathcal{G}_k(\hat{\mathcal{W}}^{\bar{C}}(\bullet)) \circ \mathfrak{ch}^*(\mathfrak{tr}_i^{\mathrm{sc}})\big) \qquad 16.39.1.$$

Proof: It follows from Lemma 15.51 that any $(\mathcal{F}^i)^{\mathrm{sc}}$*-chain* $\mathfrak{r}^i : \Delta_n \to (\mathcal{F}^i)^{\mathrm{sc}}$ can be extended to a suitable $(\mathcal{F}^i)^{\mathrm{sc}}$*-chain* $\hat{\mathfrak{r}}^i : \Delta_{n+1} \to (\mathcal{F}^i)^{\mathrm{sc}}$ such that we have

$$\mathbf{O}_p\Big(\tilde{\mathcal{F}}^i\big(\hat{\mathfrak{r}}^i(n+1)\big)\Big) = \{1\} \qquad 16.39.2$$

by applying this lemma to $\mathfrak{r}^i(n)$; then, applying the functor $\widehat{\mathfrak{aut}}_{(\mathcal{F}^i)^{\mathrm{sc}}}$ (cf. Theorem 11.32) to the $\mathfrak{ch}^*(\mathcal{F}^i)$-morphism

$$(\mathrm{id}_{\mathfrak{r}_i}, \delta^n_{n+1}) : (\hat{\mathfrak{r}}^i, \Delta_{n+1}) \longrightarrow (\mathfrak{r}^i, \Delta_n) \qquad 16.39.3,$$

we get a k^*-group isomorphism (cf. 15.51.1)

$$\hat{\mathcal{F}}^i(\hat{\mathfrak{r}}^i) \cong \hat{\mathcal{F}}^i(\mathfrak{r}^i) \qquad 16.39.4.$$

Consequently, denoting by $\mathcal{R}^i$ the set of $\mathcal{F}^i$-selfcentralizing subgroups R_i of Q_i fulfilling

$$\mathbf{O}_p\big(\tilde{\mathcal{F}}^i(R_i)\big) = \{1\} \qquad 16.39.5$$

and by $\mathfrak{tr}_i^{\mathcal{R}^i}$ the restriction of the functor $\mathfrak{tr}_i$ to the full subcategory $(\mathcal{F}^i)^{\mathcal{R}^i}$ of $(\mathcal{F}^i)^{\mathrm{sc}}$ over $\mathcal{R}^i$, it is not difficult to check that we have a canonical $\mathcal{O}\mathrm{Out}_{k^*}(\hat{G})_b$-module isomorphism

$$\begin{aligned}&\lim_{\longleftarrow}\big(\mathcal{R}_{\hat{\mathcal{W}}(\bullet)}\mathcal{G}_k(\hat{\mathcal{W}}^{\bar{C}}(\bullet)) \circ \mathfrak{ch}^*(\mathfrak{tr}_i^{\mathrm{sc}})\big)\\ &\qquad\cong \lim_{\longleftarrow}\big(\mathcal{R}_{\hat{\mathcal{W}}(\bullet)}\mathcal{G}_k(\hat{\mathcal{W}}^{\bar{C}}(\bullet)) \circ \mathfrak{ch}^*(\mathfrak{tr}_i^{\mathcal{R}^i})\big)\end{aligned} \qquad 16.39.6.$$

Similarly, it follows from Lemma 15.51 and Lemma 16.40 below that any $\mathcal{F}^{\mathrm{sc}}$*-chain* $\mathfrak{r} : \Delta_n \to \mathcal{F}^{\mathrm{sc}}$ such that $\bar{C}_{\mathfrak{r}} = \bar{C}$ (cf. 15.33) can be extended to a suitable $\mathcal{F}^{^{\rm sc}}$-*chain* $\hat{\mathfrak{r}}:\Delta_{n+1}\to\mathcal{F}^{^{\rm sc}}$ such that we have $\hat{r}(n+1)=\mathfrak{tr}_i(R_i)$ for some $R_i\in\mathcal{R}^i$ and that, applying the functor $\widehat{\mathfrak{aut}}_{\mathcal{F}^{\rm sc}}$ (cf. Theorem 11.32) to the $\mathfrak{ch}^*(\mathcal{F})$-morphism

$$(\mathrm{id}_{\mathfrak{r}_i},\delta^n_{n+1}):(\hat{\mathfrak{r}},\Delta_{n+1})\longrightarrow(\mathfrak{r},\Delta_n) \qquad 16.39.7,$$

we still get a k^*-group isomorphism

$$\hat{\mathcal{F}}(\hat{\mathfrak{r}})\cong\hat{\mathcal{F}}(\mathfrak{r}) \qquad 16.39.8.$$

As above, it is not difficult to check that, in order to compute the inverse limit of the functor $\mathcal{R}_{\hat{\mathcal{W}}(\bullet)}\mathcal{G}_k\big(\hat{\mathcal{W}}^{\bar{C}}(\bullet)\big)$, it suffices to consider its restriction to $\mathfrak{ch}^*_{\bar{C}}\big((\mathcal{W}^{\bar{C}})^{^{\mathfrak{tr}_i(\mathcal{R}^i)}}\big)$ (cf. 16.35) where $(\mathcal{W}^{\bar{C}})^{^{\mathfrak{tr}_i(\mathcal{R}^i)}}=\mathcal{F}^{^{\mathfrak{tr}_i(\mathcal{R}^i)}}$ denotes the full subcategory of $(\mathcal{W}^{\bar{C}})^{^{\rm sc}}=\mathcal{F}^{^{\rm sc}}$ over the set $\mathfrak{tr}_i(\mathcal{R}^i)$ (cf. Proposition 12.12).

Moreover, considering the obvious *image category* $\mathfrak{tr}_i\big((\mathcal{F}^i)^{^{\mathcal{R}^i}}\big)\subset\mathcal{F}^{^{\mathfrak{tr}_i(\mathcal{R}^i)}}$, it is easily checked that any $\mathcal{F}^{^{\mathfrak{tr}_i(\mathcal{R}^i)}}$-*chain* $\mathfrak{r}:\Delta_n\to\mathcal{F}^{^{\mathfrak{tr}_i(\mathcal{R}^i)}}$ such that $\bar{C}_{\mathfrak{r}}=\bar{C}$ is $\mathfrak{ch}^*\big(\mathcal{F}^{^{\mathfrak{tr}_i(\mathcal{R}^i)}}\big)$-isomorphic to a $\mathfrak{tr}_i((\mathcal{F}^i)^{^{\mathcal{R}^i}})$-*chain*, and it is clear that two $\mathfrak{tr}_i((\mathcal{F}^i)^{^{\mathcal{R}^i}})$-*chains* are $\mathfrak{ch}^*\big(\mathfrak{tr}_i((\mathcal{F}^i)^{^{\mathcal{R}^i}})\big)$-isomorphic if (and only if) they are $\mathfrak{ch}^*\big(\mathcal{F}^{^{\mathfrak{tr}_i(\mathcal{R}^i)}}\big)$-isomorphic. Then, since $\mathcal{F}(\mathfrak{r})$ acts trivially on $\mathcal{G}_k\big(\hat{\mathcal{F}}(\mathfrak{r})\big)$, it is not difficult to check that, in order to compute the inverse limit of the functor $\mathcal{R}_{\hat{\mathcal{W}}(\bullet)}\mathcal{G}_k\big(\hat{\mathcal{W}}^{\bar{C}}(\bullet)\big)$, it suffices to consider its restriction to the *whole* proper category of chains $\mathfrak{ch}^*\big(\mathfrak{tr}_i((\mathcal{F}^i)^{^{\mathcal{R}^i}})\big)$. But, it is clear that $\mathfrak{tr}_i$ induces an equivalence of categories between $(\mathcal{F}^i)^{^{\mathcal{R}^i}}$ and $\mathfrak{tr}_i((\mathcal{F}^i)^{^{\mathcal{R}^i}})$. We are done.

Lemma 16.40 *With the notation and the hypothesis above, any $\mathcal{F}$-selfcentralizing subgroup R of Q fulfilling*

$$\mathcal{F}(R)/\mathcal{W}(R)\cong\bar{C}\quad\textit{and}\quad\mathbf{O}_p\big(\tilde{\mathcal{F}}(R)\big)=\{1\} \qquad 16.40.1$$

is $\mathcal{F}$-isomorphic to $\mathfrak{tr}_i(R_i)$ where R_i is the image of R in Q_i.

Proof: For any $j\in I$, denote by R_j the image of R in Q_j and by T the direct product $\prod_{j\in I}R_j$; then, since $\mathcal{F}(R)/\mathcal{W}(R)\cong\bar{C}$, the image of $N_T(R)$ in $\mathcal{F}(R)$ is a normal p-subgroup and therefore, since we assume that $\mathbf{O}_p\big(\tilde{\mathcal{F}}(R)\big)=\{1\}$ and R is $\mathcal{F}$-selfcentralizing, we have $N_T(R)=R$ which forces $T=R$.

Moreover, if g_j is the block of $C_{\hat{G}^j}(R_j)$ such that (Q_j,f_j) contains (R_j,g_j) for any $j\in I$, and we set $g=\otimes_{j\in I}g_j$, it is clear that

$$N_{\hat{W}}(R,g)=\prod_{j\in I}N_{\hat{G}^j}(R_j,g_j) \qquad 16.40.2;$$

thus, since $N_{\hat{G}}(R,g)/N_{\hat{W}}(R,g) \cong \bar{C}$, as in 16.31.1 above we have a k^*-group isomorphism (cf. 1.32)

$$N_G(R,g) \cong N_{G^i}(R_i,g_i) \wr \bar{C} \qquad 16.40.3$$

and, since $\bar{C}$ is cyclic, it is easily checked that a complement of $N_{\hat{W}}(R,g)$ in $N_{\hat{G}}(R,g)$ is G-conjugate to the image of $\bar{C}$ by isomorphism 16.32.2; consequently, R is $\mathcal{F}$-isomorphic to $\mathfrak{tr}_i(R_i)$. We are done.

16.41 Now, let us discuss the first isomorphism in 16.36. First of all, we have to define a natural map

$$\mathfrak{g}_k \circ \widehat{\mathfrak{aut}}_{(\mathcal{F}^i)^{\mathrm{sc}}} \longrightarrow \mathcal{R}_{\hat{\mathcal{W}}(\bullet)}\mathcal{G}_k\big(\hat{\mathcal{W}}^{\bar{C}}(\bullet)\big) \circ \mathfrak{ch}^*(\mathfrak{tr}_i^{\mathrm{sc}}) \qquad 16.41.1;$$

that is to say, for any $(\mathcal{F}^i)^{\mathrm{sc}}$*-chain* $\mathfrak{r}^i : \Delta_n \to (\mathcal{F}^i)^{\mathrm{sc}}$, we have to define an $\mathcal{O}$-module homomorphism from $\mathcal{G}_k\big(\hat{\mathcal{F}}^i(\mathfrak{r}^i)\big)$ to $\mathcal{R}_{\hat{W}(\mathfrak{r})}\mathcal{G}_k\big(\hat{\mathcal{F}}(\mathfrak{r})\big)$ where we set $\mathfrak{r} = \mathfrak{tr}_i \circ \mathfrak{r}^i$. But, it is quite clear that (cf. 1.32)

$$\mathcal{F}(\mathfrak{r}) \cong \mathcal{F}^i(\mathfrak{r}^i) \wr \bar{C} \quad \text{and} \quad \hat{\mathcal{F}}(\mathfrak{r}) \cong \hat{\mathcal{W}}(\mathfrak{r}) \rtimes \bar{C} \qquad 16.41.2$$

and therefore, setting $\hat{\hat{\mathcal{F}}}(\mathfrak{r}) = \hat{\mathcal{F}}^i(\mathfrak{r}^i) \wr \bar{C}$, we get (cf. 16.32)

$$\hat{\hat{\mathcal{F}}}(\mathfrak{r})/\mathrm{Ker}(\nabla_{k^*}) \cong \hat{\mathcal{F}}(\mathfrak{r}) \quad \text{and} \quad k \otimes_{k(k^*)^I} k\hat{\hat{\mathcal{F}}}(\mathfrak{r}) \cong k_*\hat{\mathcal{F}}(\mathfrak{r}) \qquad 16.41.3.$$

Thus, as in 16.33 above, for any $j \in I$ denoting by $\mathfrak{r}^j$ the conjugate of $\mathfrak{r}^i$ *via* the corresponding element of $\bar{C}$, and setting

$$\hat{\hat{\mathcal{W}}}(\mathfrak{r}) = \prod_{j\in I} \hat{\mathcal{F}}^j(\mathfrak{r}^j) \qquad 16.41.4,$$

any $k_*\hat{\mathcal{F}}^i(\mathfrak{r}^i)$-module M_i can be considered as a $k\hat{\hat{\mathcal{W}}}(\mathfrak{r})$-module and we consider the tensor induction $\mathrm{Ten}_{\hat{\hat{\mathcal{W}}}(\mathfrak{r})}^{\hat{\hat{\mathcal{F}}}(\mathfrak{r})}(M_i)$ which becomes a $k_*\hat{\mathcal{F}}(\mathfrak{r})$-module.

Proposition 16.42 *With the notation and the hypothesis above, the correspondence mapping any* $(\mathcal{F}^i)^{\mathrm{sc}}$*-chain* $\mathfrak{r}^i : \Delta_n \to (\mathcal{F}^i)^{\mathrm{sc}}$ *and any* $k_*\hat{\mathcal{F}}^i(\mathfrak{r}^i)$*-module* M_i *on the* $k_*\hat{\mathcal{F}}(\mathfrak{tr}_i \circ \mathfrak{r}^i)$*-module* $\mathrm{Ten}_{\hat{\hat{\mathcal{W}}}(\mathfrak{tr}_i\circ\mathfrak{r}^i)}^{\hat{\hat{\mathcal{F}}}(\mathfrak{tr}_i\circ\mathfrak{r}^i)}(M_i)$ *determines a natural map*

$$\tau_i : \mathfrak{g}_k \circ \widehat{\mathfrak{aut}}_{(\mathcal{F}^i)^{\mathrm{sc}}} \longrightarrow \mathcal{R}_{\hat{\mathcal{W}}(\bullet)}\mathcal{G}_k\big(\hat{\mathcal{W}}^{\bar{C}}(\bullet)\big) \circ \mathfrak{ch}^*(\mathfrak{tr}_i^{\mathrm{sc}}) \qquad 16.42.1$$

which induces an $\mathcal{O}\mathrm{Out}_{k^*}(\hat{G})_b$*-module isomorphism*

$$\mathcal{R}\mathcal{G}_k(\bar{C}) \otimes_{\mathcal{O}} \mathcal{G}_k(\mathcal{F}^i, \widehat{\mathfrak{aut}}_{(\mathcal{F}^i)^{\mathrm{nc}}}) \cong \varprojlim \big(\mathcal{R}_{\hat{\mathcal{W}}(\bullet)}\mathcal{G}_k\big(\hat{\mathcal{W}}^{\bar{C}}(\bullet)\big) \circ \mathfrak{ch}^*(\mathfrak{tr}_i^{\mathrm{sc}})\big) \qquad 16.42.2$$

Proof: As we remark in 16.31, our arguments above apply to a more general situation; in particular, our arguments in the proof of Proposition 16.34 can be translated here proving that, for any $(\mathcal{F}^i)^{\rm sc}$*-chain* $\mathfrak{r}^i : \Delta_n \to (\mathcal{F}^i)^{\rm sc}$, we get an $\mathcal{O}$-module homomorphism

$$(\tau_i)_{\mathfrak{r}} : \mathcal{G}_k\big(\hat{\mathcal{F}}^i(\mathfrak{r}^i)\big) \longrightarrow \mathcal{R}_{\hat{\mathcal{W}}(\mathfrak{tr}_i\circ\mathfrak{r}^i)}\mathcal{G}_k\big(\hat{\mathcal{F}}(\mathfrak{tr}_i\circ \mathfrak{r}^i)\big) \qquad 16.42.3$$

which actually induces an $\mathcal{O}\mathrm{Out}_{k^*}\big(\hat{\mathcal{F}}(\mathfrak{tr}_i\circ\mathfrak{r}^i)\big)$-module isomorphism

$$\mathcal{R}\mathcal{G}_k(\bar{C})\otimes_{\mathcal{O}} \mathcal{G}_k\big(k_*\hat{\mathcal{F}}^i(\mathfrak{r}^i)\big) \cong \mathcal{R}_{\hat{\mathcal{W}}(\mathfrak{tr}_i\circ\mathfrak{r}^i)}\mathcal{G}_k\big(\hat{\mathcal{F}}(\mathfrak{tr}_i\circ \mathfrak{r}^i)\big) \qquad 16.42.4.$$

On the other hand, applying the functor $\widehat{\mathfrak{aut}}_{(\mathcal{F}^i)^{\rm sc}}$ to a $\mathfrak{ch}^*\big((\mathcal{F}^i)^{\rm sc}\big)$-morphism $(\mu^i,\delta) : (\mathfrak{t}^i,\Delta_m) \to (\mathfrak{r}^i,\Delta_n)$, we get a k^*-group homomorphism

$$\hat{\mu}^i = \widehat{\mathfrak{aut}}_{(\mathcal{F}^i)^{\rm sc}}(\nu^i,\delta) : \hat{\mathcal{F}}^i(\mathfrak{t}^i) \longrightarrow \hat{\mathcal{F}}^i(\mathfrak{r}^i) \qquad 16.42.5$$

which clearly induces $(k^*)^I$*-extension* homomorphisms (cf. 16.41)

$$\hat{\hat{\mu}} : \hat{\hat{\mathcal{F}}}(\mathfrak{tr}_i\circ\mathfrak{t}^i) \longrightarrow \hat{\hat{\mathcal{F}}}(\mathfrak{tr}_i\circ\mathfrak{r}^i) \quad \text{and} \quad \hat{\hat{\mathcal{W}}}(\mathfrak{tr}_i\circ\mathfrak{t}^i) \longrightarrow \hat{\hat{\mathcal{W}}}(\mathfrak{tr}_i\circ\mathfrak{r}^i) \qquad 16.42.6.$$

Moreover, it is easily checked that we have a canonical isomorphism

$$\mathrm{Ten}_{\hat{\hat{\mathcal{W}}}(\mathfrak{tr}_i\circ\mathfrak{t}^i)}^{\hat{\hat{\mathcal{F}}}(\mathfrak{tr}_i\circ\mathfrak{t}^i)}\big(\mathrm{Res}_{\hat{\mu}^i}(M_i)\big) \cong \mathrm{Res}_{\hat{\hat{\mu}}}\big(\mathrm{Ten}_{\hat{\hat{\mathcal{W}}}(\mathfrak{tr}_i\circ\mathfrak{r}^i)}^{\hat{\hat{\mathcal{F}}}(\mathfrak{tr}_i\circ\mathfrak{r}^i)}(M_i)\big) \qquad 16.42.7$$

and, since $\hat{\hat{\mu}}$ is clearly compatible with the k^*-group homomorphism

$$\widehat{\mathfrak{aut}}_{\mathcal{F}^{\rm sc}}(\mathfrak{tr}_i * \mu^i,\delta) : \hat{\mathcal{F}}(\mathfrak{tr}_i\circ\mathfrak{t}^i) \longrightarrow \hat{\mathcal{F}}(\mathfrak{tr}_i\circ\mathfrak{r}^i) \qquad 16.42.8,$$

isomorphism 16.42.7 shows the naturality of the correspondence mapping $\mathfrak{r}$ on $(\tau_i)_{\mathfrak{r}}$. Then, it is quite clear that the corresponding naturality of isomorphisms 16.42.4 proves isomorphism 16.42.2. We are done.

Proposition 16.43 *With the notation and the hypothesis in* 16.26 *above, assume that the group* $\bar{C} = \hat{G}/\hat{W}$ *acts regularly on* I *and let us choose an element* $i\in I$. *We have* $\mathcal{O}\mathrm{Out}_{k^*}(\hat{G})_b$*-module isomorphisms*

$$\begin{gathered}\mathcal{G}_k(\bar{C})\otimes_{\mathcal{O}}\mathcal{G}_k(\hat{G}^i,b_i)\cong\mathcal{G}_k(\hat{G},b)\\ \mathcal{G}_k(\bar{C})\otimes_{\mathcal{O}}\mathcal{G}_k(\mathcal{F}^i,\widehat{\mathfrak{aut}}_{(\mathcal{F}^i)^{\rm nc}})\cong\mathcal{G}_k(\mathcal{F},\widehat{\mathfrak{aut}}_{\mathcal{F}^{\rm nc}})\end{gathered} \qquad 16.43.1.$$

In particular, if (Q) *holds for* $(b_i,\hat{G}^i)$ *then it holds for* $(b,\hat{G})$.

Proof: For any subgroup $\bar{D}$ of $\bar{C}$, it follows from Propositions 16.34, 16.39 and 16.42 that we have $\mathcal{O}\mathrm{Out}_{k^*}(\hat{W}^{\bar{D}})_b$-module isomorphisms

$$\begin{aligned}\mathcal{R}\mathcal{G}_k(\bar{D})\otimes_{\mathcal{O}}\mathcal{G}_k(\hat{G}^i,b_i)&\cong\mathcal{R}_{\hat{W}}\mathcal{G}_k(\hat{W}^{\bar{D}},b)\\ \mathcal{R}\mathcal{G}_k(\bar{D})\otimes_{\mathcal{O}}\mathcal{G}_k(\mathcal{F}^i,\widehat{\mathfrak{aut}}_{(\mathcal{F}^i)^{\mathrm{nc}}})&\cong\mathcal{R}_{\hat{\mathcal{W}}}\mathcal{G}_k(\mathcal{W}^{\bar{D}},\widehat{\mathfrak{aut}}_{(\mathcal{W}^{\bar{D}})^{\mathrm{nc}}})\end{aligned} \qquad 16.43.2$$

where $\hat{W}^{\bar{D}}$ denotes the converse image of $\bar{D}$ in $\hat{G}$; in particular, since $\bar{C}$ acts on $\hat{W}^{\bar{D}}$ stabilizing b and we have a canonical group homomorphism

$$\mathcal{O}\mathrm{Aut}_{k^*}(\hat{G})_b\longrightarrow\mathcal{O}\mathrm{Out}_{k^*}(\hat{W}^{\bar{D}})_b \qquad 16.43.3$$

containing and normalizing the image of $\bar{C}$, we still have $\mathcal{O}\mathrm{Out}_{k^*}(\hat{G})_b$-module isomorphisms

$$\begin{aligned}\mathcal{R}\mathcal{G}_k(\bar{D})^{\bar{C}}\otimes_{\mathcal{O}}\mathcal{G}_k(\hat{G}^i,b_i)&\cong\mathcal{R}_{\hat{W}}\mathcal{G}_k(\hat{W}^{\bar{D}},b)^{\bar{C}}\\ \mathcal{R}\mathcal{G}_k(\bar{D})^{\bar{C}}\otimes_{\mathcal{O}}\mathcal{G}_k(\mathcal{F}^i,\widehat{\mathfrak{aut}}_{(\mathcal{F}^i)^{\mathrm{nc}}})&\cong\mathcal{R}_{\hat{\mathcal{W}}}\mathcal{G}_k(\mathcal{W}^{\bar{D}},\widehat{\mathfrak{aut}}_{(\mathcal{W}^{\bar{D}})^{\mathrm{nc}}})^{\bar{C}}\end{aligned} \qquad 16.43.4.$$

Then, considering the direct sum of these $\mathcal{O}\mathrm{Out}_{k^*}(\hat{G})_b$-module isomorphisms when $\bar{D}$ runs over the set of subgroups of $\bar{C}$, isomorphisms 16.43.1 follow from isomorphisms 15.23.4 and 15.38.1.

In particular, if we have an $\mathcal{O}\mathrm{Out}_{k^*}(\hat{G}^i)_{b_i}$-module isomorphism

$$\mathcal{G}_k(\mathcal{F}^i,\widehat{\mathfrak{aut}}_{(\mathcal{F}^i)^{\mathrm{nc}}})\cong\mathcal{G}_k(\hat{G}^i,b_i) \qquad 16.43.5,$$

then, from isomorphisms 16.43.1 and *via* the canonical group homomorphism $\mathrm{Out}_{k^*}(\hat{G})_b\to\mathrm{Out}_{k^*}(\hat{G}^i)_{b_i}$ (cf. 16.33.1), we get an $\mathcal{O}\mathrm{Out}_{k^*}(\hat{G})_b$-module isomorphism

$$\mathcal{G}_k(\mathcal{F},\widehat{\mathfrak{aut}}_{\mathcal{F}^{\mathrm{nc}}})\cong\mathcal{G}_k(\hat{G},b) \qquad 16.43.6.$$

16.44 In conclusion, putting all together Propositions 16.6, 16.7, 16.8, 16.9, 16.15, 16.19, 16.23, 16.25, 16.28, 16.30 and 16.43, we have proved the following reduction result.

Theorem 16.45 *Assume that any block $(c,\hat{H})$ having a normal sub-block $(d,\hat{S})$ of positive defect such that the k^*-quotient of $\hat{S}$ is simple, that we have $C_{\hat{H}}(\hat{S})=k^*$, that $\hat{H}/\hat{S}$ is a p'-group and that $\hat{H}/\hat{S}\cdot C_{\hat{H}}(Q,f)$ is cyclic where (Q,f) is a maximal Brauer $(d,\hat{S})$-pair, fulfills the following three conditions:*

16.45.1 $\mathrm{Out}(S)$ *is solvable.*

16.45.2 *Setting $\hat{K}=\hat{S}\cdot C_{\hat{H}}(Q,f)$, the canonical k^*-group homomorphism*

$$C_{\hat{S}}(Q)\longrightarrow\big(k_*\bar{C}_{\hat{S}}(Q)\bar{f}\big)^*$$

can be extended to an $\mathrm{Aut}_{k^*}(\hat{K})_{(Q,f)}$-stable k^*-group homomorphism

$$N_{\hat{K}}(Q,f)\longrightarrow\big(k_*\bar{C}_{\hat{S}}(Q)\bar{f}\big)^*.$$

16.45.3 *There is an* $\mathcal{O}\mathrm{Out}_{k^*}(\hat{H})_c$*-module isomorphism*

$$\mathcal{G}_k(\mathcal{F}_{(c,\hat{H})}, \widehat{\mathfrak{aut}}_{(\mathcal{F}_{(c,\hat{H})})^{\mathrm{nc}}}) \cong \mathcal{G}_k(\hat{H}, c)\,.$$

Then, for any block $(b, \hat{G})$ *there is an* $\mathcal{O}\mathrm{Out}_{k^*}(\hat{G})_b$*-module isomorphism*

$$\mathcal{G}_k(\mathcal{F}_{(b,\hat{G})}, \widehat{\mathfrak{aut}}_{(\mathcal{F}_{(b,\hat{G})})^{\mathrm{nc}}}) \cong \mathcal{G}_k(\hat{G}, b) \qquad 16.45.4.$$

Chapter 17

Localities associated with a Frobenius P-category

17.1 Let us come back to our abstract setting: let P be a finite p-group and $\mathcal{F}$ a P-category (cf. 2.2). In this chapter, we introduce the *localities* associated with $\mathcal{F}$ — which are *extensions* of the category $\mathcal{F}$ and generalize the *"localités"* associated with a finite group G having P as a Sylow p-subgroup, introduced in [35]. As we explain in the Introduction (cf. I41), we are mainly interested in the possible existence and uniqueness of the *perfect locality* — defined in 17.13 below — for a Frobenius P-category, which to some extent generalizes the **O**-*localité* introduced in [35, Ch. VI]. But, we find other meaningful localities canonically associated with any Frobenius P-category, as the *basic locality* in chapter 22.

17.2 A possible motivation to consider extensions of $\mathcal{F}$ could be supplied by the case where a finite group G acts on a set Ω; then, it seems reasonable to consider the category $\mathcal{F}_{G,\Omega}$ where the objects are all the p-subgroups of G and the morphisms are the pairs formed by an $\mathcal{F}_G$-morphism (cf. 1.8) and by a permutation of Ω both determined by the *same* element of G, the composition of morphisms being induced by the product in G. As usual, we identify $\mathcal{F}_{G,\Omega}$ with its equivalent full subcategory over all the subgroups of a Sylow p-subgroup P. In particular, the *regular* action of G over G defines the *transporter* category $\mathcal{T}_G = \mathcal{F}_{G,G}$.

17.3 Denote by $\kappa : \mathcal{T}_P \to \mathcal{F}$ the canonical functor determined by the conjugation (cf. 1.8). A *locality* $\mathcal{L}$ *associated with* $\mathcal{F}$ — an $\mathcal{F}$-*locality* for short — is a category where the objects are all the subgroups of P, endowed with two functors

$$\tau : \mathcal{T}_P \longrightarrow \mathcal{L} \quad \text{and} \quad \pi : \mathcal{L} \longrightarrow \mathcal{F} \qquad 17.3.1$$

which are the identity on the set of objects and fulfill $\pi \circ \tau = \kappa$, π being full; as usual, for any pair of subgroups Q and R of P, we denote by $\mathcal{L}(Q,R)$ the set of $\mathcal{L}$-morphisms from R to Q, by $\cdot$ the composition law and by

$$\tau_{Q,R} : \mathcal{T}_P(Q,R) \longrightarrow \mathcal{L}(Q,R) \quad \text{and} \quad \pi_{Q,R} : \mathcal{L}(Q,R) \longrightarrow \mathcal{F}(Q,R) \qquad 17.3.2$$

the corresponding maps; often, we will write π_x and τ_u instead of $\pi_{Q,R}(x)$ and $\tau_{Q,R}(u)$, where x and u are $\mathcal{L}$- and $\mathcal{T}_P$-morphisms respectively; moreover, we set $i_R^Q = \tau_{Q,R}(1)$ whenever $R \subset Q$ and, in all these notations, we write Q only once if $Q = R$ (cf. 1.8).

17.4 As for the P-categories, it is handy to consider *partially defined $\mathcal{F}$-localities:* if $\mathfrak{X}$ is a nonempty set of subgroups of P, which contains any subgroup Q of P such that $\mathcal{F}(Q,R) \neq \emptyset$ for some $R \in \mathfrak{X}$, an $(\mathcal{F},\mathfrak{X})$- or an $\mathcal{F}^{\mathfrak{X}}$*-locality* $\mathcal{L}^{\mathfrak{X}}$ is a category where the objects are the elements of $\mathfrak{X}$, endowed with two functors as above, fulfilling the analogous conditions where we respectively replace $\mathcal{T}_P$ and $\mathcal{F}$ by the full subcategories $\mathcal{T}_P^{\mathfrak{X}}$ of $\mathcal{T}_P$ and $\mathcal{F}^{\mathfrak{X}}$ of $\mathcal{F}$ over $\mathfrak{X}$; note that $\mathcal{L}^{\mathfrak{X}}$ can always be extended to an *ordinary* $\mathcal{F}$-locality $\mathcal{L}$ by setting $\mathcal{L}(R,T) = \mathcal{F}(R,T)$ for any pair of subgroups R and T of P such that $T \notin \mathfrak{X}$, and $x{\cdot}\psi = \pi_x \circ \psi$ for any $x \in \mathcal{L}(Q,R)$ where Q is another subgroup of P. In particular, if $\mathcal{L}$ is an $\mathcal{F}$-locality, we denote by $\mathcal{L}^{\mathrm{sc}}$ the *full* subcategory of $\mathcal{L}$ over the set of $\mathcal{F}$-selfcentralizing subgroups of P, endowed with the corresponding two functors.

17.5 Let $\mathcal{L}$ be an $\mathcal{F}$-locality, Q a subgroup of P and K a subgroup of $\mathrm{Aut}(Q)$, and assume that Q is fully K-normalized in $\mathcal{F}$ (cf. 2.6); then, denoting by $\big(N_{\mathcal{L}}^K(Q)\big)(R,T)$ the converse image of $\big(N_{\mathcal{F}}^K(Q)\big)(R,T)$ in $\mathcal{L}(R,T)$ for any pair of subgroups R and T of $N_P^K(Q)$, it is quite clear that we obtain a subcategory $N_{\mathcal{L}}^K(Q)$ of $\mathcal{L}$, where the objects are the subgroups of $N_P^K(Q)$, endowed with two functors

$$\mathcal{T}_{N_P^K(Q)} \longrightarrow N_{\mathcal{L}}^K(Q) \quad \text{and} \quad N_{\mathcal{L}}^K(Q) \longrightarrow N_{\mathcal{F}}^K(Q) \qquad 17.5.1$$

induced by the functors above; that is to say, we obtain a locality associated with the $N_P^K(Q)$-category $N_{\mathcal{F}}^K(Q)$ that we call *K-normalizer of Q in $\mathcal{L}$* — or *centralizer* and *normalizer* in the cases where $K = \{\mathrm{id}_Q\}$ and $K = \mathrm{Aut}(Q)$.

17.6 As we will see in Proposition 18.4 below, it is useful to define the $N_{\mathcal{F}}^K(Q)$-locality $N_{\mathcal{F},Q}^K(Q)$ (cf. 17.2) endowed with the functors

$$\mathcal{T}_{N_P^K(Q)} \longrightarrow N_{\mathcal{F},Q}^K(Q) \quad \text{and} \quad N_{\mathcal{F},Q}^K(Q) \longrightarrow N_{\mathcal{F}}^K(Q) \qquad 17.6.1,$$

where the objects in $N_{\mathcal{F},Q}^K(Q)$ are, once again, all the subgroups of $N_P^K(Q)$, where the morphisms are the pairs formed by an $N_{\mathcal{F}}^K(Q)$-morphism and an element of K both determined by the *same* morphism in $\mathcal{F}$, and where the composition and the two structural functors are the obvious ones. If we consider a *partially defined* $\mathcal{F}^{\mathfrak{X}}$-locality $\mathcal{L}^{\mathfrak{X}}$, it suffices that $N_P^K(Q)$ belongs to $\mathfrak{X}$ to be able to define the corresponding *K-normalizer of Q* in $\mathcal{L}^{\mathfrak{X}}$ — an $N_{\mathcal{F}}^K(Q)^{N_{\mathfrak{X}}^K(Q)}$-locality, where $N_{\mathfrak{X}}^K(Q)$ denotes the obvious set.

17.7 As in 2.3 above, we say that an $\mathcal{F}$-locality $\mathcal{L}$ is *divisible* whenever it fulfills the condition

17.7.1 *If Q, R and T are subgroups of P, for any $x \in \mathcal{L}(Q,R)$ and any $y \in \mathcal{L}(Q,T)$ such that $\pi_y(T) \subset \pi_x(R)$ there is a unique $z \in \mathcal{L}(R,T)$ such that $x{\cdot}z = y$.*

Actually, all the localities we will consider are divisible; in this case, for any $x \in \mathcal{L}(Q, R)$, the equality $|Q| = |R|$ implies that $\mathcal{L}(R, Q)$ has an element x^{-1} such that $x{\cdot}x^{-1}$ and $x^{-1}{\cdot}x$ are the unity elements in $\mathcal{L}(Q) = \mathcal{L}(Q, Q)$ and $\mathcal{L}(R)$ respectively; further, setting $T = \pi_x(R)$, we have $i_T^Q = x{\cdot}(x_*)^{-1}$ for a suitable $x_* \in \mathcal{L}(T, R)$.

17.8 Obviously, the divisibility of $\mathcal{L}$ forces the divisibility of $\mathcal{F}$ and then $\mathrm{Ker}(\pi_R)$ acts regularly on the "fibers" of the maps

$$\pi_{Q,R} : \mathcal{L}(Q, R) \longrightarrow \mathcal{F}(Q, R) \qquad 17.8.1;$$

conversely, if $\mathcal{F}$ is divisible and, for any pair of subgroups Q and R of P, $\mathrm{Ker}(\pi_R)$ acts regularly on the "fibers" of $\pi_{Q,R}$ then $\mathcal{L}$ is clearly divisible. In this case, the correspondence mapping Q on $\mathrm{Ker}(\pi_Q)$ can be extended to a *contravariant* functor

$$\mathfrak{Ker}(\pi) : \mathcal{L} \longrightarrow \mathfrak{Gr} \qquad 17.8.2$$

mapping $x \in \mathcal{L}(Q, R)$ on the group homomorphism $\mathrm{Ker}(\pi_Q) \to \mathrm{Ker}(\pi_R)$ sending $z \in \mathrm{Ker}(\pi_Q)$ to the unique element $w \in \mathrm{Ker}(\pi_R)$ fulfilling $x{\cdot}w = z{\cdot}x$ (cf. 17.7.1). It is clear that if $\mathcal{L}$ is divisible then, for any subgroup Q of P and any subgroup K of $\mathrm{Aut}(Q)$ such that Q is fully K-normalized in $\mathcal{F}$, the $N_\mathcal{F}^K(Q)$-locality $N_\mathcal{L}^K(Q)$ is divisible too.

17.9 Let $\mathcal{L}$ be a divisible $\mathcal{F}$-locality; if Q and R are subgroups of P, for any $u \in \mathcal{T}_P(Q, R)$ and any $v \in R \subset \mathcal{T}_P(R, R)$ we obviously have

$$uv = (uvu^{-1})u = \kappa_u(v)u \qquad 17.9.1,$$

so that $\tau_{Q,R}(u){\cdot}\tau_R(v) = \tau_Q\big(\kappa_u(v)\big){\cdot}\tau_{Q,R}(u)$. We say that $\mathcal{L}$ is *coherent* whenever we still have†

$$x{\cdot}\tau_R(v) = \tau_Q\big(\pi_x(v)\big){\cdot}x \qquad 17.9.2$$

for any pair of subgroups Q and R of P, any $x \in \mathcal{L}(Q, R)$ and any $v \in R$. Once again, if $\mathcal{L}$ is *coherent* then, for any subgroup Q of P and any subgroup K of $\mathrm{Aut}(Q)$ such that Q is fully K-normalized in $\mathcal{F}$, $N_\mathcal{L}^K(T)$ is also *coherent*. If $\mathcal{F}$ is a Frobenius P-category, in order to get *coherence* it suffices to restrict condition 17.9.2 to the automorphism group of suitable $\mathcal{F}$-objects; precisely, we have the following statement, which remains true for divisible $(\mathcal{F}, \mathfrak{X})$-localities whenever $\mathfrak{X}$ contains all the $\mathcal{F}$-*essential* subgroups of P (cf. 5.8).

---

† This corresponds to condition (C) in [13, Definition 1.6] which is equivalent to condition 1.8.2 in [34] for suitable groups $\mathcal{L}(Q)$, as shown in Proposition 17.10 and Remark 17.11.

Proposition 17.10 *If $\mathcal{F}$ is a Frobenius P-category, a divisible $\mathcal{F}$-locality $\mathcal{L}$ is coherent whenever we have* $\mathrm{Ker}(\pi_Q) \subset C_{\mathcal{L}(Q)}\big(\tau_Q(Q)\big)$ *for any subgroup Q of P, and if either $Q = P$ or Q is $\mathcal{F}$-essential then, for any $u \in Q$ and any $x \in \mathcal{L}(Q)$, we have*

$$x{\cdot}\tau_Q(u){\cdot}x^{-1} = \tau_Q\big(\pi_x(u)\big) \qquad 17.10.1.$$

Proof: Since for any pair of subgroups Q and R of P, for any $x \in \mathcal{T}_P(Q,R)$ and for any $v \in R$ we clearly have

$$\begin{aligned} i_Q^P{\cdot}\tau_Q\big(\pi_x(v)\big){\cdot}x &= \tau_{P,Q}\Big(\big(\pi_{P,Q}(i_Q^P{\cdot}x)\big)(v)\Big){\cdot}x \\ &= \tau_P\big(\pi_{i_Q^P\cdot x}(v)\big){\cdot}i_Q^P{\cdot}x \end{aligned} \qquad 17.10.2,$$

it suffices to prove equality 17.9.2 when $Q = P$. We argue by induction on the *length* ℓ of $\pi_{P,R}(x)$ (cf. 5.15); if $\ell = 0$ then

$$\pi_{P,R}(x) = \pi_P(n) \circ \pi_{P,R}(i_R^P) \qquad 17.10.3$$

for some $n \in \mathcal{L}(P)$, and therefore we get $x = n{\cdot}i_R^P{\cdot}z$ for some $z \in \mathcal{L}(R)$ such that $\pi_R(z) = \mathrm{id}_R$; thus, since z centralizes $\tau_R(v)$, we can write

$$\begin{aligned} x{\cdot}\tau_R(v) &= n{\cdot}i_R^P{\cdot}\tau_R\big(\pi_z(v)\big){\cdot}z = n{\cdot}\tau_P\big(\pi_{i_R^P\cdot z}(v)\big){\cdot}i_R^P{\cdot}z \\ &= \tau_P\big(\pi_x(v)\big){\cdot}x \end{aligned} \qquad 17.10.4.$$

If $\ell \geq 1$ then we have

$$\pi_{P,R}(x) = \pi_{P,Q}(i_Q^P) \circ \pi_Q(m) \circ \pi_{Q,R}(y) \qquad 17.10.5$$

for some $\mathcal{F}$-essential subgroup Q of P, some $m \in \mathcal{L}(Q)$ and some $y \in \mathcal{L}(Q,R)$ such that $\pi_{P,R}(i_Q^P{\cdot}y)$ has length $\ell - 1$ (cf. 5.15); up to a suitable choice of y, we may assume that $x = i_Q^P{\cdot}m{\cdot}y$ and, by the induction hypothesis, we still have

$$i_Q^P{\cdot}y{\cdot}\tau_R(v) = \tau_P\big(\pi_{i_Q^P\cdot y}(v)\big){\cdot}i_Q^P{\cdot}y = i_Q^P{\cdot}\tau_Q\big(\pi_y(v)\big){\cdot}y \qquad 17.10.6,$$

so that we get $y{\cdot}\tau_R(v) = \tau_Q\big(\pi_y(v)\big){\cdot}y$ (cf. 17.7.1), which finally implies that

$$\begin{aligned} x{\cdot}\tau_R(v) &= i_Q^P{\cdot}m{\cdot}\tau_Q\big(\pi_y(v)\big){\cdot}y = i_Q^P{\cdot}\tau_Q\big(\pi_{m\cdot y}(v)\big){\cdot}m{\cdot}y \\ &= \tau_P\big(\pi_x(v)\big){\cdot}x \end{aligned} \qquad 17.10.7.$$

We are done.

Remark 17.11 Note that if $\mathcal{L}$ is a coherent $\mathcal{F}$-locality, Q is a subgroup of P, R and T are subgroups of $N_P(Q)$ containing Q, and x is an element of $\mathcal{L}(R,T)$ fulfilling $\pi_x(Q) = Q$, then, on the one hand there is $y \in \mathcal{L}(Q)$ such that $i_Q^R \cdot y = x \cdot i_Q^T$ (cf. 17.7.1), so that we have $\pi_y(u) = \pi_x(u)$ for any $u \in Q$; on the other hand, for any $w \in T$ we have (cf. 17.9.2)

$$\begin{aligned} i_Q^R \cdot y \cdot \tau_Q(w) &= x \cdot \tau_T(w) \cdot i_Q^T = \tau_R\big(\pi_x(w)\big) \cdot x \cdot i_Q^T \\ &= i_Q^R \cdot \tau_Q\big(\pi_x(w)\big) \cdot y \end{aligned} \qquad 17.11.1,$$

so that we get $y \cdot \tau_Q(w) \cdot y^{-1} = \tau_Q\big(\pi_x(w)\big)$ (cf. 17.7.1); in other words,

17.11.2 *The group homomorphisms ψ in $\mathcal{F}(R,T)$ which stabilize Q induce group homomorphisms $\tau_Q(T) \to \tau_Q(R)$ already induced by elements of $\mathcal{L}(Q)$ lifting the action of ψ over Q.*

17.12 Assume that $\mathcal{F}$ is a Frobenius P-category; as a matter of fact, we are mostly interested in $\mathcal{F}$-localities which only depend on the knowledge of P and $\mathcal{F}$, and thus the groups $\mathrm{Ker}(\pi_Q)$, when Q runs over the set of subgroups of P, should only depend on P. We say that a coherent $\mathcal{F}$-locality $\mathcal{L}$ is *P-bounded* if, for any subgroup Q of P fully normalized in $\mathcal{F}$, we have

$$\mathrm{Ker}(\pi_Q) \subset \tau_Q\big(N_P(Q)\big) \qquad 17.12.1;$$

in this case, since $\mathcal{F}_P(Q)$ is a Sylow p-subgroup of $\mathcal{F}(Q)$ (cf. Proposition 2.11), *$\tau_Q\big(N_P(Q)\big)$ is a Sylow p-subgroup of $\mathcal{L}(Q)$*. It is clear that, for any subgroup P' of P and any Frobenius P'-subcategory $\mathcal{F}'$ of $\mathcal{F}$, a coherent $\mathcal{F}'$-sublocality $\mathcal{L}'$ of $\mathcal{L}$ is P'-bounded too; in particular, for any subgroup Q of P and any subgroup K of $\mathrm{Aut}(Q)$ such that Q is fully K-normalized in $\mathcal{F}$, *$N_{\mathcal{L}}^K(Q)$ is $N_P(Q)$-bounded.*

17.13 On the other hand, if $\mathrm{Ker}(\pi_Q)$ is a p-group for any subgroup Q of P, it follows from Remark 17.11 that, if Q is fully centralized in $\mathcal{F}$ and T is a subgroup of $C_P(Q)$, all the elements in $\big(C_{\mathcal{F}}(Q)\big)\big(C_P(Q),T\big)$ induce group homomorphisms from $\tau_Q(T)$ to $\tau_Q\big(C_P(Q)\big)$ which are already induced by inner automorphisms of $\mathrm{Ker}(\pi_Q)$; thus, denoting by $H_{C_{\mathcal{F}}(Q)}$ the corresponding *$C_{\mathcal{F}}(Q)$-hyperfocal subgroup* (cf. 13.2), in this case we get

$$H_{C_{\mathcal{F}}(Q)} \subset \mathrm{Ker}(\tau_Q) \qquad 17.13.1.$$

Let us say that a coherent $\mathcal{F}$-locality $\mathcal{L}$ is *perfect* when $\mathcal{L}$ is P-bounded and all these inclusions are equalities†. As above, it is not difficult to prove that if $\mathcal{L}$ is *perfect* then, for any subgroup Q of P and any subgroup K of $\mathrm{Aut}(Q)$ such that Q is fully K-normalized in $\mathcal{F}$, $N_{\mathcal{L}}^K(Q)$ is *perfect* too.

† Called a *centric linking system* in [13].

17.14 When $\mathcal{F}$ is a Frobenius P-category, we already know from Proposition 12.3 that, for any $\mathcal{F}$-*stable* subgroup U of P, there exists a U-*quotient* of $\mathcal{F}$, which is a Frobenius category $\bar{\mathcal{F}}$ over $\bar{P} = P/U$; similarly, if we have a *perfect* $\mathcal{F}$-*locality* $\mathcal{L}$, it makes sense to look for a *perfect* $\bar{\mathcal{F}}$-*locality* $\bar{\mathcal{L}}$ as a kind of U-quotient of $\mathcal{L}$. In this last part of the chapter, we describe the construction of such a quotient.

17.15 From the definition of *perfect localities*, it is quite easy *a priori* to foresee what should be the kernel $U_{\mathcal{L}}(Q)$ of the canonical group homomorphism $\mathcal{L}(Q) \to \bar{\mathcal{L}}(\bar{Q})$ for any subgroup Q of P, where $\bar{Q}$ denotes the image of Q in $\bar{P}$; indeed, since $\mathrm{Ker}(\pi_{\bar{Q}})$ should be a p-group (cf. 17.13), we have the following necessary condition

$$\bar{\mathcal{L}}(\bar{Q})/\mathbf{O}_p\big(\bar{\mathcal{L}}(\bar{Q})\big) \cong \bar{\mathcal{F}}(\bar{Q})/\mathbf{O}_p\big(\bar{\mathcal{F}}(\bar{Q})\big) \qquad 17.15.1$$

and therefore, denoting by $U_{\mathcal{F}}(Q)$ the kernel of the canonical group homomorphism $\mathcal{F}(Q) \to \bar{\mathcal{F}}(\bar{Q})$, we necessarily have

$$\mathbf{O}^p\big(U_{\mathcal{L}}(Q)\big) = \mathbf{O}^p\Big(\pi_Q^{-1}\big(U_{\mathcal{F}}(Q)\big)\Big) \qquad 17.15.2.$$

17.16 Moreover, if $\varphi : Q \to P$ is an $\mathcal{F}$-morphism such that $Q' = \varphi(Q)$ is fully normalized in $\mathcal{F}$ (cf. Proposition 2.7), Q' is also fully $U_{\mathcal{F}}(Q')$-normalized in $\mathcal{F}$ (cf. Proposition 2.11) and then it is easily checked that the Frobenius $C_{\bar{P}}(\bar{Q})$-category $C_{\bar{\mathcal{F}}}(\bar{Q}')$ is the $N_U(Q')$-quotient of $N_{\mathcal{F}}^{U_{\mathcal{F}}(Q')}(Q')$; thus, according to the perfectness condition (cf. 17.13) and to Proposition 13.10, it is quite clear that, setting

$$\mathcal{F}^{Q'} = N_{\mathcal{F}}^{U_{\mathcal{F}}(Q')}(Q') \quad \text{and} \quad P^{Q'} = N_P^{U_{\mathcal{F}}(Q')}(Q') \qquad 17.16.1,$$

the image of $N_U(Q')\cdot H_{\mathcal{F}Q'}$ in $\mathcal{L}(Q')$ should be a Sylow p-subgroup of $U_{\mathcal{L}}(Q')$.

17.17 Consequently, choosing $x \in \mathcal{L}(Q, Q')$ fulfilling $\pi_x\big(\varphi(u)\big) = u$ for any $u \in Q$, we set

$$U_{\mathcal{L}}(Q) = \mathbf{O}^p\Big(\pi_Q^{-1}\big(U_{\mathcal{F}}(Q)\big)\Big)\cdot x\cdot \tau_{Q'}\big(N_U(Q')\cdot H_{\mathcal{F}Q'}\big)\cdot x^{-1} \qquad 17.17.1$$

which does not depend on our choices of φ and x; indeed, if Q'' is a subgroup of P which is $\mathcal{F}$-isomorphic to Q and fully $U_{\mathcal{F}}(Q'')$-normalized in $\mathcal{F}$ then, according to condition 2.8.2, there is an $\mathcal{F}$-morphism

$$\zeta : N_P^{U_{\mathcal{F}}(Q')}(Q') \longrightarrow N_P^{U_{\mathcal{F}}(Q'')}(Q'') \qquad 17.17.2$$

such that $\zeta(Q') = Q''$; since Q' is fully $U_{\mathcal{F}}(Q')$-normalized in $\mathcal{F}$, we actually have

$$\zeta\big(N_P^{U_{\mathcal{F}}(Q')}(Q')\big) = N_P^{U_{\mathcal{F}}(Q'')}(Q'') \qquad 17.17.3$$

and, with the notation in 17.16.1 above, it is easily checked that the isomorphism $P^{Q'} \cong P^{Q''}$ determined by ζ is $(\mathcal{F}^{Q'}, \mathcal{F}^{Q''})$*-functorial* (cf. 12.1); hence, we still we have (cf. 13.2 and 12.2)

$$\zeta(H_{\mathcal{F}^{Q'}}) = H_{\mathcal{F}^{Q''}} \quad \text{and} \quad \zeta\big(N_U(Q')\big) = N_U(Q'') \qquad 17.17.4.$$

Note that our next result remains true for *perfect partially defined $\mathcal{F}$-localities.*

Theorem 17.18 *Assume that $\mathcal{F}$ is a Frobenius P-category. Let $\mathcal{L}$ be a perfect $\mathcal{F}$-locality and U an $\mathcal{F}$-stable subgroup of P, set $\bar{P} = P/U$ and denote by $\bar{\mathcal{F}}$ the U-quotient of $\mathcal{F}$. Then, we have a perfect $\bar{\mathcal{F}}$-locality $\bar{\mathcal{L}}$ where, for any pair of subgroups $\bar{Q} = Q/U$ and $\bar{R} = R/U$ of $\bar{P}$, the set of morphisms from $\bar{R}$ to $\bar{Q}$ is the quotient set*

$$\bar{\mathcal{L}}(\bar{Q}, \bar{R}) = \mathcal{L}(Q, R)/U_{\mathcal{L}}(R) \qquad 17.18.1$$

and the structural functors are induced by the structural functors of $\mathcal{L}$. Moreover, there is a unique functor $\mathcal{L} \to \bar{\mathcal{L}}$ compatible with the structural functors which, for any pair of subgroups Q and R of P containing U, maps $y \in \mathcal{L}(Q, R)$ on its class $\bar{y} \in \bar{\mathcal{L}}(\bar{Q}, \bar{R})$.

Proof: By the definition above it is clear that, for any pair of mutually $\mathcal{F}$-isomorphic subgroups R and R' of P, and any $x \in \mathcal{L}(R', R)$, we have

$$U_{\mathcal{L}}(R') = x{\cdot}U_{\mathcal{L}}(R){\cdot}x^{-1} \qquad 17.18.2.$$

Moreover, we claim that

17.18.3 *If Q is a subgroup of P containing R and R contains $Q \cap U$ then any $x \in U_{\mathcal{L}}(Q)$ fulfills $\pi_x(R) = R$ and the element $y \in \mathcal{L}(R)$ such that $x{\cdot}i_R^Q = i_R^Q{\cdot}y$ belongs to $U_{\mathcal{L}}(R)$.*

Indeed, arguing by induction on $|Q : R|$, we may assume that R is normal in Q; since R contains $Q \cap U$ and π_x acts trivially on $\bar{Q}$, it is clear that $\pi_x(R) = R$ and that π_y acts trivially on $\bar{R}$; furthermore, since we still have $x^n{\cdot}i_R^Q = i_R^Q{\cdot}y^n$ for any $n \in \mathbb{N}$, if x is a p'-element of $\pi_Q^{-1}\big(U_{\mathcal{F}}(Q)\big)$ then y is a p'-element of $\pi_R^{-1}\big(U_{\mathcal{F}}(R)\big)$ (cf. 17.7.1).

That is to say, denoting by $\mathcal{L}(Q)_R$ the stabilizer of R in $\mathcal{L}(Q)$, the divisibility of $\mathcal{L}$ determines a group homomorphism

$$\mathfrak{g}_R^Q : \mathcal{L}(Q)_R \longrightarrow N_{\mathcal{L}(R)}\big(\tau_R(Q)\big) \qquad 17.18.4$$

and actually we have proved above that we have

$$U_{\mathcal{L}}(Q) \subset \mathcal{L}(Q)_R \quad \text{and} \quad \mathfrak{g}_R^Q\Big(\mathbf{O}^p\big(U_{\mathcal{L}}(Q)\big)\Big) \subset \mathbf{O}^p\big(U_{\mathcal{L}}(R)\big) \qquad 17.18.5;$$

thus, in order to prove that $\mathfrak{g}_R^Q\big(U_{\mathcal{L}}(Q)\big) \subset U_{\mathcal{L}}(R)$, it suffices to prove that the image of a Sylow p-subgroup of $U_{\mathcal{L}}(Q)$ is contained in $U_{\mathcal{L}}(R)$.

It follows from Proposition 2.7 that there is an $\mathcal{F}$-morphism $\varphi : Q \to P$ such that $R' = \varphi(R)$ is fully normalized in $\mathcal{F}$; moreover, since $N_{\mathcal{F}}(R')$ is a Frobenius $N_P(R')$-category (cf. Proposition 2.16) and $Q' = \varphi(Q)$ is contained in $N_P(R')$, there is an $N_{\mathcal{F}}(R')$-morphism $\nu' : Q' \to N_P(R')$ such that $Q'' = \nu'(Q')$ is fully normalized in $N_{\mathcal{F}}(R')$ (cf. Proposition 2.7) and then, since

$$N_P^{U_{\mathcal{F}}(Q'')}(Q'') \subset N_P^{U_{\mathcal{F}}(R')}(R') = P^{R'} \qquad 17.18.6,$$

it follows from Proposition 2.11 and Lemma 2.17 that Q'' is fully $U_{\mathcal{F}}(Q'')$-normalized in $\mathcal{F}$; but, with the notation in 17.16.1, $\mathcal{F}^{Q''}$ is a subcategory of $\mathcal{F}^{R'}$ and therefore we get $H_{\mathcal{F}^{Q''}} \subset H_{\mathcal{F}^{R'}}$ (cf. 13.2); moreover, it is quite clear that $N_U(Q'') \subset N_U(R')$. Hence, we get

$$\mathfrak{g}_R^Q\Big(\tau_{Q''}\big(N_U(Q'')\cdot H_{\mathcal{F}^{Q''}}\big)\Big) \subset \tau_{R'}\big(N_U(R')\cdot H_{\mathcal{F}^{R'}}\big) \qquad 17.18.7$$

and, in order to prove our claim, it suffices to consider $z \in \mathcal{L}(Q, Q'')$ such that $\pi_z\Big(\nu'\big(\varphi(u)\big)\Big) = u$ for any $u \in Q$ (cf. 17.16).

Consequently, denoting by $\mathfrak{X}$ the set of subgroups of P containing U and by $\mathcal{L}^{\mathfrak{X}}$ the full subcategory of $\mathcal{L}$ over $\mathfrak{X}$, the correspondence $U_{\mathcal{L}}$ defined over $\mathfrak{X}$ induces a *contravariant* functor from $\mathcal{L}^{\mathfrak{X}}$ to the category $\mathfrak{Gr}$ (cf. 1.1) and therefore it is easily checked that there is a category $\bar{\mathcal{L}}$ where the objects are the subgroups of $\bar{P}$, where the set of morphisms from $\bar{R}$ to $\bar{Q}$ is the quotient set $\mathcal{L}(Q,R)/U_{\mathcal{L}}(R)$ for any $Q, R \in \mathfrak{X}$, and where the composition is induced by the composition in $\mathcal{L}$. At the same time, we get an evident functor from $\mathcal{L}^{\mathfrak{X}}$ to $\bar{\mathcal{L}}$ and the structural functors of the $\mathfrak{X}$-locality $\mathcal{L}^{\mathfrak{X}}$ determine structural functors $\bar{\pi} : \bar{\mathcal{L}} \to \bar{\mathcal{F}}$ and $\bar{\tau} : \mathcal{T}_{\bar{P}} \to \bar{\mathcal{L}}$, so that $\bar{\mathcal{L}}$ becomes an $\bar{\mathcal{F}}$-locality.

Since $\mathcal{L}$ is divisible, $\mathrm{Ker}(\pi_R)$ acts regularly on the "fibers" of the map $\pi_{Q,R}$ (cf. 17.3.2) and clearly $U_{\mathcal{F}}(R)$ acts regularly on the "fibers" of the map

$$\mathcal{F}(Q,R) \longrightarrow \bar{\mathcal{F}}(\bar{Q},\bar{R}) \qquad 17.18.8;$$

it easily follows that the kernel of the homomorphism $\bar{\pi}_{\bar{R}} : \bar{\mathcal{L}}(\bar{R}) \to \bar{\mathcal{F}}(\bar{R})$ still acts regularly on the "fibers" of the map

$$\bar{\pi}_{\bar{Q},\bar{R}} : \bar{\mathcal{L}}(\bar{Q},\bar{R}) \longrightarrow \bar{\mathcal{F}}(\bar{Q},\bar{R}) \qquad 17.18.9;$$

thus, since $\bar{\mathcal{F}}$ is divisible, $\bar{\mathcal{L}}$ is divisible too. Moreover, it is quite clear that the coherence of $\mathcal{L}^{\mathfrak{X}}$ forces the coherence of $\bar{\mathcal{L}}$.

On the other hand, since from the very definition of $U_{\mathcal{F}}(R)$ we have the isomorphism

$$\mathcal{L}(R)/\pi_R^{-1}\big(U_{\mathcal{F}}(R)\big) \cong \bar{\mathcal{F}}(\bar{R}) \qquad 17.18.10,$$

it is easily checked that $\bar{\mathcal{L}}$ is $\bar{P}$-bounded; similarly, if R is fully normalized in $\mathcal{F}$ then $\bar{R}$ is fully normalized in $\bar{\mathcal{F}}$, R is also $U_{\mathcal{F}}(R)$-normalized in $\mathcal{F}$ (cf. Proposition 2.11) and $H_{C_{\bar{\mathcal{F}}}(\bar{R})}$ is the image of $H_{\mathcal{F}^R}$ (cf. Proposition 13.10), so that we have $\mathrm{Ker}(\bar{\tau}_{\bar{R}}) = H_{C_{\bar{\mathcal{F}}}(\bar{R})}$. In conclusion, $\bar{\mathcal{L}}$ is perfect too.

Finally, we prove that the canonical functor $\mathcal{L}^{\mathfrak{x}} \to \bar{\mathcal{L}}$ has a unique extension to a functor $\lambda: \mathcal{L} \to \bar{\mathcal{L}}$ which is the identity map over the set of objects. Consider a pair of mutually $\mathcal{F}$-isomorphic subgroups Q and Q' of P, assume that Q' is fully normalized in $\mathcal{F}$, set $\hat{Q} = N_{Q\cdot U}(Q)$ and $\hat{Q}' = N_{Q'\cdot U}(Q')$, and denote by $\mathcal{F}(\hat{Q}', \hat{Q})_{Q',Q}$ the set of $\varphi \in \mathcal{F}(\hat{Q}', \hat{Q})$ such that $\varphi(Q) = Q'$ and by $\mathcal{L}(\hat{Q}', \hat{Q})_{Q',Q}$ the corresponding converse image in $\mathcal{L}(\hat{Q}', \hat{Q})$; these sets are nonempty since there is an $\mathcal{F}$-morphism $\zeta : N_P(Q) \to P$ such that $\zeta(Q) = Q'$ (cf. condition 2.8.2) and then we have $\zeta(\hat{Q}) \subset \hat{Q}'$. Once again, the divisibility of $\mathcal{L}$ determines an evident map

$$\mathfrak{g}_{Q',Q} : \mathcal{L}(\hat{Q}', \hat{Q})_{Q',Q} \longrightarrow \mathcal{L}(Q', Q) \qquad 17.18.11.$$

Moreover, since $\tau_{Q'}\big(N_P(Q')\big)$ is a Sylow p-subgroup of $\mathcal{L}(Q')$ (cf. 17.12), by the *Frattini argument* we get

$$\mathcal{L}(Q') = \mathbf{O}^p\big(U_{\mathcal{L}}(Q')\big)\cdot N_{\mathcal{L}(Q')}\Big(U_{\mathcal{L}}(Q') \cap \tau_{Q'}\big(N_P(Q')\big)\Big) \qquad 17.18.12$$

and, since U is $\mathcal{F}$-stable and $U_{\mathcal{L}}(Q')$ contains $\tau_{Q'}(\hat{Q}')$, the second factor stabilizes $\hat{Q}'$; hence, we still get

$$\mathcal{L}(Q') = \mathbf{O}^p\big(U_{\mathcal{L}}(Q')\big)\cdot N_{\mathcal{L}(Q')}\big(\tau_{Q'}(\hat{Q}')\big) \qquad 17.18.13.$$

But, since Q' is fully centralized in $\mathcal{F}$ (cf. Proposition 2.11), it follows from statement 2.10.1 that the homomorphism

$$\mathfrak{g}_{Q'}^{\hat{Q}'} : \mathcal{L}(\hat{Q}')_{Q'} \longrightarrow N_{\mathcal{L}(Q')}\big(\tau_{Q'}(\hat{Q}')\big) \qquad 17.18.14$$

is surjective. Consequently, since the map $\mathfrak{g}_{Q',Q}$ is not empty and $\mathcal{L}(Q')$ acts regularly on $\mathcal{L}(Q',Q)$, $\mathfrak{g}_{Q',Q}$ induces a surjective map (cf. 17.18.2)

$$\bar{\mathfrak{g}}_{Q',Q} : \mathcal{L}(\hat{Q}', \hat{Q})_{Q',Q} \longrightarrow \mathcal{L}(Q',Q)/\mathbf{O}^p\big(U_{\mathcal{L}}(Q)\big) \qquad 17.18.15.$$

Furthermore, for any subgroup R' of $\hat{Q}'$ containing Q', the divisibility of $\mathcal{L}$ determines as above a group homomorphism

$$\mathfrak{g}_{Q'}^{R'} : \mathcal{L}(R')_{Q'} \longrightarrow N_{\mathcal{L}(Q')}\big(\tau_{Q'}(R')\big) \qquad 17.18.16$$

and we claim that

$$(\mathfrak{g}_{Q'}^{R'})^{-1}\Big(\mathbf{O}^p\big(U_{\mathcal{L}}(Q')\big)\Big) \subset U_{\mathcal{L}}(R') \qquad 17.18.17.$$

Indeed, since the image of $U_{\mathcal{L}}(Q')$ in $\bar{\mathcal{F}}(\bar{Q}')$ is trivial and $\bar{Q}' = \bar{R}'$, it is clear that

$$(\mathfrak{g}_{Q'}^{R'})^{-1}\big(U_{\mathcal{L}}(Q')\big) \subset \pi_{R'}^{-1}\big(U_{\mathcal{F}}(R')\big) \qquad 17.18.18$$

and therefore we have

$$(\mathfrak{g}_{Q'}^{R'})^{-1}\Big(\mathbf{O}^p\big(U_{\mathcal{L}}(Q')\big)\Big) \subset \mathbf{O}^p\Big(\pi_{R'}^{-1}\big(U_{\mathcal{F}}(R')\big)\Big)\cdot\mathrm{Ker}(\mathfrak{g}_{Q'}^{R'}) \qquad 17.18.19;$$

hence, since $\mathrm{Ker}(\mathfrak{g}_{Q'}^{R'}) \subset \pi_{R'}^{-1}\big(U_{\mathcal{F}}(R')\big)$, it suffices to prove that $U_{\mathcal{L}}(R')$ contains a Sylow p-subgroup of $\mathrm{Ker}(\mathfrak{g}_{Q'}^{R'})$.

We already know that $\tau_{Q'}\big(N_P(Q')\big)$ is a Sylow p-subgroup of $\mathcal{L}(Q')$ and may assume that it contains a Sylow p-subgroup of $N_{\mathcal{L}(Q')}\big(\tau_{Q'}(R')\big)$; moreover, since $\mathcal{L}$ is perfect (cf. 17.13) and Q' is fully centralized in $\mathcal{F}$, we actually have

$$\tau_{Q'}\big(N_P(Q')\big) \cong N_P(Q')/H_{C_{\mathcal{F}}(Q')} \qquad 17.18.20;$$

then, it is easily checked that $\tau_{R'}\big(N_{H_{C_{\mathcal{F}}(Q')}}(R')\big)$ is a Sylow p-subgroup of $\mathrm{Ker}(\mathfrak{g}_{Q'}^{R'})$. But, the inclusions

$$C_P(Q') \subset N_P^{U_{\mathcal{F}}(Q')}(Q') = P^{Q'} \quad \text{and} \quad C_{\mathcal{F}}(Q') \subset \mathcal{F}^{Q'} \qquad 17.18.21$$

imply that (cf. 13.2)

$$H_{C_{\mathcal{F}}(Q')} \subset H_{\mathcal{F}^{Q'}} \qquad 17.18.22;$$

moreover, an $\mathcal{F}$-morphism $\xi : P^{Q'} \to P$ such that $R'' = \xi(R')$ is fully normalized in $\mathcal{F}$ (cf. Proposition 2.7) determines a group homomorphism from $P^{Q'}$ to $P^{R''}$ which is clearly $(\mathcal{F}^{Q'}, \mathcal{F}^{R''})$-functorial and therefore we still have

$$\xi(H_{\mathcal{F}^{Q'}}) \subset H_{\mathcal{F}^{R''}} \qquad 17.18.23.$$

Hence, since $U_{\mathcal{L}}(R'')$ contains $\tau_{R''}(H_{\mathcal{F}^{R''}})$, $U_{\mathcal{L}}(R')$ contains indeed a Sylow p-subgroup of $\mathrm{Ker}(\mathfrak{g}_{Q'}^{R'})$.

Since the equality $Q = \hat{Q}$ implies $U \subset Q$, arguing by induction on $|P:Q|$, we may assume that the functor we are looking for is already defined over $\mathcal{L}(\hat{Q}',\hat{Q})$, so that it determines a map

$$\lambda_{\hat{Q}',\hat{Q}} : \mathcal{L}(\hat{Q}',\hat{Q}) \longrightarrow \bar{\mathcal{L}}(\bar{Q}',\bar{Q}) \qquad 17.18.24$$

and then, according to the surjectivity of the map 17.18.15 and to our claim above, we get a new map

$$\lambda_{Q',Q} : \mathcal{L}(Q',Q) \longrightarrow \bar{\mathcal{L}}(\bar{Q}',\bar{Q}) \qquad 17.18.25$$

which, for any $x \in \mathcal{L}(\hat{Q}',\hat{Q})_{Q',Q}$ and any $y \in \mathbf{O}^p\big(U_{\mathcal{L}}(Q)\big)$, sends $\mathfrak{g}_{Q',Q}(x)\cdot y$ to $\lambda_{\hat{Q}',\hat{Q}}(x)$. Note that, if we already have a functor $\lambda' : \mathcal{L} \to \bar{\mathcal{L}}$ as above and we want to prove that it is unique, arguing by induction on $|P:Q|$ we may assume that $\lambda'_{\hat{Q}',\hat{Q}} = \lambda_{\hat{Q}',\hat{Q}}$ and then it is easily checked that $\lambda'_{Q',Q} = \lambda_{Q',Q}$.

Now, if Q'' is a subgroup of P which is $\mathcal{F}$-isomorphic to Q and Q', and fully normalized in $\mathcal{F}$, we have the maps $\lambda_{Q'',Q}$, $\lambda_{Q'',Q'}$ and $\lambda_{Q',Q}$, and we claim that, for any $x \in \mathcal{L}(Q',Q)$ and any $x' \in \mathcal{L}(Q'',Q')$, we get

$$\lambda_{Q'',Q}(x'\cdot x) = \lambda_{Q'',Q'}(x')\cdot\lambda_{Q',Q}(x) \qquad 17.18.26.$$

Indeed, setting $\hat{Q}'' = N_{Q''\cdot U}(Q'')$ and choosing elements $\hat{x} \in \mathcal{L}(\hat{Q}',\hat{Q})$ and $\hat{x}' \in \mathcal{L}(\hat{Q}'',\hat{Q}')$ such that

$$\begin{aligned} x^{-1}\cdot\mathfrak{g}_{Q',Q}(\hat{x}) &\in \mathbf{O}^p\big(U_{\mathcal{L}}(Q)\big) \\ x'^{-1}\cdot\mathfrak{g}_{Q'',Q'}(\hat{x}') &\in \mathbf{O}^p\big(U_{\mathcal{L}}(Q')\big) \end{aligned} \qquad 17.18.27,$$

it is easily checked that $(x'\cdot x)^{-1}\cdot\mathfrak{g}_{Q'',Q}(\hat{x}'\cdot\hat{x})$ belongs to $\mathbf{O}^p\big(U_{\mathcal{L}}(Q)\big)$ and therefore we get

$$\begin{aligned} \lambda_{Q'',Q}(x'\cdot x) &= \lambda_{\hat{Q}'',\hat{Q}}(\hat{x}'\cdot\hat{x}) = \lambda_{\hat{Q}'',\hat{Q}'}(\hat{x}')\cdot\lambda_{\hat{Q}',\hat{Q}}(\hat{x}) \\ &= \lambda_{Q'',Q'}(x')\cdot\lambda_{Q',Q}(x)\,. \end{aligned} \qquad 17.18.28.$$

In particular, if we modify our hypothesis and Q' is no more fully normalized in $\mathcal{F}$, then equality 17.18.26 supplies a definition for $\lambda_{Q',Q}$; explicitly, consider a triple of $\mathcal{F}$-isomorphic subgroups Q, Q' and Q'' of P, assume that Q'' is fully normalized in $\mathcal{F}$, and choose $x' \in \mathcal{L}(Q'',Q')$; then, we define a map

$$\lambda_{Q',Q} : \mathcal{L}(Q',Q) \longrightarrow \bar{\mathcal{L}}(\bar{Q}',\bar{Q}) \qquad 17.18.29$$

sending $x \in \mathcal{L}(Q',Q)$ to $\lambda_{Q'',Q'}(x')^{-1}\cdot\lambda_{Q'',Q}(x'\cdot x)$.

This definition does not depend on our choices since, if Q''' is a subgroup of P which is $\mathcal{F}$-isomorphic to Q and fully normalized in $\mathcal{F}$, and x'' is an element of $\mathcal{L}(Q''',Q')$, we have (cf. 17.18.26)

$$\begin{aligned}\lambda_{Q''',Q'}(x'')^{-1}\cdot\lambda_{Q''',Q}(x''\cdot x)&\\ &= \lambda_{Q''',Q'}(x'')^{-1}\cdot\lambda_{Q''',Q}\big((x''\cdot x'^{-1})\cdot(x'\cdot x)\big)\\ &= \lambda_{Q''',Q'}(x'')^{-1}\cdot\lambda_{Q''',Q''}(x''\cdot x'^{-1})\cdot\lambda_{Q'',Q}(x'\cdot x)\\ &= \lambda_{Q'',Q'}(x')^{-1}\cdot\lambda_{Q'',Q}(x'\cdot x)\,.\end{aligned} \qquad 17.18.30.$$

More generally, if we remove the hypothesis that Q'' is fully normalized, it follows from definition 17.18.29 above that, since $x''\cdot x'^{-1}$ belongs to $\mathcal{L}(Q''',Q'')$, for any $x'\in\mathcal{L}(Q'',Q')$ and any $x\in\mathcal{L}(Q',Q)$ we still have

$$\begin{aligned}&\lambda_{Q'',Q}(x'\cdot x) = \lambda_{Q''',Q''}(x''\cdot x'^{-1})^{-1}\cdot\lambda_{Q''',Q}(x''\cdot x)\\ &= \lambda_{Q''',Q''}(x''\cdot x'^{-1})^{-1}\cdot\lambda_{Q''',Q'}(x'')\cdot\lambda_{Q''',Q'}(x'')^{-1}\cdot\lambda_{Q''',Q}(x''\cdot x)\\ &= \lambda_{Q''',Q''}(x''\cdot x'^{-1})^{-1}\cdot\lambda_{Q''',Q'}\big((x''\cdot x'^{-1})\cdot x'\big)\cdot\lambda_{Q',Q}(x)\\ &= \lambda_{Q'',Q'}(x')\cdot\lambda_{Q',Q}(x)\,.\end{aligned} \qquad 17.18.31.$$

On the other hand, if R is a subgroup of Q, the compatibility with the structural functors forces

$$\lambda_{Q,R}(i^{Q}_{R}) = \bar{i}^{\bar{Q}}_{\bar{R}} = \bar{\tau}_{\bar{Q},\bar{R}}(1) \qquad 17.18.32;$$

then, the uniqueness of λ is clear and it is not difficult to see that, in order to complete its construction, it suffices to prove that, for any $x\in\mathcal{L}(Q',Q)$, setting $R'=\pi_x(R)$ and denoting by y the element of $\mathcal{L}(R',R)$ such that $x\cdot i^{Q}_{R} = i^{Q'}_{R'}\cdot y$, we have

$$\lambda_{Q',Q}(x)\cdot\bar{i}^{\bar{Q}}_{\bar{R}} = \bar{i}^{\bar{Q}'}_{\bar{R}'}\cdot\lambda_{R',R}(y) \qquad 17.18.33.$$

We argue by induction on $|P\colon Q|$ and on $|Q\colon R|$; we clearly may assume that $|Q\colon R| = p$; in this case, either $Q\cap U\subset R$ or $Q\subset\hat{R} = N_{R\cdot U}(R)$. Firstly assume that $Q\subset\hat{R}$; setting $\hat{R}' = N_{R'\cdot U}(R')$, let $\psi'\colon\hat{R}'\to P$ be an $\mathcal{F}$-morphism such that $R''=\psi'(R')$ is fully normalized in $\mathcal{F}$ (cf. Proposition 2.7), set $Q''=\psi'(Q')$, choose $x'\in\mathcal{L}(Q'',Q')$ and denote by y' the element of $\mathcal{L}(R'',R')$ such that $x'\cdot i^{Q'}_{R'} = i^{Q''}_{R''}\cdot y'$; then, it suffices to prove equality 17.18.33 for $x'\cdot x$ and for x'.

That is to say, we may assume that R' is fully normalized in $\mathcal{F}$; then, there is $\hat{y}\in\mathcal{L}(\hat{R}',\hat{R})$ such that $y^{-1}\cdot\mathfrak{g}_{R',R}(\hat{y})$ belongs to $\mathbf{O}^p\big(U_{\mathcal{L}}(R)\big)$; hence, denoting by $\hat{x}$ the element of $\mathcal{L}(Q',Q)$ such that $\hat{y}\cdot i_{Q}^{\hat{R}} = i_{Q'}^{\hat{R}'}\cdot\hat{x}$, we get (cf. 17.18.16)

$$\mathfrak{g}_{R}^{Q}(x^{-1}\cdot\hat{x}) = y^{-1}\cdot\mathfrak{g}_{R',R}(\hat{y}) \in \mathbf{O}^{p}\big(U_{\mathcal{L}}(R)\big) \qquad 17.18.34$$

and, since $U_{\mathcal{L}}(Q)$ contains $(\mathfrak{g}_{R}^{Q})^{-1}\Big(\mathbf{O}^{p}\big(U_{\mathcal{L}}(R)\big)\Big)$, $x^{-1}\cdot\hat{x}$ belongs to $U_{\mathcal{L}}(Q)$, so that $\lambda_Q(x^{-1}\cdot\hat{x})$ is the trivial element $\bar{\imath}_{\bar{Q}}^{\bar{Q}}$ of $\bar{\mathcal{L}}(\bar{Q})$. Consequently, according to our definition, we have $\lambda_{R',R}(y) = \lambda_{\hat{R}',\hat{R}}(\hat{y})$ and, since $\bar{R} = \bar{Q}$ and $\bar{R}' = \bar{Q}'$, by our induction hypothesis we may assume that $\lambda_{\hat{R}',\hat{R}}(\hat{y}) = \lambda_{Q',Q}(\hat{x})$; since $\lambda_Q(x^{-1}\cdot\hat{x}) = \bar{\imath}_{\bar{Q}}^{\bar{Q}}$, we obtain $\lambda_{R',R}(y) = \lambda_{Q',Q}(x)$.

Secondly, assume that $Q\cap U \subset R$, so that we have $Q'\cap U \subset R'$ too; let $\psi' : Q' \to P$ be an $\mathcal{F}$-morphism such that $Q'' = \psi'(Q')$ is fully normalized in $\mathcal{F}$, set $R'' = \psi'(R')$, choose $x' \in \mathcal{L}(Q'',Q')$ and denote by y' the element of $\mathcal{L}(R'',R')$ such that $x'\cdot i_{R'}^{Q'} = i_{R''}^{Q''}\cdot y'$; once again, it suffices to prove equality 17.18.33 for $x'\cdot x$ and for x'. That is to say, in this case we may assume that Q' is fully normalized in $\mathcal{F}$; then, according to our definition, there is $\hat{x} \in \mathcal{L}(\hat{Q}',\hat{Q})$ such that

$$x^{-1}\cdot\mathfrak{g}_{Q',Q}(\hat{x}) \in \mathbf{O}^{p}\big(U_{\mathcal{L}}(Q)\big) \quad \text{and} \quad \lambda_{Q',Q}(x) = \lambda_{\hat{Q}',\hat{Q}}(\hat{x}) \qquad 17.18.35.$$

Moreover, denoting by $\hat{y}$ the element of $\mathcal{L}(R',R)$ such that $\hat{x}\cdot i_{R}^{\hat{Q}} = i_{R'}^{\hat{Q}'}\cdot\hat{y}$, it is easily checked that

$$\big(x^{-1}\cdot\mathfrak{g}_{Q',Q}(\hat{x})\big)\cdot i_{R}^{Q} = i_{R}^{Q}\cdot(y^{-1}\cdot\hat{y}) \qquad 17.18.36$$

and, since $Q\cap U \subset R$ and $x^{-1}\cdot\mathfrak{g}_{Q',Q}(\hat{x}) \in U_{\mathcal{L}}(Q)$, it follows from statement 17.18.3 that $y^{-1}\cdot\hat{y}$ belongs to $U_{\mathcal{L}}(R)$ so that we get $\lambda_{R',R}(y) = \lambda_{R',R}(\hat{y})$. Consequently, by the induction hypothesis, we have

$$\lambda_{Q',Q}(x)\cdot\bar{\imath}_{\bar{R}}^{\bar{Q}} = \lambda_{\hat{Q}',\hat{Q}}(\hat{x})\cdot\bar{\imath}_{\bar{R}}^{\bar{Q}} = \bar{\imath}_{\bar{R}'}^{\bar{Q}'}\cdot\lambda_{R',R}(\hat{y}) = \bar{\imath}_{\bar{R}'}^{\bar{Q}'}\cdot\lambda_{R',R}(y) \qquad 17.18.37.$$

We are done.

Chapter 18

The localizers in a Frobenius P-category

18.1 Let P be a finite p-subgroup and $\mathcal{F}$ a Frobenius P-category. As a matter of fact, independently of the possible existence of a *perfect $\mathcal{F}$-locality $\mathcal{L}$* (cf. 17.13), the groups $\mathcal{L}(Q)$ where Q runs over the set of subgroups of P have a direct characterization in terms of $\mathcal{F}$ — we call them the *$\mathcal{F}$-localizers* — and their existence and uniqueness admits a direct proof. In this chapter we give this characterization and such a direct proof; this allows us, in the next chapter, to determine the (p-)*solvable* Frobenius P-categories. Moreover, we show the *functorial nature* of the *$\mathcal{F}$-localizers* — provided we replace $\mathcal{F}$ by the *proper category of $\mathcal{F}$-chains* $\mathfrak{ch}^*(\mathcal{F})$ of $\mathcal{F}$ (cf. A2.8) — by constructing with them a suitable functor from $\mathfrak{ch}^*(\mathcal{F})$ — the *localizing functor* of $\mathcal{F}$ — which actually holds some kind of "universal property" (cf. 18.20 below).

18.2 Let Q be a subgroup of P and K a subgroup of $\mathrm{Aut}(Q)$ such that Q is fully K-normalized in $\mathcal{F}$, so that it is fully centralized too (cf. Proposition 2.11); recall that then $N_{\mathcal{F}}^K(Q)$ is a Frobenius $N_P^K(Q)$-category (cf. Proposition 2.16). It is clear that $C_P(Q)$ is an $N_{\mathcal{F}}^K(Q)$-*stable* subgroup of $N_P^K(Q)$ (cf. 12.2) and that the Frobenius $C_P(Q)$-category $C_{\mathcal{F}}(Q)$ is *normal* in $N_{\mathcal{F}}^K(Q)$ (cf. 12.6); moreover, it follows from Proposition 13.9 that the *hyperfocal subgroup* $H_{C_{\mathcal{F}}(Q)}$ is also a $N_{\mathcal{F}}^K(Q)$-stable subgroup of $N_P^K(Q)$ and therefore, by Proposition 12.3, we have the Frobenius category $N_{\mathcal{F}}^K(Q)/H_{C_{\mathcal{F}}(Q)}$ over the p-group $N_P^K(Q)/H_{C_{\mathcal{F}}(Q)}$.

18.3 We are interested in an *extension* of this category, which actually will be a *quotient* of the $N_{\mathcal{F}}^K(Q)$-locality $N_{\mathcal{F},Q}^K(Q)$ introduced in 17.6; precisely, let us denote by $N_{\mathcal{F},Q}^K(Q)/H_{C_{\mathcal{F}}(Q)}$ the category where the objects are just the $N_{\mathcal{F}}^K(Q)/H_{C_{\mathcal{F}}(Q)}$-objects and the morphisms are the pairs formed by an $N_{\mathcal{F}}^K(Q)/H_{C_{\mathcal{F}}(Q)}$-morphism and by an element of K both determined by the *same* $N_{\mathcal{F},Q}^K(Q)$-morphism, the composition being induced by the composition in $N_{\mathcal{F}}^K(Q)/H_{C_{\mathcal{F}}(Q)}$ and the product in K. On the other hand, if $\mathcal{L}$ is an $\mathcal{F}$-locality then $\mathcal{L}(Q)$ acts on Q throughout $\mathcal{F}(Q)$ and, considering Q just as a set, recall that in 17.2 we have defined the $\mathcal{F}_{\mathcal{L}(Q)}$-locality $\mathcal{F}_{\mathcal{L}(Q),Q}$.

Proposition 18.4 *Assume that $\mathcal{L}$ is a perfect $\mathcal{F}$-locality. For any subgroup Q of P fully normalized in $\mathcal{F}$, the group homomorphisms τ_Q and π_Q induce an equivalence of categories*

$$N_{\mathcal{F},Q}(Q)/H_{C_{\mathcal{F}}(Q)} \cong \mathcal{F}_{\mathcal{L}(Q),Q} \qquad 18.4.1.$$

Proof: Since $\mathcal{F}_P(Q)$ is a Sylow p-subgroup of $\mathcal{F}(Q)$ (cf. Proposition 2.11), the image of the homomorphism τ_Q is a Sylow p-subgroup of $\mathcal{L}(Q)$ (cf. 17.12); thus, since we have (cf. 17.13.1)

$$H_{C_{\mathcal{F}}(Q)} = \mathrm{Ker}(\tau_Q) \qquad 18.4.2,$$

we may consider the Frobenius category $\mathcal{F}_{\mathcal{L}(Q)}$ associated with $\mathcal{L}(Q)$ (cf. 1.8) as a Frobenius category over the p-group $N_P(Q)/H_{C_{\mathcal{F}}(Q)}$. Then, if R and T are subgroups of $N_P(Q)$ containing $H_{C_{\mathcal{F}}(Q)}$, for any $\mathcal{F}$-morphism

$$\varphi : Q{\cdot}T \longrightarrow Q{\cdot}R \qquad 18.4.3$$

stabilizing Q and inducing some $\mathcal{F}$-morphism $\psi : T \to R$, it follows from Remark 17.11 that there is $x \in \mathcal{T}_{\mathcal{L}(Q)}\big(\tau_Q(R), \tau_Q(T)\big)$ inducing ψ and fulfilling $\pi_x(u) = \varphi(u)$ for any $u \in Q$; this proves the existence of a faithful functor

$$N_{\mathcal{F},Q}(Q)/H_{C_{\mathcal{F}}(Q)} \longrightarrow \mathcal{F}_{\mathcal{L}(Q),Q} \qquad 18.4.4.$$

Moreover, any $\mathcal{F}_{\mathcal{L}(Q),Q}$-morphism from $\tau_Q(T)$ to $\tau_Q(R)$ is determined by a suitable $x \in \mathcal{L}(Q)$ fulfilling $\tau_Q(T) \subset \tau_Q(R)^x$ and, according to statement 2.10.1, there is an $\mathcal{F}$-morphism

$$\varphi : Q{\cdot}T \longrightarrow Q{\cdot}R \qquad 18.4.5$$

extending π_x; then, it follows from Remark 17.11 that there is an element $y \in \mathcal{L}(Q)$ determining the $\mathcal{F}_{\mathcal{L}(Q),Q}$-morphism $\tau_Q(T) \to \tau_Q(Q{\cdot}R)$ induced by φ and fulfilling $\pi_y = \pi_x$; hence, $y^{-1}{\cdot}x$ belongs to (cf. 17.13.1)

$$\mathrm{Ker}(\pi_Q) = \tau_Q\big(C_P(Q)\big) \qquad 18.4.6,$$

which proves the fullness. We are done.

18.5 Let $\mathcal{L}$ be a perfect $\mathcal{F}$-locality and Q an $\mathcal{F}$-selfcentralizing subgroup of P fully normalized in $\mathcal{F}$; since $\tau_Q\big(N_P(Q)\big)$ is a Sylow p-subgroup of $\mathcal{L}(Q)$ (cf. 17.12) and we have $\mathrm{Ker}(\pi_Q) = Z(Q)$ (cf. 13.2.2 and 17.13.1), the isomorphism class of the group $\mathcal{L}(Q)$ is certainly determined by the element of $\mathbb{H}^2\big(\mathcal{F}(Q), Z(Q)\big)$ which once restricted to $N_P(Q)/Z(Q)$ coincides with the element of $\mathbb{H}^2\big(N_P(Q)/Z(Q), Z(Q)\big)$ determined by $N_P(Q)$. Conversely, this remark shows a way to construct the group $\mathcal{L}(Q)$ from the corresponding Frobenius P-category, and it turns out that this group *already* fulfills condition 17.11.2† and therefore the equivalence of categories 18.4.1. Actually, as

† In our old work [46], we proved the equivalence between this condition and the fact that a suitable 1-cocycle was a 1-coboundary; recently, we found that Jackowski-McClure proved in [32] a general result which actually implies that *all* the n-cocycles are n-coboundaries for $n{\geq}1$; but, here we give a direct proof in our situation. In an earlier version of [13], the authors proved the existence of the localizers from their *obstruction theory* and from the *same* result of Jackowski-McClure for $n{=}3$; presently, their result appears in [15].

we show in our next result, for *any* subgroup Q of P we can construct an essentially unique extension $\mathcal{L}(Q)$ — called the $\mathcal{F}$*-localizer* of Q — of $\mathcal{F}(Q)$ by $\mathfrak{c}^{\mathfrak{h}}_{\mathcal{F}}(Q)$ (cf. Proposition 13.14), fulfilling the equivalence of categories 18.4.1.

Theorem 18.6 *For any subgroup Q of P fully normalized in $\mathcal{F}$, there is a triple formed by a finite group $\mathcal{L}(Q)$ and by two group homomorphisms*

$$\tau_Q : N_P(Q) \longrightarrow \mathcal{L}(Q) \quad \textit{and} \quad \pi_Q : \mathcal{L}(Q) \longrightarrow \mathcal{F}(Q) \qquad 18.6.1$$

fulfilling the two conditions

18.6.2 $\mathrm{Ker}(\tau_Q) = H_{C_{\mathcal{F}}(Q)}$, $\mathrm{Ker}(\pi_Q) = \tau_Q\big(C_P(Q)\big) \cong \mathfrak{c}^{\mathfrak{h}}_{\mathcal{F}}(Q)$, π_Q *is surjective and we have* $\pi_Q \circ \tau_Q = \kappa_Q$.

18.6.3 τ_Q *and* π_Q *induce an equivalence of categories*

$$N_{\mathcal{F},Q}(Q)/H_{C_{\mathcal{F}}(Q)} \cong \mathcal{F}_{\mathcal{L}(Q),Q} .$$

Moreover, for another such a triple L', τ'_Q and π'_Q, there is a group isomorphism $\lambda : \mathcal{L}(Q) \cong L'$, unique up to $\mathfrak{c}^{\mathfrak{h}}_{\mathcal{F}}(Q)$-conjugation, fulfilling $\lambda \circ \tau_Q = \tau'_Q$ and $\pi'_Q \circ \lambda = \pi_Q$.

Proof: By Proposition 2.11, we already know that $\mathcal{F}_P(Q)$ is a Sylow p-subgroup of $\mathcal{F}(Q)$. Assume first that

$$Z = Z(Q) = C_P(Q) \qquad 18.6.4;$$

then we have $H_{C_{\mathcal{F}}(Q)} = \{1\}$ (cf. 13.2.2) and we claim that the element $\bar{\nu}$ of $\mathbb{H}^2\big(\mathcal{F}_P(Q), Z\big)$ induced by $N = N_P(Q)$ belongs to the image of $\mathbb{H}^2(\mathcal{F}(Q), Z)$. Indeed, according to Theorem 10.1 in [18, Ch. XII], it suffices to prove that, for any subgroup R of $N_P(Q)$ containing $Z(Q)$ and any $\varphi \in \mathcal{F}(Q)$ such that

$$\varphi \circ \mathcal{F}_R(Q) \circ \varphi^{-1} \subset \mathcal{F}_P(Q) \qquad 18.6.5,$$

the restriction $\mathrm{res}_\varphi(\bar{\nu})$ of $\bar{\nu}$ *via* the conjugation by φ coincides with the element of $\mathbb{H}^2\big(\mathcal{F}_R(Q), Z\big)$ determined by R; but, since Q is fully normalized in $\mathcal{F}$, it is also fully ${}^{\varphi}\mathcal{F}_R(Q)$-normalized (cf. 2.10) and therefore, by condition 2.8.2, there is an $\mathcal{F}$-morphism $\zeta : Q{\cdot}R \to N$ extending φ; now, the existence of a group homomorphism $R \to N$ which coincides with φ over $Z(Q)$ and with the conjugation by φ over $\mathcal{F}_R(Q)$ proves the claim.

In particular, the corresponding element of $\mathbb{H}^2\big(\mathcal{F}(Q), Z\big)$ determines a group extension

$$1 \longrightarrow Z \longrightarrow L \xrightarrow{\ \pi\ } \mathcal{F}(Q) \longrightarrow 1 \qquad 18.6.6$$

and an injective *group extension* homomorphism $\tau : N \to L$ such that $\mathrm{Im}(\tau)$ is a Sylow p-subgroup of L; moreover, since τ is injective, we may choose π in such a way that, for any $x \in L$ and any $u \in Q$, we have

$$x\tau(u)x^{-1} = \tau\Big(\big(\pi(x)\big)(u)\Big) \qquad 18.6.7;$$

then, we claim that, up to a suitable modification of our choice of τ, the group L endowed with τ fulfills condition 18.6.3.

In any case, since Q is $\mathcal{F}$-selfcentralizing, for any pair of subgroups R and T of N containing Q and any $\varphi \in \mathcal{F}(R,T)$ such that $\varphi(Q) = Q$, the element $\psi \in \mathcal{F}(Q)$ determined by φ fulfills

$$\psi \circ \mathcal{F}_T(Q) \circ \psi^{-1} \subset \mathcal{F}_R(Q) \qquad 18.6.8;$$

conversely, it follows from statement 2.10.1 that in this way we obtain *all* the elements of $\mathcal{F}(Q)$ fulfilling this inclusion. Thus, for any $x \in L$ such that $\tau(T) \subset \tau(R)^x$ there is $\varphi_x \in \mathcal{F}(R,T)$ lifting $\pi(x)$; that is to say, for any $w \in T$, we have

$$\tau\big(w\,\theta_x(w)\big) = \tau\big(\varphi_x(w)\big)^x \qquad 18.6.9$$

for a suitable $\theta_x(w) \in Z$; note that, for any $u \in Q$, we have $\theta_x(u) = 1$, and whenever $x \in \tau(R)$ we may assume that $\theta_x(w) = 1$.

Now, for any $w, w' \in T$, we get

$$\begin{aligned}\tau\big(ww'\theta_x(ww')\big) &= \tau\big(\varphi_x(ww')\big)^x = \tau\big(\varphi_x(w)\big)^x\tau\big(\varphi_x(w')\big)^x \\ &= \tau\big(w\,\theta_x(w)\big)\tau\big(w'\theta_x(w')\big) \qquad 18.6.10 \\ &= \tau\big(w\,\theta_x(w)\,w'\theta_x(w')\big) = \tau\big(ww'\theta_x(w)^{w'}\theta_x(w')\big)\end{aligned}$$

and therefore we still get

$$\theta_x(ww') = \theta_x(w)^{w'}\theta_x(w') \qquad 18.6.11;$$

hence, the map θ_x determines a Z-valued 1-*cocycle* from the image $\tilde{T}$ of T in (cf. 1.3)

$$\widetilde{\mathrm{Aut}}(Q) = \mathrm{Out}(Q) \qquad 18.6.12.$$

Actually, the *cohomology class* $\bar{\theta}_x$ of this 1-cocycle does not depend on the choice of φ_x; indeed, if another choice φ'_x determines θ'_x then, for a suitable $z \in Z$, we have (cf. Proposition 4.6)

$$\tau\big(w\,\theta'_x(w)\big) = \tau\big(\varphi'_x(w)\big)^x = \tau\big(\varphi_x(w^z)\big)^x = \tau\big(w^z\,\theta_x(w^z)\big) \qquad 18.6.13$$

for any $w \in T$, and therefore we get

$$\theta'_x(w)\theta_x(w)^{-1} = w^{-1}w^z = (z^{-1})^w z \qquad 18.6.14.$$

Consequently, this correspondence determines a map (cf. 17.2)

$$\bar\theta_{_{R,T}} : \mathcal{T}_L\big(\tau(R),\tau(T)\big) \longrightarrow \mathbb{H}^1(\tilde T, Z) \qquad 18.6.15$$

sending $x \in \mathcal{T}_L\big(\tau(R),\tau(T)\big)$ to the cohomology class $\bar\theta_x$ of θ_x .

Moreover, if U is a subgroup of N containing Q and y an element of L fulfilling $\tau(U) \subset \tau(T)^y$ then, for a choice of $\varphi_y \in \mathcal{F}(T,U)$ lifting $\pi(y)$, $\varphi_x \circ \varphi_y$ lifts $\pi(xy)$ and, for any $w \in U$, we may assume that

$$\begin{aligned}\tau\big(w\,\theta_{xy}(w)\big) &= \tau\big((\varphi_x\circ\varphi_y)(w)\big)^{xy} = \tau\Big(\varphi_y(w)\,\theta_x\big(\varphi_y(w)\big)\Big)^y \\ &= \tau\big(w\,\theta_y(w)\big)\tau\Big(\theta_x\big(\varphi_y(w)\big)\Big)^y \qquad 18.6.16; \\ &= \tau\Big(w\,\theta_y(w)\,\pi(y)^{-1}\Big(\theta_x\big(\varphi_y(w)\big)\Big)\Big)\end{aligned}$$

hence, denoting by $\tilde y$ the image of y in $\widetilde{\mathrm{Aut}}(Q) = \mathrm{Out}(Q)$ (cf. 1.3) and by $\varphi_{\tilde y} : \tilde U \to \tilde T$ and $\mathcal{Z}(\tilde y) : Z \cong Z$ the corresponding group homomorphisms, we get the 1-*cocycle condition* (cf. A3.11)

$$\bar\theta_{xy} = \bar\theta_y + \mathbb{H}^1\big(\varphi_{\tilde y}, \mathcal{Z}(\tilde y)\big)(\bar\theta_x) \qquad 18.6.17;$$

in particular, it is easily checked from this condition that $\bar\theta_x$ only depends on the class of x in the *exterior quotient* (cf. 1.3)

$$\tilde{\mathcal{T}}_L\big(\tau(R),\tau(T)\big) = \tau(R)\backslash\mathcal{T}_L\big(\tau(R),\tau(T)\big) \qquad 18.6.18.$$

Thus, respectively denoting by $\tilde L$, $\tilde R$ and $\tilde N$ the images of L, R and N in $\widetilde{\mathrm{Aut}}(Q)$, the map $\bar\theta_{_{R,T}}$ admits the factorization

$$\tilde{\bar\theta}_{_{\tilde R,\tilde T}} : \tilde{\mathcal{T}}_{\tilde L}(\tilde R,\tilde T) = \tilde R\backslash\mathcal{T}_{\tilde L}(\tilde R,\tilde T) \longrightarrow \mathbb{H}^1\big(\tilde T, Z\big) \qquad 18.6.19.$$

That is to say, considering the *exterior quotient* $\tilde{\mathcal{T}}_{\tilde L}$ (cf. 1.3) of the category $\mathcal{T}_{\tilde L}$ (cf. 17.2), it is clear that we have a *contravariant* functor $\mathfrak{h}^1_Z$ from this category to $\mathfrak{Ab}$ mapping $\tilde T$ on $\mathbb{H}^1\big(\tilde T, Z\big)$ and then, identifying the $\tilde{\mathcal{T}}_{\tilde L}$-morphism $\tilde x \in \tilde{\mathcal{T}}_{\tilde L}(\tilde R,\tilde T)$ with the obvious $\tilde{\mathcal{T}}_{\tilde L}$-chain $\Delta_1 \longrightarrow \tilde{\mathcal{T}}_{\tilde L}$, the family $\bar\theta = \{\bar\theta_{\tilde x}\}_{\tilde x}$ defines a 1-*cocycle* from $\tilde{\mathcal{T}}_{\tilde L}$ to $\mathfrak{h}^1_Z$ (cf. A3.8) since equalities 18.6.17 guarantee that the *differential* of $\bar\theta$ is zero (cf. A3.11.2).

As in [32], we claim that this 1-cocycle is a 1-coboundary; indeed, for any subgroup $\tilde R$ of $\tilde N$, choose a set of representatives $\tilde X_{\tilde R} \subset \tilde L$ for $\tilde N\backslash\tilde L/\tilde R$ and, for any $\tilde n \in \tilde X_{\tilde R}$, set $\tilde R_{\tilde n} = \tilde R \cap N^{\tilde n}$, consider the $\tilde{\mathcal{T}}_{\tilde L}$-morphisms $\tilde n : \tilde R_{\tilde n} \to \tilde N$

and $\tilde\imath^{\tilde R}_{\tilde R_{\tilde n}}:\tilde R_{\tilde n}\to\tilde R$ respectively determined by $\tilde n$ and by the trivial element of $\tilde L$ (cf. 17.2), and denote by

$$(\mathfrak{h}^1_Z)^\circ(\tilde\imath^{\tilde R}_{\tilde R_{\tilde n}}):\mathbb{H}^1\big(\tilde R_{\tilde n},Z\big)\longrightarrow\mathbb{H}^1\big(\tilde R,Z\big) \qquad 18.6.20$$

the *transfer homomorphism* (cf. §8 in [18, Ch. XII]); then, setting

$$\bar\sigma_{\tilde R}=\frac{|N|}{|L|}\cdot\sum_{\tilde n\in\tilde X_{\tilde R}}\big((\mathfrak{h}^1_Z)^\circ(\tilde\imath^{\tilde R}_{\tilde R_{\tilde n}})\big)(\bar\theta_{\tilde n}) \qquad 18.6.21,$$

we claim that, for any $\tilde x\in\tilde{\mathcal{T}}_{\tilde L}(\tilde R,\tilde T)$, we have

$$\bar\theta_{\tilde x}=\bar\sigma_{\tilde T}-\big(\mathfrak{h}^1_Z(\tilde x)\big)(\bar\sigma_{\tilde R}) \qquad 18.6.22.$$

Indeed, note that $\mathfrak{h}^1_Z(\tilde x)$ is the composition of the restriction *via* the $\tilde{\mathcal{T}}_{\tilde L}$-morphism

$$\tilde\imath^{\tilde R}_{\tilde x\tilde T\tilde x^{-1}}:\tilde x\tilde T\tilde x^{-1}\longrightarrow\tilde R \qquad 18.6.23$$

with the conjugation determined by $\tilde x$, which we denote by $\mathfrak{h}^1_Z(\tilde x_*)$; thus, by the corresponding Mackey equalities (cf. Proposition 9.1 in [18, Ch. XII]), we get

$$\begin{aligned}&\mathfrak{h}^1_Z(\tilde x)\Big(\sum_{\tilde n\in\tilde X_R}\big((\mathfrak{h}^1_Z)^\circ(\tilde\imath^{\tilde R}_{\tilde R_{\tilde n}})\big)(\bar\theta_{\tilde n})\Big)\\ &\quad=\mathfrak{h}^1_Z(\tilde x_*)\Big(\sum_{\tilde n\in\tilde X_{\tilde R}}\sum_{\tilde r\in\tilde Y_{\tilde n}}\big((\mathfrak{h}^1_Z)^\circ(\tilde\imath^{\tilde x\tilde T\tilde x^{-1}}_{\tilde N^{\tilde n\tilde r}\cap\tilde x\tilde T\tilde x^{-1}})\circ\mathfrak{h}^1_Z(\tilde r)\big)(\bar\theta_{\tilde n})\Big)\\ &\quad=\sum_{\tilde n\in\tilde X_{\tilde R}}\sum_{\tilde r\in\tilde Y_{\tilde n}}\big((\mathfrak{h}^1_Z)^\circ(\tilde\imath^{\tilde T}_{\tilde N^{\tilde n\tilde r\tilde x}\cap\tilde T})\circ\mathfrak{h}^1_Z(\tilde r\tilde x)\big)(\bar\theta_{\tilde n})\end{aligned} \qquad 18.6.24$$

where, for any $\tilde n\in X_{\tilde R}$, $\tilde Y_{\tilde n}\subset\tilde R$ is a set of representatives for $\tilde R_{\tilde n}\backslash\tilde R/\,\tilde x\tilde T\tilde x^{-1}$ and, for any $\tilde r\in\tilde Y_{\tilde n}$, we consider the $\tilde{\mathcal{T}}_{\tilde L}$-morphisms

$$\tilde r:\tilde N^{\tilde n\tilde r}\cap\tilde x\tilde T\tilde x^{-1}\longrightarrow\tilde R_{\tilde n}\quad\text{and}\quad\tilde r\tilde x:\tilde N^{\tilde n\tilde r\tilde x}\cap\tilde T\longrightarrow\tilde R_{\tilde n} \qquad 18.6.25.$$

Moreover, setting $\tilde m=\tilde n\tilde r\tilde x$ for $\tilde n\in\tilde X_{\tilde R}$ and $\tilde r\in\tilde Y_{\tilde n}$, it follows from equality 18.6.17 that

$$\big(\mathfrak{h}^1_Z(\tilde r\tilde x)\big)(\bar\theta_{\tilde n})=\bar\theta_{\tilde m}-\bar\theta_{\tilde r\tilde x}=\bar\theta_{\tilde m}-\big(\mathfrak{h}^1_Z(\tilde\imath^{\tilde T}_{\tilde T_{\tilde m}})\big)(\bar\theta_{\tilde x}) \qquad 18.6.26$$

since we assume that $\theta_{\tilde{r}} = 0$; thus, choosing $\tilde{X}_{\tilde{T}} = \bigsqcup_{\tilde{n}\in\tilde{X}_{\tilde{R}}} \tilde{n}\,\tilde{Y}_{\tilde{n}}\,\tilde{x}$, we get (cf. equality (6) in [18, Ch. XII, §8])

$$\begin{aligned}\bar{\sigma}_{\tilde{T}} - \big(\mathfrak{h}^1_Z(\tilde{x})\big)(\bar{\sigma}_{\tilde{R}}) &\\ &= \frac{|N|}{|L|}\cdot\sum_{\tilde{m}\in\tilde{X}_{\tilde{T}}} \big((\mathfrak{h}^1_Z)^\circ(\tilde{\imath}^{\tilde{T}}_{\tilde{T}_{\tilde{m}}})\big)\Big(\bar{\theta}_{\tilde{m}} - \big(\mathfrak{h}^1_Z(\tilde{r}\tilde{x})\big)(\bar{\theta}_{\tilde{n}})\Big)\\ &= \frac{|N|}{|L|}\cdot\sum_{\tilde{m}\in\tilde{X}_{\tilde{T}}} \big((\mathfrak{h}^1_Z)^\circ(\tilde{\imath}^{\tilde{T}}_{\tilde{T}_{\tilde{m}}})\big)\Big(\big(\mathfrak{h}^1_Z(\tilde{\imath}^{\tilde{T}}_{\tilde{T}_{\tilde{m}}})\big)(\bar{\theta}_{\tilde{x}})\Big) \qquad 18.6.27.\\ &= \sum_{\tilde{m}\in\tilde{X}_{\tilde{T}}} \frac{|\tilde{T}/\tilde{T}_{\tilde{m}}|}{|\tilde{L}/\tilde{N}|}\cdot\bar{\theta}_{\tilde{x}} = \bar{\theta}_{\tilde{x}}\end{aligned}$$

In particular, for any subgroup $\tilde{R}$ of $\tilde{N}$, we get

$$\bar{\sigma}_{\tilde{R}} = \big(\mathfrak{h}^1_Z(\tilde{\imath}^{\tilde{N}}_{\tilde{R}})\big)(\bar{\sigma}_{\tilde{N}}) \qquad 18.6.28$$

and the element $\bar{\sigma}_{\tilde{N}} \in \mathbb{H}^1(\tilde{N}, Z)$ can be lifted to a 1-cocycle $\sigma_{\tilde{N}} : \tilde{N} \to Z$ which determines a group automorphism $\sigma : N \cong N$ mapping $u \in N_P(Q)$ on $u\sigma_{\tilde{N}}(\tilde{u})$ where $\tilde{u}$ denotes the image of u in $\tilde{N}$; moreover, according to equality 18.6.22, in 18.6.9 we may choose

$$\theta_x(w) = \sigma_{\tilde{N}}(\tilde{w})\big(\pi(x)\big)^{-1}\Big(\sigma_{\tilde{N}}\big(\widetilde{\varphi_x(w)}\big)\Big)^{-1} \qquad 18.6.29.$$

Hence, replacing τ by $\hat{\tau} = \tau\circ\sigma$, the maps π and $\hat{\tau}$ still fulfill condition 18.6.2 and in equality 18.6.9 we get

$$\begin{aligned}\tau\big(\varphi_x(w)\big)^x &= \tau\big(w\,\theta_x(w)\big)\\ &= \tau\Big(w\big(w^{-1}\sigma(w)\big)\big(\pi(x)\big)^{-1}\Big(\varphi_x(w)^{-1}\sigma\big(\varphi_x(w)\big)\Big)^{-1}\Big)\\ &= \tau\Big(\sigma(w)\big(\pi(x)\big)^{-1}\Big(\sigma\big(\varphi_x(w)\big)^{-1}\varphi_x(w)\Big)\Big) \qquad 18.6.30,\\ &= \hat{\tau}(w)\tau\Big(\sigma\big(\varphi_x(w)\big)^{-1}\varphi_x(w)\Big)^x\\ &= \hat{\tau}(w)\hat{\tau}\big(\varphi_x(w)^{-1}\big)^x\tau\big(\varphi_x(w)\big)^x\end{aligned}$$

so that we get

$$\hat{\tau}\big(\varphi_x(w)\big)^x = \hat{\tau}(w) \quad\text{and}\quad \varphi_x(u) = \big(\pi(x)\big)(u) \text{ for any } u\in Q \qquad 18.6.31.$$

Consequently, we get a functor from $\mathcal{F}_{L,Q}$ to $N_{\mathcal{F},Q}(Q)$ (cf. 17.2 and 17.6) which, for any pair of subgroups R and T of N, maps the $\mathcal{F}_{L,Q}$-morphism from $\hat\tau(T)$ to $\hat\tau(R)$ determined by an element $x \in L$ such that $\hat\tau(T) \subset \hat\tau(R)^x$ on the $N_{\mathcal{F},Q}(Q)$-morphism from T to R determined by the corresponding $\mathcal{F}$-morphism $\varphi_x : Q{\cdot}T \to Q{\cdot}R$, and this functor is necessarily faithful. Moreover, for any $\mathcal{F}$-morphism $\psi : Q{\cdot}T \to Q{\cdot}R$ fulfilling $\psi(Q) = Q$ and $\psi(T) \subset R$, there is $y \in L$ such that $\pi(y)$ coincides with the restriction of ψ and therefore, according to Proposition 4.6, we have $\psi = \varphi_y \circ \kappa_z$ for some $z \in Z$, so that we still have $\psi = \varphi_{y\hat\tau(z)}$. In conclusion, the triple formed by L, $\hat\tau$ and π fulfills conditions 18.6.2 and 18.6.3. This completes the proof in the $\mathcal{F}$-selfcentralizing case.

Otherwise, choose an $\mathcal{F}$-morphism $\xi : Q{\cdot}C_P(Q) \to P$ such that the image $\hat Q = \xi\big(Q{\cdot}C_P(Q)\big)$ is fully normalized in $\mathcal{F}$ (cf. Proposition 2.7); thus, we have

$$^{\xi}\mathcal{F}_P\big(Q{\cdot}C_P(Q)\big) \subset \mathcal{F}_P(\hat Q) \qquad 18.6.32$$

and, in particular, it follows from statement 2.10.1 that there is an $\mathcal{F}$-morphism $\zeta : N \to P$ extending ξ and therefore fulfilling $\zeta(N) \subset N_P(\hat Q)$; moreover, we have $C_P(\hat Q) = Z(Q')$ and therefore there is a triple formed by a group $\hat L$ and by two group homomorphisms

$$\hat\tau : N_P(\hat Q) \longrightarrow \hat L \quad \text{and} \quad \hat\pi : \hat L \longrightarrow \mathcal{F}(\hat Q) \qquad 18.6.33$$

which fulfills conditions 18.6.2 and 18.6.3 with respect to $\hat Q$.

Now, setting $U = \hat\tau\big(\zeta(Q)\big)$, we claim that the triple formed by the group

$$L = N_{\hat L}(U)/\mathbf{O}^p\big(C_{\hat L}(U)\big){\cdot}\hat\tau\big(\zeta(H_{C_{\mathcal{F}}(Q)})\big) \qquad 18.6.34$$

and by the two group homomorphisms

$$\tau : N \longrightarrow L \quad \text{and} \quad \pi : L \longrightarrow \mathcal{F}(Q) \qquad 18.6.35$$

respectively mapping $u \in N$ on the image of $\hat\tau\big(\zeta(u)\big)$ in L, and the image $x \in L$ of $\hat x \in \hat L$ on the automorphism φ of Q such that $^{\zeta}\varphi = \hat\pi(\hat x)$, fulfills the conditions above with respect to Q.

Indeed, since we assume that the group $\hat L$ endowed with $\hat\tau$ and $\hat\pi$ fulfills condition 18.6.3, for any $\varphi \in \mathcal{F}(Q)$ clearly there is $\hat x \in \hat L$ such that

$$\hat\tau\Big(\zeta\big(\varphi(u)\big)\Big) = \hat x\hat\tau\big(\zeta(u)\big)\hat x^{-1} = \hat\tau\Big(\big(\hat\pi(\hat x)\big)\big(\zeta(u)\big)\Big) \qquad 18.6.36$$

for any $u \in Q$, and therefore, since $\hat\tau$ is injective, we get $\varphi = \pi(x)$ where x is the image of $\hat x$ in L. From the same condition, it is not difficult to prove that $\hat\tau\Big(\zeta\big(N_P(Q)\big)\Big)$ is a Sylow p-subgroup of $N_{\hat L}(U)$, so that $\hat\tau\Big(\zeta\big(C_P(Q)\big)\Big)$ is a Sylow p-subgroup of $C_{\hat L}(U)$, and then it is easy to check that

$$\mathbf{O}^p\big(C_{\hat L}(U)\big)\cap\hat\tau\Big(\zeta\big(C_P(Q)\big)\Big)\subset\hat\tau\big(\zeta(H_{C_{\mathcal F}(Q)})\big) \qquad 18.6.37.$$

On the other hand, the kernel of the homomorphism $N_{\hat L}(U)\to\mathcal F(Q)$ induced by ζ and $\hat\tau$ clearly coincides with $C_{\hat L}(U)$. Now, it is clear that the triple L, π and τ fulfills condition 18.6.2.

Moreover, let R and T be subgroups of N containing $H_{C_{\mathcal F}(Q)}$; it follows from Proposition 13.12 that any $\mathcal F$-morphism $\varphi:Q{\cdot}T\to Q{\cdot}R$ fulfilling $\varphi(Q)=Q$ can be extended to an $\mathcal F$-morphism

$$\psi : C_P(Q){\cdot}Q{\cdot}T \longrightarrow P \qquad 18.6.38$$

and then we have $\psi\big(C_P(Q)\big) = C_P(Q)$ since Q is fully centralized in $\mathcal F$ (cf. Proposition 2.11). Thus, setting $\hat R=\zeta(R)$ and $\hat T=\zeta(T)$, there is an $\mathcal F$-morphism

$$\hat\psi : \hat Q{\cdot}\hat T \longrightarrow \hat Q{\cdot}\hat R \qquad 18.6.39$$

such that $\hat\psi(\hat Q)=\hat Q$ and $\hat\psi\big(\zeta(w)\big)=\zeta\big(\varphi(w)\big)$ for any $w\in T$; hence, there is $\hat x\in\hat L$ such that

$$\big(\hat\pi(\hat x)\big)(\hat u)=\hat\psi(\hat u) \quad\text{and}\quad \hat x\hat\tau(\hat w)\hat x^{-1}=\hat\tau\big(\hat\psi(\hat w)\big) \qquad 18.6.40$$

for any $\hat u\in\hat Q$ and any $\hat w\in\hat T$; in particular, $\hat x$ normalizes U and, denoting by x the image of $\hat x$ in L, we get

$$\big(\pi(x)\big)(u)=\varphi(u) \quad\text{and}\quad x\tau(w)x^{-1}=\tau\big(\varphi(w)\big) \qquad 18.6.41$$

for any $u\in Q$ and any $w\in T$. That is to say, we get a functor

$$N_{\mathcal F,Q}(Q)/H_{C_{\mathcal F}(Q)} \longrightarrow \mathcal F_{L,Q} \qquad 18.6.42$$

which is necessarily faithful.

Conversely, if $x\in L$ fulfills $\tau(T)\subset\tau(R)^x$ then, since $\hat\tau(\hat R)$ and $\hat\tau(\hat T)$ are Sylow p-subgroups of the respective converse images of $\tau(R)$ and $\tau(T)$ in $\hat L$ (cf. 18.6.37), we can lift x to $\hat x\in\hat L$ such that $\hat\tau(\hat T)\subset\hat\tau(\hat R)^{\hat x}$ and therefore there is an $\mathcal F$-morphism $\hat\psi:\hat Q{\cdot}\hat T\to\hat Q{\cdot}\hat R$ still fulfilling equalities 18.6.40; now, the fullness of the functor above follows from considering the $\mathcal F$-morphism $\varphi:T\to R$ such that $\hat\psi\big(\zeta(w)\big)=\zeta\big(\varphi(w)\big)$ for any $w\in T$.

Finally, assume that there is a triple formed by a group L' and by two group homomorphisms

$$\tau' : N \longrightarrow L' \quad \text{and} \quad \pi' : L' \longrightarrow \mathcal{F}(Q) \qquad 18.6.43$$

fulfilling conditions 18.6.2 and 18.6.3; we will apply Lemma 18.8 below to L, to L', to the normal subgroup $\mathrm{Ker}(\pi') = \tau'\big(C_P(Q)\big)$ of L', and to the group homomorphism $\rho : \tau\big(N_P(Q)\big) \to L'$ mapping $\tau(u)$ on $\tau'(u)$ for any $u \in N_P(Q)$. We claim that they fulfill condition 18.8.1 below.

Indeed, if R is a subgroup of $N_P(Q)$, any $x \in L$ such that $\tau\big(N_P(Q)\big)$ contains $\tau(R)^x$ also fulfills

$$\tau\big(Q{\cdot}C_P(Q){\cdot}R\big)^x \subset \tau\big(N_P(Q)\big) \qquad 18.6.44$$

and therefore, by condition 18.6.3, there is an $\mathcal{F}$-morphism

$$\varphi : Q{\cdot}C_P(Q){\cdot}R \longrightarrow N_P(Q) \qquad 18.6.45$$

such that, for any $u \in Q$ and for any $w \in Q{\cdot}C_P(Q){\cdot}R$, we have

$$\big(\pi(x)\big)^{-1}(u) = \varphi(u) \quad \text{and} \quad x^{-1}\tau(w)x = \tau\big(\varphi(w)\big) \qquad 18.6.46;$$

in particular, we still have $\varphi(Q) = Q$ and therefore, by condition 18.6.3 again, there is an element $y \in L'$ such that we have

$$\big(\pi'(y)\big)^{-1}(u) = \varphi(u) = \big(\pi(x)\big)^{-1}(u) \qquad 18.6.47$$

for any $u \in Q$, so that $\pi'(y) = \pi(x)$, and that, for any $w \in R$, we still have

$$y^{-1}\tau'(w)y = \tau'\big(\varphi(w)\big) = \rho\Big(\tau\big(\varphi(w)\big)\Big) = \rho\big(x^{-1}\tau(w)x\big) \qquad 18.6.48.$$

Consequently, it follows from the lemma below that there is a group homomorphism $\lambda : L \to L'$ fulfilling $\lambda \circ \tau = \tau'$ and $\pi' \circ \lambda = \pi$, which is actually an isomorphism since both groups fulfill condition 18.6.2; moreover, since $\mathrm{Ker}(\pi') \subset \tau'\big(N_P(Q)\big)$, another group homomorphism $\lambda' : L \to L'$ fulfilling the corresponding equalities induces the same L-action over $\mathrm{Ker}(\pi')$ as λ and therefore, according to the lemma below, it is $Z\big(\mathrm{Ker}(\pi')\big)$-conjugate of λ. We are done.

Remark 18.7 If Q is $\mathcal{F}$-nilcentralized (cf. 4.3) then τ_Q is injective (cf. 13.2.2) and the categories $N_{\mathcal{F},Q}(Q)$ and $\mathcal{F}_{\mathcal{L}(Q),Q}$ are equivalent; conversely, in this case *the existence of an injective group homomorphism*

$$\tau_Q : N_P(Q) \longrightarrow \mathcal{L}(Q) \qquad 18.7.1$$

which induces an equivalence of categories $N_{\mathcal{F},Q}(Q) \cong \mathcal{F}_{\mathcal{L}(Q),Q}$ can replace both conditions above.

Lemma 18.8 *Let L and M be finite groups, P a Sylow p-subgroup of L, Q a normal p-subgroup of M and $\bar\sigma : L \to \bar M = M/Q$ a group homomorphism. Assume that there is a group homomorphism $\tau : P \to M$ lifting the restriction of $\bar\sigma$ and fulfilling the condition*

18.8.1 *For any subgroup R of P and any $x \in L$ such that $R^x \subset P$, there is $y \in M$ such that $\bar\sigma(x) = \bar y$ and $\tau(u^x) = \tau(u)^y$ for any $u \in R$.*

Then, there is a group homomorphism $\sigma : L \to M$ lifting $\bar\sigma$ and extending τ. Moreover, if $\sigma' : L \to M$ is a group homomorphism which lifts $\bar\sigma$, extends τ and induces the same L-action over Q as σ†, then there is $z \in Z(Q)$ such that $\sigma'(x) = \sigma(x)^z$ for any $x \in L$.

Proof: First of all, we prove the existence of σ; set $Z = Z(Q)$, $\tilde M = M/Z$ and $\tilde Q = Q/Z$; a group homomorphism τ as above determines a group homomorphism $\tilde\tau : P \to \tilde M$ which still fulfills the corresponding condition 18.8.1; thus, arguing by induction on $|Q|$, there is a group homomorphism $\tilde\sigma : L \to \tilde M$ lifting $\bar\sigma$ and extending $\tilde\tau$; in particular, $\tilde\sigma$ determines an action of L on Z and it makes sense to consider the cohomology groups $\mathbb{H}^n(L,Z)$ and $\mathbb{H}^n(P,Z)$.

But, M determines an element $\tilde\mu$ of $\mathbb{H}^2(\tilde M,Z)$ and if there is a group homomorphism $\tau : P \to M$ lifting the restriction of $\tilde\sigma$ then the corresponding image of $\tilde\mu$ in $\mathbb{H}^2(P,Z)$ has to be zero; thus, since the restriction map

$$\mathbb{H}^2(L,Z) \longrightarrow \mathbb{H}^2(P,Z) \qquad 18.8.2$$

is injective (cf. Theorem 10.1 in [18, Ch. XII]), we also get

$$\big(\mathbb{H}^2(\tilde\sigma, \mathrm{id}_Z)\big)(\tilde\mu) = 0 \qquad 18.8.3$$

and therefore there is a group homomorphism $\sigma : L \to M$ lifting $\tilde\sigma$.

At this point, the difference between τ and the restriction of σ to P defines a 1-cocycle $\theta : P \to Z$ and, for any subgroup R of P and any $x \in L$ such that $R^x \subset P$, it follows from condition 18.8.1 that, for a suitable $y \in M$ fulfilling $\bar y = \bar\sigma(x)$, for any $u \in R$ we have

$$\begin{aligned} \theta(u^x) &= \tau(u^x)^{-1}\sigma(u^x) = \tau(u^{-1})^y\sigma(u)^{\sigma(x)} \\ &= \tau(u^{-1})^y\tau(u)^{\sigma(x)}\theta(u)^{\sigma(x)} \\ &= \Big(\big(y\sigma(x)^{-1}\big)^{-1}\big(y\sigma(x)^{-1}\big)^{\tau(u)}\theta(u)\Big)^{\sigma(x)} \end{aligned} \qquad 18.8.4;$$

consequently, since the map sending $u \in R$ to $\big(y\sigma(x)^{-1}\big)^{-1}\big(y\sigma(x)^{-1}\big)^{\tau(u)}$ is a 1-coboundary, the cohomology class $\bar\theta$ of θ is *L-stable* and it follows again from Theorem 10.1 in [18, Ch. XII] that it is the restriction of a suitable element $\bar\eta \in \mathbb{H}^1(L,Z)$; then, it suffices to modify σ from a representative of $\bar\eta$ to get a new group homomorphism $\sigma' : L \to M$ lifting $\bar\sigma$ and extending τ.

† We thank Peter Symonds for helping us in the uniqueness of the noncommutative case.

Now, if $\sigma' : L \to M$ is a group homomorphism which lifts $\bar{\sigma}$, extends τ and induces the same L-action on Q as σ , the element $\sigma'(x)\sigma(x)^{-1}$ belongs to $Z(Q)$ for any $x \in L$ and thus, we get a 1-cocycle $\lambda : L \to Z(Q)$ mapping $x \in L$ on $\sigma'(x)\sigma(x)^{-1}$, which vanish over P; hence, it is a 1-coboundary (cf. Theorem 10.1 in [18, Ch. XII]) and therefore there exists $z \in Z(Q)$ such that

$$\lambda(x) = z^{-1}\sigma(x)z\sigma(x)^{-1} \qquad 18.8.5$$

so that we have $\sigma'(x) = \sigma(x)^z$ for any $x \in L$. We are done.

Remark 18.9 Whenever Q is Abelian the L-actions over Q induced by σ and σ' only depend on $\bar{\sigma}$ and therefore always coincide.

18.10 This lemma allows us to exhibit the *functorial nature* of the $\mathcal{F}$-localizers mentioned above; but, the lack of uniqueness in the general situation forces us to replace the $\mathcal{F}$-localizer $\mathcal{L}(Q)$ of a subgroup Q of P fully normalized in $\mathcal{F}$ by the quotient (cf. Lemma 13.3)

$$\bar{\mathcal{L}}(Q) = \mathcal{L}(Q)/\tau_Q\big(F_{C_{\mathcal{F}}(Q)}\big) = \mathcal{L}(Q)/\big[\mathrm{Ker}(\pi_Q), \mathrm{Ker}(\pi_Q)\big] \qquad 18.10.1;$$

coherently, we denote by

$$\bar{\tau}_Q : N_P(Q) \longrightarrow \bar{\mathcal{L}}(Q) \quad \text{and} \quad \bar{\pi}_Q : \bar{\mathcal{L}}(Q) \longrightarrow \mathcal{F}(Q) \qquad 18.10.2$$

the group homomorphisms induced by the structural ones. Note that if Q is $\mathcal{F}$-selfcentralizing then $\bar{\mathcal{L}}(Q) = \mathcal{L}(Q)$ since $\mathrm{Ker}(\pi_Q) \cong Z(Q)$ (cf. condition 18.6.2).

18.11 Recall that the objects of the *proper category of chains* $\mathfrak{ch}^*(\mathcal{F})$ of $\mathcal{F}$ (cf. A2.8) are the pairs $(\mathfrak{q}, \Delta_n)$ where $n \in \mathbb{N}$ and $\mathfrak{q} : \Delta_n \to \mathcal{F}$ is a functor — called an $(n, \mathcal{F})$-*chain* — and the morphisms from $(\mathfrak{q}, \Delta_n)$ to another object $(\mathfrak{r}, \Delta_m)$ are the pairs (ν, δ) where $\delta : \Delta_m \to \Delta_n$ is a *functor* or, equivalently, an order-preserving map, and $\nu : \mathfrak{q} \circ \delta \cong \mathfrak{r}$ is a natural isomorphism, the composition with another morphism $(\mu, \varepsilon) : (\mathfrak{r}, \Delta_m) \to (\mathfrak{t}, \Delta_\ell)$ being defined by (cf. A2.6.3)

$$(\mu, \varepsilon) \circ (\nu, \delta) = \big(\mu \circ (\nu * \varepsilon), \delta \circ \varepsilon\big) \qquad 18.11.1.$$

Then, it follows from Proposition A2.10 that we have a functor

$$\mathfrak{aut}_{\mathcal{F}} : \mathfrak{ch}^*(\mathcal{F}) \longrightarrow \mathfrak{Gr} \qquad 18.11.2$$

mapping any object $(\mathfrak{q}, \Delta_n)$ on its group of automorphisms in $\mathfrak{ch}^*(\mathcal{F})$ — denoted by $\mathcal{F}(\mathfrak{q})$ (cf. 2.19) — which is isomorphic to $\mathfrak{Nat}(\mathfrak{q}, \mathfrak{q})^*$ (cf. 1.5).

18.12 Moreover, let us denote by $\mathfrak{Loc}$ the category where the objects are the pairs (L, Q) formed by a finite group L and a normal p-subgroup Q of L and where the morphisms from (L, Q) to (L', Q') are the group homomorphisms $f : L \to L'$ fulfilling $f(Q) \subset Q'$. This category has an *inner structure* (cf. 1.3) mapping any object (L, Q) on the subgroup of the group of automorphisms determined by the Q-conjugation, and we denote by $\widetilde{\mathfrak{Loc}}$ the corresponding *exterior quotient* (cf. 1.3); that is to say, the category $\widetilde{\mathfrak{Loc}}$ has the same objects as $\mathfrak{Loc}$ and the morphisms from (L,Q) to (L',Q') are the Q-conjugacy classes of group homomorphisms $f:L \to L'$ fulfilling $f(Q) \subset Q'$. Obviously, $\mathfrak{Loc}$ and $\widetilde{\mathfrak{Loc}}$ contain $\mathfrak{Gr}$ as a full subcategory and we consider the *Levi* functor $\mathfrak{lv}:\widetilde{\mathfrak{Loc}} \to \mathfrak{Gr}$ mapping (L,Q) on L/Q. Our purpose is to exhibit a functor

$$\mathfrak{loc}_{\mathcal{F}} : \mathfrak{ch}^*(\mathcal{F}) \longrightarrow \widetilde{\mathfrak{Loc}} \qquad 18.12.1$$

— called the *$\mathcal{F}$-localizing functor* — lifting $\mathfrak{aut}_{\mathcal{F}}$ (cf. 18.11.2) and mapping the $(0,\mathcal{F})$-chain determined by a subgroup Q of P fully normalized in $\mathcal{F}$ on the pair $\big(\bar{\mathcal{L}}(Q), \mathrm{Ker}(\bar{\pi}_Q)\big)$ (cf. Theorem 18.6).

18.13 Explicitly, for any $(n,\mathcal{F})$-chain $\mathfrak{q}:\Delta_n \to \mathcal{F}$ and any $i \in \Delta_n$ we have an evident group homomorphism

$$\mathcal{F}(\mathfrak{q}) \longrightarrow \mathcal{F}\big(\mathfrak{q}(i)\big) \qquad 18.13.1$$

and actually we can identify $\mathcal{F}(\mathfrak{q})$ with its image in $\mathcal{F}\big(\mathfrak{q}(n)\big)$; if $\mathfrak{q}$ is *fully normalized* in $\mathcal{F}$ (cf. 2.18) then Theorem 18.6 supplies an essentially unique triple $\big(\tau_{\mathfrak{q}(n)}, \mathcal{L}\big(\mathfrak{q}(n)\big), \pi_{\mathfrak{q}(n)}\big)$ formed by a group $\mathcal{L}\big(\mathfrak{q}(n)\big)$ and two group homomorphisms

$$N_P\big(\mathfrak{q}(n)\big) \xrightarrow{\tau_{\mathfrak{q}(n)}} \mathcal{L}\big(\mathfrak{q}(n)\big) \xrightarrow{\pi_{\mathfrak{q}(n)}} \mathcal{F}\big(\mathfrak{q}(n)\big) \qquad 18.13.2,$$

and we denote by $\mathcal{L}(\mathfrak{q})$ the converse image of $\mathcal{F}(\mathfrak{q})$ in $\mathcal{L}\big(\mathfrak{q}(n)\big)$, by $N_P(\mathfrak{q})$ the converse image of $\mathcal{F}_P(\mathfrak{q})$ in $N_P\big(\mathfrak{q}(n)\big)$ (cf. 2.19) and by

$$N_P(\mathfrak{q}) \xrightarrow{\tau_{\mathfrak{q}}} \mathcal{L}(\mathfrak{q}) \xrightarrow{\pi_{\mathfrak{q}}} \mathcal{F}(\mathfrak{q}) \qquad 18.13.3$$

the corresponding restrictions of homomorphisms 18.13.2.

18.14 Since any $\mathcal{F}$-chain admits a naturally isomorphic chain which is *fully normalized* in $\mathcal{F}$ (cf. 2.18), in order to exhibit the *$\mathcal{F}$-localizing functor* we can restrict ourselves to the full subcategory $\mathfrak{ch}^*(\mathcal{F})^{\mathrm{fn}}$ of $\mathfrak{ch}^*(\mathcal{F})$ over the set of fully normalized chains of $\mathcal{F}$. Let $(\nu,\delta):(\mathfrak{r},\Delta_m) \to (\mathfrak{q},\Delta_n)$ be a $\mathfrak{ch}^*(\mathcal{F})^{\mathrm{fn}}$-morphism; by the very definition of the proper category of chains of $\mathcal{F}$ (cf. A2.8), we have a group isomorphism $\nu_n:\mathfrak{r}\big(\delta(n)\big) \cong \mathfrak{q}(n)$ and therefore we get a group homomorphism (cf. 18.13.1)

$$\mathcal{F}(\mathfrak{r}) \longrightarrow \mathcal{F}\Big(\mathfrak{r}\big(\delta(n)\big)\Big) \cong \mathcal{F}\big(\mathfrak{q}(n)\big) \qquad 18.14.1;$$

it is easy to check that the image of this group homomorphism is contained in $\mathcal{F}(\mathfrak{q})$ and, actually, this inclusion determines the functor 18.11.2 above

$$\mathfrak{aut}_{\mathcal{F}}(\nu,\delta) : \mathcal{F}(\mathfrak{r}) \longrightarrow \mathcal{F}(\mathfrak{q}) \qquad 18.14.2.$$

18.15 In particular, since $\mathcal{F}_P(\mathfrak{q})$ is a Sylow p-subgroup of $\mathcal{F}(\mathfrak{q})$ (cf. 2.18 and Proposition 2.11), there is $\rho \in \mathcal{F}(\mathfrak{q})$ such that

$$\big(\mathfrak{aut}_{\mathcal{F}}(\nu,\delta)\big)\big(\mathcal{F}_P(\mathfrak{r})\big) \subset \mathcal{F}_P(\mathfrak{q})^\rho \qquad 18.15.1;$$

that is to say, the composition with $\rho\circ\nu_n : \mathfrak{r}\big(\delta(n)\big) \cong \mathfrak{q}(n)$ of the inverse of the group isomorphism induced by $\mathfrak{r}(\delta(n)\bullet m)$

$$\mathfrak{r}(\delta(n)\bullet m)_* : \mathfrak{r}\big(\delta(n)\big) \cong \mathrm{Im}\big(\mathfrak{r}(\delta(n)\bullet m)\big) \qquad 18.15.2$$

maps the action of $N_P(\mathfrak{r})$ over $\mathrm{Im}\big(\mathfrak{r}(\delta(n)\bullet m)\big)$ into the action of $N_P(\mathfrak{q})$ over $\mathfrak{q}(n)$. Consequently, since $C_P\big(\mathfrak{q}(n)\big) \subset N_P(\mathfrak{q})$, it follows from Proposition 2.11 that there is an $\mathcal{F}$-morphism

$$\zeta : N_P(\mathfrak{r})\cdot\mathrm{Im}\big(\mathfrak{r}(\delta(n)\bullet m)\big) \longrightarrow N_P(\mathfrak{q})\cdot\mathfrak{q}(n) \qquad 18.15.3$$

fulfilling $\zeta\big(\mathfrak{r}(\delta(n)\bullet m)(v)\big) = \rho\big(\nu_n(v)\big)$ for any $v\in\mathfrak{r}\big(\delta(n)\big)$ and therefore, according to inclusion 18.15.1, ζ maps $N_P(\mathfrak{r})$ into $N_P(\mathfrak{q})$.

Proposition 18.16 *With the notation above, for any* $\mathfrak{ch}^*(\mathcal{F})^{^{\mathrm{fn}}}$*-morphism* $(\nu,\delta) : (\mathfrak{r},\Delta_m) \to (\mathfrak{q},\Delta_n)$, *there is a group homomorphism*

$$\lambda : \mathcal{L}(\mathfrak{r}) \longrightarrow \mathcal{L}(\mathfrak{q}) \qquad 18.16.1$$

such that

$$\pi_{\mathfrak{q}}\circ\lambda = \mathfrak{aut}_{\mathcal{F}}(\nu,\delta)\circ\pi_{\mathfrak{r}} \qquad 18.16.2$$

and that, for some $r\in\mathcal{L}(\mathfrak{q})$ *lifting* ρ *and any* $v\in N_P(\mathfrak{r})$, *we have*

$$\lambda\big(\tau_{\mathfrak{r}}(v)\big) = \tau_{\mathfrak{q}}\big(\zeta(v)\big)^r \qquad 18.16.3.$$

Moreover, these group homomorphisms induce mutually $\mathrm{Ker}(\bar\pi_{\mathfrak{q}})$*-conjugate homomorphisms*

$$\bar\lambda : \bar{\mathcal{L}}(\mathfrak{r}) \longrightarrow \bar{\mathcal{L}}(\mathfrak{q}) \qquad 18.16.4.$$

Proof: Set $Q=\mathfrak{q}(n)$, $R=\mathfrak{r}(m)$ and $\varphi = \mathfrak{r}(\delta(n)\bullet m)_*$ (cf. 18.15.2); first of all, note that we have

$$\zeta\big(C_P(R)\big)\subset C_P(Q) \quad\text{and}\quad \zeta(v) = \rho\Big(\nu_n\big(\varphi^{-1}(v)\big)\Big) \text{ if } v\in\mathrm{Im}(\varphi) \qquad 18.16.5;$$

moreover, as in the proof of Proposition 13.14, it is easily checked that ζ induces a $\big(C_{\mathcal{F}}(R), C_{\mathcal{F}}(Q)\big)$-functorial homomorphism from $C_P(R)$ to $C_P(Q)$ (cf. 12.1), so that we get (cf. 13.2)

$$\zeta(H_{C_{\mathcal{F}}(R)}) \subset H_{C_{\mathcal{F}}(Q)} \quad\text{and}\quad \zeta(F_{C_{\mathcal{F}}(R)}) \subset F_{C_{\mathcal{F}}(Q)} \qquad 18.16.6.$$

Denoting by $\kappa_\rho : \mathcal{F}(\mathfrak{q}) \cong \mathcal{F}(\mathfrak{q})$ the inner automorphism determined by ρ, on the one hand consider the composed homomorphism

$$\mathcal{L}(\mathfrak{r}) \xrightarrow{\pi_\mathfrak{r}} \mathcal{F}(\mathfrak{r}) \xrightarrow{\mathfrak{aut}_\mathcal{F}(\nu,\delta)} \mathcal{F}(\mathfrak{q}) \overset{\kappa_\rho}{\cong} \mathcal{F}(\mathfrak{q}) \qquad 18.16.7;$$

on the other hand, consider the homomorphism $\eta : \tau_\mathfrak{r}\big(N_P(\mathfrak{r})\big) \to \mathcal{L}(\mathfrak{q})$ mapping $\tau_\mathfrak{r}(v)$ on $\tau_\mathfrak{q}\big(\zeta(v)\big)$ for any $v \in N_P(\mathfrak{r})$, which makes sense by the left-hand inclusion in 18.16.6. We claim that these group homomorphisms together with the groups $\mathcal{L}(\mathfrak{r})$ and $\mathcal{L}(\mathfrak{q})$, the Sylow p-subgroup $\tau_\mathfrak{r}\big(N_P(\mathfrak{r})\big)$ of $\mathcal{L}(\mathfrak{r})$, and the normal subgroup $\mathrm{Ker}(\pi_\mathfrak{q})$ of $\mathcal{L}(\mathfrak{q})$ fulfill condition 18.8.1.

Indeed, if T is a subgroup of $N_P(\mathfrak{r})$ containing $H_{C_\mathcal{F}(R)}$ and y an element of $\mathcal{L}(\mathfrak{r})$ such that $\tau_\mathfrak{r}(T)^y \subset \tau_\mathfrak{r}\big(N_P(\mathfrak{r})\big)$, it is clear that y determines a morphism $\tau_\mathfrak{r}(T) \to \tau_\mathfrak{r}\big(N_P(\mathfrak{r})\big)$ in the $\mathcal{F}_{\mathcal{L}(R)}$-locality $\mathcal{F}_{\mathcal{L}(R),R}$ (cf. 17.11.2); then, it follows from condition 18.6.3 that there is an $\mathcal{F}$-morphism $\psi : R{\cdot}T \to R{\cdot}N_P(\mathfrak{r})$ fulfilling

$$\psi(v) = \big(\pi_R(y)\big)(v) \quad \text{and} \quad \tau_R\big(\psi(w)\big) = \tau_R(w)^y \qquad 18.16.8$$

for any $v \in R$ and any $w \in T$.

Hence, since $\pi_R(y)$ stabilizes the image in R of $\mathfrak{r}(j \bullet m)$ for any $j \in \Delta_m$, the restriction of ψ gives an $\mathcal{F}$-morphism $\mathrm{Im}(\varphi){\cdot}T \to \mathrm{Im}(\varphi){\cdot}N_P(\mathfrak{r})$ (cf. 2.4) and then, since we have $\zeta\big(\mathrm{Im}(\varphi)\big) = Q$, the "translation" *via* ζ induces an $\mathcal{F}$-morphism

$$\theta : Q{\cdot}\zeta(T) \longrightarrow Q{\cdot}N_P(\mathfrak{q}) \qquad 18.16.9$$

fulfilling $\theta(Q) = Q$ and $\theta\big(\zeta(T)\big) \subset N_P(\mathfrak{q})$ (cf. 2.4); in other words, θ determines a morphism $\zeta(T) \to N_P(\mathfrak{q})$ in the $N_\mathcal{F}(Q)$-locality $N_{\mathcal{F},Q}(Q)$ (cf. 17.6) and therefore, it follows again from condition 18.6.3 that there is $x \in \mathcal{L}(Q)$ fulfilling

$$\theta(u) = \big(\pi_Q(x)\big)(u) \quad \text{and} \quad \tau_Q\Big(\theta\big(\zeta(w)\big)\Big) = \tau_Q\big(\zeta(w)\big)^x \qquad 18.16.10$$

for any $u \in Q$ and any $w \in T$. Hence, setting $\xi = \varphi \circ (\nu_n)^{-1} \circ \rho^{-1}$, for any $u \in Q$ we get (cf. 18.15 and 18.16.5)

$$\begin{aligned} &\Big(\big(\kappa_\rho \circ \mathfrak{aut}_\mathcal{F}(\nu,\delta) \circ \pi_R\big)(y)\Big)(u) \\ &\qquad = \Big(\rho \circ \mathfrak{aut}_\mathcal{F}(\nu,\delta)\big(\pi_R(y)\big) \circ \rho^{-1}\Big)(u) \\ &\qquad = \zeta\Big(\big(\pi_R(y)\big)\big(\xi(u)\big)\Big) = \zeta\Big(\psi\big(\xi(u)\big)\Big) \\ &\qquad = \theta\Big(\zeta\big(\xi(u)\big)\Big) = \theta(u) = \big(\pi_Q(x)\big)(u) \end{aligned} \qquad 18.16.11,$$

so that $x \in \mathcal{L}(\mathfrak{q})$ is a lifting of the image of y by homomorphism 18.16.7.

Moreover, for any $w \in T$ we still get

$$\begin{aligned}\eta\big(\tau_{R}(w)^y\big) &= \eta\Big(\tau_{\mathfrak{r}}\big(\psi(w)\big)\Big) = \tau_{\mathfrak{q}}\Big(\zeta\big(\psi(w)\big)\Big)\\ &= \tau_{\mathfrak{q}}\Big(\theta\big(\zeta(w)\big)\Big) = \tau_{Q}\big(\zeta(w)\big)^x\end{aligned} \qquad 18.16.12.$$

At this point, it follows from Lemma 18.8 that there is a group homomorphism $\lambda : \mathcal{L}(\mathfrak{r}) \to \mathcal{L}(\mathfrak{q})$ fulfilling

$$\pi_{\mathfrak{q}} \circ \lambda = \kappa_\rho \circ \mathfrak{aut}_{\mathcal{F}}(\nu,\delta) \circ \pi_{\mathfrak{r}} \quad \text{and} \quad \lambda\big(\tau_{\mathfrak{r}}(v)\big) = \tau_{\mathfrak{q}}\big(\zeta(v)\big) \qquad 18.16.13$$

for any $v \in N_P(\mathfrak{r})$; in particular, for any $r \in \mathcal{L}(\mathfrak{q})$ lifting ρ, the group homomorphism $\lambda^r : \mathcal{L}(\mathfrak{r}) \to \mathcal{L}(\mathfrak{q})$ fulfills the announced condition.

Thus, it follows from the right-hand inclusion in 18.16.6 that

$$\lambda^r\big(\tau_{\mathfrak{r}}(F_{C_{\mathcal{F}}(R)})\big) \subset \tau_{\mathfrak{q}}(F_{C_{\mathcal{F}}(Q)}) \qquad 18.16.14$$

and therefore λ^r induces a group homomorphism

$$\overline{\lambda^r} : \bar{\mathcal{L}}(\mathfrak{r}) \longrightarrow \bar{\mathcal{L}}(\mathfrak{q}) \qquad 18.16.15.$$

Similarly, it is clear that any group homomorphism $\lambda' : \mathcal{L}(\mathfrak{r}) \to \mathcal{L}(\mathfrak{q})$ fulfilling condition 18.16.13 still fulfills inclusion 18.16.14 and therefore it induces also a group homomorphism $\overline{(\lambda')^r} : \bar{\mathcal{L}}(\mathfrak{r}) \to \bar{\mathcal{L}}(\mathfrak{q})$; then, it follows from Lemma 18.8 and Remark 18.9 that $\overline{\lambda^r}$ and $\overline{(\lambda')^r}$ are $\mathrm{Ker}(\bar{\pi}_{\mathfrak{q}})$-conjugate. We are done.

Remark 18.17 Our condition above does not depend on the choices of ρ and ζ. Indeed, for another $\rho' \in \mathcal{F}(\mathfrak{q})$ fulfilling inclusion 18.15.1, it is clear that $\rho' \circ \rho^{-1}$ maps the action of $\zeta\big(N_P(\mathfrak{r})\big)$ over $\mathfrak{q}(n)$ into the action of $N_P(\mathfrak{q})$ and therefore it follows again from Proposition 2.11 that there is an $\mathcal{F}$-morphism

$$\theta : \mathfrak{q}(n)\cdot\zeta\big(N_P(\mathfrak{r})\big) \longrightarrow \mathfrak{q}(n)\cdot N_P(\mathfrak{q}) \qquad 18.17.1$$

extending $\rho' \circ \rho^{-1}$; at this point, it follows from Theorem 18.6 that there is $t \in \mathcal{L}\big(\mathfrak{q}(n)\big)$ fulfilling

$$\pi_{\mathfrak{q}(n)}(t) = \rho' \circ \rho^{-1} \quad \text{and} \quad \tau_{\mathfrak{q}}\Big(\theta\big(\zeta(v)\big)\Big) = \tau_{\mathfrak{q}}\big(\zeta(v)\big)^t \text{ if } v \in N_P(\mathfrak{r}) \qquad 18.17.2;$$

then, setting $r' = tr$ and $\zeta' = \theta \circ \zeta_*$ where ζ_* denotes the $\mathcal{F}$-isomorphism induced by ζ, we still have $\lambda_{(\nu,\delta)}\big(\tau_{\mathfrak{r}}(u)\big) = \tau_{\mathfrak{q}}\big(\zeta'(v)\big)^{r'}$ for any $v \in N_P(\mathfrak{r})$. Similarly, for another choice ζ' of ζ, we still can define an $\mathcal{F}$-morphism θ fulfilling $\zeta' = \theta \circ \zeta_*$ but this time θ induces the identity on $\mathfrak{q}(n)$, so that t belongs to $\mathrm{Ker}(\pi_{\mathfrak{q}})$.

18.18 Thus, with the notation above, it follows from this proposition and from Lemma 18.8 and Remark 18.9 that, for any $\mathfrak{ch}^*(\mathcal{F})^{\mathrm{fn}}$-morphism $(\nu,\delta):(\mathfrak{r},\Delta_m)\to(\mathfrak{q},\Delta_n)$, there is a unique $\widetilde{\mathfrak{Loc}}$-morphism

$$\mathfrak{loc}_{(\nu,\delta)}:\big(\bar{\mathcal{L}}(\mathfrak{r}),\mathrm{Ker}(\bar{\pi}_{\mathfrak{r}})\big)\longrightarrow\big(\bar{\mathcal{L}}(\mathfrak{q}),\mathrm{Ker}(\bar{\pi}_{\mathfrak{q}})\big) \qquad 18.18.1$$

admitting a representative $\bar{\lambda}:\bar{\mathcal{L}}(\mathfrak{r})\to\bar{\mathcal{L}}(\mathfrak{q})$ which, for some $\bar{r}\in\bar{\mathcal{L}}(\mathfrak{q})$ lifting ρ (cf. 18.15), fulfills

$$\bar{\pi}_{\mathfrak{q}}\circ\bar{\lambda}=\mathfrak{aut}_{\mathcal{F}}(\nu,\delta)\circ\bar{\pi}_{\mathfrak{r}} \quad\text{and}\quad \bar{\lambda}\big(\bar{\tau}_{\mathfrak{r}}(v)\big)=\bar{\tau}_{\mathfrak{q}}\big(\zeta(v)\big)^{\bar{r}} \qquad 18.18.2$$

for any $v\in N_P(\mathfrak{r})$. Recall that we denote by $\mathfrak{lv}:\widetilde{\mathfrak{Loc}}\to\mathfrak{Gr}$ the functor mapping any $\widetilde{\mathfrak{Loc}}$-object (L,Q) on L/Q (cf. 18.12).

Proposition 18.19 *There is a unique isomorphism class of functors*

$$\mathfrak{loc}_{\mathcal{F}}:\mathfrak{ch}^*(\mathcal{F})\longrightarrow\widetilde{\mathfrak{Loc}} \qquad 18.19.1$$

mapping any $\mathfrak{ch}^*(\mathcal{F})^{\mathrm{fn}}$*-object* $(\mathfrak{q},\Delta_n)$ *on* $\big(\bar{\mathcal{L}}(\mathfrak{q}),\mathrm{Ker}(\bar{\pi}_{\mathfrak{q}})\big)$ *and any* $\mathfrak{ch}^*(\mathcal{F})^{\mathrm{fn}}$*-morphism* $(\nu,\delta):(\mathfrak{r},\Delta_m)\to(\mathfrak{q},\Delta_n)$ *on* $\mathfrak{loc}_{(\nu,\delta)}$. *In particular, we have*

$$\mathfrak{lv}\circ\mathfrak{loc}_{\mathcal{F}}\cong\mathfrak{aut}_{\mathcal{F}} \qquad 18.19.2.$$

Proof: We already know that we can restrict ourselves to the full subcategory $\mathfrak{ch}^*(\mathcal{F})^{\mathrm{fn}}$; thus, let $(\eta,\varepsilon):(\mathfrak{t},\Delta_\ell)\to(\mathfrak{r},\Delta_m)$ be a second $\mathfrak{ch}^*(\mathcal{F})^{\mathrm{fn}}$-morphism and denote by $\bar{\mu}:\bar{\mathcal{L}}(\mathfrak{t})\to\bar{\mathcal{L}}(\mathfrak{r})$ a representative of $\mathfrak{loc}_{(\eta,\varepsilon)}$. Setting $Q=\mathfrak{q}(n)$, $R=\mathfrak{r}(m)$, $T=\mathfrak{t}(\ell)$, $\varphi=\mathfrak{r}(\delta(n)\bullet m)_*$ and $\psi=\mathfrak{t}(\varepsilon(m)\bullet\ell)_*$ (cf. 18.15.2), we already know that (cf. 18.18)

$$\bar{\pi}_{\mathfrak{r}}\circ\bar{\mu}=\mathfrak{aut}_{\mathcal{F}}(\eta,\varepsilon)\circ\bar{\pi}_{\mathfrak{t}} \qquad 18.19.3$$

and that there are an $\mathcal{F}$-morphism (cf. 18.15)

$$\xi:\mathrm{Im}(\psi)\cdot N_P(\mathfrak{t})\longrightarrow R\cdot N_P(\mathfrak{r}) \qquad 18.19.4$$

and an element $\bar{s}$ in $\bar{\mathcal{L}}(\mathfrak{r})$ fulfilling

$$\bar{\mu}\big(\bar{\tau}_{\mathfrak{t}}(u)\big)=\bar{\tau}_{\mathfrak{r}}\big(\xi(u)\big)^{\bar{s}} \quad\text{and}\quad \xi\big(\psi(v)\big)=\big(\bar{\pi}_{\mathfrak{r}}(\bar{s})\big)\big(\eta_m(v)\big) \qquad 18.19.5$$

for any $u\in N_P(\mathfrak{t})$ and any $v\in\mathfrak{t}\big(\varepsilon(m)\big)$. Since $(\nu,\delta)\circ(\eta,\varepsilon)=\big(\nu\circ(\eta*\delta),\varepsilon\circ\delta\big)$ (cf. A2.6.3), it is clear that

$$\begin{aligned}\mathfrak{aut}_{\mathcal{F}}\big(\nu\circ(\eta*\delta),\varepsilon\circ\delta\big)\circ\bar{\pi}_{\mathfrak{t}}&=\mathfrak{aut}_{\mathcal{F}}(\nu,\delta)\circ\mathfrak{aut}_{\mathcal{F}}(\eta,\varepsilon)\circ\bar{\pi}_{\mathfrak{t}}\\&=\mathfrak{aut}_{\mathcal{F}}(\nu,\delta)\circ\bar{\pi}_{\mathfrak{r}}\circ\bar{\mu}\\&=\bar{\pi}_{\mathfrak{q}}\circ(\bar{\lambda}\circ\bar{\mu})\end{aligned} \qquad 18.19.6.$$

On the other hand, considering the $\mathcal{F}$-morphisms

$$\begin{array}{ccccc} \mathrm{Im}(\psi)\cdot N_P(\mathfrak{t}) & \xrightarrow{\xi} & R\cdot N_P(\mathfrak{r}) & & \\ & & \cup & & \\ & & \mathrm{Im}(\varphi)\cdot N_P(\mathfrak{r}) & \xrightarrow{\zeta} & Q\cdot N_P(\mathfrak{q}) \end{array} \qquad 18.19.7$$

and setting $\omega = \mathfrak{t}\big((\varepsilon\circ\delta)(n)\bullet\ell\big)_*$ (cf. 18.15.2), we get the $\mathcal{F}$-morphism

$$\theta : \mathrm{Im}(\omega)\cdot N_P(\mathfrak{t}) \longrightarrow Q\cdot N_P(\mathfrak{q}) \qquad 18.19.8$$

mapping $w\in \mathrm{Im}(\omega)\cdot N_P(\mathfrak{t})$ on $\zeta\big(\xi(w)\big)$. In particular, since we have

$$\begin{aligned} \mathfrak{t}\big((\varepsilon\circ\delta)(n)\bullet\ell\big) &= \mathfrak{t}(\varepsilon(m)\bullet\ell)\circ\mathfrak{t}\big((\varepsilon\circ\delta)(n)\bullet\varepsilon(m)\big) \\ &= \mathfrak{t}(\varepsilon(m)\bullet\ell)\circ(\eta_m)^{-1}\circ\mathfrak{r}(\delta(n)\bullet m)\circ\eta_{\delta(n)} \end{aligned} \qquad 18.19.9,$$

if $v\in\mathfrak{t}\big((\varepsilon\circ\delta)(n)\big)$ setting $w=\varphi\big(\eta_{\delta(n)}(v)\big)$ then we get

$$\omega(v) = \psi\big((\eta_{\varepsilon(m)})^{-1}(w)\big) \qquad 18.19.10$$

and therefore, setting $\sigma = \bar\pi_{\mathfrak{r}}(\bar s)$ we still have

$$\zeta\Big(\xi\big(\omega(v)\big)\Big) = \zeta\bigg(\xi\Big(\psi\big((\eta_{\varepsilon(m)})^{-1}(w)\big)\Big)\bigg) = \zeta\big(\sigma(w)\big) \qquad 18.19.11;$$

thus, denoting by $\sigma_{\delta(n)}\in\mathcal{F}\Big(\mathfrak{r}\big(\delta(n)\big)\Big)$ the element induced by σ, for any $v\in\mathfrak{t}\big((\varepsilon\circ\delta)(n)\big)$ from the naturality of η we get (cf. 18.15)

$$\begin{aligned} \theta\big(\omega(v)\big) &= \zeta\Big(\varphi\big((\sigma_{\delta(n)}\circ\eta_{\delta(n)})(v)\big)\Big) \\ &= \rho\Big(\nu_n\big((\sigma_{\delta(n)}\circ\eta_{\delta(n)})(v)\big)\Big) \\ &= \big(\rho\circ\mathfrak{aut}_{\mathcal{F}}(\nu,\delta)(\sigma)\big)\big((\nu_n\circ\eta_{\delta(n)})(v)\big) \\ &= \Big(\big(\rho\circ\mathfrak{aut}_{\mathcal{F}}(\nu,\delta)(\sigma)\big)\circ\big(\nu\circ(\eta*\delta)\big)_n\Big)(v) \end{aligned} \qquad 18.19.12\,.$$

That is to say, since we have (cf. 18.15.1)

$$\begin{aligned} \big(\mathfrak{aut}_{\mathcal{F}}(\nu,\delta)\circ\mathfrak{aut}_{\mathcal{F}}(\eta,\varepsilon)\big)\big(\mathcal{F}_P(\mathfrak{t})\big) &\subset \mathfrak{aut}_{\mathcal{F}}(\nu,\delta)\big(\mathcal{F}_P(\mathfrak{r})^\sigma\big) \\ = \mathfrak{aut}_{\mathcal{F}}(\nu,\delta)\big(\mathcal{F}_P(\mathfrak{r})\big)^{\mathfrak{aut}_{\mathcal{F}}(\nu,\delta)(\sigma)} &\subset \mathcal{F}_P(\mathfrak{q})^{\rho\circ\mathfrak{aut}_{\mathcal{F}}(\nu,\delta)(\sigma)} \end{aligned} \qquad 18.19.13,$$

the $\mathcal{F}$-morphism θ fulfills the last condition in 18.15 with respect to the element $\rho\circ\mathfrak{aut}_{\mathcal{F}}(\nu,\delta)(\sigma)\in\mathcal{F}(\mathfrak{q})$; moreover, it is quite clear that, for any $u\in N_P(\mathfrak{t})$, we get (cf. 18.19.5 and 18.18.2)

$$(\bar\lambda\circ\bar\mu)\big(\bar\tau_{\mathfrak{t}}(u)\big) = \bar\lambda\Big(\bar\tau_{\mathfrak{r}}\big(\xi(u)\big)^{\bar s}\Big) = \bar\tau_{\mathfrak{q}}\big(\theta(u)\big)^{\bar r\bar\lambda(\bar s)} \qquad 18.19.14.$$

Consequently, $\lambda \circ \mu$ fulfills condition 18.18.2 with respect to the $\mathfrak{ch}^*(\mathcal{F})$-morphism $\big(\nu\circ(\omega*\delta), \varepsilon\circ\delta\big)$ and therefore it is a representative of the $\widetilde{\mathfrak{Loc}}$-morphism $\mathfrak{loc}_{(\nu\circ(\omega*\delta),\varepsilon\circ\delta)}$, so that

$$\mathfrak{loc}_{\mathcal{F}}\big((\nu,\delta)\circ(\eta,\varepsilon)\big) = \mathfrak{loc}_{\mathcal{F}}(\nu,\delta)\circ\mathfrak{loc}_{\mathcal{F}}(\eta,\varepsilon) \qquad 18.19.15.$$

We are done.

18.20 Finally, let us discuss a particular kind of "universality" of the $\mathcal{F}$*-localizing functor* $\mathfrak{loc}_{\mathcal{F}}$. Let $\hat{\mathcal{L}}$ be a coherent $\mathcal{F}$-locality, denote by

$$\hat{\tau} : \mathcal{T}_P \longrightarrow \hat{\mathcal{L}} \quad \text{and} \quad \hat{\pi} : \hat{\mathcal{L}} \longrightarrow \mathcal{F} \qquad 18.20.1$$

the structural functors and assume that, for any subgroup Q of P, $\mathrm{Ker}(\hat{\pi}_Q)$ is an Abelian p-group; it follows from Proposition A2.10 that we have a functor

$$\mathfrak{aut}_{\hat{\mathcal{L}}} : \mathfrak{ch}^*(\hat{\mathcal{L}}) \longrightarrow \mathfrak{Gr} \qquad 18.20.2$$

mapping any $\mathfrak{ch}^*(\hat{\mathcal{L}})$-object $(\hat{\mathfrak{q}}, \Delta_n)$ on the group of all the $\mathfrak{ch}^*(\hat{\mathcal{L}})$-automorphisms of $(\hat{\mathfrak{q}}, \Delta_n)$; since $\hat{\mathcal{L}}$ is divisible (cf. 17.7), this group can be identified with the converse image of $\mathcal{F}(\hat{\pi}\circ\hat{\mathfrak{q}})$ in $\mathcal{L}\big(\hat{\mathfrak{q}}(n)\big)$ — denoted by $\hat{\mathcal{L}}(\hat{\mathfrak{q}})$. Then, it is clear that $\mathfrak{aut}_{\hat{\mathcal{L}}}$ induces an obvious functor

$$\mathfrak{loc}_{\hat{\mathcal{L}}} : \mathfrak{ch}^*(\mathcal{F}) \longrightarrow \widetilde{\mathfrak{Loc}} \qquad 18.20.3$$

mapping any $\mathfrak{ch}^*(\mathcal{F})$-object $(\mathfrak{q}, \Delta_n)$ on $\big(\hat{\mathcal{L}}(\hat{\mathfrak{q}}), \mathrm{Ker}(\hat{\pi}_{\mathfrak{q}})\big)$, where $\hat{\mathfrak{q}} : \Delta_n \to \hat{\mathcal{L}}$ is a functor lifting $\mathfrak{q} : \Delta_n \to \mathcal{F}$ and $\hat{\pi}_{\mathfrak{q}}$ is the restriction of $\hat{\pi}_{\mathfrak{q}(n)}$ to $\hat{\mathcal{L}}(\hat{\mathfrak{q}})$; note that $\hat{\mathcal{L}}(\hat{\mathfrak{q}})$ does not depend on the choice of $\hat{\mathfrak{q}}$. In particular, we have (cf. 18.12)

$$\mathfrak{lo}\circ\mathfrak{loc}_{\hat{\mathcal{L}}} = \mathfrak{aut}_{\mathcal{F}} \qquad 18.20.4.$$

Proposition 18.21 *With the notation above, there is a unique natural map*

$$\lambda_{\hat{\mathcal{L}}} : \mathfrak{loc}_{\mathcal{F}} \longrightarrow \mathfrak{loc}_{\hat{\mathcal{L}}} \qquad 18.21.1$$

such that $\mathfrak{lo}*\lambda_{\hat{\mathcal{L}}} = \mathrm{id}_{\mathfrak{aut}_{\mathcal{F}}}$ *and that, for any chain* $\mathfrak{q} : \Delta_n \to \mathcal{F}$ *fully normalized in* $\mathcal{F}$, *we have* $(\lambda_{\hat{\mathcal{L}}})_{(\mathfrak{q},\Delta_n)}\circ\bar{\tau}_{\mathfrak{q}} = \hat{\tau}_{\mathfrak{q}}$.

Proof: First of all, for any subgroup Q of P fully normalized in $\mathcal{F}$, choose an $\mathcal{F}$-localizer $\mathcal{L}(Q)$ of Q and denote by

$$\tau_Q : N_P(Q) \longrightarrow \mathcal{L}(Q) \quad \text{and} \quad \pi_Q : \mathcal{L}(Q) \longrightarrow \mathcal{F}(Q) \qquad 18.21.2$$

the structural homomorphisms; we will apply Lemma 18.8 to the groups $\bar{\mathcal{L}}(Q)$ (cf. 18.10) and $\hat{\mathcal{L}}(Q)$, to the quotient $\mathcal{F}(Q)$ of both, and to the group homomorphism

$$\eta : \bar{\tau}_Q\big(N_P(Q)\big) \longrightarrow \hat{\mathcal{L}}(Q) \qquad 18.21.3$$

mapping $\bar{\tau}_Q(u)$ on $\hat{\tau}_Q(u)$ for any $u \in N_P(Q)$, which makes sense since $\mathrm{Ker}(\hat{\pi}_Q)$ is an Abelian p-group, so that $F_{C_{\mathcal{F}}(Q)} \subset \mathrm{Ker}(\hat{\tau}_Q)$ (cf. Lemma 13.3 and 17.13.1)

We know that $\bar{\tau}_Q\big(N_P(Q)\big)$ is a Sylow p-subgroup of $\bar{\mathcal{L}}(Q)$ (cf. 17.12) and it is clear that the restriction of $\bar{\pi}_Q$ to $\bar{\tau}_Q\big(N_P(Q)\big)$ coincides with the composition $\hat{\pi}_Q \circ \eta$. Let R be a subgroup of $N_P(Q)$ and x an element of $\mathcal{L}(Q)$ such that $\tau_Q(R)^x \subset \tau_Q\big(N_P(Q)\big)$; since we still have

$$\tau_Q(Q{\cdot}R)^x \subset \tau_Q\big(N_P(Q)\big) \qquad 18.21.4,$$

it follows from Theorem 18.6 that there is an $\mathcal{F}$-morphism $\varphi : Q{\cdot}R \to N_P(Q)$ such that we have

$$\varphi(u) = \pi_x(u) \quad \text{and} \quad \tau_Q\big(\varphi(v)\big) = \tau_Q(v)^x \qquad 18.21.5$$

for any $u \in Q$ and any $v \in Q{\cdot}R$.

At this point, it follows from statement 17.11.2 that there is $\hat{x} \in \hat{\mathcal{L}}(Q)$ fulfilling

$$\varphi(u) = \hat{\pi}_{\hat{x}}(u) \quad \text{and} \quad \hat{\tau}_Q\big(\varphi(v)\big) = \hat{\tau}_Q(v)^{\hat{x}} \qquad 18.21.6$$

for any $u \in Q$ and any $v \in Q{\cdot}R$. Consequently, denoting by $\bar{x}$ the image of x in $\bar{\mathcal{L}}(Q)$, we get

$$\hat{\pi}_Q(\hat{x}) = \bar{\pi}_Q(\bar{x}) \quad \text{and} \quad \eta\big(\bar{\tau}_Q(v)^{\bar{x}}\big) = \hat{\tau}_Q(v)^{\hat{x}} = \eta\big(\bar{\tau}_Q(v)\big)^{\hat{x}} \qquad 18.21.7$$

for any $v \in Q{\cdot}R$. Now, it follows from Lemma 18.8 that there is a unique $\mathrm{Ker}(\hat{\pi}_Q)$-conjugacy class of group homomorphisms

$$\lambda_Q : \bar{\mathcal{L}}(Q) \longrightarrow \hat{\mathcal{L}}(Q) \qquad 18.21.8$$

fulfilling $\hat{\pi}_Q \circ \lambda_Q = \bar{\pi}_Q$ and $\lambda_Q \circ \bar{\tau}_Q = \hat{\tau}_Q$.

Then, it is clear that, for any chain $\mathfrak{q} : \Delta_n \to \mathcal{F}$ fully normalized in $\mathcal{F}$, the corresponding group homomorphism

$$\lambda_{\mathfrak{q}(n)} : \bar{\mathcal{L}}\big(\mathfrak{q}(n)\big) \longrightarrow \hat{\mathcal{L}}\big(\mathfrak{q}(n)\big) \qquad 18.21.9$$

maps $\bar{\mathcal{L}}(\mathfrak{q})$ into $\hat{\mathcal{L}}(\mathfrak{q}) = (\hat{\pi}_{\mathfrak{q}})^{-1}\big(\mathcal{F}(\mathfrak{q})\big)$ and $\mathrm{Ker}(\bar{\pi}_{\mathfrak{q}})$ into $\mathrm{Ker}(\hat{\pi}_{\mathfrak{q}})$, and therefore determines a unique $\widetilde{\mathfrak{Loc}}$-morphism

$$\tilde{\lambda}_{\mathfrak{q}} : \mathfrak{loc}_{\mathcal{F}}(\mathfrak{q}, \Delta_n) \longrightarrow \mathfrak{loc}_{\hat{\mathcal{L}}}(\mathfrak{q}, \Delta_n) \qquad 18.21.10.$$

Now, we claim that the correspondence $\lambda_{\hat{\mathcal{L}}}$ which maps any $\mathfrak{ch}^*(\mathcal{F})^{\mathrm{fn}}$-object $(\mathfrak{q}, \Delta_n)$ on $\tilde{\lambda}_{\mathfrak{q}}$ defines a natural map over $\mathfrak{ch}^*(\mathcal{F})^{\mathrm{fn}}$ and therefore, since this subcategory is equivalent to $\mathfrak{ch}^*(\mathcal{F})$, it can be uniquely extended to the announced natural map.

Actually, up to equivalence, we can restrict ourselves to the full subcategory $\mathfrak{ch}^*(\mathcal{F})^{\mathrm{fni}}$ of $\mathfrak{ch}^*(\mathcal{F})^{\mathrm{fn}}$ over the $\mathfrak{ch}^*(\mathcal{F})^{\mathrm{fn}}$-objects $(\mathfrak{q}, \Delta_n)$ such that we have $\mathfrak{q}(\ell) \subset \mathfrak{q}(n)$ and $\mathfrak{q}(\ell \bullet n) = \iota_{\mathfrak{q}(i)}^{\mathfrak{q}(n)}$ for any $\ell \in \Delta_n$; in this case, we denote by $\hat{\mathfrak{q}} : \Delta_n \to \hat{\mathcal{L}}$ the $\hat{\mathcal{L}}$-*chain* mapping $\ell \in \Delta_n$ on $\mathfrak{q}(\ell)$ and $\ell \bullet n$ on $\hat{\imath}_{\mathfrak{q}(\ell)}^{\mathfrak{q}(n)} = \hat{\tau}_{\mathfrak{q}(n),\mathfrak{q}(\ell)}(1)$. Let $(\nu, \delta) : (\mathfrak{r}, \Delta_m) \to (\mathfrak{q}, \Delta_n)$ be a $\mathfrak{ch}^*(\mathcal{F})^{\mathrm{fni}}$-morphism and denote by $(\hat{\nu}, \delta) : (\hat{\mathfrak{r}}, \Delta_m) \to (\hat{\mathfrak{q}}, \Delta_n)$ a lifting of (ν, δ) to the corresponding full subcategory of $\mathfrak{ch}^*(\hat{\mathcal{L}})$; it is quite clear that the group homomorphism (cf. 18.20.2)

$$\mathfrak{aut}_{\hat{\mathcal{L}}}(\hat{\nu}, \delta) : \hat{\mathcal{L}}(\hat{\mathfrak{r}}) \longrightarrow \hat{\mathcal{L}}(\hat{\mathfrak{q}}) \subset \hat{\mathcal{L}}\big(\mathfrak{q}(n)\big) \qquad 18.21.11$$

is induced by the composition

$$\hat{\mathcal{L}}(\hat{\mathfrak{r}}) \longrightarrow \hat{\mathcal{L}}\Big(\mathfrak{r}\big(\delta(n)\big)\Big) \cong \hat{\mathcal{L}}\big(\mathfrak{q}(n)\big) \qquad 18.21.12$$

where the right isomorphism is determined by $\hat{\nu}_n : \mathfrak{r}\big(\delta(n)\big) \cong \mathfrak{q}(n)$ and the left homomorphism maps $\hat{x} \in \hat{\mathcal{L}}(\hat{\mathfrak{r}}) \subset \hat{\mathcal{L}}\big(\mathfrak{r}(m)\big)$ on the element $\hat{y} \in \hat{\mathcal{L}}\Big(\mathfrak{r}\big(\delta(n)\big)\Big)$ fulfilling

$$\hat{\imath}_{\mathfrak{r}(\delta(n))}^{\mathfrak{r}(m)} \cdot \hat{y} = \hat{x} \cdot \hat{\imath}_{\mathfrak{r}(\delta(n))}^{\mathfrak{r}(m)} \qquad 18.21.13;$$

thus, for any $v \in N_P(\hat{\mathfrak{r}})$, this homomorphism maps $\hat{\tau}_{\mathfrak{r}(m)}(v)$ on $\hat{\tau}_{\mathfrak{r}(\delta(n))}(v)$.

On the other hand, if we have

$$\nu_n \circ \mathcal{F}_P(\mathfrak{r}) \circ (\nu_n)^{-1} \subset \mathcal{F}_P(\mathfrak{q})^\rho \qquad 18.21.14$$

for some $\rho \in \mathcal{F}(\mathfrak{q})$ (cf. 18.15) then, since $\mathfrak{q}(n)$ is fully centralized in $\mathcal{F}$ (cf. Proposition 2.11), there is an $\mathcal{F}$-morphism (cf. condition 2.10.1)

$$\zeta : \mathfrak{r}\big(\delta(n)\big) \cdot N_P(\mathfrak{r}) \longrightarrow \mathfrak{q}(n) \cdot N_P(\mathfrak{q}) \qquad 18.21.15$$

fulfilling $\zeta(w) = (\rho \circ \nu_n)(w)$ for any $w \in \mathfrak{r}\big(\delta(n)\big)$; moreover, since we have $C_P\big(\mathfrak{q}(n)\big) \subset N_P(\mathfrak{q})$, it is easily checked that $\zeta\big(N_P(\mathfrak{r})\big) \subset N_P(\mathfrak{q})$. Hence, there are $\hat{z} \in \hat{\mathcal{L}}\Big(\mathfrak{q}(n), \mathfrak{r}\big(\delta(n)\big)\Big)$ lifting the element of $\mathcal{F}\Big(\mathfrak{q}(n), \mathfrak{r}\big(\delta(n)\big)\Big)$ determined by ζ and $\bar{r} \in \bar{\mathcal{L}}(\mathfrak{q})$ lifting ρ such that, for any $v \in N_P(\mathfrak{r})$, we have

$$\hat{z} = \lambda_{\mathfrak{q}}(\bar{r}) \cdot \hat{\nu}_n \quad \text{and} \quad \big(\mathfrak{aut}_{\hat{\mathcal{L}}}(\hat{\nu}, \delta)\big)\big(\hat{\tau}_{\mathfrak{r}}(v)\big) = \hat{\tau}_{\mathfrak{q}}\big(\zeta(v)\big)^{\lambda_{\mathfrak{q}}(\bar{r})} \qquad 18.21.16.$$

But, according to Proposition 18.16, we have a group homomorphism

$$\bar\lambda_{(\nu,\delta)} : \bar{\mathcal{L}}(\mathfrak{r}) \longrightarrow \bar{\mathcal{L}}(\mathfrak{q}) \qquad 18.21.17$$

such that $\bar\pi_{\mathfrak{q}} \circ \bar\lambda_{(\nu,\delta)} = \mathfrak{aut}_{\mathcal{F}}(\nu,\delta) \circ \bar\pi_{\mathfrak{r}}$ and that, for some $\bar r' \in \mathcal{L}(\mathfrak{q})$ lifting ρ, we have $\bar\lambda_{(\nu,\delta)}\big(\bar\tau_{\mathfrak{r}}(v)\big) = \bar\tau_{\mathfrak{q}}\big(\zeta(v)\big)^{\bar r'}$ for any $v \in N_P(\mathfrak{r})$; moreover, we always can modify our lifting $\hat\nu$ of ν in such a way that $\bar r = \bar r'$. In this case, for any $v \in N_P(\mathfrak{r})$ we get

$$\big(\mathfrak{aut}_{\hat{\mathcal{L}}}(\hat\nu,\delta)\big)\big(\hat\tau_{\mathfrak{r}}(v)\big) = \hat\tau_{\mathfrak{q}}\big(\zeta(v)\big)^{\lambda_{\mathfrak{q}}(\bar r)} = \lambda_{\mathfrak{q}}\Big(\bar\lambda_{(\nu,\delta)}\big(\bar\tau_{\mathfrak{r}}(v)\big)\Big) \qquad 18.21.18;$$

that is to say, the two group homomorphisms $\mathfrak{aut}_{\hat{\mathcal{L}}}(\hat\nu,\delta) \circ \lambda_{\mathfrak{r}}$ and $\lambda_{\mathfrak{q}} \circ \bar\lambda_{(\nu,\delta)}$ from $\bar{\mathcal{L}}(\mathfrak{r})$ to $\hat{\mathcal{L}}(\mathfrak{q})$ coincide over $\bar\tau_{\mathfrak{r}}\big(N_P(\mathfrak{r})\big)$ which is a Sylow p-subgroup of $\bar{\mathcal{L}}(\mathfrak{r})$, and induce the same homomorphism from $\bar{\mathcal{L}}(\mathfrak{r})$ to $\mathcal{F}(\mathfrak{q})$.

Consequently, it follows from the uniqueness part of Lemma 18.8 that the group homomorphisms $\mathfrak{aut}_{\hat{\mathcal{L}}}(\hat\nu,\delta) \circ \lambda_{\mathfrak{r}}$ and $\lambda_{\mathfrak{q}} \circ \bar\lambda_{(\nu,\delta)}$ are mutually $\mathrm{Ker}(\hat\pi_{\mathfrak{q}})$-conjugate and therefore we obtain the following equality of $\widetilde{\mathfrak{Loc}}$-morphisms

$$\mathfrak{loc}_{\hat{\mathcal{L}}}(\nu,\delta) \circ \tilde\lambda_{\mathfrak{r}} = \tilde\lambda_{\mathfrak{q}} \circ \mathfrak{loc}_{\mathcal{F}}(\nu,\delta) \qquad 18.21.19$$

which proves the announced naturality. Finally, the uniqueness of the natural map $\lambda_{\hat{\mathcal{L}}}$ follows from the uniqueness of the $\mathrm{Ker}(\hat\pi_Q)$-conjugacy classes of the group homomorphisms 18.21.8 above. We are done.

18.22 Naturally, this proposition raises a question: does there exist an $\mathcal{F}$-sublocality $\mathcal{L}$ of $\hat{\mathcal{L}}$ such that the corresponding subfunctor $\mathfrak{loc}_{\mathcal{L}} \subset \mathfrak{loc}_{\hat{\mathcal{L}}}$ is just the image of $\lambda_{\hat{\mathcal{L}}}$? Note that, restricting our attention to the *full* subcategories $\mathcal{L}^{^{\mathrm{sc}}}$ and $\hat{\mathcal{L}}^{^{\mathrm{sc}}}$ over the set of $\mathcal{F}$-selfcentralizing subgroups Q of P and assuming that the restricted functor $\hat\tau^{\mathrm{sc}} : \mathcal{T}_P^{^{\mathrm{sc}}} \to \hat{\mathcal{L}}^{^{\mathrm{sc}}}$ (cf. 17.4) is *faithful*, in this case $\mathcal{L}^{^{\mathrm{sc}}}$ is a *perfect* $\mathcal{F}^{^{\mathrm{sc}}}$*-locality* since $\mathcal{L}(Q)$ is indeed the $\mathcal{F}$*-localizer* of Q (cf. 17.13 and 18.5). As a matter of fact, these questions on the *full* subcategory $\hat{\mathcal{L}}^{^{\mathrm{sc}}}$ — on this existence and on the corresponding uniqueness — admit *cohomological answers* in terms of two *stable* $\widetilde{\mathfrak{Ker}}(\hat\pi^{\mathrm{sc}})/\bar{\tilde{\mathfrak{z}}}_{\mathcal{F}^{\mathrm{sc}}}$*-valued* 2- *and* 1-*cocycles over* $\tilde{\mathcal{F}}^{^{\mathrm{sc}}}$ (cf. A3.18), where $\hat\pi^{\mathrm{sc}} : \hat{\mathcal{L}}^{^{\mathrm{sc}}} \to \mathcal{F}^{^{\mathrm{sc}}}$ denotes the restriction of $\hat\pi$, $\widetilde{\mathfrak{Ker}}(\hat\pi^{\mathrm{sc}})$ is the factorization throughout $\tilde{\mathcal{F}}^{^{\mathrm{sc}}}$ of the *contravariant* functor $\mathfrak{Ker}(\hat\pi^{\mathrm{sc}}) : \hat{\mathcal{L}}^{^{\mathrm{sc}}} \to \mathfrak{Ab}$ (cf. 17.8 and Proposition 17.10), and $\bar{\tilde{\mathfrak{z}}}_{\mathcal{F}^{\mathrm{sc}}} : \tilde{\mathcal{F}}^{^{\mathrm{sc}}} \to \mathfrak{Ab}$ is the corresponding image of the *contravariant* functor defined by the *centers* (cf. 13.13).

18.23 Let us exhibit these *cocycles*. First of all note that, denoting by $\bar{\hat{\mathcal{L}}}$ the obvious quotient $\mathcal{F}^{^{\mathrm{sc}}}$-locality $\hat{\mathcal{L}}^{^{\mathrm{sc}}}/\mathfrak{z}_{\mathcal{F}^{\mathrm{sc}}}$ and setting $\bar{\tilde{\mathfrak{k}}} = \widetilde{\mathfrak{Ker}}(\hat\pi^{\mathrm{sc}})/\bar{\tilde{\mathfrak{z}}}_{\mathcal{F}^{\mathrm{sc}}}$, for any $\mathcal{F}$-selfcentralizing subgroup Q of P we have

$$\bar{\hat{\mathcal{L}}}(Q) \cong \bar{\tilde{\mathfrak{k}}}(Q) \rtimes \mathcal{F}(Q) \qquad 18.23.1$$

since, identifying the $\mathcal{F}$-localizer $\mathcal{L}(Q)$ to its image in $\hat{\mathcal{L}}(Q)$ by a representative of $(\lambda_{\hat{\mathcal{L}}})_Q$ (cf. Proposition 18.21), we have

$$\hat{\mathcal{L}}(Q) = \mathrm{Ker}(\pi_Q)\cdot\mathcal{L}(Q) \qquad 18.23.2.$$

Secondly, any $\mathcal{F}^{^{\rm sc}}$-morphism $\varphi : R \to Q$ clearly determines a $(1, \mathcal{F}^{^{\rm sc}})$-*chain* $\mathfrak{o}_\varphi : \Delta_1 \to \mathcal{F}^{^{\rm sc}}$ mapping 0 on R, 1 on Q and the Δ_1-morphism $0 \bullet 1$ on φ, and then we consider the obvious $\mathfrak{ch}^*(\mathcal{F}^{^{\rm sc}})$-morphism

$$(\mathrm{id}_R, \delta_1^0) : (\mathfrak{o}_\varphi, \Delta_1) \longrightarrow (\mathfrak{o}_R, \Delta_0) \qquad 18.23.3,$$

where $\mathfrak{o}_R : \Delta_0 \to \mathcal{F}^{^{\rm sc}}$ maps 0 on R; the functor $\mathfrak{aut}_{\mathcal{F}^{\rm sc}}$ (cf. Proposition A2.10) maps this $\mathfrak{ch}^*(\mathcal{F}^{^{\rm sc}})$-morphism on a group homomorphism

$$\mathfrak{aut}_{\mathcal{F}^{\rm sc}}(\mathrm{id}_R, \delta_1^0) : \mathcal{F}(Q)_{R'} \longrightarrow \mathcal{F}(R) \qquad 18.23.4,$$

where we set $R' = \varphi(R)$ and, as usual, $\mathcal{F}(Q)_{R'}$ denotes the stabilizer of R' in $\mathcal{F}(Q)$.

18.24 Moreover, the functor $\bar{\hat{\pi}} : \bar{\hat{\mathcal{L}}} \to \mathcal{F}^{^{\rm sc}}$ induced by $\hat{\pi}^{\rm sc}$ determines a *natural map* (cf. Proposition A2.10)

$$\mathfrak{aut}_{\bar{\hat{\pi}}} : \mathfrak{aut}_{\bar{\hat{\mathcal{L}}}} \longrightarrow \mathfrak{aut}_{\mathcal{F}^{\rm sc}} \qquad 18.24.1$$

and actually, identifying $\mathfrak{Gr}$ with the *full subcategory* of $\widetilde{\mathfrak{Loc}}$ over the objects $(G, \{1\})$ (cf. 18.12) where G runs over the finite groups, it still determines a *natural map* (cf. 18.20)

$$\mathfrak{loc}_{\bar{\hat{\pi}}} : \mathfrak{loc}_{\bar{\hat{\mathcal{L}}}} \longrightarrow \mathfrak{aut}_{\mathcal{F}^{\rm sc}} \qquad 18.24.2;$$

then, this natural map and the natural map $\bar{\lambda}_{\hat{\mathcal{L}}} : \mathfrak{loc}_{\mathcal{F}^{\rm sc}} \to \mathfrak{loc}_{\bar{\hat{\mathcal{L}}}}$ determined by $\lambda_{\hat{\mathcal{L}}}$ (cf. 18.21.1), both applied to the $\mathfrak{ch}^*(\mathcal{F}^{^{\rm sc}})$-morphism 18.23.3, yield the commutative $\widetilde{\mathfrak{Loc}}$-diagram

$$\begin{array}{ccccc}
\mathfrak{loc}_{\mathcal{F}^{\rm sc}}(R) & \xrightarrow{(\bar{\lambda}_{\hat{\mathcal{L}}})_R} & \mathfrak{loc}_{\bar{\hat{\mathcal{L}}}}(R) & \xrightarrow{(\mathfrak{loc}_{\bar{\hat{\pi}}})_R} & \mathfrak{aut}_{\mathcal{F}^{\rm sc}}(R) \\
{\scriptstyle \mathfrak{loc}_{\mathcal{F}^{\rm sc}}(\mathrm{id}_R,\delta_1^0)}\uparrow & & {\scriptstyle \mathfrak{loc}_{\bar{\hat{\mathcal{L}}}}(\mathrm{id}_R,\delta_1^0)}\uparrow & & {\scriptstyle \mathfrak{aut}_{\mathcal{F}^{\rm sc}}(\mathrm{id}_R,\delta_1^0)}\uparrow \\
\mathfrak{loc}_{\mathcal{F}^{\rm sc}}(\mathfrak{o}_\varphi) & \longrightarrow & \mathfrak{loc}_{\bar{\hat{\mathcal{L}}}}(\mathfrak{o}_\varphi) & \longrightarrow & \mathfrak{aut}_{\mathcal{F}^{\rm sc}}(\mathfrak{o}_\varphi)
\end{array} \qquad 18.24.3.$$

18.25 In this diagram, since the right-hand vertical arrow is simply a group homomorphism, fixing a representative λ_R for $(\bar{\lambda}_{\hat{\mathcal{L}}})_R$ we can choose a representative for the middle vertical arrow in such a way that we still have the commutative diagram of *group homomorphisms*

$$\begin{array}{ccccc}
\mathcal{L}(R) & \xrightarrow{\lambda_R} & \bar{\hat{\mathcal{L}}}(R) & \xrightarrow{(\mathfrak{aut}_{\bar{\hat{\pi}}})_R} & \mathcal{F}(R) \\
\uparrow & & \uparrow & & {\scriptstyle \mathfrak{aut}_{\mathcal{F}^{\rm sc}}(\mathrm{id}_R,\delta_1^0)}\uparrow \\
\mathcal{L}(Q)_{R'} & \longrightarrow & \bar{\hat{\mathcal{L}}}(Q)_{R'} & \longrightarrow & \mathcal{F}(Q)_{R'}
\end{array} \qquad 18.25.1.$$

That is to say, by the very definition of $\mathfrak{loc}_{\bar{\hat{\mathcal{L}}}}(\mathrm{id}_R, \delta_1^0)$ (cf. 18.20), we have found $\bar{\hat{x}}_\varphi \in \bar{\hat{\mathcal{L}}}(Q,R)$ which lifts $\varphi \in \mathcal{F}(Q,R)$; more precisely, considering as above the $(1,\bar{\hat{\mathcal{L}}})$- and the $(0,\bar{\hat{\mathcal{L}}})$-*chains* $\bar{\hat{\mathfrak{o}}}_\varphi : \Delta_1 \to \bar{\hat{\mathcal{L}}}$ and $\bar{\hat{\mathfrak{o}}}_R : \Delta_0 \to \bar{\hat{\mathcal{L}}}$ respectively defined by $\bar{\hat{x}}_\varphi$ and R, and the obvious $\mathfrak{ch}^*(\bar{\hat{\mathcal{L}}})$-morphism

$$\big(\bar{\hat{\tau}}_R(1), \delta_1^0\big) : (\bar{\hat{\mathfrak{o}}}_\varphi, \Delta_1) \longrightarrow (\bar{\hat{\mathfrak{o}}}_R, \Delta_0) \qquad 18.25.2,$$

we are just putting $\mathfrak{aut}_{\hat{\mathcal{L}}^{\mathrm{sc}}}\big(\hat{\tau}_R(1), \delta_1^0\big)$ in the middle vertical arrow of diagram 18.25.1.

18.26 In particular, note that the commutativity of the left-hand square in diagram 18.25.1 implies that $\mathfrak{aut}_{\hat{\mathcal{L}}^{\mathrm{sc}}}\big(\hat{\tau}_R(1), \delta_1^0\big)$ maps the image in $\bar{\hat{\mathcal{L}}}(Q)_{R'}$ of $\mathcal{L}(Q)_{\mathcal{R}'}$ on the image in $\bar{\hat{\mathcal{L}}}(R)$ of $\mathcal{L}(R)$; but, by the very definition of isomorphisms 18.23.1, these images respectively coincide with the images of $\mathcal{F}(Q)_{R'}$ and $\mathcal{F}(R)$; consequently, considering the action of $\mathcal{F}(Q) \times \mathcal{F}(R)$ on $\bar{\hat{\mathcal{L}}}(Q,R)$ defined by the composition on the left and on the right *via* isomorphisms 18.23.1, we claim that the stabilizer $\big(\mathcal{F}(Q) \times \mathcal{F}(R)\big)_{\bar{\hat{x}}_\varphi}$ of $\bar{\hat{x}}_\varphi$ coincides with the stabilizer $\big(\mathcal{F}(Q) \times \mathcal{F}(R)\big)_\varphi$ of φ with respect to the corresponding action on $\mathcal{F}(Q,R)$.

18.27 Indeed, since $\bar{\hat{x}}_\varphi$ lifts φ, the inclusion

$$\big(\mathcal{F}(Q) \times \mathcal{F}(R)\big)_{\bar{\hat{x}}_\varphi} \subset \big(\mathcal{F}(Q) \times \mathcal{F}(R)\big)_\varphi \qquad 18.27.1$$

is clear; conversely, for any $(\sigma,\tau) \in \big(\mathcal{F}(Q) \times \mathcal{F}(R)\big)_\varphi$, we have $\sigma \circ \varphi = \varphi \circ \tau$ and, in particular, σ belongs to $\mathcal{F}(Q)_{R'}$ and we get

$$\big(\mathfrak{aut}_{\mathcal{F}^{\mathrm{sc}}}(\mathrm{id}_R, \delta_1^0)\big)(\sigma) = \tau \qquad 18.27.2;$$

hence, by our remark above and the commutativity of the right-hand square in diagram 18.25.1, $\mathfrak{aut}_{\hat{\mathcal{L}}^{\mathrm{sc}}}\big(\hat{\tau}_R(1), \delta_1^0\big)$ still maps the image of σ on the image of τ and we still have $\sigma\cdot\bar{\hat{x}}_\varphi = \bar{\hat{x}}_\varphi\cdot\tau$. In conclusion, we can choose the lifting $\bar{\hat{x}}_\varphi \in \bar{\hat{\mathcal{L}}}(Q,R)$ of $\varphi \in \mathcal{F}(Q,R)$ in such a way that

18.27.3 *We have* $\bar{\hat{x}}_{\sigma\circ\varphi\circ\tau} = \sigma\cdot\bar{\hat{x}}_\varphi\cdot\tau$ *for any* $\sigma \in \mathcal{F}(Q)$ *and any* $\tau \in \mathcal{F}(R)$.

Now, considering the *full* subcategory $\mathcal{F}^{\mathfrak{X}}$ of $\mathcal{F}^{\mathrm{sc}}$ over a set of representatives $\mathfrak{X}$ for the set of isomorphism classes of $\mathcal{F}^{\mathrm{sc}}$-objects, for all the $\mathcal{F}^{\mathfrak{X}}$-morphisms we make a choice as above and then, for any $\mathcal{F}^{\mathrm{sc}}$-morphism $\varphi' : R' \to Q'$, we define $\bar{\hat{x}}_{\varphi'} = \sigma'\cdot\bar{\hat{x}}_\varphi\cdot\tau'$ where $\sigma' : Q \cong Q'$ and $\tau' : R' \cong R$ are $\mathcal{F}$-isomorphisms, Q and R belong to $\mathfrak{X}$ and $\varphi' = \sigma' \circ \varphi \circ \tau'$.

Proposition 18.28 *With the notation above, the correspondence mapping any pair of $\mathcal{F}^{sc}$-morphisms $\psi:T \to R$ and $\varphi:R \to Q$ on the element $\bar{\bar{k}}_{\varphi,\psi}$ of $\bar{\bar{\mathfrak{k}}}(T) = \mathrm{Ker}(\pi_T)/Z(T)$ such that*

$$\bar{\hat{x}}_\varphi\cdot\bar{\hat{x}}_\psi = \bar{\hat{x}}_{\varphi\circ\psi}\cdot\bar{\bar{k}}_{\varphi,\psi} \qquad 18.28.1$$

determines a stable $\bar{\bar{\mathfrak{k}}}$-valued 2-cocycle over $\tilde{\mathcal{F}}^{sc}$. Moreover, this 2-cocycle is a 2-coboundary if and only if the functor $\bar{\hat{\pi}}:\bar{\hat{\mathcal{L}}} \to \mathcal{F}^{sc}$ admits a functorial section. In this case, there is an $\mathcal{F}^{sc}$-sublocality $\mathcal{L}^{sc}$ of $\hat{\mathcal{L}}^{sc}$ such that $\mathfrak{loc}_{\mathcal{L}^{sc}}$ is the image of $\lambda_{\hat{\mathcal{L}}^{sc}}$.

Proof: Since $\bar{\hat{x}}_\varphi\cdot\bar{\hat{x}}_\psi$ and $\bar{\hat{x}}_{\varphi\circ\psi}$ have the same image in $\mathcal{F}(Q,T)$, the divisibility of $\bar{\hat{\mathcal{L}}}$ guarantees the existence and the uniqueness of $\bar{\bar{k}}_{\varphi,\psi} \in \bar{\bar{\mathfrak{k}}}(T)$; that is to say, we have a correspondence mapping any $(2,\mathcal{F}^{sc})$-*chain* $\mathfrak{q}:\Delta_2 \to \mathcal{F}^{sc}$ on $\bar{\bar{k}}_{\mathfrak{q}(0\bullet 1),\mathfrak{q}(1\bullet 2)}$. Moreover, for any $\mathcal{F}$-isomorphisms $\sigma:Q \cong Q'$, $\tau:R \cong R'$ and $\omega:T \cong T'$, setting $\varphi' = \sigma\circ\varphi\circ\tau^{-1}$ and $\psi' = \tau\circ\psi\circ\omega^{-1}$ and assuming that Q, R and T belong to $\mathfrak{X}$, we get

$$\begin{aligned}\bar{\hat{x}}_{\varphi'\circ\psi'}\cdot\bar{\bar{k}}_{\varphi',\psi'} &= \bar{\hat{x}}_{\varphi'}\cdot\bar{\hat{x}}_{\psi'} = (\sigma\cdot\bar{\hat{x}}_\varphi\cdot\tau^{-1})\cdot(\tau\cdot\bar{\hat{x}}_\psi\cdot\omega^{-1})\\ &= \sigma\cdot(\bar{\hat{x}}_{\varphi\circ\psi}\cdot\bar{\bar{k}}_{\varphi,\psi})\cdot\omega^{-1} = \sigma\cdot\bar{\hat{x}}_{\varphi\circ\psi}\cdot\omega^{-1}\cdot\big(\bar{\bar{\mathfrak{k}}}(\tilde{\omega})\big)\big(\bar{\bar{k}}_{\varphi,\psi}\big) \qquad 18.28.2\\ &= \bar{\hat{x}}_{\varphi'\circ\psi'}\cdot\big(\bar{\bar{k}}(\tilde{\omega})\big)\big(\bar{\bar{k}}_{\varphi,\psi}\big)\end{aligned}$$

and therefore we obtain $\bar{\bar{k}}_{\varphi',\psi'} = \big(\bar{\bar{\mathfrak{k}}}(\tilde{\omega})\big)\big(\bar{\bar{k}}_{\varphi,\psi}\big)$.

Hence, the same equality is true for *any* choice of the $\mathcal{F}$-isomorphisms $\sigma:Q \cong Q'$, $\tau:R \cong R'$ and $\omega:T \cong T'$, proving the *stability* of the above correspondence (cf. A3.18). In particular note that, for any $u \in Q$ and any $v \in R$, we obtain

$$\bar{\bar{k}}_{\kappa_Q(u)\circ\varphi,\kappa_R(v)\circ\psi} = \bar{\bar{k}}_{\varphi,\psi} \qquad 18.28.3,$$

proving that $\bar{\bar{k}}_{\varphi,\psi}$ only depends on the classes $\tilde{\varphi}$ and $\tilde{\psi}$ in $\tilde{\mathcal{F}}^{sc}$ of φ and ψ; that is to say, the above correspondence factorizes throughout the set of $(2,\tilde{\mathcal{F}}^{sc})$-*chains* $\tilde{\mathfrak{q}}:\Delta_2 \to \tilde{\mathcal{F}}^{sc}$.

Moreover, for a third $\mathcal{F}^{sc}$-morphism $\eta:U \to T$, it is clear that

$$\begin{aligned}\bar{\hat{x}}_\varphi\cdot\bar{\hat{x}}_\psi\cdot\bar{\hat{x}}_\eta &= \bar{\hat{x}}_{\varphi\circ\psi}\cdot\bar{\bar{k}}_{\varphi,\psi}\cdot\bar{\hat{x}}_\eta = \bar{\hat{x}}_{\varphi\circ\psi}\cdot\bar{\hat{x}}_\eta\cdot\big(\bar{\bar{\mathfrak{k}}}(\tilde{\eta})\big)\big(\bar{\bar{k}}_{\varphi,\psi}\big)\\ &= \bar{\hat{x}}_{\varphi\circ\psi\circ\eta}\cdot\bar{\bar{k}}_{\varphi\circ\psi,\eta}\cdot\big(\bar{\bar{\mathfrak{k}}}(\tilde{\eta})\big)\big(\bar{\bar{k}}_{\varphi,\psi}\big)\end{aligned} \qquad 18.28.4$$

and, similarly, we have

$$\bar{\hat{x}}_\varphi\cdot\bar{\hat{x}}_\psi\cdot\bar{\hat{x}}_\eta = \bar{\hat{x}}_{\varphi\circ\psi\circ\eta}\cdot\bar{\bar{k}}_{\varphi,\psi\circ\eta}\cdot\bar{\bar{k}}_{\psi,\eta} \qquad 18.28.5;$$

hence, we get the 2-*cocycle* condition

$$\bar{\bar{k}}_{\varphi\circ\psi,\eta}\cdot\big(\bar{\bar{\mathfrak{k}}}(\tilde{\eta})\big)(\bar{\bar{k}}_{\varphi,\psi}) = \bar{\bar{k}}_{\varphi,\psi\circ\eta}\cdot\bar{\bar{k}}_{\psi,\eta} \tag{18.28.6.}$$

In conclusion, the above correspondence determines indeed a stable $\bar{\bar{\mathfrak{k}}}$-valued 2-cocycle over $\tilde{\mathcal{F}}^{sc}$.

On the other hand, if the functor $\bar{\bar{\pi}}:\bar{\hat{\mathcal{L}}} \to \mathcal{F}^{sc}$ admits a functorial section, it is clear that such a section provides a choice fulfilling $\bar{\bar{k}}_{\varphi,\psi} = 1$ for any φ and ψ. Conversely, if there exists a family of elements $\bar{\bar{\ell}}_{\tilde{\varphi}} \in \bar{\bar{\mathfrak{k}}}(R)$, where $\tilde{\varphi}:R \to Q$ runs over the set of $\tilde{\mathcal{F}}^{sc}$-morphisms, fulfilling

$$\bar{\bar{k}}_{\varphi,\psi} = \big(\bar{\bar{\mathfrak{k}}}(\tilde{\psi})\big)(\bar{\bar{\ell}}_{\tilde{\varphi}})\cdot(\bar{\bar{\ell}}_{\tilde{\varphi}\circ\tilde{\psi}})^{-1}\cdot\bar{\bar{\ell}}_{\tilde{\psi}} \tag{18.28.7}$$

then equality 18.28.1 becomes

$$\big(\bar{\hat{x}}_{\varphi}\cdot(\bar{\bar{\ell}}_{\tilde{\varphi}})^{-1}\big)\cdot\big(\bar{\hat{x}}_{\psi}\cdot(\bar{\bar{\ell}}_{\tilde{\psi}})^{-1}\big) = \bar{\hat{x}}_{\varphi\circ\psi}\cdot(\bar{\bar{\ell}}_{\tilde{\varphi}\circ\tilde{\psi}})^{-1} \tag{18.28.8.}$$

Thus, we can modify our starting choice of liftings, obtaining the announced functorial section; the converse image in $\hat{\mathcal{L}}^{sc}$ of the image of this section yields the announced $\mathcal{F}^{sc}$-sublocality. We are done.

Proposition 18.29 *With the notation above, assume that there are two functorial sections* $\mathfrak{s}$ *and* $\mathfrak{s}'$ *of* $\bar{\bar{\pi}}:\bar{\hat{\mathcal{L}}} \to \mathcal{F}^{sc}$ *such that* $\mathfrak{s}'\big(\mathcal{F}(Q)\big) = \mathfrak{s}\big(\mathcal{F}(Q)\big)$ *for any* $\mathcal{F}^{sc}$*-object* Q. *Then, the correspondence mapping any* $\mathcal{F}^{sc}$*-morphism* $\varphi:R \to Q$ *on the element* $\bar{\bar{\ell}}_{\varphi}$ *of* $\bar{\bar{\mathfrak{k}}}(T) = \mathrm{Ker}(\pi_T)/Z(T)$ *such that*

$$\mathfrak{s}'(\varphi) = \mathfrak{s}(\varphi)\cdot\bar{\bar{\ell}}_{\varphi} \tag{18.29.1}$$

determines a stable $\bar{\bar{\mathfrak{k}}}$*-valued* 1*-cocycle over* $\tilde{\mathcal{F}}^{sc}$. *Moreover, this* 1*-cocycle is a* 1*-coboundary if and only if we have a natural isomorphism* $\mathfrak{s}' \cong \mathfrak{s}$.

Proof: Since $\mathfrak{s}'(\varphi)$ and $\mathfrak{s}(\varphi)$ have the same image in $\mathcal{F}(Q,R)$, the divisibility of $\bar{\hat{\mathcal{L}}}$ guarantees the existence and the uniqueness of $\bar{\bar{\ell}}_{\varphi} \in \bar{\bar{\mathfrak{k}}}(R)$. Moreover, for any $\mathcal{F}$-isomorphisms $\sigma:Q \cong Q'$ and $\tau:R \cong R'$, setting $\varphi' = \sigma\circ\varphi\circ\tau^{-1}$ we have

$$\begin{aligned}\mathfrak{s}(\sigma)\cdot\mathfrak{s}(\varphi)\cdot\mathfrak{s}(\tau)^{-1}\cdot\bar{\bar{\ell}}_{\varphi'} = \mathfrak{s}(\varphi')\cdot\bar{\bar{\ell}}_{\varphi'} = \mathfrak{s}'(\varphi') = \mathfrak{s}'(\sigma)\cdot\mathfrak{s}'(\varphi)\cdot\mathfrak{s}'(\tau)^{-1} \\ = \mathfrak{s}'(\sigma)\cdot\mathfrak{s}(\varphi)\cdot\bar{\bar{\ell}}_{\varphi}\cdot\mathfrak{s}'(\tau)^{-1}\end{aligned} \tag{18.29.2;}$$

but, since $\mathfrak{s}'\big(\mathcal{F}(Q)\big) = \mathfrak{s}\big(\mathcal{F}(Q)\big)$, $\mathfrak{s}(\sigma)^{-1}\cdot\mathfrak{s}'(\sigma)$ belongs to $\mathfrak{s}\big(\mathcal{F}(Q)\big)$ and therefore we easily get $\mathfrak{s}'(\sigma) = \mathfrak{s}(\sigma)$ (cf. 18.23.1); similarly, we get $\mathfrak{s}'(\tau) = \mathfrak{s}(\tau)$

and thus equality 18.29.2 forces $\bar{\bar{\ell}}_{\varphi'} = \big(\bar{\bar{\mathfrak{k}}}(\tilde{\tau})\big)(\bar{\bar{\ell}}_\varphi)$, proving the *stability* of the above correspondence. In particular note that, for any $u \in Q$, we obtain

$$\bar{\bar{\ell}}_{\kappa_Q(u)\circ\varphi} = \bar{\bar{\ell}}_\varphi \qquad 18.29.3,$$

proving that $\bar{\bar{\ell}}_\varphi$ only depends on the class $\tilde{\varphi}$ in $\tilde{\mathcal{F}}^{^{sc}}$ of φ; thus, the above correspondence factorizes throughout the set of $\tilde{\mathcal{F}}^{^{sc}}$-*chains* $\tilde{\mathfrak{r}} : \Delta_1 \to \tilde{\mathcal{F}}^{^{sc}}$.

Moreover, for a second $\mathcal{F}^{^{sc}}$-morphism $\psi : T \to R$, we have

$$\begin{aligned}\mathfrak{s}(\varphi)\cdot\mathfrak{s}(\psi)\cdot\bar{\bar{\ell}}_{\varphi\circ\psi} &= \mathfrak{s}(\varphi\circ\psi)\cdot\bar{\bar{\ell}}_{\varphi\circ\psi} = \mathfrak{s}'(\varphi\circ\psi) = \mathfrak{s}'(\varphi)\cdot\mathfrak{s}'(\psi)\\ &= \mathfrak{s}(\varphi)\cdot\bar{\bar{\ell}}_\varphi\cdot\mathfrak{s}(\psi)\cdot\bar{\bar{\ell}}_\psi = \mathfrak{s}(\varphi)\cdot\mathfrak{s}(\psi)\cdot\big(\bar{\bar{\mathfrak{k}}}(\tilde{\psi})\big)(\bar{\bar{\ell}}_\varphi)\cdot\bar{\bar{\ell}}_\psi\end{aligned} \qquad 18.29.4$$

and therefore we get the 1-*cocycle* condition

$$\bar{\bar{\ell}}_{\varphi\circ\psi} = \big(\bar{\bar{\mathfrak{k}}}(\tilde{\psi})\big)(\bar{\bar{\ell}}_\varphi)\cdot\bar{\bar{\ell}}_\psi \qquad 18.29.5.$$

In conclusion, the above correspondence determines indeed a stable $\bar{\bar{\mathfrak{k}}}$-valued 1-cocycle over $\tilde{\mathcal{F}}^{^{sc}}$.

On the other hand, if we have a natural isomorphism $\nu : \mathfrak{s}' \cong \mathfrak{s}$ then the naturality of ν yields $\nu_Q\cdot\mathfrak{s}'(\varphi) = \mathfrak{s}(\varphi)\cdot\nu_R$; but, according to decomposition 18.23.1, we have $\nu_Q = \mathfrak{s}(\sigma_Q)\cdot\bar{\bar{m}}_Q$ and $\nu_R = \mathfrak{s}(\sigma_R)\cdot\bar{\bar{m}}$ for suitable $\sigma_Q \in \mathcal{F}(Q)$, $\sigma_R \in \mathcal{F}(R)$, $\bar{\bar{m}} \in \bar{\bar{\mathfrak{k}}}(Q)$ and $\bar{\bar{m}} \in \bar{\bar{\mathfrak{k}}}(R)$; consequently, we get

$$\begin{aligned}\nu_Q\cdot\mathfrak{s}'(\varphi) &= \mathfrak{s}(\sigma_Q)\cdot\bar{\bar{m}}\cdot\mathfrak{s}'(\varphi) = \mathfrak{s}(\sigma_Q\circ\varphi)\cdot\big(\bar{\bar{\mathfrak{k}}}(\tilde{\varphi})\big)(\bar{\bar{m}})\cdot\bar{\bar{\ell}}_\varphi\\ \mathfrak{s}(\varphi)\cdot\nu_R &= \mathfrak{s}(\varphi)\cdot\mathfrak{s}(\sigma_R)\cdot\bar{\bar{m}} = \mathfrak{s}(\varphi\circ\sigma_R)\cdot\bar{\bar{m}}\end{aligned} \qquad 18.29.6$$

and therefore, since the image by $\bar{\bar{\pi}}$ implies that $\sigma_Q\circ\varphi = \varphi\circ\sigma_R$, we still get (cf. 18.23.1)

$$\bar{\bar{\ell}}_\varphi = \big(\bar{\bar{\mathfrak{k}}}(\tilde{\varphi})\big)(\bar{\bar{m}})^{-1}\cdot\bar{\bar{m}} \qquad 18.29.7.$$

Conversely, it is clear that a correspondence which maps any $\mathcal{F}^{^{sc}}$-object Q on $\bar{\bar{m}} \in \bar{\bar{\mathfrak{k}}}(Q)$ and fulfills equality 18.29.7 for any $\mathcal{F}^{^{sc}}$-morphism φ, determines a natural isomorphism $\mathfrak{s}' \cong \mathfrak{s}$ (with $\sigma_Q = \mathrm{id}_Q$ for any $\mathcal{F}^{^{sc}}$-object Q). We are done.

Chapter 19

Solvability for Frobenius P-categories

19.1 As we explain in the Introduction (cf. I28), one of the positive features of the Frobenius categories over finite p-groups is the fact that they admit a reasonable definition of *solvability* which forces a *solvable one* to come from a finite group, showing that our abstract setting lives not so far from the *true* finite groups. In this chapter we expose these facts as a consequence of our results in chapters 12, 13 and 18.

19.2 Let P be a finite p-group and $\mathcal{F}$ a Frobenius P-category; consider the following two sequences of $\mathcal{F}$-stable subgroups P_i of P and of *normal* Frobenius P_i-subcategories $\mathcal{F}_i$ of $\mathcal{F}$ (cf. 12.6); we set $P_0 = P$ and $\mathcal{F}_0 = \mathcal{F}$, and for any $i \in \mathbb{N}$ we inductively define (cf. Corollary 12.17 and 13.8)

$$\begin{aligned} P_{2i+1} &= P_{2i} &\quad\text{and}\quad \mathcal{F}_{2i+1} &= (\mathcal{F}_{2i})^{\mathfrak{a}} \\ P_{2i+2} &= H_{\mathcal{F}_{2i+1}} &\quad\text{and}\quad \mathcal{F}_{2i+2} &= (\mathcal{F}_{2i+1})^{\mathfrak{h}} \end{aligned} \qquad 19.2.1;$$

indeed, assuming that P_i is $\mathcal{F}$-stable and $\mathcal{F}_i$ is normal in $\mathcal{F}$, it follows either from Corollary 12.17 or from Proposition 13.9 that P_{i+1} is $\mathcal{F}$-stable and $\mathcal{F}_{i+1}$ is normal in $\mathcal{F}$. Then, we say that $\mathcal{F}$ is (p-)*solvable* if, for some $n \in \mathbb{N}$, we have $P_n = \{1\}$.

19.3 From Corollary 12.17 and Proposition 13.9, it is not difficult to see that this definition does not change if we consider the sequences of $\mathcal{F}$-stable subgroups P'_i of P and normal Frobenius P'_i-subcategories $\mathcal{F}'_i$ of $\mathcal{F}$ starting on P and $\mathcal{F}$, and exchanging the indices $^{\mathfrak{a}}$ and $^{\mathfrak{h}}$ in equalities 19.2.1; indeed, from those statements, for any $i \in \mathbb{N}$ we get

$$\mathcal{F}'_{i+1} \subset \mathcal{F}_i \quad\text{and}\quad \mathcal{F}_{i+1} \subset \mathcal{F}'_i \qquad 19.3.1.$$

By similar arguments, if $\mathcal{F}$ is solvable then, for any $\mathcal{F}$-stable subgroup P' of P, any *normal* Frobenius P'-subcategory $\mathcal{F}'$ of $\mathcal{F}$ is also solvable and the P'-*quotient* of $\mathcal{F}$ is solvable too; moreover, in this case the groups $\mathcal{F}(Q)$ are p-solvable for any subgroup Q of P (cf. Corollary 12.17 and Theorem 13.6).

19.4 In order to give an alternative definition of *solvability* for $\mathcal{F}$, recall that a subgroup N of P is *normal in* $\mathcal{F}$ or $\mathcal{F}$-*normal* whenever it is fully normalized in $\mathcal{F}$ and we have $N_{\mathcal{F}}(N) = \mathcal{F}$ (cf. 12.6); according to the next proposition, there is a *biggest* $\mathcal{F}$-normal subgroup of P — denoted by $\mathbf{O}_p(\mathcal{F})$.

Proposition 19.5 *A subgroup N of P is $\mathcal{F}$-normal if and only if it is $\mathcal{F}(P)$-stable and, for any $\mathcal{F}$-essential subgroup Q of P, Q contains N and $\mathcal{F}(Q)$ stabilizes N. Moreover, if N is $\mathcal{F}$-normal and $\mathcal{F}$-nilcentralized then $\mathcal{F}$ coincides with the Frobenius category $\mathcal{F}_L$ associated with the $\mathcal{F}$-localizer L of N.*

Proof: If N is fully normalized in $\mathcal{F}$ and we have $N_{\mathcal{F}}(N) = \mathcal{F}$ then N is clearly $\mathcal{F}(P)$-stable and, for any $\mathcal{F}$-essential subgroup Q of P (cf. 5.8), it follows from the very definition of $N_{\mathcal{F}}(N)$ (cf. 2.14) that any element of $\mathcal{F}(Q)$ is the restriction of an element of $\mathcal{F}(N{\cdot}Q)$ stabilizing N; in particular, $\mathcal{F}_N(Q)$ is normal in $\mathcal{F}(Q)$ and therefore it follows from Theorem 5.11 that $\mathcal{F}_N(Q)$ is contained in $\mathcal{F}_Q(Q)$, and more precisely, that $N_N(Q) \subset Q$. Consequently, we get $N \subset Q$ and $\mathcal{F}(Q)$ stabilizes N.

Conversely, it suffices to prove that, for any subgroup R of P, any $\mathcal{F}$-morphism $\psi : R \to P$ can be extended to $N{\cdot}R$, stabilizing N; we argue by induction on the *length* $\ell(\psi)$ of ψ (cf. 5.15). If ψ belongs to $\mathcal{F}(P) \circ \iota_R^P$, then the statement is clear; otherwise, we have $\psi = \iota_Q^P \circ \sigma \circ \eta$ where Q is an $\mathcal{F}$-essential subgroup of P, σ an element of $X_{\mathcal{F}}(Q)$, $\eta : R \to Q$ an $\mathcal{F}$-morphism and we have $\ell(\iota_Q^P \circ \eta) = \ell(\psi) - 1$ (cf. 5.15); hence, by the induction hypothesis, there is an $\mathcal{F}$-morphism $\hat{\eta} : N{\cdot}R \to P$ extending $\iota_Q^P \circ \eta$ and fulfilling $\hat{\eta}(N) = N \subset Q$, so that $\hat{\eta}$ defines an $\mathcal{F}$-morphism $\mu : N{\cdot}R \to Q$ extending η and fulfilling $\mu(N) = N$; then, $\hat{\psi} = \iota_Q^P \circ \sigma \circ \mu$ extends ψ and fulfills $\hat{\psi}(N) = N$.

Moreover, if N is $\mathcal{F}$-nilcentralized then we have

$$H_{C_{\mathcal{F}}(N)} = \{1\} \quad \text{and} \quad N_{\mathcal{F},N}(N) = N_{\mathcal{F}}(N) \qquad 19.5.1;$$

similarly, denoting by L the $\mathcal{F}$-localizer of N (cf. 18.5), we get $\mathcal{F}_{L,N} = \mathcal{F}_L$; now, the last statement follows from Theorem 18.6 and Remark 18.7.

Corollary 19.6 *If P' is an $\mathcal{F}$-stable subgroup of P, $\mathcal{F}'$ a normal Frobenius P'-subcategory of $\mathcal{F}$ and N an $\mathcal{F}$-normal subgroup of P', then N is normal in $\mathcal{F}'$. Moreover, $\mathbf{O}_p(\mathcal{F}')$ is normal in $\mathcal{F}$.*

Proof: Since we assume that $N_{\mathcal{F}}(N) = \mathcal{F}$, $\mathcal{F}(P')$ stabilizes N and, for any subgroup Q of P, $\mathcal{F}_N(Q)$ is normal in $\mathcal{F}(Q)$; in particular, if Q is an $\mathcal{F}'$-essential subgroup of P', the image of $N_N(Q)$ in $\mathcal{F}'(Q)$ is contained in $\mathcal{F}_Q(Q)$ by condition 5.11.2, and therefore Q contains $N_N(Q)$ by condition 5.11.1; thus, Q contains N and then N is $\mathcal{F}(Q)$-stable; at this point it suffices to apply Proposition 19.5 to prove the first statement.

Moreover, set $T = \mathbf{O}_p(\mathcal{F}')$; since $\mathcal{F}(P')$ stabilizes $\mathcal{F}'$, it stabilizes T and, in particular, $\mathcal{F}(P)$ still stabilizes T (cf. Proposition 12.8); on the other hand, if R is an $\mathcal{F}$-essential subgroup of P, any $\sigma \in \mathcal{F}(R)$ stabilizes $R' = R \cap P'$ and, denoting by $\varphi : R' \to P'$ the homomorphism mapping $u \in R'$ on $\sigma(u)$, it follows from Proposition 12.8 that $\varphi = \tau \circ \varphi'$ for suitable $\tau \in \mathcal{F}(P')$ and $\varphi' \in \mathcal{F}'(P', R')$. Then, since T is normal in $\mathcal{F}'$, there exists an $\mathcal{F}'$-morphism $\psi' : T{\cdot}R' \to P'$ fulfilling $\psi'(T) = T$ and inducing φ' on R' and therefore we get an $\mathcal{F}$-morphism $\psi : T{\cdot}R' \to P'$ mapping $u \in T{\cdot}R'$ on $\tau\big(\psi'(u)\big)$; thus, ψ fulfills $\psi(T) = T$ and extends φ.

Consequently, there exists a group automorphism $\eta : T{\cdot}R \cong T{\cdot}R$ mapping $v{\cdot}u$ on $\psi(v){\cdot}\sigma(u)$ for any $v \in T$ and any $u \in R$; then, since R is $\mathcal{F}$-selfcentralizing, it follows from statement 2.10.1 and from the existence of η that σ can be extended to an $\mathcal{F}$-automorphism of $N_T(R){\cdot}R$ and, in particular, σ normalizes the image of $N_T(R){\cdot}R$ in $\mathcal{F}(R)$. That is to say, $\mathcal{F}(R)$ normalizes its own p-subgroup $\mathcal{F}_{T\cdot R}(R)$; then it follows from Theorem 5.11 that $\mathcal{F}_{T\cdot R}(R) = \mathcal{F}_R(R)$ and therefore that $N_{T\cdot R}(R) = R$, which forces $T \subset R$; now, the last statement follows from Proposition 19.5. We are done.

19.7 We are ready to prove our main result on *solvability*. Set $\overline{P}^0 = P$, $\overline{\mathcal{F}}^0 = \mathcal{F}$ and, for any $i \in \mathbb{N}$, inductively define

$$\overline{P}^{i+1} = \overline{P}^i/\mathbf{O}_p(\overline{\mathcal{F}}^i) \quad \text{and} \quad \overline{\mathcal{F}}^{i+1} = \overline{\mathcal{F}}^i/\mathbf{O}_p(\overline{\mathcal{F}}^i) \qquad 19.7.1$$

where $\overline{\mathcal{F}}^i/\mathbf{O}_p(\overline{\mathcal{F}}^i)$ denotes the $\mathbf{O}_p(\overline{\mathcal{F}}^i)$-quotient of $\overline{\mathcal{F}}^i$.

Theorem 19.8 *The following statements on the Frobenius P-category $\mathcal{F}$ are mutually equivalent:*

19.8.1 *$\mathcal{F}$ is solvable.*

19.8.2 *There is $n \in \mathbb{N}$ such that $\overline{P}^n = \{1\}$.*

19.8.3 *There is a chain $\{Q_i\}_{0\le i\le n}$ of $\mathcal{F}$-stable subgroups of P such that $Q_0 = \{1\}$, $Q_n = P$ and, for any $1 \le i \le n$, Q_i/Q_{i-1} is normal in the Q_{i-1}-quotient of $\mathcal{F}$.*

19.8.4 *$\mathcal{F} = \mathcal{F}_G$ for a p-solvable finite group G.*

19.8.5 *There is an $\mathcal{F}$-normal $\mathcal{F}$-selfcentralizing subgroup Q of P such that the group $\mathcal{F}(Q)$ is p-solvable.*

Moreover, in this case $G/\mathbf{O}_{p'}(G)$ is isomorphic to the $\mathcal{F}$-localizer of $\mathbf{O}_p(\mathcal{F})$.

Proof: We may assume that $P \neq \{1\}$. If $\mathcal{F}$ is solvable and $n \in \mathbb{N}$ is the first integer fulfilling $P_{n+1} = \{1\}$ (cf. 19.2) then n is odd and $\mathcal{F}_n$ necessarily coincides with the Frobenius category $\mathcal{F}_{P_n}$ associated with the group P_n; but, P_n is clearly $\mathcal{F}(P)$-stable (cf. 19.2); moreover, for any $\mathcal{F}$-essential subgroup Q of P, since $\mathcal{F}_n = \mathcal{F}_{P_n}$ is normal in $\mathcal{F}$ (cf. 19.2), we have

$$\mathcal{F}_{P_n}(Q) \lhd \mathcal{F}(Q) \qquad 19.8.6$$

which forces $N_{P_n}(Q) \subset Q$ (cf. Theorem 5.11), and therefore P_n is contained in Q. That is to say, P_n is $\mathcal{F}$-normal and, in particular, $\mathbf{O}_p(\mathcal{F})$ is not trivial; then, since the $\mathbf{O}_p(\mathcal{F})$-quotient of $\mathcal{F}$ is also solvable (cf. 19.3), to get statement 19.8.2 it suffices to argue by induction on $|P|$.

It is obvious that statement 19.8.2 implies statement 19.8.3. Statement 19.8.4 implies statement 19.8.5 since it is easily checked from Theorem 3.2 in [28, Ch. 6] that, in statement 19.8.4, $\mathbf{O}_p(\mathcal{F})$ coincides with a

Sylow p-subgroup of $\mathbf{O}_{p',p}(G)$ and that it is $\mathcal{F}$-selfcentralizing; moreover, in this case Proposition 19.5 implies the last statement. Similarly, applying Proposition 19.5 again, statement 19.8.5 implies that we have $\mathcal{F} = \mathcal{F}_L$ for an $\mathcal{F}$-localizer L of Q (cf. 18.5) and, since L is then p-solvable, it is very easy to prove statement 19.8.1.

From now on, assume that statement 19.8.3 holds and set

$$P' = \mathbf{O}_p(\mathcal{F})\cdot C_P\big(\mathbf{O}_p(\mathcal{F})\big) \quad \text{and} \quad U = \mathcal{F}_{\mathbf{O}_p(\mathcal{F})}\big(\mathbf{O}_p(\mathcal{F})\big) \qquad 19.8.7;$$

it is not difficult to check that P' is $\mathcal{F}$-stable (cf. Corollary 12.5) and that the Frobenius P'-subcategory

$$\mathcal{F}' = \mathbf{O}_p(\mathcal{F})\cdot C_{\mathcal{F}}\big(\mathbf{O}_p(\mathcal{F})\big) = N_{\mathcal{F}}^U\big(\mathbf{O}_p(\mathcal{F})\big) \qquad 19.8.8$$

is normal in $\mathcal{F}$; it is easily checked from Corollary 19.6 that $\mathcal{F}'$ still fulfills the corresponding statement 19.8.3. Hence, arguing by induction on $|P|$ and on $\prod_Q |\mathcal{F}(P,Q)|$ where Q runs over the set of subgroups of P, if either $P' \neq P$, or $P' = P$ and $\mathcal{F}' \neq \mathcal{F}$, then $\mathcal{F}'$ is the Frobenius category $\mathcal{F}_{G'}$ associated with a p-solvable finite group G', and we may assume that $\mathbf{O}_{p'}(G') = 1$; thus, according to Corollary 19.6, $\mathbf{O}_p(\mathcal{F}')$ is normal in $\mathcal{F}$, so that it is contained in $\mathbf{O}_p(\mathcal{F})$, and, conversely, $\mathbf{O}_p(\mathcal{F})$ is normal in $\mathcal{F}'$, so that we get the equality $\mathbf{O}_p(\mathcal{F}') = \mathbf{O}_p(\mathcal{F})$; moreover, since $\mathcal{F}' = \mathcal{F}_{G'}$, it follows from Theorem 3.2 in [28, Ch. 6] that $\mathbf{O}_p(G') = \mathbf{O}_p(\mathcal{F}')$ and then, by the very definition of $\mathcal{F}'$, that

$$C_{\mathcal{F}}\big(\mathbf{O}_p(\mathcal{F})\big) = C_{\mathcal{F}'}\big(\mathbf{O}_p(\mathcal{F})\big) = C_{\mathcal{F}'}\big(\mathbf{O}_p(G')\big) = \mathcal{F}_{Z(\mathbf{O}_p(G'))} \qquad 19.8.9.$$

In this case, by the very definition of $\mathcal{F}'$, we simply have $G' = \mathbf{O}_p(\mathcal{F})$ and therefore $\mathbf{O}_p(\mathcal{F})$ is $\mathcal{F}$-selfcentralizing; hence, according to Proposition 19.5, $\mathcal{F}$ is the Frobenius category associated with the $\mathcal{F}$-localizer L of $\mathbf{O}_p(\mathcal{F})$; moreover, it is quite clear that $\bar{\mathcal{F}} = \mathcal{F}/\mathbf{O}_p(\mathcal{F})$ still fulfills condition 19.8.3 and therefore, by the induction hypothesis, $\bar{\mathcal{F}}$ is the Frobenius category associated with a p-solvable finite group $\bar{G}$; in particular, $\bar{\mathcal{F}}$ is solvable which clearly implies that $\mathcal{F}$ is also solvable, so that $\mathcal{F}\big(\mathbf{O}_p(\mathcal{F})\big)$ is a p-solvable group (cf. 19.3); thus, since $L/\mathbf{O}_p(\mathcal{F}) \cong \mathcal{F}\big(\mathbf{O}_p(\mathcal{F})\big)$ the group L is p-solvable too.

Finally, assume that $P' = P$ and $\mathcal{F}' = \mathcal{F}$; in this case, we prove that

$$P = \mathbf{O}_p(\mathcal{F}) \quad \text{and} \quad \mathcal{F} = \mathcal{F}_{\mathbf{O}_p(\mathcal{F})} \qquad 19.8.10.$$

Clearly, the Q_1-quotient $\bar{\mathcal{F}}$ of $\mathcal{F}$ over $\bar{P} = P/Q_1$ still fulfills the corresponding statement 19.8.3 and therefore, arguing by induction on $|P|$, $\bar{\mathcal{F}}$ is the Frobenius category $\mathcal{F}_{\bar{G}}$ associated with a finite p-solvable group $\bar{G}$, and we may assume that $\mathbf{O}_{p'}(\bar{G}) = 1$. Then, we claim that the converse image T of $\mathbf{O}_p(\bar{G})$ in P is normal in $\mathcal{F}$.

It follows from Corollary 12.5 that T is $\mathcal{F}(P)$-stable. On the other hand, let R be an $\mathcal{F}$-essential subgroup of P fully normalized in $\mathcal{F}$ and denote by $\bar{R}$ its image in $\bar{P}$; since $Q_1 \subset \mathbf{O}_p(\mathcal{F}) \subset R$, by the very definition of $\mathcal{F}'$ we have

$$\mathcal{F}(Q_1) = \mathcal{F}'(Q_1) = \mathcal{F}_{\mathbf{O}_p(\mathcal{F})}(Q_1) \qquad 19.8.11$$

and therefore any p'-subgroup K of $\mathcal{F}'(R) = \mathcal{F}(R)$ centralizes Q_1; thus, if moreover K centralizes $\bar{R}$, it is trivial (cf. Theorem 7.2 in [28, Ch. 5]). Hence, the kernel of the canonical homomorphism $\mathcal{F}(R) \to \bar{\mathcal{F}}(\bar{R})$ is a p-group and therefore it is contained in $\mathcal{F}_R(R)$ (cf. Theorem 5.11); in particular, the converse image of $C_{\bar{P}}(\bar{R})$ in P is contained in R and therefore $\bar{R}$ is $\bar{\mathcal{F}}$-selfcentralizing. Moreover, since

$$Q_1 \subset \mathbf{O}_p(\mathcal{F}) \subset R \qquad 19.8.12,$$

the homomorphism $\mathcal{F}(R) \to \bar{\mathcal{F}}(\bar{R})$ is surjective (cf. Proposition 12.3). Consequently, it follows from Theorem 5.11 that $\bar{R}$ is $\bar{\mathcal{F}}$-essential and therefore, according to Proposition 19.5, it contains $\mathbf{O}_p(\bar{\mathcal{F}}) = \mathbf{O}_p(\bar{G})$ which implies that R contains T.

At this point, it follows from Proposition 19.5 again that T is $\mathcal{F}$-normal and therefore it is contained in $\mathbf{O}_p(\mathcal{F})$; but, the image of $\mathbf{O}_p(\mathcal{F})$ in $\bar{P}$ is clearly normal in $\bar{\mathcal{F}}$ and, since its normalizer in $\bar{\mathcal{F}}$ is the Frobenius category associated with its normalizer in $\bar{G}$ (cf. Corollary 3.6), it follows from the equality (cf. Theorem 3.2 in [28, Ch. 6])

$$C_{\bar{G}}\big(\mathbf{O}_p(\bar{G})\big) = Z\big(\mathbf{O}_p(\bar{G})\big) \qquad 19.8.13$$

that this image is normal in $\bar{G}$; hence, we get $T = \mathbf{O}_p(\mathcal{F})$ and therefore we have

$$\mathcal{F}(T) = \mathcal{F}'(T) = \mathcal{F}_{\mathbf{O}_p(\mathcal{F})}(T) \qquad 19.8.14,$$

so that any p'-subgroup of $\bar{G}$ centralizes $\mathbf{O}_p(\bar{G})$; by equality 19.8.13, this forces $\bar{G} = \mathbf{O}_p(\bar{G})$ and therefore $\mathcal{F}$ is the Frobenius category associated with $\mathbf{O}_p(\mathcal{F})$. We are done.

19.9 This result leads us to consider the following class of subgroups of P. We say that a subgroup Q of P is *$\mathcal{F}$-solvcentralized* whenever, for some $\varphi \in \mathcal{F}(P,Q)$ such that $\varphi(Q)$ is fully centralized in $\mathcal{F}$, the Frobenius $C_P\big(\varphi(Q)\big)$-category $C_{\mathcal{F}}\big(\varphi(Q)\big)$ is *solvable;* in this case, it is clear that the Frobenius $C_P\big(\psi(Q)\big)$-category $C_{\mathcal{F}}\big(\psi(Q)\big)$ is solvable for any $\psi \in \mathcal{F}(P,Q)$ such that $\psi(Q)$ is fully centralized in $\mathcal{F}$. Note that

19.9.1 *If Q contains a $\mathcal{F}$-solvcentralized subgroup R of P then it is $\mathcal{F}$-solvcentralized too.*

Indeed, we may assume that $R \triangleleft Q$ and then that both are fully centralized in $\mathcal{F}$ (cf. Corollary 2.21); in this case, Q is also fully centralized in the solvable Frobenius $Q{\cdot}C_P(R)$-category

$$Q{\cdot}C_{\mathcal{F}}(R) = N_{\mathcal{F}}^{\mathcal{F}_Q(R)}(R) \qquad 19.9.2,$$

we clearly have

$$C_P(Q) = C_{Q\cdot C_P(R)}(Q) \qquad 19.9.3$$

and the Frobenius categories $C_{\mathcal{F}}(Q)$ and $C_{Q\cdot C_{\mathcal{F}}(R)}(Q)$ coincide. Our last result in this chapter gives several characterizations of the $\mathcal{F}$-solvcentralized subgroups of P; as usual, if R is a subgroup of Q, we denote by $\mathcal{F}(Q)_R$ the stabilizer of R in $\mathcal{F}(Q)$.

Proposition 19.10 *Let Q be a subgroup of P fully centralized in $\mathcal{F}$. Then the following statements on Q are mutually equivalent:*

19.10.1 *Q is $\mathcal{F}$-solvcentralized.*

19.10.2 *$U = Q{\cdot}\mathbf{O}_p\big(C_{\mathcal{F}}(Q)\big)$ is $\mathcal{F}$-selfcentralizing and the kernel of the restriction map $\mathcal{F}(U)_Q \to \mathcal{F}(Q)$ is p-solvable.*

19.10.3 *There exists an $\mathcal{F}$-selfcentralizing subgroup R of $Q{\cdot}C_P(Q)$ such that we have $Q \subset R \triangleleft Q{\cdot}C_{\mathcal{F}}(Q)$ and that the kernel of the restriction map $\mathcal{F}(R)_Q \to \mathcal{F}(Q)$ is p-solvable.*

Moreover, if $\mathcal{F}\big(Q{\cdot}C_P(Q)\big)_Q$ stabilizes R, the restriction map $\mathcal{F}(R)_Q \to \mathcal{F}(Q)$ is surjective.

Proof: If Q is $\mathcal{F}$-solvcentralized then $\mathbf{O}_p\big(C_{\mathcal{F}}(Q)\big)$ is $C_{\mathcal{F}}(Q)$-selfcentralizing by statement 19.8.5; moreover, according to Corollary 2.21, we may assume that $U = Q{\cdot}\mathbf{O}_p\big(C_{\mathcal{F}}(Q)\big)$ is also fully centralized in $\mathcal{F}$; then, since $C_P(U)$ is contained in $C_P(Q)$ and centralizes $\mathbf{O}_p\big(C_{\mathcal{F}}(Q)\big)$, U is $\mathcal{F}$-selfcentralizing. On the other hand, we are assuming that the Frobenius $Q{\cdot}C_P(Q)$-category $\mathcal{H} = Q{\cdot}C_{\mathcal{F}}(Q)$ is solvable and, by the very definition of $C_{\mathcal{H}}(Q)$ (cf. 2.14), the kernel of the restriction map $\mathcal{F}(U)_Q \to \mathcal{F}(Q)$ coincides with $\big(C_{\mathcal{H}}(Q)\big)(U)$.

It is clear that statement 19.10.2 implies statement 19.10.3. Assume that statement 19.10.3 holds; since R is also $\mathcal{H}$-selfcentralizing and the kernel of the restriction homomorphism $\mathcal{F}(R)_Q \to \mathcal{F}(Q)$ coincides with $\big(C_{\mathcal{H}}(Q)\big)(R)$ (cf. 2.14), statement 19.10.1 follows from 19.8.5. Moreover, since we have

$$\mathcal{F}_{Q\cdot C_P(Q)}(Q) = \mathcal{F}_Q(Q) \qquad 19.10.4,$$

it follows from statement 2.10.1 that any $\varphi \in \mathcal{F}(P,Q)$ can be extended to $Q{\cdot}C_P(Q)$; in particular, any $\sigma \in \mathcal{F}(Q)$ can be lifted to some element τ in $\mathcal{F}\big(Q{\cdot}C_P(Q)\big)_Q$ and if $\mathcal{F}\big(Q{\cdot}C_P(Q)\big)_Q$ stabilizes R then τ induces $\rho \in \mathcal{F}(R)$ lifting σ. We are done.

Remark 19.11 From the arguments in [35, Ch. V], it is not difficult to prove that if $p \neq 2$ and the subgroup $L(P)$ of P defined there is *not* normal in $\mathcal{F}$ then there is some $\mathcal{F}$-essential subgroup Q of P such that $\mathcal{F}(Q)$ is of *even* order — the proof of this statement actually need not quote Thompson's Classification of the so-called *Quadratic pairs*. In other words, we have

19.11.1 *If $p \neq 2$ and $\mathcal{F}(Q)$ has odd order for any subgroup Q of P then $\mathcal{F}$ is solvable.*

In general, assuming that $p \geq 5$ and quoting Thompson's Classification of the *Quadratic pairs* (cf. [58]), the arguments developed in [35, Ch. V] remain essentially valid, providing a step towards an eventual classification of the *simple* Frobenius P-categories defined in 12.19.

Chapter 20

A perfect $\mathcal{F}$-locality from a perfect $\mathcal{F}^{sc}$-locality

20.1 Let P be a finite p-group and $\mathcal{F}$ a Frobenius P-category; recall that we denote by $\mathcal{F}^{sc}$ the full subcategory of $\mathcal{F}$ over the set of $\mathcal{F}$-selfcentralizing subgroups of P (cf. 6.1) and that in chapter 17 we have introduced the $\mathcal{F}$- and $\mathcal{F}^{sc}$-localities (cf. 17.4), introducing in 17.13 the *perfect ones*. In this chapter, we prove that any *perfect* $\mathcal{F}^{sc}$*-locality* $\mathcal{L}^{sc}$ can be extended to a unique *perfect* $\mathcal{F}$*-locality* $\mathcal{L}$. Our argument depends on a precise consequence of Theorem 17.18 (cf. 20.9); the remainder of our long proof is just routine, yet we have not found a shorter method. As Theorem 4.12, this result illustrates the fact that the full subcategory $\mathcal{F}^{sc}$ "determines" what happens in $\mathcal{F}$.

20.2 From now on, we assume that there exists a perfect $\mathcal{F}^{sc}$-locality $\mathcal{L}^{sc}$ and denote by

$$\tau^{sc} : \mathcal{T}_P^{sc} \longrightarrow \mathcal{L}^{sc} \quad \text{and} \quad \pi^{sc} : \mathcal{L}^{sc} \longrightarrow \mathcal{F}^{sc} \qquad 20.2.1$$

the structural functors (cf. 17.3 and 17.4); recall that, for any pair of subgroups Q and R of P such that $R \subset Q$, we set $i_R^Q = (\tau^{sc})_R^Q(1)$ (cf. 17.3). In order to construct $\mathcal{L}$, we start by defining the $\mathcal{L}$*-isomorphisms* and the *restriction maps* between them. Explicitly, let Q and Q' be $\mathcal{F}$-isomorphic subgroups of P, R a subgroup of Q and R' a subgroup of Q', and let us assume that the set $\mathcal{F}(Q',Q)_{R',R}$ of $\varphi \in \mathcal{F}(Q',Q)$ fulfilling $\varphi(R) = R'$ is not empty; since $\mathcal{F}$ is divisible, there is a unique *restriction map*

$$\mathfrak{f}_{R',R}^{Q',Q} : \mathcal{F}(Q',Q)_{R',R} \longrightarrow \mathcal{F}(R',R) \qquad 20.2.2$$

sending $\varphi \in \mathcal{F}(Q',Q)_{R',R}$ to $\psi \in \mathcal{F}(R',R)$ such that $\iota_{R'}^{Q'} \circ \psi = \varphi \circ \iota_R^Q$, and we start by defining the sets $\mathcal{L}(Q',Q)_{R',R}$ and $\mathcal{L}(R',R)$, and by lifting $\mathfrak{f}_{R',R}^{Q',Q}$ to a map

$$\mathfrak{g}_{R',R}^{Q',Q} : \mathcal{L}(Q',Q)_{R',R} \longrightarrow \mathcal{L}(R',R) \qquad 20.2.3.$$

Whenever all these subgroups are $\mathcal{F}$*-selfcentralizing*, the existence of all the items in 20.2.3 will be an easy consequence of the existence of $\mathcal{L}^{sc}$; in a second step, we will consider the case where all these subgroups are *fully centralized* in $\mathcal{F}$; finally, in the general case, we argue by induction on $|P\!:\!Q|$.

20.3 If Q and Q' are $\mathcal{F}$-selfcentralizing, we simply set

$$\mathcal{L}(Q',Q) = \mathcal{L}^{sc}(Q',Q) \qquad 20.3.1$$

and denote by $\mathcal{L}(Q',Q)_{R',R}$ the converse image of $\mathcal{F}(Q',Q)_{R',R}$ in $\mathcal{L}(Q',Q)$ *via* the structural homomorphisms $(\pi^{\mathrm{sc}})_{Q',Q}$. Moreover, if R and R' are $\mathcal{F}$-selfcentralizing too, it follows from the *divisibility* of $\mathcal{L}^{^{\mathrm{sc}}}$ that there exists a map $\mathfrak{g}^{Q',Q}_{R',R}$ from $\mathcal{L}(Q',Q)_{R',R}$ to $\mathcal{L}(R',R)$ such that, for any $x \in \mathcal{L}(Q',Q)_{R',R}$, we have

$$i^{Q'}_{R'}\cdot\mathfrak{g}^{Q',Q}_{R',R}(x) = x\cdot i^{Q}_{R} \qquad 20.3.2;$$

actually, it follows from Proposition 4.6 and from the injectivity of $(\tau^{\mathrm{sc}})_R$ (cf. 17.13) that $\mathfrak{g}^{Q',Q}_{R',R}$ is injective. In particular, for any $x \in \mathcal{L}(Q',Q)_{R',R}$ and any $u \in \mathcal{T}_P(Q',Q) \cap \mathcal{T}_P(R',R)$, we have

$$\begin{aligned}(\pi^{\mathrm{sc}})_{Q',Q}\big(\mathfrak{g}^{Q',Q}_{R',R}(x)\big) &= \mathfrak{f}^{Q',Q}_{R',R}\big((\pi^{\mathrm{sc}})_{R',R}(x)\big)\\ \mathfrak{g}^{Q',Q}_{R',R}\big((\tau^{\mathrm{sc}})_{Q',Q}(u)\big) &= (\tau^{\mathrm{sc}})_{R',R}(u)\end{aligned} \qquad 20.3.3$$

and we simply set $\pi_{Q',Q} = (\pi^{\mathrm{sc}})_{Q',Q}$ and $\tau_{Q',Q} = (\tau^{\mathrm{sc}})_{Q',Q}$.

20.4 Note that the family of these maps, for the different choices of $\mathcal{F}$-selfcentralizing subgroups, agrees with the composition of $\mathcal{L}^{^{\mathrm{sc}}}$-morphisms and fulfills a *transitivity* condition. Namely, if Q'' is an $\mathcal{F}$-selfcentralizing subgroup of P which is $\mathcal{F}$-isomorphic to Q and Q', R'' is an $\mathcal{F}$-selfcentralizing subgroup of Q'', and x' is an element of $\mathcal{L}(Q'',Q')_{R'',R'}$, then we clearly have

$$\mathfrak{g}^{Q'',Q}_{R'',R}(x'\cdot x) = \mathfrak{g}^{Q'',Q'}_{R'',R'}(x')\cdot\mathfrak{g}^{Q',Q}_{R',R}(x) \qquad 20.4.1.$$

Moreover, if T and T' are $\mathcal{F}$-selfcentralizing subgroups of P which are respectively contained in R and R', and $x \in \mathcal{L}(Q',Q)_{R',R} \cap \mathcal{L}(Q',Q)_{T',T}$, then it is easily checked that

$$\mathfrak{g}^{Q',Q}_{T',T}(x) = \mathfrak{g}^{R',R}_{T',T}\big(\mathfrak{g}^{Q',Q}_{R',R}(x)\big) \qquad 20.4.2.$$

20.5 Now, assume that Q is *fully centralized* in $\mathcal{F}$ and set $\hat{Q} = Q\cdot C_P(Q)$ which is clearly $\mathcal{F}$-selfcentralizing; in particular, $\mathcal{L}(\hat{Q}) = \mathcal{L}(\hat{Q},\hat{Q})$ is already defined and the structural functor $\tau^{\mathrm{sc}} : \mathcal{T}_P^{^{\mathrm{sc}}} \to \mathcal{L}^{^{\mathrm{sc}}}$ determines a group homomorphism $\tau_{\hat{Q}} : N_P(\hat{Q}) \to \mathcal{L}(\hat{Q})$; then, we need to consider the following subgroup of $\mathcal{L}(\hat{Q})$ (cf. 13.2)

$$O(Q) = \mathbf{O}^p\Big(C_{\mathcal{L}(\hat{Q})}\big(\tau_{\hat{Q}}(Q)\big)\Big)\cdot\tau_{\hat{Q}}(H_{C_{\mathcal{F}}(Q)}) \qquad 20.5.1.$$

If Q' is also fully centralized in $\mathcal{F}$ and we set $\hat{Q}' = Q'\cdot C_P(Q')$, it follows from statement 2.10.1 that any $\varphi \in \mathcal{F}(Q',Q)$ can be extended to a suitable $\hat{\varphi} \in \mathcal{F}(\hat{Q}',\hat{Q})$; thus, the map

$$\mathfrak{f}^{\hat{Q}',\hat{Q}}_{Q',Q} : \mathcal{F}(\hat{Q}',\hat{Q})_{Q',Q} \longrightarrow \mathcal{F}(Q',Q) \qquad 20.5.2$$

is surjective.

20.6 Moreover, in this case, $\mathcal{L}(\hat{Q}',\hat{Q})$ is already defined and the groups $\mathcal{L}(\hat{Q})$ and $\mathcal{L}(\hat{Q}')$ act on this set by composition on the right and on the left respectively; moreover, it is quite clear that the respective subgroups $O(Q)$ and $O(Q')$ stabilize $\mathcal{L}(\hat{Q}',\hat{Q})_{Q',Q}$ and that the corresponding quotient sets $\mathcal{L}(\hat{Q}',\hat{Q})_{Q',Q}/O(Q)$ and $O(Q')\backslash\mathcal{L}(\hat{Q}',\hat{Q})_{Q',Q}$ coincide. Thus, we define

$$\mathcal{L}(Q',Q) = \mathcal{L}(\hat{Q}',\hat{Q})_{Q',Q}/O(Q) = O(Q')\backslash\mathcal{L}(\hat{Q}',\hat{Q})_{Q',Q} \qquad 20.6.1$$

and denote by

$$\mathfrak{g}_{Q',Q}^{\hat{Q}',\hat{Q}} : \mathcal{L}(\hat{Q}',\hat{Q})_{Q',Q} \longrightarrow \mathcal{L}(Q',Q) \qquad 20.6.2$$

the canonical map.

20.7 It is clear that there is a unique map

$$\pi_{Q',Q} : \mathcal{L}(Q',Q) \longrightarrow \mathcal{F}(Q',Q) \qquad 20.7.1$$

which, for any $x \in \mathcal{L}(\hat{Q}',\hat{Q})_{Q',Q}$, fulfills

$$\pi_{Q',Q}\big(\mathfrak{g}_{Q',Q}^{\hat{Q}',\hat{Q}}(x)\big) = \mathfrak{f}_{Q',Q}^{\hat{Q}',\hat{Q}}\big(\pi_{\hat{Q}',\hat{Q}}(x)\big) \qquad 20.7.2;$$

similarly, if u belongs to $\mathcal{T}_P(Q',Q)$ then it belongs to $\mathcal{T}_P(\hat{Q}',\hat{Q})$ too, and we consider the map $\tau_{Q',Q} : \mathcal{T}_P(Q',Q) \to \mathcal{L}(Q',Q)$ defined by

$$\tau_{Q',Q}(u) = \mathfrak{g}_{Q',Q}^{\hat{Q}',\hat{Q}}\big(\tau_{\hat{Q}',\hat{Q}}(u)\big) \qquad 20.7.3.$$

Furthermore, it is clear that the composition in $\mathcal{L}^{sc}$ defines a compatible *composition* between these morphisms; explicitly, if Q'' is a subgroup of P which is $\mathcal{F}$-isomorphic to Q and Q', and is fully centralized in $\mathcal{F}$, setting $\hat{Q}'' = Q''{\cdot}C_P(Q'')$ we get a *composition map*

$$\mathcal{L}(Q'',Q') \times \mathcal{L}(Q',Q) \longrightarrow \mathcal{L}(Q'',Q) \qquad 20.7.4$$

— denoted by $\cdot$ — that for any $x \in \mathcal{L}(\hat{Q}',\hat{Q})_{Q',Q}$ and any $x' \in \mathcal{L}(\hat{Q}'',\hat{Q}')_{Q'',Q'}$ fulfills

$$\mathfrak{g}_{Q'',Q}^{\hat{Q}'',\hat{Q}}(x'{\cdot}x) = \mathfrak{g}_{Q'',Q'}^{\hat{Q}'',\hat{Q}'}(x'){\cdot}\mathfrak{g}_{Q',Q}^{\hat{Q}',\hat{Q}}(x) \qquad 20.7.5$$

and the associativity of the composition in $\mathcal{L}^{sc}$ forces the obvious *associativity* here.

20.8 Assume that R and R' are also fully centralized in $\mathcal{F}$ and normal in Q and Q' respectively; thus, $\hat{Q}$ and $\hat{Q}'$ respectively normalize R and R', and we set

$$\hat{T} = \hat{Q}{\cdot}C_P(R) = Q{\cdot}\hat{R} \quad \text{and} \quad \hat{T}' = \hat{Q}'{\cdot}C_P(R') = Q'{\cdot}\hat{R}' \qquad 20.8.1$$

which respectively coincide with the converse images of $\mathcal{F}_Q(R)$ in $N_P(R)$ and of $\mathcal{F}_{Q'}(R')$ in $N_P(R')$; since R is also fully $\mathcal{F}_Q(R)$-normalized in $\mathcal{F}$ (cf. 2.10) and we have $N_P^{\mathcal{F}_Q(R)}(R) = \hat{T}$, we get the Frobenius $\hat{T}$-category and the associated partial perfect locality (cf. 17.4 and 17.5)

$$\mathcal{F}^{^R} = N_{\mathcal{F}}^{\mathcal{F}_Q(R)}(R) \quad \text{and} \quad \mathcal{L}^{^R} = N_{\mathcal{L}^{\rm sc}}^{\mathcal{F}_Q(R)}(R) \qquad 20.8.2$$

defined over the set $\mathfrak{X}^{^R}$ of $\mathcal{F}$-selfcentralizing subgroups of P contained in $\hat{T}$; we identify $\mathcal{F}^{^R}$ and $\mathcal{L}^{^R}$ with their canonical image in $\mathcal{F}^{^{\rm sc}}$ and $\mathcal{L}^{^{\rm sc}}$ respectively.

20.9 Then, since $H_{\mathcal{F}^R}$ is an $\mathcal{F}^{^R}$-stable subgroup of $\hat{T}$ (cf. Lemma 13.3), it follows from Theorem 17.18 that we have the *quotient* partial perfect $\mathcal{F}^{^R}/H_{\mathcal{F}^R}$-locality $\mathcal{L}^{^R}/H_{\mathcal{F}^R}$; but, it is easily checked from Lemma 13.3 that $\mathcal{L}^{^R}/H_{\mathcal{F}^R}$ can be identified to the *full* subcategory of $\mathcal{T}_{\hat{T}/H_{\mathcal{F}^R}}$ (cf. 17.2) over the image in $\hat{T}/H_{\mathcal{F}^R}$ of the set $\mathfrak{X}^{^R}$; hence, we have a canonical functor

$$\mathfrak{t}^R : \mathcal{L}^{^R} \longrightarrow \mathcal{T}_{\hat{T}/H_{\mathcal{F}^R}} \qquad 20.9.1$$

compatible with the structural functors, and, since $H_{\mathcal{F}^R}$ centralizes R, it follows from Theorem 17.18 that, for any $x \in \mathcal{L}^{^R}(\hat{T},\hat{Q})$ fulfilling $\mathfrak{t}^R(x) = 1$, π_x acts trivially on R.

Proposition 20.10 *With the notation and the hypothesis above, for any element* $\hat{x} \in \mathcal{L}(\hat{Q}',\hat{Q})_{Q',Q} \cap \mathcal{L}(\hat{Q}',\hat{Q})_{R',R}$ *there is* $\hat{w} \in \mathcal{L}(\hat{T}',\hat{T})_{R',R}$ *such that*

$$\hat{w}^{-1}\cdot i_{\hat{Q}'}^{\hat{T}'}\cdot\hat{x} \in \mathcal{L}^{^R}(\hat{T},\hat{Q}) \quad \textit{and} \quad \mathfrak{t}^R(\hat{w}^{-1}\cdot i_{\hat{Q}'}^{\hat{T}'}\cdot\hat{x}) = 1 \qquad 20.10.1.$$

Moreover, the correspondence sending $\mathfrak{g}_{Q',Q}^{\hat{Q}',\hat{Q}}(\hat{x})$ *to* $\mathfrak{g}_{R',R}^{\hat{R}',\hat{R}}\big(\mathfrak{g}_{\hat{R}',\hat{R}}^{\hat{T}',\hat{T}}(\hat{w})\big)$ *determines a unique map*

$$\mathfrak{g}_{R',R}^{Q',Q} : \mathcal{L}(Q',Q)_{R',R} \longrightarrow \mathcal{L}(R',R) \qquad 20.10.2.$$

Proof: By the very definition of $\mathcal{L}(Q',Q)$ (cf. 20.6.1), it is clear that

$$\mathfrak{g}_{Q',Q}^{\hat{Q}',\hat{Q}}\big(\mathcal{L}(\hat{Q}',\hat{Q})_{Q',Q} \cap \mathcal{L}(\hat{Q}',\hat{Q})_{R',R}\big) = \mathcal{L}(Q',Q)_{R',R} \qquad 20.10.3;$$

moreover, denoting by $\psi \in \mathcal{F}(R',R)$ the element fulfilling $\pi_{\hat{x}} \circ \iota_R^{\hat{Q}} = \iota_{R'}^{\hat{Q}'} \circ \psi$, we clearly have ${}^{\psi}\big(\mathcal{F}_Q(R)\big) = \mathcal{F}_{Q'}(R')$; then, since R is normal in $\hat{T}$ and we have $\mathcal{F}_{\hat{T}}(R) = \mathcal{F}_Q(R)$, it follows from statement 2.10.1 that ψ can be extended to an $\mathcal{F}$-morphism $\omega : \hat{T} \to P$, and, since $\hat{T}'$ coincides with the converse image of $\mathcal{F}_{Q'}(R')$ in $N_P(R')$, we clearly have $\omega(\hat{T}) = \hat{T}'$; in particular, there is $\hat{w} \in \mathcal{L}(\hat{T}',\hat{T})_{R',R}$ such that $\hat{w}^{-1}{\cdot}i_{\hat{Q}'}^{\hat{T}'}{\cdot}\hat{x}$ belongs to $\mathcal{L}^{^R}(\hat{T},\hat{Q})$, and actually we can modify our choice of $\hat{w}$ in such a way that

$$\mathfrak{t}^R(\hat{w}^{-1}{\cdot}i_{\hat{Q}'}^{\hat{T}'}{\cdot}\hat{x}) = 1 \qquad 20.10.4.$$

Moreover, we claim that $\mathfrak{g}_{R',R}^{\hat{R}',\hat{R}}(\mathfrak{g}_{\hat{R}',\hat{R}}^{\hat{T}',\hat{T}}(\hat{w}))$ only depends on $x = \mathfrak{g}_{Q',Q}^{\hat{Q}',\hat{Q}}(\hat{x})$, defining the announced map. Indeed, if $\hat{x}' \in \mathcal{L}(\hat{Q}',\hat{Q})_{Q',Q}$ is another lifting of x then, by definition 20.6.1 we have $\hat{x}' = \hat{x}{\cdot}z{\cdot}\tau_{\hat{Q}}(u)$ for suitable elements

$$z \in \mathbf{O}^p\Big(C_{\mathcal{L}(\hat{Q})}\big(\tau_{\hat{Q}}(Q)\big)\Big) \quad \text{and} \quad u \in H_{C_{\mathcal{F}}(Q)} \qquad 20.10.5;$$

in particular, note that z and $\tau_{\hat{Q}}(u)$ belong to $\mathcal{L}^{^R}(\hat{Q}) = N_{\mathcal{L}(\hat{Q})}^{\mathcal{F}_Q(R)}\big(\tau_{\hat{Q}}(R)\big)$; but, the inclusion $R \subset Q$ forces

$$\mathbf{O}^p\Big(C_{\mathcal{L}(\hat{Q})}\big(\tau_{\hat{Q}}(Q)\big)\Big) \subset \mathbf{O}^p\Big(C_{\mathcal{L}(\hat{Q})}\big(\tau_{\hat{Q}}(R)\big)\Big) \qquad 20.10.6;$$

hence, we have $\mathfrak{t}^R(z) = 1$. Similarly, the inclusions

$$C_P(Q) \subset C_P(R) \quad \text{and} \quad C_{\mathcal{F}}(Q) \subset C_{\mathcal{F}}(R) \qquad 20.10.7$$

force $H_{C_{\mathcal{F}}(R)} \subset H_{C_{\mathcal{F}}(R)}$ (cf. Proposition 13.9) and therefore we still have $\mathfrak{t}^R\big(\tau_{\hat{Q}}(u)\big) = 1$. Consequently, for $\hat{x}'$ we can make the same choice $\hat{w}$.

On the other hand, if $\hat{w}' \in \mathcal{L}(\hat{T}',\hat{T})_{R',R}$ is another choice such that

$$\hat{w}'^{-1}{\cdot}i_{\hat{Q}'}^{\hat{T}'}{\cdot}\hat{x} \in \mathcal{L}^{^R}(\hat{T},\hat{Q}) \quad \text{and} \quad \mathfrak{t}^R(\hat{w}'^{-1}{\cdot}i_{\hat{Q}'}^{\hat{T}'}{\cdot}\hat{x}) = 1 \qquad 20.10.8,$$

then the difference $\hat{w}^{-1}{\cdot}\hat{w}'$ belongs to $\mathcal{L}^{^R}(\hat{T})$ and we have $\mathfrak{t}^R(\hat{w}^{-1}{\cdot}\hat{w}') = 1$; but, according to Theorem 17.18 and definition 20.8.2, it is easily checked that the kernel of the homomorphism $\mathcal{L}^{^R}(\hat{T}) \to \hat{T}/H_{\mathcal{F}^R}$ determined by the functor $\mathfrak{t}^R$ is equal to

$$\mathbf{O}^p\big(\mathcal{L}^{^R}(\hat{T})\big){\cdot}(\tau^R)_{\hat{T}}(H_{\mathcal{F}^R}) = \mathbf{O}^p\Big(C_{\mathcal{L}(\hat{T})}\big(\tau_{\hat{T}}(R)\big)\Big){\cdot}\tau_{\hat{T}}(H_{C_{\mathcal{F}}(R)}) \qquad 20.10.9$$

where $\tau^{^R} : \mathcal{T}_{\hat{T}}^{^R} \to \mathcal{L}^{^R}$ denotes the structural functor; consequently, since

$$\begin{gathered}\mathfrak{g}_{\hat{R},\hat{R}}^{\hat{T},\hat{T}}\bigg(\mathbf{O}^p\Big(C_{\mathcal{L}(\hat{T})}\big(\tau_{\hat{T}}(R)\big)\Big)\bigg) \subset \mathbf{O}^p\Big(C_{\mathcal{L}(\hat{R})}\big(\tau_{\hat{R}}(R)\big)\Big) \\ \mathfrak{g}_{\hat{R},\hat{R}}^{\hat{T},\hat{T}}\big(\tau_{\hat{T}}(H_{C_{\mathcal{F}}(R)})\big) = \tau_{\hat{R}}(H_{C_{\mathcal{F}}(R)})\end{gathered} \qquad 20.10.10,$$

the element $\mathfrak{g}_{\hat{R},\hat{R}}^{\hat{T},\hat{T}}(\hat{w}^{-1}{\cdot}\hat{w}')$ belongs to $O(R)$. We are done.

Lemma 20.11 *With the notation and the hypothesis above, we have*

$$\begin{aligned}\pi_{R',R}\big(\mathfrak{g}^{Q',Q}_{R',R}(x)\big) &= \mathfrak{f}^{Q',Q}_{R',R}\big(\pi_{Q',Q}(x)\big)\\ \mathfrak{g}^{Q',Q}_{R',R}\big(\tau_{Q',Q}(u)\big) &= \tau_{R',R}(u)\end{aligned} \qquad 20.11.1.$$

for any $x \in \mathcal{L}(Q',Q)_{R',R}$ *and any* $u \in \mathcal{T}_P(Q',Q)\cap\mathcal{T}_P(R',R)$.

Proof: With the notation and the hypothesis above, let $\hat{x}$ be an element of $\mathcal{L}(\hat{Q}',\hat{Q})_{Q',Q}$ fulfilling $x = \mathfrak{g}^{\hat{Q}',\hat{Q}}_{Q',Q}(\hat{x})$ (cf. 20.5.2), and $\hat{w}$ an element of $\mathcal{L}(\hat{T}',\hat{T})_{R',R}$ such that $\hat{w}^{-1}\cdot i^{\hat{T}'}_{\hat{Q}'}\cdot\hat{x}$ belongs to $\mathcal{L}^{R}(\hat{T},\hat{Q})$ and fulfills the equality $\mathfrak{t}^R(\hat{w}^{-1}\cdot i^{\hat{T}'}_{\hat{Q}'}\cdot\hat{x}) = 1$; in particular, note that $\pi_{\hat{T},\hat{Q}}(\hat{w}^{-1}\cdot i^{\hat{T}'}_{\hat{Q}'}\cdot\hat{x})$ acts trivially on R (cf. definition 20.8.2). Then, according to the definition of $\mathfrak{g}^{Q',Q}_{R',R}$ above, we have

$$\mathfrak{g}^{Q',Q}_{R',R}(x) = \mathfrak{g}^{\hat{R}',\hat{R}}_{R',R}\big(\mathfrak{g}^{\hat{T}',\hat{T}}_{\hat{R}',\hat{R}}(\hat{w})\big) \qquad 20.11.2$$

and, since $\pi_{\hat{T},\hat{Q}}(\hat{w}^{-1}\cdot i^{\hat{T}'}_{\hat{Q}'}\cdot\hat{x})$ acts trivially on R, it follows from equalities 20.7.2 and 20.10.3 that we still have

$$\begin{aligned}\pi_{R',R}\big(\mathfrak{g}^{Q',Q}_{R',R}(x)\big) &= \pi_{R',R}\Big(\mathfrak{g}^{\hat{R}',\hat{R}}_{R',R}\big(\mathfrak{g}^{\hat{T}',\hat{T}}_{\hat{R}',\hat{R}}(\hat{w})\big)\Big)\\ &= \mathfrak{f}^{\hat{R}',\hat{R}}_{R',R}\Big(\pi_{\hat{R}',\hat{R}}\big(\mathfrak{g}^{\hat{T}',\hat{T}}_{\hat{R}',\hat{R}}(\hat{w})\big)\Big) = \mathfrak{f}^{\hat{T}',\hat{T}}_{R',R}\big(\pi_{\hat{T}',\hat{T}}(\hat{w})\big)\\ &= \mathfrak{f}^{\hat{Q}',\hat{Q}}_{R',R}\big(\pi_{\hat{Q}',\hat{Q}}(\hat{x})\big) = \mathfrak{f}^{Q',Q}_{R',R}\big(\pi_{Q',Q}(x)\big)\end{aligned} \qquad 20.11.3.$$

Similarly, by equalities 20.3.3 and 20.7.3, since u still belongs to $\mathcal{T}_P(\hat{T}',\hat{T})$ and we have $\tau_{Q',Q}(u) = \mathfrak{g}^{\hat{Q}',\hat{Q}}_{Q',Q}(\tau_{\hat{Q}',\hat{Q}}(u))$, by the proposition above we get

$$\mathfrak{g}^{Q',Q}_{R',R}\big(\tau_{Q',Q}(u)\big) = \mathfrak{g}^{\hat{R}',\hat{R}}_{R',R}\big(\mathfrak{g}^{\hat{T}',\hat{T}}_{\hat{R}',\hat{R}}(\tau_{\hat{T}',\hat{T}}(u))\big) = \tau_{R',R}(u) \qquad 20.11.4.$$

We are done.

20.12 Now, we claim that the family of maps $\mathfrak{g}^{Q',Q}_{R',R}$ obtained in Proposition 20.10, for the different choices of the subgroups of P fully centralized in $\mathcal{F}$, is compatible with the *composition maps* defined in 20.7.4 and fulfills the corresponding *transitivity condition*.

Lemma 20.13 *With the notation and hypothesis above, let* Q'' *be a subgroup of* P *which is* $\mathcal{F}$*-isomorphic to* Q *and* Q', *and* R'' *a normal subgroup of* Q''. *Assume that both are fully centralized in* $\mathcal{F}$. *Then, for any* $(x',x) \in \mathcal{L}(Q'',Q')_{R'',R'} \times \mathcal{L}(Q',Q)_{R',R}$, *we have*

$$\mathfrak{g}^{Q'',Q}_{R'',R}(x'\cdot x) = \mathfrak{g}^{Q'',Q'}_{R'',R'}(x')\cdot\mathfrak{g}^{Q',Q}_{R',R}(x) \qquad 20.13.1.$$

Proof: With the notation above, set $\hat{T}'' = Q''\cdot\hat{R}''$ and choose an element $\hat{x}'$ in $\mathcal{L}(\hat{Q}'',\hat{Q}')_{Q'',Q'}$ fulfilling $x' = \mathfrak{g}_{Q'',Q'}^{\hat{Q}'',\hat{Q}'}(\hat{x}')$, and an element $\hat{w}'$ in $\mathcal{L}(\hat{T}'',\hat{T}')_{R'',R'}$ such that

$$\hat{w}'^{-1}\cdot i_{\hat{Q}''}^{\hat{T}''}\cdot\hat{x}' \in \mathcal{L}^{R}(\hat{T}',\hat{Q}') \quad \text{and} \quad \mathfrak{t}^{R}(\hat{w}'^{-1}\cdot i_{\hat{Q}''}^{\hat{T}''}\cdot\hat{x}') = 1 \qquad 20.13.2;$$

then, the product $\hat{x}'' = \hat{x}'\cdot\hat{x}$ lifts $x'' = x'\cdot x$ and, setting $\hat{w}'' = \hat{w}'\cdot\hat{w}$, we claim that

$$\hat{w}''^{-1}\cdot i_{\hat{Q}''}^{\hat{T}''}\cdot\hat{x}'' \in \mathcal{L}^{R}(\hat{T},\hat{Q}) \quad \text{and} \quad \mathfrak{t}^{R}(\hat{w}''^{-1}\cdot i_{\hat{Q}''}^{\hat{T}''}\cdot\hat{x}'') = 1 \qquad 20.13.3.$$

We argue by induction on the *length* ℓ of $\pi_{\hat{T}',\hat{Q}'}(\hat{w}'^{-1}\cdot i_{\hat{Q}''}^{\hat{T}''}\cdot\hat{x}')$ as an $\mathcal{F}^{R}$*-morphism* (cf. 5.15 and 20.8.2); if $\ell = 0$ then we have

$$\hat{w}'^{-1}\cdot i_{\hat{Q}''}^{\hat{T}''}\cdot\hat{x}' = \hat{y}\cdot i_{\hat{Q}'}^{\hat{T}'} \qquad 20.13.4$$

for some $\hat{y} \in \mathcal{L}^{R'}(\hat{T}')$ fulfilling $\mathfrak{t}^{R'}(\hat{y}) = 1$ and therefore we easily get

$$\hat{w}''^{-1}\cdot i_{\hat{Q}''}^{\hat{T}''}\cdot\hat{x}'' = (\hat{w}^{-1}\cdot\hat{y}\cdot\hat{w})\cdot(\hat{w}^{-1}\cdot i_{\hat{Q}'}^{\hat{T}'}\cdot\hat{x}) \qquad 20.13.5;$$

since $\hat{w}^{-1}\cdot\hat{y}\cdot\hat{w} \in \mathcal{L}^{R}(\hat{T})$ and $\mathfrak{t}^{R}(\hat{w}^{-1}\cdot\hat{y}\cdot\hat{w}) = 1$, in this case we are done.

If $\ell \geq 1$ then we have

$$\hat{w}'^{-1}\cdot i_{\hat{Q}''}^{\hat{T}''}\cdot\hat{x}' = i_{\hat{U}'}^{\hat{T}'}\cdot\hat{y}\cdot\hat{v} \qquad 20.13.6$$

for some $\mathcal{F}^{R'}$-essential subgroup $\hat{U}'$ of $\hat{T}'$, some p'-element $\hat{y}$ of the converse image of $X_{\mathcal{F}^{R'}}(\hat{U}')$ in $\mathcal{L}^{R'}(\hat{U}')$ and some element $\hat{v} \in \mathcal{L}^{R'}(\hat{U}',\hat{Q}')$, in such a way that $\pi_{\hat{T}',\hat{Q}'}(i_{\hat{U}'}^{\hat{T}'}\cdot\hat{v})$ has length $\ell - 1$ (cf. 5.15); thus, by the induction hypothesis, we already know that

$$\hat{w}^{-1}\cdot(i_{\hat{U}'}^{\hat{T}'}\cdot\hat{v})\cdot\hat{x} \in \mathcal{L}^{R}(\hat{T},\hat{Q}) \quad \text{and} \quad \mathfrak{t}^{R}\big(\hat{w}^{-1}\cdot(i_{\hat{U}'}^{\hat{T}'}\cdot\hat{v})\cdot\hat{x}\big) = 1 \qquad 20.13.7$$

and therefore, setting $\hat{U} = (\pi_{\hat{w}})^{-1}(\hat{U}')$ and denoting by $\hat{u} \in \mathcal{L}(\hat{U}',\hat{U})$ the element fulfilling $\hat{w}^{-1}\cdot i_{\hat{U}'}^{\hat{T}'} = i_{\hat{U}}^{\hat{T}}\cdot\hat{u}^{-1}$, the divisibility in $\mathcal{L}^{R}$ implies that $\hat{u}^{-1}\cdot\hat{v}\cdot\hat{x}$ belongs to $\mathcal{L}^{R}(\hat{U},\hat{Q})$ and then we have $\mathfrak{t}^{R}\big(\hat{u}^{-1}\cdot\hat{v}\cdot\hat{x}\big) = 1$; consequently, we still get

$$\hat{w}''^{-1}\cdot i_{\hat{Q}''}^{\hat{T}''}\cdot\hat{x}'' = \hat{w}^{-1}\cdot(i_{\hat{U}'}^{\hat{T}'}\cdot\hat{y}\cdot\hat{v})\cdot\hat{x} = i_{\hat{U}}^{\hat{T}}\cdot(\hat{u}^{-1}\cdot\hat{y}\cdot\hat{u})\cdot(\hat{u}^{-1}\cdot\hat{v}\cdot\hat{x}) \qquad 20.13.8$$

and, since $\hat{u}^{-1}\cdot\hat{y}\cdot\hat{u}$ is a p'-element of $\mathcal{L}^{R}(\hat{U})$, we have $\mathfrak{t}^{R}(\hat{u}^{-1}\cdot\hat{y}\cdot\hat{u}) = 1$, which proves the claim.

Now, according to 20.4, to 20.7 and to Proposition 20.10, we have

$$\begin{aligned}\mathfrak{g}_{R'',R}^{Q'',Q}(x'') &= \mathfrak{g}_{R'',R}^{Q'',Q}\big(\mathfrak{g}_{Q'',Q}^{\hat{Q}'',\hat{Q}}(\hat{x}'')\big) = \mathfrak{g}_{R'',R}^{\hat{R}'',\hat{R}}\big(\mathfrak{g}_{\hat{R}'',\hat{R}}^{\hat{T}'',\hat{T}}(\hat{w}'')\big)\\ &= \mathfrak{g}_{R'',R}^{\hat{R}'',\hat{R}}\big(\mathfrak{g}_{\hat{R}'',\hat{R}'}^{\hat{T}'',\hat{T}'}(\hat{w}')\cdot\mathfrak{g}_{\hat{R}',\hat{R}}^{\hat{T}',\hat{T}}(\hat{w})\big)\\ &= \mathfrak{g}_{R'',R'}^{\hat{R}'',\hat{R}'}\big(\mathfrak{g}_{\hat{R}'',\hat{R}'}^{\hat{T}'',\hat{T}'}(\hat{w}')\big)\cdot\mathfrak{g}_{R',R}^{\hat{R}',\hat{R}}\big(\mathfrak{g}_{\hat{R}',\hat{R}}^{\hat{T}',\hat{T}}(\hat{w})\big)\\ &= \mathfrak{g}_{R'',R'}^{Q'',Q'}(x')\cdot\mathfrak{g}_{R',R}^{Q',Q}(x)\end{aligned} \qquad 20.13.9.$$

We are done.

Lemma 20.14 *With the notation and the hypothesis above, let U be a normal subgroup of Q which contains R and is fully centralized in $\mathcal{F}$. Then, for any $x \in \mathcal{L}(Q',Q)_{R',R}$, setting $U' = \pi_x(U)$ we have*

$$\mathfrak{g}_{R',R}^{U',U}\big(\mathfrak{g}_{U',U}^{Q',Q}(x)\big) = \mathfrak{g}_{R',R}^{Q',Q}(x) \qquad 20.14.1.$$

Proof: As above, $\hat{U} = U{\cdot}C_P(U)$ normalizes R and we set

$$\hat{W} = \hat{U}{\cdot}C_P(R) = U{\cdot}\hat{R} \qquad 20.14.2$$

which clearly is contained in $\hat{T}$; moreover, since $\hat{Q}$ normalizes U, it normalizes $\hat{U}$ and we analogously set

$$\hat{V} = Q{\cdot}\hat{U} = \hat{Q}{\cdot}C_P(U) \qquad 20.14.3;$$

mutatis mutandis, we also consider $\hat{W}' = U'{\cdot}\hat{R}'$ and $\hat{V}' = Q'{\cdot}\hat{U}'$. If $\hat{x}$ is an element of $\mathcal{L}(\hat{Q}',\hat{Q})_{Q',Q}$ lifting x then there is $\hat{v} \in \mathcal{L}(\hat{V}',\hat{V})_{U',U}$ such that

$$\hat{v}^{-1}{\cdot}i_{\hat{Q}'}^{\hat{V}'}{\cdot}\hat{x} \in \mathcal{L}^{U}(\hat{V},\hat{Q}) \quad \text{and} \quad \mathfrak{t}^{U}(\hat{v}^{-1}{\cdot}i_{\hat{Q}'}^{\hat{V}'}{\cdot}\hat{x}) = 1 \qquad 20.14.4$$

and moreover we have (cf. Proposition 20.10)

$$\mathfrak{g}_{U',U}^{Q',Q}(x) = \mathfrak{g}_{U',U}^{\hat{U}',\hat{U}}\big(\mathfrak{g}_{\hat{U}',\hat{U}}^{\hat{V}',\hat{V}}(\hat{v})\big) \qquad 20.14.5.$$

Similarly, since we have $\hat{T} = \hat{V}{\cdot}C_P(R)$ and $\hat{v}^{-1}{\cdot}i_{\hat{Q}'}^{\hat{V}'}{\cdot}\hat{x}$ centralizes $R \subset U$ (cf. definition 20.8.2), so that $\hat{v}$ also belongs to $\mathcal{L}(\hat{V}',\hat{V})_{R',R}$, there exists an element $\hat{t} \in \mathcal{L}(\hat{T}',\hat{T})_{R',R}$ such that

$$\hat{t}^{-1}{\cdot}i_{\hat{V}'}^{\hat{T}'}{\cdot}\hat{v} \in \mathcal{L}^{R}(\hat{T},\hat{V}) \quad \text{and} \quad \mathfrak{t}^{R}(\hat{t}^{-1}{\cdot}i_{\hat{V}'}^{\hat{T}'}{\cdot}\hat{v}) = 1 \qquad 20.14.6.$$

On the other hand, with the notation in 20.8 above applied to R and to U in Q, it is clear that the inclusion

$$\hat{V} = N_P^{\mathcal{F}_Q(U)}(U) \subset \hat{T} = N_P^{\mathcal{F}_Q(R)}(R) \qquad 20.14.7$$

is $(\mathcal{F}^{^U},\mathcal{F}^{^R})$-functorial; thus, we have an obvious faithful functor and an inclusion (cf. Proposition 13.9)

$$\mathcal{L}^{^U} \longrightarrow \mathcal{L}^{^R} \quad \text{and} \quad H_{\mathcal{F}^U} \subset H_{\mathcal{F}^R} \qquad 20.14.8.$$

Moreover, we claim that we have the commutative diagram

$$\begin{array}{ccc} \mathcal{L}^{^R} & \xrightarrow{\mathfrak{t}^R} & \mathcal{T}_{\hat{T}/H_{\mathcal{F}^R}} \\ \uparrow & & \uparrow \\ \mathcal{L}^{^U} & \xrightarrow{\mathfrak{t}^U} & \mathcal{T}_{\hat{V}/H_{\mathcal{F}^U}} \end{array} \qquad 20.14.9$$

Indeed, it is clear that $\hat{V}\cap H_{\mathcal{F}^R}$ is an $\mathcal{F}^{^U}$-stable subgroup of $\hat{V}$ (cf. 13.4) and then, it follows from the uniqueness of the functor in Theorem 17.18 that the composition of $\mathfrak{t}^U$ with the canonical functor

$$\mathcal{T}_{\hat{V}/H_{\mathcal{F}^U}} \longrightarrow \mathcal{T}_{\hat{V}/\hat{V}\cap H_{\mathcal{F}^R}} \qquad 20.14.10$$

coincides with the functor $\mathcal{L}^{^U} \to \mathcal{T}_{\hat{V}/\hat{V}\cap H_{\mathcal{F}^R}}$ induced by the composition of the functors $\mathcal{L}^{^U} \to \mathcal{L}^{^R}$ and $\mathfrak{t}^R$. In particular,

$$\hat{v}^{-1}\cdot i_{\hat{Q}'}^{\hat{V}'}\cdot\hat{x} \in \mathcal{L}^{^R}(\hat{V},\hat{Q}) \quad \text{and} \quad \mathfrak{t}^R(\hat{v}^{-1}\cdot i_{\hat{Q}'}^{\hat{V}'}\cdot\hat{x}) = 1 \qquad 20.14.11;$$

consequently, the composition $(\hat{t}^{-1}\cdot i_{\hat{V}'}^{\hat{T}'}\cdot\hat{v})\cdot(\hat{v}^{-1}\cdot i_{\hat{Q}'}^{\hat{V}'}\cdot\hat{x}) = \hat{t}^{-1}\cdot i_{\hat{Q}'}^{\hat{T}'}\cdot\hat{x}$ belongs to $\mathcal{L}^{^R}(\hat{T},\hat{Q})$ and fulfills $\mathfrak{t}^R(\hat{t}^{-1}\cdot i_{\hat{Q}'}^{\hat{T}'}\cdot\hat{x}) = 1$, and therefore we get

$$\mathfrak{g}_{R',R}^{Q',Q}(x) = \mathfrak{g}_{R',R}^{\hat{R}',\hat{R}}\big(\mathfrak{g}_{\hat{R}',\hat{R}}^{\hat{T}',\hat{T}}(\hat{t})\big) \qquad 20.14.12.$$

Finally, by equality 20.14.5, $\hat{u} = \mathfrak{g}_{\hat{U}',\hat{U}}^{\hat{V}',\hat{V}}(\hat{v})$ is a lifting of $\mathfrak{g}_{U',U}^{Q',Q}(x)$ in $\mathcal{L}(\hat{U}',\hat{U})_{U',U}$; thus, if we choose $\hat{w} \in \mathcal{L}(\hat{W}',\hat{W})_{R',R}$ such that $\hat{w}^{-1}\cdot i_{\hat{U}'}^{\hat{W}'}\cdot\hat{u}$ belongs to $\mathcal{L}^{^R}(\hat{W},\hat{U})$ and fulfills $\mathfrak{t}^R(\hat{w}^{-1}\cdot i_{\hat{U}'}^{\hat{W}'}\cdot\hat{u}) = 1$, we have

$$\mathfrak{g}_{R',R}^{U',U}\big(\mathfrak{g}_{U',U}^{Q',Q}(x)\big) = \mathfrak{g}_{R',R}^{\hat{R}',\hat{R}}\big(\mathfrak{g}_{\hat{R}',\hat{R}}^{\hat{W}',\hat{W}}(\hat{w})\big) \qquad 20.14.13.$$

Moreover, we still have (cf. 20.3.2)

$$\begin{aligned}(\hat{t}^{-1}\cdot i_{\hat{V}'}^{\hat{T}'}\cdot\hat{v})\cdot i_{\hat{U}}^{\hat{V}} &= (\hat{t}^{-1}\cdot i_{\hat{V}'}^{\hat{T}'})\cdot(\hat{v}\cdot i_{\hat{U}}^{\hat{V}}) = \hat{t}^{-1}\cdot i_{\hat{U}'}^{\hat{T}'}\cdot\hat{u} \\ &= (\hat{t}^{-1}\cdot i_{\hat{W}'}^{\hat{T}'}\cdot\hat{w})\cdot(\hat{w}^{-1}\cdot i_{\hat{U}'}^{\hat{W}'}\cdot\hat{u})\end{aligned} \qquad 20.14.14$$

and therefore, since $\hat t^{-1}{\cdot}i^{\hat T'}_{\hat W'}{\cdot}\hat w$ belongs to $\mathcal{L}^{^R}(\hat T,\hat W)$ and we already know that

$$\mathfrak{t}^R(\hat w^{-1}{\cdot}i^{\hat W'}_{\hat U'}{\cdot}\hat u) = 1 = \mathfrak{t}^R(\hat t^{-1}{\cdot}i^{\hat T'}_{\hat V'}{\cdot}\hat v) \qquad 20.14.15,$$

we get $\mathfrak{t}^R(\hat t^{-1}{\cdot}i^{\hat T'}_{\hat W'}{\cdot}\hat w) = 1$.

Now, we claim that the element

$$y = \mathfrak{g}^{\hat T',\hat T}_{\hat R',\hat R}(\hat t)^{-1}{\cdot}\mathfrak{g}^{\hat W',\hat W}_{\hat R',\hat R}(\hat w) \qquad 20.14.16$$

belongs to $O(R)$; indeed, since we have (cf. 20.3.2)

$$(\hat t^{-1}{\cdot}i^{\hat T'}_{\hat W'}{\cdot}\hat w){\cdot}i^{\hat W}_{\hat R} = \hat t^{-1}{\cdot}i^{\hat T'}_{\hat R'}{\cdot}\mathfrak{g}^{\hat W',\hat W}_{\hat R',\hat R}(\hat w) = i^{\hat T}_{\hat R}{\cdot}y \qquad 20.14.17,$$

y belongs to $\mathcal{L}^{^R}(\hat R) = N^{\mathcal{F}_Q(R)}_{\mathcal{L}(\hat R)}(R)$ and we still have $\mathfrak{t}^R(y) = 1$; but, it is easily checked that

$$\mathbf{O}^p\big(O(R)\big) = \mathbf{O}^p\big(\mathcal{L}^{^R}(\hat R)\big) \quad \text{and} \quad O(R)\cap \tau_{\hat R}(\hat T) = \tau_{\hat R}(H_{\mathcal{F}^R}) \qquad 20.14.18$$

which proves the claim. Consequently, we obtain

$$\mathfrak{g}^{Q',Q}_{R',R}(x) = \mathfrak{g}^{\hat R',\hat R}_{R',R}\big(\mathfrak{g}^{\hat T',\hat T}_{\hat R',\hat R}(\hat t)\big) = \mathfrak{g}^{\hat R',\hat R}_{R',R}\big(\mathfrak{g}^{\hat W',\hat W}_{\hat R',\hat R}(\hat w)\big) = \mathfrak{g}^{U',U}_{R',R}\big(\mathfrak{g}^{Q',Q}_{U',U}(x)\big) \quad 20.14.19.$$

We are done.

20.15 We are ready to define the set $\mathcal{L}(Q',Q)$ for any pair of $\mathcal{F}$-isomorphic subgroups Q and Q' of P; we proceed by induction on $|P\,{:}\,Q|$ and, obviously, our definition will extend the definitions in the previous situations; moreover, since $\mathcal{L}(P)$ is already defined, we may assume that $Q \neq P$. Then, we have $N = N_P(Q) \neq Q$ and it follows from Corollary 2.21 that there is an $\mathcal{F}$-morphism $\nu : N \to P$ such that $\nu(N)$ and $\nu(Q)$ are both fully centralized in $\mathcal{F}$; in this situation, according to our induction hypothesis, we may assume that the set $\mathcal{L}\big(\nu(N),N\big)$ and the canonical map

$$\pi_{\nu(N),N} : \mathcal{L}\big(\nu(N),N\big) \longrightarrow \mathcal{F}\big(\nu(N),N\big) \qquad 20.15.1$$

are already defined and may choose $n \in \mathcal{L}\big(\nu(N),N\big)$ lifting ν. That is to say, we may assume that

20.15.2 *There is a pair (N,n) formed by a subgroup N of P which strictly contains and normalizes Q, and by $n \in \mathcal{L}\big(\nu(N),N\big)$ lifting ν for a suitable $\mathcal{F}$-morphism $\nu : N \to P$ such that $\nu(N)$ and $\nu(Q)$ are both fully centralized in $\mathcal{F}$.*

20.16 We denote by $\mathfrak{N}(Q)$ the set of such pairs and often we write n instead of (N,n), setting ${}^nN = \nu(N)$ and $\pi_n = \pi_{_{\nu(N),N}}(n)$. For another pair $(\bar N,\bar n)\in\mathfrak{N}(Q)$, note that the element $\bar n{\cdot}n^{-1}$ belongs to $\mathcal{L}({}^{\bar n}\bar N,{}^nN)$, it fulfills $\pi_{\bar n\cdot n^{-1}}({}^nQ) = {}^{\bar n}Q$ and, since ${}^nQ$ and ${}^{\bar n}Q$ are fully centralized in $\mathcal{F}$, we get the element

$$g_{\bar n,n} = \mathfrak{g}^{{}^{\bar n}\bar N,{}^nN}_{{}^{\bar n}Q,{}^nQ}(\bar n{\cdot}n^{-1})\in\mathcal{L}({}^{\bar n}Q,{}^nQ) \qquad 20.16.1.$$

Then, for any pair $(\bar{\bar N},\bar{\bar n})\in\mathfrak{N}(Q)$, it follows from Lemma 20.13 that

$$g_{\bar{\bar n},\bar n}{\cdot}g_{\bar n,n} = g_{\bar{\bar n},n} \qquad 20.16.2.$$

On the other hand, if Q is fully centralized in $\bar{\mathcal{F}}$ then $N = N_P(Q)$ is $\mathcal{F}$-selfcentralizing, so that it is fully centralized too, and therefore the pair (N, i^{N}_{N}) belongs to $\mathfrak{N}(Q)$.

20.17 Now, for any pair of mutually $\mathcal{F}$-isomorphic subgroups Q and Q' of P, we denote by $\mathcal{L}(Q',Q)$ the subset of

$$\hat{\mathcal{L}}(Q',Q) = \prod_{n\in\mathfrak{N}(Q)}\ \prod_{n'\in\mathfrak{N}(Q')}\mathcal{L}({}^{n'}Q',{}^nQ) \qquad 20.17.1$$

formed by the families $\{x_{n',n}\}_{n\in\mathfrak{N}(Q),n'\in\mathfrak{N}(Q')}$ fulfilling

$$g_{\bar n',n'}{\cdot}x_{n',n} = x_{\bar n',\bar n}{\cdot}g_{\bar n,n} \qquad 20.17.2.$$

In other words, the set $\mathcal{L}(Q',Q)$ is the *inverse limit* of the family formed by the sets $\mathcal{L}\big({}^{n'}Q',{}^nQ\big)$ and by the bijections between them induced by $g_{\bar n,n}$ and $g_{\bar n',n'}$.

20.18 Note that, according to equalities 20.16.2, the *projection map* onto the factor labelled by the pair $\big((N,n),(N',n')\big)$ induces a bijection

$$\mathfrak{n}_{n',n} : \mathcal{L}(Q',Q)\cong\mathcal{L}\big({}^{n'}Q',{}^nQ\big) \qquad 20.18.1;$$

moreover, we have a map

$$\pi_{_{Q',Q}} : \mathcal{L}(Q',Q)\longrightarrow\mathcal{F}(Q',Q) \qquad 20.18.2$$

sending $x\in\mathcal{L}(Q',Q)$ to (cf. 20.5)

$$\pi_{_{Q',Q}}(x) = \mathfrak{f}^{{}^{n'}N',N'}_{{}^{n'}Q',Q'}(\pi_{n'})^{-1}\circ\pi_{_{{}^{n'}Q',{}^nQ}}\big(\mathfrak{n}_{n',n}(x)\big)\circ\mathfrak{f}^{{}^nN,N}_{{}^nQ,Q}(\pi_n) \qquad 20.18.3;$$

then, from equality 20.17.2, it is not difficult to prove that this map does not depend on the choice of (N,n) and (N',n'). In particular, if Q and Q' are already fully centralized in $\mathcal{F}$, setting $N = N_P(Q)$ and $N' = N_P(Q')$, $\mathcal{L}(Q',Q)$ can be identified with the factor labelled by the pair $\big((N',i^{N'}_{N'}),(N,i^{N}_{N})\big)$ throughout $\mathfrak{n}_{i^{N'}_{N'},i^{N}_{N}}$.

20.19 On the other hand, if Q'' is a subgroup of P, $\mathcal{F}$-isomorphic to Q and Q', the *composition map* in 20.7.4 determines a new *composition map*

$$\mathcal{L}(Q'',Q')\times\mathcal{L}(Q',Q)\longrightarrow\mathcal{L}(Q'',Q) \qquad 20.19.1$$

sending $(x',x)\in\mathcal{L}(Q'',Q')\times\mathcal{L}(Q',Q)$ to

$$x'\cdot x=(\mathfrak{n}_{n'',n})^{-1}\big(\mathfrak{n}_{n'',n'}(x')\cdot\mathfrak{n}_{n',n}(x)\big) \qquad 20.19.2$$

for a choice of $(N,n)\in\mathfrak{N}(Q)$, $(N',n')\in\mathfrak{N}(Q')$ and $(N'',n'')\in\mathfrak{N}(Q'')$; indeed, for another choice of pairs $(\bar N,\bar n)\in\mathfrak{N}(Q)$, $(\bar N',\bar n')\in\mathfrak{N}(Q')$ and $(\bar N'',\bar n'')\in\mathfrak{N}(Q'')$, we get (cf. 20.16.2)

$$\begin{aligned}g_{n'',\bar n''}\cdot\big(\mathfrak{n}_{\bar n'',\bar n'}(x')\cdot\mathfrak{n}_{\bar n',\bar n}(x)\big)&=\mathfrak{n}_{n'',n'}(x')\cdot g_{n',\bar n}\cdot\mathfrak{n}_{\bar n',\bar n}(x)\\&=\mathfrak{n}_{n'',n'}(x')\cdot\mathfrak{n}_{n',n}(x)\cdot g_{n,\bar n}=\mathfrak{n}_{n'',n}(x'\cdot x)\cdot g_{n,\bar n}\end{aligned} \qquad 20.19.3.$$

If Q''' is a subgroup of P, $\mathcal{F}$-isomorphic to Q, Q' and Q'', and x'' an element of $\mathcal{L}(Q''',Q'')$, it is quite clear that

$$(x''\cdot x')\cdot x=x''\cdot(x'\cdot x) \qquad 20.19.4.$$

20.20 From now on, for any pair of subgroups R of Q and R' of Q' such that $\mathcal{F}(Q',Q)_{R',R}$ is not empty we will construct the *restriction map*

$$\mathfrak{g}^{Q',Q}_{R',R}:\mathcal{L}(Q',Q)_{R',R}\longrightarrow\mathcal{L}(R',R) \qquad 20.20.1$$

announced in 20.2. Firstly, we assume that R is normal in Q and may assume that $R\neq Q$; then, it follows from Corollary 2.21 that there exists an $\mathcal{F}$-morphism $\nu:Q\to P$ such that the subgroups $\nu(Q)$ and $\nu(R)$ are fully centralized in $\mathcal{F}$ and we may choose $n\in\mathcal{L}\big(\nu(Q),Q\big)$ lifting ν (cf. 20.18.2); that is to say, we have the pair (Q,n) in $\mathfrak{N}(R)$ (cf. 20.16) and, since ${}^nR$ is fully centralized in $\mathcal{F}$, we can consider the element

$$n_R=\big(\mathfrak{n}_{i^{{}^nQ}_{{}^nQ},n}\big)^{-1}\big(i^{{}^nR}_{{}^nR}\big)\in\mathcal{L}({}^nR,R) \qquad 20.20.2.$$

Note that, up to our identification, for another pair $(Q,\bar n)\in\mathfrak{N}(R)$ we have (cf. Proposition 20.10 and 20.16.1)

$$g_{n,\bar n}=\mathfrak{g}^{{}^nQ,{}^{\bar n}Q}_{{}^nR,{}^{\bar n}R}(n\cdot\bar n^{-1})=n_R\cdot(\bar n_R)^{-1} \qquad 20.20.3;$$

indeed, according to equality 20.17.2 and definition 20.19.2, we have

$$\begin{aligned}\mathfrak{n}_{i^{{}^nQ}_{{}^nQ},i^{{}^{\bar n}Q}_{{}^{\bar n}Q}}\big(n_R\cdot(\bar n_R)^{-1}\big)&=\mathfrak{n}_{i^{{}^nQ}_{{}^nQ},n}(n_R)\cdot\mathfrak{n}_{n,i^{{}^{\bar n}Q}_{{}^{\bar n}Q}}\big((\bar n_R)^{-1}\big)\\&=\mathfrak{n}_{n,i^{{}^{\bar n}Q}_{{}^{\bar n}Q}}\big((\bar n_R)^{-1}\big)=g_{n,\bar n}\cdot\mathfrak{n}_{\bar n,i^{{}^{\bar n}Q}_{{}^{\bar n}Q}}\big((\bar n_R)^{-1}\big)=g_{n,\bar n}\end{aligned} \qquad 20.20.4.$$

We are ready to define the map $\mathfrak{g}_{R',R}^{Q',Q}$ in the normal case.

Lemma 20.21 *With the notation and the hypothesis above, the restriction map $\mathfrak{f}_{R',R}^{Q',Q}$ admits a lifting*

$$\mathfrak{g}_{R',R}^{Q',Q} : \mathcal{L}(Q',Q)_{R',R} \longrightarrow \mathcal{L}(R',R) \qquad 20.21.1$$

sending $x \in \mathcal{L}(Q',Q)_{R',R}$ to the element

$$\mathfrak{g}_{R',R}^{Q',Q}(x) = (n'_{R'})^{-1}\cdot\mathfrak{g}_{{}^{n'}R',{}^{n}R}^{{}^{n'}Q',{}^{n}Q}(n'\cdot x\cdot n^{-1})\cdot n_R \qquad 20.21.2$$

for any choice of pairs (Q',n') in $\mathfrak{N}(R')$ and (Q,n) in $\mathfrak{N}(R)$. Moreover, for any subgroup Q'' of P $\mathcal{F}$-isomorphic to Q and Q', any normal subgroup R'' of Q'' such that $\mathcal{F}(Q'',Q')_{R'',R'}$ is not empty and any $x' \in \mathcal{L}(Q'',Q')_{R'',R'}$ we have

$$\mathfrak{g}_{R'',R}^{Q'',Q}(x'\cdot x) = \mathfrak{g}_{R'',R'}^{Q'',Q'}(x')\cdot\mathfrak{g}_{R',R}^{Q',Q}(x) \qquad 20.21.3.$$

Proof: We firstly prove that our definition does not depend on the choices of (Q,n) and (Q',n'); indeed, if $(Q,\bar n) \in \mathfrak{N}(R)$, we have (cf. Lemma 20.13 and equality 20.20.3)

$$\begin{aligned}
&(n'_{R'})^{-1}\cdot\mathfrak{g}_{{}^{n'}R',{}^{\bar n}R}^{{}^{n'}Q',{}^{\bar n}Q}(n'\cdot x\cdot\bar n^{-1})\cdot\bar n_R\\
&\quad= (n'_{R'})^{-1}\cdot\mathfrak{g}_{{}^{n'}R',{}^{\bar n}R}^{{}^{n'}Q',{}^{\bar n}Q}\big((u'\cdot x\cdot n^{-1})\cdot(n\cdot\bar n^{-1})\big)\cdot\bar n_R\\
&\quad= (n'_{R'})^{-1}\cdot\mathfrak{g}_{{}^{n'}R',{}^{n}R}^{{}^{n'}Q',{}^{u}Q}(n'\cdot x\cdot n^{-1})\cdot g_{n\cdot\bar n}\cdot\bar u_R\\
&\quad= (n'_{R'})^{-1}\cdot\mathfrak{g}_{{}^{n'}R',{}^{n}R}^{{}^{n'}Q',{}^{n}Q}(n'\cdot x\cdot n^{-1})\cdot n_R
\end{aligned} \qquad 20.21.4.$$

Similarly, if $(Q',\bar n') \in \mathfrak{N}(R')$, we get the “symmetric” equalities.

Now, we prove equality 20.21.3; according to 20.21.2 and Lemma 20.13, choosing $(Q,n) \in \mathfrak{N}(R)$, $(Q',n') \in \mathfrak{N}(R')$ and $(Q'',n'') \in \mathfrak{N}(R'')$, we get

$$\begin{aligned}
&\mathfrak{g}_{R'',R'}^{Q'',Q'}(x')\cdot\mathfrak{g}_{R',R}^{Q',Q}(x)\\
&= (n''_{R'})^{-1}\cdot\mathfrak{g}_{{}^{n''}R'',{}^{n'}R'}^{{}^{n''}Q'',{}^{n'}Q'}(n''\cdot x'\cdot n'^{-1})\cdot\mathfrak{g}_{{}^{n'}R',{}^{n}R}^{{}^{n'}Q',{}^{n}Q}(n'\cdot x\cdot n^{-1})\cdot n_R\\
&= (n''_{R''})^{-1}\cdot\mathfrak{g}_{{}^{n''}R'',{}^{n'}R'}^{{}^{n''}Q'',{}^{n'}Q'}(n''\cdot x'\cdot x\cdot n^{-1})\cdot n_R = \mathfrak{g}_{R'',R}^{Q'',Q}(x'\cdot x)\,.
\end{aligned} \qquad 20.21.5.$$

We are done.

20.22 Moreover, for any mutually $\mathcal{F}$-isomorphic subgroups Q and Q' of P, the elements considered in 20.20.2 allow us to define a map

$$\tau_{_{Q',Q}} : \mathcal{T}_P(Q',Q) \longrightarrow \mathcal{L}(Q',Q) \tag{20.22.1}$$

extending the corresponding definition in 20.7 when Q and Q' are fully centralized in $\mathcal{F}$. For any $v \in \mathcal{T}_P(Q',Q)$ and any pair (N,n) in $\mathfrak{N}(Q)$, setting $N' = {}^vN$, $\hat{N} = {}^nN$ and $\hat{N}' = {}^v\hat{N}$, it is obvious that v still belongs to the intersection

$$\mathcal{T}_P(N',N) \cap \mathcal{T}_P(\hat{N}',\hat{N}) \tag{20.22.2}$$

and, arguing by induction on $|P\!:\!Q|$, we may assume that the maps $\tau_{_{N',N}}$ and $\tau_{_{\hat{N}',\hat{N}}}$ are already defined; then, setting $n' = \tau_{_{\hat{N}',\hat{N}}}(v){\cdot}n{\cdot}\tau_{_{N',N}}(v)^{-1}$ it is quite clear that (N',n') belongs to $\mathfrak{N}(Q')$ and that v still belongs to $\mathcal{T}_P({}^{n'}\!Q',{}^nQ)$; thus, since $\tau_{_{n'Q',nQ}}(v)$ is already defined (cf. definition 20.7.3), we set

$$\tau_{_{Q',Q}}(v) = (n_{_{Q'}})^{-1}{\cdot}\tau_{_{n'Q',nQ}}(v){\cdot}n_{_Q} \tag{20.22.3}$$

and it is easily checked from Lemma 20.11 and equality 20.20.3 that $\tau_{_{Q',Q}}(v)$ does not depend on our choice of (N,n). Note that, always arguing by induction on $|P\!:\!Q|$, these maps are clearly compatible with the *composition maps.* We are ready to define $\mathfrak{g}^{Q',Q}_{R',R}$ in the general case.

Proposition 20.23 *For any pair of mutually $\mathcal{F}$-isomorphic subgroups Q and Q' of P, any subgroup R of Q and any subgroup R' of Q' such that $\mathcal{F}(Q',Q)_{R',R} \neq \emptyset$, there exists a unique map*

$$\mathfrak{g}^{Q',Q}_{R',R} : \mathcal{L}(Q',Q)_{R',R} \longrightarrow \mathcal{L}(R',R) \tag{20.23.1}$$

lifting $\mathfrak{f}^{Q',Q}_{R',R}$ and fulfilling the two conditions

20.23.2 *For any subgroup T of R, any subgroup T' of R' such that the set $\mathcal{F}(R',R)_{T',T}$ is not empty, and any $x \in \mathcal{L}(Q',Q)_{R',R} \cap \mathcal{L}(Q',Q)_{T',T}$, we have*

$$\mathfrak{g}^{R',R}_{T',T}\big(\mathfrak{g}^{Q',Q}_{R',R}(x)\big) = \mathfrak{g}^{Q',Q}_{T',T}(x)\,.$$

20.23.3 *If R and R' are respectively normal in Q and Q' then, for any $x \in \mathcal{L}(Q',Q)_{R',R}$ and any pairs $(Q,n) \in \mathfrak{N}(R)$ and $(Q',n') \in \mathfrak{N}(R')$, we have*

$$\mathfrak{g}^{Q',Q}_{R',R}(x) = (n'_{_{R'}})^{-1}{\cdot}\mathfrak{g}^{n'Q',nQ}_{n'R',nR}\big(n'{\cdot}x{\cdot}n^{-1}\big){\cdot}n_{_R}\,.$$

Moreover, the following two statements hold:

20.23.4 *For any subgroup Q'' of P which is $\mathcal{F}$-isomorphic to Q and Q', any subgroup R'' of Q'' such that $\mathcal{F}(Q'',Q')_{R'',R'} \neq \emptyset$, any $x' \in \mathcal{L}(Q'',Q')_{R'',R'}$ and any $x \in \mathcal{L}(Q',Q)_{R',R}$, we have*

$$\mathfrak{g}_{R'',R}^{Q'',Q}(x'\cdot x) = \mathfrak{g}_{R'',R'}^{Q'',Q'}(x')\cdot\mathfrak{g}_{R',R}^{Q',Q}(x)\,.$$

20.23.5 *For any $x \in \mathcal{L}(Q',Q)_{R',R}$ and any $v \in \mathcal{T}_P(Q',Q) \cap \mathcal{T}_P(R',R)$, we have*

$$\pi_{R',R}\big(\mathfrak{g}_{R',R}^{Q',Q}(x)\big) = \mathfrak{f}_{R',R}^{Q',Q}\big(\pi_{Q',Q}(x)\big)$$
$$\mathfrak{g}_{R',R}^{Q',Q}\big(\tau_{Q',Q}(v)\big) = \tau_{R',R}(v)\,.$$

Proof: We firstly consider all the pairs Q and R where R is a normal proper subgroup of Q; then R' is normal in Q' and we have all the liftings defined above, which fulfill conditions 20.23.4 and 20.23.3; we claim that they also fulfill conditions 20.23.2 and 20.23.5. Indeed, in this case, it follows from Lemma 20.11 that

$$\begin{aligned}\pi_{R',R}\big(\mathfrak{g}_{R',R}^{Q',Q}(x)\big) &= \pi_{R',R}\big((n'_{R'})^{-1}\cdot\mathfrak{g}_{{}^{n'}R',{}^{n}R}^{{}^{n'}Q',{}^{n}Q}(n'\cdot x\cdot n)\cdot n_R\big)\\ &= \pi_{R',{}^{n'}R'}(n'_{R'})^{-1}\circ\mathfrak{f}_{{}^{n'}R',{}^{n}R}^{{}^{n'}Q',{}^{n}Q}\big(\pi_{{}^{n'}Q',{}^{n}Q}(n'\cdot x\cdot n)\big)\circ\pi_{{}^{n}R,R}(n_R)\\ &= (\delta_{n'})^{-1}\circ\mathfrak{f}_{R',R}^{Q',Q}\big(\pi_{Q',Q}(x)\big)\circ\delta_n\end{aligned} \qquad 20.23.6$$

where we set

$$\begin{aligned}\delta_n &= \mathfrak{f}_{{}^{n}R,R}^{{}^{n}Q,Q}\big(\pi_{{}^{n}Q,Q}(n)\big)^{-1}\cdot\pi_{{}^{n}R,R}(n_R)\\ \delta_{n'} &= \mathfrak{f}_{{}^{n'}R',R'}^{{}^{n'}Q',Q'}\big(\pi_{{}^{n'}Q',Q'}(n')\big)^{-1}\cdot\pi_{{}^{n'}R',R'}(n_{R'})\end{aligned} \qquad 20.23.7;$$

but, according to equality 20.18.3, we get

$$\begin{aligned}\pi_{{}^{n}R,R}(n_R) &= \mathfrak{f}_{{}^{n}R,{}^{n}R}^{{}^{n}Q,{}^{n}Q}\big(\pi_{{}^{n}Q,{}^{n}Q}(1)\big)\circ\pi_{{}^{n}R,{}^{n}R}(1)\circ\mathfrak{f}_{{}^{n}R,R}^{{}^{n}Q,Q}\big(\pi_{{}^{n}Q,Q}(n)\big)\\ &= \mathfrak{f}_{{}^{n}R,R}^{{}^{n}Q,Q}\big(\pi_{{}^{n}Q,Q}(n)\big)\end{aligned} \qquad 20.23.8,$$

so that $\delta_n = \mathrm{id}_R$ and similarly $\delta_{n'} = \mathrm{id}_{R'}$.

Let T and T' be respective proper subgroups of R and R', normal in Q and Q', and assume that $\mathcal{L}(Q',Q)_{R',R} \cap \mathcal{L}(Q',Q)_{T',T}$ is not empty; then, for any $x \in \mathcal{L}(Q',Q)_{R',R} \cap \mathcal{L}(Q',Q)_{T',T}$, we claim that

$$\mathfrak{g}_{T',T}^{R',R}\big(\mathfrak{g}_{R',R}^{Q',Q}(x)\big) = \mathfrak{g}_{T',T}^{Q',Q}(x) \qquad 20.23.9;$$

indeed, it follows from Corollary 2.21 that we can choose pairs $(Q,n) \in \mathfrak{N}(R)$ and $(Q',n') \in \mathfrak{N}(R')$ which still belong to $\mathfrak{N}(T)$ and $\mathfrak{N}(T')$ respectively; hence, we have (cf. 20.20.2 and 20.23.3)

$$\mathfrak{g}_{R',R}^{Q',Q}(x) = (n'_{R'})^{-1}\cdot\mathfrak{g}_{n'R',{}^nR}^{n'Q',{}^nQ}(n'\cdot x\cdot n^{-1})\cdot n_R \\ \mathfrak{g}_{T',T}^{Q',Q}(x) = (n'_{T'})^{-1}\cdot\mathfrak{g}_{n'T',{}^nT}^{n'Q',{}^nQ}(n'\cdot x\cdot n^{-1})\cdot n_T \qquad 20.23.10;$$

but, according to equality 20.23.8, we have ${}^{n_R}R = {}^nR$ and ${}^{n_R}T = {}^nT$, and therefore the pair (R,n_R) belongs to $\mathfrak{N}(T)$ and it is easily checked that $(n_R)_T = n_T$ (cf. 20.20.2); similarly, $(R',n'_{R'})$ belongs to $\mathfrak{N}(R')$ and we have $(n_{R'})_{T'} = n_{T'}$ (cf. 20.20.2); consequently, we still have

$$\mathfrak{g}_{T',T}^{R',R}\big(\mathfrak{g}_{R',R}^{Q',Q}(x)\big) = (n'_{T'})^{-1}\cdot\mathfrak{g}_{n'T',{}^nT}^{n'R',{}^nR}\big(n'_{R'}\cdot\mathfrak{g}_{R',R}^{Q',Q}(x)\cdot(n_R^{-1})\big)\cdot n_T \qquad 20.23.11$$

and then it follows from equalities 20.23.10 and Lemma 20.14 that

$$\begin{aligned}\mathfrak{g}_{T',T}^{R',R}\big(\mathfrak{g}_{R',R}^{Q',Q}(x)\big) &= (n'_{T'})^{-1}\cdot\mathfrak{g}_{n'T',{}^nT}^{n'R',{}^nR}\big(\mathfrak{g}_{n'R',{}^nR}^{n'Q',{}^nQ}(n'\cdot x\cdot n^{-1})\big)\cdot n_T \\ &= (n'_{T'})^{-1}\cdot\mathfrak{g}_{n'T',{}^nT}^{n'Q',{}^nQ}(n'\cdot x\cdot n^{-1})\cdot n_T \\ &= \mathfrak{g}_{T',T}^{Q',Q}(x)\end{aligned} \qquad 20.23.12.$$

Now, we can complete the proof of condition 20.23.5 in the normal case; according to Lemma 20.11, to 20.20 and to statement 20.23.3, we have

$$\begin{aligned}&\pi_{R',R}\big(\mathfrak{g}_{R',R}^{Q',Q}(x)\big) \\ &= \pi_{n'R',R'}(n'_{R'})^{-1}\circ\pi_{n'R',{}^nR}\big(\mathfrak{g}_{n'R',{}^nR}^{n'Q',{}^nQ}(n'\cdot x\cdot n^{-1})\big)\circ\pi_{{}^nR,R}(n_R) \\ &= \pi_{n'R',R'}(n'_{R'})^{-1}\circ\mathfrak{f}_{n'R',{}^nR}^{n'Q',{}^nQ}\big(\pi_{n'Q',{}^nQ}(n'\cdot x\cdot n^{-1})\big)\circ\pi_{{}^nR,R}(n_R) \\ &= \mathfrak{f}_{R',R}^{Q',Q}\big(\pi_{n'Q',Q'}(n')^{-1}\circ\pi_{n'Q',{}^nQ}(n'\cdot x\cdot n^{-1})\circ\pi_{{}^nQ,Q}(n)\big) \\ &= \mathfrak{f}_{R',R}^{Q',Q}\big(\pi_{Q',Q}(x)\big)\end{aligned} \qquad 20.23.13;$$

similarly, according to Lemma 20.11, to definition 20.22.3 and to statement 20.23.3, we still have

$$\begin{aligned}\mathfrak{g}_{R',R}^{Q',Q}\big(\tau_{Q',Q}(v)\big) &= (n'_{R'})^{-1}\cdot\mathfrak{g}_{n'R',{}^nR}^{n'Q',{}^nQ}\big(\tau_{n'Q',{}^nQ}(v)\big)\cdot n_R \\ &= (n'_{R'})^{-1}\cdot\tau_{n'R',{}^nR}(v)\cdot n_R \\ &= \tau_{R',R}(v)\end{aligned} \qquad 20.23.14.$$

In the not normal case, we argue by induction on $|Q:R|$; it is clear that, setting $N = N_Q(R)$ and $N' = N_{Q'}(R')$, we may assume that $N \neq Q$ and $N' \neq Q'$, and therefore that the maps $\mathfrak{g}^{Q',Q}_{N',N}$ and $\mathfrak{g}^{N',N}_{R',R}$ are already defined; then, for any element $x \in \mathcal{L}(Q',Q)_{R',R}$, we define

$$\mathfrak{g}^{Q',Q}_{R',R}(x) = \mathfrak{g}^{N',N}_{R',R}\big(\mathfrak{g}^{Q',Q}_{N',N}(x)\big) \qquad 20.23.15;$$

note that, in the case that R is normal in Q, $\mathfrak{g}^{Q',Q}_{R',R} = \mathfrak{g}^{N',N}_{R',R}$ is already defined and $\mathfrak{g}^{Q',Q}_{N',N} = \mathrm{id}_{\mathcal{L}(Q',Q)}$, so that equality 20.23.15 still holds. In order to prove statement 20.23.3, let T and T' be respectively subgroups of R and R', assume that $\mathcal{F}(R',R)_{T',T} \neq \emptyset$ and set $M = N_Q(T)$ and $M' = N_{Q'}(T')$. For any element $x \in \mathcal{L}(Q',Q)_{R',R} \cap \mathcal{L}(Q',Q)_{T',T}$, we argue by induction on $|Q:R||Q:T|$. First of all, if $M \neq Q$, we get

$$\begin{aligned}\mathfrak{g}^{R',R}_{T',T}\big(\mathfrak{g}^{Q',Q}_{R',R}(x)\big) &= \mathfrak{g}^{M'\cap R',M\cap R}_{T',T}\Big(\mathfrak{g}^{R',R}_{M'\cap R',M\cap R}\big(\mathfrak{g}^{Q',Q}_{R',R}(x)\big)\Big)\\ &= \mathfrak{g}^{M'\cap R',M\cap R}_{T',T}\big(\mathfrak{g}^{Q',Q}_{M'\cap R',M\cap R}(x)\big)\\ &= \mathfrak{g}^{M'\cap R',M\cap R}_{T',T}\Big(\mathfrak{g}^{M',M}_{M'\cap R',M\cap R}\big(\mathfrak{g}^{Q',Q}_{M',M}(x)\big)\Big)\\ &= \mathfrak{g}^{M',M}_{T',T}\big(\mathfrak{g}^{Q',Q}_{M',M}(x)\big) = \mathfrak{g}^{Q',Q}_{T',T}(x)\end{aligned} \qquad 20.23.16.$$

Secondly, if $M = Q$ we may assume that $N \neq Q$ and we still get (cf. 20.23.15)

$$\begin{aligned}\mathfrak{g}^{R',R}_{T',T}\big(\mathfrak{g}^{Q',Q}_{R',R}(x)\big) &= \mathfrak{g}^{R',R}_{T',T}\Big(\mathfrak{g}^{N',N}_{R',R}\big(\mathfrak{g}^{Q',Q}_{N',N}(x)\big)\Big)\\ &= \mathfrak{g}^{N',N}_{T',T}\big(\mathfrak{g}^{Q',Q}_{N',N}(x)\big) = \mathfrak{g}^{Q',Q}_{T',T}(x)\end{aligned} \qquad 20.23.17.$$

Finally, statements 20.23.4 and 20.23.5 follow from statement 20.23.2, from our induction hypothesis and from the corresponding statement in the normal case (cf. Lemmas 20.11 and 20.13). We are done.

Theorem 20.24 *The partial perfect $\mathcal{F}^{sc}$-locality $\mathcal{L}^{sc}$ can be extended to a unique perfect $\mathcal{F}$-locality $\mathcal{L}$.*

Proof: For any pair of subgroups Q and R of P, we define the set of morphisms in $\mathcal{L}$ from R to Q by the disjoint union

$$\mathcal{L}(Q,R) = \bigsqcup_{R'} \mathcal{L}(R',R) \qquad 20.24.1$$

where R' runs over the set of subgroups of Q which are $\mathcal{F}$-isomorphic to R. If x is an $\mathcal{L}$-morphism from R to Q, T a subgroup of P and y an $\mathcal{L}$-morphism from T to R, then, respectively denoting by R' and T' the subgroups of Q and R labelling the corresponding terms and by x_* and y_* the corresponding elements in $\mathcal{L}(R',R)$ and $\mathcal{L}(T',T)$, and setting $T'' = \big(\pi_{R',R}(x)\big)(T')$, the composition $x{\cdot}y$ of x and y in $\mathcal{L}$ is defined by

$$(x{\cdot}y)_* = \mathfrak{g}_{T'',T'}^{R',R}(x_*){\cdot}y_* \qquad 20.24.2.$$

This makes sense since all the sets $\mathcal{L}(R',R)$ are already defined, together with the maps $\pi_{R',R}:\mathcal{L}(R',R) \to \mathcal{F}(R',R)$, and therefore x_* obviously belongs to $\mathcal{L}(R',R)_{T'',T'}$, so that, by Proposition 20.23, we get the element $\mathfrak{g}_{T'',T'}^{R',R}(x_*)$ in $\mathcal{L}(T'',T')$; moreover, in 20.19 we already have defined the *composition* of $\mathfrak{g}_{T'',T'}^{R',R}(x_*)$ and y_*. Since $\mathfrak{g}_{R',R}^{R',R} = \mathrm{id}_{\mathcal{L}(R',R)}$ (cf. Proposition 20.10 and statement 20.23.3), the new composition extends the *composition maps* 20.19.1.

In order to prove the associativity of the new composition, consider a subgroup U of P and an $\mathcal{L}$-morphism z from U to T; that is to say, denoting by U' the subgroup of T labelling the corresponding term and by z_* the corresponding element in $\mathcal{L}(U',U)$, and analogously setting $U'' = \pi_y(U')$ and $U''' = \pi_x(U'')$, we have

$$\begin{aligned}\big((x{\cdot}y){\cdot}z\big)_* &= \mathfrak{g}_{U''',U''}^{T'',T'}\big((x{\cdot}y)_*\big){\cdot}z_* = \mathfrak{g}_{U''',U''}^{T'',T'}\big(\mathfrak{g}_{T'',T'}^{R',R}(x_*){\cdot}y_*\big){\cdot}z_*\\ &= \mathfrak{g}_{U''',U''}^{R',R}(x_*){\cdot}\mathfrak{g}_{U''',U''}^{T'',T'}(y_*){\cdot}z_* = \mathfrak{g}_{U''',U''}^{R',R}(x_*){\cdot}(y{\cdot}z)_*\\ &= \big(x{\cdot}(y{\cdot}z)\big)_*\end{aligned} \qquad 20.24.3.$$

We have an evident functor $\pi:\mathcal{L} \to \mathcal{F}$ which is the identity over the objects and, for any pair of subgroups Q and R of P, defines the map

$$\pi_{Q,R}:\mathcal{L}(Q,R) \longrightarrow \mathcal{F}(Q,R) \qquad 20.24.4$$

sending $x \in \mathcal{L}(Q,R)$ to $\iota_{R'}^{Q}\circ\pi_{R',R}(x_*)$, where R' is the subgroup of Q labelling the corresponding term (cf. 20.18.2) and x_* is the corresponding element of $\mathcal{L}(R',R)$; indeed, with the notation above, we get (cf. Lemma 20.11)

$$\begin{aligned}\pi_{Q,T}(x{\cdot}y) &= \iota_{T''}^{Q}\circ\pi_{T'',T}\big(\mathfrak{g}_{T'',T'}^{R',R}(x_*){\cdot}y_*\big)\\ &= \iota_{T''}^{Q}\circ\mathfrak{f}_{T'',T'}^{R',R}\big(\pi_{R',R}(x_*)\big)\circ\pi_{T',T}(y_*)\\ &= \big(\iota_{R'}^{Q}\circ\pi_{R',R}(x_*)\big)\circ\big(\iota_{T'}^{R}\circ\pi_{T',T}(y_*)\big)\\ &= \pi_{Q,R}(x)\circ\pi_{R,T}(y)\end{aligned} \qquad 20.24.5.$$

We also have a canonical functor $\tau : \mathcal{T}_P \to \mathcal{L}$ which is the identity over the objects and defines the map

$$\tau_{Q,R} : \mathcal{T}_P(Q,R) \to \mathcal{L}(Q,R) \qquad 20.24.6$$

sending $v \in \mathcal{T}_P(Q,R)$ to the element x in $\mathcal{L}(Q,R)$ fulfilling $x_* = \tau_{{}^vR,R}(v)$; indeed, with the notation above, if $w \in \mathcal{T}_P(R,T)$ then, setting $R' = {}^vR$, $T' = {}^wT$ and $T'' = {}^vT'$, we have (cf. 20.23.3)

$$\begin{aligned}\big(\tau_{Q,R}(v)\cdot\tau_{R,T}(w)\big)_* &= \mathfrak{g}^{R',R}_{T'',T'}\big(\tau_{R',R}(v)\big)\cdot\tau_{T',T}(w)\\ &= \tau_{T'',T'}(v)\cdot\tau_{T',T}(w) = \tau_{T'',T}(vw)\\ &= \big(\tau_{Q,T}(vw)\big)_*\end{aligned} \qquad 20.24.7.$$

Now, the category $\mathcal{L}$ endowed with the functors $\tau : \mathcal{T}_P \to \mathcal{L}$ and $\pi : \mathcal{L} \to \mathcal{F}$ is clearly an $\mathcal{F}$-locality (cf. 17.3). We claim that $\mathcal{L}$ is divisible; indeed, if Q, R and T are subgroups of P, x is an element of $\mathcal{L}(Q,R)$ and $y \in \mathcal{L}(Q,T)$ fulfills $T' = \pi_y(T) \subset \pi_x(R) = R'$, then x_* belongs to $\mathcal{L}(R',R)$, y_* belongs to $\mathcal{L}(T',T)$ and, setting $T'' = (\pi_x)^{-1}(T')$, the element $\mathfrak{g}^{R',R}_{T',T''}(x_*)$ belongs to $\mathcal{L}(T',T'')$; moreover, the composition $g^{R',R}_{T',T''}(x_*)^{-1}\cdot y_*$ belongs to $\mathcal{L}(T'',T)$ and determines an element $z \in \mathcal{L}(R,T)$ such that $x\cdot z = y$; thus, $\mathcal{L}$ is indeed divisible (cf. 17.7).

Moreover, since $\mathcal{L}^{^{sc}}$ is coherent and any $\mathcal{F}$-essential subgroup of P is $\mathcal{F}$-selfcentralizing, it is clear that $\mathcal{L}$ still fulfills the conditions in Proposition 17.10 and therefore it is coherent too. Finally, by definition 20.6.1, for any subgroup Q of P fully centralized in $\mathcal{F}$, we have

$$\mathrm{Ker}(\tau_Q) = H_{C_\mathcal{F}(Q)} \qquad 20.24.8,$$

so that $\mathcal{L}$ is perfect. The uniqueness of $\mathcal{L}$ follows easily from the uniqueness of our construction and from Lemma 20.21.

Chapter 21

Frobenius P-categories: the second definition

21.1 Let P be a finite p-group; in this chapter we give an equivalent definition of a *Frobenius P-category* in terms of a *$P\times P$-set*; as usual, we call *$P\times P$-set* a set Ω endowed with a $P\times P$-action, but we write $u{\cdot}\omega{\cdot}v$ instead of $(u,v){\cdot}\omega$ for any $(u,v)\in P\times P$ and any $\omega\in\Omega$. Although this definition is surprisingly simple — at least from our point of view — it does not reflect our motivation for considering such an object; however, this new definition is very suitable for the construction of *localities* (cf. 17.3) associated with a Frobenius P-category, as we show in chapters 22 and 23 below.

21.2 Let Ω be a finite nonempty $P\times P$-set such that $\{1\}\times P$ acts freely on Ω and assume that

$$\Omega^\circ\cong\Omega \quad\text{and}\quad |\Omega|/|P|\not\equiv 0 \pmod p \qquad 21.2.1$$

where Ω° denotes the *symmetric* $P\times P$-set, namely the $P\times P$-set obtained exchanging both factors; denote by G the *automorphism group* of $\mathrm{Res}_{\{1\}\times P}(\Omega)$ or, equivalently, the centralizer of $\{1\}\times P$ in the *symmetric group* $\mathfrak{S}_\Omega$ of Ω (cf. 1.32), and identify P with the image of $P\times\{1\}$ in G. It is clear that, for any pair of subgroups Q and R of P, the image of $\mathcal{T}_G(Q,R)$ (cf. 17.2) in $\mathrm{Hom}(R,Q)$ — denoted by $\mathcal{F}^\Omega(Q,R)$ — is determined by the set of group homomorphisms $\varphi:R\to P$ such that

$$\varphi(R)\subset Q \quad\text{and}\quad \mathrm{Res}_{\varphi\times\mathrm{id}_P}(\Omega)\cong\mathrm{Res}_{\iota_R^P\times\mathrm{id}_P}(\Omega) \qquad 21.2.2$$

and moreover that this correspondence defines a *divisible P-category* (cf. 2.3) — denoted by $\mathcal{F}^\Omega$.

21.3 On the other hand, according to our hypothesis, the stabilizer in $P\times P$ of any element of Ω intersects $P\times\{1\}$ and $\{1\}\times P$ trivially and therefore it has the form

$$\Delta_\varphi(Q)=\big\{\big(u,\varphi(u)\big)\big\}_{u\in Q} \qquad 21.3.1$$

for some subgroup Q of P and a suitable injective group homomorphism $\varphi:Q\to P$; with this in mind, for any subgroup Q of P, let us denote by $\mathrm{Hom}^\Omega(Q,P)$ the set of injective group homomorphisms $\varphi:Q\to P$ such that the set of fixed points $\Omega^{\Delta_\varphi(Q)}$ of $\Delta_\varphi(Q)$ in Ω is nonempty; note that condition 21.2.1 forces $\mathrm{Hom}^\Omega(P,P)\neq\emptyset$. Then, for any $\psi\in\mathcal{F}^\Omega(Q,R)$, it is quite clear that

$$\mathrm{Hom}^\Omega(Q,P)\circ\psi\subset\mathrm{Hom}^\Omega(R,P) \qquad 21.3.2$$

and, since $\mathrm{Hom}^\Omega(P,P)\neq\emptyset$, we get

$$|\mathcal{F}^{\Omega}(P,Q)|\leq|\mathrm{Hom}^\Omega(Q,P)| \qquad 21.3.3;$$

more precisely, for any $\sigma\in\mathrm{Hom}^\Omega(P,P)$, we still get

$$\mathcal{F}^{\Omega}(P,Q)\subset\sigma^{-1}\circ\mathrm{Hom}^\Omega(Q,P)=\mathrm{Hom}^{\mathrm{Res}_{\mathrm{id}_P\times\sigma}(\Omega)}(Q,P) \qquad 21.3.4.$$

21.4 We are ready to make our definition. We say that a $P\times P$-set Ω is *basic* if $\{1\}\times P$ acts freely on Ω, Ω fulfills condition 21.2.1 and, for any subgroup Q of P, we have

$$\mathcal{F}^{\Omega}(P,Q)=\mathrm{Hom}^\Omega(Q,P) \qquad 21.4.1;$$

moreover, we say that a *basic* $P\times P$-set is *thick* if the multiplicity of the indecomposable $P\times P$-set $(P\times P)/\Delta_\varphi(Q)$ is at least two for any $\varphi\in\mathcal{F}^{\Omega}(P,Q)$.

21.5 Note that, if $\mathcal{F}$ is a divisible P-category and Ω is a $P\times P$-set fulfilling all the conditions in 21.2 in such a way that, for any subgroup Q of P, we have

$$\mathrm{Hom}^\Omega(Q,P)\subset\mathcal{F}(P,Q)\quad\text{and}\quad\mathcal{F}(P,Q)\subset\mathcal{F}^{\Omega}(P,Q) \qquad 21.5.1,$$

then it follows from inequality 21.3.3 that Ω is *basic* and that $\mathcal{F}=\mathcal{F}^{\Omega}$; in this case we say that Ω is an *$\mathcal{F}$-basic $P\times P$-set*. We always can embed the disjoint union of two $\mathcal{F}$-basic $P\times P$-sets Ω and Ω' into a third $\mathcal{F}$-basic $P\times P$-set Ω'', namely the disjoint union of Ω with p copies of Ω'. Actually, we prove below that $\mathcal{F}$ admits an $\mathcal{F}$-basic $P\times P$-set if and only if $\mathcal{F}$ is a Frobenius P-category.

21.6 Equivalently, a $P\times P$-set Ω is *$\mathcal{F}$-basic* if and only if, setting

$$\Delta_{\varphi,\varphi'}(Q)=\{(\varphi(u),\varphi'(u))\}_{u\in Q} \qquad 21.6.1$$

for any subgroup Q of P, it fulfills condition 21.2.1 and the statement:

21.6.2 *The stabilizer of any element of Ω coincides with $\Delta_{\psi,\psi'}(R)$ for some subgroup R of P and suitable $\psi,\psi'\in\mathcal{F}(P,R)$, and we have*

$$|\Omega^{\Delta_{\varphi,\varphi'}(Q)}|=|\Omega^{\Delta(Q)}|$$

for any subgroup Q of P and any $\varphi,\varphi'\in\mathcal{F}(P,Q)$.

Indeed, it easily follows from its very definition that an $\mathcal{F}$-basic $P\times P$-set fulfills this statement; conversely, if Ω fulfills this statement then it is clear that $\{1\}\times P$ acts freely on Ω and that we have $\mathrm{Hom}^\Omega(Q,P)\subset\mathcal{F}(P,Q)$ for any subgroup Q of P; moreover, for any $\varphi\in\mathcal{F}(P,Q)$ and any subgroup W of $Q\times P$, $\mathrm{Res}_{\varphi\times\mathrm{id}_P}(\Omega)^W$ and Ω^W are empty unless $W=\Delta_\psi(R)$ for some subgroup R of Q and a suitable $\psi\in\mathcal{F}(P,R)$, and then we have

$$\left|\mathrm{Res}_{\varphi\times\mathrm{id}_P}(\Omega)^{\Delta_\psi(R)}\right|=|\Omega^{\Delta_{\varphi',\psi}(R)}|=|\Omega^{\Delta(R)}|=|\Omega^{\Delta_\psi(R)}| \qquad 21.6.3,$$

where $\varphi' \in \mathcal{F}(P,R)$ is the restriction of φ, which proves that

$$\mathrm{Res}_{\varphi\times \mathrm{id}_P}(\Omega) \cong \mathrm{Res}_{\iota_Q^P\times \mathrm{id}_P}(\Omega) \qquad 21.6.4,$$

so that $\mathcal{F}(P,Q) \subset \mathcal{F}^{\Omega}(P,Q)$.

21.7 This leads us to introduce the following *relative* situation which is useful in some inductive arguments. If $\mathfrak{X}$ is a set of subgroups of P such that any subgroup Q of P fulfilling $\mathcal{F}(Q,R) \neq \emptyset$ for some $R \in \mathfrak{X}$ belongs to $\mathfrak{X}$, let us say that a $P\times P$-set Ω is $\mathcal{F}^{\mathfrak{X}}$*-basic* if it is either empty or fulfills condition 21.2.1 and the statement

21.7.1 *The stabilizer of any element of Ω coincides with $\Delta_{\psi,\psi'}(R)$ for some $R \in \mathfrak{X}$ and suitable $\psi, \psi' \in \mathcal{F}(P,R)$, and we have*

$$|\Omega^{\Delta_{\varphi,\varphi'}(Q)}| = |\Omega^{\Delta(Q)}|$$

for any $Q \in \mathfrak{X}$ and any $\varphi, \varphi' \in \mathcal{F}(P,Q)$.

Moreover, we say that an $\mathcal{F}$-basic $P\times P$-set Ω is *thick outside* $\mathfrak{X}$ if the multiplicity of the indecomposable $P\times P$-set $(P\times P)/\Delta_\varphi(Q)$ is at least 2 for any subgroup $Q \notin \mathfrak{X}$ of P and any $\varphi \in \mathcal{F}(P,Q)$.

Lemma 21.8 *If Ω is a basic $P\times P$-set then, for any subgroup Q of P and any $\varphi \in \mathcal{F}^{\Omega}(P,Q)$, the number of $Q\times P$-orbits isomorphic to $(Q\times P)/\Delta_\varphi(Q)$ in Ω is equal to $|\Omega^{\Delta(Q)}|/|C_P(\varphi(Q))|$. Moreover, if $|C_P(\varphi(Q))|$ is maximal in $\{|C_P(\varphi'(Q))|\}_{\varphi'\in\mathcal{F}^{\Omega}(P,Q)}$ then we have*

$$|\Omega^{\Delta(Q)}|/|C_P(\varphi(Q))| \not\equiv 0 \pmod p \qquad 21.8.1.$$

Proof: For any $\varphi \in \mathrm{Hom}^{\Omega}(P,Q)$, it is easily checked that the group $\Delta_\varphi(Q)$ has $|C_P(\varphi(Q))|$ fixed elements in $(Q\times P)/\Delta_\varphi(Q)$ and no fixed element in the $Q\times P$-orbits of Ω which are *not* isomorphic to $(Q\times P)/\Delta_\varphi(Q)$; thus, the number k_φ of $Q\times P$-orbits in Ω isomorphic to $(Q\times P)/\Delta_\varphi(Q)$ is clearly equal to $|\Omega^{\Delta_\varphi(Q)}|/|C_P(\varphi(Q))|$. But, since we are assuming that

$$\mathrm{Hom}^{\Omega}(P,Q) = \mathcal{F}^{\Omega}(P,Q) \qquad 21.8.2,$$

it follows from the *symmetry* of Ω and the very definition of $\mathcal{F}^{\Omega}$ that we have

$$\mathrm{Res}_{\iota_Q^P,\iota_Q^P}(\Omega) \cong \mathrm{Res}_{\iota_Q^P,\varphi}(\Omega) \qquad 21.8.3$$

and therefore we get $|\Omega^{\Delta_\varphi(Q)}| = |\Omega^{\Delta(Q)}|$, so that

$$k_\varphi = |\Omega^{\Delta(Q)}|/|C_P(\varphi(Q))| \qquad 21.8.4.$$

Moreover, if $|C_P(\varphi(Q))|$ is maximal in $\{|C_P(\varphi'(Q))|\}_{\varphi'\in\mathcal{F}^\Omega(P,Q)}$ then pk_φ obviously divides $k_{\varphi'}$ for any $\varphi'\in\mathcal{F}^\Omega(P,Q)$ such that $|C_P(\varphi'(Q))|$ is not maximal; since all the other $Q\times P$-orbits in Ω have a cardinal divisible by $p|P|$, we easily get the announced congruence from the congruence in condition 21.2.1.

Proposition 21.9 *If Ω is a basic $P\times P$-set, $\mathcal{F}^\Omega$ is a Frobenius P-category.*

Proof: The group $\mathcal{F}_P(P)$ of inner automorphisms of P clearly acts on the set $\mathrm{Hom}^\Omega(P,P)$ and there is a natural bijection between the quotient set $\mathcal{F}_P(P)\backslash\mathrm{Hom}^\Omega(P,P)$ and the set of isomorphism classes of $P\times P$-orbits of cardinal $|P|$ in Ω; hence, since $\mathcal{F}^\Omega(P)=\mathrm{Hom}^\Omega(P,P)$, it follows from isomorphism 21.2.2 that all these $P\times P$-orbits have the same multiplicity in Ω. Moreover, by condition 21.2.1, the number of $P\times P$-orbits of cardinal $|P|$ in Ω is prime to p; consequently, $|\mathcal{F}^\Omega(P)/\mathcal{F}_P(P)|$ is also prime to p.

Let Q be an $\mathcal{F}^\Omega$-selfcentralizing subgroup of P; since $\mathcal{F}^\Omega$ is divisible (cf. 21.2), according to Theorem 4.12 it suffices to prove that, for any $\mathcal{F}$-morphism $\varphi:Q\to P$ and any subgroup R of $N_P(Q)$ such that

$$Q\subset R \quad\text{and}\quad {}^{\varphi}\mathcal{F}_R(Q)\subset\mathcal{F}_P(\varphi(Q)) \qquad 21.9.1,$$

there is $\rho\in\mathcal{F}^\Omega(P,R)$ such that $\rho(u)=\varphi(u)$ for any $u\in Q$. Note that an element $(v,w)\in P\times P$ normalizes $\Delta_\varphi(Q)$ if and only if we have $\varphi(u)^w=\varphi(u^v)$ for any $u\in Q$ and therefore R is contained in the image T of $N_{P\times P}(\Delta_\varphi(Q))$ in $N_P(Q)$ by the first projection.

On the other hand, since $C_P(\varphi(Q))=Z(\varphi(Q))$ (cf. 4.8) and the group $\{1\}\times Z(\varphi(Q))$ is normal in the normalizer $N_{P\times P}(\Delta_\varphi(Q))$, it is clear that this normalizer acts on the quotient set $\Omega^{\Delta_\varphi(Q)}/Z(\varphi(Q))$ which, according to Lemma 21.8, has a cardinal prime to p; consequently, the group $N_{P\times P}(\Delta_\varphi(Q))$ stabilizes $\omega\cdot Z(\varphi(Q))$ for some $\omega\in\Omega^{\Delta_\varphi(Q)}$, and therefore the stabilizer of ω in $N_{P\times P}(\Delta_\varphi(Q))$ has the form $\Delta_\rho(T)$ for some $\rho\in\mathcal{F}^\Omega(P,T)$; since $\Delta_\varphi(Q)\subset\Delta_\rho(T)$, for any $u\in Q\subset T$, we get $\varphi(u)=\rho(u)$ and, since $\mathcal{F}^\Omega$ is divisible, the restriction of ρ to R belongs to $\mathcal{F}^\Omega(P,R)$. We are done.

Remark 21.10 Once we have this proposition, it follows from statement 2.10.1 that our maximality condition in Lemma 21.8 amounts to saying that Q is fully centralized in $\mathcal{F}^\Omega$.

Proposition 21.11 *Assume that Ω is a basic $P\times P$-set. Let Q be a subgroup of P and K a subgroup of* $\mathrm{Aut}(Q)$ *such that Q is fully K-normalized in $\mathcal{F}^\Omega$. Then, the union $\Omega_Q^K=\bigcup_{\chi\in K}\Omega^{\Delta_\chi(Q)}$ is a basic $N_P^K(Q)\times N_P^K(Q)$-set and we have $\mathcal{F}^{\Omega_Q^K}=N_{\mathcal{F}^\Omega}^K(Q)$.*

Proof: Let us firstly prove that $N_P^K(Q) \times N_P^K(Q)$ stabilizes Ω_Q^K ; indeed, if $\chi \in K$, $\omega \in \Omega^{\Delta_\chi(Q)}$ and $v, w \in N_P^K(Q)$ then, for any $u \in Q$, we have

$$\begin{aligned} v^{-1}{\cdot}\omega{\cdot}w &= v^{-1}{\cdot}u{\cdot}\omega{\cdot}\chi(u)^{-1}{\cdot}w = (u^v){\cdot}(v^{-1}{\cdot}\omega{\cdot}w){\cdot}\chi(u^{-1})^w \\ &= (u^v){\cdot}(v^{-1}{\cdot}\omega{\cdot}w){\cdot}\chi'(u^v)^{-1} \end{aligned} \qquad 21.11.1$$

where $\chi' = \kappa_{w^{-1}} \circ \chi \circ \kappa_{v^{-1}}$ still belongs to K .

Moreover, for any $\chi \in K$, we know that the stabilizer in $P \times P$ of any element $\omega \in \Omega^{\Delta_\chi(Q)}$ coincides with $\Delta_\psi(R)$ for some subgroup R of P and some $\psi \in \mathcal{F}(P,R)$, so that R contains Q and $\psi(u) = \chi(u)$ for any $u \in Q$. Thus, on the one hand, ω determines χ and therefore the union

$$\Omega_Q^K = \bigcup_{\chi \in K} \Omega^{\Delta_\chi(Q)} \qquad 21.11.2$$

is a disjoint union; on the other hand, the stabilizer of ω in $N_P^K(Q) \times N_P^K(Q)$ is equal to $\Delta_{\psi_Q}\big(N_R^K(Q)\big)$, where $\psi_Q : N_R^K(Q) \to N_P^K(Q)$ is the $\mathcal{F}^\Omega$-morphism determined by ψ which actually is an $\big(N_{\mathcal{F}^\Omega}^K(Q)\big)$-morphism (cf. 2.14); but, it is easily checked that

$$\big(N_{\mathcal{F}^\Omega}^K(Q)\big)\big(N_P^K(Q), N_R^K(Q)\big) \subset \mathcal{F}^{\Omega_Q^K}\big(N_P^K(Q), N_R^K(Q)\big) \qquad 21.11.3.$$

If $s:\Omega \cong \Omega$ is a bijection such that, for any $\omega \in \Omega$ and any $u, v \in P$, we have $s(u{\cdot}\omega{\cdot}v^{-1}) = v{\cdot}s(\omega){\cdot}u^{-1}$, then s maps $\Omega^{\Delta_\chi(Q)}$ on $\Omega^{\Delta_{\chi^{-1}}(Q)}$ for any $\chi \in K$ and therefore it stabilizes Ω_Q^K . Finally, it follows from Lemma 21.8 that

$$\Big|\bigcup_{\chi \in K} \Omega^{\Delta_\chi(Q)}\Big| = |K \cap \mathcal{F}^\Omega(Q)||\Omega^{\Delta(Q)}| \qquad 21.11.4$$

and it is quite clear that $|N_P^K(Q)| = |K \cap \mathcal{F}_P(Q)||C_P(Q)|$; but, according to Proposition 21.9, $\mathcal{F}^\Omega$ is a Frobenius P-category and therefore, by Proposition 2.11, $K \cap \mathcal{F}_P(Q)$ is a Sylow p-subgroup of $K \cap \mathcal{F}^\Omega(Q)$. Consequently, Ω_Q^K also fulfills condition 21.2.1. We are done.

Proposition 21.12 *Let $\mathcal{F}$ be a Frobenius P-category, $\mathfrak{X}$ a set of subgroups of P such that any subgroup Q of P fulfilling $\mathcal{F}(Q,R) \neq \emptyset$ for some $R \in \mathfrak{X}$ belongs to $\mathfrak{X}$, and Ω an $\mathcal{F}^\mathfrak{X}$-basic $P\times P$-set. There is an $\mathcal{F}$-basic $P\times P$-set $\widehat{\Omega}$ containing Ω which is thick outside $\mathfrak{X}$ and fulfills*

$$\widehat{\Omega}^{\Delta_\varphi(Q)} = \Omega^{\Delta_\varphi(Q)} \qquad 21.12.1$$

for any $Q \in \mathfrak{X}$ and any $\varphi \in \mathcal{F}(P,Q)$.

Proof:† Let $\{\widehat{\Omega}_n\}_{n\in\mathbb{N}}$ be a family of $P\times P$-sets inductively defined as follows; if $\mathfrak{X}\neq\emptyset$ then we set $\widehat{\Omega}_0=\Omega$; if $\mathfrak{X}=\emptyset$ then Ω is also empty and, choosing $k_0\in\mathbb{N}-\{0,1\}$ prime to p, the $P\times P$-set $\widehat{\Omega}_0$ is the disjoint union of k_0 copies of the $P\times P$-sets $(P\times P)/\Delta_\sigma(P)$ where σ runs over a set of representatives for $\mathcal{F}(P)/\mathcal{F}_P(P)$ in $\mathcal{F}(P)$; note that, for any $Q\in\mathfrak{X}\cup\{P\}$ and any elements $\varphi,\varphi'\in\mathcal{F}(P,Q)$, we have

$$\left|(\widehat{\Omega}_0)^{\Delta(Q)}\right|=\left|(\widehat{\Omega}_0)^{\Delta_{\varphi,\varphi'}(Q)}\right| \tag{21.12.2}$$

and that the $P\times P$-sets $\widehat{\Omega}_0$ and $(\widehat{\Omega}_0)^\circ$ are isomorphic.

For any $n\in\mathbb{N}$, we choose a set of representatives $\mathfrak{R}_n$ for the $\mathcal{F}$-isomorphism classes of subgroups of P of index p^n which do *not* belong to $\mathfrak{X}$ in such a way that any element of $\mathfrak{R}_n$ is fully normalized in $\mathcal{F}$; we argue by induction on n, so that, if $n\geq 1$ and $m<n$, we assume that the $P\times P$-sets $\widehat{\Omega}_m$ and $(\widehat{\Omega}_m)^\circ$ are isomorphic, that $\{1\}\times P$ acts freely on $\widehat{\Omega}_m$ and that, for any $Q\in\mathfrak{X}\cup\mathfrak{R}_m$ and any $\varphi,\varphi'\in\mathcal{F}(P,Q)$, we have

$$\left|(\widehat{\Omega}_m)^{\Delta(Q)}\right|=\left|(\widehat{\Omega}_m)^{\Delta_{\varphi,\varphi'}(Q)}\right| \tag{21.12.3}$$

Now, choose $k_R\in\mathbb{N}-\{0,1\}$ such that

$$\begin{aligned}&\left|(\widehat{\Omega}_{n-1})^{\Delta(R)}\right|+k_R\left|\big((P\times P)/\Delta(R)\big)^{\Delta(R)}\right|\\ &\qquad\geq\left|(\widehat{\Omega}_{n-1})^{\Delta_{\psi,\psi'}(R)}\right|+2\left|\big((P\times P)/\Delta_{\psi,\psi'}(R)\big)^{\Delta_{\psi,\psi'}(R)}\right|\end{aligned} \tag{21.12.4}$$

for any $\psi,\psi'\in\mathcal{F}(P,R)$, and denote by $k_{\psi,\psi'}$ the positive rational number fulfilling

$$\begin{aligned}&\left|(\widehat{\Omega}_{n-1})^{\Delta(R)}\right|+k_R\left|\big((P\times P)/\Delta(R)\big)^{\Delta(R)}\right|\\ &\qquad=\left|(\widehat{\Omega}_{n-1})^{\Delta_{\psi,\psi'}(R)}\right|+k_{\psi,\psi'}\left|\big((P\times P)/\Delta_{\psi,\psi'}(R)\big)^{\Delta_{\psi,\psi'}(R)}\right|\end{aligned} \qquad 21.12.5;$$

then, provided we prove that $k_{\psi,\psi'}$ is an *integer* for any $\psi,\psi'\in\mathcal{F}(P,R)$, we will choose $\widehat{\Omega}_n$ as the disjoint union of $\widehat{\Omega}_{n-1}$ with the disjoint union of $k_{\psi,\psi'}$ copies of $(P\times P)/\Delta_{\psi,\psi'}(R)$ where R runs over $\mathfrak{R}_n$ and ψ and ψ' run over $\mathcal{F}(P,R)$; at that point, it will be clear that

$$|(\widehat{\Omega}_n)^{\Delta(Q)}|=|(\widehat{\Omega}_n)^{\Delta_{\varphi,\varphi'}(Q)}| \quad\text{and}\quad \widehat{\Omega}_n\cong(\widehat{\Omega}_n)^\circ \tag{21.12.6}$$

In order to prove that $k_{\psi,\psi'}$ is an integer, note that

$$\left|\big((P\times P)/\Delta_{\psi,\psi'}(R)\big)^{\Delta_{\psi,\psi'}(R)}\right|=\left|\bar{N}_{P\times P}\big(\Delta_{\psi,\psi'}(R)\big)\right| \tag{21.12.7}$$

† This proof follows the arguments of the proof of Proposition 5.5 in [13].

and that the quotient $\bar{N}_{P\times P}\big(\Delta_{\psi,\psi'}(R)\big)$ acts on $(\widehat{\Omega}_{n-1})^{\Delta_{\psi,\psi'}(R)}$; the key point of the argument in [13] is that this quotient still acts on $(\widehat{\Omega}_{n-1})^{\Delta(R)}$ and that the "difference" between these two $\bar{N}_{P\times P}\big(\Delta_{\psi,\psi'}(R)\big)$-sets is a multiple of the regular orbit.

Indeed, since R is fully normalized in $\mathcal{F}$, considering the inverse of the isomorphism $\psi : R \cong \psi(R)$ determined by ψ, it follows from condition 2.8.2 that there are an $\mathcal{F}$-morphism $\zeta : N_P\big(\psi(R)\big) \to P$ and an element $\chi \in \mathcal{F}(R)$ such that $\zeta\big(\psi(v)\big) = \chi(v)$ for any $v \in R$; moreover, according to Proposition 2.11, R is also fully centralized in $\mathcal{F}$ and therefore, considering the group homomorphism

$$\eta : \psi'(R) \longrightarrow P \qquad 21.12.8$$

mapping $\psi'(v)$ on $\chi(v)$ for any $v \in R$, and the image

$$T = \pi'\Big(N_{P\times P}\big(\Delta_{\psi,\psi'}(R)\big)\Big) \subset N_P\big(\psi'(R)\big) \qquad 21.12.9$$

where $\pi' : P \times P \to P$ is the second projection, there is $\xi \in \mathcal{F}(P,T)$ extending η and, in particular, fulfilling $\xi\big(\psi'(v)\big) = \chi(v)$ for any $v \in R$ (cf. statement 2.10.1); thus, $\zeta \times \xi$ determines a group homomorphism

$$N_{P\times P}\big(\Delta_{\psi,\psi'}(R)\big) \longrightarrow P \times P \qquad 21.12.10$$

which maps $\Delta_{\psi,\psi'}(R)$ onto $\Delta(R)$ and therefore it fulfills

$$(\zeta \times \xi)\Big(N_{P\times P}\big(\Delta_{\psi,\psi'}(R)\big)\Big) \subset N_{P\times P}\big(\Delta(R)\big) \qquad 21.12.11.$$

Hence, the homomorphism $\zeta \times \xi$ induces an action of $\bar{N}_{P\times P}\big(\Delta_{\psi,\psi'}(R)\big)$ on $(\widehat{\Omega}_{n-1})^{\Delta(R)}$ and from equality 21.12.7 we get

$$\begin{aligned}&\big|\big((P\times P)/\Delta(R)\big)^{\Delta(R)}\big| \\ &\quad = \big|\big((P\times P)/\Delta_{\psi,\psi'}(R)\big)^{\Delta_{\psi,\psi'}(R)}\big|\frac{\big|N_{P\times P}\big(\Delta(R)\big)\big|}{\big|N_{P\times P}\big(\Delta_{\psi,\psi'}(R)\big)\big|}\end{aligned} \qquad 21.12.12.$$

Moreover, at this point it follows from our induction hypothesis that we have

$$\big|\big((\widehat{\Omega}_{n-1})^{\Delta(R)}\big)^{\bar{Q}}\big| = \big|\big((\widehat{\Omega}_{n-1})^{\Delta_{\psi,\psi'}(R)}\big)^{\bar{Q}}\big| \qquad 21.12.13$$

for any nontrivial subgroup $\bar{Q}$ of $\bar{N}_{P\times P}\big(\Delta_{\psi,\psi'}(R)\big)$; this is equivalent to saying that the two $\bar{N}_{P\times P}\big(\Delta_{\psi,\psi'}(R)\big)$-sets obtained removing all the regular orbits from $(\widehat{\Omega}_{n-1})^{\Delta(R)}$ and $(\widehat{\Omega}_{n-1})^{\Delta_{\psi,\psi'}(R)}$ are isomorphic, which proves our claim.

Finally, we claim that the union $\widehat{\Omega} = \bigcup_{n\in\mathbb{N}} \widehat{\Omega}_n$ is an $\mathcal{F}$-basic $P \times P$-set; indeed, it is clear that

$$|\widehat{\Omega}|/|P| \equiv |\widehat{\Omega}_0|/|P| \pmod{p} \qquad 21.12.14$$

and we have either $\widehat{\Omega}_0 = \Omega$ or $|\widehat{\Omega}_0| = k_0|\mathcal{F}(P)||Z(P)|$, so that $|\widehat{\Omega}|/|P|$ is prime to p; now, it easily follows from 21.12.6 that $\widehat{\Omega}$ fulfills conditions 21.2.1 and 21.6.2; moreover, for any $R \in \bigcup_{n\in\mathbb{N}} \mathfrak{R}_n$, it is clear that the multiplicity $k_{\psi,\psi'}$ of $(P \times P)/\Delta_{\psi,\psi'}(R)$ is at least 2, so that $\widehat{\Omega}$ is *thick outside* $\mathfrak{X}$, and that $\Delta(Q)$ fixes no element in $(P\times P)/\Delta_{\psi,\psi'}(R)$ for any $Q \in \mathfrak{X}$ which implies equalities 21.12.1. We are done.

Chapter 22

The basic $\mathcal{F}$-locality

22.1 Let P be a finite p-group, Ω a thick basic $P \times P$-set (cf. 21.4) and $\mathcal{F} = \mathcal{F}^{\Omega}$ the corresponding Frobenius P-category (cf. Proposition 21.9). In this chapter we exhibit a *coherent* $\mathcal{F}$*-locality* $\mathcal{L}^{b}$ — called the *basic* $\mathcal{F}$*-locality* — which, although constructed from this $P \times P$-set Ω, actually does not depend on the chosen thick $\mathcal{F}$-basic $P \times P$-set (cf. Proposition 22.12 below).

22.2 Denote by G the group of automorphisms of $\mathrm{Res}_{\{1\}\times P}(\Omega)$ and identify the p-group P with the image of $P \times \{1\}$ in G, so that from now on P acts freely on Ω (cf. 21.2). In order to construct $\mathcal{L}^{b}$, we firstly have to discuss the behaviour of the centralizers in G of the subgroups Q of P; as a matter of fact, the *thickness* condition will guarantee a "stable" description — with respect to the different choices of Ω such that $\mathcal{F}^{\Omega} = \mathcal{F}$ — of these centralizers and of their Abelian quotients.

22.3 Let Q be a subgroup of P; clearly, the centralizer $C_G(Q)$ coincides with the group of automorphisms of $\mathrm{Res}_{Q\times P}(\Omega)$; moreover, any orbit of this $Q \times P$-set is isomorphic to the quotient set $(Q \times P)/\Delta_\eta(T)$ (cf. 21.3) for some subgroup T of Q and some $\eta \in \mathcal{F}(P,T)$ and, by the thickness, *all* these $Q \times P$-sets are isomorphic to orbits of $\mathrm{Res}_{Q\times P}(\Omega)$ (cf. 21.4). Note that, denoting by $\mathrm{Aut}\big((Q \times P)/\Delta_\eta(T)\big)$ the group of automorphisms of this $Q \times P$-set, the multiplication by $N_{Q\times P}\big(\Delta_\eta(T)\big)$ on the right induces a group isomorphism (cf. 1.9)

$$\mathrm{Aut}\big((Q \times P)/\Delta_\eta(T)\big) \cong \bar{N}_{Q\times P}\big(\Delta_\eta(T)\big) \qquad 22.3.1$$

and that this group acts regularly on $\big((Q \times P)/\Delta_\eta(T)\big)^{\Delta_\eta(T)}$.

22.4 Actually, the isomorphism class of this $Q \times P$-set $(Q \times P)/\Delta_\eta(T)$ only depends on the conjugacy class of T in Q and on the class $\tilde{\eta}$ of η in $\tilde{\mathcal{F}}(P,T)$ (cf. 1.3); in other words, by the very definition of the *product cover* $\mathfrak{pc}(\tilde{\mathcal{F}})$ (cf. A2.7) of the *exterior quotient* $\tilde{\mathcal{F}}$ of $\mathcal{F}$ (cf. 1.3), the pair (Q,P) defines the $\mathfrak{pc}(\tilde{\mathcal{F}})$-object $Q\tilde{\times}P$ (cf. A2.7) and the set of isomorphism classes of $Q \times P$-orbits in $\mathrm{Res}_{Q\times P}(\Omega)$ corresponds bijectively with the set of isomorphism classes of objects in the *full subcategory* $\tilde{\mathcal{F}}_{Q\tilde{\times}P}$ of $\mathfrak{pc}(\tilde{\mathcal{F}})_{Q\tilde{\times}P}$ (cf. 1.7) — the full subcategory of $\mathfrak{pc}(\tilde{\mathcal{F}})_{Q\tilde{\times}P}$ over the set of $\mathfrak{pc}(\tilde{\mathcal{F}})$-morphisms $\tilde{\eta}: T \to Q\tilde{\times}P$ when T runs over the set of subgroups of P.

22.5 Let us denote by $\mathfrak{O}_Q$ a set of representatives for the set of isomorphism classes of $\tilde{\mathcal{F}}_{Q\tilde{\times}P}$-objects; actually, we may assume that, for any

$\mathfrak{pc}(\tilde{\mathcal{F}})$-morphism $\tilde{\eta}: T \to Q\tilde{\times}P$ in $\mathfrak{O}_Q$, T is contained in Q and the composition of $\tilde{\eta}$ with the structural $\mathfrak{pc}(\tilde{\mathcal{F}})$-morphism $Q\tilde{\times}P \to Q$ coincides with $\tilde{\iota}_T^Q$ (cf. 1.9); then, we denote by $\hat{\tilde{\eta}}$ the composition of $\tilde{\eta}$ with the structural $\mathfrak{pc}(\tilde{\mathcal{F}})$-morphism $Q\tilde{\times}P \to P$. Moreover, we denote by $Q \times_{\tilde{\eta}} P$ the corresponding *indecomposable* $Q \times P$-set, by $k_{\tilde{\eta}}$ the multiplicity of this $Q \times P$-set in Ω and by $\mathfrak{S}_{k_{\tilde{\eta}}}$ the corresponding *symmetric group* (cf. 1.32). With all this notation, it is quite clear that we have a canonical $\widetilde{\mathfrak{Gr}}$-isomorphism (cf. 1.3 and 1.32)

$$\tilde{\omega}_Q : C_G(Q) \cong \prod_{\tilde{\eta}\in\mathfrak{O}_Q} \mathrm{Aut}(Q \times_{\tilde{\eta}} P) \wr \mathfrak{S}_{k_{\tilde{\eta}}} \qquad 22.5.1.$$

22.6 Let us exhibit a representative of $\tilde{\omega}_Q$; it is clear that the correspondence mapping any injective $Q \times P$-set homomorphism

$$h : Q \times_{\tilde{\eta}} P \longrightarrow \mathrm{Res}_{Q\times P}(\Omega) \qquad 22.6.1$$

on its image induces a bijection between the set of $\mathrm{Aut}(Q \times_{\tilde{\eta}} P)$-orbits in the set of these injective $Q \times P$-set homomorphisms, and the set of $Q \times P$-orbits isomorphic to $Q \times_{\tilde{\eta}} P$ in $\mathrm{Res}_{Q\times P}(\Omega)$; thus, denoting by $\mathcal{K}_{\tilde{\eta}}$ a set of representatives for the first set of orbits, we have $|\mathcal{K}_{\tilde{\eta}}| = k_{\tilde{\eta}}$ and we can identify $\mathfrak{S}_{k_{\tilde{\eta}}}$ with the group of permutations $\mathfrak{S}_{\mathcal{K}_{\tilde{\eta}}}$ of $\mathcal{K}_{\tilde{\eta}}$ (cf. 1.32); up to this identification, for any $z \in C_G(Q)$ and any $h \in \mathcal{K}_{\tilde{\eta}}$, there are unique elements $\sigma_z \in \mathfrak{S}_{\mathcal{K}_{\tilde{\eta}}}$ and $a_{z,h} \in \mathrm{Aut}(Q \times_{\tilde{\eta}} P)$ fulfilling

$$z \circ h = \sigma_z(h) \circ a_{z,h} \qquad 22.6.2$$

and a representative θ_Q of the $\widetilde{\mathfrak{Gr}}$-isomorphism 22.5.1 maps $z \in C_G(Q)$ on the element of the direct product having $\big((a_{z,h})_{h\in\mathcal{K}_{\tilde{\eta}}}, \sigma_z\big)$ in its $\tilde{\eta}$-component.

Proposition 22.7 *For any subgroup Q of P, denote by $\mathfrak{S}(Q)$ the minimal normal subgroup of $C_G(Q)$ containing $(\omega_Q)^{-1}\big(\prod_{\tilde{\eta}\in\mathfrak{O}_Q} \mathfrak{S}_{k_{\tilde{\eta}}}\big)$ for a representative ω_Q of $\tilde{\omega}_Q$. Then $\tilde{\omega}_Q$ induces an $\mathfrak{Ab}$-isomorphism*

$$\bar{\omega}_Q : C_G(Q)/\mathfrak{S}(Q) \cong \prod_{\tilde{\eta}\in\mathfrak{O}_Q} \mathfrak{ab}\big(\mathrm{Aut}(Q \times_{\tilde{\eta}} P)\big) \qquad 22.7.1.$$

Proof: It follows from the thickness condition and from the general lemma below.

Lemma 22.8 *If $n \in \mathbb{N} - \{0,1\}$, H is a group and N is the minimal normal subgroup containing $\mathfrak{S}_n$ in the wreath product $H \wr \mathfrak{S}_n$, then we have*

$$(H \wr \mathfrak{S}_n)/N \cong \mathfrak{ab}(H) \qquad 22.8.1.$$

Proof: Firstly assume that H is Abelian; then, we have an obvious group homomorphism $H \wr \mathfrak{S}_n \to H$ (cf. 1.32) mapping an element

$$\big((h_i)_{1\le i\le n}, \sigma\big) \in \big(\prod_{i=1}^n H\big) \rtimes \mathfrak{S}_n \qquad 22.8.2$$

on the product $\prod_{i=1}^n h_i$ in H, which is clearly surjective; moreover, denoting by $h^{(i)}$ the element of $\prod_{i=1}^n H$ having $h \in H$ in the entry i and 1 elsewhere, and by τ_i the transposition $(i, i+1)$ for any $1 \le i < n$, the kernel of this homomorphism is generated by the elements

$$(h^{(i+1)})^{-1} h^{(i)} = [\tau_i, h^{(i)}] \qquad 22.8.3.$$

In the general case, for any $h, g \in H$, the elements $(h^{(i+1)})^{-1}h^{(i)}$, $(g^{(i+1)})^{-1}g^{(i)}$ and $\big((hg)^{(i+1)}\big)^{-1}(hg)^{(i)}$ still belong to N and therefore the product

$$(h^{(i)})^{-1}h^{(i+1)}(g^{(i)})^{-1}g^{(i+1)}\big((hg)^{(i+1)}\big)^{-1}(hg)^{(i)} = [h^{(i)}, g^{(i)}] \qquad 22.8.4$$

also belongs to N; hence, the n-direct product $\prod_{i=1}^n [H, H]$ is contained in N. We are done.

22.9 Let Q and R be subgroups of P such that $R \subset Q$; in order to determine the inclusion $C_G(Q) \subset C_G(R)$ in terms of the right-hand members of the corresponding isomorphisms 22.5.1, we need some general notation. If $m \le n$, let us denote by $\tilde{\iota}_m^n : \mathfrak{S}_m \to \mathfrak{S}_n$ the *natural* $\widetilde{\mathfrak{Gr}}$*-inclusion*; it is clear that, if $\sum_{i\in I} m_i \le n$ then we can find representatives of the $\widetilde{\mathfrak{Gr}}$-morphisms $\tilde{\iota}_{m_i}^n$ (cf. 1.3), where i runs over I, determining a unique *injective* $\widetilde{\mathfrak{Gr}}$-morphism

$$\tilde{\iota}_{\{m_i\}_{i\in I}}^n : \prod_{i\in I} \mathfrak{S}_{m_i} \longrightarrow \mathfrak{S}_n \qquad 22.9.1.$$

In the case where $m_i = m$ for any $i \in I$ and $n = |I|m$, we denote by

$$\tilde{\delta}_m^{|I|} : \mathfrak{S}_m \longrightarrow \mathfrak{S}_{|I|m} \qquad 22.9.2$$

the composition of the diagonal map $\mathfrak{S}_m \to \prod_{i\in I} \mathfrak{S}_m$ with the corresponding $\widetilde{\mathfrak{Gr}}$-morphism 22.9.1; more generally, if $\sum_{i\in I} \ell_i m_i \le n$ then we denote by $\tilde{\iota}_{\{(\ell_i, m_i)\}_{i\in I}}^n$ the composition

$$\prod_{i\in I} \mathfrak{S}_{m_i} \xrightarrow{\tilde{\iota}_{\{\ell_i m_i\}_{i\in I}}^n} \prod_{i\in I} \mathfrak{S}_{\ell_i m_i} \xrightarrow{\prod_{i\in I} \tilde{\delta}_{m_i}^{\ell_i}} \mathfrak{S}_n \qquad 22.9.3.$$

22.10 Moreover, for any $\tilde{\eta} \in \mathfrak{O}_Q$ and any $\tilde{\nu} \in \mathfrak{O}_R$, denote by $k_{\tilde{\nu}}^{\tilde{\eta}}$ the multiplicity of $R \times_{\tilde{\nu}} P$ in $\mathrm{Res}_{R\times P}(Q \times_{\tilde{\eta}} P)$ and choose a set of representatives $\mathcal{K}_{\tilde{\nu}}^{\tilde{\eta}}$ for the set of $\mathrm{Aut}(R \times_{\tilde{\nu}} P)$-orbits of injective $R \times P$-set homomorphisms

$$g : R \times_{\tilde{\nu}} P \longrightarrow \mathrm{Res}_{R\times P}(Q \times_{\tilde{\eta}} P) \qquad 22.10.1,$$

so that $|\mathcal{K}_{\tilde{\nu}}^{\tilde{\eta}}| = k_{\tilde{\nu}}^{\tilde{\eta}}$; note that the composition induces a bijection

$$\mathcal{K}_{\tilde{\eta}} \times \mathcal{K}_{\tilde{\nu}}^{\tilde{\eta}} \cong \mathcal{K}_{\tilde{\eta}} \circ \mathcal{K}_{\tilde{\nu}}^{\tilde{\eta}} \qquad 22.10.2$$

and, since $\{\mathrm{Im}(h\circ g)\}_{h\in\mathcal{K}_{\tilde{\eta}},\, g\in\mathcal{K}_{\tilde{\nu}}^{\tilde{\eta}}}$ is a family of mutually disjoint subsets of Ω, we may assume that $\mathcal{K}_{\tilde{\nu}}$ coincides with the disjoint union $\bigsqcup_{\tilde{\eta}\in\mathfrak{O}_Q} \mathcal{K}_{\tilde{\eta}} \circ \mathcal{K}_{\tilde{\nu}}^{\tilde{\eta}}$, for any representative $\tilde{\nu} \in \mathfrak{O}_R$,

Proposition 22.11 *With the notation above, let Q and R be subgroups of P such that $R \subset Q$. Then, the inclusion $C_G(Q) \subset C_G(R)$ and the $\widetilde{\mathfrak{Gr}}$-morphisms $(\tilde{\omega}_Q)^{-1}$, $(\tilde{\omega}_R)^{-1}$ and*

$$\tilde{\iota}^{k_{\tilde{\nu}}}_{\{(k_{\tilde{\eta}},k_{\tilde{\nu}}^{\tilde{\eta}})\}_{\tilde{\eta}\in\mathfrak{O}_Q}} : \prod_{\tilde{\eta}\in\mathfrak{O}_Q} \mathfrak{S}_{k_{\tilde{\eta}}} \longrightarrow \mathfrak{S}_{k_{\tilde{\nu}}} \qquad 22.11.1$$

where $\tilde{\nu}$ runs over $\mathfrak{O}_R$, determine the following commutative $\widetilde{\mathfrak{Gr}}$-diagram

$$\begin{array}{ccc} C_G(Q) & \longrightarrow & C_G(R) \\ \uparrow & & \uparrow \\ \prod_{\tilde{\eta}\in\mathfrak{O}_Q} \mathfrak{S}_{k_{\tilde{\eta}}} & \longrightarrow & \prod_{\tilde{\nu}\in\mathfrak{O}_R} \mathfrak{S}_{k_{\tilde{\nu}}} \end{array} \qquad 22.11.2.$$

In particular, we have $\mathfrak{S}(Q) \subset \mathfrak{S}(R)$.

Proof: First of all note that, considering the decomposition of $\mathrm{Res}_{Q\times P}(\Omega)$ in $R \times P$-orbits, for any $\tilde{\nu} \in \mathfrak{O}_R$ we clearly have

$$k_{\tilde{\nu}} = \sum_{\tilde{\eta}\in\mathfrak{O}_Q} k_{\tilde{\nu}}^{\tilde{\eta}}\, k_{\tilde{\eta}} \qquad 22.11.3.$$

Moreover, for any $\tilde{\eta} \in \mathfrak{O}_Q$, any $\sigma \in \mathfrak{S}_{k_{\tilde{\eta}}}$ and any $h \in \mathcal{K}_{\tilde{\eta}}$, let us identify, up to the choice of a representative ω_Q of $\tilde{\omega}_Q$, $\mathfrak{S}_{k_{\tilde{\eta}}}$ with $\mathfrak{S}_{\mathcal{K}_{\tilde{\eta}}}$ acting on Ω just permuting the $Q \times P$-orbits isomorphic to $Q \times_{\tilde{\eta}} P$; then, according to our definition in 22.6.2, we have $(\omega_Q)^{-1}(\sigma) \circ h = \sigma(h)$ and therefore, considering $(\omega_Q)^{-1}(\sigma)$ as an element of $C_G(R)$, for any $g \in \mathcal{K}_{\tilde{\nu}}^{\tilde{\eta}}$ our definition in 22.6.2 simply becomes

$$(\omega_Q)^{-1}(\sigma) \circ (h \circ g) = \sigma(h) \circ g \qquad 22.11.4.$$

Now, it is straightforward to check that the map sending σ to the permutation over $\mathcal{K}_{\tilde\eta} \circ \mathcal{K}_{\tilde\nu}^{\tilde\eta}$ mapping $h \circ g$ on $\sigma(h) \circ g$ coincides with the group homomorphism

$$\mathfrak{S}_{\mathcal{K}_{\tilde\eta}} \longrightarrow \mathfrak{S}_{\mathcal{K}_{\tilde\eta} \times \mathcal{K}_{\tilde\nu}^{\tilde\eta}} \cong \mathfrak{S}_{\mathcal{K}_{\tilde\eta} \circ \mathcal{K}_{\tilde\nu}^{\tilde\eta}} \qquad 22.11.5$$

coming from the evident diagonal map

$$\mathfrak{S}_{\mathcal{K}_{\tilde\eta}} \longrightarrow \prod_{g \in \mathcal{K}_{\tilde\nu}^{\tilde\eta}} \mathfrak{S}_{\mathcal{K}_{\tilde\eta} \times \{g\}} \subset \mathfrak{S}_{\mathcal{K}_{\tilde\eta} \times \mathcal{K}_{\tilde\nu}^{\tilde\eta}} \qquad 22.11.6;$$

then, assuming that $\mathcal{K}_{\tilde\nu} = \bigsqcup_{\tilde\eta \in \mathfrak{O}_Q} \mathcal{K}_{\tilde\eta} \circ \mathcal{K}_{\tilde\nu}^{\tilde\eta}$, it is quite clear that the direct product of these maps when $\tilde\eta$ runs over $\mathfrak{O}_Q$ coincides with the $\widetilde{\mathfrak{Gr}}$-morphism 22.11.1.

Finally, for any $\tilde\nu \in \mathfrak{O}_R$, any $\tau \in \mathfrak{S}_{k_{\tilde\nu}}$ and any $\ell \in \mathcal{K}_{\tilde\nu}$, identifying as above $\mathfrak{S}_{k_{\tilde\nu}}$ with $\mathfrak{S}_{\mathcal{K}_{\tilde\nu}}$ acting on Ω just permuting the $R \times P$-orbits isomorphic to $Q \times_{\tilde\nu} P$, we still have $(\omega_R)^{-1}(\tau) \circ \ell = \tau(\ell)$ by our definition 22.6.2. We are done.

Proposition 22.12 *The correspondence mapping any pair of subgroups Q and R of P on the quotient set*

$$\overset{\mathrm{b}}{\mathcal{L}}(Q,R) = \mathcal{T}_G(Q,R)/\mathfrak{S}(R) \qquad 22.12.1,$$

endowed with the natural maps

$$\tau^{\mathrm{b}}_{Q,R} : \mathcal{T}_P(Q,R) \to \overset{\mathrm{b}}{\mathcal{L}}(Q,R) \quad \textit{and} \quad \pi^{\mathrm{b}}_{Q,R} : \overset{\mathrm{b}}{\mathcal{L}}(Q,R) \to \mathcal{F}(Q,R) \qquad 22.12.2,$$

defines a coherent $\mathcal{F}$-locality $(\tau^{\mathrm{b}}, \overset{\mathrm{b}}{\mathcal{L}}, \pi^{\mathrm{b}})$ which does not depend on the chosen thick $\mathcal{F}$-basic $P \times P$-set Ω.

Proof: It follows from the last statement in Proposition 22.11 that, for any triple of subgroups Q, R and T of P, the product in G induces a map

$$\overset{\mathrm{b}}{\mathcal{L}}(Q,R) \times \overset{\mathrm{b}}{\mathcal{L}}(R,T) \longrightarrow \overset{\mathrm{b}}{\mathcal{L}}(Q,T) \qquad 22.12.3;$$

then, it is quite clear that these maps determine a *composition* in $\overset{\mathrm{b}}{\mathcal{L}}$ † and that the natural maps in 22.12.2 define the structural functors (cf. 17.3)

$$\tau^{\mathrm{b}} : \mathcal{T}_P \longrightarrow \overset{\mathrm{b}}{\mathcal{L}} \quad \text{and} \quad \pi^{\mathrm{b}} : \overset{\mathrm{b}}{\mathcal{L}} \longrightarrow \mathcal{F} \qquad 22.12.4;$$

moreover, the *divisibility* (cf. 17.7) and the *coherence* (cf. 17.9) of $\overset{\mathrm{b}}{\mathcal{L}}$ are easy consequences of the fact that G is a group.

---

† It corresponds to the very definition of *"localité"* in [35, Ch.I]

On the other hand, for another choice of a thick $\mathcal{F}$-basic $P \times P$-set Ω', we already know that we can embed their *disjoint union* in a third thick $\mathcal{F}$-basic $P \times P$-set Ω'' (cf. 21.5); thus, setting

$$G' = \mathrm{Aut}_{\{1\}\times P}(\Omega') \quad \text{and} \quad G'' = \mathrm{Aut}_{\{1\}\times P}(\Omega'') \qquad 22.12.5$$

and denoting by $(G'')_{\Omega,\Omega'}$ the stabilizer in G'' of the images of Ω and Ω', we have surjective canonical group homomorphisms

$$G \longleftarrow (G'')_{\Omega,\Omega'} \longrightarrow G' \qquad 22.12.6$$

mapping $P \subset (G'')_{\Omega,\Omega'}$ onto both $P \subset G$ and $P \subset G'$.

More precisely, for any pair of subgroups Q and R of P, it is quite clear that we still have surjective maps (cf. 17.2)

$$\mathcal{T}_G(Q,R) \longleftarrow \mathcal{T}_{(G'')_{\Omega,\Omega'}}(Q,R) \longrightarrow \mathcal{T}_{G'}(Q,R) \qquad 22.12.7;$$

then, denoting by $\mathfrak{S}'(R)$ and $\mathfrak{S}''(R)$ the respective subgroups of $C_{G'}(R)$ and $C_{G''}(R)$ analogous to $\mathfrak{S}(R)$, it is easily checked that we get

$$\mathcal{T}_{G''}(Q,R) = \mathcal{T}_{(G'')_{\Omega,\Omega'}}(Q,R)\cdot\mathfrak{S}''(R) \qquad 22.12.8$$

and that this equality, together with the surjective maps 22.12.7, induces bijections

$$\mathcal{T}_G(Q,R)/\mathfrak{S}(R) \cong \mathcal{T}_{G''}(Q,R)/\mathfrak{S}''(R) \cong \mathcal{T}_{G'}(Q,R)/\mathfrak{S}'(R) \qquad 22.12.9$$

which are clearly compatible with the corresponding natural maps 22.12.2. Consequently, the three $\mathcal{F}$-localities we obtain are mutually equivalent. We are done.

22.13 Since the *basic* $\mathcal{F}$-locality is *divisible* and, for any subgroup Q of P, the group $\mathrm{Ker}(\pi^{\mathrm{b}}_Q) = C_G(Q)/\mathfrak{S}(Q)$ is Abelian, $\mathfrak{Ker}(\pi^{\mathrm{b}})$ induces a *contravariant* functor (cf. 17.8)

$$\tilde{\mathfrak{k}} : \tilde{\mathcal{F}} \longrightarrow \mathfrak{Ab} \qquad 22.13.1$$

which, according to Proposition 22.7, maps Q on

$$\tilde{\mathfrak{k}}(Q) = \prod_{\tilde\eta \in \mathfrak{O}_Q} \mathfrak{ab}\big(\mathrm{Aut}(Q \times_{\tilde\eta} P)\big) \qquad 22.13.2.$$

In particular, for any $\tilde{\mathcal{F}}$-morphism $\tilde\varphi : R \to Q$, it is clear that we have

$$\tilde{\mathfrak{k}}(\tilde\varphi) = \bigoplus_{\tilde\eta \in \mathfrak{O}_Q,\, \tilde\nu \in \mathfrak{O}_R} \tilde{\mathfrak{k}}(\tilde\varphi)^{\tilde\eta}_{\tilde\nu} \qquad 22.13.3$$

for suitable group homomorphisms

$$\tilde{\mathfrak{k}}(\tilde{\varphi})^{\tilde{\eta}}_{\tilde{\nu}} : \mathfrak{ab}\big(\mathrm{Aut}(Q \times_{\tilde{\eta}} P)\big) \longrightarrow \mathfrak{ab}\big(\mathrm{Aut}(R \times_{\tilde{\nu}} P)\big) \qquad 22.13.4$$

and we will make them explicit.

22.14 It suffices to consider the case where R is a subgroup of Q and $\tilde{\varphi}$ the class of the inclusion map, and then we write $\tilde{\mathfrak{k}}^{\tilde{\eta}}_{\tilde{\nu}}$ instead of $\tilde{\mathfrak{k}}(\tilde{\iota}^Q_R)^{\tilde{\eta}}_{\tilde{\nu}}$. For any $\tilde{\eta} \in \mathfrak{O}_Q$ and any $\tilde{\nu} \in \mathfrak{O}_R$, it is clear that $\mathrm{Aut}(Q \times_{\tilde{\eta}} P)$ acts on the set of $R \times P$-orbits in $Q \times_{\tilde{\eta}} P$, preserving their isomorphism class, and therefore, for any identification of $\mathfrak{S}_{k^{\tilde{\eta}}_{\tilde{\nu}}}$ with $\mathfrak{S}_{\mathcal{K}^{\tilde{\eta}}_{\tilde{\nu}}}$, we get a group homomorphism

$$\alpha^{\tilde{\eta}}_{\tilde{\nu}} : \mathrm{Aut}(Q \times_{\tilde{\eta}} P) \longrightarrow \mathfrak{S}_{k^{\tilde{\eta}}_{\tilde{\nu}}} \qquad 22.14.1$$

which, for any $a \in \mathrm{Aut}(Q \times_{\tilde{\eta}} P)$ and any $g \in \mathcal{K}^{\tilde{\eta}}_{\tilde{\nu}}$, fulfills

$$a \circ g = \big(\alpha^{\tilde{\eta}}_{\tilde{\nu}}(a)\big)(g) \circ b_{a,g} \qquad 22.14.2$$

for a suitable $b_{a,g} \in \mathrm{Aut}(R \times_{\tilde{\nu}} P)$; it is clear that the set of these group homomorphisms form a unique $\widetilde{\mathfrak{Gr}}$-morphism $\tilde{\alpha}^{\tilde{\eta}}_{\tilde{\nu}}$.

Proposition 22.15 *With the notation above, let Q and R be subgroups of P such that $R \subset Q$. Then, the inclusion $C_G(Q) \subset C_G(R)$ and the $\widetilde{\mathfrak{Gr}}$-morphisms $(\tilde{\omega}_Q)^{-1}$, $\tilde{\omega}_R$ and $\tilde{\alpha}^{\tilde{\eta}}_{\tilde{\nu}}$, where $\tilde{\eta}$ and $\tilde{\nu}$ respectively run over $\mathfrak{O}_Q$ and $\mathfrak{O}_R$, determine the commutative $\widetilde{\mathfrak{Gr}}$-diagram*

$$\begin{array}{ccc} C_G(Q) & \longrightarrow & C_G(R) \\ \uparrow & & \downarrow \\ \prod_{\tilde{\eta}\in\mathfrak{O}_Q} \prod_{i=1}^{k_{\tilde{\eta}}} \mathrm{Aut}(Q \times_{\tilde{\eta}} P) & \longrightarrow & \prod_{\tilde{\nu}\in\mathfrak{O}_R} \mathfrak{S}_{k_{\tilde{\nu}}} \end{array} \qquad 22.15.1.$$

Proof: With the notation in 22.6 above, identify $\prod_{i=1}^{k_{\tilde{\eta}}} \mathrm{Aut}(Q \times_{\tilde{\eta}} P)$ with $\prod_{h\in\mathcal{K}_{\tilde{\eta}}} \mathrm{Aut}(Q \times_{\tilde{\eta}} P)$ and, for any $h \in \mathcal{K}_{\tilde{\eta}}$, identify any $a \in \mathrm{Aut}(Q \times_{\tilde{\eta}} P)$ in the factor indexed by h with its image in $C_G(Q)$ *via* a representative ω_Q of $\tilde{\omega}_Q$; then, it is clear that $a \circ h = h \circ a$ (cf. 22.6.2); thus, since we may assume that $h \circ g \in \mathcal{K}_{\tilde{\nu}}$ for any $g \in \mathcal{K}^{\tilde{\eta}}_{\tilde{\nu}}$ (cf. 22.10.2), from definition 22.14.2 above we get

$$a \circ (h \circ g) = h \circ \big(\alpha^{\tilde{\eta}}_{\tilde{\nu}}(a)\big)(g) \circ b_{a,g} \qquad 22.15.2$$

for a suitable $b_{a,g} \in \mathrm{Aut}(R \times_{\tilde{\nu}} P)$.

For any $\tilde\nu \in \mathfrak{O}_R$, on the one hand we consider the direct product of the homomorphisms $\prod_{h\in\mathcal{K}_{\tilde\eta}} \alpha_{\tilde\nu}^{\tilde\eta}$ when $\tilde\eta$ runs over $\mathfrak{O}_Q$ (cf. 22.14.1)

$$\prod_{\tilde\eta\in\mathfrak{O}_Q}\prod_{h\in\mathcal{K}_{\tilde\eta}} \mathrm{Aut}(Q\times_{\tilde\eta} P) \longrightarrow \prod_{\tilde\eta\in\mathfrak{O}_Q}\prod_{h\in\mathcal{K}_{\tilde\eta}} \mathfrak{S}_{k_{\tilde\nu}^{\tilde\eta}} \qquad 22.15.3;$$

on the other hand, we consider the family of integers

$$\mathfrak{K}_{\tilde\nu} = \{k_{\tilde\nu}^{(\tilde\eta,h)}\}_{(\tilde\eta,h)\in\mathfrak{O}_Q\times\mathcal{K}_{\tilde\eta}} \qquad 22.15.4$$

where $k_{\tilde\nu}^{(\tilde\eta,h)} = k_{\tilde\nu}^{\tilde\eta}$ for any $h\in\mathcal{K}_{\tilde\eta}$, and the $\widetilde{\mathfrak{Gr}}$-morphism (cf. 22.9.1 and 22.10)

$$\tilde\iota_{\mathfrak{K}_{\tilde\nu}}^{k_{\tilde\nu}} : \prod_{\tilde\eta\in\mathfrak{O}_Q}\prod_{h\in\mathcal{K}_{\tilde\eta}} \mathfrak{S}_{k_{\tilde\nu}^{\tilde\eta}} \longrightarrow \mathfrak{S}_{k_{\tilde\nu}} \qquad 22.15.5;$$

thus, the composition of the $\widetilde{\mathfrak{Gr}}$-morphisms 22.15.3 and 22.15.5 determines a $\widetilde{\mathfrak{Gr}}$-morphism

$$\prod_{\tilde\eta\in\mathfrak{O}_Q}\prod_{h\in\mathcal{K}_{\tilde\eta}} \mathrm{Aut}(Q\times_{\tilde\eta} P) \longrightarrow \mathfrak{S}_{k_{\tilde\nu}} \qquad 22.15.6.$$

Finally, from equality 22.15.2 it is easy to check that the $\widetilde{\mathfrak{Gr}}$-morphism

$$\prod_{\tilde\eta\in\mathfrak{O}_Q}\prod_{h\in\mathcal{K}_{\tilde\eta}} \mathrm{Aut}(Q\times_{\tilde\eta} P) \longrightarrow \prod_{\tilde\nu\in\mathfrak{O}_R} \mathfrak{S}_{k_{\tilde\nu}} \qquad 22.15.7$$

obtained from the $\widetilde{\mathfrak{Gr}}$-morphisms 22.15.6 when $\tilde\nu$ runs over $\mathfrak{O}_R$ yields the commutativity of diagram 22.15.1. We are done.

22.16 Now, for any pair of subgroups Q and R of P such that $R\subset Q$, any $\tilde\eta\in\mathfrak{O}_Q$ and any $\tilde\nu\in\mathfrak{O}_R$, we are ready to describe the group homomorphism above

$$\tilde{\mathfrak{k}}_{\tilde\nu}^{\tilde\eta} : \mathfrak{ab}\big(\mathrm{Aut}(Q\times_{\tilde\eta} P)\big) \longrightarrow \mathfrak{ab}\big(\mathrm{Aut}(R\times_{\tilde\nu} P)\big) \qquad 22.16.1.$$

It is clear that the direct product $\mathrm{Aut}(Q\times_{\tilde\eta} P)\times\mathrm{Aut}(R\times_{\tilde\nu} P)$ acts, by composition on the left and on the right, on the set of injective $R\times P$-set homomorphisms

$$g : R\times_{\tilde\nu} P \longrightarrow \mathrm{Res}_{R\times P}(Q\times_{\tilde\eta} P) \qquad 22.16.2$$

and let us consider a set of representatives $\mathcal{I}_{\tilde\nu}^{\tilde\eta}$ for the corresponding set of orbits; moreover, for any $g\in\mathcal{I}_{\tilde\nu}^{\tilde\eta}$, denote by $\mathrm{Aut}(Q\times_{\tilde\eta} P)_g$ the stabilizer of $g\circ\mathrm{Aut}(R\times_{\tilde\nu} P)$ in $\mathrm{Aut}(Q\times_{\tilde\eta} P)$ and by

$$\iota_g : \mathrm{Aut}(Q\times_{\tilde\eta} P)_g \longrightarrow \mathrm{Aut}(Q\times_{\tilde\eta} P) \qquad 22.16.3$$

the inclusion map; then, we clearly have an injective group homomorphism

$$\beta_g : \mathrm{Aut}(Q \times_{\tilde\eta} P)_g \longrightarrow \mathrm{Aut}(R \times_{\tilde\nu} P) \qquad 22.16.4$$

such that $a \circ g = g \circ \beta_g(a)$ for any $a \in \mathrm{Aut}(Q \times_{\tilde\eta} P)_g$.

Proposition 22.17 *With the notation above, for any pair of subgroups Q and R of P such that $R \subset Q$, any $\tilde\eta \in \mathfrak{O}_Q$ and any $\tilde\nu \in \mathfrak{O}_R$, we have*

$$\mathfrak{k}_{\tilde\nu}^{\tilde\eta} = \prod_{g \in \mathcal{I}_{\tilde\nu}^{\tilde\eta}} \mathfrak{ab}(\beta_g) \circ \mathfrak{ab}^\circ(\iota_g) \qquad 22.17.1.$$

Proof: As above, let us identify $\mathfrak{S}_{k_{\tilde\eta}}$ with $\mathfrak{S}_{\mathcal{K}_{\tilde\eta}}$ and $\mathfrak{S}_{k_{\tilde\nu}}$ with $\mathfrak{S}_{\mathcal{K}_{\tilde\nu}}$; it follows from Propositions 22.11 and 22.15 that the inclusion $C_G(Q) \subset C_G(R)$ maps an element (cf. 22.5.1)

$$\big((a_h)_{h \in \mathcal{K}_{\tilde\eta}}, \sigma_{\tilde\eta}\big) \in \Big(\prod_{h \in \mathcal{K}_{\tilde\eta}} \mathrm{Aut}(Q \times_{\tilde\eta} P)\Big) \rtimes \mathfrak{S}_{\mathcal{K}_{\tilde\eta}} \qquad 22.17.2$$

on the element

$$\big\{\big((b_\ell)_{\ell \in \mathcal{K}_{\tilde\nu}}, \tau_{\tilde\nu}\big)\big\}_{\tilde\nu \in \mathfrak{O}_R} \in \prod_{\tilde\nu \in \mathfrak{O}_R} \Big(\prod_{\ell \in \mathcal{K}_{\tilde\nu}} \mathrm{Aut}(R \times_{\tilde\nu} P)\Big) \rtimes \mathfrak{S}_{\mathcal{K}_{\tilde\nu}} \qquad 22.17.3$$

where, identifying the set $\mathcal{K}_{\tilde\nu}^{\tilde\eta} \times \mathcal{K}_{\tilde\eta}$ with the corresponding subset of $\mathcal{K}_{\tilde\nu}$ (cf. 22.10), $\tau_{\tilde\nu}$ maps $(g,h) \in \mathcal{K}_{\tilde\nu}^{\tilde\eta} \times \mathcal{K}_{\tilde\eta}$ on $\Big(\big(\alpha_{\tilde\nu}^{\tilde\eta}(a_h)\big)(g), \sigma_{\tilde\eta}(h)\Big)$ and fixes all the other elements in $\mathcal{K}_{\tilde\nu}$, and where $b_{h\circ g}$ fulfills

$$\big(\alpha_{\tilde\nu}^{\tilde\eta}(a_h)\big)(g) \circ b_{h \circ g} = a_h \circ g \qquad 22.17.4,$$

whereas $b_\ell = \{1\}$ otherwise. Then, denoting by $\bar a_h$ and $\bar b_\ell$ the respective classes of a_h in $\mathfrak{ab}\big(\mathrm{Aut}(Q \times_{\tilde\eta} P)\big)$ and of b_ℓ in $\mathfrak{ab}\big(\mathrm{Aut}(R \times_{\tilde\nu} P)\big)$, it is clear that $\mathfrak{k}_{\tilde\nu}^{\tilde\eta}$ maps $\prod_{h \in \mathcal{K}_{\tilde\eta}} \bar a_h$ on $\prod_{\ell \in \mathcal{K}_{\tilde\nu}} \bar b_\ell$.

More explicitly, we may choose the set of representatives $\mathcal{I}_{\tilde\nu}^{\tilde\eta}$ inside the set of representatives $\mathcal{K}_{\tilde\nu}^{\tilde\eta}$; moreover, since $\mathrm{Aut}(Q \times_{\tilde\eta} P)$ acts *freely* on $Q \times_{\tilde\eta} P$, it still acts freely on the set of injective $Q \times P$-set homomorphisms 22.16.2 and therefore, for any $g \in \mathcal{I}_{\tilde\nu}^{\tilde\eta}$, choosing a set of representatives $A_g \subset \mathrm{Aut}(Q \times_{\tilde\eta} P)$ for $\mathrm{Aut}(Q \times_{\tilde\eta} P)/\mathrm{Aut}(Q \times_{\tilde\eta} P)_g$, we may assume that

$$\mathcal{K}_{\tilde\nu}^{\tilde\eta} = \bigsqcup_{g \in \mathcal{I}_{\tilde\nu}^{\tilde\eta}} A_g \circ g \qquad 22.17.5.$$

With these choices in hand, for any $g \in \mathcal{I}_{\tilde\nu}^{\tilde\eta}$ and any $a_g \in A_g$, equality 22.17.4 becomes

$$\big(\alpha_{\tilde\nu}^{\tilde\eta}(a_h)\big)(a_g \circ g) \circ b_{h\circ a_g\circ g} = a_h \circ a_g \circ g \qquad 22.17.6$$

and, since the element $\big(\alpha_{\tilde\nu}^{\tilde\eta}(a_h)\big)(a_g\circ g)$ of $\mathcal{K}_{\tilde\nu}^{\tilde\eta}$ belongs to the $\mathrm{Aut}(Q\times_{\tilde\eta}P)$-orbit of g, for some $\hat a_g \in A_g$ we have

$$\big(\alpha_{\tilde\nu}^{\tilde\eta}(a_h)\big)(a_g \circ g) = \hat a_g \circ g \qquad 22.17.7;$$

now, equality 22.17.6 implies that

$$g \circ b_{h\circ a_g\circ g} = (\hat a_g)^{-1} \circ a_h \circ a_g \circ g \qquad 22.17.8$$

and therefore, by the very definition of β_g, finally we get

$$b_{h\circ a_g\circ g} = \beta_g\big((\hat a_g)^{-1} \circ a_h \circ a_g\big) \qquad 22.17.9.$$

Consequently, it follows from the argument above and from all the equalities 22.17.9 that for any $a \in \mathrm{Aut}(Q \times_{\tilde\eta} P)$ we have

$$\begin{aligned}\alpha_{\tilde\nu}^{\tilde\eta}(\bar a) &= \prod_{g\in\mathcal{I}_{\tilde\nu}^{\tilde\eta}} \prod_{a_g\in A_g} \overline{\beta_g\big((\hat a_g)^{-1} \circ a \circ a_g\big)} \\ &= \prod_{g\in\mathcal{I}_{\tilde\nu}^{\tilde\eta}} \mathfrak{ab}(\beta_g)\Big(\prod_{a_g\in A_g} \overline{(\hat a_g)^{-1} \circ a \circ a_g}\Big)\end{aligned} \qquad 22.17.10$$

where $\bar a$ is the image of a in $\mathfrak{ab}\big(\mathrm{Aut}(Q \times_{\tilde\eta} P)\big)$ and, for any $a_g \in A_g$, we denote by $\overline{\beta_g(a_g)}$ the image of $\beta_g(a_g)$ in $\mathfrak{ab}\big(\mathrm{Aut}(R \times_{\tilde\nu} P)\big)$ and by $\hat a_g$ the element of A_g such that $a \circ a_g$ belongs to $\hat a_g \circ \mathrm{Aut}(Q \times_{\tilde\eta} P)_g$ (cf. 22.17.7). But, by the very definition of $\mathfrak{ab}^\circ(\iota_g)$, for any $g \in \mathcal{I}_{\tilde\nu}^{\tilde\eta}$ we have

$$\big(\mathfrak{ab}^\circ(\iota_g)\big)(\bar a) = \prod_{a_g\in A_g} \overline{(\hat a_g)^{-1} \circ a \circ a_g} \qquad 22.17.11.$$

We are done.

22.18 Finally, let us be more explicit on homomorphisms 22.16.2. For any subgroup Q of P and any $\mathfrak{pc}(\tilde{\mathcal{F}})$-morphism $\tilde\eta : T \to Q\tilde\times P$ in $\mathfrak{O}_Q$, choose a representative η of $\hat{\tilde\eta}$; then, for any $\mathcal{F}$-morphism $\varphi : R \to Q$, any $\mathfrak{pc}(\tilde{\mathcal{F}})$-morphism $\tilde\nu : U \to R\tilde\times P$ in $\mathfrak{O}_R$ and any representative ν of $\hat{\tilde\nu}$, it is clear that an injective $R \times P$-set homomorphism

$$g : (R \times P)/\Delta_\nu(U) \longrightarrow \mathrm{Res}_{\varphi\times\mathrm{id}_P}\big((Q \times P)/\Delta_\eta(T)\big) \qquad 22.18.1$$

is determined by the image $(w,\hat{w})\Delta_\eta(T)$ of the class $(1,1)\Delta_\nu(U)$ where $(w,\hat{w})$ is an element of $Q\times P$ fulfilling

$$\varphi(U)^w = \varphi(R)^w \cap T \quad \text{and} \quad \eta\big(\varphi(v)^w\big) = \nu(v)^{\hat{w}} \qquad 22.18.2$$

for any $v \in U$; in this case, we denote by $\psi_w : U \to T$ the $\mathcal{F}$-morphism mapping $v\in U$ on $\varphi(v)^w$.

Lemma 22.19 *With the notation above, the correspondence mapping any injective $R\times P$-set homomorphism*

$$g : (R\times P)/\Delta_\nu(U) \longrightarrow \mathrm{Res}_{\varphi\times \mathrm{id}_P}\big((Q\times P)/\Delta_\eta(T)\big) \qquad 22.19.1$$

such that $g\big((1,1)\Delta_\nu(U)\big) = (w,\hat{w})\Delta_\eta(T)$ on $\varphi(R)wT$ induces an injective map to $\varphi(R)\backslash Q/T$ from the set of $\mathrm{Aut}(R\times_{\hat{\nu}} P)$-orbits in the set of these homomorphisms. Moreover, a double class in $\varphi(R)\backslash Q/T$ belongs to the image of this map if and only if it admits a representative $w\in Q$ fulfilling

$$\varphi(U)^w = \varphi(R)^w \cap T \quad \text{and} \quad \hat{\tilde{\eta}}\circ\tilde{\psi}_w = \hat{\tilde{\nu}} \qquad 22.19.2.$$

Proof: Let g' be such an injective $R\times P$-set homomorphism, set

$$g'\big((1,1)\Delta_\nu(U)\big) = (w',\hat{w}')\Delta_\eta(T) \qquad 22.19.3$$

and, with the notation above, assume that $w' = \varphi(r)wt$ for some $r\in R$ and some $t\in T$; thus, w' and $\hat{w}'$ fulfill equalities 20.18.2 and in particular we have

$$\varphi(U) = \varphi(R)\cap T^{w'^{-1}} = \big(\varphi(R)\cap T^{w^{-1}}\big)^{\varphi(r)^{-1}} = \varphi(U^{r^{-1}}) \qquad 22.19.4,$$

so that r normalizes U.

Moreover, for any $v\in U$ we still have

$$\nu(v)^{\hat{w}'} = \eta\big(\varphi(v)^{w'}\big) = \eta\big(\varphi(v^r)^w\big)^{\eta(t)} = \nu(v^r)^{\hat{w}\eta(t)} \qquad 22.19.5,$$

so that the element $(r,\hat{w}'\eta(t)^{-1}\hat{w}^{-1})$ normalizes $\Delta_v(U)$; hence, since

$$(w',\hat{w}') = \big(\varphi(r),\hat{w}'\eta(t)^{-1}\hat{w}^{-1}\big)(w,\hat{w})\big(t,\eta(t)\big) \qquad 22.19.6,$$

g' belongs to the $\mathrm{Aut}(R\times_{\hat{\nu}} P)$-orbit of g (cf. 22.3.1).

Conversely, if g' belongs to the $\mathrm{Aut}(R\times_{\hat{\nu}} P)$-orbit of g, then there are $(r,\hat{r})\in N_{R\times P}\big(\Delta_\nu(U)\big)$ and $t\in T$ fulfilling

$$(w',\hat{w}') = \big(\varphi(r),\hat{r}\big)(w,\hat{w})\big(t,\eta(t)\big) \qquad 22.19.7,$$

so that $w' = \varphi(r)wt$.

Finally, it is clear that equalities 22.18.2 imply equalities 22.19.2; conversely, the equality $\hat{\tilde{\eta}}\circ\tilde{\psi}_w = \hat{\tilde{\nu}}$ forces the existence of $\hat{w}\in P$ which fulfills equalities 22.18.2 and, in particular, defines an injective $Q\times P$-set homomorphism g as above mapping on w. We are done.

Chapter 23

Narrowing the basic $\mathcal{F}^{sc}$-locality

23.1 Let P be a finite p-group and $\mathcal{F}$ a Frobenius P-category. As in chapter 4, denote by $\mathcal{F}^{sc}$ the full subcategory of $\mathcal{F}$ over the set of $\mathcal{F}$-selfcentralizing subgroups of P and by $\tilde{\mathcal{F}}^{sc}$ its *exterior quotient* (cf. 1.3). In this chapter, we exhibit some significant quotients of the full subcategory $\mathcal{L}^{b,sc}$ of $\mathcal{L}^{b}$ over the set of $\mathcal{F}$-selfcentralizing subgroups of P — namely, the *polycentral* and the *reduced $\mathcal{F}^{sc}$-localities* — which are *necessary* candidates to contain any *perfect $\mathcal{F}^{sc}$-locality*, as we show in the last chapter; in particular, we give a sufficient condition for the existence of a *perfect $\mathcal{F}^{sc}$-locality* (cf. 17.4 and 17.13). We freely use the notation introduced in the previous chapter and abundantly employ the terminology and the results of chapter 6.

23.2 Let Ω be a thick $\mathcal{F}$-basic $P \times P$-set and set $G = \mathrm{Aut}_{\{1\}\times P}(\Omega)$ (cf. 21.2); recall that, for any subgroup Q of P, we denote by $\mathfrak{O}_Q$ a set of representatives for the set of isomorphism classes of objects in the full subcategory $\tilde{\mathcal{F}}_{Q\tilde{\times}P}$ of $\mathfrak{pc}(\tilde{\mathcal{F}})_{Q\tilde{\times}P}$ over the set of $\mathfrak{pc}(\tilde{\mathcal{F}})$-morphisms $\tilde{\eta}:T \to Q\tilde{\times}P$ where T is a subgroup of P (cf. 22.5). For any $\tilde{\eta}:T \to Q\tilde{\times}P$ in $\mathfrak{O}_Q$, we assume that T is contained in Q and that the composition of $\tilde{\eta}$ with the structural $\mathfrak{pc}(\tilde{\mathcal{F}})$-morphism $Q\tilde{\times}P \to Q$ coincides with $\tilde{\iota}_T^Q$ (cf. 1.9). Then, denoting by $\tilde{\mathfrak{k}}$ the factorization of $\mathfrak{Ker}(\pi^{b})$ throughout $\tilde{\mathcal{F}}$ (cf. 22.13.1), we have (cf. isomorphism 22.7.1)

$$\tilde{\mathfrak{k}}(Q) = C_G(Q)/\mathfrak{S}(Q) \cong \prod_{\tilde{\eta}\in\mathfrak{O}_Q} \mathfrak{ab}\big(\mathrm{Aut}(Q \times_{\tilde{\eta}} P)\big) \qquad 23.2.1$$

and, for any $\tilde{\mathcal{F}}$-morphism $\tilde{\varphi}:R \to Q$, any $\tilde{\eta} \in \mathfrak{O}_Q$ and any $\tilde{\nu} \in \mathfrak{O}_R$, the evident homomorphisms (cf. 22.13.4)

$$\tilde{\mathfrak{k}}(\tilde{\varphi})_{\tilde{\nu}}^{\tilde{\eta}} : \mathfrak{ab}\big(\mathrm{Aut}(Q \times_{\tilde{\eta}} P)\big) \longrightarrow \mathfrak{ab}\big(\mathrm{Aut}(R \times_{\tilde{\nu}} P)\big) \qquad 23.2.2$$

are trivial unless $\tilde{\mathfrak{k}}(\tilde{\varphi})_{\tilde{\nu}}^{\tilde{\eta}} \neq 0$ which forces $\tilde{\mathcal{F}}(T,U) \neq \emptyset$ (cf. 22.18).

23.3 Thus, if $\mathfrak{X}$ is a nonempty set of subgroups of P containing any subgroup Q such that $\mathcal{F}(Q,R) \neq \emptyset$ for some $R \in \mathfrak{X}$ then, with the notation above, we have

23.3.1 $\tilde{\mathfrak{k}}(\tilde{\varphi})_{\tilde{\nu}}^{\tilde{\eta}} \neq 0$ *and* $U \in \mathfrak{X}$ *implies that* $T \in \mathfrak{X}$.

That is to say, denoting by $\mathfrak{O}_Q^{\mathfrak{X}}$ the set of $\mathfrak{pc}(\tilde{\mathcal{F}})$-morphisms $\tilde{\eta}:T \to Q\tilde{\times}P$ in $\mathfrak{O}_Q$ such that $T \in \mathfrak{X}$, and by $\mathfrak{S}^{\mathfrak{X}}(Q)$ the subgroup of $C_G(Q)$ defined by

$$\mathfrak{S}^{\mathfrak{X}}(Q)/\mathfrak{S}(Q) \cong \prod_{\tilde{\eta}\in\mathfrak{O}_Q-\mathfrak{O}_Q^{\mathfrak{X}}} \mathfrak{ab}\big(\mathrm{Aut}(Q \times_{\tilde{\eta}} P)\big) \qquad 23.3.2,$$

so that we have

$$C_G(Q)/\mathfrak{S}^{\mathfrak{X}}(Q) \cong \prod_{\tilde\eta\in\mathfrak{O}_Q^{\mathfrak{X}}} \mathfrak{ab}\big(\mathrm{Aut}(Q\times_{\tilde\eta}P)\big) \qquad 23.3.3,$$

it is easily checked from this remark that, for any $\mathcal{F}$-morphism $\varphi: R\to Q$, we still have $\mathfrak{S}^{\mathfrak{X}}(Q)^x \subset \mathfrak{S}^{\mathfrak{X}}(R)$ for any $x\in G$ inducing φ.

23.4 Consequently, respectively denoting by $\mathcal{F}^{\mathfrak{X}}$ and $\mathcal{L}^{\mathrm{b},\mathfrak{X}}$ the full subcategories over $\mathfrak{X}$ of $\mathcal{F}$ and $\mathcal{L}^{\mathrm{b}}$, we obtain a new *coherent* $\mathcal{F}^{\mathfrak{X}}$*-locality* $\bar{\mathcal{L}}^{\mathrm{b},\mathfrak{X}}$ (cf. 17.4) which is the *quotient* of $\mathcal{L}^{\mathrm{b},\mathfrak{X}}$ defined by

$$\bar{\mathcal{L}}^{\mathrm{b},\mathfrak{X}}(Q,R) = \mathcal{T}_G(Q,R)/\mathfrak{S}^{\mathfrak{X}}(R) \qquad 23.4.1$$

for any pair of elements Q and R of $\mathfrak{X}$, together with the induced functors

$$\bar\tau^{\mathrm{b},\mathfrak{X}}:\mathcal{T}_P^{\mathfrak{X}}\to\bar{\mathcal{L}}^{\mathrm{b},\mathfrak{X}} \quad\text{and}\quad \bar\pi^{\mathrm{b},\mathfrak{X}}:\bar{\mathcal{L}}^{\mathrm{b},\mathfrak{X}}\to\mathcal{F}^{\mathfrak{X}} \qquad 23.4.2.$$

We are particularly interested in the case where $\mathfrak{X}$ is the set of $\mathcal{F}$-selfcentralizing subgroups of P and then we denote by $(\bar\tau^{\mathrm{b,sc}},\bar{\mathcal{L}}^{\mathrm{b,sc}},\bar\pi^{\mathrm{b,sc}})$ the corresponding $\mathcal{F}^{\mathrm{sc}}$-locality.

23.5 First of all, let us give an explicit description of the *contravariant* functor

$$\tilde{\mathfrak{k}}^{\mathrm{sc}}:\tilde{\mathcal{F}}^{\mathrm{sc}}\longrightarrow\mathfrak{Ab} \qquad 23.5.1$$

induced by $\mathfrak{Ker}(\bar\pi^{\mathrm{b,sc}})$ (cf. 17.8). Consider the *additive cover* $\mathfrak{ac}(\tilde{\mathcal{F}}^{\mathrm{sc}})$ of $\tilde{\mathcal{F}}^{\mathrm{sc}}$ (cf. 6.2) which coincides with $\mathfrak{pc}\big((\tilde{\mathcal{F}}^{\mathrm{sc}})^\circ\big)^\circ$ (cf. A2.7.3); the existence of the *exterior intersection* in $\mathfrak{ac}(\tilde{\mathcal{F}}^{\mathrm{sc}})$ (cf. 6.8) allows us to define a *distributive* functor

$$\mathfrak{pc}(\tilde{\mathcal{F}}^{\mathrm{sc}})\longrightarrow\mathfrak{ac}(\tilde{\mathcal{F}}^{\mathrm{sc}}) \qquad 23.5.2$$

which is the identity over the *full* subcategory $\tilde{\mathcal{F}}^{\mathrm{sc}}$ of both covers and, for any pair of $\tilde{\mathcal{F}}^{\mathrm{sc}}$-objects Q and R, maps $Q\tilde\times R$ on $Q\tilde\cap R$ (cf. 22.4); the fact that the *exterior intersection* is a *direct product* in $\mathfrak{ac}(\tilde{\mathcal{F}}^{\mathrm{sc}})$ (cf. Proposition 6.14) guarantees the existence and uniqueness of such a functor.

23.6 Actually, this functor induces an equivalence of categories between the subcategories $(\tilde{\mathcal{F}}^{\mathrm{sc}})_{Q\tilde\times R}$ of $\mathfrak{pc}(\tilde{\mathcal{F}}^{\mathrm{sc}})_{Q\tilde\times R}$ and $(\tilde{\mathcal{F}}^{\mathrm{sc}})_{Q\tilde\cap R}$ of $\mathfrak{ac}(\tilde{\mathcal{F}}^{\mathrm{sc}})_{Q\tilde\cap R}$ (cf. 23.2); coherently, we denote by $\mathfrak{O}^{Q,\mathrm{sc}}$ the corresponding set of representatives for the set of isomorphism classes of objects in $(\tilde{\mathcal{F}}^{\mathrm{sc}})_{Q\tilde\cap P}$ and, for any $(\tilde{\mathcal{F}}^{\mathrm{sc}})_{Q\tilde\cap P}$-object $\tilde\eta: T\to Q\tilde\cap P$, we still denote by $Q\times_{\tilde\eta}P$ the corresponding $Q\times P$-set. Moreover, let us denote by $\Omega^{Q,\mathrm{sc}}$ the union of all the $Q\times P$-orbits in Ω which are isomorphic to $Q\times_{\tilde\eta}P$ for some $\tilde\eta\in\mathfrak{O}^{Q,\mathrm{sc}}$; note that, since all the $Q\times P$-orbits outside $\Omega^{Q,\mathrm{sc}}$ have a cardinal divisible by $p|P|$, we still have

$$|\Omega^{Q,\mathrm{sc}}|/|P| \equiv |\Omega|/|P| \neq 0 \pmod p \qquad 23.6.1.$$

23.7 More explicitly, for any $\mathcal{F}$-selfcentralizing subgroup Q of P, choose a set of representatives $\tilde{\mathcal{S}}_Q$ for the set of Q-conjugacy classes of the $\mathcal{F}$-selfcentralizing subgroups Q' of Q such that (cf. 6.4.1)

$$\tilde{\mathcal{F}}(P,Q')_{\tilde{\iota}^Q_{Q'}} \neq \emptyset \qquad 23.7.1$$

and, for any $Q' \in \tilde{\mathcal{S}}_Q$, choose a set of representatives $\tilde{\mathcal{M}}^Q_{Q'}$ for the set of $\tilde{\mathcal{F}}_Q(Q')$-orbits in $\tilde{\mathcal{F}}(P,Q')_{\tilde{\iota}^Q_{Q'}}$. Then, according to our definition of the *exterior intersection* (cf. 6.11), we may assume that

$$Q\,\tilde{\cap}\,P = \bigoplus_{Q'\in\tilde{\mathcal{S}}_Q}\ \bigoplus_{\tilde{\eta}'\in\tilde{\mathcal{M}}^Q_{Q'}} Q' \qquad 23.7.2;$$

in particular any $Q' \in \tilde{\mathcal{S}}_Q$ and any $\tilde{\eta}' \in \tilde{\mathcal{M}}^Q_{Q'}$ determine an $(\tilde{\mathcal{F}}^{^{\mathrm{sc}}})_{Q\,\tilde{\cap}\,P}$-object

$$\tilde{\theta}_{Q',\tilde{\eta}'} : Q' \longrightarrow Q\,\tilde{\cap}\,P \qquad 23.7.3$$

which is clearly *maximal* (cf. A5.1); these $(\tilde{\mathcal{F}}^{^{\mathrm{sc}}})_{Q\,\tilde{\cap}\,P}$-objects are mutually nonisomorphic and we denote by $\mathfrak{O}^{Q,\mathrm{sc}}_{\max} \subset \mathfrak{O}^{Q,\mathrm{sc}}$ the subset of them.

23.8 Conversely, any $(\tilde{\mathcal{F}}^{^{\mathrm{sc}}})_{Q\,\tilde{\cap}\,P}$-object $\tilde{\eta}: T \to Q\,\tilde{\cap}\,P$ determines some $Q' \in \tilde{\mathcal{S}}_Q$ and some $\tilde{\eta}' \in \tilde{\mathcal{M}}^Q_{Q'}$ (cf. 23.7.2); moreover, up to an $(\tilde{\mathcal{F}}^{^{\mathrm{sc}}})_{Q\,\tilde{\cap}\,P}$-isomorphism, we may assume that Q' contains T and that $\tilde{\iota}^Q_T$ and $\tilde{\eta}'\circ\tilde{\iota}^{Q'}_T$ coincide with the composition of $\tilde{\eta}$ with the respective structural $\mathfrak{ac}(\tilde{\mathcal{F}}^{^{\mathrm{sc}}})$-morphisms

$$Q\,\tilde{\cap}\,P \longrightarrow Q \quad \text{and} \quad Q\,\tilde{\cap}\,P \longrightarrow P \qquad 23.8.1;$$

we choose a representative η' of $\tilde{\eta}'\circ\tilde{\iota}^{Q'}_T$ and set $\eta = \eta'\circ\iota^{Q'}_T$. Then, it is easily checked that we have

$$N_{Q\times P}\big(\Delta_\eta(T)\big) = \big(Z(T)\times\{1\}\big)\rtimes\Delta_{\eta'}\big(N_{Q'}(T)\big) \qquad 23.8.2$$

and that $\bar{N}_{Q'}(T)$ is isomorphic to the group of $(\tilde{\mathcal{F}}^{^{\mathrm{sc}}})_{Q\,\tilde{\cap}\,P}$-automorphisms of the $(\tilde{\mathcal{F}}^{^{\mathrm{sc}}})_{Q\,\tilde{\cap}\,P}$-object $\tilde{\eta}: T \to Q\,\tilde{\cap}\,P$, which actually coincides with the intersection (cf. Proposition 6.14)

$$\mathrm{Aut}(\tilde{\eta}) = (\tilde{\mathcal{F}}^{^{\mathrm{sc}}})_{Q\,\tilde{\cap}\,P}(\tilde{\eta}) = \tilde{\mathcal{F}}_Q(T)\cap{}^{\eta^*}\tilde{\mathcal{F}}_P\big(\eta(T)\big) \qquad 23.8.3$$

where η^* denotes the inverse of the isomorphism $T\cong\eta(T)$ induced by η; in particular, setting $Z(\tilde{\eta}) = Z(T)$ and considering the evident action of $\mathrm{Aut}(\tilde{\eta})$ on $Z(\tilde{\eta})$, from isomorphism 22.3.1 we get

$$\mathrm{Aut}(Q\times_{\tilde{\eta}} P) \cong Z(\tilde{\eta})\rtimes\mathrm{Aut}(\tilde{\eta}) \qquad 23.8.4$$

and therefore, setting $\bar{Z}(\tilde{\eta}) = Z(\tilde{\eta})/\big[Z(\tilde{\eta}),\mathrm{Aut}(\tilde{\eta})\big]$, we still get

$$\mathfrak{ab}\big(\mathrm{Aut}(Q\times_{\tilde{\eta}} P)\big) \cong \bar{Z}(\tilde{\eta})\times\mathfrak{ab}\big(\mathrm{Aut}(\tilde{\eta})\big) \qquad 23.8.5.$$

23.9 At this point, for any $\mathcal{F}$-selfcentralizing subgroup Q of P, from isomorphism 23.3.3 we get

$$\tilde{\mathfrak{k}}^{^{\rm sc}}(Q) = C_G(Q)/\mathfrak{S}^{^{\rm sc}}(Q) \cong \prod_{\tilde\eta \in \mathfrak{O}^{Q,{\rm sc}}} \bar Z(\tilde\eta) \times \mathfrak{ab}\big({\rm Aut}(\tilde\eta)\big) \qquad 23.9.1.$$

Moreover, for any $\mathcal{F}$-selfcentralizing subgroup R of Q, any $\tilde\eta \in \mathfrak{O}^{^{Q,{\rm sc}}}$ and any $\tilde\nu \in \mathfrak{O}^{^{R,{\rm sc}}}$, it follows from Proposition 22.17 that the group homomorphism

$$\tilde{\mathfrak{k}}^{^{\rm sc}}(\tilde\iota_R^Q) : \bar Z(\tilde\eta) \times \mathfrak{ab}\big({\rm Aut}(\tilde\eta)\big) \longrightarrow \bar Z(\tilde\nu) \times \mathfrak{ab}\big({\rm Aut}(\tilde\nu)\big) \qquad 23.9.2$$

is equal to

$$\tilde{\mathfrak{k}}_{\tilde\nu}^{\tilde\eta} = \prod_{g \in \mathcal{I}_{\tilde\nu}^{\tilde\eta}} \mathfrak{ab}(\beta_g) \circ \mathfrak{ab}^\circ(\iota_g) \qquad 23.9.3,$$

where $\mathcal{I}_{\tilde\nu}^{\tilde\eta}$ is a set of representatives for the set of orbits of the direct product ${\rm Aut}(Q \times_{\tilde\eta} P) \times {\rm Aut}(R \times_{\tilde\nu} P)$ over the set of injective $Q \times P$-set homomorphisms

$$g : R \times_{\tilde\nu} P \longrightarrow {\rm Res}_{R\times P}(Q \times_{\tilde\eta} P) \qquad 23.9.4,$$

and where we denote by ${\rm Aut}(Q \times_{\tilde\eta} P)_g$ the stabilizer of $g \circ {\rm Aut}(R \times_{\tilde\nu} P)$ in ${\rm Aut}(Q \times_{\tilde\eta} P)$, by

$$\iota_g : {\rm Aut}(Q \times_{\tilde\eta} P)_g \longrightarrow {\rm Aut}(Q \times_{\tilde\eta} P) \qquad 23.9.5$$

the inclusion map and by

$$\beta_g : {\rm Aut}(Q \times_{\tilde\eta} P)_g \longrightarrow {\rm Aut}(R \times_{\tilde\nu} P) \qquad 23.9.6$$

the group homomorphism fulfilling $a \circ g = g \circ \beta_g(a)$ for any $a \in {\rm Aut}(Q \times_{\tilde\eta} P)_g$. We are ready to exhibit the *polycentral* $\mathcal{F}^{^{\rm sc}}$*-locality* $(\tau^{\rm c,sc}, \mathcal{L}^{^{\rm c,sc}}, \pi^{\rm c,sc})$.

Proposition 23.10 *With the notation above, we have a natural direct decomposition* $\tilde{\mathfrak{k}}^{^{\rm sc}} \cong \hat{\hat{\mathfrak{z}}} \times \hat{\hat{\mathfrak{k}}}$ *where* $\hat{\hat{\mathfrak{z}}}$ *and* $\hat{\hat{\mathfrak{k}}}$ *are contravariant functors from* $\tilde{\mathcal{F}}^{^{\rm sc}}$ *to* $\mathfrak{Ab}$ *mapping any* $\mathcal{F}$*-selfcentralizing subgroup* Q *of* P *on*

$$\hat{\hat{\mathfrak{z}}}(Q) = \prod_{\tilde\eta \in \mathfrak{O}^{Q,{\rm sc}}} \bar Z(\tilde\eta) \quad \text{and} \quad \hat{\hat{\mathfrak{k}}}(Q) = \prod_{\tilde\eta \in \mathfrak{O}^{Q,{\rm sc}}} \mathfrak{ab}\big({\rm Aut}(\tilde\eta)\big) \qquad 23.10.1.$$

In particular, there is a coherent $\mathcal{F}^{^{\rm sc}}$*-locality* $(\tau^{\rm c,sc}, \mathcal{L}^{^{\rm c,sc}}, \pi^{\rm c,sc})$ *where*

$$\mathcal{L}^{^{\rm c,sc}}(Q,R) = \bar{\mathcal{L}}^{^{\rm b,sc}}(Q,R)/\,\hat{\hat{\mathfrak{k}}}(R) \qquad 23.10.2$$

for any pair of $\mathcal{F}$*-selfcentralizing subgroups* Q *and* R *of* P, *and where the structural functors are induced by* $\bar\tau^{\rm b,sc}$ *and* $\bar\pi^{\rm b,sc}$.

Proof: With the notation in 23.9 above, it is quite clear that it suffices to show that the group homomorphism $\tilde{\mathfrak{k}}_{\tilde\nu}^{\tilde\eta}$ in 23.9.2 maps $\bar Z(\tilde\eta)$ on $\bar Z(\tilde\nu)$ and $\mathfrak{ab}\big({\rm Aut}(\tilde\eta)\big)$ on $\mathfrak{ab}\big({\rm Aut}(\tilde\nu)\big)$.

First of all, it is easily checked that the image of $Z(\tilde\eta)$ in $\mathrm{Aut}(Q\times_{\tilde\eta}P)$ stabilizes $g\circ\mathrm{Aut}(R\times_{\tilde\nu}P)$ and therefore, denoting by $\mathrm{Aut}(\tilde\eta)_g$ the corresponding stabilizer in $\mathrm{Aut}(\tilde\eta)$, we have

$$\mathrm{Aut}(Q\times_{\tilde\eta}P)_g\cong Z(\tilde\eta)\rtimes\mathrm{Aut}(\tilde\eta)_g \qquad 23.10.3;$$

then, it is quite clear that ι_g decomposes on the direct product of the identity over $Z(\tilde\eta)$ and the inclusion map

$$\hat\iota_g:\mathrm{Aut}(\tilde\eta)_g\longrightarrow\mathrm{Aut}(\tilde\eta) \qquad 23.10.4;$$

coherently, it is easily checked that $\mathfrak{ab}^\circ(\iota_g)$ still decomposes on the direct product of the homomorphism $\mathfrak{ab}^\circ(\hat\iota_g)$ and the group homomorphism

$$\bar Z(\tilde\eta)\longrightarrow Z(\tilde\eta)/\big[\mathrm{Aut}(\tilde\eta)_g,Z(\tilde\eta)\big] \qquad 23.10.5$$

mapping the class of $z\in Z(\tilde\eta)$ in the left-hand term on the class of $\prod_u uzu^{-1}$ in the right-hand term where, with the notation in 23.7 above, $u\in N_{Q'}(T)$ runs over a set of representatives for $\mathrm{Aut}(\tilde\eta)/\mathrm{Aut}(\tilde\eta)_g$.

On the other hand, denoting by T and U the respective origins of $\tilde\eta$ and $\tilde\nu$, and by $\psi_g:U\to T$ the group homomorphism induced by g (cf. 22.18), we already know that $\tilde\iota_T^Q\circ\tilde\psi_g=\tilde\iota_U^Q$ and $\hat{\tilde\eta}\circ\tilde\psi_g=\hat{\tilde\nu}$ (cf. 22.19.2) and therefore we get $\tilde\eta\circ\tilde\psi_g=(\tilde\iota_R^Q\,\tilde\cap\,\tilde{\mathrm{id}}_P)\circ\tilde\nu$; that is to say, with the notation in 23.7 above, $\tilde\nu:U\to R\,\tilde\cap\,P$ determines some $R'\in\tilde{\mathcal{S}}_R$ and some $\tilde\nu'\in\tilde{\mathcal{M}}_{R'}^R$, and then $\tilde\psi_g:U\to T$ induces an $\tilde{\mathcal{F}}$-morphism $\tilde\psi'_g:R'\to Q'$ fulfilling

$$\tilde\iota_{Q'}^Q\circ\tilde\psi'_g=\tilde\iota_{R'}^Q\quad\text{and}\quad\tilde\eta'\circ\tilde\psi'_g=\tilde\nu' \qquad 23.10.6;$$

thus, choosing a coherent family of representatives η' of $\tilde\eta'\circ\tilde\iota_T^{Q'}$ and ν' of $\tilde\nu'\circ\tilde\iota_U^{R'}$, and setting $\eta=\eta'\circ\iota_T^{Q'}$ and $\nu=\nu'\circ\iota_U^{R'}$, we may assume that we have $\eta'\big(\psi'_g(v')\big)=\nu'(v')$ for any $v'\in N_{R'}(U)$, and then we still have

$$(\psi'_g\times\mathrm{id}_P)^{-1}\Big(\Delta_{\eta'}\big(N_{Q'}(T)\big)\Big)\subset\Delta_{\nu'}\big(N_{R'}(U)\big) \qquad 23.10.7.$$

Consequently, up to a suitable choice of isomorphisms 23.10.3, β_g induces an injective group homomorphism

$$\hat\beta_g:\mathrm{Aut}(\tilde\eta)_g\longrightarrow\mathrm{Aut}(\tilde\nu) \qquad 23.10.8$$

mapping any element of $\mathrm{Aut}(\tilde\eta)_g$, which is necessarily determined by $\Delta_{\eta'}(u')$ for some $u'\in N_{Q'}(T)$ in the image of ψ'_g, on the element of $\mathrm{Aut}(\tilde\nu)$ determined by the element $v'\in N_{R'}(U)$ fulfilling $\psi'_g(v')=u'$; with this choice, throughout isomorphism 23.10.3 β_g also decomposes on the direct product of $\mathfrak{z}_{\mathcal{F}^{\rm sc}}(\psi_g):Z(T)\to Z(U)$ (cf. 13.13.2) and $\hat\beta_g$. Finally, independently of any choice, $\mathfrak{ab}(\beta_g)$ decomposes on the direct product of $\mathfrak{ab}(\hat\beta_g)$ and the group homomorphism

$$Z(\tilde\eta)/\big[\mathrm{Aut}(\tilde\eta)_g,Z(\tilde\eta)\big]\longrightarrow\bar Z(\tilde\nu) \qquad 23.10.9$$

induced by $\mathfrak{z}_{\mathcal{F}^{\rm sc}}(\psi_g):Z(\tilde\eta)\to Z(\tilde\nu)$. We are done.

23.11 As a matter of fact, for a suitable *representation* $\tilde{\mathcal{F}}^{sc} \to \mathfrak{CC}$ of $\tilde{\mathcal{F}}^{sc}$ endowed with a suitable *weak structure* (cf. A2.3), the corresponding *stable adjoin image* introduced in A2.14 supplies a good description of the functor $\hat{\hat{\mathfrak{z}}}$ above; then, this description provides a significant *vanishing cohomology property* (cf. Theorem 23.20 and equality 23.23.5 below). Explicitly, let us consider the functor

$$\mathfrak{int}_P : \tilde{\mathcal{F}}^{sc} \longrightarrow \mathfrak{ac}(\tilde{\mathcal{F}}^{sc}) \qquad 23.11.1$$

mapping any $\tilde{\mathcal{F}}^{sc}$-object Q on $Q \tilde{\cap} P$ and any $\tilde{\mathcal{F}}^{sc}$-morphism $\tilde{\varphi}$ on $\tilde{\varphi} \tilde{\cap} \widetilde{\mathrm{id}}_P$; then, the *representation* of $\tilde{\mathcal{F}}^{sc}$ we have to consider is the composed functor (cf. A2.2 and A2.4)

$$\mathfrak{rn} = \mathfrak{rg}_{\mathfrak{ac}(\tilde{\mathcal{F}}^{sc})} \circ \mathfrak{int}_P : \tilde{\mathcal{F}}^{sc} \longrightarrow \mathfrak{ac}(\tilde{\mathcal{F}}^{sc}) \longrightarrow \mathfrak{CC} \qquad 23.11.2$$

which maps any $\tilde{\mathcal{F}}^{sc}$-object Q on the category $\mathfrak{ac}(\tilde{\mathcal{F}}^{sc})_{Q \tilde{\cap} P}$, and the *weak structure* (cf. A2.3) we need

$$\mathfrak{rn}^{\mathfrak{i}} = \mathfrak{rg}^{\mathfrak{i}}_{\mathfrak{ac}(\tilde{\mathcal{F}}^{sc})} \circ \mathfrak{int}_P : \tilde{\mathcal{F}}^{sc} \longrightarrow \mathfrak{CC} \qquad 23.11.3$$

maps any $\tilde{\mathcal{F}}^{sc}$-object Q on the subcategory $\mathfrak{ac}(\tilde{\mathcal{F}}^{sc})^{\mathfrak{i}}_{Q \tilde{\cap} P}$ of $\mathfrak{ac}(\tilde{\mathcal{F}}^{sc})_{Q \tilde{\cap} P}$ with the *same* objects and only with the $\mathfrak{ac}(\tilde{\mathcal{F}}^{sc})_{Q \tilde{\cap} P}$-morphisms defined by an injective map and $\tilde{\mathcal{F}}^{sc}$-isomorphisms (cf. 1.7 and 6.2).

23.12 The point is that, for any $\tilde{\mathcal{F}}^{sc}$-morphism $\tilde{\varphi} : R \to Q$, the functor (cf. 1.7)

$$\mathfrak{ac}(\tilde{\mathcal{F}}^{sc})_{\tilde{\varphi} \tilde{\cap} \widetilde{\mathrm{id}}_P} : \mathfrak{ac}(\tilde{\mathcal{F}}^{sc})_{R \tilde{\cap} P} \longrightarrow \mathfrak{ac}(\tilde{\mathcal{F}}^{sc})_{Q \tilde{\cap} P} \qquad 23.12.1$$

admits a *right adjoin*

$$\big(\mathfrak{ac}(\tilde{\mathcal{F}}^{sc})_{\tilde{\varphi} \tilde{\cap} \widetilde{\mathrm{id}}_P}\big)^{\mathfrak{a}} : \mathfrak{ac}(\tilde{\mathcal{F}}^{sc})_{Q \tilde{\cap} P} \longrightarrow \mathfrak{ac}(\tilde{\mathcal{F}}^{sc})_{R \tilde{\cap} P} \qquad 23.12.2$$

defined by the $\mathfrak{ac}(\tilde{\mathcal{F}}^{sc})$*-pull-backs* (cf. Proposition 6.21)

$$\begin{array}{ccc} R \tilde{\cap} P & \xrightarrow{\tilde{\varphi} \tilde{\cap} \widetilde{\mathrm{id}}_P} & Q \tilde{\cap} P \\ \tilde{\nu} \uparrow & & \uparrow \tilde{\eta} \\ U & \xrightarrow{\tilde{\psi}} & T \end{array} \qquad 23.12.3$$

and by the $\mathfrak{ac}(\tilde{\mathcal{F}}^{sc})$-morphisms between them

$$\begin{array}{ccccccc} U' & & & \xrightarrow{\tilde{\psi}'} & & & T' \\ & \searrow^{\tilde{\nu}'} & & & & {}^{\tilde{\eta}'}\swarrow & \\ \tilde{\omega}\uparrow & & R \tilde{\cap} P & \longrightarrow & Q \tilde{\cap} P & & \uparrow\tilde{\theta} \\ & \nearrow_{\tilde{\nu}} & & & & {}_{\tilde{\eta}}\nwarrow & \\ U & & & \xrightarrow{\tilde{\psi}} & & & T \end{array} \qquad 23.12.4;$$

namely, $\big(\mathfrak{ac}(\tilde{\mathcal{F}}^{sc})_{\tilde\varphi\,\tilde\cap\,\widetilde{\mathrm{id}}_P}\big)^{\mathfrak{a}}$ sends the $\mathfrak{ac}(\tilde{\mathcal{F}}^{sc})_{Q\,\tilde\cap\,P}$-object $\tilde\eta$ in diagram 23.12.3 to the $\mathfrak{ac}(\tilde{\mathcal{F}}^{sc})_{R\,\tilde\cap\,P}$-object $\tilde\nu$, and the $\mathfrak{ac}(\tilde{\mathcal{F}}^{sc})_{Q\,\tilde\cap\,P}$-morphism $\tilde\theta$ in diagram 23.12.4 to the $\mathfrak{ac}(\tilde{\mathcal{F}}^{sc})_{R\,\tilde\cap\,P}$-morphism $\tilde\omega$. Moreover, it is easily checked that $\big(\mathfrak{ac}(\tilde{\mathcal{F}}^{sc})_{\tilde\varphi\,\tilde\cap\,\widetilde{\mathrm{id}}_P}\big)^{\mathfrak{a}}$ preserves the *weak structure*

$$\big(\mathfrak{ac}(\tilde{\mathcal{F}}^{sc})_{\tilde\varphi\,\tilde\cap\,\widetilde{\mathrm{id}}_P}\big)^{\mathfrak{a}}\big(\mathfrak{ac}(\tilde{\mathcal{F}}^{sc})^{\mathfrak{i}}_{Q\,\tilde\cap\,P}\big)\subset \mathfrak{ac}(\tilde{\mathcal{F}}^{sc})^{\mathfrak{i}}_{R\,\tilde\cap\,P} \qquad 23.12.5.$$

23.13 Consequently, according to A2.14, we have the $\mathfrak{rn}^{\mathfrak{i}}$-*stable adjoin image* functor

$$(\mathfrak{p}_{\mathfrak{rn},\mathfrak{rn}^{\mathfrak{i}}})_{\mathfrak{a}} : \mathfrak{Fct}\big((\mathfrak{rn}\rtimes\tilde{\mathcal{F}}^{sc})^{\circ},\mathfrak{Ab}\big)\longrightarrow \mathfrak{Fct}\big((\tilde{\mathcal{F}}^{sc})^{\circ},\mathfrak{Ab}\big) \qquad 23.13.1;$$

moreover, for any $\tilde{\mathcal{F}}^{sc}$-object Q, it is clear that the category $\mathfrak{ac}(\tilde{\mathcal{F}}^{sc})_{Q\,\tilde\cap\,P}$ has a final object, namely the $\mathfrak{ac}(\tilde{\mathcal{F}}^{sc})$-morphism $\widetilde{\mathrm{id}}_{Q\,\tilde\cap\,P}$; hence, according to A2.17, we still have another functor and a natural map

$$\begin{aligned}\mathfrak{o}_{\mathfrak{rn}} &: \mathfrak{Fct}\big((\mathfrak{rn}\rtimes\tilde{\mathcal{F}}^{sc})^{\circ},\mathfrak{Ab}\big)\longrightarrow \mathfrak{Fct}\big((\tilde{\mathcal{F}}^{sc})^{\circ},\mathfrak{Ab}\big)\\ \omega_{\mathfrak{rn},\mathfrak{rn}^{\mathfrak{i}}} &: \mathfrak{o}_{\mathfrak{rn}}\longrightarrow (\mathfrak{p}_{\mathfrak{rn},\mathfrak{rn}^{\mathfrak{i}}})_{\mathfrak{a}}\end{aligned} \qquad 23.13.2.$$

23.14 More explicitly, note that for any $\tilde{\mathcal{F}}^{sc}$-morphism $\tilde\varphi : R\to Q$, the structural $\mathfrak{ac}(\tilde{\mathcal{F}}^{sc})$-morphisms $R\,\tilde\cap\,P\to R$ and $Q\,\tilde\cap\,P\to Q$ yield the $\mathfrak{ac}(\tilde{\mathcal{F}}^{sc})$-*pull-back*

$$\begin{array}{ccc} R & \xrightarrow{\ \tilde\varphi\ } & Q\\ \uparrow & & \uparrow\\ R\,\tilde\cap\,P & \xrightarrow{\ \tilde\varphi\,\tilde\cap\,\widetilde{\mathrm{id}}_P\ } & Q\,\tilde\cap\,P\end{array} \qquad 23.14.1;$$

hence, for any $\mathfrak{ac}(\tilde{\mathcal{F}}^{sc})$-morphism $\tilde\eta : T\to Q\,\tilde\cap\,P$ in $\mathfrak{O}^{Q,sc}$ — where we are assuming that $T\subset Q$ and that the composition of $\tilde\eta$ with the first structural map is defined by this inclusion (cf. 23.8) — it follows from the *transitivity* of *pull-backs* and from Proposition 6.19 that the $\mathfrak{ac}(\tilde{\mathcal{F}}^{sc})$-*pull-back* 23.12.3 is given by

$$\begin{array}{ccc} R\,\tilde\cap\,P & \xrightarrow{\ \tilde\varphi\,\tilde\cap\,\widetilde{\mathrm{id}}_P\ } & Q\,\tilde\cap\,P\\ {\scriptstyle(\tilde\nu_w)_{w\in W_{\tilde\eta}}}\uparrow & & \uparrow{\scriptstyle\tilde\eta}\\ \bigoplus_{w\in W_{\tilde\eta}} U_w & \xrightarrow{\ (\tilde\psi_w)_{w\in W_{\tilde\eta}}\ } & T\end{array} \qquad 23.14.2$$

where, for a choice of $\varphi\in\tilde\varphi$, $W_{\tilde\eta}$ is a set of representatives for the set of double classes $\varphi(R)wT\in\varphi(R)\backslash Q/T$ such that $U_w=\varphi(R)^w\cap T$ is $\mathcal{F}$-selfcentralizing, $\tilde\psi_w : U_w\to T$ is the $\tilde{\mathcal{F}}^{sc}$-morphism induced by the inclusion, and $\tilde\nu_w$ is defined by the $\mathcal{F}$-morphism $U_w\to R$ sending $\varphi(u)^w\in U_w$ to $u\in R$ and by the composition of $\tilde\eta\circ\tilde\psi_w$ with the structural $\tilde{\mathcal{F}}^{sc}$-morphism $Q\,\tilde\cap\,P\to P$.

23.15 Moreover, for any *contravariant* functor

$$\mathfrak{b} : \mathfrak{tn} \rtimes \tilde{\mathcal{F}}^{^{\mathrm{sc}}} \longrightarrow \mathfrak{Ab} \tag{23.15.1}$$

and any $\mathfrak{tn} \rtimes \tilde{\mathcal{F}}^{^{\mathrm{sc}}}$-object $(\tilde\eta, Q)$ (cf. A2.7), an automorphism $\tilde\sigma : \tilde\eta \cong \tilde\eta$ of the $\mathfrak{ac}(\tilde{\mathcal{F}}^{^{\mathrm{sc}}})_{Q\,\tilde\cap\,P}$-object $\tilde\eta$ determines the $\mathfrak{tn} \rtimes \tilde{\mathcal{F}}^{^{\mathrm{sc}}}$-automorphism $(\tilde\sigma, \widetilde{\mathrm{id}}_Q)$ of $(\tilde\eta, Q)$ and therefore $\mathrm{Aut}(\tilde\eta)$ acts on $\mathfrak{b}(\tilde\eta, Q)$ (cf. 23.8.3). In order to compute $\big((\mathfrak{p}_{\mathfrak{tn},\mathfrak{tn}^{\mathfrak{i}}})_{\mathfrak{a}}(\mathfrak{b})\big)(Q)$ (cf. 23.13.1) we have to consider the quotient

$$\bar{\mathfrak{b}}(\tilde\eta, Q) = \mathfrak{b}(\tilde\eta, Q)/\big[\mathfrak{b}(\tilde\eta, Q), \mathrm{Aut}(\tilde\eta)\big] \tag{23.15.2}$$

and note that, for any $\mathfrak{ac}(\tilde{\mathcal{F}}^{^{\mathrm{sc}}})_{Q\,\tilde\cap\,P}$-object $\tilde\eta' : T \to Q\,\tilde\cap\,P$ isomorphic to $\tilde\eta$, we may identify $\bar{\mathfrak{b}}(\tilde\eta', Q)$ with $\bar{\mathfrak{b}}(\tilde\eta, Q)$ since any $\mathfrak{ac}(\tilde{\mathcal{F}}^{^{\mathrm{sc}}})_{Q\,\tilde\cap\,P}$-isomorphism between them determines the *same* isomorphism $\bar{\mathfrak{b}}(\tilde\eta', Q) \cong \bar{\mathfrak{b}}(\tilde\eta, Q)$.

23.16 Now, for any $\tilde{\mathcal{F}}^{^{\mathrm{sc}}}$-object Q, we claim that

$$\big((\mathfrak{p}_{\mathfrak{tn},\mathfrak{tn}^{\mathfrak{i}}})_{\mathfrak{a}}(\mathfrak{b})\big)(Q) \cong \prod_{\tilde\eta \in \mathfrak{O}^{Q,\mathrm{sc}}} \bar{\mathfrak{b}}(\tilde\eta, Q) \tag{23.16.1;}$$

indeed, denoting by $(\mathfrak{b}_Q)^{\mathfrak{i}}$ the restriction of the *contravariant* functor $\mathfrak{b}_Q$ to the subcategory $\mathfrak{ac}(\tilde{\mathcal{F}}^{^{\mathrm{sc}}})^{\mathfrak{i}}_{Q\,\tilde\cap\,P}$ (cf. A2.15), it is quite clear that the *direct limit* $\underrightarrow{\lim}(\mathfrak{b}_Q)^{\mathfrak{i}}$ is a quotient of the direct sum $\bigoplus_{\tilde\eta} \bar{\mathfrak{b}}(\tilde\eta, Q)$ where $\tilde\eta$ runs over a set of representatives for the set of isomorphism classes of $\mathfrak{ac}(\tilde{\mathcal{F}}^{^{\mathrm{sc}}})_{Q\,\tilde\cap\,P}$-objects; then, for any $\mathfrak{ac}(\tilde{\mathcal{F}}^{^{\mathrm{sc}}})_{Q\,\tilde\cap\,P}$-object

$$\tilde\eta : \bigoplus_{i\in I} T_i \longrightarrow Q\,\tilde\cap\,P \tag{23.16.2,}$$

it suffices to consider the $\mathfrak{ac}(\tilde{\mathcal{F}}^{^{\mathrm{sc}}})^{\mathfrak{i}}_{Q\,\tilde\cap\,P}$-morphisms obtained from the obvious $\mathfrak{ac}(\tilde{\mathcal{F}}^{^{\mathrm{sc}}})$-morphisms $T_i \to \bigoplus_{i'\in I} T_{i'}$ where i runs over I, to easily get isomorphism 23.16.1.

23.17 Similarly, for any $\tilde{\mathcal{F}}^{^{\mathrm{sc}}}$-morphism $\tilde\varphi : R \to Q$, choose $\varphi \in \tilde\varphi$ and a set of representatives $W_{\tilde\eta}$ for the set of double classes $\varphi(R)wT \in \varphi(R)\backslash Q/T$ such that $U_w = \varphi(R)^w \cap T$ is $\mathcal{F}$-selfcentralizing; then, for any $\mathfrak{ac}(\tilde{\mathcal{F}}^{^{\mathrm{sc}}})$-morphism $\tilde\eta : T \to Q\,\tilde\cap\,P$ in $\mathfrak{O}^{Q,\mathrm{sc}}$, with the notation in 23.14 it is easily checked from definition A2.16.2 that the group homomorphism

$$\big((\mathfrak{p}_{\mathfrak{tn},\mathfrak{tn}^{\mathfrak{i}}})_{\mathfrak{a}}(\mathfrak{b})\big)(\tilde\varphi) : \prod_{\tilde\eta \in \mathfrak{O}^{Q,\mathrm{sc}}} \bar{\mathfrak{b}}(\tilde\eta, Q) \longrightarrow \prod_{\tilde\nu \in \mathfrak{O}^{R,\mathrm{sc}}} \bar{\mathfrak{b}}(\tilde\nu, R) \tag{23.17.1}$$

maps $\bar{b} \in \bar{\mathfrak{b}}(\tilde{\eta}, Q)$ on

$$\Big(\big((\mathfrak{p}_{\mathfrak{rn},\mathfrak{rn}^{i}})_{\mathfrak{a}}(\mathfrak{b})\big)(\tilde{\varphi})\Big)(b) = \sum_{w \in W_{\tilde{\eta}}} \overline{\big(\mathfrak{b}(\tilde{\psi}_w, \tilde{\varphi})\big)(b)} \qquad 23.17.2$$

where $b \in \mathfrak{b}(\tilde{\eta}, Q)$ is a representative for $\bar{b}$ and $\overline{\big(\mathfrak{b}(\tilde{\psi}_w, \tilde{\varphi})\big)(b)}$ is the image in $\bar{\mathfrak{b}}(\tilde{\nu}_w, R)$ of $\big(\mathfrak{b}(\tilde{\psi}_w, \tilde{\varphi})\big)(b) \in \mathfrak{b}(\tilde{\nu}_w, R)$. Finally, for any $\tilde{\mathcal{F}}^{sc}$-object Q, it follows from our definition in A2.17 that we have the equality

$$\big(\mathfrak{o}_{\mathfrak{rn}}(\mathfrak{b})\big)(Q) = \mathfrak{b}(\widetilde{\mathrm{id}}_{Q \,\tilde{\cap}\, P}, Q) \qquad 23.17.3$$

and a natural group homomorphism (cf. 23.13.2)

$$\omega_{\mathfrak{rn},\mathfrak{rn}^{i}}(\mathfrak{b})_Q : \mathfrak{b}(\widetilde{\mathrm{id}}_{Q \,\tilde{\cap}\, P}, Q) \longrightarrow \prod_{\tilde{\eta} \in \mathfrak{D}^{Q,sc}} \bar{\mathfrak{b}}(\tilde{\eta}, Q) \qquad 23.17.4$$

determined by the $\mathfrak{rn} \rtimes \tilde{\mathcal{F}}^{sc}$-morphisms $(\tilde{\eta}, \widetilde{\mathrm{id}}_Q) : (\tilde{\eta}, Q) \to (\widetilde{\mathrm{id}}_{Q \,\tilde{\cap}\, P}, Q)$ when $\tilde{\eta}$ runs over $\mathfrak{D}^{Q,sc}$.

23.18 In order to state the announced *vanishing cohomology criterion*, we have to consider the functor

$$\mathfrak{im} : \mathfrak{rg}_{\mathfrak{ac}(\tilde{\mathcal{F}}^{sc})} \rtimes \mathfrak{ac}(\tilde{\mathcal{F}}^{sc}) \longrightarrow \mathfrak{rg}_{\mathfrak{ac}(\tilde{\mathcal{F}}^{sc})} \rtimes \mathfrak{ac}(\tilde{\mathcal{F}}^{sc}) \qquad 23.18.1$$

defined as follows. Recall that an $\mathfrak{rg}_{\mathfrak{ac}(\tilde{\mathcal{F}}^{sc})} \rtimes \mathfrak{ac}(\tilde{\mathcal{F}}^{sc})$-object is a pair $(\tilde{\eta}, Q)$ formed by an $\mathfrak{ac}(\tilde{\mathcal{F}}^{sc})$-object $Q = \bigoplus_{i \in I} Q_i$ and an $\mathfrak{ac}(\tilde{\mathcal{F}}^{sc})$-morphism (cf. A2.7)

$$\tilde{\eta} : T = \bigoplus_{\ell \in L} T_\ell \longrightarrow Q = \bigoplus_{i \in I} Q_i \qquad 23.18.2$$

given by a map $g : L \to I$ and by $\tilde{\mathcal{F}}^{sc}$-morphisms $\tilde{\eta}_\ell : T_\ell \to Q_{g(\ell)}$ where ℓ runs over L (cf. 6.2); then, the functor $\mathfrak{im}$ maps $(\tilde{\eta}, Q)$ on the pair formed by the $\mathfrak{ac}(\tilde{\mathcal{F}}^{sc})$-object $Q' = \bigoplus_{i \in g(L)} Q_i$ and by the $\mathfrak{ac}(\tilde{\mathcal{F}}^{sc})$-morphism $\tilde{\eta}' : T \to Q'$ given by the induced map and the *same* $\tilde{\mathcal{F}}^{sc}$-morphisms. Similarly, the functor $\mathfrak{im}$ maps any $\mathfrak{rg}_{\mathfrak{ac}(\tilde{\mathcal{F}}^{sc})} \rtimes \mathfrak{ac}(\tilde{\mathcal{F}}^{sc})$-morphism

$$(\tilde{\psi}, \tilde{\varphi}) : (\tilde{\nu}, R) \longrightarrow (\tilde{\eta}, Q) \qquad 23.18.3$$

on the $\mathfrak{rg}_{\mathfrak{ac}(\tilde{\mathcal{F}}^{sc})} \rtimes \mathfrak{ac}(\tilde{\mathcal{F}}^{sc})$-morphism

$$(\tilde{\psi}, \tilde{\varphi}') : (\tilde{\nu}', R') \longrightarrow (\tilde{\eta}', Q') \qquad 23.18.4$$

where $\tilde{\varphi}'$ is given by the induced map and the same $\tilde{\mathcal{F}}^{sc}$-morphisms. Note that we have an evident *natural map*

$$\iota : \mathfrak{im} \to \mathfrak{id}_{\mathfrak{rg}_{\mathfrak{ac}(\tilde{\mathcal{F}}^{sc})} \rtimes \mathfrak{ac}(\tilde{\mathcal{F}}^{sc})} \qquad 23.18.5$$

and that we still have

$$\mathfrak{im} \circ \mathfrak{im} = \mathfrak{im} \quad \text{and} \quad \iota * \mathfrak{im} = \mathrm{id}_{\mathfrak{im}} = \mathfrak{im} * \iota \qquad 23.18.6;$$

moreover, recall that, according to Proposition A2.20, we have a functor

$$\mathrm{id}_{\mathfrak{tn}} \rtimes \mathfrak{int}_P : \mathfrak{tn} \rtimes \tilde{\mathcal{F}}^{^{\mathrm{sc}}} \longrightarrow \mathfrak{tg}_{\mathfrak{ac}(\tilde{\mathcal{F}}^{\mathrm{sc}})} \rtimes \mathfrak{ac}(\tilde{\mathcal{F}}^{^{\mathrm{sc}}}) \qquad 23.18.7.$$

23.19 To prove our announced *vanishing cohomology criterion*, we still need the following notation. For any $\tilde{\mathcal{F}}^{^{\mathrm{sc}}}$*-chain* $\mathfrak{q} : \Delta_n \to \tilde{\mathcal{F}}^{^{\mathrm{sc}}}$ (cf. A2.8), let us denote by $X_{\mathfrak{q}}$ the set of pairs $(\mu, \mathfrak{q}')$ formed by an $\tilde{\mathcal{F}}^{^{\mathrm{sc}}}$*-chain* $\mathfrak{q}' : \Delta_n \to \tilde{\mathcal{F}}^{^{\mathrm{sc}}}$ such that $\mathfrak{q}'(i-1) \subset \mathfrak{q}'(i)$ and that $\mathfrak{q}'(i-1 \bullet i) = \tilde{\iota}_{\mathfrak{q}'(i-1)}^{\mathfrak{q}'(i)}$ for any $1 \le i \le n$, and by a natural map $\mu : \mathfrak{q}' \to \mathfrak{q}$ such that $\tilde{\iota}_{\mathfrak{q}'(i)}^{P}$ belongs to $\tilde{\mathcal{F}}\big(P, \mathfrak{q}'(i)\big)_{\tilde{\mu}_i}$ for any $i \in \Delta_n$ (cf. 6.4.1); then, if $u \in P$, we denote by $(\mu, \mathfrak{q}')^u$ the element of $X_{\mathfrak{q}}$ formed by the $\tilde{\mathcal{F}}^{^{\mathrm{sc}}}$*-chain* and by the natural map respectively sending $i \in \Delta_n$ to (cf. 1.8)

$$\mathfrak{q}'^u(i) = \mathfrak{q}'(i)^u \quad \text{and} \quad (\mu^u)_i = \tilde{\mu}_i \circ \tilde{\kappa}_{\mathfrak{q}'(i), \mathfrak{q}'(i)^u}(u) \qquad 23.19.1.$$

Recall that we denote by $\mathbb{H}_*^n(\tilde{\mathcal{F}}^{^{\mathrm{sc}}}, \bullet)$ the *stable n-cohomology groups* of $\tilde{\mathcal{F}}^{^{\mathrm{sc}}}$ (cf. A3.17).

Theorem 23.20 *With all the notation above, choose* $\mathfrak{b} = \mathfrak{a} \circ \mathfrak{im} \circ (\mathrm{id}_{\mathfrak{tn}} \rtimes \mathfrak{int}_P)$ *for a contravariant functor*

$$\mathfrak{a} : \mathfrak{tg}_{\mathfrak{ac}(\tilde{\mathcal{F}}^{\mathrm{sc}})} \rtimes \mathfrak{ac}(\tilde{\mathcal{F}}^{^{\mathrm{sc}}}) \longrightarrow \mathfrak{Ab} \qquad 23.20.1.$$

Then, for any $n \ge 1$ *we have* $\mathbb{H}_*^n\big(\tilde{\mathcal{F}}^{^{\mathrm{sc}}}, (\mathfrak{p}_{\mathfrak{tn}, \mathfrak{tn}^{\mathfrak{i}}})_{\mathfrak{a}}(\mathfrak{b})\big) = \{0\}$.

Proof: For any $\tilde{\mathcal{F}}^{^{\mathrm{sc}}}$*-chain* $\mathfrak{q} : \Delta_n \to \tilde{\mathcal{F}}^{^{\mathrm{sc}}}$ and any pair $(\mu, \mathfrak{q}')$ in the set $X_{\mathfrak{q}}$ defined above, let us consider the $\tilde{\mathcal{F}}^{^{\mathrm{sc}}}$*-chain* $\mathfrak{q}'^P : \Delta_{n+1} \to \tilde{\mathcal{F}}^{^{\mathrm{sc}}}$ defined by

$$\mathfrak{q}'^P \circ \delta_{n+1}^n = \mathfrak{q}' \quad , \quad \mathfrak{q}'^P(n+1) = P \quad \text{and} \quad \mathfrak{q}'^P(n \bullet n{+}1) = \tilde{\iota}_{\mathfrak{q}'(n)}^P \qquad 23.20.2$$

and moreover, for any $\ell \in \Delta_n$, the $\tilde{\mathcal{F}}^{^{\mathrm{sc}}}$*-chain* $\mathfrak{h}_\ell^n(\mu) : \Delta_{n+1} \to \tilde{\mathcal{F}}^{^{\mathrm{sc}}}$ defined in Lemma A4.2. Set $\hat{\mathfrak{b}} = (\mathfrak{p}_{\mathfrak{tn}, \mathfrak{tn}^{\mathfrak{i}}})_{\mathfrak{a}}(\mathfrak{b})$ and let $b = \{b_{\mathfrak{q}}\}_{\mathfrak{q} \in \mathfrak{Fct}(\Delta_n, \tilde{\mathcal{F}}^{\mathrm{sc}})}$ be a *stable* $\hat{\mathfrak{b}}$*-valued n-cocycle* (cf. A3.18); thus, considering the *components* of the *differential* of b labelled by $\mathfrak{q}'^P$ and by $\mathfrak{h}_\ell^n(\mu)$ for any $\ell \in \Delta_n$, we get (cf. A3.11.2)

$$\begin{aligned}
0 &= \big(\hat{\mathfrak{b}}(\mathfrak{q}'(0\bullet 1))\big)(b_{\mathfrak{q}'^P \circ \delta_0^n}) + \sum_{i=1}^{n} b_{\mathfrak{q}'^P \circ \delta_i^n} + (-1)^{n+1} b_{\mathfrak{q}'} \\
0 &= \big(\hat{\mathfrak{b}}(\mathfrak{q}'(0\bullet 1))\big)(b_{\mathfrak{h}_\ell^n(\mu) \circ \delta_0^n}) + \sum_{i=1}^{n+1} (-1)^i b_{\mathfrak{h}_\ell^n(\mu) \circ \delta_i^n} \quad \text{if } 1 \le \ell \le n \qquad 23.20.3. \\
0 &= \big(\hat{\mathfrak{b}}(\tilde{\mu}_0)\big)(b_{\mathfrak{q}}) + \sum_{i=1}^{n+1} (-1)^i b_{\mathfrak{h}_0^n(\mu) \circ \delta_i^n}
\end{aligned}$$

Then, setting

$$a_{(\mu,\mathfrak{q}'),i} = \sum_{\ell=0}^{n-1} (-1)^\ell b_{\mathfrak{h}_\ell^{n-1}(\mu*\delta_i^{n-1})} + (-1)^n b_{(\mathfrak{q}'\circ\delta_i^{n-1})^P} \qquad 23.20.4$$

which belongs to $\hat{\mathfrak{b}}\big(\mathfrak{q}'(1)\big)$ if $i = 0$ and to $\hat{\mathfrak{b}}\big(\mathfrak{q}'(0)\big)$ if $1 \le i \le n$, it follows from Lemma A4.2 that the "alternating sum" of equalities 23.20.3 yields

$$\big(\hat{\mathfrak{b}}(\tilde{\mu}_0)\big)(b_\mathfrak{q}) = \big(\hat{\mathfrak{b}}(\mathfrak{q}'(0\bullet 1))\big)(a_{(\mu,\mathfrak{q}'),0}) + \sum_{i=1}^{n} (-1)^i a_{(\mu,\mathfrak{q}'),i} \qquad 23.20.5;$$

indeed, the last terms in the top equality and in the n-th middle equality cancel with each other in the alternating sum; the same happens with the second terms in the first middle equality and in the bottom equality, and with the $(\ell+2)$-th terms in the ℓ-th and $(\ell+1)$-th middle equalities. Moreover, since $(\mu, \mathfrak{q}')$ belongs to $X_\mathfrak{q}$, it is quite clear that, for any $0 \le i \le n$, the pair $(\mu * \delta_i^{n-1}, \mathfrak{q}' \circ \delta_i^{n-1})$ belongs to $X_{\mathfrak{q}\circ\delta_i^{n-1}}$ and that we have

$$\mathfrak{q}'^P \circ \delta_i^n = (\mathfrak{q}' \circ \delta_i^{n-1})^P \qquad 23.20.6.$$

On the other hand, for any $\tilde{\mathcal{F}}^{^{sc}}$-object Q, recall that (cf. isomorphism 23.16.1)

$$\hat{\mathfrak{b}}(Q) = \prod_{\tilde{\gamma}\in\mathfrak{O}^{Q,sc}} \bar{\mathfrak{b}}(\tilde{\gamma}, Q) \qquad 23.20.7$$

and let us denote by $x_{\tilde{\gamma}}$ the $\tilde{\gamma}$-component of $x \in \hat{\mathfrak{b}}(Q)$ for any $\tilde{\gamma} \in \mathfrak{O}^{Q,sc}$; thus, equality 23.20.5 becomes

$$\big(\hat{\mathfrak{b}}(\tilde{\mu}_0)\big)(b_\mathfrak{q})_{\tilde{\eta}'} = \big(\hat{\mathfrak{b}}(\mathfrak{q}'(0\bullet 1))\big)(a_{(\mu,\mathfrak{q}'),0})_{\tilde{\eta}'} + \sum_{i=1}^{n} (-1)^i (a_{(\mu,\mathfrak{q}'),i})_{\tilde{\eta}'} \qquad 23.20.8$$

for any $\tilde{\eta}' \in \mathfrak{O}^{\mathfrak{q}'(0),sc}$. Moreover, setting (cf. Proposition 6.14)

$$\tilde{\eta} = (\tilde{\mu}_0 \tilde{\cap} \widetilde{\mathrm{id}}_P) \circ \tilde{\eta}' : T \longrightarrow \mathfrak{q}'(0) \tilde{\cap} P \longrightarrow \mathfrak{q}(0) \tilde{\cap} P \qquad 23.20.9,$$

we claim that $\mathrm{Aut}(\tilde{\eta}') = \mathrm{Aut}(\tilde{\eta})$.

Indeed, according to definition 23.20.9, it is quite clear that $\mathrm{Aut}(\tilde{\eta}')$ is contained in $\mathrm{Aut}(\tilde{\eta})$ and, since $\tilde{\imath}^P_{\mathfrak{q}'(0)}$ belongs to $\tilde{\mathcal{F}}\big(\mathfrak{q}(0), \mathfrak{q}'(0)\big)_{\tilde{\mu}_0}$, from statement 6.5.1 it is not difficult to check that we get the equality; actually, it follows from Lemma 23.21 below that any $\mathfrak{ac}(\tilde{\mathcal{F}}^{^{sc}})$-morphism $\tilde{\eta} : T \to \mathfrak{q}(0) \tilde{\cap} P$ determines a P-orbit of pairs $(\mathfrak{q}', \mu) \in X_\mathfrak{q}$ such that $\tilde{\eta}$ factorizes as in 23.20.9 for a suitable $\tilde{\eta}' \in \mathfrak{O}^{\mathfrak{q}'(0),sc}$, and that we have

$$\mathfrak{im}\big(\tilde{\eta}, \mathfrak{q}(0) \tilde{\cap} P\big) = \big(\tilde{\eta}', \mathfrak{q}'(0)\big) = \mathfrak{im}\big(\tilde{\eta}', \mathfrak{q}'(0) \tilde{\cap} P\big) \qquad 23.20.10,$$

so that $\tilde{\eta}$ and $\tilde{\eta}'$ mutually determine each other. Consequently, it follows from 23.18.6 that we have (cf. 23.15.2)

$$\bar{\mathfrak{b}}(\tilde{\eta}, \mathfrak{q}(0)) = \bar{\mathfrak{b}}(\tilde{\eta}', \mathfrak{q}'(0)) \qquad 23.20.11$$

and that $\mathfrak{b}(\widetilde{\mathrm{id}}_T, \tilde{\mu}_0)$ is equal to the identity on it.

We are ready to show that b is a *stable* $\hat{\mathfrak{b}}$*-valued* n*-coboundary* (cf. A3.18); assuming that for any $\tilde{\mathcal{F}}^{\mathrm{sc}}$*-chain* $\mathfrak{q}:\Delta_n \to \tilde{\mathcal{F}}_{\mathrm{sc}}$ and any $(\tilde{\mathcal{F}}^{\mathrm{sc}})_{\mathfrak{q}(0)\,\tilde{\cap}\,P}$-object $\tilde{\varepsilon}:U \to \mathfrak{q}(0)\,\tilde{\cap}\,P$ in $\mathfrak{O}^{\mathfrak{q}(0),\mathrm{sc}}$ such that $|P:U| < p^m$ we have $(b_{\mathfrak{q}})_{\tilde{\varepsilon}} = 0$, we argue by induction on $m \in \mathbb{N}$. Denoting by $\{d_n\}_{n\in\mathbb{N}}$ the *differential map* determined by the functor $(\mathfrak{p}_{\mathfrak{ch}^*(\tilde{\mathcal{F}}^{\mathrm{sc}})})_*(\hat{\mathfrak{b}} \circ \mathfrak{v}_{\tilde{\mathcal{F}}^{\mathrm{sc}}})$ (cf. A3.11 and A3.18), it suffices to exhibit a *stable* $\hat{\mathfrak{b}}$*-valued* n*-coboundary* $d_{n-1}(a)$ fulfilling

$$(d_{n-1}(a)_{\mathfrak{q}})_{\tilde{\varepsilon}} = 0 \quad \text{and} \quad (\hat{b}_{\mathfrak{q}})_{\tilde{\eta}} - (d_{n-1}(a)_{\mathfrak{q}})_{\tilde{\eta}} = 0 \qquad 23.20.12$$

for any $(\tilde{\mathcal{F}}^{\mathrm{sc}})_{\mathfrak{q}(0)\,\tilde{\cap}\,P}$-objects $\tilde{\varepsilon}:U \to \mathfrak{q}(0)\,\tilde{\cap}\,P$ such that $|P:U| < p^m$ and $\tilde{\eta}:T \to \mathfrak{q}(0)\,\tilde{\cap}\,P$ such that $|P:T| = p^m$, both in $\mathfrak{O}^{\mathfrak{q}(0),\mathrm{sc}}$; indeed, the element $b - d_{n-1}(a)$ is then a *stable* $\hat{\mathfrak{b}}$*-valued* n*-coboundary* by our induction hypothesis.

In order to exhibit a, for any $\tilde{\mathcal{F}}^{\mathrm{sc}}$*-chain* $\mathfrak{r}:\Delta_{n-1} \to \tilde{\mathcal{F}}^{\mathrm{sc}}$, let us consider the element $a_{\mathfrak{r}} \in \hat{\mathfrak{b}}(\mathfrak{r}(0))$ which, for any $(\tilde{\mathcal{F}}^{\mathrm{sc}})_{\mathfrak{r}(0)\,\tilde{\cap}\,P}$-object $\tilde{\varepsilon}:U \to \mathfrak{r}(0)\,\tilde{\cap}\,P$ in $\mathfrak{O}^{\mathfrak{r}(0),\mathrm{sc}}$, has the following $\tilde{\varepsilon}$*-component* (cf. 23.20.7 and 23.20.11)

$$(a_{\mathfrak{r}})_{\tilde{\varepsilon}} = \sum_{\ell=0}^{n-1} (-1)^{\ell} (b_{\mathfrak{h}_{\ell}^{n-1}(\nu)})_{\tilde{\varepsilon}'} + (-1)^n (b_{\mathfrak{r}'^P})_{\tilde{\varepsilon}'} \qquad 23.20.13,$$

where $(\nu, \mathfrak{r}')$ and $\tilde{\varepsilon}':U \to \mathfrak{r}'(0)\,\tilde{\cap}\,P$ are respectively a pair in $X_{\mathfrak{r}}$ and an $(\tilde{\mathcal{F}}^{\mathrm{sc}})_{\mathfrak{r}'(0)\,\tilde{\cap}\,P}$-object determined, up to a P-action, by $\tilde{\varepsilon}$ (cf. Lemma 23.21 below); in particular, we have $\mathrm{Aut}(\tilde{\varepsilon}') = \mathrm{Aut}(\tilde{\varepsilon})$ and, for any $0 \le i \le n$, considering the pair $(\mu * \delta_i^{n-1}, \mathfrak{q}' \circ \delta_i^{n-1})$, we get (cf. 23.20.4)

$$(a_{\mathfrak{q}\circ\delta_i^{n-1}})_{\tilde{\varepsilon}} = (a_{(\mu,\mathfrak{q}'),i})_{\tilde{\varepsilon}'} \qquad 23.20.14.$$

Finally, we set

$$a = \{a_{\mathfrak{r}}\}_{\mathfrak{r}\in\mathfrak{Fct}(\Delta_{n-1},\tilde{\mathcal{F}}^{\mathrm{sc}})} \qquad 23.20.15;$$

note that our definition does not depend on the choice of the pair $(\nu, \mathfrak{r}')$ in its P-orbit since, for any $u \in P$, considering $(\nu, \mathfrak{r}')^u = (\nu \circ \kappa_u, \mathfrak{r}'^u)$ (cf. 23.19), the corresponding $\tilde{\mathcal{F}}^{\mathrm{sc}}$*-chains* $(\mathfrak{r}'^u)^P$ and $\mathfrak{h}_{\ell}^{n-1}(\nu \circ \kappa_u)$ are respectively $\mathfrak{ch}^*(\tilde{\mathcal{F}}^{\mathrm{sc}})$-isomorphic to $\mathfrak{r}'^P$ and to $\mathfrak{h}_{\ell}^{n-1}(\nu)$ for any $\ell \in \Delta_{n-1}$, and then it suffices to quote the *stability* of b (cf. A3.18).

We claim that the family a is also *stable* by $\mathfrak{ch}^*(\tilde{\mathcal{F}}^{^{\rm sc}})$-isomorphisms; indeed, if $\bar{\mathfrak{r}}:\Delta_{n-1}\to\tilde{\mathcal{F}}^{^{\rm sc}}$ is an $\tilde{\mathcal{F}}^{^{\rm sc}}$*-chain* $\mathfrak{ch}^*(\tilde{\mathcal{F}}^{^{\rm sc}})$-isomorphic to $\mathfrak{r}$ and $\lambda:\mathfrak{r}\cong\bar{\mathfrak{r}}$ is a $\mathfrak{ch}^*(\tilde{\mathcal{F}}^{^{\rm sc}})$-isomorphism then, for the $(\tilde{\mathcal{F}}^{^{\rm sc}})_{\bar{\mathfrak{r}}(0)\,\tilde{\cap}\,P}$-object

$$(\tilde{\lambda}_0\,\tilde{\cap}\,\widetilde{\rm id}_P)\circ\tilde{\varepsilon}:U\longrightarrow\mathfrak{r}(0)\,\tilde{\cap}\,P\longrightarrow\bar{\mathfrak{r}}(0)\,\tilde{\cap}\,P \qquad 23.20.16,$$

we can define the $(\tilde{\lambda}_0\,\tilde{\cap}\,\widetilde{\rm id}_P)\circ\tilde{\varepsilon}$-component of $a_{\bar{\mathfrak{r}}}$ from the pair $(\lambda\circ\nu,\mathfrak{r}')$; in this case, the $\tilde{\mathcal{F}}^{^{\rm sc}}$*-chain* $\mathfrak{r}'^P$ does not change and, for any $\ell\in\Delta_{n-1}$, the corresponding $\tilde{\mathcal{F}}^{^{\rm sc}}$*-chain* $\mathfrak{h}_\ell^{n-1}(\lambda\circ\nu)$ is $\mathfrak{ch}^*(\tilde{\mathcal{F}}^{^{\rm sc}})$-isomorphic to $\mathfrak{h}_\ell^{n-1}(\nu)$; it suffices again to quote the *stability* of b.

Let $\mathfrak{q}:\Delta_n\to\tilde{\mathcal{F}}^{^{\rm sc}}$ be an $\tilde{\mathcal{F}}^{^{\rm sc}}$*-chain* and $\tilde{\varepsilon}:U\to\mathfrak{q}(0)\,\tilde{\cap}\,P$ an $(\tilde{\mathcal{F}}^{^{\rm sc}})_{\mathfrak{q}(0)\,\tilde{\cap}\,P}$-object such that $|P:U|<p^m$; by the very definition of the differential map, we have (cf. A3.11)

$$(d_{n-1}(a)_\mathfrak{q})_{\tilde{\varepsilon}}=\big(\hat{\mathfrak{b}}(\mathfrak{q}(0\bullet 1))\big)(a_{\mathfrak{q}\circ\delta_0^{n-1}})_{\tilde{\varepsilon}}+\sum_{i=1}^{n}(-1)^i(a_{\mathfrak{q}\circ\delta_i^{n-1}})_{\tilde{\varepsilon}} \qquad 23.20.17.$$

But, according to equality 23.20.14 and to definition 23.20.13, we obtain

$$\begin{aligned}(a_{\mathfrak{q}\circ\delta_0^{n-1}})_{\tilde{\gamma}}&=(a_{(\mu,\mathfrak{q}'),0})_{\tilde{\gamma}'}=0\\(a_{\mathfrak{q}\circ\delta_i^{n-1}})_{\tilde{\varepsilon}}&=(a_{(\mu,\mathfrak{q}'),i})_{\tilde{\varepsilon}'}=0\end{aligned} \qquad 23.20.18$$

for any $(\tilde{\mathcal{F}}^{^{\rm sc}})_{\mathfrak{q}(1)\,\tilde{\cap}\,P}$-object $\tilde{\gamma}:U\to\mathfrak{q}(1)\,\tilde{\cap}\,P$ and any $1\le i\le n$; moreover, it is clear that

$$\big(\hat{\mathfrak{b}}(\mathfrak{q}(0\bullet 1))\big)(a_{\mathfrak{q}\circ\delta_0^{n-1}})_{\tilde{\varepsilon}}=\sum_{\tilde{\beta}}\big(\hat{\mathfrak{b}}(\mathfrak{q}(0\bullet 1))\big)\big((a_{\mathfrak{q}\circ\delta_0^{n-1}})_{\tilde{\beta}}\big)_{\tilde{\varepsilon}} \qquad 23.20.19$$

where $\tilde{\beta}:V\to\mathfrak{q}(1)\,\tilde{\cap}\,P$ runs over the set of elements of $\mathfrak{O}^{\mathfrak{q}(1),\rm sc}$ such that $|U|\le|V|$ (cf. 23.17). Consequently, this sum is also equal to zero and we get indeed $(d_{n-1}(a)_\mathfrak{q})_{\tilde{\varepsilon}}=0$.

On the other hand, let $\tilde{\eta}:T\to\mathfrak{q}(0)\,\tilde{\cap}\,P$ be an $(\tilde{\mathcal{F}}^{^{\rm sc}})_{\mathfrak{q}(0)\,\tilde{\cap}\,P}$-object such that $|P:T|=p^m$; once again, it follows from Lemma 23.21 below that $\tilde{\eta}$ determines a P-orbit of pairs $(\mu,\mathfrak{q}')\in X_\mathfrak{q}$ and also an $(\tilde{\mathcal{F}}^{^{\rm sc}})_{q'(0)\,\tilde{\cap}\,P}$-object $\tilde{\eta}':T\to\mathfrak{q}'(0)\,\tilde{\cap}\,P$ such that $\tilde{\eta}=(\mu_0\,\tilde{\cap}\,\widetilde{\rm id}_P)\circ\tilde{\eta}'$. As above, we have

$$\big(\hat{\mathfrak{b}}(\tilde{\mu}_0)\big)(b_\mathfrak{q})_{\tilde{\eta}'}=\sum_{\tilde{\gamma}\in\mathfrak{O}^{\mathfrak{q}(0),\rm sc}}\big(\hat{\mathfrak{b}}(\tilde{\mu}_0)\big)(b_{\mathfrak{q},\tilde{\gamma}})_{\tilde{\eta}'} \qquad 23.20.20$$

where $\tilde{\gamma}:V\to\mathfrak{q}(0)\,\tilde{\cap}\,P$ runs over the set of elements in $\mathfrak{O}^{\mathfrak{q}(0),\rm sc}$ admitting an $\tilde{\mathcal{F}}^{^{\rm sc}}$-morphism $\tilde{\psi}:T\to V$ such that the following $\mathfrak{ac}(\tilde{\mathcal{F}}^{^{\rm sc}})$-diagram is commutative (cf. 23.17)

$$\begin{array}{ccc}V&\xrightarrow{\ \tilde{\gamma}\ }&\mathfrak{q}(0)\,\tilde{\cap}\,P\\ {\scriptstyle\tilde{\psi}}\uparrow&&\uparrow{\scriptstyle\tilde{\mu}_0\,\tilde{\cap}\,\widetilde{\rm id}_P}\\ T&\xrightarrow{\ \tilde{\eta}'\ }&\mathfrak{q}'(0)\,\tilde{\cap}\,P\end{array} \qquad 23.20.21;$$

consequently, all the terms in this sum are equal to zero unless $\tilde\psi$ is an $\tilde{\mathcal{F}}^{^{sc}}$-isomorphism and then we necessarily have $V = T$ and $\tilde\gamma = \tilde\eta$; thus, we get (cf. 23.20.11)

$$\big(\hat{\mathfrak{b}}(\tilde\mu_0)\big)(b_{\mathfrak{q}})_{\tilde\eta'} = (b_{\mathfrak{q}})_{\tilde\eta} \qquad 23.20.22.$$

Moreover, according to Lemma 23.21 below, for any $1 \le i \le n$, the P-orbit of $(\mu * \delta_i^{n-1}, \mathfrak{q}' \circ \delta_i^{n-1})$ in $X_{\mathfrak{q}\circ\delta_i^{n-1}}$ actually coincides with the P-orbit determined by $\tilde\eta$, whereas the composition $\mathfrak{q}(0\bullet 1)\circ\tilde\eta$ determines the P-orbit of $(\mu * \delta_0^{n-1}, \mathfrak{q}' \circ \delta_0^{n-1})$ in $X_{\mathfrak{q}\circ\delta_0^{n-1}}$, together with the $(\tilde{\mathcal{F}}^{^{sc}})_{\mathfrak{q}'(1)\,\tilde\cap\, P}$-object

$$\mathfrak{q}(0\bullet 1)\circ\tilde\eta' : T \longrightarrow \mathfrak{q}'(1)\,\tilde\cap\, P \qquad 23.20.23;$$

hence, arguing as in 23.20.19 above, we still have (cf. 23.20.14)

$$\begin{aligned}\big(\hat{\mathfrak{b}}(\mathfrak{q}(0\bullet 1))\big)(a_{\mathfrak{q}\circ\delta_0^{n-1}})_{\tilde\eta} &= (a_{\mathfrak{q}\circ\delta_0^{n-1}})_{\mathfrak{q}(0\bullet 1)\circ\tilde\eta} = (a_{(\mu,\mathfrak{q}'),0})_{\mathfrak{q}(0\bullet 1)\circ\tilde\eta'} \\ &= \big(\hat{\mathfrak{b}}(\mathfrak{q}'(0\bullet 1))\big)(a_{(\mu,\mathfrak{q}'),0})_{\tilde\eta'}\end{aligned} \qquad 23.20.24.$$

In conclusion, from equalities 23.20.8, 23.20.14, 23.20.22 and 23.20.24, we get

$$\begin{aligned}(b_{\mathfrak{q}})_{\tilde\eta} &= \big(\hat{\mathfrak{b}}(\tilde\mu_0)\big)(b_{\mathfrak{q}})_{\tilde\eta'} \\ &= \big(\hat{\mathfrak{b}}(\mathfrak{q}'(0\bullet 1))\big)(a_{(\mu,\mathfrak{q}'),0})_{\tilde\eta'} + \sum_{i=1}^{n}(-1)^i(a_{(\mu,\mathfrak{q}'),i})_{\tilde\eta'} \\ &= \big(\hat{\mathfrak{b}}(\mathfrak{q}(0\bullet 1))\big)(a_{\mathfrak{q}\circ\delta_0^{n-1}})_{\tilde\eta} + \sum_{i=1}^{n}(-1)^i(a_{\mathfrak{q}\circ\delta_i^{n-1}})_{\tilde\eta} \\ &= (d_{n-1}(b)_{\mathfrak{q}})_{\tilde\eta}\end{aligned} \qquad 23.20.25$$

which completes our induction argument. We are done.

Lemma 23.21 *With the notation above, for any $\tilde{\mathcal{F}}^{^{sc}}$-chain $\mathfrak{q}:\Delta_n \to \tilde{\mathcal{F}}^{^{sc}}$ we have*

$$\mathfrak{q}(0)\,\tilde\cap\, P \cong \bigoplus_{(\mu,\mathfrak{q}')} \mathfrak{q}'(0) \qquad 23.21.1$$

where the pair $(\mu,\mathfrak{q}')$ runs over a set of representatives in $X_{\mathfrak{q}}$ for the set of P-orbits. In particular, for any $(\tilde{\mathcal{F}}^{^{sc}})_{\mathfrak{q}(0)\,\tilde\cap\, P}$-object $\tilde\eta : T \to \mathfrak{q}(0)\,\tilde\cap\, P$, there is a unique P-orbit of pairs $(\mu,\mathfrak{q}')$ in $X_{\mathfrak{q}}$ such that $\tilde\eta$ factorizes throughout the $\mathfrak{ac}(\tilde{\mathcal{F}}^{^{sc}})$-morphism $\mathfrak{q}'(0) \to \mathfrak{q}(0)\,\tilde\cap\, P$ determined by $\tilde\mu_0 : \mathfrak{q}'(0) \to \mathfrak{q}(0)$ and $\tilde\iota^P_{\mathfrak{q}'(0)} : \mathfrak{q}'(0) \to P$.

Proof: Recall that, for any $\mathcal{F}$-selfcentralizing subgroup Q of P, we may assume that (cf 23.17.2)

$$Q\,\tilde\cap\, P = \bigoplus_{Q' \in \tilde{\mathcal{S}}_Q}\ \bigoplus_{\tilde\delta' \in \tilde{\mathcal{M}}^Q_{Q'}} Q' \qquad 23.21.2$$

where $\tilde{\mathcal{S}}_Q$ is a set of representatives for the set of Q-conjugacy classes of the $\mathcal{F}$-selfcentralizing subgroups Q' of Q such that (cf. 6.4.1)

$$\tilde{\mathcal{F}}(P,Q')_{\tilde{\iota}^Q_{Q'}} \neq \emptyset \qquad 23.21.3$$

and, for any $Q' \in \tilde{\mathcal{S}}_Q$, $\tilde{\mathcal{M}}^Q_{Q'}$ is a set of representatives for the set of $\tilde{\mathcal{F}}_Q(Q')$-orbits in $\tilde{\mathcal{F}}(P,Q')_{\tilde{\iota}^Q_{Q'}}$; but, according to statement 6.5.2, $\tilde{\delta}' \in \tilde{\mathcal{F}}(Q,Q')_{\tilde{\iota}^P_{Q'}}$ is equivalent to $\tilde{\iota}^P_{Q'} \in \tilde{\mathcal{F}}(P,Q')_{\tilde{\delta}'}$ and we know that an $\mathfrak{ac}(\tilde{\mathcal{F}}^{^{sc}})$-morphism $\tilde{\eta}:T \to Q \tilde{\cap} P$ is defined by an $\tilde{\mathcal{F}}^{^{sc}}$-morphism $\tilde{\eta}':T \to Q'$ for some $Q' \in \tilde{\mathcal{S}}_Q$ and some $\delta' \in \tilde{\mathcal{M}}^Q_{Q'}$ (cf. 6.2). This proves the lemma for $n = 0$.

Assume that $n \geq 1$; for any $(\mu, \mathfrak{q}') \in X_{\mathfrak{q}}$ it is clear that $(\mu * \delta^n_0, \mathfrak{q}' \circ \delta^n_0)$ still belongs to $X_{\mathfrak{q}\circ\delta^n_0}$; but, for any $(\nu, \mathfrak{r}') \in X_{\mathfrak{q}\circ\delta^n_0}$, by definition 6.18.2, we have

$$\mathfrak{q}(0)\,\tilde{\cap}_{\mathfrak{q}(1)}\,\mathfrak{r}'(0) \cong \bigoplus_{(\tilde{\varepsilon}',R',\tilde{\iota}^{\mathfrak{r}'(0)}_{R'})} R' \qquad 23.21.4$$

where R' runs over a set of representatives for the set of $\mathfrak{r}'(0)$-conjugacy classes of $\mathcal{F}$-selfcentralizing subgroups of $\mathfrak{r}'(0)$ such that

$$\tilde{\mathcal{F}}\big(\mathfrak{q}(0),R'\big)_{\tilde{\iota}^{\mathfrak{r}'(0)}_{R'}} \neq \emptyset \quad \text{and} \quad \mathfrak{q}(0\bullet 1)\circ\tilde{\gamma}' = \tilde{\nu}_0 \circ \tilde{\iota}^{\mathfrak{r}'(0)}_{R'} \qquad 23.21.5$$

for some $\tilde{\gamma}' \in \tilde{\mathcal{F}}\big(\mathfrak{q}(0),R'\big)_{\tilde{\iota}^{\mathfrak{r}'(0)}_{R'}}$, and, for such a R', $\tilde{\varepsilon}'$ runs over a set of representatives for $\tilde{\mathcal{F}}\big(\mathfrak{q}(0),R'\big)_{\tilde{\iota}^{\mathfrak{r}'(0)}_{R'}}/\tilde{\mathcal{F}}_{\mathfrak{r}'(0)}(R')$ fulfilling

$$\mathfrak{q}(0\bullet 1)\circ\tilde{\varepsilon}' = \tilde{\nu}_0 \circ \tilde{\iota}^{\mathfrak{r}'(0)}_{R'} \qquad 23.21.6.$$

Moreover, for such a triple $(\tilde{\varepsilon}',R',\tilde{\iota}^{\mathfrak{r}'(0)}_{R'})$, it is easily checked that the pair $(\mu,\mathfrak{q}')$ formed by the functor $\mathfrak{q}':\Delta_n \to \tilde{\mathcal{F}}^{^{sc}}$ defined by

$$\mathfrak{q}'(0) = R' \quad , \quad \mathfrak{q}'\circ\delta^n_0 = \mathfrak{r}' \quad \text{and} \quad \mathfrak{q}'(0\bullet 1) = \tilde{\iota}^{\mathfrak{r}'(0)}_{R'} \qquad 23.21.7$$

and by the natural map $\mu:\mathfrak{q}' \to \mathfrak{q}$ fulfilling

$$\tilde{\mu}_0 = \tilde{\varepsilon}' \quad \text{and} \quad \mu * \delta^n_0 = \nu \qquad 23.21.8,$$

belongs to $X_{\mathfrak{q}}$.

On the other hand, arguing by induction on n, we may assume that

$$\mathfrak{q}(1)\,\tilde{\cap}\,P \cong \bigoplus_{(\nu,\mathfrak{r}')} \mathfrak{q}'(1) \qquad 23.21.9$$

where the pair $(\nu, \mathfrak{t}')$ runs over a set of representatives in $X_{\mathfrak{q}\circ\delta_0^n}$ for the set of P-orbits; but, it easily follows from Propositions 6.14 and 6.21 that we have

$$\mathfrak{q}(0)\,\tilde{\cap}\,P \cong \mathfrak{q}(0)\,\tilde{\cap}_{\mathfrak{q}(1)}\big(\mathfrak{q}(1)\,\tilde{\cap}\,P\big) = \bigoplus_{(\nu,\mathfrak{t}')} \mathfrak{q}(0)\,\tilde{\cap}_{\mathfrak{q}(1)}\,\mathfrak{q}'(1) \qquad 23.21.10$$

where $(\nu, \mathfrak{t}')$ runs over the same set of representatives; consequently, it suffices to apply isomorphism 23.21.4 to any summand, and then the argument above proves isomorphism 23.21.1.

Finally, for any $\mathcal{F}$-selfcentralizing subgroup T of P and any $\mathfrak{ac}(\tilde{\mathcal{F}}^{^{\mathrm{sc}}})$-morphism $\tilde{\eta}\colon T \to \mathfrak{q}(0)\,\tilde{\cap}\,P$, it follows from our induction hypothesis that the composed morphism

$$\big(\mathfrak{q}(0\bullet 1)\,\tilde{\cap}\,\widetilde{\mathrm{id}}_P\big)\circ\tilde{\eta} : T \longrightarrow \mathfrak{q}(0)\,\tilde{\cap}\,P \longrightarrow \mathfrak{q}(1)\,\tilde{\cap}\,P \qquad 23.21.11$$

determines the P-orbit of a suitable pair $(\nu, \mathfrak{t}')$ in isomorphism 23.21.9 above in such a way that the $(\tilde{\mathcal{F}}^{^{\mathrm{sc}}})_{\mathfrak{q}(1)\,\tilde{\cap}\,P}$-object $\big(\mathfrak{q}(0\bullet 1)\,\tilde{\cap}\,\widetilde{\mathrm{id}}_P\big)\circ\tilde{\eta}$ has the announced factorization; but, $\tilde{\eta}$ clearly determines a term of the sum in the right-hand member of equality 23.21.10 (cf. 6.2), and then a term of the sum in the right-hand member of the corresponding isomorphism 23.21.4 (cf. 6.2); once again, by the argument above we get the announced factorization. We are done.

23.22 We are ready to exhibit our announced description of the functor $\hat{\tilde{\mathfrak{z}}}$ (cf. Proposition 23.10). Setting $\tilde{\mathfrak{z}} = \tilde{\mathfrak{z}}_{\tilde{\mathcal{F}}^{\mathrm{sc}}}$ (cf. 13.18) and considering the *covering functor* (cf. A2.18)

$$\mathfrak{c}_{\mathfrak{ac}(\tilde{\mathcal{F}}^{\mathrm{sc}})} : \mathfrak{rg}_{\mathfrak{ac}(\tilde{\mathcal{F}}^{\mathrm{sc}})} \rtimes \mathfrak{ac}(\tilde{\mathcal{F}}^{^{\mathrm{sc}}}) \longrightarrow \mathfrak{ac}(\tilde{\mathcal{F}}^{^{\mathrm{sc}}}) \qquad 23.22.1$$

and the *natural transformation* $(\mathfrak{int}_P, \mathrm{id}_{\mathfrak{rn}})$ from the pair $(\tilde{\mathcal{F}}^{^{\mathrm{sc}}}, \mathfrak{rn})$ to the pair $\big(\mathfrak{ac}(\tilde{\mathcal{F}}^{^{\mathrm{sc}}}), \mathfrak{rg}_{\mathfrak{ac}(\tilde{\mathcal{F}}^{\mathrm{sc}})}\big)$ (cf. A2.19), we get the *contravariant* functor (cf. 23.18.7 and A4.10)

$$\mathfrak{rn}\rtimes\tilde{\mathcal{F}}^{^{\mathrm{sc}}} \xrightarrow{\mathrm{id}_{\mathfrak{rn}}\rtimes\mathfrak{int}_P} \mathfrak{rg}_{\mathfrak{ac}(\tilde{\mathcal{F}}^{\mathrm{sc}})} \rtimes \mathfrak{ac}(\tilde{\mathcal{F}}^{^{\mathrm{sc}}}) \xrightarrow{\mathfrak{c}_{\mathfrak{ac}(\tilde{\mathcal{F}}^{\mathrm{sc}})}} \mathfrak{ac}(\tilde{\mathcal{F}}^{^{\mathrm{sc}}}) \xrightarrow{\mathfrak{ac}(\tilde{\mathfrak{z}})} \mathfrak{Ab} \qquad 23.22.2$$

and then it is easy to check that

$$\hat{\tilde{\mathfrak{z}}} = (\mathfrak{p}_{\mathfrak{rn},\mathfrak{rn}^{\mathfrak{i}}})_{\mathfrak{a}}\big(\mathfrak{ac}(\tilde{\mathfrak{z}})\circ\mathfrak{c}_{\mathfrak{ac}(\tilde{\mathcal{F}}^{\mathrm{sc}})}\circ(\mathrm{id}_{\mathfrak{rn}}\rtimes\mathfrak{int}_P)\big) \qquad 23.22.3.$$

Moreover, it follows from our definition of $\mathfrak{im}$ (cf. 23.18) that we have

$$\mathfrak{c}_{\mathfrak{ac}(\tilde{\mathcal{F}}^{\mathrm{sc}})}\circ\mathfrak{im} = \mathfrak{c}_{\mathfrak{ac}(\tilde{\mathcal{F}}^{\mathrm{sc}})} \qquad 23.22.4$$

and therefore we still have

$$\hat{\tilde{\mathfrak{z}}} = (\mathfrak{p}_{\mathfrak{rn},\mathfrak{rn}^{\mathfrak{i}}})_{\mathfrak{a}}\big(\mathfrak{ac}(\tilde{\mathfrak{z}})\circ\mathfrak{c}_{\mathfrak{ac}(\tilde{\mathcal{F}}^{\mathrm{sc}})}\circ\mathfrak{im}\circ(\mathrm{id}_{\mathfrak{rn}}\rtimes\mathfrak{int}_P)\big) \qquad 23.22.5,$$

so that Theorem 23.20 guarantees that the positive *stable* $\hat{\tilde{\mathfrak{z}}}$*-valued cohomology* of $\tilde{\mathcal{F}}^{^{\rm sc}}$ vanish. But, for instance, in the last member of these equalities we can replace $\mathfrak{c}_{\mathfrak{ac}(\tilde{\mathcal{F}}^{\rm sc})}$ by $\mathfrak{p}_{\mathfrak{ac}(\tilde{\mathcal{F}}^{\rm sc})}$ (cf. A2.18) obtaining a new vanishing result.

23.23 More precisely, we are especially interested in the vanishing result corresponding to the following quotient of $\mathfrak{ac}(\tilde{\mathfrak{z}})\circ\mathfrak{c}_{\mathfrak{ac}(\tilde{\mathcal{F}}^{\rm sc})}$. Consider the natural map defined in A2.14

$$\kappa_{\mathfrak{ac}(\tilde{\mathcal{F}}^{\rm sc})} : \mathfrak{c}_{\mathfrak{ac}(\tilde{\mathcal{F}}^{\rm sc})} \longrightarrow \mathfrak{p}_{\mathfrak{ac}(\tilde{\mathcal{F}}^{\rm sc})} \qquad 23.23.1$$

which determines a new *natural map*

$$\mathfrak{ac}(\tilde{\mathfrak{z}}) * \kappa_{\mathfrak{ac}(\tilde{\mathcal{F}}^{\rm sc})} : \mathfrak{ac}(\tilde{\mathfrak{z}}) \circ \mathfrak{p}_{\mathfrak{ac}(\tilde{\mathcal{F}}^{\rm sc})} \longrightarrow \mathfrak{ac}(\tilde{\mathfrak{z}}) \circ \mathfrak{c}_{\mathfrak{ac}(\tilde{\mathcal{F}}^{\rm sc})} \qquad 23.23.2,$$

and, setting $\overline{\mathfrak{ac}(\tilde{\mathfrak{z}}) \circ \mathfrak{p}_{\mathfrak{ac}(\tilde{\mathcal{F}}^{\rm sc})}} = \mathrm{Im}\big(\mathfrak{ac}(\tilde{\mathfrak{z}}) * \kappa_{\mathfrak{ac}(\tilde{\mathcal{F}}^{\rm sc})}\big)$, we consider the quotient

$$\big(\mathfrak{ac}(\tilde{\mathfrak{z}}) \circ \mathfrak{c}_{\mathfrak{ac}(\tilde{\mathcal{F}}^{\rm sc})}\big)/\big(\overline{\mathfrak{ac}(\tilde{\mathfrak{z}}) \circ \mathfrak{p}_{\mathfrak{ac}(\tilde{\mathcal{F}}^{\rm sc})}}\big) : \mathfrak{rg}_{\mathfrak{ac}(\tilde{\mathcal{F}}^{\rm sc})} \rtimes \mathfrak{ac}(\tilde{\mathcal{F}}^{^{\rm sc}}) \longrightarrow \mathfrak{Ab} \qquad 23.23.3.$$

In order to simplify notation, let us set

$$\begin{aligned} \mathfrak{m} &= \mathfrak{ac}(\tilde{\mathfrak{z}}) \circ \mathfrak{c}_{\mathfrak{ac}(\tilde{\mathcal{F}}^{\rm sc})} \circ (\mathrm{id}_{\mathfrak{rn}} \rtimes \mathfrak{int}_P) : \mathfrak{rn} \rtimes \tilde{\mathcal{F}}^{^{\rm sc}} \longrightarrow \mathfrak{Ab} \\ \mathfrak{n} &= \big(\overline{\mathfrak{ac}(\tilde{\mathfrak{z}}) \circ \mathfrak{p}_{\mathfrak{ac}(\tilde{\mathcal{F}}^{\rm sc})}}\big) \circ \mathfrak{im} \circ (\mathrm{id}_{\mathfrak{rn}} \rtimes \mathfrak{int}_P) : \mathfrak{rn} \rtimes \tilde{\mathcal{F}}^{^{\rm sc}} \longrightarrow \mathfrak{Ab} \end{aligned} \qquad 23.23.4,$$

so that we have $\hat{\tilde{\mathfrak{z}}} = (\mathfrak{p}_{\mathfrak{rn},\mathfrak{rn}^{\mathrm{i}}})_{\mathfrak{a}}(\mathfrak{m})$ and we will consider $\tilde{\mathfrak{y}} = (\mathfrak{p}_{\mathfrak{rn},\mathfrak{rn}^{\mathrm{i}}})_{\mathfrak{a}}(\mathfrak{n})$; then Theorem 23.20 still guarantees that, for any $n \geq 1$, we have

$$\mathbb{H}^n_*(\tilde{\mathcal{F}}^{^{\rm sc}}, \hat{\tilde{\mathfrak{z}}}/\tilde{\mathfrak{y}}) = \{0\} = \mathbb{H}^n_*(\tilde{\mathcal{F}}^{^{\rm sc}}, \tilde{\mathfrak{y}}) \qquad 23.23.5.$$

Corollary 23.24 *With the notation above, there exists an* $\mathcal{F}^{^{\rm sc}}$*-sublocality* $(\tau^{\rm d,sc}, \mathcal{L}^{^{\rm d,sc}}, \pi^{\rm d,sc})$ *of* $\mathcal{L}^{^{\rm c,sc}}$, *unique up to natural isomorphisms, such that*

$$\mathfrak{Ker}(\pi^{\rm d,sc}) = \tilde{\mathfrak{y}} \circ \tilde{\pi}^{\rm c,sc} \qquad 23.24.1$$

where $\tilde{\pi}^{\rm c,sc} : \mathcal{L}^{^{\rm c,sc}} \to \tilde{\mathcal{F}}^{^{\rm sc}}$ *denotes the composition of* $\pi^{\rm c,sc}$ *with the canonical functor* $\mathcal{F}^{^{\rm sc}} \to \tilde{\mathcal{F}}^{^{\rm sc}}$.

Proof: We consider the evident *quotient* $\mathcal{F}^{^{\rm sc}}$*-locality* $\mathcal{L}^{^{\rm c,sc}}/(\tilde{\mathfrak{y}} \circ \tilde{\pi}^{\rm c,sc})$ and claim that the canonical functor

$$\bar{\pi}^{\rm c,sc} : \bar{\mathcal{L}}^{^{\rm c,sc}} = \mathcal{L}^{^{\rm c,sc}}/(\tilde{\mathfrak{y}} \circ \tilde{\pi}^{\rm c,sc}) \longrightarrow \mathcal{F}^{^{\rm sc}} \qquad 23.24.2$$

admits a *functorial section* $\mathfrak{s}$, unique up to natural isomorphisms; then, the converse image in $\mathcal{L}^{^{\mathrm{c,sc}}}$ of the image of $\mathfrak{s}$ supplies the announced $\mathcal{F}^{^{\mathrm{sc}}}$-sublocality $\mathcal{L}^{^{\mathrm{d,sc}}}$. The existence and the uniqueness of $\mathfrak{s}$ is a consequence of the equalities (cf. 23.23.5)

$$\mathbb{H}^2_*(\tilde{\mathcal{F}}^{^{\mathrm{sc}}},\hat{\tilde{\mathfrak{z}}}/\tilde{\mathfrak{y}}) = \{0\} = \mathbb{H}^1_*(\tilde{\mathcal{F}}^{^{\mathrm{sc}}},\hat{\tilde{\mathfrak{z}}}/\tilde{\mathfrak{y}}) \qquad 23.24.3.$$

Indeed, first of all note that the obvious inclusion $Z(Q) \subset C_G(Q)$ for any $\tilde{\mathcal{F}}^{^{\mathrm{sc}}}$-object Q, determines injective natural maps from $\tilde{\mathfrak{z}}$ to $\tilde{\mathfrak{k}}^{^{\mathrm{sc}}}$ and to $\hat{\tilde{\mathfrak{z}}}$ (cf. Proposition 23.10), and that its image in $\hat{\tilde{\mathfrak{z}}}$ is contained in $\tilde{\mathfrak{y}}$; moreover, since $\mathrm{Ker}(\pi_Q^{\mathrm{c,sc}})$ is an Abelian p-group, Proposition 18.21 applies to the $\mathcal{F}^{^{\mathrm{sc}}}$-locality $\mathcal{L}^{^{\mathrm{c,sc}}}$ and in particular we have (cf. 18.23.1)

$$\bar{\mathcal{L}}^{^{\mathrm{c,sc}}}(Q) \cong (\hat{\tilde{\mathfrak{z}}}/\tilde{\mathfrak{y}})(Q) \rtimes \mathcal{F}(Q) \qquad 23.24.4.$$

Consequently, it follows from 18.27 that there exists a correspondence which maps any $\mathcal{F}^{^{\mathrm{sc}}}$-morphism $\varphi : R \to Q$ on an element $\bar{x}_\varphi \in \bar{\mathcal{L}}^{^{\mathrm{c,sc}}}(Q,R)$ in such a way that, for any $\mathcal{F}$-isomorphisms $\sigma : Q \cong Q'$ and $\tau : R \cong R'$, we have

$$\bar{x}_{\sigma\circ\varphi\circ\tau^{-1}} = \sigma\cdot\bar{x}_\varphi\cdot\tau^{-1} \qquad 23.24.5$$

where $\mathcal{F}(Q)$ and $\mathcal{F}(R)$ act on $\bar{\mathcal{L}}^{^{\mathrm{c,sc}}}(Q,R)$ *via* isomorphisms 23.23.4.

Now, it follows from Proposition 18.28 that the correspondence mapping any pair of $\mathcal{F}^{^{\mathrm{sc}}}$-morphisms $\psi : T \to R$ and $\varphi : R \to Q$ on the element $\tilde{z}_{\varphi,\psi}$ of $(\hat{\tilde{\mathfrak{z}}}/\tilde{\mathfrak{y}})(T)$ such that

$$\bar{x}_\varphi\cdot\bar{x}_\psi = \bar{x}_{\varphi\circ\psi}\cdot\tilde{z}_{\varphi,\psi} \qquad 23.24.6$$

determines a stable $\hat{\tilde{\mathfrak{z}}}/\tilde{\mathfrak{y}}$-valued 2-cocycle over $\tilde{\mathcal{F}}^{^{\mathrm{sc}}}$, which is a 2-coboundary since $\mathbb{H}^2_*(\tilde{\mathcal{F}}^{^{\mathrm{sc}}},\hat{\tilde{\mathfrak{z}}}/\tilde{\mathfrak{y}}) = \{0\}$; as in Proposition 18.28, this easily proves the existence of the announced functorial section $\mathfrak{s}$.

Similarly, if we have two functorial sections $\mathfrak{s}$ and $\mathfrak{s}'$ from $\mathcal{F}^{^{\mathrm{sc}}}$ to $\bar{\mathcal{L}}^{^{\mathrm{c,sc}}}$, it follows from the uniqueness part of Proposition 18.21 that we may assume that $\mathfrak{s}'\big(\mathcal{F}(Q)\big) = \mathfrak{s}\big(\mathcal{F}(Q)\big)$ for any $\mathcal{F}^{^{\mathrm{sc}}}$-object Q. Then, it follows from Proposition 18.29 that the correspondence mapping any $\mathcal{F}^{^{\mathrm{sc}}}$-morphism $\varphi : R \to Q$ on the element $\tilde{\ell}_\varphi \in (\hat{\tilde{\mathfrak{z}}}/\tilde{\mathfrak{y}})(T)$ fulfilling

$$\mathfrak{s}'(\varphi) = \mathfrak{s}(\varphi)\cdot\tilde{\ell}_\varphi \qquad 23.24.7$$

determines a stable $\hat{\tilde{\mathfrak{z}}}/\tilde{\mathfrak{y}}$-valued 1-cocycle over $\tilde{\mathcal{F}}^{^{\mathrm{sc}}}$, which is a 1-coboundary since $\mathbb{H}^1_*(\tilde{\mathcal{F}}^{^{\mathrm{sc}}},\hat{\tilde{\mathfrak{z}}}/\tilde{\mathfrak{y}}) = \{0\}$; as in Proposition 18.29, this easily proves that $\mathfrak{s}$ and $\mathfrak{s}'$ are naturally isomorphic. We are done.

23.25 This $\mathcal{F}^{sc}$-locality $(\tau^{d,sc}, \mathcal{L}^{d,sc}, \pi^{d,sc})$ still can be "narrowed" till the *reduced $\mathcal{F}^{sc}$-locality* $(\tau^{r,sc}, \mathcal{L}^{r,sc}, \pi^{r,sc})$ below where the *contravariant* functor

$$\mathfrak{o}_{\mathfrak{r}\mathfrak{n}}(\mathfrak{n}) = \mathfrak{ac}(\tilde{\mathfrak{z}}) \circ \mathfrak{int}_P : \tilde{\mathcal{F}}^{sc} \longrightarrow \mathfrak{Ab} \qquad 23.25.1$$

determines the *kernel* of $\pi^{r,sc}$. For any $\tilde{\mathcal{F}}^{sc}$-object Q, recall that (cf. 23.7)

$$\mathfrak{int}_P(Q) = Q \,\tilde{\cap}\, P = \bigoplus_{Q' \in \tilde{\mathcal{S}}_Q} \bigoplus_{\tilde{\eta}' \in \tilde{\mathcal{M}}^Q_{Q'}} Q' \qquad 23.25.2$$

and that any $Q' \in \tilde{\mathcal{S}}_Q$ and any $\tilde{\eta}' \in \tilde{\mathcal{M}}^Q_{Q'}$ determine a maximal $(\tilde{\mathcal{F}}^{sc})_{Q \,\tilde{\cap}\, P}$-object $\tilde{\theta}_{Q',\tilde{\eta}'} : Q' \to Q \,\tilde{\cap}\, P$ (cf. 23.7.3), getting a bijection

$$\bigsqcup_{Q' \in \tilde{\mathcal{S}}_Q} \tilde{\mathcal{M}}^Q_{Q'} \cong \mathfrak{D}^{Q,sc}_{\max} \qquad 23.25.3$$

which, with evident notation, allows us to write

$$\mathfrak{int}_P(Q) = Q \,\tilde{\cap}\, P = \bigoplus_{\tilde{\theta} \in \mathfrak{D}^{Q,sc}_{\max}} Q_{\tilde{\theta}} \qquad 23.25.4;$$

moreover, any $\tilde{\eta} \in \mathfrak{D}^{Q,sc}$ determines an element $\tilde{\theta}_{\tilde{\eta}} : Q_{\tilde{\theta}_{\tilde{\eta}}} \to Q \,\tilde{\cap}\, P$ in $\mathfrak{D}^{Q,sc}_{\max}$ and, in particular, we have (cf. 23.18)

$$Q_{\tilde{\theta}_{\tilde{\eta}}} = \mathfrak{p}_{\mathfrak{ac}(\tilde{\mathcal{F}}^{sc})}\big(\mathfrak{im}(\tilde{\eta}, Q \,\tilde{\cap}\, P)\big) \qquad 23.25.5.$$

Hence, for any $\tilde{\eta} : T \to Q \,\tilde{\cap}\, P$ in $\mathfrak{D}^{Q,sc}$, setting $Z(\tilde{\eta}) = Z(T)$ and (cf. definition 23.15.2)

$$\overline{Z(\tilde{\theta}_{\tilde{\eta}})}^{\tilde{\eta}} = Z(\tilde{\theta}_{\tilde{\eta}}) / \big(Z(\tilde{\theta}_{\tilde{\eta}}) \cap [Z(T), \mathrm{Aut}(\tilde{\eta})]\big) \qquad 23.25.6,$$

since $\mathrm{Aut}(\tilde{\theta}) = \{\widetilde{\mathrm{id}}_{Q_{\tilde{\theta}}}\}$ for any $\tilde{\theta} \in \mathfrak{D}^{Q,sc}_{\max}$ (cf. 23.8) we get (cf. 23.16.1)

$$Z(Q \,\tilde{\cap}\, P) = \prod_{\tilde{\theta} \in \mathfrak{D}^{Q,sc}_{\max}} Z(\tilde{\theta}) \quad \text{and} \quad \tilde{\mathfrak{y}}(Q) = \prod_{\tilde{\eta} \in \mathfrak{D}^{Q,sc}} \overline{Z(\tilde{\theta}_{\tilde{\eta}})}^{\tilde{\eta}} \qquad 23.25.7$$

and the natural map in 23.17.4 applied to $\mathfrak{n}$ gives a natural map sending Q to the group homomorphism

$$\omega_{\mathfrak{r},\mathfrak{r}^{\mathfrak{i}}}(\mathfrak{n})_Q : \prod_{\tilde{\theta} \in \mathfrak{D}^{Q,sc}_{\max}} Z(\tilde{\theta}) \longrightarrow \prod_{\tilde{\eta} \in \mathfrak{D}^{Q,sc}} \overline{Z(\tilde{\theta}_{\tilde{\eta}})}^{\tilde{\eta}} \qquad 23.25.8$$

determined by all the canonical homomorphisms $Z(\tilde\theta_{\tilde\eta}) \to \overline{Z(\tilde\theta_{\tilde\eta})}^{\tilde\eta}$; it is clear that the projection over the factors in $\mathfrak{O}^{Q,sc}_{\max}$ defines a section of $\omega_{\mathfrak{rn},\mathfrak{rn}^{\mathfrak{i}}}(\mathfrak{n})_Q$ but these sections need *not* define a *natural* section of $\omega_{\mathfrak{rn},\mathfrak{rn}^{\mathfrak{i}}}(\mathfrak{n})$; however, we have the following result.

Proposition 23.26 *With the notation above, for any $n \in \mathbb{N}$ the correspondence sending any $\tilde{\mathcal{F}}^{sc}$-object Q to the k-module homomorphism*

$$k \otimes_{\mathbb{Z}} \mathfrak{h}(Q)^{p^n} \longrightarrow k \otimes_{\mathbb{Z}} Z(Q \,\tilde\cap\, P)^{p^n} \qquad 23.26.1$$

defined by the projection over the factors in $\mathfrak{O}^{Q,sc}_{\max} \subset \mathfrak{O}^{Q,sc}$ is a natural section of $k \otimes_{\mathbb{Z}} \omega_{\mathfrak{rn},\mathfrak{rn}^{\mathfrak{i}}}(\mathfrak{n}^{p^n})$.

Proof: It suffices to prove that, for any $\tilde{\mathcal{F}}^{sc}$-morphism $\tilde\varphi : R \to Q$, any $\tilde\eta$ in $\mathfrak{O}^{Q,sc} - \mathfrak{O}^{Q,sc}_{\max}$ and any $n \in \mathbb{N}$, $\mathfrak{h}(\tilde\varphi) = \big((\mathfrak{p}_{\mathfrak{rn},\mathfrak{rn}^{\mathfrak{i}}})_{\mathfrak{a}}(\mathfrak{n})\big)(\tilde\varphi)$ (cf. 23.23) maps the elements of $\overline{Z(\tilde\theta_{\tilde\eta})^{p^n}}^{\tilde\eta}$ on the elements of the product

$$\prod_{\tilde\varpi \in \mathfrak{O}^{R,sc}_{\max}} Z(\tilde\varpi)^{p^{n+1}} \times \prod_{\tilde\nu \in \mathfrak{O}^{R,sc} - \mathfrak{O}^{R,sc}_{\max}} \overline{Z(\tilde\varpi_{\tilde\nu})^{p^n}}^{\tilde\nu} \qquad 23.26.2$$

where $\tilde\varpi_{\tilde\nu}$ denotes the element of $\mathfrak{O}^{R,sc}_{\max}$ determined by $\tilde\nu \in \mathfrak{O}^{R,sc}$.

But it follows from 23.14 and 23.17 that, for any $z \in Z(\tilde\theta_{\tilde\eta})^{p^n}$, $\mathfrak{h}(\tilde\varphi)$ maps its image $\bar z$ in $\overline{Z(\tilde\theta_{\tilde\eta})^{p^n}}^{\tilde\eta}$ on

$$\Big(\big((\mathfrak{p}_{\mathfrak{rn},\mathfrak{rn}^{\mathfrak{i}}})_{\mathfrak{a}}(\mathfrak{n})\big)(\tilde\varphi)\Big)(\bar z) = \sum_{w \in W_{\tilde\eta}} \overline{\big(\mathfrak{n}(\tilde\psi_w, \tilde\varphi)\big)(z)} \qquad 23.26.3$$

where $\overline{\big(\mathfrak{n}(\tilde\psi_w, \tilde\varphi)\big)(z)}$ denotes the image in $\overline{Z(\tilde\varpi_{\tilde\nu})^{p^n}}^{\tilde\nu}$ of $\big(\mathfrak{n}(\tilde\psi_w, \tilde\varphi)\big)(z)$; moreover, for any $w \in W_{\tilde\eta}$, it follows from our definition of $\mathfrak{n}$ (cf. 23.23.4) that, denoting by $\tilde\varphi_w$ the $\tilde{\mathcal{F}}^{sc}$-morphism determined by $\mathfrak{im}(\tilde\psi_w, \tilde\varphi \,\tilde\cap\, \mathrm{id}_P)$ between the origins of $\tilde\varpi_{\tilde\nu_w}$ and $\tilde\theta_{\tilde\eta}$, and setting $Z(\tilde\varphi_w) = \big(\mathfrak{ac}(\mathfrak{z})\big)(\tilde\varphi_w)$, we have

$$\mathfrak{n}(\tilde\psi_w, \tilde\varphi) = Z(\tilde\varphi_w) : Z(\tilde\theta_{\tilde\eta}) \longrightarrow Z(\tilde\varpi_{\tilde\nu_w}) \qquad 23.26.4.$$

On the other hand, respectively denoting by T and Q' the origins of $\tilde\eta$ and $\tilde\theta_{\tilde\eta}$, recall that $\mathrm{Aut}(\tilde\eta) \cong \bar N_{Q'}(T)$ (cf. 23.8) and therefore, since $\tilde\eta \notin \mathfrak{O}^{Q,sc}_{\max}$, we have $T \neq Q'$ and in particular $\mathrm{Aut}(\tilde\eta)$ is a nontrivial p-group. Moreover, with our notation in 23.14, it is clear that $N_{Q'}(T)$ acts on the set of double classes $\varphi(R)\backslash Q/T$ by the multiplication on the right and therefore it acts on the set of representatives $W_{\tilde\eta}$; but, for any $w \in W_{\tilde\eta}$ and any $u' \in N_{Q'}(T)$, since

$$U_{wu'} = \varphi(R)^{wu'} \cap T = (U_w)^{u'} \qquad 23.26.5,$$

it is clear that the elements of $\mathfrak{O}^{R,\mathrm{sc}}$ which are $(\tilde{\mathcal{F}}^{\mathrm{sc}})_{R\tilde{\cap}P}$-isomorphic to

$$\tilde{\nu}_{wu'} : U_{wu'} \longrightarrow R\tilde{\cap}P \quad \text{or} \quad \tilde{\nu}_w : U_w \longrightarrow R\tilde{\cap}P \qquad 23.26.6$$

determine the *same* element $\tilde{\varpi} \in \mathfrak{O}^{R,\mathrm{sc}}_{\max}$; finally, denoting by R' the origin of $\tilde{\varpi}$, we may assume that $\varphi(R')^w \subset Q'$ and that the corresponding group homomorphism determines $\tilde{\varphi}_w$; consequently, since u' acts trivially on $Z(\tilde{\theta}_{\tilde{\eta}}) = Z(Q')$, we get $Z(\tilde{\varphi}_{wu'}) = Z(\tilde{\varphi}_w)$.

In conclusion, for any orbit O of $\mathrm{Aut}(\tilde{\eta}) \cong \bar{N}_{Q'}(T)$ in $W_{\tilde{\eta}}$ and any $w \in O$, we actually get

$$\sum_{w'\in O} \overline{\big(Z(\tilde{\varphi}_{w'})\big)(z)}^{\tilde{\nu}_{w'}} = |O|\cdot\overline{\big(Z(\tilde{\varphi}_w)\big)(z)}^{\tilde{\nu}_w} \qquad 23.26.7$$

and therefore if $O \neq \{w\}$ then this sum belongs to $\overline{Z(R')^{p^{n+1}}}^{\tilde{\nu}_w}$; thus, if $N_{Q'}(T)$ has no fixed points in $W_{\tilde{\eta}}$, we are done. But, if $N_{Q'}(T)$ fixes some $w \in W_{\tilde{\eta}}$ then, for any $u' \in N_{Q'}(T)$, we have $wu' = \varphi(v)wt$ for suitable $v \in R$ and $t \in T$, and therefore $u't^{-1}$ belongs to

$$\varphi(R)^w \cap Q' = \varphi(R')^w \qquad 23.26.8$$

and normalizes $\varphi(R)^w \cap T = U_w$, namely $u't^{-1}$ determines an element of $\mathrm{Aut}(\tilde{\nu}_w) \cong \bar{N}_{\varphi(R')^w}(U_w)$; finally, we get a group homomorphism

$$\mathrm{Aut}(\tilde{\eta}) \longrightarrow \mathrm{Aut}(\tilde{\nu}_w) \qquad 23.26.9$$

which is injective by Corollary 4.9 and, in particular, it forces $\mathrm{Aut}(\tilde{\nu}_w) \neq \{1\}$, so that $\tilde{\nu}_w \notin \mathfrak{O}^{R,\mathrm{sc}}_{\max}$. We are done.

Corollary 23.27 *With the notation above, for any* $m \in \mathbb{N}$ *and any* $n \geq 1$*, we have*

$$\begin{aligned} \mathbb{H}^n_*\big(\tilde{\mathcal{F}}^{\mathrm{sc}}, k\otimes_{\mathbb{Z}} (\mathfrak{ac}(\tilde{\mathfrak{z}})\circ \mathfrak{int}_P)^{p^m}\big) &= \{0\} \\ \mathbb{H}^n_*\big(\tilde{\mathcal{F}}^{\mathrm{sc}}, k\otimes_{\mathbb{Z}} \tilde{\mathfrak{h}}^{p^m}/(\mathfrak{ac}(\tilde{\mathfrak{z}})\circ \mathfrak{int}_P)^{p^m}\big) &= \{0\} \end{aligned} \qquad 23.27.1.$$

Proof: It follows from Proposition 23.26 that, for any $m \in \mathbb{N}$, we have

$$k\otimes_{\mathbb{Z}} \tilde{\mathfrak{h}}^{p^m} \cong \big(k\otimes_{\mathbb{Z}} (\mathfrak{ac}(\tilde{\mathfrak{z}})\circ \mathfrak{int}_P)^{p^m}\big) \oplus \big(k\otimes_{\mathbb{Z}} \tilde{\mathfrak{h}}^{p^m}/(\mathfrak{ac}(\tilde{\mathfrak{z}})\circ \mathfrak{int}_P)^{p^m}\big) \quad 23.27.2;$$

but, according to Theorem 23.20 applied to the *contravariant* functor

$$k\otimes_{\mathbb{Z}} \big(\overline{\mathfrak{ac}(\tilde{\mathfrak{z}})\circ \mathfrak{p}_{\mathfrak{ac}(\tilde{\mathcal{F}}^{\mathrm{sc}})}}\big)^{p^m} : \mathfrak{tg}_{\mathfrak{ac}(\tilde{\mathcal{F}}^{\mathrm{sc}})} \rtimes \mathfrak{ac}(\tilde{\mathcal{F}}^{\mathrm{sc}}) \longrightarrow k\text{-}\mathfrak{mod} \qquad 23.27.3,$$

for any $n \geq 1$ we still have (cf. 23.23)

$$\mathbb{H}^n_*(\tilde{\mathcal{F}}^{\mathrm{sc}}, k\otimes_{\mathbb{Z}} \tilde{\mathfrak{h}}^{p^m}) = \{0\} \qquad 23.27.4;$$

hence, we get equalities 23.27.1.

Corollary 23.28 *With the notation above, there exists an $\mathcal{F}^{sc}$-sublocality $(\tau^{r,sc}, \mathcal{L}^{r,sc}, \pi^{r,sc})$ of $\mathcal{L}^{d,sc}$, unique up to natural isomorphisms, such that*

$$\mathfrak{Ker}(\pi^{r,sc}) = \mathfrak{ac}(\mathfrak{z}) \circ \mathfrak{int}_P \circ \tilde\pi^{d,sc} \qquad 23.28.1$$

where $\tilde\pi^{d,sc} : \mathcal{L}^{d,sc} \to \tilde{\mathcal{F}}^{sc}$ denotes the composition of $\pi^{d,sc}$ with the canonical functor $\mathcal{F}^{sc} \to \tilde{\mathcal{F}}^{sc}$.

Proof: The proof follows the same pattern as the proof of Corollary 23.24 except that the *graded equalities* 23.27.1 force us to argue by induction on m. Set $(\tau_0^{d,sc}, \mathcal{L}_0^{d,sc}, \pi_0^{d,sc}) = (\tau^{d,sc}, \mathcal{L}^{d,sc}, \pi^{d,sc})$, $\tilde{\mathfrak{a}}_0 = \mathfrak{ac}(\mathfrak{z}) \circ \mathfrak{int}_P$ and $\tilde{\mathfrak{h}}_0 = \mathfrak{h}$; for any $1 \le i \le m$, inductively define $\tilde{\mathfrak{a}}_i$ and $\tilde{\mathfrak{h}}_i$ as the respective p-th powers of $\tilde{\mathfrak{a}}_{i-1}$ and $\tilde{\mathfrak{h}}_{i-1}$ and assume that there exists a coherent $\mathcal{F}^{sc}$-sublocality $(\tau_i^{d,sc}, \mathcal{L}_i^{d,sc}, \pi_i^{d,sc})$ of $(\tau_{i-1}^{d,sc}, \mathcal{L}_{i-1}^{d,sc}, \pi_{i-1}^{d,sc})$ fulfilling

$$(\tilde{\mathfrak{h}}_i + \tilde{\mathfrak{a}}_0) \circ \tilde\pi_i^{d,sc} = \mathfrak{Ker}(\pi_i^{d,sc}) \quad \text{and} \quad \tilde{\mathfrak{h}}_i \cap \tilde{\mathfrak{a}}_0 = \tilde{\mathfrak{a}}_i \qquad 23.28.2.$$

Then, considering the obvious *surjective* functor

$$\mathfrak{r}_m : \hat{\mathcal{L}}_m = \mathcal{L}_m^{d,sc} / \big((\tilde{\mathfrak{h}}_{m+1} + \tilde{\mathfrak{a}}_m) \circ \tilde\pi_m^{d,sc}\big) \longrightarrow \mathcal{L}_m^{d,sc} / (\tilde{\mathfrak{h}}_m \circ \tilde\pi_m^{d,sc}) = \mathcal{L}_m \qquad 23.28.3,$$

it is clear that

$$\mathfrak{Ker}(\mathfrak{r}_m) = \big(\tilde{\mathfrak{h}}_m / (\tilde{\mathfrak{h}}_{m+1} + \tilde{\mathfrak{a}}_m)\big) \circ \tilde\pi_m^{d,sc} \qquad 23.28.4.$$

But, setting $\bar{\tilde{\mathfrak{h}}}_m = \tilde{\mathfrak{h}}_m / (\tilde{\mathfrak{h}}_{m+1} + \tilde{\mathfrak{a}}_m)$, it follows from Corollary 23.27 that, for any $n \ge 1$, we have

$$\mathbb{H}_*^n(\tilde{\mathcal{F}}^{sc}, \bar{\tilde{\mathfrak{h}}}_m) = \{0\} \qquad 23.28.5.$$

At this point, we claim that the particular equalities

$$\mathbb{H}_*^2(\tilde{\mathcal{F}}^{sc}, \bar{\tilde{\mathfrak{h}}}_m) = \{0\} = \mathbb{H}_*^1(\tilde{\mathcal{F}}^{sc}, \bar{\tilde{\mathfrak{h}}}_m) \qquad 23.28.6$$

force the existence of a *functorial section* $\mathfrak{s}_m$ of $\mathfrak{r}_m$, unique up to natural isomorphisms; then, the *converse image* in $\mathcal{L}_m^{d,sc}$ of the *image* of $\mathfrak{s}_m$ in $\hat{\mathcal{L}}_m$ defines a coherent $\mathcal{F}^{sc}$-sublocality $(\tau_{m+1}^{d,sc}, \mathcal{L}_{m+1}^{d,sc}, \pi_{m+1}^{d,sc})$ of $(\tau_m^{d,sc}, \mathcal{L}_m^{d,sc}, \pi_m^{d,sc})$ fulfilling condition 23.28.2.

Indeed, for any $\tilde{\mathcal{F}}^{sc}$-object Q, since $\mathrm{Ker}(\pi_Q^{d,sc})$ is an Abelian p-group, Proposition 18.21 applies to the $\mathcal{F}^{sc}$-locality $\hat{\mathcal{L}}_m$ and in particular we get

$$\hat{\mathcal{L}}_m(Q) \cong \bar{\tilde{\mathfrak{h}}}_m(Q) \rtimes \mathcal{L}_m(Q) \qquad 23.28.7.$$

Consequently, arguing as in 18.23, 18.24, 18.25, 18.26 and 18.27 above, it is not difficult to prove that there exists a correspondence which maps any

$\mathcal{L}_m$-morphism $x:R \to Q$ on an element $\hat{x} \in \hat{\mathcal{L}}_m(Q,R)$ in such a way that, for any $\mathcal{L}_m$-isomorphisms $s:Q \cong Q'$ and $t:R' \cong R$, we have

$$\widehat{s\cdot x\cdot t} = s\cdot\hat{x}\cdot t \qquad 23.28.8$$

where $\mathcal{L}_m(Q)$ and $\mathcal{L}_m(R)$ act on $\hat{\mathcal{L}}_m(Q,R)$ *via* isomorphisms 23.27.7.

Now, arguing as in Proposition 18.28 it is easy to check that the correspondence mapping any pair of $\mathcal{L}_m$-morphisms $y:T \to R$ and $x:R \to Q$ on the element $\tilde{a}_{x,y}$ of $\bar{\tilde{\mathfrak{h}}}_m(T)$ such that

$$\hat{x}\cdot\hat{y} = \widehat{x\cdot y}\cdot\tilde{a}_{x,y} \qquad 23.28.9$$

determines a stable $\bar{\tilde{\mathfrak{h}}}_m$-valued 2-cocycle over $\tilde{\mathcal{L}}_m = \tilde{\mathcal{F}}^{sc}$ (cf. 1.3), which is a 2-coboundary since $\mathbb{H}^2_*(\tilde{\mathcal{F}}^{sc},\bar{\tilde{\mathfrak{h}}}_m) = \{0\}$; as in Proposition 18.28, this easily proves the existence of the announced functorial section $\mathfrak{s}_m$.

Similarly, if we have two functorial sections $\mathfrak{s}_m$ and $\mathfrak{s}'_m$ from $\mathcal{L}_m$ to $\hat{\mathcal{L}}_m$, it follows from the uniqueness part of Proposition 18.21 that we may assume that

$$\mathfrak{s}'_m\big(\mathcal{L}_m(Q)\big) = \mathfrak{s}_m\big(\mathcal{L}_m(Q)\big) \qquad 23.28.10$$

for any $\mathcal{L}_m$-object Q. Then, arguing as in Proposition 18.29 it is easily checked that the correspondence mapping any $\mathcal{L}_m$-morphism $x:R \to Q$ on the element $\tilde{b}_x \in \bar{\tilde{\mathfrak{h}}}_m(T)$ fulfilling

$$\mathfrak{s}'_m(x) = \mathfrak{s}_m(x)\cdot\tilde{b}_x \qquad 23.28.11$$

determines a stable $\bar{\tilde{\mathfrak{h}}}_m$-valued 1-cocycle over $\tilde{\mathcal{L}}_m = \tilde{\mathcal{F}}^{sc}$ (cf. 1.3), which is a 1-coboundary since $\mathbb{H}^1_*(\tilde{\mathcal{F}}^{sc},\bar{\tilde{\mathfrak{h}}}_m) = \{0\}$; as in Proposition 18.29, this easily proves that $\mathfrak{s}_m$ and $\mathfrak{s}'_m$ are naturally isomorphic. This completes our induction argument.

But, for m big enough, the *contravariant* functor $\tilde{\mathfrak{h}}_m$ is trivial, so that we have $\mathcal{L}^{d,sc}_m = \mathcal{L}^{d,sc}_{m+1}$; in this situation, it follows from condition 23.28.2 that $(\tau^{d,sc}_m, \mathcal{L}^{d,sc}_m, \pi^{d,sc}_m)$ fulfills condition 23.28.1; moreover, the uniqueness follows from the uniqueness in each step of our induction. We are done.

23.29 At this point, the main difficulty to obtain a *perfect $\mathcal{F}^{sc}$-locality* as a quotient of $\mathcal{L}^{r,sc}$ is the lack of naturality of the splitting of the exact sequences

$$1 \longrightarrow Z(Q) \longrightarrow Z(Q\,\tilde{\cap}\,P) \longrightarrow Z(Q\,\tilde{\cap}\,P)/Z(Q) \longrightarrow 1 \qquad 23.29.1$$

when Q runs over the set of $\tilde{\mathcal{F}}^{sc}$-objects. More precisely, respectively denoting by

$$\mathfrak{j}_{\tilde{\mathcal{F}}^{sc}} : \tilde{\mathcal{F}}^{sc} \to \mathfrak{ac}(\tilde{\mathcal{F}}^{sc}) \quad \text{and} \quad \gamma_P : \mathfrak{int}_P \to \mathfrak{j}_{\tilde{\mathcal{F}}^{sc}} \qquad 23.29.2$$

the canonical functor and the natural map obtained from the first structural morphism (cf. 23.11.1), the corresponding *injective* natural map

$$\mathrm{id}_{\mathfrak{ac}(\tilde{\mathfrak{z}})} * \gamma_P : \tilde{\mathfrak{z}} = \mathfrak{ac}(\tilde{\mathfrak{z}}) \circ \mathfrak{j}_{\tilde{\mathcal{F}}^{\mathrm{sc}}} \longrightarrow \mathfrak{ac}(\tilde{\mathfrak{z}}) \circ \mathfrak{int}_P \qquad 23.29.3$$

only admits an "almost" natural section in the sense of the following result. For this result, we need the functor we have defined in 13.18

$$\tilde{\mathfrak{z}}^{\circ} = \tilde{\mathfrak{z}}^{\circ}_{\tilde{\mathcal{F}}^{\mathrm{sc}}} : \tilde{\mathcal{F}}^{\mathrm{sc}} \longrightarrow \mathfrak{Ab} \qquad 23.29.4.$$

Proposition 23.30 *For any $\mathcal{F}$-selfcentralizing subgroup Q, the structural injection $Z(Q) \to Z(Q \tilde{\cap} P)$ raised to a suitable power $c_Q \equiv |\tilde{\mathcal{F}}(P)|$ (mod p) admits a section*

$$\sigma_Q : Z(Q \tilde{\cap} P) = \bigoplus_{Q' \in \tilde{\mathcal{S}}_Q} \bigoplus_{\tilde{\theta} \in \tilde{\mathcal{M}}^Q_{Q'}} Z(Q') \longrightarrow Z(Q) \qquad 23.30.1$$

sending $z' \in Z(Q')$ to $\big(\tilde{\mathfrak{z}}^{\circ}(\tilde{\iota}^Q_{Q'})\big)(z')$ for any $Q' \in \tilde{\mathcal{S}}_Q$ and any $\tilde{\theta} \in \tilde{\mathcal{M}}^Q_{Q'}$. Moreover, for any $\tilde{\mathcal{F}}^{\mathrm{sc}}$-morphism $\tilde{\varphi} : R \to Q$, denote by $\overline{Z(R)}$ the quotient of $Z(R)$ by the sum

$$\sum_{(\varphi,Q')} \sum_{Q''} \big(\tilde{\mathfrak{z}}^{\circ}(\tilde{\delta}_{\varphi}) \circ \tilde{\mathfrak{z}}(\tilde{\iota}^{Q''}_{Q''\cap\varphi(R)}) \circ \tilde{\mathfrak{z}}^{\circ}(\tilde{\iota}^{Q''}_{Q'})\big)\big(Z(Q')\big) \qquad 23.30.2$$

where $\varphi \in \tilde{\varphi}$ and $Q' \in \tilde{\mathcal{S}}_Q$ run over the pairs such that $Q' \cap \varphi(R)$ is not $\mathcal{F}$-selfcentralizing and where, for such a pair, Q'' runs over the set of minimal subgroups of Q containing Q' such that $Q'' \cap \varphi(R)$ is $\mathcal{F}$-selfcentralizing, and $\delta_{\varphi} : Q'' \cap \varphi(R) \to R$ maps $\varphi(v)$ on $v \in R$. Then, we have the commutative diagram

$$\begin{array}{ccc} Z(Q \tilde{\cap} P) & \xrightarrow{\sigma_Q} & Z(Q) \\ {\scriptstyle (\mathfrak{ac}(\tilde{\mathfrak{z}})\circ\mathfrak{int}_P)(\tilde{\varphi})} \downarrow & & \downarrow {\scriptstyle \overline{\tilde{\mathfrak{z}}(\tilde{\varphi})}} \\ Z(R \tilde{\cap} P) & \xrightarrow{\bar{\sigma}_R} & \overline{Z(R)} \end{array} \qquad 23.30.3.$$

Proof: It follows from Lemma 13.19 that σ_Q maps the image in $Z(Q \tilde{\cap} P)$ of $z \in Z(Q)$ on $\prod_{Q' \in \tilde{\mathcal{S}}_Q} \prod_{\tilde{\theta} \in \tilde{\mathcal{M}}^Q_{Q'}} z^{|Q:Q'|}$, and from Proposition 6.7 that

$$\sum_{Q' \in \tilde{\mathcal{S}}_Q} \sum_{\tilde{\theta} \in \tilde{\mathcal{M}}^Q_{Q'}} |Q : Q'| \equiv |\tilde{\mathcal{F}}(P, Q)| \equiv |\tilde{\mathcal{F}}(P)| \not\equiv 0 \ (\mathrm{mod}\ p) \qquad 23.30.4.$$

Let $\tilde\varphi : R \to Q$ be an $\tilde{\mathcal{F}}^{sc}$-morphism; it follows from 23.14 that, for any $(\tilde{\mathcal{F}}^{sc})_{Q\,\tilde\cap\, P}$-object $\tilde\theta : Q' \to Q\,\tilde\cap\, P$ in $\mathfrak{O}^{Q,sc}_{\max}$, we have the $\mathfrak{ac}(\tilde{\mathcal{F}}^{sc})$*-pull-back*

$$\begin{array}{ccc} Q' & \xrightarrow{\tilde\theta} & Q\,\tilde\cap\, P \\ {\scriptstyle \oplus_{w\in W_{\tilde\theta}} \tilde\varphi'_w}\uparrow & & \uparrow{\scriptstyle \tilde\varphi\,\tilde\cap\,\widetilde{\mathrm{id}}_P} \\ \bigoplus_{w\in W_{\tilde\theta}} R'_w & \xrightarrow{\oplus_{w\in W_{\tilde\theta}} \tilde\varpi_w} & R\,\tilde\cap\, P \end{array} \qquad 23.30.5$$

for a choice of an element $\varphi \in \tilde\varphi$ and of a set of representatives $W_{\tilde\theta}$ in Q for $\varphi(R)\backslash Q/Q'$, in such a way that for any $w \in W_{\tilde\theta}$ we have

$$\varphi(R'_w)^w = \varphi(R)^w \cap Q' \qquad 23.30.6;$$

then, it is quite clear that the $\mathfrak{ac}(\tilde{\mathcal{F}}^{sc})$*-pull-back* 23.14.1 forces

$$\bigoplus_{\tilde\theta\in\mathfrak{O}^{Q,sc}_{\max}} \bigoplus_{w\in W_{\tilde\theta}} R'_w \cong R\,\tilde\cap\, P \qquad 23.30.7$$

or, equivalently, up to suitable choices it implies that

$$\bigsqcup_{\tilde\theta\in\mathfrak{O}^{Q,sc}_{\max}} \{\tilde\varpi_w\}_{w\in W_{\tilde\theta}} = \mathfrak{O}^{R,sc}_{\max} \qquad 23.30.8.$$

Then, we claim that the following $\mathfrak{ac}(\tilde{\mathcal{F}}^{sc})$-diagram is commutative

$$\begin{array}{ccccc} Z(Q\,\tilde\cap\, P) & = & \prod_{\tilde\theta\in\mathfrak{O}^{Q,sc}_{\max}} Z(\tilde\theta) & \xrightarrow{\sigma_Q} & Z(Q) \\ \downarrow{\scriptstyle (\mathfrak{ac}(\mathfrak{z}))(\tilde\varphi\,\tilde\cap\,\widetilde{\mathrm{id}}_P)} & & & & \downarrow{\scriptstyle \overline{\mathfrak{z}(\tilde\varphi)}} \\ Z(R\,\tilde\cap\, P) & = & \prod_{\tilde\varpi\in\mathfrak{O}^{R,sc}_{\max}} Z(\tilde\varpi) & \xrightarrow{\bar\sigma_R} & \overline{Z(R)} \end{array} \qquad 23.30.9;$$

indeed, it is quite clear that it suffices to prove the commutativity of the restricted $\mathfrak{ac}(\tilde{\mathcal{F}}^{sc})$-diagrams

$$\begin{array}{ccc} Z(Q') & \xrightarrow{\mathfrak{z}^\circ(\tilde\iota^{Q}_{Q'})} & Z(Q) \\ {\scriptstyle \mathfrak{z}(\tilde\varphi_w)}\downarrow & & \downarrow{\scriptstyle \overline{\mathfrak{z}(\tilde\varphi)}} \\ \prod_{w\in W_{\tilde\theta}} Z(R'_w) & \xrightarrow{\prod_{w\in W_{\tilde\theta}} \overline{\mathfrak{z}^\circ(\tilde\iota^{R}_{R'_w})}} & \overline{Z(R)} \end{array} \qquad 23.30.10;$$

but, since we have (cf. Proposition 6.19 and equality 23.30.6)

$$\prod_{w\in W_{\tilde\theta}} Z(R'_w) \cong \prod_{w\in W_{\tilde\theta}} Z\big(\varphi(R)^w \cap Q'\big) = Z(R\,\tilde\cap_Q\, Q') \qquad 23.30.11,$$

the commutativity follows from Lemma 13.19. We are done.

23.31 A sufficient (strong!) condition to overcome this difficulty is to assume that

23.31.1 *For every $\mathcal{F}$-selfcentralizing subgroup Q of P and every nontrivial subgroup $\bar{R}$ of $\bar{N}_P(Q)$, the $k\bar{R}$-module $k \otimes_{\mathbb{Z}} Z(Q)$ has no nonzero projective factors.*

This condition is fulfilled if P has no p-elementary Abelian subgroups of order p^p which, as a bound, is stronger than the bounds p^{p^3} and p^{p^2} respectively guaranteeing the existence and the uniqueness of the *perfect* $\mathcal{F}$-locality in [13]; but, condition 23.31.1 can be fulfilled independently of such a bound.

Corollary 23.32 *With the notation above, assume that for every $\mathcal{F}$-selfcentralizing subgroup Q of P and every nontrivial subgroup $\bar{R}$ of $\bar{N}_P(Q)$, the $k\bar{R}$-module $k \otimes_{\mathbb{Z}} Z(Q)$ has no nonzero projective factors. Then, for any $m \in \mathbb{N}$, the correspondence sending any $\mathcal{F}$-selfcentralizing subgroup Q of P to the k-module homomorphism*

$$k \otimes_{\mathbb{Z}} Z(Q \tilde{\cap} P)^{p^m} \longrightarrow k \otimes_{\mathbb{Z}} Z(Q)^{p^m} \qquad 23.32.1$$

induced by σ_Q is a natural map. In particular, for any $n \geq 1$ we have

$$\begin{aligned} \mathbb{H}^n_*\big(\tilde{\mathcal{F}}^{sc}, k \otimes_{\mathbb{Z}} (\mathfrak{ac}(\tilde{\mathfrak{z}}) \circ \mathfrak{int}_P)^{p^m}/\tilde{\mathfrak{z}}^{p^m}\big) &= \{0\} \\ \mathbb{H}^n_*\big(\tilde{\mathcal{F}}^{sc}, k \otimes_{\mathbb{Z}} \tilde{\mathfrak{z}}^{p^m}\big) &= \{0\} \end{aligned} \qquad 23.32.2.$$

Proof: Since we have an obvious surjective $k\bar{N}_P(Q)$-module homomorphism from $k \otimes_{\mathbb{Z}} Z(Q)$ to $k \otimes_{\mathbb{Z}} Z(Q)^{p^m}$, for every nontrivial subgroup $\bar{N}$ of $\bar{N}_P(Q)$ the $k\bar{N}$-module $k \otimes_{\mathbb{Z}} Z(Q)^{p^m}$ has no nonzero projective factors and therefore, according to Lemma 23.33 below, we have

$$\mathrm{Tr}_1^{\bar{N}}\big(Z(Q)^{p^m}\big) \subset Z(Q)^{p^{m+1}} \qquad 23.32.3.$$

Consequently, for any $\tilde{\mathcal{F}}^{sc}$-morphism $\tilde{\varphi}: R \to Q$, since in the sum 23.30.2 we always have $Q' \neq Q''$, we get

$$\big(\tilde{\mathfrak{z}}^\circ(\tilde{\iota}_{Q'}^{Q''})\big)\big(Z(Q')^{p^m}\big) \subset Z(Q')^{p^{m+1}} \qquad 23.32.4$$

and therefore the corresponding sum (cf. 23.30.2)

$$\sum_{(\varphi,Q')} \sum_{Q''} \big(\tilde{\mathfrak{z}}^\circ(\tilde{\delta}_\varphi) \circ \tilde{\mathfrak{z}}(\tilde{\iota}_{Q''\cap\varphi(R)}^{Q''}) \circ \tilde{\mathfrak{z}}^\circ(\tilde{\iota}_{Q'}^{Q''})\big)\big(Z(Q')^{p^m}\big) \qquad 23.32.5$$

is contained in $Z(Q')^{p^{m+1}}$; thus, we have the commutative diagram

$$\begin{array}{ccc} k \otimes_{\mathbb{Z}} Z(Q \tilde{\cap} P)^{p^m} & \xrightarrow{k\otimes_{\mathbb{Z}}\sigma_Q} & k \otimes_{\mathbb{Z}} Z(Q)^{p^m} \\ {\scriptstyle k\otimes_{\mathbb{Z}}(\mathfrak{ac}(\tilde{\mathfrak{z}})\circ\mathfrak{int}_P)(\tilde{\varphi})} \downarrow & & \downarrow {\scriptstyle k\otimes_{\mathbb{Z}}\tilde{\mathfrak{z}}(\tilde{\varphi})} \\ k \otimes_{\mathbb{Z}} Z(R \tilde{\cap} P)^{p^m} & \xrightarrow{k\otimes_{\mathbb{Z}}\sigma_R} & k \otimes_{\mathbb{Z}} Z(R)^{p^m} \end{array} \qquad 23.32.6.$$

Now, according to Proposition 23.30, we have

$$k\otimes_{\mathbb{Z}}(\mathfrak{ac}(\tilde{\mathfrak{z}})\circ \mathfrak{int}_P)^{p^m} \cong (k\otimes_{\mathbb{Z}}\tilde{\mathfrak{z}}^{p^m}) \oplus \big(k\otimes_{\mathbb{Z}}(\mathfrak{ac}(\tilde{\mathfrak{z}})\circ \mathfrak{int}_P)^{p^m}/\tilde{\mathfrak{z}}^{p^m}\big) \qquad 23.32.7$$

and therefore the last equalities are a consequence of the top equality in 23.27.1. We are done.

Lemma 23.33 *Let H be a finite group and M a kH-module. The restriction from M to $M_1^H = \mathrm{Tr}_1^H(M)$ induces a surjective homomorphism*

$$\mathrm{End}_k(M)_1^H \longrightarrow \mathrm{End}_k(M_1^H) \qquad 23.33.1.$$

In particular, if M has no nonzero projective factors then $M_1^H = \{0\}$.

Proof: Setting $A = \mathrm{End}_k(M)$, it is easily checked that $A_1^H{\cdot}M^H = M_1^H$; in particular, the restriction from M to M_1^H induces a homomorphism from A_1^H to $\mathrm{End}_k(M_1^H)$. Moreover, if $g \in \mathrm{End}_k(M_1^H)$ and $\{m_i\}_{i\in I}$ is a basis of M_1^H, it is quite clear that, completing $\{m_i\}_{i\in I}$ till to a basis of M, we can define a k-linear map $f: M \to M$ fulfilling

$$\mathrm{Tr}_1^H\big(f(m_i)\big) = g(m_i) \qquad 23.33.2$$

for any $i \in I$ and then homomorphism 23.33.1 sends $\mathrm{Tr}_1^H(f)$ to g; that is to say, we get a surjective homomorphism $A_1^H \to \mathrm{End}_k(M_1^H)$.

In particular, by the so-called *Higman Criterion* (cf. [29]), if M has no nonzero projective factors then A_1^H has no nonzero idempotents and therefore it is *nilpotent* (cf. Corollary 2.14 in [51]), which forces M_1^H to be zero.

Corollary 23.34 *Assume that for every $\mathcal{F}$-selfcentralizing subgroup Q of P and every nontrivial subgroup $\bar{N}$ of $\bar{N}_P(Q)$, the $k\bar{N}$-module $k\otimes_{\mathbb{Z}} Z(Q)$ has no nonzero projective factors. Then, there is a perfect $\mathcal{F}^{sc}$-sublocality $\mathcal{L}^{sc}$ of $\mathcal{L}^{r,sc}$, unique up to conjugation by an element of $\varprojlim(\pi^{t,sc})$.*

Proof: The proof follows the same pattern as in the proof of Corollary 23.28, replacing equalities 23.27.1 by equalities 23.32.2.

Remark 23.35 According to our hypothesis, it follows from Theorem 4.13 in [52] that $\mathbb{H}^n(\tilde{\mathcal{F}}^{sc},\tilde{\mathfrak{z}}) = \{0\}$ for any $n \geq 1$ and therefore Proposition 3.1 in [13] already proves the existence and the uniqueness of a partial perfect $\mathcal{F}^{sc}$-sublocality. Here we provide another proof of this fact since this corollary guarantees the existence and the uniqueness of $\mathcal{L}^{r,sc}$ inside $\mathcal{L}^{r,sc}$ and in the next chapter we show that any perfect $\mathcal{F}^{sc}$-locality lies inside $\mathcal{L}^{r,sc}$.

Chapter 24

Looking for a perfect $\mathcal{F}^{sc}$-locality

24.1 Let P be a finite p-group and $\mathcal{F}$ a Frobenius P-category. In the previous chapter, we already have given a sufficient condition for the existence and the uniqueness of a *perfect* $\mathcal{F}^{sc}$*-locality* $\mathcal{L}^{sc}$ as a *sublocality* of the *reduced* $\mathcal{F}^{sc}$*-locality* $\mathcal{L}^{r,sc}$; as we point out in Remark 23.35, in this case Proposition 3.1 in [13] already provides a complete proof of the uniqueness of $\mathcal{L}^{sc}$. In this last chapter, we prove that *any* perfect $\mathcal{F}^{sc}$-locality has to be a *sublocality* of $\mathcal{L}^{r,sc}$ independently of any condition; in particular, when our sufficient condition is fulfilled, this provides another proof of the uniqueness of $\mathcal{L}^{sc}$. From now on, $(\tau^{sc}, \mathcal{L}^{sc}, \pi^{sc})$ is a *perfect* $\mathcal{F}^{sc}$*-locality* $\mathcal{L}^{sc}$; recall that the structural functor $\pi^{sc} : \mathcal{L}^{sc} \to \mathcal{F}^{sc}$ induces an equivalence of categories

$$\tilde{\mathcal{L}}^{sc} \cong \tilde{\mathcal{F}}^{sc} \qquad 24.1.1$$

between their respective *exterior quotients* (cf. 1.3); we write τ and π for short. We will consider the *additive cover* $\mathfrak{ac}(\mathcal{L}^{sc})$ of $\mathcal{L}^{sc}$ (cf. A2.7 and A4.10).

Proposition 24.2 *Any morphism* $x : R \to Q$ *in the category* $\mathcal{L}^{sc}$ *is a monomorphism and an epimorphism.*

Proof: Let T be an $\mathcal{F}$-selfcentralizing subgroup of P and $y, y' \in \mathcal{L}^{sc}(R,T)$ two morphisms such that $x{\cdot}y = x{\cdot}y'$; in particular, we have $\pi_x \circ \pi_y = \pi_x \circ \pi_{y'}$ and therefore, since π_x is injective, we get $\pi_y = \pi_{y'}$; thus, we have $y' = y{\cdot}z$ for some $z \in \tau_T\big(Z(T)\big)$ and therefore we still have $x{\cdot}y = x{\cdot}y' = x{\cdot}y{\cdot}z$; now, the divisibility of $\mathcal{L}^{sc}$ forces $z = \tau_T(1)$, so that $y' = y$.

On the other hand, if $w, w' \in \mathcal{L}^{sc}(P,Q)$ fulfill $w{\cdot}x = w'{\cdot}x$, we have again $\pi_w \circ \pi_x = \pi_{w'} \circ \pi_x$ and therefore, by Proposition 4.6, we get $\pi_{w'} = \pi_w \circ \kappa_{\pi_x(z)}$ for some $z \in Z(R)$; since $Z(Q) \subset \pi_x\big(Z(R)\big)$, up to a suitable choice of z , we actually may assume that $w' = w{\cdot}\tau_Q\big(\pi_x(z)\big)$ and then, by the coherence in $\mathcal{L}^{sc}$, we still get (cf. 17.9.2)

$$w{\cdot}x = w'{\cdot}x = w{\cdot}\tau_Q\big(\pi_x(z)\big){\cdot}x = w{\cdot}x{\cdot}\tau_R(z) \qquad 24.2.1;$$

once again the divisibility of $\mathcal{L}^{sc}$, together with the injectivity of τ_R (cf. 17.13 and 13.2.2), forces $z = 1$, so that $w' = w$.

24.3 The fact that any morphism in $\mathcal{L}^{sc}$ is an epimorphism allows us to develop in this category the same constructions as, in chapter 6, we do in $\tilde{\mathcal{F}}^{sc}$. Explicitly, for any triple of $\mathcal{F}$-selfcentralizing subgroups Q , R and T , any morphism $x:Q \to R$ in $\mathcal{L}^{^{sc}}$ induces an injective map from $\mathcal{L}^{^{sc}}(T,R)$ to $\mathcal{L}^{^{sc}}(T,Q)$ and, as in 6.4.1, we set

$$\mathcal{L}^{^{sc}}(T,Q)_x = \mathcal{L}^{^{sc}}(T,Q) - \bigcup_{z'} \mathcal{L}^{^{sc}}(T,Q')\cdot z' \qquad 24.3.1$$

where z' runs over the set of *nonisomorphisms* $z':Q \to Q'$ from Q in $\mathcal{L}^{^{sc}}$ — the set of *nonfinal objects* in $(\mathcal{L}^{^{sc}})^Q$ (cf. 1.7) — fulfilling $x'.z' = x$ for some $x' \in \mathcal{L}^{^{sc}}(R,Q')$; then, x' is uniquely determined by this equality and we simply say that z' *divides* x setting $x' = x/z'$. Note that the existence of x' for some $z' \in \mathcal{L}^{^{sc}}(Q',Q)$ is equivalent to the existence of a subgroup of R which is $\mathcal{F}$-isomorphic to Q' and contains $\pi_x(Q)$. Thus, it is quite clear that

24.3.2 $\mathcal{L}^{^{sc}}(T,Q)_x$ *is the converse image of* $\tilde{\mathcal{F}}(T,Q)_{\tilde{\pi}_x}$ *in* $\mathcal{L}^{^{sc}}(T,Q)$.

In particular, we have $\mathcal{L}^{^{sc}}(T,Q)_x = \mathcal{L}^{^{sc}}(T,Q)$ if and only if π_x is an isomorphism, and moreover, for any $y \in \mathcal{L}^{^{sc}}(T,Q)$ we have (cf. 6.5.2)

24.3.3 $y \in \mathcal{L}^{^{sc}}(T,Q)_x$ *is equivalent to* $x \in \mathcal{L}^{^{sc}}(R,Q)_y$.

Proposition 24.4 *For any triple of $\mathcal{F}$-selfcentralizing subgroups Q, R and T of P and any $x \in \mathcal{L}^{^{sc}}(R,Q)$, we have*

$$\mathcal{L}^{^{sc}}(T,Q) = \bigsqcup_{z'} \mathcal{L}^{^{sc}}(T,Q')_{x/z'}\cdot z' \qquad 24.4.1$$

where $z':Q \to Q'$ runs over a set of representatives for the isomorphism classes of objects in $(\mathcal{L}^{^{sc}})^Q$ dividing x.

Proof: It easily follows from Proposition 6.7 and statement 24.3.2 above.

24.5 Let us consider the *additive cover* $\mathfrak{ac}(\mathcal{L}^{^{sc}})$ of $\mathcal{L}^{^{sc}}$ (cf. A4.10); recall that the $\mathfrak{ac}(\mathcal{L}^{^{sc}})$-objects are the finite sequences $Q = \{Q_i\}_{i\in I}$ — denoted by $\bigoplus_{i\in I} Q_i$ — of $\mathcal{F}$-selfcentralizing subgroups Q_i of P, and an $\mathfrak{ac}(\mathcal{L}^{^{sc}})$-morphism from another $\mathfrak{ac}(\mathcal{L}^{^{sc}})$-object $R = \bigoplus_{j\in J} R_j$ to $Q = \bigoplus_{i\in I} Q_i$ is a pair $(\mathfrak{x}, f)$ formed by a map $f: J \to I$ and a family $\mathfrak{x} = \{x_j\}_{j\in J}$ of $\mathcal{L}^{^{sc}}$-morphisms $x_j : R_j \to Q_{f(j)}$; the composition of $(\mathfrak{x}, f)$ with another $\mathfrak{ac}(\mathcal{L}^{^{sc}})$-morphism

$$(\mathfrak{y}, g) : T = \bigoplus_{\ell\in L} T_\ell \longrightarrow R \qquad 24.5.1$$

is the pair $\big((\mathfrak{x} * g)\circ \mathfrak{y}, f\circ g\big)$ (cf. A2.6.3) where $(\mathfrak{x} * g)\circ \mathfrak{y}$ is the family of $\mathcal{L}^{^{sc}}$-morphisms

$$x_{g(\ell)} \circ y_\ell : T_\ell \longrightarrow R_{g(\ell)} \longrightarrow Q_{(f\circ g)(\ell)} \qquad 24.5.2$$

where ℓ runs over L.

24.6 Now, the decomposition 24.4.1 leads us to consider in $\mathfrak{ac}(\mathcal{L}^{sc})$ the $\mathcal{L}^{sc}$*-intersection* of two $\mathcal{F}$-selfcentralizing subgroups of P; explicitly, if R and T are two $\mathcal{F}$-selfcentralizing subgroups of P, we consider the set $\mathfrak{Y}_{R,T}$ of triples (x, Q, y) where Q is an $\mathcal{F}$-selfcentralizing subgroup of P, x belongs to $\mathcal{L}(R,Q)_y$ and y belongs to $\mathcal{L}^{sc}(T,Q)_x$ or, equivalently, we have (cf. 6.5.1)

$$
{}^{(\pi_x)^*}\mathcal{F}_R\big(\pi_x(Q)\big) \cap {}^{(\pi_y)^*}\mathcal{F}_T\big(\pi_y(Q)\big) = \mathcal{F}_Q(Q) \qquad 24.6.1
$$

where $(\pi_x)^* : \pi_x(Q) \cong Q$ and $(\pi_y)^* : \pi_y(Q) \cong Q$ are the inverse of the respective isomorphisms determined by π_x and π_y; note that, in our present situation, we can identify $\mathcal{L}^{sc}(R,Q)$ and $\mathcal{L}^{sc}(T,Q)$ with the respective subsets

$$
\tau_{P,R}(1)\cdot\mathcal{L}^{sc}(R,Q) \subset \mathcal{L}^{sc}(P,Q) \supset \tau_{P,T}(1)\cdot\mathcal{L}^{sc}(T,Q) \qquad 24.6.2.
$$

It is clear that, for any $v \in R$ and any $w \in T$, the triple

$$
v\cdot(x,Q,y)\cdot w^{-1} = \big(\tau_R(v)\cdot x, Q, \tau_T(w)\cdot y\big) \qquad 24.6.3
$$

still belongs to $\mathfrak{Y}_{R,T}$ and that the quotient set $(R\times T)\backslash\mathfrak{Y}_{R,T}$ coincides with the set of triples $\mathfrak{T}_{R,T}$ introduced in 6.9.

24.7 As in 6.9, we say that two triples (x, Q, y) and (x', Q', y') are *equivalent* if there exists an $\mathcal{L}^{sc}$-isomorphism $z : Q \cong Q'$ fulfilling

$$
x'\cdot z = x \quad \text{and} \quad y'\cdot z = y \qquad 24.7.1;
$$

then, we denote by $\widetilde{\mathfrak{Y}}_{R,T}$ the set of equivalence classes of such triples; since $\mathcal{L}^{sc}$ is divisible, such an $\mathcal{L}^{sc}$-isomorphism z is unique; in particular, we always may assume that $Q \subset R$ and that $x = \tau_{R,Q}(1)$, and then $\big(\tau_{R,Q}(1), Q, y\big)$ is the unique element of this form in its equivalence class. As above, note that the quotient set $(R\times T)\backslash\widetilde{\mathfrak{Y}}_{R,T}$ coincides with the set of classes $\widetilde{\mathfrak{T}}_{R,T}$ in 6.9. Then, in $\mathfrak{ac}(\mathcal{L}^{sc})$ we set

$$
R \cap^{\mathcal{L}^{sc}} T = \bigoplus_{(x,Q,y)\in\breve{\mathfrak{Y}}_{R,T}} Q \qquad 24.7.2
$$

for a choice of a set of representatives $\breve{\mathfrak{Y}}_{R,T}$ of $\widetilde{\mathfrak{Y}}_{R,T}$ in $\mathfrak{Y}_{R,T}$, and we have canonical $\mathfrak{ac}(\mathcal{L}^{sc})$-morphisms

$$
R \longleftarrow R\cap^{\mathcal{L}^{sc}} T \longrightarrow T \qquad 24.7.3
$$

respectively determined by $x : Q \to R$ and $y : Q \to T$. Note that, for another choice of the set of representatives, we get an isomorphic object and a *unique* $\mathfrak{ac}(\mathcal{L}^{sc})$-isomorphism which is compatible with the canonical morphisms.

Proposition 24.8 *With the notation above, the category* $\mathfrak{ac}(\mathcal{L}^{^{sc}})$ *admits a distributive direct product mapping any pair of selfcentralizing subgroups* R *and* T *of* P *on their* $\mathcal{L}^{^{sc}}$*-intersection* $R\cap^{\mathcal{L}^{^{sc}}} T$.

Proof: With the notation above, in order to discuss the functorial nature of the $\mathcal{L}^{^{sc}}$-intersection, consider an $\mathcal{F}$-selfcentralizing subgroup U of P and two morphisms $a\in\mathcal{L}^{^{sc}}(R,U)$ and $b\in\mathcal{L}^{^{sc}}(T,U)$; it follows from Proposition 24.4 that b determines an isomorphism class of $(\mathcal{L}^{^{sc}})^U$-objects $z'\colon U\to U'$ dividing a such that, setting $a'=a/z'$, we have $b=b'\cdot z'$ for a suitable $b'\in\mathcal{L}^{^{sc}}(T,U)_{a'}$ and, once again, b' is uniquely determined; that is to say, the pair (a,b) determines an equivalence class of triples in $\widetilde{\mathfrak{Y}}_{R,T}$ and, once we have chosen a set of representatives $\check{\mathfrak{Y}}_{R,T}$, it determines a unique triple (a',U',b') and a unique $\mathcal{L}^{^{sc}}$-morphism $z'\colon U\to U'$ fulfilling $a=a'\cdot z'$ and $b=b'\cdot z'$; hence, the map

$$\bigsqcup_{(x,Q,y)\in\check{\mathfrak{Y}}_{R,T}}\mathcal{L}^{^{sc}}(Q,U)\longrightarrow\mathcal{L}^{^{sc}}(R,U)\times\mathcal{L}^{^{sc}}(T,U) \qquad 24.8.1.$$

sending the element $z\in\bigsqcup_{(x,Q,y)\in\check{\mathfrak{Y}}_{R,T}}\mathcal{L}^{^{sc}}(Q,U)$ in the term labelled by (x,Q,y) to $(x\cdot z,y\cdot z)$ is bijective.

In particular, considering two $\mathcal{L}^{^{sc}}$-morphisms $a\colon R'\to R$ and $b\colon T'\to T$ and a triple $(x',Q',y')\in\mathfrak{Y}'_{R',T'}$, we get the $\mathcal{L}^{^{sc}}$-morphisms

$$R\xleftarrow{\;a\cdot x'\;}Q'\xrightarrow{\;b\cdot y'\;}T \qquad 24.8.2$$

and therefore we still get a triple (x,Q,y) in $\check{\mathfrak{Y}}_{R,T}$ and an $\mathcal{L}^{^{sc}}$-morphism $z\colon Q\to Q'$ fulfilling

$$a\cdot x'=x\cdot z\quad\text{and}\quad b\cdot y'=y\cdot z \qquad 24.8.3.$$

In conclusion, we have obtained a map

$$\widetilde{\mathfrak{Y}}_{a,b}:\widetilde{\mathfrak{Y}}_{R',T'}\longrightarrow\widetilde{\mathfrak{Y}}_{R,T} \qquad 24.8.4$$

and, for any respective choices of sets of representatives $\check{\mathfrak{Y}}_{R,T}$ and $\check{\mathfrak{Y}}_{R',T'}$ for $\widetilde{\mathfrak{Y}}_{R,T}$ and $\widetilde{\mathfrak{Y}}_{R',T'}$, this map induces a new map

$$\check{\mathfrak{Y}}_{a,b}:\check{\mathfrak{Y}}_{R',T'}\longrightarrow\check{\mathfrak{Y}}_{R,T} \qquad 24.8.5$$

which, together with the $\mathcal{L}^{^{sc}}$-morphisms $z\colon Q\to Q'$ and the isomorphisms coming from the equivalences, determines an $\mathfrak{ac}(\mathcal{L}^{^{sc}})$-morphism

$$a\cap^{\mathcal{L}^{^{sc}}}b:R'\cap^{\mathcal{L}^{^{sc}}}T'\longrightarrow R\cap^{\mathcal{L}^{^{sc}}}T \qquad 24.8.6.$$

Finally, always from the bijection 24.8.1, it is not difficult to check that, for two other $\mathcal{L}^{^{sc}}$-morphisms $a'\colon R''\to R'$ and $b'\colon T''\to T'$, we get

$$(a\cap^{\mathcal{L}^{^{sc}}}b)\cdot(a'\cap^{\mathcal{L}^{^{sc}}}b')=(a\cdot a')\cap^{\mathcal{L}^{^{sc}}}(b\cdot b') \qquad 24.8.7.$$

By distributivity, we can extend the $\mathcal{L}^{sc}$-intersection to the whole category $\mathfrak{ac}(\mathcal{L}^{sc})$, namely if $R = \bigoplus_{i\in I} R_i$ and $T = \bigoplus_{j\in J} T_j$ are two objects in this category, where R_i and T_j are $\mathcal{F}$-selfcentralizing subgroups of P, then we set

$$R \cap^{\mathcal{L}^{sc}} T = \bigoplus_{(i,j)\in I\times J} R_i \cap^{\mathcal{L}^{sc}} T_j \qquad 24.8.8.$$

Similarly, if we have two $\mathfrak{ac}(\mathcal{L}^{sc})$-morphisms

$$(\mathfrak{a}, f) : R \longrightarrow R' = \bigoplus_{i'\in I'} R'_{i'} \quad \text{and} \quad (\mathfrak{b}, g) : T \longrightarrow T' = \bigoplus_{j'\in J'} T'_{j'} \qquad 24.8.9$$

where $f : I \to I'$ and $g : J \to J'$ are maps and where $\mathfrak{a}$ and $\mathfrak{b}$ are respective families of $\mathcal{L}^{sc}$-morphisms $a_i : R_i \to R'_{f(i)}$ and $b_j : T_j \to T'_{g(j)}$, i and j respectively running over I and J, then we have $\mathfrak{ac}(\mathcal{L}^{sc})$-morphisms

$$a_i \cap^{\mathcal{L}^{sc}} b_j : R_i \cap^{\mathcal{L}^{sc}} T_j \longrightarrow R'_{f(i)} \cap^{\mathcal{L}^{sc}} T'_{g(j)} \qquad 24.8.10$$

which, together with the map $f \times g : I \times J \to I' \times J'$, clearly define a new $\mathfrak{ac}(\mathcal{L}^{sc})$-morphism

$$(\mathfrak{a}, f) \cap^{\mathcal{L}^{sc}} (\mathfrak{b}, g) : R \cap^{\mathcal{L}^{sc}} T \longrightarrow R' \cap^{\mathcal{L}^{sc}} T' \qquad 24.8.11.$$

Finally, it is quite clear that the bijections 24.8.1 imply the bijections

$$\big(\mathfrak{ac}(\mathcal{L}^{sc})\big)(R \cap^{\mathcal{L}^{sc}} T, U) \cong \big(\mathfrak{ac}(\mathcal{L}^{sc})\big)(R, U) \times \big(\mathfrak{ac}(\mathcal{L}^{sc})\big)(T, U) \qquad 24.8.12$$

for any $\mathfrak{ac}(\mathcal{L}^{sc})$-object U; this proves that the $\mathcal{L}^{sc}$-intersection is a *direct product* in the category $\mathfrak{ac}(\mathcal{L}^{sc})$. We are done.

24.9 Here, we are particularly interested in the $\mathcal{L}^{sc}$-intersection of P with itself; more explicitly, denoting by Ω the set of pairs (Q, y) formed by an $\mathcal{F}$-selfcentralizing subgroup Q of P and an element y of $\mathcal{L}^{sc}(P, Q)_{\tau_{P,Q}(1)}$, we have (cf. 24.7)

$$P \cap^{\mathcal{L}^{sc}} P = \bigoplus_{(Q,y)\in\Omega} Q \qquad 24.9.1$$

since the set of triples $\{(\tau_{P,Q}(1), Q, y)\}_{(Q,y)\in\Omega}$ is a set of representatives for $\tilde{\mathfrak{Y}}_{P,P}$; moreover, since $P \times P$ acts on $\tilde{\mathfrak{Y}}_{P,P}$ (cf. 24.6 and 24.7), this group acts on Ω; explicitly, for any $u, v \in P$, we have

$$u\cdot(\tau_{P,Q}(1), Q, y)\cdot v^{-1} = (\tau_{P,Q}(u), Q, \tau_P(v)\cdot y) \qquad 24.9.2$$

which is equivalent to the triple

$$\big(\tau_{P,Q^{u^{-1}}}(1), Q^{u^{-1}}, \tau_P(v)\cdot y\cdot \tau_{Q,Q^{u^{-1}}}(u^{-1})\big) \qquad 24.9.3;$$

that is to say, we have proved that

24.9.4 $(u,v) \in P \times P$ *maps* $(Q,y) \in \Omega$ *on* $\big(Q^{u^{-1}}, \tau_P(v)\cdot y\cdot\tau_{Q,Q^{u^{-1}}}(u^{-1})\big)$.

In particular, $\{1\} \times P$ acts *freely* on Ω. On the other hand, it is clear that the map sending a triple (x,Q,y) in $\mathfrak{Y}_{P,P}$ to (y,Q,x) induces a $P \times P$-set isomorphism $\Omega \cong \Omega^\circ$.

Proposition 24.10 *With the notation above, the stabilizer of $(Q,y) \in \Omega$ in $P \times P$ coincides with $\Delta_{\pi_y}(Q)$. In particular, we have*

$$|\Omega|/|P| \not\equiv 0 \pmod{p} \qquad 24.10.1.$$

Proof: On the one hand, if $(u,v) \in \Delta_{\pi_y}(Q)$ then $u \in Q$, so that $Q^{u^{-1}} = Q$, and from the coherence of $\mathcal{L}^{\mathrm{sc}}$ (cf. 17.9.2) we get

$$y\cdot\tau_{Q,Q^{u^{-1}}}(u^{-1}) = \tau_P(v^{-1})\cdot y \qquad 24.10.2,$$

so that $\Delta_{\pi_y}(Q)$ fixes (Q,y).

On the other hand, if (u,v) fixes (Q,y) then it follows from statement 24.9.4 that u normalizes Q and that $y = \tau_P(v)\cdot y\cdot\tau_Q(u^{-1})$; in particular, v normalizes $\pi_y(Q)$ and π_y maps the action of u over Q on the action of v over $\pi_y(Q)$. But, since $y \in \mathcal{L}^{\mathrm{sc}}(P,Q)_{\tau_{P,Q}(1)}$ (cf. 24.9), we have (cf. equality 6.5.1 and statement 24.3.2)

$$\tilde{\mathcal{F}}_P(Q) \cap {}^{(\pi_y)^*}\tilde{\mathcal{F}}_P\big(\pi_y(Q)\big) = \{1\} \qquad 24.10.3.$$

Consequently, u belongs to Q and, once again from the coherence of $\mathcal{L}^{\mathrm{sc}}$ (cf. 17.9.2), we get

$$\begin{aligned} y = \tau_P(v)\cdot y\cdot\tau_Q(u^{-1}) &= \tau_P(v)\cdot\tau_P\big(\pi_y(u^{-1})\big)\cdot y \\ &= \tau_P\big(v\pi_y(u^{-1})\big)\cdot y \end{aligned} \qquad 24.10.4;$$

then, Proposition 24.2 and the injectivity of τ_P (cf. 17.13 and 13.2.2) force $v = \pi_y(u)$, so that $(u,v) \in \Delta_{\pi_y}(Q)$.

In particular, if $Q \neq P$ then $p|P|$ divides the cardinal of the $P \times P$-orbit of (Q,y) and therefore, considering the subset Ω' of Ω determined by the pairs (P,x) where x runs over $\mathcal{L}^{\mathrm{sc}}(P)$, it is quite clear that

$$|\Omega|/|P| \equiv |\Omega'|/|P| = |\mathcal{L}^{\mathrm{sc}}(P)|/|P| \pmod{p} \qquad 24.10.5;$$

but, we know that $\tau_P(P) \cong P$ is a Sylow p-subgroup of $\mathcal{L}^{\mathrm{sc}}(P)$ (cf. 17.12). We are done.

Corollary 24.11 *With the notation above, for any* $\mathcal{F}$*-selfcentralizing subgroup* Q *of* P *and any* $\varphi, \varphi' \in \mathcal{F}(P,Q)$, *we have*

$$|\Omega^{\Delta_{\varphi,\varphi'}(Q)}| = |Z(Q)| = |\Omega^{\Delta(Q)}| \qquad 24.11.1.$$

In particular, there is an $\mathcal{F}$*-basic* $P\times P$*-set* $\widehat{\Omega}$ *which contains* Ω, *fulfills*

$$\widehat{\Omega}^{\Delta_\varphi(Q)} = \Omega^{\Delta_\varphi(Q)} \qquad 24.11.2$$

for any $\mathcal{F}$*-selfcentralizing subgroup* Q *of* P *and any* φ *in* $\mathcal{F}(P,Q)$, *and is thick outside the set of them.*

Proof: According to Proposition 24.10, an element (R,y) of Ω belongs to $\Omega^{\Delta_{\varphi,\varphi'}(Q)}$ if and only if $\Delta_{\varphi,\varphi'}(Q)$ is contained in $\Delta_{\pi_y}(R)$, namely if and only if $\varphi(Q)\subset R$ and $\pi_y\big(\varphi(u)\big) = \varphi'(u)$ for any $u\in Q$; then, the element $x = y\cdot\tau_{R,\varphi(Q)}(1)$ still fulfills $\pi_x\big(\varphi(u)\big) = \varphi'(u)$ for any $u\in Q$. Conversely, any $x\in\mathcal{L}^{sc}\big(P,\varphi(Q)\big)$ fulfilling this equality, together with $\tau_{P,\varphi(Q)}(1)$, determines an $\mathfrak{ac}(\mathcal{L}^{sc})$-morphism (cf. Proposition 24.8)

$$\varphi(Q) \longrightarrow P\cap^{\mathcal{L}^{sc}} P \qquad 24.11.3;$$

that is to say, it determines $(R,y)\in\Omega$ such that R contains $\varphi(Q)$ and $x = y\cdot\tau_{R,Q}(1)$ (cf. definition 24.9.1); hence, we get

$$|\Omega^{\Delta_{\varphi,\varphi'}(Q)}| = |Z\big(\varphi(Q)\big)| = |Z(Q)| \qquad 24.11.4.$$

Then, the last statement follows from Proposition 21.12.

Remark 24.12 Actually, this argument and Proposition 24.4 determine a canonical bijection

$$\mathcal{L}^{sc}(P,Q) \cong \bigsqcup_R \bigsqcup_{\psi\in\mathcal{F}(P,R)_{\iota_R^P}} \Omega^{\Delta_\psi(R)} \qquad 24.12.1$$

where R runs over the subgroups of P containing Q.

24.13 Thus, the $P\times P$-set Ω is contained in an $\mathcal{F}$-basic $P\times P$-set $\widehat{\Omega}$ fulfilling $\widehat{\Omega}^{\Delta(Q)} = \Omega^{\Delta(Q)}$ for any $\mathcal{F}$-selfcentralizing subgroup Q of P; since we cannot guarantee the thickness, we consider the disjoint union $\widehat{\widehat{\Omega}}$ of $\widehat{\Omega}$ with $|P|$ copies of a *thick* $\mathcal{F}$-basic $P\times P$-set Ω'. Now, $\widehat{\widehat{\Omega}}$ is thick and, as in 21.2, we denote by G the automorphism group of $\mathrm{Res}_{\{1\}\times P}(\widehat{\widehat{\Omega}})$ and identify P with the image of $P\times\{1\}$ in G; recall that, from the group G, we can construct the *basic* $\mathcal{F}$*-locality* $\mathcal{L}^{b}$ (cf. Proposition 22.12) and then the $\mathcal{F}^{sc}$*-localities* $\mathcal{L}^{c,sc}$ (cf. Proposition 23.10), $\mathcal{L}^{d,sc}$ (cf. Corollary 23.24) and $\mathcal{L}^{r,sc}$ (cf. Corollary 23.28).

24.14 On the other hand, for any $\mathcal{F}$-selfcentralizing subgroup Q of P, it follows from Proposition 24.8 that the inclusion $Q \subset P$ determines an $\mathfrak{ac}(\mathcal{L}^{^{sc}})$-morphism

$$\tau_{P,Q}(1) \cap^{\mathcal{L}^{^{sc}}} \tau_P(1) : Q \cap^{\mathcal{L}^{^{sc}}} P \longrightarrow P \cap^{\mathcal{L}^{^{sc}}} P \qquad 24.14.1$$

and in particular, denoting by Ω_Q the set of indices in the structural decomposition of the left-hand term, it determines a map $\Omega_Q \to \Omega$; actually, according to 24.7, we may identify Ω_Q with the set of pairs (T,z) formed by an $\mathcal{F}$-selfcentralizing subgroup T of Q and an element z of $\mathcal{L}^{^{sc}}(P,T)_{\tau_{Q,T}(1)}$.

Proposition 24.15 *The map $\Omega_Q \to \Omega$ determined by the $\mathfrak{ac}(\mathcal{L}^{^{sc}})$-morphism $\tau_{P,Q}(1) \cap^{\mathcal{L}^{^{sc}}} \tau_P(1)$ sends $(T,z) \in \Omega_Q$ to $(R,y) \in \Omega$ if and only if we have $T = Q \cap R$ and $z = y{\cdot}\tau_{R,T}(1)$. In particular, this map is injective and induces an injective $Q \times P$-set homomorphism*

$$\Omega_Q \longrightarrow \mathrm{Res}_{Q\times P}(\widehat{\Omega}) \qquad 24.15.1.$$

Proof: If the map associated with $\tau_{P,Q}(1) \cap^{\mathcal{L}^{^{sc}}} \tau_P(1)$ sends the pair (T,z) in Ω_Q to $(R,y) \in \Omega$, then it induces an $\mathcal{L}^{^{sc}}$-morphism $x : T \to R$ fulfilling

$$\tau_{P,T}(1) = \tau_{P,R}(1){\cdot}x \quad \text{and} \quad z = y{\cdot}x \qquad 24.15.2$$

and therefore, by the divisibility of $\mathcal{L}^{^{sc}}$, we have $x = \tau_{R,T}(1)$, so that T is contained in $Q \cap R$ and $z = y{\cdot}\tau_{R,T}(1)$; in particular, $y{\cdot}\tau_{R,Q\cap R}(1)$ extends z, which forces $T = Q \cap R$ since z belongs to $\mathcal{L}^{^{sc}}(P,T)_{\tau_{Q,T}(1)}$ (cf. 24.3.1); the converse is clear.

In particular, if the map associated to $\tau_{P,Q}(1) \cap^{\mathcal{L}^{^{sc}}} \tau_P(1)$ also sends the pair $(T',z') \in \Omega_Q$ to (R,y), we have $T' = Q \cap R = T$ and $z' = y{\cdot}\tau_{R,T}(1) = z$. We are done.

24.16 Now, for any $\mathcal{L}^{^{sc}}$-isomorphism $x : Q \cong Q'$, it follows again from Proposition 24.8 that we have an $\mathfrak{ac}(\mathcal{L}^{^{sc}})$-isomorphism

$$x \cap^{\mathcal{L}^{^{sc}}} \tau_P(1) : Q \cap^{\mathcal{L}^{^{sc}}} P \cong Q' \cap^{\mathcal{L}^{^{sc}}} P \qquad 24.16.1$$

and therefore a bijection between the sets of indices Ω_Q and $\Omega_{Q'}$, which is compatible *via* π_x with the actions of $Q \times P$ and $Q' \times P$ on both sets; that is to say, we get $Q \times P$-set isomorphism

$$f_x : \Omega_Q \cong \mathrm{Res}_{\pi_x \times \mathrm{id}_P}(\Omega_{Q'}) \qquad 24.16.2.$$

Proposition 24.17 *For any* $\mathcal{L}^{\mathrm{sc}}$*-isomorphism* $x : Q \cong Q'$*, the* $Q \times P$*-set isomorphism*

$$f_x : \Omega_Q \cong \mathrm{Res}_{\pi_x \times \mathrm{id}_P}(\Omega_{Q'}) \qquad 24.17.1$$

can be extended to an element $\hat{\hat{f}}_x$ *of* $\mathcal{T}_G(Q', Q)$ *which stabilizes any copy of* Ω' *and induces the same automorphism in all of them. Moreover, the image of* $\hat{\hat{f}}_x$ *in* $\mathcal{L}^{\mathrm{c,sc}}(Q', Q)$ *is uniquely determined by* x*.*

Proof: Since the $Q \times P$-sets $\mathrm{Res}_{Q\times P}(\widehat{\Omega})$ and $\mathrm{Res}_{\pi_x\times \mathrm{id}_P}\big(\mathrm{Res}_{Q'\times P}(\widehat{\Omega})\big)$ are isomorphic to each other (cf. 21.2.2 and 21.4), and the $Q \times P$- and $Q' \times P$-set homomorphisms

$$\Omega_Q \longrightarrow \mathrm{Res}_{Q\times P}(\widehat{\Omega}) \quad \text{and} \quad \Omega_{Q'} \longrightarrow \mathrm{Res}_{Q'\times P}(\widehat{\Omega}) \qquad 24.17.2$$

are injective (cf. Proposition 24.15), f_x can be extended to a $Q \times P$-set isomorphism

$$\hat{f}_x : \mathrm{Res}_{Q\times P}(\widehat{\Omega}) \cong \mathrm{Res}_{\pi_x\times \mathrm{id}_P}\big(\mathrm{Res}_{Q'\times P}(\widehat{\Omega})\big) \qquad 24.17.3;$$

moreover, since the $Q \times P$-sets $\mathrm{Res}_{Q\times P}(\Omega')$ and $\mathrm{Res}_{\pi_x\times \mathrm{id}_P}\big(\mathrm{Res}_{Q'\times P}(\Omega')\big)$ are mutually isomorphic too, $\hat{f}_x$ can be extended to a $Q \times P$-set isomorphism

$$\hat{\hat{f}}_x : \mathrm{Res}_{Q\times P}(\widehat{\widehat{\Omega}}) \cong \mathrm{Res}_{\pi_x\times \mathrm{id}_P}\big(\mathrm{Res}_{Q'\times P}(\widehat{\widehat{\Omega}})\big) \qquad 24.17.4$$

which stabilizes any copy of Ω' and induces the same automorphism in all of them; that is to say, we get an element $\hat{\hat{f}}_x$ of $\mathcal{T}_G(Q', Q)$.

Then, we claim that the image of $\hat{\hat{f}}_x$ in $\mathcal{L}^{\mathrm{c,sc}}(Q', Q)$ is independent of our choices; indeed, for another choice $\hat{\hat{g}}_x \in \mathcal{T}_G(Q', Q)$ fulfilling the above conditions, the composed map $(\hat{\hat{f}}_x)^{-1} \circ \hat{\hat{g}}_x$ belongs to $C_G(Q)$ and induces the identity in Ω_Q; but, it follows from Corollary 24.11 that, for any $\mathcal{F}$-selfcentralizing subgroup R of P and any $\psi \in \mathcal{F}(P, R)$, we have

$$\Omega^{\Delta_\psi(R)} = \widehat{\Omega}^{\Delta_\psi(R)} \qquad 24.17.5$$

in particular, for any $\mathfrak{pc}(\tilde{\mathcal{F}}^{\mathrm{sc}})$-morphism $\tilde{\eta} : T \to Q\tilde{\times}P$ in $\mathfrak{O}_Q$ (cf. 22.5) such that T is $\mathcal{F}$-selfcentralizing, any $Q \times P$-orbit Γ in $\mathrm{Res}_{Q\times P}(\widehat{\Omega})$ isomorphic to $Q \times_{\tilde{\eta}} P$ is contained in Ω since, for any representative η of the composition $\hat{\tilde{\eta}}$ of $\tilde{\eta}$ with the structural $\mathfrak{pc}(\tilde{\mathcal{F}})$-morphism $Q\tilde{\times}P \to P$ (cf. 22.5), $\Delta_\eta(T)$ has a fixed point in Γ. Moreover, it follows from Propositions 24.10 and 24.15 that Γ is actually contained in Ω_Q.

Consequently, since we have (cf. 22.5.1)

$$C_G(Q) \cong \prod_{\tilde{\eta}\in\mathfrak{O}_Q} \Big(\prod_{\Gamma_{\tilde{\eta}}} \mathrm{Aut}(\Gamma_{\tilde{\eta}}) \times \prod_{i=1}^{|P|} \prod_{\Gamma'_{\tilde{\eta}}} \mathrm{Aut}(\Gamma'_{\tilde{\eta}}) \Big) \rtimes \mathfrak{S}_{k_{\tilde{\eta}}} \qquad 24.17.6$$

where $\Gamma_{\tilde\eta}$ and $\Gamma'_{\tilde\eta}$ respectively run over the set of $Q\times P$-orbits isomorphic to $Q\times_{\tilde\eta} P$ in $\mathrm{Res}_{Q\times P}(\widehat\Omega)$ and in $\mathrm{Res}_{Q\times P}(\Omega')$, and, according to our choices, $(\hat{\hat f}_x)^{-1}\circ\hat{\hat g}_x$ stabilizes $\widehat\Omega$ and any copy of Ω', this element $(\hat{\hat f}_x)^{-1}\circ\hat{\hat g}_x$ has a trivial image in (cf. 23.9)

$$\tilde{\mathfrak{k}}^{^{\mathrm{sc}}}(Q)=\prod_{\tilde\eta\in\mathfrak{O}^{Q,\mathrm{sc}}}\mathfrak{ab}\big(\mathrm{Aut}(Q\times_{\tilde\eta}P)\big) \qquad 24.17.7.$$

Indeed, for any $\tilde\eta\in\mathfrak{O}^{Q,\mathrm{sc}}$, this element acts trivially on the union of the $Q\times P$-orbits $\Gamma_{\tilde\eta}$ above and therefore, denoting by $h'_{\tilde\eta}$ the automorphism of the union of the $Q\times P$-orbits $\Gamma'_{\tilde\eta}$ above induced by $(\hat{\hat f}_x)^{-1}\circ\hat{\hat g}_x$, its action on the union of all the $Q\times P$-orbits in $\widehat{\widehat\Omega}$ isomorphic to $Q\times_{\tilde\eta}P$ coincides with the product $\prod_{i=1}^{|P|}h'_{\tilde\eta}$; but, according to our choices, all the factors have the same image in $\mathfrak{ab}\big(\mathrm{Aut}(Q\times_{\tilde\eta}P)\big)$ and we clearly have (cf. 22.3.1)

$$\mathfrak{ab}\big(\mathrm{Aut}(Q\times_{\tilde\eta}P)\big)^{|P|}=\{1\} \qquad 24.17.8.$$

We are done.

Corollary 24.18 *There is a faithful functor* $\lambda:\mathcal{L}^{^{\mathrm{sc}}}\to\mathcal{L}^{^{\mathrm{r,sc}}}$ *which is compatible with the structural functors and, up to a suitable choice of the inclusion* $\mathcal{L}^{^{\mathrm{r,sc}}}\subset\mathcal{L}^{^{\mathrm{c,sc}}}$*, maps any* $\mathcal{L}^{^{\mathrm{sc}}}$*-isomorphism* $x:Q\cong Q'$ *on the image of* $\hat{\hat f}_x$ *in* $\mathcal{L}^{^{\mathrm{c,sc}}}(Q',Q)$*, and* $\tau^{\mathrm{sc}}_{P,Q}(1)$ *on* $\tau^{\mathrm{c,sc}}_{P,Q}(1)$.

Proof: Let us denote by $\lambda(x)$ the image of $\hat{\hat f}_x$ in $\mathcal{L}^{^{\mathrm{c,sc}}}(Q',Q)$; first of all, let $x':Q'\cong Q''$ be a second $\mathcal{L}^{^{\mathrm{sc}}}$-isomorphism; it is clear that the automorphism $\mathrm{Res}_{\pi_x\times\mathrm{id}_P}(\hat{\hat f}_{x'})\circ\hat{\hat f}_x$ of $\widehat{\widehat\Omega}$ extends $\mathrm{Res}_{\pi_x\times\mathrm{id}_P}(f_{x'})\circ f_x$ and stabilizes any copy of Ω', inducing the same automorphism in all of them; consequently, by the argument above, we get

$$\lambda(x'{\cdot}x)=\lambda(x'){\cdot}\lambda(x) \qquad 24.18.1.$$

Secondly, by the divisibility of $\mathcal{L}^{^{\mathrm{sc}}}$, any $\mathcal{L}^{^{\mathrm{sc}}}$-morphism $z:R\to Q$ is the composition of $\tau^{\mathrm{sc}}_{Q,R'}(1)$ with an $\mathcal{L}^{^{\mathrm{sc}}}$-isomorphism $z_*:R\cong R'=\pi_z(R)$ and we simply define

$$\lambda(z)=\tau^{\mathrm{r,sc}}_{Q,R'}(1){\cdot}\lambda(z_*) \qquad 24.18.2.$$

Now, in order to prove that this correspondence defines a functor, it suffices to show that, for any $\mathcal{L}^{^{\mathrm{sc}}}$-isomorphism $x:Q\cong Q'$ and any subgroup R of Q, setting $R' = \pi_x(R)$ and denoting by $y\colon R \cong R'$ the $\mathcal{L}^{sc}$-isomorphism induced by x (cf. 17.7), we still have

$$\lambda(x)\cdot\tau^{c,sc}_{Q,R}(1) = \tau^{c,sc}_{Q',R'}(1)\cdot\lambda(y) \qquad 24.18.3.$$

But, it is quite clear that the commutative $\mathfrak{ac}(\mathcal{L}^{sc})$-diagram (cf. Proposition 24.8)

$$\begin{array}{ccc} R\cap^{\mathcal{L}^{sc}} P & \xrightarrow{\tau_{Q,R}(1)\cap^{\mathcal{L}^{sc}}\tau_P(1)} & Q\cap^{\mathcal{L}^{sc}} P \\ {\scriptstyle y\cap^{\mathcal{L}^{sc}}\tau_P(1)}\downarrow & & \downarrow{\scriptstyle x\cap^{\mathcal{L}^{sc}}\tau_P(1)} \\ R'\cap^{\mathcal{L}^{sc}} P & \xrightarrow{\tau_{Q',R'}(1)\cap^{\mathcal{L}^{sc}}\tau_P(1)} & Q'\cap^{\mathcal{L}^{sc}} P \end{array} \qquad 24.18.4$$

determines a commutative diagram of $R\times P$-sets

$$\begin{array}{ccc} \Omega_R & \longrightarrow & \mathrm{Res}^{Q\times P}_{R\times P}(\Omega_Q) \\ {\scriptstyle f_y}\downarrow & & \downarrow{\scriptstyle \mathrm{Res}^{Q\times P}_{R\times P}(f_x)} \\ \mathrm{Res}_{\pi_y\times \mathrm{id}_P}(\Omega_{R'}) & \longrightarrow & \mathrm{Res}^{Q\times P}_{R\times P}\big(\mathrm{Res}_{\pi_x\times\mathrm{id}_P}(\Omega_{Q'})\big) \end{array} \qquad 24.18.5.$$

Consequently, we can choose $\hat{\hat{f}}_y = \hat{\hat{f}}_x$. Moreover, it is easily checked that this functor is faithful.

In particular, it follows from Proposition A2.10 that we have a *natural map*

$$\mathfrak{aut}_\lambda : \mathfrak{aut}_{\mathcal{L}^{sc}} \longrightarrow \mathfrak{aut}_{\mathcal{L}^{c,sc}} \qquad 24.18.6$$

which, according to 18.20 and to a suitable *partial version* of Propositions 18.19 and 18.21, induces a new *natural map*

$$\mathfrak{loc}_\lambda : \mathfrak{loc}_{\mathcal{F}^{sc}} \cong \mathfrak{loc}_{\mathcal{L}^{sc}} \longrightarrow \mathfrak{loc}_{\mathcal{L}^{c,sc}} \qquad 24.18.7.$$

Furthermore, it follows again from this *partial version* of Proposition 18.21 that we have a natural map

$$\lambda_{\mathcal{L}^{r,sc}} : \mathfrak{loc}_{\mathcal{F}^{sc}} \longrightarrow \mathfrak{loc}_{\mathcal{L}^{r,sc}} \qquad 24.18.8$$

and that, considering the *inclusion* functor $\mathfrak{i}^{c}_{r}\colon \mathcal{L}^{r,sc} \to \mathcal{L}^{c,sc}$, the natural map $\mathfrak{loc}_{\mathfrak{i}^c_r}\circ\lambda_{\mathcal{L}^{r,sc}}$ coincides with $\mathfrak{loc}_\lambda$; that is to say, up to a suitable choice of the $\mathcal{F}^{sc}$-sublocality $\mathcal{L}^{r,sc}$ of $\mathcal{L}^{c,sc}$, we may assume that $\lambda\big(\mathcal{L}^{sc}(Q)\big)\subset\mathcal{L}^{r,sc}(Q)$ for any $\mathcal{F}$-selfcentralizing subgroup Q of P, and then it follows from Corollary 5.14 that the image of the functor λ is contained in $\mathcal{L}^{r,sc}$. We are done.

Corollary 24.19 *Assume that for every $\mathcal{F}$-selfcentralizing subgroup Q of P and every nontrivial subgroup $\bar{R}$ of $\bar{N}_P(Q)$, the $k\bar{R}$-module $k\otimes_{\mathbb{Z}} Z(Q)$ has no nonzero projective factors. Then, there is a unique perfect $\mathcal{F}^{sc}$-locality $\mathcal{L}^{sc}$.*

Proof: It is an immediate consequence of Corollaries 23.34 and 24.18.

Appendix

Simpliciality

In this Appendix we collect all the generalities on categories and on their cohomological groups we need in this book. As we already mention in chapter 1, in general we only consider the *small* categories, namely the categories over a *set* of objects. Note that all the results below are well-known and we have no claim of originality; our purpose in stating them is both, to provide a handy reference for the reader and to develop our own point of view and our *unified* notation.

When dealing with categories and their cohomology, it is a wide experience that, although sometimes quite long, many proofs are straightforward. Thus, without pushing the argument as far as Serge Lang did in his well-known book *Algebra*†, we partially follow Lang's point of view and state our results without proofs when considering that they are straightforward.

A1 On the 2-categories

A1.1 Recall that a 2-*category* is a category $\mathfrak{X}$ where, for any pair of objects X and Y, the set of $\mathfrak{X}$-morphisms from X to Y is endowed with a structure of *category* — denoted by $\mathfrak{X}(Y,X)$ (cf. 1.2) — and where, for any triple of $\mathfrak{X}$-objects X, Y and Z, the *left-hand composition* is extended to a functor

$$\mathfrak{l}^X_{Y,Z} : \mathfrak{X}(Z,Y) \longrightarrow \mathfrak{Fct}\big(\mathfrak{X}(Y,X), \mathfrak{X}(Z,X)\big) \qquad \text{A1.1.1}$$

in such a way that, for any quadruple of $\mathfrak{X}$-objects X, Y, Z and U, any pair of pairs of $\mathfrak{X}$-morphisms g, g' from Y to Z, and h, h' from Z to U, and any respective $\mathfrak{X}(Z,Y)$- and $\mathfrak{X}(U,Z)$-morphisms $\beta : g \to g'$ and $\gamma : h \to h'$, it fulfills the *associative* conditions

$$\mathfrak{l}^X_h \circ \mathfrak{l}^X_g = \mathfrak{l}^X_{h\circ g} \quad , \quad \mathfrak{l}^X_\gamma * \mathfrak{l}^X_g = \mathfrak{l}^X_{\gamma * g} \quad \text{and} \quad \mathfrak{l}^X_h * \mathfrak{l}^X_\beta = \mathfrak{l}^X_{h*\beta} \qquad \text{A1.1.2}$$

where $*$ denotes the "composition" between functors and natural maps and we set $\mathfrak{l}^X_g = \mathfrak{l}^X_{Y,Z}(g)$, $\mathfrak{l}^X_\beta = \mathfrak{l}^X_{Y,Z}(\beta)$, $\gamma * g = \mathfrak{l}^Y_\gamma(g)$ and $h * \beta = \mathfrak{l}^Y_h(\beta)$.

A1.2 To avoid confusion, if f and f' are two $\mathfrak{X}$-morphisms from X to Y, we denote by $\mathrm{Nat}_{\mathfrak{X}}(f', f)$ the set of $\mathfrak{X}(Y,X)$-morphisms from f to f', and call them $\mathfrak{X}$-*natural maps*. We denote by $\mathfrak{X}^\circ$ the 2-category over the opposite category, by $\mathfrak{X}^\star$ the 2-subcategory of $\mathfrak{X}$ formed by the same objects, the same morphisms and the *natural* $\mathfrak{X}$-*isomorphisms*, and by $\mathfrak{X}^\mathfrak{c}$ the underlying category.

† In EXERCISES at the end of Chapter IV, Serge Lang write *"Take any book on homological algebra, and prove all the theorems without looking at the proofs given in that book."*

A1.3 Note that, if f is an $\mathfrak{X}$-morphism from X to Y, the left-hand equality in A1.1.2 implies the usual associative condition

$$h \circ (g \circ f) = (h \circ g) \circ f \qquad \text{A1.3.1;}$$

more generally, if f' is a second $\mathfrak{X}$-morphism from X to Y and $\alpha : f \to f'$ a $\mathfrak{X}$-natural map, the left-hand equality in A1.1.2 forces

$$h * (g * \alpha)) = (h \circ g) * \alpha \qquad \text{A1.3.2;}$$

moreover, if f'' is a third $\mathfrak{X}$-morphism from X to Y and $\alpha' \in \mathrm{Nat}_{\mathfrak{X}}(f'', f')$, the functoriality of $\mathfrak{l}_g^X$ implies that

$$g * (\alpha' \circ \alpha) = (g * \alpha') \circ (g * \alpha) \qquad \text{A1.3.3.}$$

A1.4 Similarly, if g' is a second $\mathfrak{X}$-morphism from Y to Z and β a $\mathfrak{X}$-natural map from g to g', the right-hand equality in A1.1.2 forces

$$h * (\beta * f) = (h * \beta) * f \qquad \text{A1.4.1;}$$

moreover, if g'' is a third $\mathfrak{X}$-morphism from Y to Z and $\beta' \in \mathrm{Nat}_{\mathfrak{X}}(g'', g')$, the functoriality of $\mathfrak{l}_{Y,Z}^X$ implies that

$$(\beta' \circ \beta) * f = (\beta' * f) \circ (\beta * f) \qquad \text{A1.4.2;}$$

finally, the middle equality in A1.1.2 implies that

$$\gamma * (g \circ f) = (\gamma * g) * f \qquad \text{A1.4.3.}$$

A1.5 Moreover, the naturality of $\mathfrak{l}_\beta^Y : \mathfrak{l}_g^Y \to \mathfrak{l}_{g'}^Y$ applied to the $\mathfrak{X}$-natural map $\alpha : f \to f'$ implies that we have

$$(g' * \alpha) \circ (\beta * f) = (\beta * f') \circ (g * \alpha) \qquad \text{A1.5.1.}$$

Conversely, it is not difficult to see that all these associative and distributive conditions on the composition of morphisms and natural maps allow the (re)construction of functors $\mathfrak{l}_{Y,Z}^X$ as above fulfilling conditions A1.1.2. Note that, since they are *symmetric*, they still allow us to construct the corresponding *right-hand composition* functors $\mathfrak{r}_{X,Y}^Z$, which actually *commute* with the left-hand ones.

A1.6 The standard example of a 2-category is the *category of small categories* $\mathfrak{CC}$ where the objects are the *small categories* (cf. 1.2) and the morphisms between two small categories $\mathfrak{C}$ and $\mathfrak{D}$ are *all* the functors $\mathfrak{Fct}(\mathfrak{C}, \mathfrak{D})$ from $\mathfrak{C}$ to $\mathfrak{D}$, which form a *set*†. Indeed, $\mathfrak{Fct}(\mathfrak{C}, \mathfrak{D})$ has a natural structure of small category since between two functors $\mathfrak{f}$ and $\mathfrak{g}$ from $\mathfrak{C}$ to $\mathfrak{D}$ we still have the set of *natural maps* $\mathfrak{Nat}(\mathfrak{f}, \mathfrak{g})$, and the natural maps can be composed.

† *A contrario*, there is no *category of categories*.

A1.7 Besides $\mathfrak{CC}$, the 2-category which motivates our interest is the *simplicial* 2-*category* $\boldsymbol{\Delta}$, usually considered as a simple category in the literature. The $\boldsymbol{\Delta}$-objects are the ordered sets

$$\Delta_n = \{i \mid 0 \le i \le n\} \qquad \text{A1.7.1}$$

for any $n \in \mathbb{N}$; then, for any $n, m \in \mathbb{N}$, the $\boldsymbol{\Delta}$-morphisms from Δ_n to Δ_m, are the order-preserving maps with the usual composition; moreover, for any pair of morphisms $\delta : \Delta_n \to \Delta_m$ and $\varepsilon : \Delta_n \to \Delta_m$ in $\boldsymbol{\Delta}$, we denote by $\mathrm{Nat}_{\boldsymbol{\Delta}}(\varepsilon, \delta)$ a set with one element $\delta \bullet \varepsilon$ if $\delta(i) \le \varepsilon(i)$ for any $i \in \Delta_n$ and the empty set otherwise, and we have an obvious composition of these elements.

A1.8 For any 2-category $\mathfrak{X}$, we consider the *category of connected components* $\pi_0(\mathfrak{X}) = \overline{\mathfrak{X}}$ of $\mathfrak{X}$ †, namely the *quotient category* of $\mathfrak{X}$ by the minimal equivalence relation between the $\mathfrak{X}$-morphisms such that, for any pair of $\mathfrak{X}$-objects X and Y, and any pair of $\mathfrak{X}$-morphisms f and g from X to Y, if $\mathrm{Nat}_{\mathfrak{X}}(f, g) \neq \emptyset$ then f and g are equivalent. The existence of the functors $\mathfrak{l}^X_{Y,Z}$ fulfilling conditions A1.1.2 guarantees that the composition in $\mathfrak{X}$ induces indeed a composition in $\overline{\mathfrak{X}}$. Note that any category $\mathfrak{C}$ admits a trivial 2-category structure where, for any pair of objects A and B, the set $\mathfrak{C}(A, B)$ has the *trivial* category structure, namely it has no morphism different from the *identity* of any object.

A1.9 If $\mathfrak{Y}$ is another 2-category, recall that a 2-*functor* $\mathfrak{x} : \mathfrak{X} \to \mathfrak{Y}$ is a functor between the underlying categories, together with a *natural* family of functors

$$\mathfrak{x}_{X',X} : \mathfrak{X}(X', X) \longrightarrow \mathfrak{Y}\big(\mathfrak{x}(X'), \mathfrak{x}(X)\big) \qquad \text{A1.9.1}$$

where X and X' run over the collection of $\mathfrak{X}$-objects, which coincide with $\mathfrak{x}$ over the corresponding morphisms and which, for any triple of $\mathfrak{X}$-objects X, X' and X'', any pair of $\mathfrak{X}$-morphisms f, f' from X to X', any pair of $\mathfrak{X}$-morphisms g, g' from X' to X'', and any *natural maps* $\alpha \in \mathrm{Nat}_{\mathfrak{X}}(f', f)$ and $\beta \in \mathrm{Nat}_{\mathfrak{X}}(g', g)$, fulfill

$$\begin{aligned} \mathfrak{x}_{X'',X}(g \circ f) &= \mathfrak{x}(g) \circ \mathfrak{x}(f) \\ \mathfrak{x}_{X'',X}(g * \alpha) &= \mathfrak{x}(g) * \mathfrak{x}_{X',X}(\alpha) \\ \mathfrak{x}_{X'',X}(\beta * f) &= \mathfrak{x}_{X'',X'}(\beta) * \mathfrak{x}(f) \end{aligned} \qquad \text{A1.9.2.}$$

Moreover, in this case $\mathfrak{x}$ induces a functor $\bar{\mathfrak{x}} : \overline{\mathfrak{X}} \to \overline{\mathfrak{Y}}$. We denote by $\mathfrak{id}_{\mathfrak{X}}$ the identity 2-functor of $\mathfrak{X}$.

A1.10 If $\mathfrak{Z}$ is a third 2-category and $\mathfrak{u} : \mathfrak{Y} \to \mathfrak{Z}$ is a 2-functor, it is easily checked that we have a 2-functor $\mathfrak{u} \circ \mathfrak{x}$ from $\mathfrak{X}$ to $\mathfrak{Z}$ such that its underlying

† Alberto Arabia helped us to understand the interest of this definition.

functor coincides with the composition of the underlying functors of $\mathfrak{x}$ and $\mathfrak{u}$, and that it fulfills

$$(\mathfrak{u}\circ\mathfrak{x})_{X',X} = \mathfrak{u}_{\mathfrak{x}(X'),\mathfrak{x}(X)}\circ\mathfrak{x}_{X',X} \qquad \text{A1.10.1}$$

for any pair of $\mathfrak{X}$-objects X and X; then, we have $\overline{\mathfrak{u}\circ\mathfrak{x}} = \bar{\mathfrak{u}}\circ\bar{\mathfrak{x}}$.

A1.11 Recall that if $\mathfrak{y}:\mathfrak{X}\to\mathfrak{Y}$ is a second 2-functor, a (strict) 2-*natural map* $\mu:\mathfrak{x}\to\mathfrak{y}$ is a correspondence mapping any $\mathfrak{X}$-object X on a $\mathfrak{Y}$-morphism $\mu_X:\mathfrak{x}(X)\to\mathfrak{y}(X)$ in such a way that, for any pair of $\mathfrak{X}$-objects X and X' we have (cf. A1.5)

$$\mathfrak{l}^{\mathfrak{x}(X)}_{\mu_{X'}}\circ\mathfrak{x}_{X',X} = \mathfrak{r}^{\mathfrak{y}(X')}_{\mu_X}\circ\mathfrak{y}_{X',X} \qquad \text{A1.11.1;}$$

in this case, μ clearly induces a natural map $\bar\mu:\bar{\mathfrak{x}}\to\bar{\mathfrak{y}}$. If $\mathfrak{t}:\mathfrak{X}\to\mathfrak{Y}$ is a third 2-functor and $\nu:\mathfrak{y}\to\mathfrak{t}$ a 2-natural map, we have a *composed* 2-natural map $\nu\circ\mu$ from $\mathfrak{x}$ to $\mathfrak{t}$.

A1.12 It is clear that the *composition* of 2-natural maps is associative and that if $\mathfrak{X}$, $\mathfrak{Y}$ and $\mathfrak{Z}$ are 2-categories, $\mathfrak{x}:\mathfrak{X}\to\mathfrak{Y}$, $\mathfrak{x}':\mathfrak{X}\to\mathfrak{Y}$, $\mathfrak{y}:\mathfrak{Y}\to\mathfrak{Z}$ and $\mathfrak{y}':\mathfrak{Y}\to\mathfrak{Z}$ are 2-functors, and $\mu:\mathfrak{x}\to\mathfrak{x}'$ and $\nu:\mathfrak{y}\to\mathfrak{y}'$ are 2-natural maps, then we have 2-natural maps

$$\nu*\mathfrak{x}:\mathfrak{y}\circ\mathfrak{x}\longrightarrow\mathfrak{y}'\circ\mathfrak{x} \quad\text{and}\quad \mathfrak{y}*\mu:\mathfrak{y}\circ\mathfrak{x}\longrightarrow\mathfrak{y}\circ\mathfrak{x}' \qquad \text{A1.12.1.}$$

With the same notation, if $\mathfrak{x}'':\mathfrak{X}\to\mathfrak{Y}$ and $\mathfrak{y}'':\mathfrak{Y}\to\mathfrak{Z}$ are 2-functors, and $\mu':\mathfrak{x}'\to\mathfrak{x}''$ and $\nu':\mathfrak{y}'\to\mathfrak{y}''$ are 2-natural maps, then we have

$$(\nu'\circ\nu)*\mathfrak{x} = (\nu'*\mathfrak{x})\circ(\nu*\mathfrak{x}) \quad\text{and}\quad \mathfrak{y}*(\mu'\circ\mu) = (\mathfrak{y}*\mu')\circ(\mathfrak{y}*\mu) \qquad \text{A1.12.2.}$$

Moreover, if $\mathfrak{U}$ is a 2-category, $\mathfrak{t}:\mathfrak{Z}\to\mathfrak{U}$ and $\mathfrak{t}':\mathfrak{Z}\to\mathfrak{U}$ are 2-functors and $\eta:\mathfrak{t}\to\mathfrak{t}'$ is a 2-natural map, then we still have

$$(\eta*\mathfrak{y})*\mathfrak{x} = \eta*(\mathfrak{y}\circ\mathfrak{x}) \quad\text{and}\quad \mathfrak{t}*(\mathfrak{y}*\mu) = (\mathfrak{t}\circ\mathfrak{y})*\mu \qquad \text{A1.12.3.}$$

A1.13 A meaningful example of a 2-functor and a 2-natural map comes from the simplicial 2-category $\boldsymbol{\Delta}$. The *shifting functor*

$$\mathfrak{sh}:\boldsymbol{\Delta}\longrightarrow\boldsymbol{\Delta} \qquad \text{A1.13.1}$$

is a 2-functor mapping any Δ_n on Δ_{n+1}, mapping any order-preserving map $\delta:\Delta_n\to\Delta_m$ on the order-preserving map $\Delta_{n+1}\to\Delta_{m+1}$ sending 0 to 0 and $i\in\Delta_{n+1}-\{0\}$ to $\delta(n-1)+1$ and, for any order-preserving map $\varepsilon:\Delta_n\to\Delta_m$ such that $\delta(i)\le\varepsilon(i)$ for any $i\in\Delta_n$, mapping the natural map $\delta\bullet\varepsilon$ on the natural map $\mathfrak{sh}(\delta)\bullet\mathfrak{sh}(\varepsilon)$, which makes sense. The *shifting natural map*

$$\varsigma:\mathfrak{id}_{\boldsymbol{\Delta}}\longrightarrow\mathfrak{sh} \qquad \text{A1.13.2}$$

is a 2-natural map which sends Δ_n to the order-preserving map $\Delta_n\to\Delta_{n+1}$ sending $i\in\Delta_n$ to $i+1$.

A1.14 Although barely needed in this book, let us recall the definition of the *supranatural maps.* If $\mathfrak{X}$ and $\mathfrak{Y}$ are 2-categories, $\mathfrak{x}:\mathfrak{X}\to\mathfrak{Y}$ and $\mathfrak{y}:\mathfrak{X}\to\mathfrak{Y}$ are 2-functors and $\lambda:\mathfrak{x}\to\mathfrak{y}$ and $\mu:\mathfrak{x}\to\mathfrak{y}$ are 2-natural maps, a *supranatural map* $\sigma:\lambda\to\mu$ is a correspondence mapping any $\mathfrak{X}$-object X on a $\mathfrak{Y}$-natural map $\sigma_X\in \mathrm{Nat}_{\mathfrak{Y}}(\mu_X,\lambda_X)$ in such a way that any $\mathfrak{X}$-morphism $f:X\to X'$ fulfills the equality

$$\mathfrak{y}(f)*\sigma_X=\sigma_{X'}*\mathfrak{x}(f) \qquad \text{A1.14.1.}$$

Moreover, if $\nu:\mathfrak{x}\to\mathfrak{y}$ is a third 2-natural map and $\tau:\mu\to\nu$ a supranatural map, the correspondence mapping X on $\tau_X\circ\sigma_X$ defines a supranatural map $\tau\circ\sigma:\lambda\to\nu$; this *composition* is associative.

A1.15 Finally, if $\mathfrak{t}:\mathfrak{X}\to\mathfrak{Y}$ is a 2-functor, $\lambda':\mathfrak{y}\to\mathfrak{t}$ and $\mu':\mathfrak{y}\to\mathfrak{t}$ are 2-natural maps, and $\sigma':\lambda'\to\mu'$ is a supranatural map, then we have the supranatural maps

$$\lambda'*\sigma:\lambda'\circ\lambda\longrightarrow\lambda'\circ\mu \quad\text{and}\quad \sigma'*\lambda:\lambda'\circ\lambda\longrightarrow\mu'\circ\lambda \qquad \text{A1.15.1.}$$

Similarly, if $\mathfrak{Z}$ is a third 2-category, $\mathfrak{t}:\mathfrak{Y}\to\mathfrak{Z}$ and $\mathfrak{u}:\mathfrak{Y}\to\mathfrak{Z}$ are 2-functors, $\nu:\mathfrak{t}\to\mathfrak{u}$ and $\eta:\mathfrak{t}\to\mathfrak{u}$ are 2-natural maps, and $\tau:\nu\to\eta$ a supranatural map, then we have the supranatural maps

$$\mathfrak{t}\diamond\sigma:\mathfrak{t}*\lambda\longrightarrow\mathfrak{t}*\mu \quad\text{and}\quad \tau\diamond\mathfrak{x}:\nu*\mathfrak{x}\longrightarrow\eta*\mathfrak{x} \qquad \text{A1.15.2.}$$

A2 Representations and semidirect products

A2.1 The formalism of the *semidirect product*† we introduce here unifies several constructions of categories employed in this book just by mimicking the definition of the semidirect product of a group by one of its linear representations.

A2.2 Let $\mathfrak{X}$ be a 2-category; we call *representation of* $\mathfrak{X}$ a 2-functor $\mathfrak{m}:\mathfrak{X}\to\mathfrak{CC}$ from $\mathfrak{X}$ to the *category of small categories* $\mathfrak{CC}$ (cf. A1.6); for any item $\bullet$ in $\mathfrak{X}$, let us denote by ${}^{\mathfrak{m}}\bullet$ its image. Our main example is the *standard representation* of the *simplicial* 2-*category* (cf. A1.7)

$$\mathfrak{st}:\mathbf{\Delta}\longrightarrow\mathfrak{CC} \qquad \text{A2.2.1}$$

mapping any *n-simplex* Δ_n on the category — still denoted by Δ_n — where the objects are all the elements $i\in\Delta_n$ and the set of $\mathbf{\Delta}$-morphisms from $i\le n$ to $j\le n$ is not empty only if $i\le j$, and then has a unique element, denoted by $i\bullet j$; moreover, $\mathfrak{st}$ maps any $\mathbf{\Delta}$-morphism from Δ_n to Δ_m on the obvious functor and any $\mathbf{\Delta}$-natural map on the obvious natural map.

† Some people pointed out to us that, although with another name, this construction already has been introduced by Alexander Grothendieck.

A2.3 Since any *set* can be considered as a *category* with no other morphisms than the *identity* for any element, and then the *maps* become *functors*, denoting by $\aleph$ the category of finite sets (cf. 1.1), we have the *identity representation*

$$\mathfrak{i} : \aleph \longrightarrow \mathfrak{CC} \qquad \text{A2.3.1.}$$

Note that any representation $\mathfrak{m}$ of $\mathfrak{X}$ induces functors $\mathfrak{m}^*$ and $\mathfrak{m}^{\mathfrak{o}}$ from the underlying category $\mathfrak{X}^{\mathfrak{c}}$ to $\mathfrak{CC}$ respectively mapping any $\mathfrak{X}$-object X on the subcategories $({}^{\mathfrak{m}}X)^*$ and $({}^{\mathfrak{m}}X)^{\mathfrak{o}}$ of ${}^{\mathfrak{m}}X$ (cf. 1.5), where the respective morphisms are the ${}^{\mathfrak{m}}X$-isomorphisms and the ${}^{\mathfrak{m}}X$*-identity* of any object (cf. A2.2), and any $\mathfrak{X}$-morphism $\mathfrak{f}: X \to Y$ on the corresponding functors

$$({}^{\mathfrak{m}}\mathfrak{f})^* : ({}^{\mathfrak{m}}X)^* \longrightarrow ({}^{\mathfrak{m}}Y)^* \quad \text{and} \quad ({}^{\mathfrak{m}}\mathfrak{f})^{\mathfrak{o}} : ({}^{\mathfrak{m}}X)^{\mathfrak{o}} \longrightarrow ({}^{\mathfrak{m}}Y)^{\mathfrak{o}} \qquad \text{A2.3.2.}$$

More generally, we call *weak structure* of $\mathfrak{m}$ a *subfunctor* $\mathfrak{m}^w$ of $\mathfrak{m}$ mapping any $\mathfrak{X}$-object X on a subcategory ${}^{\mathfrak{m}^w}X$ of ${}^{\mathfrak{m}}X$ with the same objects.

A2.4 Any small category $\mathfrak{C}$ considered as a 2-category admits the *regular representation*

$$\mathfrak{rg}_{\mathfrak{C}} : \mathfrak{C} \longrightarrow \mathfrak{CC} \qquad \text{A2.4.1}$$

defined as follows. Firstly, for any $\mathfrak{C}$-object C, we consider the category $\mathfrak{C}_C$ (cf. 1.7); recall that the $\mathfrak{C}_C$-objects are the $\mathfrak{C}$-morphisms $f: A \to C$, the $\mathfrak{C}_C$-morphisms from $f: A \to C$ to another $\mathfrak{C}_C$-object $f': A' \to C$ are the $\mathfrak{C}$-morphisms $h: A \to A'$ fulfilling $f' \circ h = f$, and the composition comes from the composition in $\mathfrak{C}$. Secondly, for any $\mathfrak{C}$-morphism $\ell: C \to C'$, we have a functor

$$\mathfrak{C}_\ell : \mathfrak{C}_C \longrightarrow \mathfrak{C}_{C'} \qquad \text{A2.4.2}$$

mapping any $\mathfrak{C}_C$-object $f: A \to C$ on the composition $\ell \circ f: A \to C'$, which is indeed a $\mathfrak{C}_{C'}$-object, and any $\mathfrak{C}_C$-morphism h from $f: A \to C$ to another $\mathfrak{C}_C$-object $f': A' \to C$ — namely any $\mathfrak{C}$-morphism from A to A' fulfilling the equality $f' \circ h = f$ in $\mathfrak{C}$ — on the *same* h considered as a $\mathfrak{C}_C$-morphism from $\ell \circ f: A \to C'$ to $\ell \circ f': A' \to C'$, since we still have $\ell \circ f' \circ h = \ell \circ f$. Then, the *regular representation* $\mathfrak{rg}_{\mathfrak{C}}$ maps any $\mathfrak{C}$-object C on the category $\mathfrak{C}_C$ and any $\mathfrak{C}$-morphism $\ell: C \to C'$ on the functor $\mathfrak{C}_\ell : \mathfrak{C}_C \to \mathfrak{C}_{C'}$.

A2.5 Moreover, for any small category $\mathfrak{C}$, we have the $\mathfrak{C}$-*dual* of a representation $\mathfrak{m}$ of $\mathfrak{X}$, which is a representation of $\mathfrak{X}^\circ$

$$\mathfrak{Fct}(\mathfrak{m}, \mathfrak{C}) : \mathfrak{X}^\circ \longrightarrow \mathfrak{CC} \qquad \text{A2.5.1}$$

mapping any $\mathfrak{X}$-object X on the category of functors $\mathfrak{Fct}({}^{\mathfrak{m}}X, \mathfrak{C})$, any $\mathfrak{X}$-morphism $\mathfrak{f}: Y \to X$ on the functor

$$\mathfrak{Fct}({}^{\mathfrak{m}}\mathfrak{f}, \mathfrak{C}) : \mathfrak{Fct}({}^{\mathfrak{m}}X, \mathfrak{C}) \longrightarrow \mathfrak{Fct}({}^{\mathfrak{m}}Y, \mathfrak{C}) \qquad \text{A2.5.2}$$

determined by the composition with ${}^{\mathfrak{m}}\mathfrak{f}\colon {}^{\mathfrak{m}}Y \to {}^{\mathfrak{m}}X$, and any $\mathfrak{X}$-natural map $\mu\colon \mathfrak{f} \to \mathfrak{g}$ on the natural map

$$\mathfrak{Fct}({}^{\mathfrak{m}}\mu, \mathfrak{C}) : \mathfrak{Fct}({}^{\mathfrak{m}}\mathfrak{f}, \mathfrak{C}) \longrightarrow \mathfrak{Fct}({}^{\mathfrak{m}}\mathfrak{g}, \mathfrak{C}) \qquad \text{A2.5.3}$$

sending any functor $\mathfrak{h}\colon {}^{\mathfrak{m}}X \to \mathfrak{C}$ to the natural map

$$\mathfrak{h} * {}^{\mathfrak{m}}\mu : \mathfrak{h} \circ {}^{\mathfrak{m}}\mathfrak{f} \longrightarrow \mathfrak{h} \circ {}^{\mathfrak{m}}\mathfrak{g} \qquad \text{A2.5.4.}$$

The *weak structure* $\mathfrak{Fct}^*(\mathfrak{m}, \mathfrak{C})$ of $\mathfrak{Fct}(\mathfrak{m}, \mathfrak{C})$ maps any $\mathfrak{X}$-object X on the category of functors $\mathfrak{Fct}({}^{\mathfrak{m}}X, \mathfrak{C})^*$ with natural isomorphisms.

Proposition A2.6 *For any 2-category $\mathfrak{X}$ and any representation $\mathfrak{m}\colon \mathfrak{X} \to \mathfrak{CC}$ of $\mathfrak{X}$, there is a 2-category $\mathfrak{m} \rtimes \mathfrak{X}$ fulfilling:*

A2.6.1 *The $\mathfrak{m} \rtimes \mathfrak{X}$-objects are the pairs (A, X) formed by an $\mathfrak{X}$-object X and a ${}^{\mathfrak{m}}X$-object A.*

A2.6.2 *The $\mathfrak{m} \rtimes \mathfrak{X}$-morphisms from (A, X) to another $\mathfrak{m} \rtimes \mathfrak{X}$-object (B, Y) are the pairs $(f, \mathfrak{f})$ formed by an $\mathfrak{X}$-morphism $\mathfrak{f}\colon X \to Y$ and a ${}^{\mathfrak{m}}Y$-morphism*

$$f : {}^{\mathfrak{m}}\mathfrak{f}\,(A) \longrightarrow B\,.$$

A2.6.3 *The composition of the $\mathfrak{m} \rtimes \mathfrak{X}$-morphism $(f, \mathfrak{f})\colon (A, X) \to (B, Y)$ with another $\mathfrak{m} \rtimes \mathfrak{X}$-morphism $(g, \mathfrak{g})\colon (B, Y) \to (C, Z)$ is given by*

$$(g, \mathfrak{g}) \circ (f, \mathfrak{f}) = \bigl(g \circ {}^{\mathfrak{m}}\mathfrak{g}\,(f), \mathfrak{g} \circ \mathfrak{f}\bigr)\,.$$

A2.6.4 *The $\mathfrak{m} \rtimes \mathfrak{X}$-natural maps from $(f, \mathfrak{f})$ to another $\mathfrak{m} \rtimes \mathfrak{X}$-morphism $(f', \mathfrak{f}')\colon (A, X) \to (B, Y)$ are the $\mathfrak{X}$-natural maps $\alpha\colon \mathfrak{f} \to \mathfrak{f}'$ fulfilling*

$$f' \circ ({}^{\mathfrak{m}}\alpha)_A = f\,.$$

A2.6.5 *The compositions between $\mathfrak{m} \rtimes \mathfrak{X}$-morphisms and $\mathfrak{m} \rtimes \mathfrak{X}$-natural maps, and between $\mathfrak{m} \rtimes \mathfrak{X}$-natural maps are induced by the corresponding compositions in $\mathfrak{X}$.*

A2.7 We call *semidirect product of $\mathfrak{X}$ by $\mathfrak{m}$* the 2-category $\mathfrak{m} \rtimes \mathfrak{X}$; note that we have an obvious 2-functor

$$\mathfrak{p}_{\mathfrak{X},\mathfrak{m}} : \mathfrak{m} \rtimes \mathfrak{X} \longrightarrow \mathfrak{X} \qquad \text{A2.7.1;}$$

moreover, a *weak structure* $\mathfrak{m}^w$ of $\mathfrak{m}$ determines a subcategory $\mathfrak{m}^w \rtimes \mathfrak{X}$ of $\mathfrak{m} \rtimes \mathfrak{X}$ with the same objects. Our first example is the *product cover* of a small category: for any small category $\mathfrak{C}$, we consider the $\mathfrak{C}$-dual representation $\mathfrak{Fct}(\mathfrak{i}, \mathfrak{C})$ (cf. A2.5) of the *identity representation* of $\aleph$ (cf. A2.3) and then the *product cover* of $\mathfrak{C}$ is the category

$$\mathfrak{pc}(\mathfrak{C}) = \mathfrak{Fct}(\mathfrak{i}, \mathfrak{C}) \rtimes \aleph^\circ \qquad \text{A2.7.2.}$$

Explicitly, a $\mathfrak{pc}(\mathfrak{C})$-object is just a finite sequence $\{C_i\}_{i\in I}$ of objects in $\mathfrak{C}$ — denoted by $\prod_{i\in I} C_i$, or $\widetilde{\prod}_{i\in I} C_i$ to avoid confusion with a possible direct product in $\mathfrak{C}$ — and a $\mathfrak{pc}(\mathfrak{C})$-morphism from $\prod_{i\in I} C_i$ to another $\mathfrak{pc}(\mathfrak{C})$-object $\prod_{j\in J} D_j$ is formed by a map $f : J \to I$ and by a $\mathfrak{C}$-morphism $\alpha_j : C_{f(j)} \to D_j$ for any $j \in J$. In this book, we widely employ the *additive cover* of $\mathfrak{C}$, introduced by S. Jackowski and J. McClure in [32], which coincides with (cf. A4.10 below)

$$\mathfrak{ac}(\mathfrak{C}) = \mathfrak{pc}(\mathfrak{C}^\circ)^\circ \qquad \text{A2.7.3.}$$

A2.8 Our second example of a semidirect product is the *category of chains* of a small category: for any small category $\mathfrak{C}$, we consider the $\mathfrak{C}$-dual representation $\mathfrak{Fct}(\mathfrak{st}, \mathfrak{C})$ (cf. A2.5) of the *standard representation* $\mathfrak{st}$ of $\boldsymbol{\Delta}$ (cf. A2.2) and then the *category of $\mathfrak{C}$-chains* is the 2-category

$$\mathfrak{ch}(\mathfrak{C}) = \mathfrak{Fct}(\mathfrak{st}, \mathfrak{C}) \rtimes \boldsymbol{\Delta}^\circ \qquad \text{A2.8.1.}$$

Actually, in this book we are only interested in the 2-subcategory $\mathfrak{ch}^*(\mathfrak{C})$ of $\mathfrak{ch}(\mathfrak{C})$ — called *proper category of $\mathfrak{C}$-chains* — coming from the *weak structure* $\mathfrak{Fct}(\mathfrak{st}, \mathfrak{C})^*$ of $\mathfrak{Fct}(\mathfrak{st}, \mathfrak{C})$ (cf. A2.5); explicitly, $\mathfrak{ch}^*(\mathfrak{C})$ is formed by the same objects $(\mathfrak{q}, \Delta_n)$, where $\mathfrak{q} : \Delta_n \to \mathfrak{C}$ is a functor, by the morphisms

$$(\nu, \delta) : (\mathfrak{r}, \Delta_m) \longrightarrow (\mathfrak{q}, \Delta_n) \qquad \text{A2.8.2,}$$

where $\delta : \Delta_m \to \Delta_n$ is an order-preserving map and where — writing δ for ${}^{\mathfrak{st}}\delta$ — $\nu : \mathfrak{r} \cong \mathfrak{q} \circ \delta$ is a *natural isomorphism* instead of a simple natural map, and by the same $\mathfrak{ch}(\mathfrak{C})$-natural maps between these $\mathfrak{ch}(\mathfrak{C})$-morphisms.

A2.9 Note that $\mathfrak{ch}^*(\mathfrak{C})$ need *not* be the semidirect product of $\boldsymbol{\Delta}^\circ$ by the representation of $\boldsymbol{\Delta}^\circ$ mapping Δ_n on the category $\mathfrak{Fct}(\Delta_n, \mathfrak{C})^*$, where the morphisms are the natural isomorphisms, since the $\mathfrak{ch}^*(\mathfrak{C})$-natural maps need *not* be isomorphisms; this is meaningful when dealing with the *category of connected components* $\overline{\mathfrak{ch}^*(\mathfrak{C})}$ (cf. Proposition A5.4). The main reason to consider the *proper category of chains* comes from the following fact which has to be understood as a generalization of the "barycentric decomposition".

Proposition A2.10 *For any small category $\mathfrak{C}$, the correspondence mapping any $\mathfrak{ch}^*(\mathfrak{C})$-object $(\mathfrak{q}, \Delta_n)$, where $\mathfrak{q} : \Delta_n \to \mathfrak{C}$ is a functor, on its $\mathfrak{ch}^*(\mathfrak{C})$-automorphism group $\mathfrak{Nat}(\mathfrak{q}, \mathfrak{q})^*$ can be extended to a functor*

$$\mathfrak{aut}_{\mathfrak{C}} : \mathfrak{ch}^*(\mathfrak{C}) \longrightarrow \mathfrak{Gr} \qquad \text{A2.10.1}$$

mapping any $\mathfrak{ch}^*(\mathfrak{C})$-morphism $(\nu, \delta) : (\mathfrak{r}, \Delta_m) \to (\mathfrak{q}, \Delta_n)$ on the group homomorphism sending $\rho \in \mathfrak{Nat}(\mathfrak{r}, \mathfrak{r})^*$ to $\nu \circ (\rho * \delta) \circ \nu^{-1} \in \mathfrak{Nat}(\mathfrak{q}, \mathfrak{q})^*$.

A2.11 Let us illustrate the usefulness of the *semidirect product* in the usual construction of the *direct image* of an $\mathfrak{Ab}$-valued *contravariant* functor (cf. 1.6). Let $\mathfrak{B}$ and $\mathfrak{C}$ be small categories and $\mathfrak{f}:\mathfrak{B}\to\mathfrak{C}$; the starting point is the *graph representation* of $\mathfrak{f}$ (cf. 1.5)

$$\mathfrak{gr}_{\mathfrak{f}} : \mathfrak{C}^\circ \times \mathfrak{B} \longrightarrow \mathfrak{CC} \qquad \text{A2.11.1}$$

mapping any $\mathfrak{C}^\circ \times \mathfrak{B}$-object (C,B) on the set $\mathfrak{C}\big(C,\mathfrak{f}(B)\big)$ considered as a trivial category and any $\mathfrak{C}^\circ \times \mathfrak{B}$-morphism on the obvious map; it is quite clear that we can identify this functor with a new functor

$$\mathfrak{gr}_{\mathfrak{f}} : \mathfrak{C}^\circ \longrightarrow \mathfrak{Fct}(\mathfrak{B},\mathfrak{CC}) \qquad \text{A2.11.2.}$$

A2.12 That is to say, for any $\mathfrak{C}$-object C we get a *representation*

$$\mathfrak{gr}_{\mathfrak{f}}(C) : \mathfrak{B} \longrightarrow \mathfrak{CC} \qquad \text{A2.12.1}$$

of $\mathfrak{B}$ and therefore we still get a category $\mathfrak{gr}_{\mathfrak{f}}(C)\rtimes\mathfrak{B}$, actually endowed with a structural functor (cf. A2.7)

$$\mathfrak{p}_C = \mathfrak{p}_{\mathfrak{B},\mathfrak{gr}_{\mathfrak{f}}(C)} : \mathfrak{gr}_{\mathfrak{f}}(C)\rtimes\mathfrak{B} \longrightarrow \mathfrak{B} \qquad \text{A2.12.2.}$$

Analogously, for any $\mathfrak{C}$-morphism $g:C\to C'$ we have a natural map

$$\mathfrak{gr}_{\mathfrak{f}}(g) : \mathfrak{gr}_{\mathfrak{f}}(C') \longrightarrow \mathfrak{gr}_{\mathfrak{f}}(C) \qquad \text{A2.12.3}$$

and therefore we still have a functor (cf. Proposition 2.20 below)

$$\mathfrak{gr}_{\mathfrak{f}}(g)\rtimes\mathfrak{id}_{\mathfrak{B}} : \mathfrak{gr}_{\mathfrak{f}}(C')\rtimes\mathfrak{B} \longrightarrow \mathfrak{gr}_{\mathfrak{f}}(C)\rtimes\mathfrak{B} \qquad \text{A2.12.4;}$$

moreover, it is clear that

$$\mathfrak{p}_C \circ (\mathfrak{gr}_{\mathfrak{f}}(g)\rtimes\mathfrak{id}_{\mathfrak{B}}) = \mathfrak{p}_{C'} \qquad \text{A2.12.5.}$$

A2.13 At this point, with evident terminology, it is clear that we have obtained a *pointed representation* of $\mathfrak{C}^\circ$ (cf. 1.7)

$$\mathfrak{pr}_{\mathfrak{f}} : \mathfrak{C}^\circ \longrightarrow (\mathfrak{CC})_{\mathfrak{B}} \qquad \text{A2.13.1}$$

mapping any $\mathfrak{C}$-object C on the $(\mathfrak{CC})_{\mathfrak{B}}$-object $\mathfrak{p}_C$ defined in A2.12.2 above and any $\mathfrak{C}$-morphism $g:C\to C'$ on the $(\mathfrak{CC})_{\mathfrak{B}}$-morphism $\mathfrak{gr}_{\mathfrak{f}}(g)\rtimes\mathfrak{id}_{\mathfrak{B}}$. On the other hand, we know that any *contravariant* functor $\mathfrak{a}:\mathfrak{B}\to\mathfrak{Ab}$ and the *inverse limit* define two functors (cf. 1.7)

$$(\mathfrak{CC})_{\mathfrak{a}} : (\mathfrak{CC})_{\mathfrak{B}} \longrightarrow (\mathfrak{CC})_{\mathfrak{Ab}^\circ} \quad \text{and} \quad \varinjlim : (\mathfrak{CC})_{\mathfrak{Ab}^\circ} \longrightarrow \mathfrak{Ab} \qquad \text{A2.13.2.}$$

Finally, it is easily checked that

$$\mathfrak{f}_*(\mathfrak{a}) = \varinjlim \circ (\mathfrak{CC})_{\mathfrak{a}} \circ \mathfrak{pr}_{\mathfrak{f}} \qquad \text{A2.13.3.}$$

A2.14 But in this book, we are also interested in another kind of *image* involving the *semidirect product*. Namely, we are interested in the particular situation where, for a representation $\mathfrak{m}$ of a 2-category $\mathfrak{X}$, endowed with a *weak structure* $\mathfrak{m}^w$, the functor ${}^{\mathfrak{m}}\mathfrak{f}\colon {}^{\mathfrak{m}}X \to {}^{\mathfrak{m}}Y$ admits a *right adjoin* $({}^{\mathfrak{m}}\mathfrak{f})^{\mathfrak{a}}$ (cf. 1.6) such that

$$({}^{\mathfrak{m}}\mathfrak{f})^{\mathfrak{a}}({}^{\mathfrak{m}^w}X) \subset {}^{\mathfrak{m}^w}Y \qquad \text{A2.14.1}$$

for any $\mathfrak{X}$-morphism $\mathfrak{f}\colon X \to Y$. In this case, let us denote by

$$\alpha_{\mathfrak{f}} : {}^{\mathfrak{m}}\mathfrak{f} \circ ({}^{\mathfrak{m}}\mathfrak{f})^{\mathfrak{a}} \longrightarrow \mathfrak{id}_{{}^{\mathfrak{m}}Y} \qquad \text{A2.14.2}$$

the corresponding *adjoinness natural map* and by

$$({}^{\mathfrak{m}}\mathfrak{f})^{\mathfrak{a},w} : {}^{\mathfrak{m}^w}X \to {}^{\mathfrak{m}^w}Y \qquad \text{A2.14.3}$$

the corresponding restriction. The point is the existence of a functor

$$(\mathfrak{p}_{\mathfrak{m},\mathfrak{m}^w})_{\mathfrak{a}} : \mathfrak{Fct}\big((\mathfrak{m} \rtimes \mathfrak{X})^\circ, \mathfrak{Ab}\big) \longrightarrow \mathfrak{Fct}(\mathfrak{X}^\circ, \mathfrak{Ab}) \qquad \text{A2.14.4}$$

— the $\mathfrak{m}^w$*-stable adjoin image* — defined as follows.

A2.15 First of all, for any *contravariant* functor

$$\mathfrak{b} : \mathfrak{m} \rtimes \mathfrak{X} \longrightarrow \mathfrak{Ab} \qquad \text{A2.15.1}$$

and any $\mathfrak{X}$-object X, we consider the *contravariant* functor

$$\mathfrak{b}_X : {}^{\mathfrak{m}}X \longrightarrow \mathfrak{Ab} \qquad \text{A2.15.2}$$

mapping any ${}^{\mathfrak{m}}X$-object A on $\mathfrak{b}(A,X)$ and any ${}^{\mathfrak{m}}X$-morphism $f\colon A \to A'$ on $\mathfrak{b}(f,\mathrm{id}_X)$, and denote by $(\mathfrak{b}_X)^w$ the restriction of $\mathfrak{b}_X$ to ${}^{\mathfrak{m}^w}X$; note that, for any $\mathfrak{X}$-morphism $\mathfrak{f}\colon X \to Y$, the *adjoinness* $\alpha_{\mathfrak{f}}$ determines a *natural map*

$$\mathfrak{b}_{\mathfrak{f}} : \mathfrak{b}_Y \longrightarrow \mathfrak{b}_X \circ ({}^{\mathfrak{m}}\mathfrak{f})^{\mathfrak{a}} \qquad \text{A2.15.3}$$

sending any ${}^{\mathfrak{m}}Y$-object B to $\mathfrak{b}\big((\alpha_{\mathfrak{f}})_B,\mathfrak{f}\big)$ (cf. A2.6.2). Moreover, for any natural map $\mu\colon \mathfrak{b} \to \mathfrak{b}'$ from $\mathfrak{b}$ to another *contravariant* functor $\mathfrak{b}'\colon \mathfrak{m} \rtimes \mathfrak{X} \to \mathfrak{Ab}$ and any $\mathfrak{X}$-object X, we have an obvious natural map $\mu_X : \mathfrak{b}_X \to \mathfrak{b}'_X$.

A2.16 Then, $(\mathfrak{p}_{\mathfrak{m},\mathfrak{m}^w})_{\mathfrak{a}}$ maps $\mathfrak{b}$ on the *contravariant* functor

$$(\mathfrak{p}_{\mathfrak{m},\mathfrak{m}^w})_{\mathfrak{a}}(\mathfrak{b}) : \mathfrak{X} \longrightarrow \mathfrak{Ab} \qquad \text{A2.16.1}$$

sending X to the *direct limit* $\varinjlim (\mathfrak{b}_X)^w$ and any $\mathfrak{X}$-morphism $\mathfrak{f}\colon X \to Y$ to the group homomorphism

$$\varinjlim (\mathfrak{b}_Y)^w \longrightarrow \varinjlim \big((\mathfrak{b}_X)^w \circ ({}^{\mathfrak{m}}\mathfrak{f})^{\mathfrak{a},w}\big) \longrightarrow \varinjlim (\mathfrak{b}_X)^w \qquad \text{A2.16.2}$$

which is the composition of $\varinjlim \mathfrak{b}_{\mathfrak{f}}$ with the corresponding canonical homomorphism. Moreover, $(\mathfrak{p}_{\mathfrak{m},\mathfrak{m}^w})_{\mathfrak{a}}$ maps $\mu:\mathfrak{b}\to\mathfrak{b}'$ on the natural map

$$(\mathfrak{p}_{\mathfrak{m},\mathfrak{m}^w})_{\mathfrak{a}}(\mu) : (\mathfrak{p}_{\mathfrak{m},\mathfrak{m}^w})_{\mathfrak{a}}(\mathfrak{b}) \longrightarrow (\mathfrak{p}_{\mathfrak{m},\mathfrak{m}^w})_{\mathfrak{a}}(\mathfrak{b}') \qquad \text{A2.16.3}$$

sending any $\mathfrak{X}$-object X to

$$\varinjlim \mu_X : \varinjlim (\mathfrak{b}_X)^w \longrightarrow \varinjlim (\mathfrak{b}'_X)^w \qquad \text{A2.16.4.}$$

A2.17 In this situation, note that if the category ${}^{\mathfrak{m}}X$ has a final object O_X for any $\mathfrak{X}$-object X then we have $({}^{\mathfrak{m}}\mathfrak{f})^{\mathfrak{a}}(O_Y) = O_X$ for any $\mathfrak{X}$-morphism $\mathfrak{f}:X\to Y$ and therefore we get a *contravariant* functor and a *natural map*

$$\begin{aligned} \mathfrak{o}_{\mathfrak{m}} &: \mathfrak{Fct}\big((\mathfrak{m}\rtimes\mathfrak{X})^\circ,\mathfrak{Ab}\big) \longrightarrow \mathfrak{Fct}(\mathfrak{X}^\circ,\mathfrak{Ab}) \\ \omega_{\mathfrak{m},\mathfrak{m}^w} &: \mathfrak{o}_{\mathfrak{m}} \longrightarrow (\mathfrak{p}_{\mathfrak{m},\mathfrak{m}^w})_{\mathfrak{a}} \end{aligned} \qquad \text{A2.17.1}$$

defined as follows; the functor $\mathfrak{o}_{\mathfrak{m}}$ maps $\mathfrak{b}$ on the *contravariat* functor sending any $\mathfrak{X}$-object X to $\mathfrak{b}(O_X,X)$ and any $\mathfrak{X}$-morphism $\mathfrak{f}:X\to Y$ to $\mathfrak{b}(f,\mathfrak{f})$ where $f:{}^{\mathfrak{m}}\mathfrak{f}(O_X)\to O_Y$ is the *unique* ${}^{\mathfrak{m}}Y$-morphism, and $\mu:\mathfrak{b}\to\mathfrak{b}'$ on the natural map $\mathfrak{o}_{\mathfrak{m}}(\mu):\mathfrak{o}_{\mathfrak{m}}(\mathfrak{b})\to\mathfrak{o}_{\mathfrak{m}}(\mathfrak{b}')$ sending X to the group homomorphism

$$\mu_{(O_X,X)} : \mathfrak{b}(O_X,X) \longrightarrow \mathfrak{b}'(O_X,X) \qquad \text{A2.17.2;}$$

the natural map $\omega_{\mathfrak{m},\mathfrak{m}^w}$ sends $\mathfrak{b}$ to the natural map

$$\omega_{\mathfrak{m},\mathfrak{m}^w}(\mathfrak{b}) : \mathfrak{o}_{\mathfrak{m}}(\mathfrak{b}) \longrightarrow (\mathfrak{p}_{\mathfrak{m},\mathfrak{m}^w})_{\mathfrak{a}}(\mathfrak{b}) \qquad \text{A2.17.3}$$

sending X to the obvious group homomorphism

$$\mathfrak{b}(O_X,X) \longrightarrow \varinjlim (\mathfrak{b}_X)^w \qquad \text{A2.17.4.}$$

A2.18 Let us come back to the general situation. For any small category $\mathfrak{C}$ considered as a 2-category, it is also useful to consider the semidirect product $\mathfrak{rg}_{\mathfrak{C}}\rtimes\mathfrak{C}$ of $\mathfrak{C}$ by its regular representation; in this case, besides the canonical functor $\mathfrak{p}_{\mathfrak{C}}:\mathfrak{rg}_{\mathfrak{C}}\rtimes\mathfrak{C}\to\mathfrak{C}$ mentioned in A2.7 above, we have the *covering functor* and the *covering natural map*

$$\mathfrak{c}_{\mathfrak{C}} : \mathfrak{rg}_{\mathfrak{C}}\rtimes\mathfrak{C} \longrightarrow \mathfrak{C} \quad \text{and} \quad \kappa_{\mathfrak{C}} : \mathfrak{c}_{\mathfrak{C}} \longrightarrow \mathfrak{p}_{\mathfrak{C}} \qquad \text{A2.18.1}$$

sending any pair (f,C), where C is a $\mathfrak{C}$-object and $f:A\to C$ a $\mathfrak{C}$-morphism, to A and to $f:A\to C$ respectively, and any $\mathfrak{rg}_{\mathfrak{C}}\rtimes\mathfrak{C}$-morphism

$$(h,\ell) : (f,C) \longrightarrow (f',C') \qquad \text{A2.18.2,}$$

where C' is a $\mathfrak{C}$-object and $f':A'\to C'$ a $\mathfrak{C}$-morphism fulfilling $f'\circ h = \ell\circ f$, to $h:A\to A'$.

A2.19 Let $\mathfrak{X}$ and $\mathfrak{Y}$ be two 2-categories and consider representations

$$\mathfrak{m} : \mathfrak{X} \longrightarrow \mathfrak{CC} \quad \text{and} \quad \mathfrak{n} : \mathfrak{Y} \longrightarrow \mathfrak{CC} \tag{A2.19.1}$$

of $\mathfrak{X}$ and $\mathfrak{Y}$; we call *natural transformation from* $(\mathfrak{X},\mathfrak{m})$ *to* $(\mathfrak{Y},\mathfrak{n})$ any pair $(\mathfrak{r},\alpha)$ formed by a 2-functor and a 2-natural map

$$\mathfrak{r} : \mathfrak{X} \longrightarrow \mathfrak{Y} \quad \text{and} \quad \alpha : \mathfrak{m} \longrightarrow \mathfrak{n}\circ\mathfrak{r} \tag{A2.19.2;}$$

in particular, we have the natural transformation

$$(\mathfrak{r}, \mathrm{id}_{\mathfrak{n}\circ\mathfrak{r}}) : (\mathfrak{X}, \mathfrak{n}\circ\mathfrak{r}) \longrightarrow (\mathfrak{Y},\mathfrak{n}) \tag{A2.19.3.}$$

Analogously, for another representation $\mathfrak{m}' : \mathfrak{X} \to \mathfrak{CC}$ and any natural map $\mu : \mathfrak{m} \to \mathfrak{m}'$, we get the natural transformation

$$(\mathfrak{id}_{\mathfrak{X}}, \mu) : (\mathfrak{X},\mathfrak{m}) \longrightarrow (\mathfrak{X},\mathfrak{m}') \tag{A2.19.4.}$$

For instance, it is quite clear that any functor $\mathfrak{f} : \mathfrak{C} \to \mathfrak{C}'$ between two categories induces a 2-natural map

$$\mathfrak{Fct}(\mathfrak{st},\mathfrak{f}) : \mathfrak{Fct}(\mathfrak{st},\mathfrak{C}) \longrightarrow \mathfrak{Fct}(\mathfrak{st},\mathfrak{C}') \tag{A2.19.5}$$

and therefore we get the natural transformation $\big(\mathfrak{id}_{\Delta^\circ}, \mathfrak{Fct}(\mathfrak{st},\mathfrak{f})\big)$; thus, according to the next proposition, we have a 2-functor

$$\mathfrak{ch}(\mathfrak{f}) = \mathfrak{Fct}(\mathfrak{st},\mathfrak{f}) \rtimes \mathfrak{id}_{\Delta^\circ} : \mathfrak{ch}(\mathfrak{C}) \longrightarrow \mathfrak{ch}(\mathfrak{C}') \tag{A2.19.6.}$$

Proposition A2.20 *Let* $\mathfrak{X}$ *and* $\mathfrak{Y}$ *be two 2-categories,* $\mathfrak{m} : \mathfrak{X} \to \mathfrak{CC}$ *and* $\mathfrak{n} : \mathfrak{Y} \to \mathfrak{CC}$ *respective representations of* $\mathfrak{X}$ *and* $\mathfrak{Y}$, *and* $(\mathfrak{r},\alpha)$ *a natural transformation from* $(\mathfrak{X},\mathfrak{m})$ *to* $(\mathfrak{Y},\mathfrak{n})$. *Then, we have a 2-functor*

$$\alpha \rtimes \mathfrak{r} : \mathfrak{m} \rtimes \mathfrak{X} \longrightarrow \mathfrak{n} \rtimes \mathfrak{Y} \tag{A2.20.1}$$

mapping

A2.20.2 *any* $\mathfrak{m} \rtimes \mathfrak{X}$*-object* (A,X) *on the* $\mathfrak{n} \rtimes \mathfrak{Y}$*-object* $\big(\alpha_X(A), \mathfrak{r}(X)\big)$,

A2.20.3 *any* $\mathfrak{m} \rtimes \mathfrak{X}$*-morphism* $(f,\mathfrak{f}) : (A,X) \to (B,Y)$ *on the* $\mathfrak{n} \rtimes \mathfrak{Y}$*-morphism*

$$\big(\alpha_Y(f) \circ \mathfrak{r}_{X,Y}(\mathfrak{f})_A^{-1}, \mathfrak{r}(\mathfrak{f})\big) : \big(\alpha_X(A), \mathfrak{r}(X)\big) \longrightarrow \big(\alpha_Y(B), \mathfrak{r}(Y)\big),$$

A2.20.4 *any* $\mathfrak{m} \rtimes \mathfrak{X}$*-natural map* $\mu : (f,\mathfrak{f}) \to (f',\mathfrak{f}')$, *where* $\mu \in \mathrm{Nat}_{\mathfrak{X}}(\mathfrak{f}',\mathfrak{f})$, *on the* $\mathfrak{n} \rtimes \mathfrak{Y}$*-natural map determined by* $\mathfrak{r}_{Y,X}(\mu)$.

A2.21 With the same notation, if $\mathfrak{Z}$ is a third 2-category and $\mathfrak{p}: \mathfrak{Z} \to \mathfrak{CC}$ a representation of $\mathfrak{Z}$, any natural transformation $(\mathfrak{y}, \beta): (\mathfrak{Y}, \mathfrak{n}) \to (\mathfrak{Z}, \mathfrak{p})$ can be *composed* with $(\mathfrak{x}, \alpha): (\mathfrak{X}, \mathfrak{m}) \to (\mathfrak{Y}, \mathfrak{n})$ to get the natural transformation

$$\big(\mathfrak{y} \circ \mathfrak{x}, (\beta * \mathfrak{x}) \circ \alpha\big) : (\mathfrak{X}, \mathfrak{m}) \longrightarrow (\mathfrak{Z}, \mathfrak{p}) \qquad \text{A2.21.1}$$

and we have

$$\big((\beta * \mathfrak{x}) \circ \alpha\big) \rtimes (\mathfrak{y} \circ \mathfrak{x}) = (\beta \rtimes \mathfrak{y}) \circ (\alpha \rtimes \mathfrak{x}) \qquad \text{A2.21.2.}$$

Moreover, if $(\mathfrak{x}', \alpha'): (\mathfrak{X}, \mathfrak{m}) \to (\mathfrak{Y}, \mathfrak{n})$ is another natural transformation, we call *supranatural transformation* from $(\mathfrak{x}, \alpha)$ to $(\mathfrak{x}', \alpha')$ any pair (θ, σ) formed by a 2-natural map and a *supranatural* map

$$\theta : \mathfrak{x} \longrightarrow \mathfrak{x}' \quad \text{and} \quad \sigma : (\mathfrak{n} * \theta) \circ \alpha \longrightarrow \alpha' \qquad \text{A2.21.3.}$$

For instance, it is quite clear that any natural map $\mu: \mathfrak{f} \to \mathfrak{f}'$, where $\mathfrak{f}$ and $\mathfrak{f}'$ are two functors from a category $\mathfrak{C}$ to a category $\mathfrak{C}'$, induces a supranatural map

$$\mathfrak{Fct}(\mathfrak{st}, \mu) : \mathfrak{Fct}(\mathfrak{st}, \mathfrak{f}) \longrightarrow \mathfrak{Fct}(\mathfrak{st}, \mathfrak{f}') \qquad \text{A2.21.4}$$

sending any $\mathfrak{C}$-*chain* $\mathfrak{q}: \Delta_n \to \mathfrak{C}$ to the natural map $\mu * \mathfrak{q}: \mathfrak{f} \circ \mathfrak{q} \to \mathfrak{f}' \circ \mathfrak{q}$; thus, according to the next proposition, we have a 2-natural map

$$\mathfrak{ch}(\mu) = \mathfrak{Fct}(\mathfrak{st}, \mu) \rtimes \mathrm{id}_{\mathfrak{i}\partial_{\Delta^\circ}} : \mathfrak{ch}(\mathfrak{f}) \longrightarrow \mathfrak{ch}(\mathfrak{f}') \qquad \text{A2.21.5.}$$

Proposition A2.22 *Let $(\mathfrak{x}, \alpha)$ and $(\mathfrak{x}', \alpha')$ be natural transformations from $(\mathfrak{X}, \mathfrak{m})$ to $(\mathfrak{Y}, \mathfrak{n})$. A supranatural transformation (θ, σ) from $(\mathfrak{x}, \alpha)$ to $(\mathfrak{x}', \alpha')$ determines a 2-natural map*

$$\sigma \rtimes \theta : \alpha \rtimes \mathfrak{x} \longrightarrow \alpha' \rtimes \mathfrak{x}' \qquad \text{A2.22.1}$$

mapping any $\mathfrak{m} \rtimes \mathfrak{X}$-object (A, X) on the $\mathfrak{n} \rtimes \mathfrak{Y}$-morphism

$$((\sigma_X)_A, \theta_X) : \big(\alpha_X(A), \mathfrak{x}(X)\big) \longrightarrow \big(\alpha'_X(A), \mathfrak{x}'(X)\big) \qquad \text{A2.22.2}$$

and any $\mathfrak{m} \rtimes \mathfrak{X}$-morphism $(f, \mathfrak{f}): (A, X) \to (B, Y)$ on the $\mathfrak{n} \rtimes \mathfrak{Y}$-natural isomorphism

$$\begin{aligned}\theta_{Y,X}(\mathfrak{f}) : \big((\sigma_Y)_B, \theta_Y\big) &\circ \big(\alpha_Y(f) \circ \mathfrak{x}_{Y,X}(\mathfrak{f})_A^{-1}, \mathfrak{x}(\mathfrak{f})\big) \\ &\cong \big(\alpha'_Y(f) \circ \mathfrak{x}'_{Y,X}(\mathfrak{f})_A^{-1}, \mathfrak{x}'(\mathfrak{f})\big) \circ \big((\sigma_X)_A, \theta_X\big)\end{aligned} \qquad \text{A2.22.3.}$$

Corollary A2.23 *With the same notation, let $(\alpha'', \mathfrak{x}'')$ natural transformations from $(\mathfrak{X}, \mathfrak{m})$ to $(\mathfrak{Y}, \mathfrak{n})$ and $(\theta', \sigma'): (\mathfrak{x}', \alpha') \to (\mathfrak{x}'', \alpha'')$ a supranatural transformation. Then, the pair $\Big(\theta' \circ \theta, \sigma' \circ \big((\mathfrak{n} * \theta') * \sigma\big)\Big)$ is a supranatural transformation from $(\mathfrak{x}, \alpha)$ to $(\mathfrak{x}'', \alpha'')$ and we have*

$$(\sigma' \rtimes \theta') \circ (\sigma \rtimes \theta) = \Big(\sigma' \circ \big((\mathfrak{n} * \theta') * \sigma\big)\Big) \rtimes (\theta' \circ \theta) \qquad \text{A2.23.1.}$$

Moreover, let $\mathfrak{Z}$ *be a 2-category,* $\mathfrak{p}:\mathfrak{Z}\to\mathfrak{CC}$ *a representation of* $\mathfrak{Z}$, $(\mathfrak{y},\beta)$ *and* $(\mathfrak{y}',\beta')$ *natural transformations from* $(\mathfrak{Y},\mathfrak{n})$ *to* $(\mathfrak{Z},\mathfrak{p})$, *and* (ω,τ) *a supranatural transformation from* $(\mathfrak{y},\beta)$ *to* $(\mathfrak{y}',\beta')$. *Then, we have*

$$\begin{aligned}(\beta\rtimes\mathfrak{y})*(\sigma\rtimes\theta) &= \Big(\big((\beta*\mathfrak{x}')*\sigma\big)\circ\big((\beta\diamond\theta)^{-1}*\alpha\big)\Big)\rtimes(\mathfrak{y}*\theta)\\ (\tau\rtimes\omega)*(\alpha\rtimes\mathfrak{x}) &= \big((\tau\diamond\mathfrak{x})*\alpha\big)\rtimes(\omega*\mathfrak{x})\end{aligned} \qquad \text{A2.23.2.}$$

A2.24 Although not needed in this book, let us mention the existence of *semidirect products* between representations, between natural transformations and between supranatural transformations. Let $\mathfrak{X}$ be a 2-category, $\mathfrak{m}:\mathfrak{X}\to\mathfrak{CC}$ a representation of $\mathfrak{X}$ and $\mathfrak{q}:\mathfrak{m}\rtimes\mathfrak{X}\to\mathfrak{CC}$ a representation of $\mathfrak{m}\rtimes\mathfrak{X}$; we obviously can consider the semidirect product $\mathfrak{q}\rtimes(\mathfrak{m}\rtimes\mathfrak{X})$ and the point is that we can define another representation of $\mathfrak{X}$

$$\mathfrak{q}\rtimes\mathfrak{m}:\mathfrak{X}\longrightarrow\mathfrak{CC} \qquad \text{A2.24.1}$$

— called *semidirect product* of $\mathfrak{m}$ by $\mathfrak{q}$ — in such a way that we have a natural isomorphism

$$(\mathfrak{q}\rtimes\mathfrak{m})\rtimes\mathfrak{X}\cong\mathfrak{q}\rtimes(\mathfrak{m}\rtimes\mathfrak{X}) \qquad \text{A2.24.2.}$$

A2.25 Similarly, let $\mathfrak{Y}$ be another 2-category, $\mathfrak{n}:\mathfrak{Y}\to\mathfrak{CC}$ a representation of $\mathfrak{Y}$ and $\mathfrak{r}:\mathfrak{n}\rtimes\mathfrak{Y}\to\mathfrak{CC}$ a representation of $\mathfrak{n}\rtimes\mathfrak{Y}$; we already know that a natural transformation $(\mathfrak{x},\alpha)$ from $(\mathfrak{X},\mathfrak{m})$ to $(\mathfrak{Y},\mathfrak{n})$ determines a 2-functor $\alpha\rtimes\mathfrak{x}$ from $\mathfrak{m}\rtimes\mathfrak{X}$ to $\mathfrak{n}\rtimes\mathfrak{Y}$ and then, a 2-natural map $\tau:\mathfrak{q}\to\mathfrak{r}\circ(\alpha\rtimes\mathfrak{x})$ builds a new natural transformation $(\alpha\rtimes\mathfrak{x},\tau)$ from $(\mathfrak{m}\rtimes\mathfrak{X},\mathfrak{q})$ to $(\mathfrak{n}\rtimes\mathfrak{Y},\mathfrak{r})$ and therefore determines a 2-functor

$$\tau\rtimes(\alpha\rtimes\mathfrak{x}):\mathfrak{q}\rtimes(\mathfrak{m}\rtimes\mathfrak{X})\longrightarrow\mathfrak{r}\rtimes(\mathfrak{n}\rtimes\mathfrak{Y}) \qquad \text{A2.25.1.}$$

Once again, the point is that we can define a 2-natural map

$$\tau\rtimes\alpha:\mathfrak{q}\rtimes\mathfrak{m}\longrightarrow(\mathfrak{r}\rtimes\mathfrak{n})\circ\mathfrak{x} \qquad \text{A2.25.2}$$

— called *semidirect product* of α by τ — defining a natural transformation $(\mathfrak{x},\tau\rtimes\alpha)$ from $(\mathfrak{X},\mathfrak{q}\rtimes\mathfrak{m})$ to $(\mathfrak{Y},\mathfrak{r}\rtimes\mathfrak{n})$ in such a way that the 2-functors $\tau\rtimes(\alpha\rtimes\mathfrak{x})$ and $(\tau\rtimes\alpha)\rtimes\mathfrak{x}$ coincide.

A2.26 Finally, with the same notation, if we have a supranatural transformation (θ,σ) from $(\mathfrak{x},\alpha)$ to another natural transformation $(\mathfrak{x}',\alpha')$ from $(\mathfrak{X},\mathfrak{m})$ to $(\mathfrak{Y},\mathfrak{n})$, we get a 2-natural map (cf. A2.22.1)

$$\sigma\rtimes\theta:\alpha\rtimes\mathfrak{x}\longrightarrow\alpha'\rtimes\mathfrak{x}' \qquad \text{A2.26.1;}$$

mutatis mutandis, a supranatural transformation $(\sigma\rtimes\theta,\omega)$ from $(\alpha\rtimes\mathfrak{x},\tau)$ to a second natural transformation $(\alpha'\rtimes\mathfrak{x}',\tau')$ from $(\mathfrak{m}\rtimes\mathfrak{X},\mathfrak{q})$ to $(\mathfrak{n}\rtimes\mathfrak{Y},\mathfrak{r})$ still determines a 2-natural map (cf. A2.20.1)

$$\omega\rtimes(\sigma\rtimes\theta):\tau\rtimes(\alpha\rtimes\mathfrak{x})\longrightarrow\tau'\rtimes(\alpha'\rtimes\mathfrak{x}') \qquad \text{A2.26.2.}$$

Once again, we can define a supranatural map

$$\omega \rtimes \sigma : \big((\mathfrak{r} \rtimes \mathfrak{n}) * \theta\big) \circ (\tau \rtimes \alpha) \longrightarrow \tau' \rtimes \alpha' \qquad \text{A2.26.3}$$

in such a way that, up to suitable identifications, we have

$$\omega \rtimes (\sigma \rtimes \theta) = (\omega \rtimes \sigma) \rtimes \theta \qquad \text{A2.26.4.}$$

A3 Cohomology of the small categories

A3.1 As usual, in the simplicial 2-category $\mathbf{\Delta}$ (cf. A1.7) we denote by $\delta_i^n : \Delta_n \to \Delta_{n+1}$ the injective order-preserving map which does not cover i in Δ_{n+1}, and by $\varepsilon_i^{n+1} : \Delta_{n+1} \to \Delta_n$ the surjective order-preserving map which covers i twice in Δ_n; it is well-known that all the nonidentity morphisms in $\mathbf{\Delta}$ are suitable compositions of these morphisms and that, for any $i, j \in \Delta_n$, we have

$$\delta_j^{n-1} \circ \varepsilon_i^n = \begin{cases} \varepsilon_{i+1}^{n+1} \circ \delta_j^n & \text{if } j \le i \\ \varepsilon_i^{n+1} \circ \delta_{j+1}^n & \text{if } i < j \end{cases} \qquad \text{A3.1.1.}$$
$$\varepsilon_i^{n+1} \circ \delta_i^n = \mathrm{id}_{\Delta_n} = \varepsilon_i^{n+1} \circ \delta_{i+1}^n$$

A3.2 Recall that, for any functor $\mathfrak{a} : \mathbf{\Delta} \to \mathfrak{Ab}$ from the simplicial 2-category to the category of Abelian groups $\mathfrak{Ab}$, the *graded homology group of* $\mathfrak{a}$

$$\mathbb{H}(\mathfrak{a}) = \bigoplus_{n \in \mathbb{N}} \mathbb{H}_n(\mathfrak{a}) \qquad \text{A3.2.1}$$

is the graded homology group of the *differential complex* formed by the abelian groups $\mathfrak{a}(\Delta_n)$ and the *differential maps*

$$d_n^{\mathfrak{a}} = \sum_{i=0}^{n+1} (-1)^i \, \mathfrak{a}(\delta_i^n) : \mathfrak{a}(\Delta_n) \longrightarrow \mathfrak{a}(\Delta_{n+1}) \qquad \text{A3.2.2}$$

where n runs over $\mathbb{N}$; explicitly, setting $d_{-1}^{\mathfrak{a}} : \{0\} \to \mathfrak{a}(\Delta_0)$, we have

$$\mathbb{H}_n(\mathfrak{a}) = \mathrm{Ker}(d_n^{\mathfrak{a}}) / \mathrm{Im}(d_{n-1}^{\mathfrak{a}}) \qquad \text{A3.2.3.}$$

A3.3 Note that if $\mathfrak{b} : \mathbf{\Delta} \to \mathfrak{Ab}$ is another functor and $\mu : \mathfrak{a} \to \mathfrak{b}$ a natural map, for any $n \in \mathbb{N}$ we get a group homomorphism

$$\mathbb{H}_n(\mu) : \mathbb{H}_n(\mathfrak{a}) \longrightarrow \mathbb{H}_n(\mathfrak{b}) \qquad \text{A3.3.1.}$$

and then, if $\mathfrak{c} : \mathbf{\Delta} \to \mathfrak{Ab}$ is a third functor and $\lambda : \mathfrak{b} \to \mathfrak{c}$ a natural map, we obviously have

$$\mathbb{H}_n(\lambda \circ \mu) = \mathbb{H}_n(\lambda) \circ \mathbb{H}_n(\mu) \qquad \text{A3.3.2.}$$

Moreover, it is well-known that whenever we have an exact sequence of such functors

$$\mathcal{E} : 0 \longrightarrow \mathfrak{a} \xrightarrow{\mu} \mathfrak{b} \xrightarrow{\lambda} \mathfrak{c} \longrightarrow 0 \qquad \text{A3.3.3,}$$

setting $\mathbb{H}_{-n}(\mathfrak{a}) = \mathbb{H}_{-n}(\mathfrak{b}) = \mathbb{H}_{-n}(\mathfrak{c}) = \{0\}$ for any $n > 0$, there is a 1-*graded* $\mathbb{Z}$-module homomorphism

$$\delta^{\mathcal{E}} : \bigoplus_{n\in\mathbb{Z}} \mathbb{H}_n(\mathfrak{c}) \longrightarrow \bigoplus_{n\in\mathbb{Z}} \mathbb{H}_n(\mathfrak{a}) \qquad \text{A3.3.4}$$

— called a *connecting homomorphism* — such that we have the following *exact triangle* (cf. §2 in [18, Ch. III] and Theorem 1.1 in [18, Ch. IV])

$$\begin{array}{ccc} \bigoplus_{n\in\mathbb{Z}} \mathbb{H}_n(\mathfrak{c}) & \xrightarrow{\delta^{\mathcal{E}}} & \bigoplus_{n\in\mathbb{Z}} \mathbb{H}_n(\mathfrak{a}) \\ {\scriptstyle\oplus_{n\in\mathbb{Z}}\mathbb{H}_n(\lambda)} \nwarrow & & \swarrow {\scriptstyle\oplus_{n\in\mathbb{Z}}\mathbb{H}_n(\mu)} \\ & \bigoplus_{n\in\mathbb{Z}} \mathbb{H}_n(\mathfrak{b}) & \end{array} \qquad \text{A3.3.5.}$$

In order to exploit this exactness, it is useful to employ the following lemma which is an easy consequence of equalities A3.1.1.

Lemma A3.4 *Let* $\mathfrak{a} : \mathbf{\Delta} \to \mathfrak{Ab}$ *be a functor. For any* $n \in \mathbb{N}$ *we have*

$$\mathrm{id}_{\mathfrak{a}(\Delta_{n+1})} = \mathfrak{a}(\varepsilon_0^{n+2}) \circ d_n^{\mathfrak{a}\circ\mathfrak{sh}} + d_{n-1}^{\mathfrak{a}\circ\mathfrak{sh}} \circ \mathfrak{a}(\varepsilon_0^{n+1}) \qquad \text{A3.4.1}$$

where we set $d_{-1}^{\mathfrak{a}\circ\mathfrak{sh}} = \mathfrak{a}(\delta_1^0)$. *In particular, we have*

$$\mathbb{H}_0(\mathfrak{a}\circ\mathfrak{sh}) = \mathrm{Im}\big(\mathfrak{a}(\delta_1^0)\big) \quad \textit{and} \quad \mathbb{H}_n(\mathfrak{a}\circ\mathfrak{sh}) = \{0\} \ \textit{for any}\ n \geq 1 \qquad \text{A3.4.2.}$$

A3.5 An immediate consequence of this lemma is:

A3.5.1 *Any functor* $\mathfrak{i} : \mathbf{\Delta} \to \mathfrak{Ab}$ *which is an injective object in the category of functors* $\mathfrak{Fct}(\mathbf{\Delta}, \mathfrak{Ab})$ *fulfills* $\mathbb{H}^n(\mathfrak{i}) = \{0\}$ *for any* $n \geq 1$.

Indeed, since $\mathfrak{i}(\varepsilon_0^{n+1}) \circ \mathfrak{i}(\delta_0^n) = \mathrm{id}_{\mathfrak{i}(\Delta_n)}$, the natural map (cf. A1.13)

$$\mathfrak{i} * \varsigma : \mathfrak{i} \longrightarrow \mathfrak{i}\circ\mathfrak{sh} \qquad \text{A3.5.2}$$

is injective, so that $\mathfrak{i}$ is a direct summand of $\mathfrak{i}\circ\mathfrak{sh}$ and it suffices to apply Lemma A3.4.

A3.6 From equality A3.2.3 it is easily checked that $\mathbb{H}_0(\mathfrak{a})$ coincides with the *inverse limit* $\varprojlim \mathfrak{a}$ and then, this remark and the exact triangle A3.3.5 shows that $\mathbb{H}_n$ is the *n-th right derived functor* (cf. §3 in [18, Ch. V]) of the functor

$$\varprojlim : \mathfrak{Fct}(\mathbf{\Delta}, \mathfrak{Ab}) \longrightarrow \mathfrak{Ab} \qquad \text{A3.6.1.}$$

On the other hand, denoting by $\mathfrak{t}$ the *trivial* functor from $\mathbf{\Delta}$ to $\mathfrak{Ab}$ mapping the $\mathbf{\Delta}$-objects on $\mathbb{Z}$ and the $\mathbf{\Delta}$-morphisms on $\mathrm{id}_{\mathbb{Z}}$, it is clear that

$$\varprojlim \mathfrak{a} \cong \mathfrak{Nat}(\mathfrak{t}, \mathfrak{a}) \qquad \text{A3.6.2}$$

and therefore, for any *projective resolution* of $\mathfrak{t}$

$$\ldots \to \mathfrak{p}_{n+1} \to \mathfrak{p}_n \to \ldots \to \mathfrak{p}_1 \to \mathfrak{t} \to 0 \qquad \text{A3.6.3,}$$

$\mathbb{H}(\mathfrak{a})$ still coincides with the graded homology group of the differential complex (cf. Proposition 1.1 in [18, Ch. VI])

$$\ldots \longleftarrow \mathfrak{Nat}(\mathfrak{p}_{n+1}, \mathfrak{a}) \longleftarrow \mathfrak{Nat}(\mathfrak{p}_n, \mathfrak{a}) \longleftarrow \ldots \longleftarrow \mathfrak{Nat}(\mathfrak{p}_1, \mathfrak{a}) \qquad \text{A3.6.4.}$$

A3.7 Let $\mathfrak{C}$ be a small category; in order to recall the definition of the cohomology groups of a *contravariant* functor from $\mathfrak{C}$ to $\mathfrak{Ab}$, we have to consider the following subcategory — called the *naive category of chains* of $\mathfrak{C}$ — of the category of chains $\mathfrak{ch}(\mathfrak{C})$ (cf. A2.2, A2.3, A2.5 and A2.7)

$$\mathfrak{ch}^{\mathrm{o}}(\mathfrak{C}) = \mathfrak{Fct}(\mathfrak{st}, \mathfrak{C})^{\circ} \rtimes \mathbf{\Delta}^{\circ} \qquad \text{A3.7.1}$$

with its structural functor (cf. A2.7.1)

$$\mathfrak{l}^{\mathrm{o}}_{\mathfrak{C}} = \mathfrak{p}_{\mathbf{\Delta}^{\circ}, \mathfrak{Fct}(\mathfrak{st},\mathfrak{C})^{\circ}} : \mathfrak{ch}^{\mathrm{o}}(\mathfrak{C}) \longrightarrow \mathbf{\Delta}^{\circ} \qquad \text{A3.7.2,}$$

and the *evaluation functor*

$$\mathfrak{v}_{\mathfrak{C}} : \mathfrak{ch}(\mathfrak{C}) \longrightarrow \mathfrak{C} \qquad \text{A3.7.3}$$

mapping any $\mathfrak{ch}(\mathfrak{C})$-object $(\mathfrak{q}, \Delta_n)$ on the $\mathfrak{C}$-object $\mathfrak{q}(0)$, and any $\mathfrak{ch}(\mathfrak{C})$-morphism $(\mu, \delta) : (\mathfrak{r}, \Delta_m) \to (\mathfrak{q}, \Delta_n)$ on the composition of $\mathfrak{C}$-morphisms

$$\mathfrak{r}(0) \xrightarrow{\mathfrak{r}(0 \bullet \delta(0))} \mathfrak{r}\big(\delta(0)\big) \xrightarrow{\mu_0} \mathfrak{q}(0) \qquad \text{A3.7.4;}$$

note that this functor factorizes throughout $\overline{\mathfrak{ch}(\mathfrak{C})} = \pi_0\big(\mathfrak{ch}(\mathfrak{C})\big)$ (cf. A1.8). We denote by $\mathfrak{v}^*_{\mathfrak{C}}$ and $\mathfrak{v}^{\mathrm{o}}_{\mathfrak{C}}$ the respective restrictions of $\mathfrak{v}_{\mathfrak{C}}$ to $\mathfrak{ch}^*(\mathfrak{C})$ and $\mathfrak{ch}^{\mathrm{o}}(\mathfrak{C})$.

A3.8 But, as explained in the Introduction (cf. I33), in this book we have to consider slightly more general cohomology groups, namely the cohomology groups of *contravariant* functors from $\mathfrak{ch}^*(\mathfrak{C})$ to $\mathfrak{Ab}$. More generally, any *contravariant* functor $\mathfrak{a}^{\mathrm{o}} : \mathfrak{ch}^{\mathrm{o}}(\mathfrak{C}) \to \mathfrak{Ab}$ determines a functor from $\mathbf{\Delta}$ to $\mathfrak{Ab}$ *via* the *direct image* throughout $\mathfrak{l}^{\mathrm{o}}_{\mathfrak{C}}$ (cf. 1.6 and A2.13.3) and for any $n \in \mathbb{N}$ we simply define

$$\mathbb{H}^n_{\mathrm{o}}(\mathfrak{C}, \mathfrak{a}^{\mathrm{o}}) = \mathbb{H}_n\big((\mathfrak{l}^{\mathrm{o}}_{\mathfrak{C}})_*(\mathfrak{a}^{\mathrm{o}})\big) \qquad \text{A3.8.1}$$

— called the *naive n-cohomology group of* $\mathfrak{C}$ *over* $\mathfrak{a}^{\mathrm{o}}$. Since $\mathbb{H}_n$ and $(\mathfrak{l}^{\mathrm{o}}_{\mathfrak{C}})_*$ are true functors

$$\mathfrak{Fct}\big(\mathfrak{ch}^{\mathrm{o}}(\mathfrak{C})^{\circ}, \mathbf{\Delta}\big) \xrightarrow{(\mathfrak{l}^{\mathrm{o}}_{\mathfrak{C}})_*} \mathfrak{Fct}(\mathbf{\Delta}, \mathfrak{Ab}) \xrightarrow{\mathbb{H}_n} \mathfrak{Ab} \qquad \text{A3.8.2,}$$

we actually get a functor

$$\mathbb{H}^n_{\mathrm{o}}(\mathfrak{C}, \bullet) : \mathfrak{Fct}\big(\mathfrak{ch}^{\mathrm{o}}(\mathfrak{C})^{\circ}, \mathfrak{Ab}\big) \longrightarrow \mathfrak{Ab} \qquad \text{A3.8.3.}$$

A3.9 Usually, we only consider *contravariant* functors $\mathfrak{a}^* : \mathfrak{ch}^*(\mathfrak{C}) \to \mathfrak{Ab}$ which *factorize* throughout $\overline{\mathfrak{ch}^*(\mathfrak{C})}$ (cf. A1.8) and then, denoting by $\mathfrak{a}^\circ$ the restriction of $\mathfrak{a}^*$ to $\mathfrak{ch}^\circ(\mathfrak{C})$, it is not difficult to check that we have a natural isomorphism

$$\lim_{\longleftarrow} \mathfrak{a}^* \cong \mathbb{H}^0_\circ(\mathfrak{C}, \mathfrak{a}^\circ) \qquad \text{A3.9.1.}$$

Moreover, if we start from an ordinary *contravariant* functor $\mathfrak{a} : \mathfrak{C} \to \mathfrak{Ab}$, in order to apply our previous definition it suffices to consider the composed functor $\mathfrak{a} \circ \mathfrak{v}^\circ_{\mathfrak{C}} : \mathfrak{ch}^\circ(\mathfrak{C}) \to \mathfrak{C} \to \mathfrak{Ab}$ and, for any $n \in \mathbb{N}$, we simply set

$$\mathbb{H}^n(\mathfrak{C}, \mathfrak{a}) = \mathbb{H}^n_\circ(\mathfrak{C}, \mathfrak{a} \circ \mathfrak{v}^\circ_{\mathfrak{C}}) = \mathbb{H}_n\big((\mathfrak{l}^\circ_{\mathfrak{C}})_*(\mathfrak{a} \circ \mathfrak{v}^\circ_{\mathfrak{C}})\big) \qquad \text{A3.9.2}$$

— called the *n-cohomology group of* $\mathfrak{C}$ *over* $\mathfrak{a}$. Once again, we get a functor

$$\mathbb{H}^n(\mathfrak{C}, \bullet) : \mathfrak{Fct}(\mathfrak{C}^\circ, \mathfrak{Ab}) \longrightarrow \mathfrak{Ab} \qquad \text{A3.9.3.}$$

A3.10 More generally, the correspondence mapping any small category $\mathfrak{C}$ on the small category $\mathfrak{Fct}\big(\mathfrak{ch}^\circ(\mathfrak{C})^\circ, \mathfrak{Ab}\big)$, mapping any functor $\mathfrak{f} : \mathfrak{C} \to \mathfrak{C}'$ between two small categories on the obvious functor

$$\mathfrak{Fct}\big(\mathfrak{ch}^\circ(\mathfrak{f}), \mathfrak{Ab}\big) : \mathfrak{Fct}\big(\mathfrak{ch}^\circ(\mathfrak{C}')^\circ, \mathfrak{Ab}\big) \longrightarrow \mathfrak{Fct}\big(\mathfrak{ch}^\circ(\mathfrak{C})^\circ, \mathfrak{Ab}\big) \qquad \text{A3.10.1}$$

which sends any *contravariant* functor $\mathfrak{a}^\circ : \mathfrak{ch}^\circ(\mathfrak{C}') \to \mathfrak{Ab}$ to $\mathfrak{a}^\circ \circ \mathfrak{ch}^\circ(\mathfrak{f})$, and mapping any natural map $\mu : \mathfrak{f} \to \mathfrak{f}'$ between two such functors on the natural map sending $\mathfrak{a}^\circ$ to $\mathfrak{a}^\circ * \mathfrak{ch}^\circ(\mu)$, actually defines a *representation* of $(\mathfrak{CC})^\circ$

$$\mathfrak{Fct}\big(\mathfrak{ch}^\circ(\bullet), \mathfrak{Ab}^\circ\big) : (\mathfrak{CC})^\circ \longrightarrow \mathfrak{CC} \qquad \text{A3.10.2}$$

and then it is easily checked that the *naive n-cohomology groups* define a functor

$$\mathbb{H}^n_\circ : \mathfrak{Fct}\big(\mathfrak{ch}^\circ(\bullet), \mathfrak{Ab}^\circ\big) \rtimes (\mathfrak{CC})^\circ \longrightarrow \mathfrak{Ab} \qquad \text{A3.10.3;}$$

mutatis mutandis, the ordinary *n-cohomology groups* define a functor

$$\mathbb{H}^n : \mathfrak{Fct}(\bullet, \mathfrak{Ab}^\circ) \rtimes (\mathfrak{CC})^\circ \longrightarrow \mathfrak{Ab} \qquad \text{A3.10.4.}$$

A3.11 Explicitly, the direct image $(\mathfrak{l}^\circ_{\mathfrak{C}})_*(\mathfrak{a}^\circ)$ (cf. 1.6) maps Δ_n on

$$\big((\mathfrak{l}^\circ_{\mathfrak{C}})_*(\mathfrak{a}^\circ)\big)(\Delta_n) = \prod_{\mathfrak{q} \in \mathfrak{Fct}(\Delta_n, \mathfrak{C})} \mathfrak{a}^\circ(\mathfrak{q}, \Delta_n) \qquad \text{A3.11.1}$$

and, according to definition A3.2.2, the corresponding *differential map* sends an element $a = (a_\mathfrak{q})_{\mathfrak{q} \in \mathfrak{Fct}(\Delta_n, \mathfrak{C})}$ of $\big((\mathfrak{l}^\circ_{\mathfrak{C}})_*(\mathfrak{a}^\circ)\big)(\Delta_n)$ on

$$d_n^{(\mathfrak{l}^\circ_{\mathfrak{C}})_*(\mathfrak{a}^\circ)}(a) = \Big(\sum_{i=0}^{n+1} (-1)^i \big(\mathfrak{a}^\circ(\mathrm{id}_{\mathfrak{r} \circ \delta^n_i}, \delta^n_i)\big)(a_{\mathfrak{r} \circ \delta^n_i})\Big)_{\mathfrak{r} \in \mathfrak{Fct}(\Delta_{n+1}, \mathfrak{C})} \qquad \text{A3.11.2.}$$

In particular, $(\mathfrak{l}^{\circ}_{\mathfrak{C}})_*$ maps an exact sequence of *contravariant* functors from $\mathfrak{ch}^{\circ}(\mathfrak{C})$ to $\mathfrak{Ab}$

$$\mathcal{E}^{\circ} : 0 \longrightarrow \mathfrak{a}^{\circ} \xrightarrow{\mu^{\circ}} \mathfrak{b}^{\circ} \xrightarrow{\lambda^{\circ}} \mathfrak{c}^{\circ} \longrightarrow 0 \qquad \text{A3.11.3}$$

on an exact sequence of functors $(\mathfrak{l}^{\circ}_{\mathfrak{C}})_*(\mathcal{E}^{\circ})$ from $\mathbf{\Delta}$ to $\mathfrak{Ab}$ and therefore, setting again $\mathbb{H}^{-n}_{\circ}(\mathfrak{C}, \mathfrak{a}^{\circ}) = \mathbb{H}^{-n}_{\circ}(\mathfrak{C}, \mathfrak{b}^{\circ}) = \mathbb{H}^{-n}_{\circ}(\mathfrak{C}, \mathfrak{c}^{\circ}) = \{0\}$ for any $n > 0$, we have the *exact triangle* (cf. A3.3.5)

$$\begin{array}{ccc} \bigoplus_{n\in\mathbb{Z}} \mathbb{H}^{n}_{\circ}(\mathfrak{C}, \mathfrak{c}^{\circ}) & \xrightarrow{\delta^{(\mathfrak{l}^{\circ}_{\mathfrak{C}})_*(\mathcal{E}^{\circ})}} & \bigoplus_{n\in\mathbb{Z}} \mathbb{H}^{n}_{\circ}(\mathfrak{C}, \mathfrak{a}^{\circ}) \\ \nwarrow & & \swarrow \\ & \bigoplus_{n\in\mathbb{Z}} \mathbb{H}^{n}_{\circ}(\mathfrak{C}, \mathfrak{b}^{\circ}) & \end{array} \qquad \text{A3.11.4.}$$

A3.12 As above, this exact triangle shows that $\mathbb{H}^{n}(\mathfrak{C}, \bullet)$ is the *n-th right derived functor* of the functor

$$\lim_{\longleftarrow} : \mathfrak{Fct}(\mathfrak{C}^{\circ}, \mathfrak{Ab}) \longrightarrow \mathfrak{Ab} \qquad \text{A3.12.1,}$$

provided we prove a statement analogous to A3.5.1. This depends on the *covering functor* $\mathfrak{c}_{\mathfrak{C}}$ and the *covering natural map* $\kappa_{\mathfrak{C}}$ introduced in A2.18; for any *contravariant* functor $\mathfrak{a} : \mathfrak{C} \to \mathfrak{Ab}$, we have to consider the new *contravariant* functor

$$(\mathfrak{p}_{\mathfrak{C}})_*(\mathfrak{a} \circ \mathfrak{c}_{\mathfrak{C}}) : \mathfrak{C} \longrightarrow \mathfrak{Ab} \qquad \text{A3.12.2}$$

and the *injective* natural map

$$(\mathfrak{p}_{\mathfrak{C}})_*(\mathfrak{a} * \kappa_{\mathfrak{C}}) \circ \iota_{\mathfrak{a}} : \mathfrak{a} \longrightarrow (\mathfrak{p}_{\mathfrak{C}})_*(\mathfrak{a} \circ \mathfrak{p}_{\mathfrak{C}}) \longrightarrow (\mathfrak{p}_{\mathfrak{C}})_*(\mathfrak{a} \circ \mathfrak{c}_{\mathfrak{C}}) \qquad \text{A3.12.3}$$

where $\iota_{\mathfrak{a}}$ denotes the image of $\mathrm{id}_{\mathfrak{a}\circ\mathfrak{p}_{\mathfrak{C}}}$ by the *adjoinness* isomorphism (cf. 1.6)

$$\mathfrak{Nat}(\mathfrak{a} \circ \mathfrak{p}_{\mathfrak{C}}, \mathfrak{a} \circ \mathfrak{p}_{\mathfrak{C}}) \cong \mathfrak{Nat}\big(\mathfrak{a}, (\mathfrak{p}_{\mathfrak{C}})_*(\mathfrak{a} \circ \mathfrak{p}_{\mathfrak{C}})\big) \qquad \text{A3.12.4.}$$

Indeed, with them we can develop a similar argument since we have the following result (cf. Lemma A3.4 and definition A3.8.1).

Proposition A3.13 *With the notation above, we have a natural isomorphism*

$$(\mathfrak{l}^{\circ}_{\mathfrak{C}})_*\big((\mathfrak{p}_{\mathfrak{C}})_*(\mathfrak{a} \circ \mathfrak{c}_{\mathfrak{C}}) \circ \mathfrak{v}^{\circ}_{\mathfrak{C}}\big) \cong (\mathfrak{l}^{\circ}_{\mathfrak{C}})_*(\mathfrak{a} \circ \mathfrak{v}^{\circ}_{\mathfrak{C}}) \circ \mathfrak{sh} \qquad \text{A3.13.1.}$$

In particular, we have $\mathbb{H}^{n}\big(\mathfrak{C}, (\mathfrak{p}_{\mathfrak{C}})_*(\mathfrak{a} \circ \mathfrak{c}_{\mathfrak{C}})\big) = \{0\}$ *for any* $n \geq 1$.

A3.14 As above, an immediate consequence of this proposition is:

A3.14.1 *Any contravariant functor* $\mathfrak{i}:\mathfrak{C}\to\mathfrak{Ab}$ *which is an injective object in the category of functors* $\mathfrak{Fct}(\mathfrak{C}^\circ,\mathfrak{Ab})$ *fulfills* $\mathbb{H}^n(\mathfrak{C},\mathfrak{i})=\{0\}$ *for any* $n\geq 1$.
Moreover, it is easy to see that any functor $\mathfrak{f}:\mathfrak{B}\to\mathfrak{C}$ such that the direct image $\mathfrak{f}_*$ (cf. 1.6) preserves monomorphisms, composed with an injective object in the category $\mathfrak{Fct}(\mathfrak{C}^\circ,\mathfrak{Ab})$, becomes an injective object in $\mathfrak{Fct}(\mathfrak{B}^\circ,\mathfrak{Ab})$; hence, from the *Isomorphism Criterion* (cf. Proposition 4.4 in [18, Ch. V]) we get the following result.

Proposition A3.15 *For any pair of small categories* $\mathfrak{B}$ *and* $\mathfrak{C}$, *any contravariant functor* $\mathfrak{a}:\mathfrak{C}\to\mathfrak{Ab}$ *and any functor* $\mathfrak{f}:\mathfrak{B}\to\mathfrak{C}$, *if* $\mathfrak{f}_*$ *preserves monomorphisms and* $\mathfrak{f}$ *induces a group isomorphism* $\mathbb{H}^0(\mathfrak{C},\mathfrak{a})\cong\mathbb{H}^0(\mathfrak{B},\mathfrak{a}\circ\mathfrak{f})$ *then, for any* $n\geq 1$, $\mathfrak{f}$ *induces a group isomorphism*

$$\mathbb{H}^n(\mathfrak{C},\mathfrak{a})\cong\mathbb{H}^n(\mathfrak{B},\mathfrak{a}\circ\mathfrak{f}) \qquad \text{A3.15.1.}$$

Corollary A3.16 *For any small category* $\mathfrak{C}$, *any contravariant functor* $\mathfrak{a}:\mathfrak{C}\to\mathfrak{Ab}$ *and any* $n\in\mathbb{N}$, *the evaluation functor* $\mathfrak{v}_{\mathfrak{C}}:\mathfrak{ch}(\mathfrak{C})\to\mathfrak{C}$ *induces a group isomorphism*

$$\mathbb{H}^n(\mathfrak{C},\mathfrak{a})\cong\mathbb{H}^n\big(\overline{\mathfrak{ch}(\mathfrak{C})},\mathfrak{a}\circ\bar{\mathfrak{v}}_{\mathfrak{C}}\big) \qquad \text{A3.16.1.}$$

A3.17 We also have to consider the *stable cohomology groups of* $\mathfrak{C}$ or, more generally, the $\mathfrak{D}$*-stable cohomology groups of* $\mathfrak{C}$ for a subcategory $\mathfrak{D}$ of $\mathfrak{C}$, with the same set of objects, over a *contravariant* functor

$$\mathfrak{a}^*:\mathfrak{ch}^*(\mathfrak{C})\longrightarrow\mathfrak{Ab} \qquad \text{A3.17.1.}$$

In order to define it, we consider the *weak structure* of $\mathfrak{Fct}(\mathfrak{st},\mathfrak{C})$ (cf. A2.3 and A2.5)

$$\mathfrak{Fct}(\mathfrak{st},\mathfrak{C})^*_{\mathfrak{D}}:\mathbf{\Delta}^\circ\longrightarrow\mathfrak{CC} \qquad \text{A3.17.2}$$

mapping Δ_n on the subcategory $\mathfrak{Fct}(\Delta_n,\mathfrak{C})^*_{\mathfrak{D}}$ of the category of functors $\mathfrak{Fct}(\mathfrak{st},\mathfrak{C})$ formed by all the functors and, for any pair of them $\mathfrak{q}:\Delta_n\to\mathfrak{C}$ and $\mathfrak{q}':\Delta_n\to\mathfrak{C}$, by the natural maps $\nu:\mathfrak{q}\to\mathfrak{q}'$ such that $\tilde{\nu}_i:\mathfrak{q}(i)\to\mathfrak{q}'(i)$ is a $\mathfrak{D}$-isomorphism; we consider the *semidirect product* (cf. A2.7)

$$\mathfrak{ch}^*_{\mathfrak{D}}(\mathfrak{C})=\mathfrak{Fct}(\mathfrak{st},\mathfrak{C})^*_{\mathfrak{D}}\rtimes\mathbf{\Delta}^\circ \qquad \text{A3.17.3}$$

which is clearly contained in the *proper category of* $\mathfrak{C}$*-chains* $\mathfrak{ch}^*(\mathfrak{C})$ (cf. A2.8) and contains the *naive category of* $\mathfrak{C}$*-chains* $\mathfrak{ch}^\circ(\mathfrak{C})$ (cf. A3.7).

A3.18 Then, denoting by $\mathfrak{l}^{\mathfrak{D}}_{\mathfrak{C}}:\mathfrak{ch}^*_{\mathfrak{D}}(\mathfrak{C})\to\mathbf{\Delta}^\circ$ the restriction to $\mathfrak{ch}^*_{\mathfrak{D}}(\mathfrak{C})$ of the structural functor $\mathfrak{p}_{\mathbf{\Delta}^\circ,\mathfrak{Fct}(\mathfrak{st},\mathfrak{C})}$ (cf. A2.7.1) and by $\mathfrak{a}^{\mathfrak{D}}$ the restriction of $\mathfrak{a}^*$ to $\mathfrak{ch}^*_{\mathfrak{D}}(\mathfrak{C})$, for any $n\in\mathbb{N}$ we also consider

$$\mathbb{H}^n_{\mathfrak{D}}(\mathfrak{C},\mathfrak{a}^*)=\mathbb{H}_n\big((\mathfrak{l}^{\mathfrak{D}}_{\mathfrak{C}})_*(\mathfrak{a}^{\mathfrak{D}})\big) \qquad \text{A3.18.1}$$

— called the $\mathfrak{D}$*-stable* n*-cohomology group of* $\mathfrak{C}$ *over* $\mathfrak{a}^*$; if $\mathfrak{D} = \mathfrak{C}$, we write $\mathfrak{l}^*_{\mathfrak{C}}$ and $\mathbb{H}^n_*(\mathfrak{C}, \mathfrak{a}^*)$ instead of $\mathfrak{l}^{\mathfrak{D}}_{\mathfrak{C}}$ and $\mathbb{H}^n_{\mathfrak{D}}(\mathfrak{C}, \mathfrak{a}^*)$ respectively, and call *stable* n*-cohomology group of* $\mathfrak{C}$ *over* $\mathfrak{a}^*$ this group $\mathbb{H}^n_*(\mathfrak{C}, \mathfrak{a}^*)$. More explicitly, the direct image $(\mathfrak{l}^{\mathfrak{D}}_{\mathfrak{C}})_*(\mathfrak{a}^{\mathfrak{D}})$ (cf. 1.6) maps Δ_n on the subgroup

$$\big((\mathfrak{l}^{\mathfrak{D}}_{\mathfrak{C}})_*(\mathfrak{a}^{\mathfrak{D}})\big)(\Delta_n) \subset \prod_{\mathfrak{q} \in \mathfrak{Fct}(\Delta_n, \mathfrak{C})} \mathfrak{a}^*(\mathfrak{q}, \Delta_n) \qquad \text{A3.18.2}$$

formed by the $\mathfrak{D}$*-stable elements,* namely the elements $a = (a_{\mathfrak{q}})_{\mathfrak{q} \in \mathfrak{Fct}(\Delta_n, \mathfrak{C})}$ fulfilling $a_{\mathfrak{q}'} = \big(\mathfrak{a}^*(\nu, \mathrm{id}_{\Delta_n})\big)(a_{\mathfrak{q}})$ for any $\mathfrak{ch}(\mathfrak{D})$-isomorphism $(\nu, \mathrm{id}_{\Delta_n})$ between two $\mathfrak{ch}(\mathfrak{C})$-objects $(\mathfrak{q}, \Delta_n)$ and $(\mathfrak{q}', \Delta_n)$. Since the direct image $(\mathfrak{l}^{\mathfrak{D}}_{\mathfrak{C}})_*$ (cf. 1.6) need not preserve the surjectivity, these groups need not provide any (exact) triangle (cf. A3.11.4).

A4 Basic results on the cohomology of small categories

A4.1 Let $\mathfrak{C}$ be a small category; here we collect some basic results on the *naive* and the *stable* cohomology of $\mathfrak{C}$ over a *contravariant* functor

$$\mathfrak{a}^* : \mathfrak{ch}^*(\mathfrak{C}) \longrightarrow \mathfrak{Ab} \qquad \text{A4.1.1}$$

and we denote by $\mathfrak{a}^{\circ}$ the restriction of $\mathfrak{a}^*$ to $\mathfrak{ch}^{\circ}(\mathfrak{C})$. Most of these results depend on the following lemma and proposition†.

Lemma A4.2 *Let* $\mathfrak{q}$ *and* $\mathfrak{r}$ *be functors from* Δ_n *to* $\mathfrak{C}$ *and* $\mu : \mathfrak{q} \to \mathfrak{r}$ *a natural map. For any* $i \in \Delta_n$, *denote by* $\mathfrak{h}^n_i(\mu) : \Delta_{n+1} \to \mathfrak{C}$ *the functor which coincides with* $\mathfrak{q}$ *over* Δ_i, *coincides with* $\mathfrak{r} \circ \varepsilon^{n+1}_i$ *over* $\Delta_{n+1} - \Delta_i$, *and maps* $i \bullet (i+1)$ *on* $\mu_i : \mathfrak{q}(i) \to \mathfrak{r}(i)$, *and by*

$$\kappa^n_i(\mu) : \mathfrak{q} \circ \varepsilon^{n+1}_i \longrightarrow \mathfrak{h}^n_i(\mu) \qquad \text{A4.2.1}$$

the natural map sending $j \le i$ *to* $\mathrm{id}_{\mathfrak{q}(j)}$, *and* $i < j \le n+1$ *to* μ_{j-1}. *Then, for any* $i \in \Delta_n$ *and any* $j \in \Delta_{n-1}$, *we have*

$$\begin{aligned}
\mathfrak{h}^n_{j+1}(\mu) \circ \delta^n_{j+1} &= \mathfrak{h}^n_j(\mu) \circ \delta^n_{j+1} \\
\kappa^n_{j+1}(\mu) * \delta^n_{j+1} &= \kappa^n_j(\mu) * \delta^n_{j+1} \\
\mathfrak{h}^{n-1}_j(\mu * \delta^{n-1}_i) &= \begin{cases} \mathfrak{h}^n_{j+1}(\mu) \circ \delta^n_i & \text{if } i \le j \\ \mathfrak{h}^n_j(\mu) \circ \delta^n_{i+1} & \text{if } j < i\,. \end{cases} \\
\kappa^{n-1}_j(\mu * \delta^{n-1}_i) &= \begin{cases} \kappa^n_{j+1}(\mu) * \delta^n_i & \text{if } i \le j \\ \kappa^n_j(\mu) * \delta^n_{i+1} & \text{if } j < i \end{cases}
\end{aligned} \qquad \text{A4.2.2.}$$

† We learned them from Alberto Arabia.

Proof: Indeed, $\mathfrak{h}^n_{j+1}(\mu)\circ\delta^n_{j+1}$ and $\mathfrak{h}^n_j(\mu)\circ\delta^n_{j+1}$ coincide with $\mathfrak{q}'$ over Δ_i and with $\mathfrak{q}$ over $\Delta_n-\Delta_i$, and both map $j\bullet(j{+}1)$ on

$$\mu_{j+1}\circ\mathfrak{q}'(j\bullet(j{+}1))=\mathfrak{q}(j\bullet(j{+}1))\circ\mu_i \qquad \text{A4.2.3;}$$

then, $\kappa^n_{j+1}(\mu)*\delta^n_{j+1}$ and $\kappa^n_j(\mu)*\delta^n_{j+1}$ coincide with the natural map from $\mathfrak{q}$ to the functor $\mathfrak{h}^n_j(\mu)\circ\delta^n_{j+1}$, sending $j\le i$ to $\mathrm{id}_{\mathfrak{q}(j)}$, and $i<j\le n$ to μ_j . Moreover, for any $\ell\le j$ we have

$$\big(\mathfrak{h}^{n-1}_j(\mu*\delta^{n-1}_i)\big)(\ell)=(\mathfrak{q}'\circ\delta^{n-1}_i)(\ell) \qquad \text{A4.2.4}$$

and therefore we get

$$\big(\mathfrak{h}^{n-1}_j(\mu*\delta^{n-1}_i)\big)(\ell)=\begin{cases}\mathfrak{q}'(\ell{+}1)=\big(\mathfrak{h}^n_{j+1}(\mu)\circ\delta^n_i\big)(\ell) & \text{if } i\le\ell\\ \mathfrak{q}'(\ell)=\big(\mathfrak{h}^n_{j+1}(\mu)\circ\delta^n_i\big)(\ell) & \text{if } \ell<i\le j\\ \mathfrak{q}'(\ell)=\big(\mathfrak{h}^n_j(\mu)\circ\delta^n_{i+1}\big)(\ell) & \text{if } j<i\end{cases} \qquad \text{A4.2.5.}$$

On the contrary, if $j<\ell\le n$ then we have

$$\big(\mathfrak{h}^{n-1}_j(\mu*\delta^{n-1}_i)\big)(\ell)=(\mathfrak{q}\circ\delta^{n-1}_i)(\ell{-}1) \qquad \text{A4.2.6}$$

and therefore we get

$$\big(\mathfrak{h}^{n-1}_j(\mu*\delta^{n-1}_i)\big)(\ell)=\begin{cases}\mathfrak{q}(\ell)=\big(\mathfrak{h}^n_{j+1}(\mu)\circ\delta^n_i\big)(\ell) & \text{if } i\le j\\ \mathfrak{q}(\ell)=\big(\mathfrak{h}^n_j(\mu)\circ\delta^n_{i+1}\big)(\ell) & \text{if } j<i<\ell\\ \mathfrak{q}(\ell{-}1)=\big(\mathfrak{h}^n_j(\mu)\circ\mathfrak{d}^n_{i+1}\big)(\ell) & \text{if } \ell\le i\end{cases} \qquad \text{A4.2.7.}$$

More precisely, it is easily checked that the announced equalities still hold over the Δ_i- and the "$\Delta_n-\Delta_i$"-morphisms; finally, we have

$$\begin{aligned}\big(\mathfrak{h}^{n-1}_j(\mu*\delta^{n-1}_i)\big)(j\bullet j{+}1)&=\mu_{\delta^{n-1}_i(j)}\\ &=\begin{cases}\big(\mathfrak{h}^n_{j+1}(\mu)\circ\delta^n_i\big)(j\bullet j{+}1) & \text{if } i\le j\\ \big(\mathfrak{h}^n_j(\mu)\circ\delta^n_{i+1}\big)(j\bullet j{+}1) & \text{if } j<i\end{cases}\end{aligned} \qquad \text{A4.2.8.}$$

Mutatis mutandis, for any $\ell\le j$ we have

$$\big(\kappa^{n-1}_j(\mu*\delta^{n-1}_i)\big)(\ell)=\mathrm{id}_{(\mathfrak{q}\circ\delta^{n-1}_i)(\ell)} \qquad \text{A4.2.9}$$

and therefore we get

$$\big(\kappa^{n-1}_j(\mu*\delta^{n-1}_i)\big)(\ell)=\begin{cases}\mathrm{id}_{\mathfrak{q}(\ell+1)}=\big(\kappa^n_{j+1}(\mu)*\delta^n_i\big)(\ell) & \text{if } i\le\ell\\ \mathrm{id}_{\mathfrak{q}(\ell)}=\big(\kappa^n_{j+1}(\mu)*\delta^n_i\big)(\ell) & \text{if } \ell<i\le j\\ \mathrm{id}_{\mathfrak{q}(\ell)}=\big(\kappa^n_j(\mathfrak{q})*\delta^n_{i+1}\big)(\ell) & \text{if } j<i\end{cases} \qquad \text{A4.2.10.}$$

On the contrary, if $j < \ell \le n$ then we have

$$\big(\kappa_j^{n-1}(\mu * \delta_i^{n-1})\big)(\ell) = \mu_{\delta_j^{n-1}(\ell-1)} \qquad \text{A4.2.11}$$

and therefore we get

$$\big(\kappa_j^{n-1}(\mu * \delta_i^{n-1})\big)(\ell) = \begin{cases} \mu_\ell = \big(\kappa_{j+1}^n(\mu) * \delta_i^n\big)(\ell) & \text{if } i \le j \\ \mu_\ell = \big(\kappa_j^n(\mu) * \delta_{i+1}^n\big)(\ell) & \text{if } j < i < \ell \\ \mu_{\ell-1} = \big(\kappa_j^n(\mu) * \delta_{i+1}^n\big)(\ell) & \text{if } \ell \le i \end{cases} \qquad \text{A4.2.12.}$$

We are done.

A4.3 Let $\mathfrak{B}$ be a second small category and $\mathfrak{f}:\mathfrak{B} \to \mathfrak{C}$ a functor, so that we have a *contravariant* functor

$$\mathfrak{a}^* \circ \mathfrak{ch}^*(\mathfrak{f}) : \mathfrak{ch}^*(\mathfrak{B}) \longrightarrow \mathfrak{Ab} \qquad \text{A4.3.1,}$$

where $\mathfrak{ch}^*(\mathfrak{f})$ denotes the functor $\mathfrak{ch}^*(\mathfrak{B}) \to \mathfrak{ch}^*(\mathfrak{C})$ obtained by the restriction of $\mathfrak{ch}(\mathfrak{f})$ (cf. A2.19.6); recall that we set (cf. A3.7.2)

$$\mathfrak{l}_{\mathfrak{C}}^{\mathrm{o}} = \mathfrak{p}_{\Delta^\circ,\mathfrak{Fct}(\mathfrak{st},\mathfrak{C})^\circ} \quad \text{and} \quad \mathfrak{l}_{\mathfrak{B}}^{\mathrm{o}} = \mathfrak{p}_{\Delta^\circ,\mathfrak{Fct}(\mathfrak{st},\mathfrak{B})^\circ} \qquad \text{A4.3.2}$$

and, denoting by $\mathfrak{ch}^{\mathrm{o}}(\mathfrak{f})$ the corresponding restriction of $\mathfrak{ch}(\mathfrak{f})$, it is clear that we have $\mathfrak{l}_{\mathfrak{B}}^{\mathrm{o}} = \mathfrak{l}_{\mathfrak{C}}^{\mathrm{o}} \circ \mathfrak{ch}^{\mathrm{o}}(\mathfrak{f})$. Then, considering the natural map

$$(\mathfrak{l}_{\mathfrak{C}}^{\mathrm{o}})_*(\mathfrak{a}^{\mathrm{o}}) \circ \mathfrak{l}_{\mathfrak{C}}^{\mathrm{o}} \longrightarrow \mathfrak{a}^{\mathrm{o}} \qquad \text{A4.3.3}$$

determined by $\mathrm{id}_{(\mathfrak{l}_{\mathfrak{C}}^{\mathrm{o}})_*(\mathfrak{a}^{\mathrm{o}})}$ and by the *adjoinness natural isomorphism* (cf. 1.6), the composition with $\mathfrak{ch}^{\mathrm{o}}(\mathfrak{f})$ yields a new natural map

$$(\mathfrak{l}_{\mathfrak{C}}^{\mathrm{o}})_*(\mathfrak{a}^{\mathrm{o}}) \circ \mathfrak{l}_{\mathfrak{B}}^{\mathrm{o}} \longrightarrow \mathfrak{a}^{\mathrm{o}} \circ \mathfrak{ch}^{\mathrm{o}}(\mathfrak{f}) \qquad \text{A4.3.4}$$

and finally, by applying again the *adjoinness natural isomorphism*, we get the natural map

$$\omega_{\mathfrak{f}} : (\mathfrak{l}_{\mathfrak{C}}^{\mathrm{o}})_*(\mathfrak{a}^{\mathrm{o}}) \longrightarrow (\mathfrak{l}_{\mathfrak{B}}^{\mathrm{o}})_*\big(\mathfrak{a}^{\mathrm{o}} \circ \mathfrak{ch}^{\mathrm{o}}(\mathfrak{f})\big) \qquad \text{A4.3.5;}$$

for any $n \in \mathbb{N}$ note that, by its very definition, $\mathbb{H}_n(\omega_{\mathfrak{f}})$ coincides with the image — denoted by $\mathbb{H}_{\mathrm{o}}^n(\mathfrak{f}, \mathfrak{a}^*)$ — of the $\mathfrak{Fct}(\mathfrak{ch}^{\mathrm{o}}(\bullet), \mathfrak{Ab}^\circ) \rtimes (\mathfrak{CC})^\circ$-morphism $(\mathrm{id}_{\mathfrak{a}^{\mathrm{o}} \circ \mathfrak{ch}^{\mathrm{o}}(\mathfrak{f})}, \mathfrak{f})$ by the functor $\mathbb{H}_{\mathrm{o}}^n$ (cf. A3.10.3).

A4.4 Let $\mathfrak{g}:\mathfrak{B} \to \mathfrak{C}$ be a second functor and $\mu:\mathfrak{f} \to \mathfrak{g}$ a natural map; on the one hand, we have a second natural map

$$\omega_{\mathfrak{g}} : (\mathfrak{l}_{\mathfrak{C}}^{\mathrm{o}})_*(\mathfrak{a}^{\mathrm{o}}) \longrightarrow (\mathfrak{l}_{\mathfrak{B}}^{\mathrm{o}})_*\big(\mathfrak{a}^{\mathrm{o}} \circ \mathfrak{ch}^{\mathrm{o}}(\mathfrak{g})\big) \qquad \text{A4.4.1.}$$

On the other hand, assume either that μ is a natural isomorphism — so that $\mathfrak{a}^* * \mathfrak{ch}(\mu)$ makes sense — or that $\mathfrak{a}^*$ can be extended to a *contravariant* functor $\mathfrak{a}$ from $\mathfrak{ch}(\mathfrak{C})$ to $\mathfrak{Ab}$ — so that $\mathfrak{a} * \mathfrak{ch}(\mu)$ is defined; in both cases, we get a natural map

$$\mathfrak{a}^{?} * \mathfrak{ch}(\mu) : \mathfrak{a}^{\circ} \circ \mathfrak{ch}^{\circ}(\mathfrak{g}) \longrightarrow \mathfrak{a}^{\circ} \circ \mathfrak{ch}^{\circ}(\mathfrak{f}) \qquad \text{A4.4.2}$$

where $\mathfrak{a}^{?}$ coincides with $\mathfrak{a}^*$ or $\mathfrak{a}$ accordingly, which determines a third natural map

$$\omega_\mu : (\mathfrak{l}^{\circ}_{\mathfrak{B}})_*\big(\mathfrak{a}^{\circ} \circ \mathfrak{ch}^{\circ}(\mathfrak{g})\big) \longrightarrow (\mathfrak{l}^{\circ}_{\mathfrak{B}})_*\big(\mathfrak{a}^{\circ} \circ \mathfrak{ch}^{\circ}(\mathfrak{f})\big) \qquad \text{A4.4.3}$$

and in the next proposition we prove that the corresponding triangle is *homotopically* commutative. For any $n \in \mathbb{N}$, let us set (cf. A3.11)

$$(d^{\,\mathfrak{a}^*})_n = d_n^{(\mathfrak{l}^{\circ}_{\mathfrak{C}})_*(\mathfrak{a}^{\circ})} \quad \text{and} \quad (d^{\,\mathfrak{a}^*,\mathfrak{f}})_n = d_n^{(\mathfrak{l}^{\circ}_{\mathfrak{B}})_*(\mathfrak{a}^{\circ}\circ\mathfrak{ch}^{\circ}(\mathfrak{f}))} \qquad \text{A4.4.4}$$

and let us consider the group homomorphism

$$h_n : (\mathfrak{l}^{\circ}_{\mathfrak{C}})_*(\Delta_{n+1}) \longrightarrow (\mathfrak{l}^{\circ}_{\mathfrak{B}})_*(\Delta_n) \qquad \text{A4.4.5}$$

mapping $a = (a_{\mathfrak{r}})_{\mathfrak{r}\in\mathfrak{Fct}(\Delta_{n+1},\mathfrak{C})} \in (\mathfrak{l}^{\circ}_{\mathfrak{C}})_*(\Delta_{n+1})$ on (cf. Lemma A4.2)

$$h_n(a) = \Big(\sum_{i=0}^{n} (-1)^i \big(\mathfrak{a}^{?}(\kappa_i^n(\mu * \mathfrak{q}), \varepsilon_i^{n+1})\big)\big(a_{\mathfrak{h}_i^n(\mu*\mathfrak{q})}\big)\Big)_{\mathfrak{q}\in\mathfrak{Fct}(\boldsymbol{\Delta}_n,\mathfrak{B})} \qquad \text{A4.4.6.}$$

Proposition A4.5 *Let $\mathfrak{B}$ and $\mathfrak{C}$ be two small categories, $\mathfrak{f}:\mathfrak{B} \to \mathfrak{C}$ and $\mathfrak{g}:\mathfrak{B} \to \mathfrak{C}$ two functors, $\mu:\mathfrak{f} \to \mathfrak{g}$ a natural map and $\mathfrak{a}^* : \mathfrak{ch}^*(\mathfrak{C}) \to \mathfrak{Ab}$ a contravariant functor. With the notation and the hypothesis above, for any $n \in \mathbb{N}$ we have*

$$(\omega_{\mathfrak{f}} - \omega_\mu \circ \omega_{\mathfrak{g}})_{\Delta_n} = (d^{\,\mathfrak{a}^*,\mathfrak{f}})_{n-1} \circ h_{n-1} + h_n \circ (d^{\,\mathfrak{a}^*})_n \qquad \text{A4.5.1.}$$

In particular, the following diagram is commutative:

$$\begin{array}{ccc} \mathbb{H}^n_{\circ}\big(\mathfrak{B}, \mathfrak{a}^{?} \circ \mathfrak{ch}^*(\mathfrak{g})\big) & \xrightarrow{\mathbb{H}^n_{\circ}(\mathfrak{B},\mathfrak{a}^{?}*\mathfrak{ch}(\mu))} & \mathbb{H}^n_{\circ}\big(\mathfrak{B}, \mathfrak{a}^{?} \circ \mathfrak{ch}^*(\mathfrak{f})\big) \\ {\scriptstyle \mathbb{H}^n_{\circ}(\mathfrak{g},\mathfrak{a}^*)} \nwarrow & & \nearrow {\scriptstyle \mathbb{H}^n_{\circ}(\mathfrak{f},\mathfrak{a}^*)} \\ & \mathbb{H}^n_{\circ}(\mathfrak{C}, \mathfrak{a}^*) & \end{array} \qquad \text{A4.5.2.}$$

Proof: For any $a = (a_{\mathfrak{q}})_{\mathfrak{q}\in\mathfrak{Fct}(\Delta_n,\mathfrak{C})}$ in $(\mathfrak{l}^{\circ}_{\mathfrak{C}})_*(\Delta_n)$ we have

$$\begin{aligned} &(d^{\,\mathfrak{a}^*,\mathfrak{f}})_{n-1}\big(h_{n-1}(a)\big) \\ &= \sum_{i=0}^{n-1} (-1)^i (d^{\,\mathfrak{a}^*,\mathfrak{f}})_{n-1}\Big(\big(\mathfrak{a}^{?}(\kappa_i^{n-1}(\mu * \mathfrak{t}), \varepsilon_i^n)\big)\big(a_{\mathfrak{h}_i^{n-1}(\mu*\mathfrak{t})}\big)\Big)_{\mathfrak{t}} \qquad \text{A4.5.3,} \\ &= \Big(\sum_{j=0}^{n}\sum_{i=0}^{n-1} (-1)^{i+j} \big(\mathfrak{a}^{?}(\kappa_i^{n-1}(\mu * \mathfrak{q}_j), \delta_j^{n-1} \circ \varepsilon_i^n)\big)\big(a_{\mathfrak{h}_i^{n-1}(\mu*\mathfrak{q}_j)}\big)\Big)_{\mathfrak{q}} \end{aligned}$$

where $\mathfrak{t}$ and $\mathfrak{q}$ respectively run over $\mathfrak{Fct}(\Delta_{n-1}, \mathfrak{B})$ and $\mathfrak{Fct}(\Delta_n, \mathfrak{C})$, and we set $\mathfrak{q}_j = \mathfrak{q} \circ \delta_j^{n-1}$; similarly, the second term of the right-hand member maps a on

$$h_n\big((d^{\mathfrak{a}^*})_n(a)\big) = \sum_{j=0}^{n+1}(-1)^j h_n\Big(\big(\mathfrak{a}^?(\mathrm{id}_{\mathfrak{r}\circ\delta_j^n}, \delta_j^n)\big)(a_{\mathfrak{r}\circ\delta_j^n})\Big)_{\mathfrak{r}}$$

$$= \Big(\sum_{j=0}^{n+1}\sum_{i=0}^{n}(-1)^{i+j}\big(\mathfrak{a}^?(\kappa_i^n(\mu * \mathfrak{q}) \circ \delta_j^n, \varepsilon_i^{n+1} \circ \delta_j^n)\big)(a_{\mathfrak{h}_i^n(\mu*\mathfrak{q})\circ\delta_j^n})\Big)_{\mathfrak{q}} \qquad \text{A4.5.4,}$$

where $\mathfrak{r}$ and $\mathfrak{q}$ respectively run over $\mathfrak{Fct}(\Delta_{n+1}, \mathfrak{C})$ and $\mathfrak{Fct}(\Delta_n, \mathfrak{C})$.

But, according to Lemma A4.2, for any $j \in \Delta_n$ and any $i \in \Delta_{n-1}$ we know that

$$\begin{aligned}
\mathfrak{h}_{i+1}^n(\mu * \mathfrak{q}) \circ \delta_{i+1}^n &= \mathfrak{h}_i^n(\mu * \mathfrak{q}) \circ \delta_{i+1}^n \\
\kappa_{i+1}^n(\mu * \mathfrak{q}) * \delta_{i+1}^n &= \kappa_i^n(\mu * \mathfrak{q}) * \delta_{i+1}^n \\
\mathfrak{h}_i^{n-1}(\mu * \mathfrak{q}_j) &= \begin{cases} \mathfrak{h}_{i+1}^n(\mu * \mathfrak{q}) \circ \delta_j^n & \text{if } j \le i \\ \mathfrak{h}_i^n(\mu * \mathfrak{q}) \circ \delta_{j+1}^n & \text{if } i < j\,. \end{cases} \\
\kappa_i^{n-1}(\mu * \mathfrak{q}_j) &= \begin{cases} \kappa_{i+1}^n(\mu * \mathfrak{q}) * \delta_j^n & \text{if } j \le i \\ \kappa_i^n(\mu * \mathfrak{q}) * \delta_{j+1}^n & \text{if } i < j \end{cases}
\end{aligned} \qquad \text{A4.5.5.}$$

Consequently, in the bottom member of equality A4.5.4, the terms where $j = i$ and $j = i + 1$ cancel with each other for any $1 \le j \le n$; moreover, the term (i, j) in the bottom member of equality 4.5.3 cancels either with the term $(i + 1, j)$ if $1 \le j \le i \le n - 1$, or with the term $(i, j + 1)$ in the bottom member of equality 4.5.4 if $0 \le i < j \le n$. Finally, the term $(i, 0)$ in the bottom member of equality 4.5.3 cancels with the term $(i + 1, 0)$ in the bottom member of equality 4.5.4 for any $0 \le i \le n - 1$, whereas the terms $(0, 0)$ and $(n, n + 1)$ respectively coincide with $-(\omega_\mu \circ \omega_{\mathfrak{g}})_{\Delta_n}(a)$ and $(\omega_{\mathfrak{f}})_{\Delta_n}(a)$ which proves equality A4.5.1. We are done.

Corollary A4.6 *With the notation and the hypothesis above, for any* $n \in \mathbb{N}$, h_n *preserves stability sending* $(\mathfrak{l}_{\mathfrak{C}}^*)_*(\Delta_{n+1})$ *to* $(\mathfrak{l}_{\mathfrak{B}}^{\circ})_*(\Delta_n)$. *In particular, we have*

$$\mathbb{H}_*^n(\mathfrak{f}, \mathfrak{a}^*) = \mathbb{H}_*^n\big(\mathfrak{B}, \mathfrak{a}^? * \mathfrak{ch}(\mu)\big) \circ \mathbb{H}_*^n(\mathfrak{g}, \mathfrak{a}^*) \qquad \text{A4.6.1.}$$

Proof: For any $a = (a_{\mathfrak{r}})_{\mathfrak{r}\in\mathfrak{Fct}(\Delta_{n+1},\mathfrak{C})} \in (\mathfrak{l}_{\mathfrak{C}}^*)_*(\Delta_{n+1})$, any isomorphic functors $\mathfrak{q}$ and $\mathfrak{q}'$ from Δ_n to $\mathfrak{B}$, any natural isomorphism $\nu : \mathfrak{q} \cong \mathfrak{q}'$ and any $i \in \Delta_n$, it is quite clear that we have a natural isomorphism

$$\mathfrak{h}_i^n(\mu, \nu) : \mathfrak{h}_i^n(\mu * \mathfrak{q}) \cong \mathfrak{h}_i^n(\mu * \mathfrak{q}') \qquad \text{A4.6.2}$$

sending $j \in \Delta_{n+1}$ to

$$\mathfrak{h}_i^n(\mu,\nu)_j = \begin{cases} \mathfrak{f}(\nu_j) : (\mathfrak{f}\circ\mathfrak{q})(j) \cong (\mathfrak{f}\circ\mathfrak{q}')(j) & \text{if } j \leq i \\ \mathfrak{g}(\nu_j) : (\mathfrak{g}\circ\mathfrak{q})(j-1) \cong (\mathfrak{g}\circ\mathfrak{q}')(j-1) & \text{if } j > i \end{cases} \qquad \text{A4.6.3}$$

and therefore we get $a_{\mathfrak{h}_i^n(\mu*\mathfrak{q}')} = \mathfrak{a}^?\big(\mathfrak{h}_i^n(\mu,\nu), \mathrm{id}_{\mathfrak{i}\mathfrak{d}_{\Delta_{n+1}}}\big)(a_{\mathfrak{h}_i^n(\mu*\mathfrak{q})})$, so that

$$\begin{aligned} &\big(\mathfrak{a}^?(\kappa_i^n(\mu*\mathfrak{q}'),\varepsilon_i^{n+1})\big)(a_{\mathfrak{h}_i^n(\mu*\mathfrak{q}')}) \\ &\quad = \Big(\mathfrak{a}^?\big(\mathfrak{h}_i^n(\mu,\nu)*\kappa_i^n(\mu*\mathfrak{q}'),\varepsilon_i^{n+1}\big)\Big)(a_{\mathfrak{h}_i^n(\mu*\mathfrak{q})}) \\ &\quad = \Big(\mathfrak{a}^?\big(\kappa_i^n(\mu*\mathfrak{q})\circ(\mathfrak{f}*\nu),\varepsilon_i^{n+1}\big)\Big)(a_{\mathfrak{h}_i^n(\mu*\mathfrak{q})}) \\ &\quad = \mathfrak{a}^?(\mathfrak{f}*\nu,\mathrm{id}_{\mathfrak{i}\mathfrak{d}_{\Delta_n}})\Big(\big(\mathfrak{a}^?(\kappa_i^n(\mu*\mathfrak{q}),\varepsilon_i^{n+1})\big)(a_{\mathfrak{h}_i^n(\mu*\mathfrak{q})})\Big) \end{aligned} \qquad 4.6.4.$$

Consequently, it follows from the very definition of h_n (cf. A4.4.6) that $h_n(a)$ belongs to $(\mathfrak{l}_{\mathfrak{B}}^{\mathfrak{o}})_*(\Delta_n)$ and then equality 4.6.1 follows from equality 4.5.1.

Proposition A4.7 *Let $\mathfrak{B}$ and $\mathfrak{C}$ be two small categories, $\mathfrak{f}:\mathfrak{B}\to\mathfrak{C}$ and $\mathfrak{g}:\mathfrak{C}\to\mathfrak{B}$ two functors, $\mu:\mathfrak{i}\mathfrak{d}_{\mathfrak{C}}\to\mathfrak{f}\circ\mathfrak{g}$ and $\nu:\mathfrak{i}\mathfrak{d}_{\mathfrak{B}}\to\mathfrak{g}\circ\mathfrak{f}$ two natural maps such that $\mu*\mathfrak{f}=\mathfrak{f}*\nu$, and $\mathfrak{a}^*:\mathfrak{ch}^*(\mathfrak{C})\to\mathfrak{Ab}$ a contravariant functor. Assume either that μ and ν are natural isomorphisms, or that $\mathfrak{a}^*$ can be extended to a contravariant functor $\mathfrak{a}$ from $\mathfrak{ch}(\mathfrak{C})$ to $\mathfrak{Ab}$, and denote by $\mathfrak{a}^?$ either $\mathfrak{a}^*$ or $\mathfrak{a}$ accordingly. Then, for any $n\in\mathbb{N}$, we have the group isomorphisms*

$$\begin{aligned} \mathbb{H}_{\mathfrak{o}}^n(\mathfrak{f},\mathfrak{a}^*) &: \mathbb{H}_{\mathfrak{o}}^n(\mathfrak{C},\mathfrak{a}^*) \cong \mathbb{H}_{\mathfrak{o}}^n\big(\mathfrak{B},\mathfrak{a}^?\circ\mathfrak{ch}(\mathfrak{f})\big) \\ \mathbb{H}_*^n(\mathfrak{f},\mathfrak{a}^*) &: \mathbb{H}_*^n(\mathfrak{C},\mathfrak{a}^*) \cong \mathbb{H}_*^n\big(\mathfrak{B},\mathfrak{a}^?\circ\mathfrak{ch}(\mathfrak{f})\big) \end{aligned} \qquad \text{A4.7.1.}$$

Proof: Applying Proposition A4.5 and Corollary A4.6 to the functors $\mathfrak{i}\mathfrak{d}_{\mathfrak{C}}$ and $\mathfrak{f}\circ\mathfrak{g}$, for any $n\in\mathbb{N}$ we get

$$\begin{aligned} \mathbb{H}_{\mathfrak{o}}^n\big(\mathfrak{C},\mathfrak{a}^?*\mathfrak{ch}(\mu)\big)\circ\mathbb{H}_{\mathfrak{o}}^n(\mathfrak{f}\circ\mathfrak{g},\mathfrak{a}^*) &= \mathbb{H}_{\mathfrak{o}}^n(\mathfrak{i}\mathfrak{d}_{\mathfrak{C}},\mathfrak{a}^*) = \mathrm{id}_{\mathbb{H}_{\mathfrak{o}}^n(\mathfrak{C},\mathfrak{a}^*)} \\ \mathbb{H}_*^n\big(\mathfrak{C},\mathfrak{a}^?*\mathfrak{ch}(\mu)\big)\circ\mathbb{H}_*^n(\mathfrak{f}\circ\mathfrak{g},\mathfrak{a}^*) &= \mathbb{H}_*^n(\mathfrak{i}\mathfrak{d}_{\mathfrak{C}},\mathfrak{a}^*) = \mathrm{id}_{\mathbb{H}_*^n(\mathfrak{C},\mathfrak{a}^*)} \end{aligned} \qquad \text{A4.7.2}$$

and therefore the group homomorphisms (cf. A3.10)

$$\begin{aligned} \mathbb{H}_{\mathfrak{o}}^n(\mathfrak{f}\circ\mathfrak{g},\mathfrak{a}^*) &= \mathbb{H}_{\mathfrak{o}}^n\big(\mathfrak{g},\mathfrak{a}^?\circ\mathfrak{ch}(\mathfrak{f})\big)\circ\mathbb{H}_{\mathfrak{o}}^n(\mathfrak{f},\mathfrak{a}^*) \\ \mathbb{H}_*^n(\mathfrak{f}\circ\mathfrak{g},\mathfrak{a}^*) &= \mathbb{H}_*^n\big(\mathfrak{g},\mathfrak{a}^?\circ\mathfrak{ch}(\mathfrak{f})\big)\circ\mathbb{H}_*^n(\mathfrak{f},\mathfrak{a}^*) \end{aligned} \qquad \text{A4.7.3}$$

are injective, so that $\mathbb{H}_{\mathfrak{o}}^n(\mathfrak{f},\mathfrak{a}^*)$ and $\mathbb{H}_*^n(\mathfrak{f},\mathfrak{a}^*)$ are injective too.

Mutatis mutandis, applying Proposition A4.5 and Corollary A4.6 to the functors $\mathfrak{id}_{\mathfrak{B}}$ and $\mathfrak{g} \circ \mathfrak{f}$, for any $n \in \mathbb{N}$ we get

$$\begin{aligned} &\mathbb{H}^n_{\mathfrak{o}}\big(\mathfrak{B}, (\mathfrak{a}^? \circ \mathfrak{ch}(\mathfrak{f})) * \mathfrak{ch}(\nu)\big) \circ \mathbb{H}^n_{\mathfrak{o}}\big(\mathfrak{g} \circ \mathfrak{f}, \mathfrak{a}^? \circ \mathfrak{ch}(\mathfrak{f})\big) \\ &\qquad = \mathrm{id}_{\mathbb{H}^n_{\mathfrak{o}}(\mathfrak{B}, \mathfrak{a}^? \circ \mathfrak{ch}(\mathfrak{f}))} \\ &\mathbb{H}^n_{*}\big(\mathfrak{B}, (\mathfrak{a}^? \circ \mathfrak{ch}(\mathfrak{f})) * \mathfrak{ch}(\nu)\big) \circ \mathbb{H}^n_{*}\big(\mathfrak{g} \circ \mathfrak{f}, \mathfrak{a}^? \circ \mathfrak{ch}(\mathfrak{f})\big) \\ &\qquad = \mathrm{id}_{\mathbb{H}^n_{*}(\mathfrak{B}, \mathfrak{a}^? \circ \mathfrak{ch}(\mathfrak{f}))} \end{aligned} \qquad \text{A4.7.4,}$$

so that the group homomorphisms (cf. A3.10)

$$\begin{aligned} &\mathbb{H}^n_{\mathfrak{o}}\big(\mathfrak{B}, (\mathfrak{a}^? \circ \mathfrak{ch}(\mathfrak{f})) * \mathfrak{ch}(\nu)\big) \circ \mathbb{H}^n_{\mathfrak{o}}\big(\mathfrak{f}, \mathfrak{a}^? \circ \mathfrak{ch}(\mathfrak{f} \circ \mathfrak{g})\big) \\ &\mathbb{H}^n_{*}\big(\mathfrak{B}, (\mathfrak{a}^? \circ \mathfrak{ch}(\mathfrak{f})) * \mathfrak{ch}(\nu)\big) \circ \mathbb{H}^n_{*}\big(\mathfrak{f}, \mathfrak{a}^? \circ \mathfrak{ch}(\mathfrak{f} \circ \mathfrak{g})\big) \end{aligned} \qquad \text{A4.7.5}$$

are surjective; but, since we are assuming that $\mu * \mathfrak{f} = \mathfrak{f} * \nu$, we get (cf. A3.10)

$$\begin{aligned} &\mathbb{H}^n_{\mathfrak{o}}(\mathfrak{f}, \mathfrak{a}^*) \circ \mathbb{H}^n_{\mathfrak{o}}\big(\mathfrak{C}, \mathfrak{a}^? * \mathfrak{ch}(\mu)\big) \\ &\qquad = \mathbb{H}^n_{\mathfrak{o}}\big(\mathfrak{B}, (\mathfrak{a}^? \circ \mathfrak{ch}(\mathfrak{f})) * \mathfrak{ch}(\nu)\big) \circ \mathbb{H}^n_{\mathfrak{o}}\big(\mathfrak{f}, \mathfrak{a}^? \circ \mathfrak{ch}(\mathfrak{f} \circ \mathfrak{g})\big) \\ &\mathbb{H}^n_{*}(\mathfrak{f}, \mathfrak{a}^*) \circ \mathbb{H}^n_{*}\big(\mathfrak{C}, \mathfrak{a}^? * \mathfrak{ch}(\mu)\big) \\ &\qquad = \mathbb{H}^n_{*}\big(\mathfrak{B}, (\mathfrak{a}^? \circ \mathfrak{ch}(\mathfrak{f})) * \mathfrak{ch}(\nu)\big) \circ \mathbb{H}^n_{*}\big(\mathfrak{f}, \mathfrak{a}^? \circ \mathfrak{ch}(\mathfrak{f} \circ \mathfrak{g})\big) \end{aligned} \qquad \text{A4.7.6;}$$

hence, $\mathbb{H}^n_{\mathfrak{o}}(\mathfrak{f}, \mathfrak{a}^*)$ and $\mathbb{H}^n_{*}(\mathfrak{f}, \mathfrak{a}^*)$ are surjective too. We are done.

Corollary A4.8 *Let $\mathfrak{C}$ be a small category with a final object and $\mathfrak{a} : \mathfrak{C} \to \mathfrak{Ab}$ a contravariant functor. For any $n \geq 1$ we have $\mathbb{H}^n(\mathfrak{C}, \mathfrak{a}) = \{0\}$.*

Proof: We apply Proposition A4.7 to the subcategory $\mathfrak{B}$ of $\mathfrak{C}$ over the final object, to the inclusion functor $\mathfrak{i} : \mathfrak{B} \to \mathfrak{C}$, to the unique functor $\mathfrak{j} : \mathfrak{C} \to \mathfrak{B}$, to the unique natural map $\mu : \mathfrak{id}_{\mathfrak{C}} \to \mathfrak{i} \circ \mathfrak{j}$, which actually fulfills $\mu * \mathfrak{i} = \mathrm{id}_{\mathfrak{i}}$, and to the contravariant functor $\mathfrak{a} \circ \mathfrak{v}_{\mathfrak{C}} : \mathfrak{ch}(\mathfrak{C}) \to \mathfrak{Ab}$; then, the corollary follows.

Corollary A4.9 *Let $\mathfrak{B}$ and $\mathfrak{C}$ be two small categories, $\mathfrak{f} : \mathfrak{B} \to \mathfrak{C}$ an equivalence of categories and $\mathfrak{a}^* : \mathfrak{ch}^*(\mathfrak{C}) \to \mathfrak{Ab}$ a contravariant functor. Then, for any $n \in \mathbb{N}$, $\mathbb{H}^n_{\mathfrak{o}}(\mathfrak{f}, \mathfrak{a}^*)$ and $\mathbb{H}^n_{*}(\mathfrak{f}, \mathfrak{a}^*)$ are group isomorphisms.*

Proof: It is well-known that there is a functor $\mathfrak{g} : \mathfrak{C} \to \mathfrak{B}$ such that we have two natural isomorphisms

$$\mu : \mathfrak{id}_{\mathfrak{C}} \cong \mathfrak{f} \circ \mathfrak{g} \quad \text{and} \quad \nu : \mathfrak{id}_{\mathfrak{B}} \cong \mathfrak{g} \circ \mathfrak{f} \qquad \text{A4.9.1}$$

and then the corollary follows from Proposition A4.7.

A4.10 Consider the *additive cover* $\mathfrak{ac}(\mathfrak{C}) = \mathfrak{pc}(\mathfrak{C}^\circ)^\circ$ of $\mathfrak{C}$ defined in A2.7.3 above. That is to say, according to Proposition A2.6, the $\mathfrak{ac}(\mathfrak{C})$-objects are the finite sequences $\{C_i\}_{i\in I}$ — denoted by $C = \bigoplus_{i\in I} C_i$ — of $\mathfrak{C}$-objects and an $\mathfrak{ac}(\mathfrak{C})$-morphism from another object $D = \bigoplus_{j\in J} D_j$ to $C = \bigoplus_{i\in I} C_i$ is a pair (α, f) formed by a map $f : J \to I$ and by a family $\alpha = \{\alpha_j\}_{j\in J}$ of $\mathfrak{C}$-morphisms $\alpha_j : D_j \to C_{f(j)}$; the composition of (α, f) with another morphism

$$(\beta, g) : E = \bigoplus_{\ell\in L} E_\ell \longrightarrow D \qquad \text{A4.10.1,}$$

is the pair $\big((\alpha * g)\circ\beta, f\circ g\big)$ where, if β is the family $\{\beta_\ell\}_{\ell\in L}$, $(\alpha * g)\circ\beta$ is the family $\{\alpha_{g(\ell)}\circ\beta_\ell\}_{\ell\in L}$ of composed $\mathfrak{C}$-morphisms

$$\alpha_{g(\ell)}\circ\beta_\ell : E_\ell \longrightarrow D_{g(\ell)} \longrightarrow C_{(f\circ g)(\ell)} \qquad \text{A4.10.2.}$$

In particular, we have an obvious functor $\mathfrak{j}_\mathfrak{C} : \mathfrak{C} \to \mathfrak{ac}(\mathfrak{C})$, and any *contravariant* functor $\mathfrak{a} : \mathfrak{C} \to \mathfrak{Ab}$ can be extended to a *contravariant* functor

$$\mathfrak{ac}(\mathfrak{a}) : \mathfrak{ac}(\mathfrak{C}) \longrightarrow \mathfrak{Ab} \qquad \text{A4.10.3}$$

mapping C on $\bigoplus_{i\in I}\mathfrak{a}(C_i)$ and (α, f) on the group homomorphism

$$\bigoplus_{i\in I}\mathfrak{a}(C_i) \longrightarrow \bigoplus_{j\in J}\mathfrak{a}(D_j) \qquad \text{A4.10.4}$$

sending $\sum_{i\in I} a_i$, where $a_i \in \mathfrak{a}(\mathfrak{C}_i)$, to $\sum_{j\in J}\big(\mathfrak{a}(\alpha_j)\big)(a_{f(j)})$. In [32], S. Jackowski and J. McClure prove the following result.

Proposition A4.11 *With the notation above, $\mathfrak{ac}(\mathfrak{a})$ is the direct image of $\mathfrak{a}$ throughout $\mathfrak{j}_\mathfrak{C}$ and, for any $n \in \mathbb{N}$, $\mathfrak{j}_\mathfrak{C}$ induces a group isomorphism*

$$\mathbb{H}^n\big(\mathfrak{ac}(\mathfrak{C}), \mathfrak{ac}(\mathfrak{a})\big) \cong \mathbb{H}^n(\mathfrak{C}, \mathfrak{a}) \qquad \text{A4.11.1.}$$

Proof: For any *contravariant* functor $\mathfrak{b} : \mathfrak{ac}(\mathfrak{C}) \to \mathfrak{Ab}$ and any natural map $\mu : \mathfrak{b}\circ\mathfrak{j}_\mathfrak{C} \to \mathfrak{a}$, we have an obvious natural map

$$\mathfrak{ac}(\mu) : \mathfrak{ac}(\mathfrak{b}\circ\mathfrak{j}_\mathfrak{C}) \longrightarrow \mathfrak{ac}(\mathfrak{a}) \qquad \text{A4.11.2.}$$

Moreover, for any $\mathfrak{ac}(\mathfrak{C})$-object $C = \bigoplus_{\subset\in I} C_i$, the evident $\mathfrak{ac}(\mathfrak{C})$-morphisms $C_i \to C$ determine group homomorphisms $\mathfrak{b}(C) \to \mathfrak{b}(C_i)$ for any $i \in I$ and therefore we get a group homomorphism $\mathfrak{b}(C) \to \big(\mathfrak{ac}(\mathfrak{a})\big)(C)$; as a matter of fact, this correspondence defines a natural map $\lambda_\mathfrak{b} : \mathfrak{b} \to \mathfrak{ac}(\mathfrak{b}\circ\mathfrak{j}_\mathfrak{C})$ which fulfills $\lambda_{\mathfrak{b}} * \mathfrak{j}_{\mathfrak{C}} = \mathrm{id}_{\mathfrak{b}\circ\mathfrak{j}_{\mathfrak{C}}}$. Consequently, we have obtained an injective group homomorphism

$$\mathfrak{Nat}(\mathfrak{b}\circ\mathfrak{j}_{\mathfrak{C}},\mathfrak{a}) \longrightarrow \mathfrak{Nat}\big(\mathfrak{b},\mathfrak{ac}(\mathfrak{a})\big) \qquad \text{A4.11.3}$$

mapping μ on $\mathfrak{ac}(\mu)\circ\lambda_{\mathfrak{b}}$, which is easily checked to be surjective, proving that $(\mathfrak{j}_{\mathfrak{C}})_*(\mathfrak{a}) = \mathfrak{ac}(\mathfrak{a})$.

Moreover, we have $\mathfrak{ac}(\mathfrak{a})\circ\mathfrak{j}_{\mathfrak{C}} = \mathfrak{a}$, $\mathfrak{ac}$ clearly preserves the monomorphisms and $\mathfrak{j}_{\mathfrak{C}}$ induces a group isomorphism

$$\mathbb{H}^0\big(\mathfrak{ac}(\mathfrak{C}),\mathfrak{ac}(\mathfrak{a})\big) \cong \varprojlim \mathfrak{ac}(\mathfrak{a}) \cong \varprojlim \mathfrak{a} \cong \mathbb{H}^0(\mathfrak{C},\mathfrak{a}) \qquad \text{A4.11.4;}$$

hence, isomorphism A4.11.1 follows from Proposition A3.15.

A4.12 Finally, for any small category $\mathfrak{C}$, any subcategory $\mathfrak{D}$ over the same set of objects and any *contravariant* functor $\mathfrak{a}^* : \mathfrak{ch}^*(\mathfrak{C}) \to \mathfrak{Ab}$, let us discuss the relationship between the groups $\mathbb{H}^n_{\mathfrak{o}}(\mathfrak{C},\mathfrak{a}^*)$ and $\mathbb{H}^n_{\mathfrak{D}}(\mathfrak{C},\mathfrak{a}^*)$ (cf. A3.18). Since we have a natural inclusion (cf. A3.7, A3.11.1 and A3.18)

$$\iota(\mathfrak{a}^{\mathfrak{D}}) : (\mathfrak{l}^{\mathfrak{D}}_{\mathfrak{C}})_*(\mathfrak{a}^{\mathfrak{D}}) \longrightarrow (\mathfrak{l}^{\mathfrak{o}}_{\mathfrak{C}})_*(\mathfrak{a}^{\mathfrak{o}}) \qquad \text{A4.12.1,}$$

we always have a group homomorphism $\mathbb{H}^n_{\mathfrak{D}}(\mathfrak{C},\mathfrak{a}^*) \to \mathbb{H}^n_{\mathfrak{o}}(\mathfrak{C},\mathfrak{a}^*)$ for any $n\in\mathbb{N}$ and, for $n=0$, it is easily checked that

$$\mathbb{H}^0_{\mathfrak{D}}(\mathfrak{C},\mathfrak{a}^*) \cong \mathbb{H}^0_{\mathfrak{o}}(\mathfrak{C},\mathfrak{a}^*) \qquad \text{A4.12.2.}$$

We are interested in the case where the group of $\mathfrak{D}$-automorphisms of any object is *finite* and has an *order* which is *invertible* in a suitable ring $\mathcal{R}$, and we consider *contravariant* functors from $\mathfrak{ch}^*(\mathfrak{C})$ to the category $\mathcal{R}$-$\mathfrak{mod}$ of finitely generated $\mathcal{R}$-modules.

Proposition A4.13 *Let $\mathfrak{C}$ be a small category, $\mathfrak{D}$ a subcategory over the same set of objects, where the group of $\mathfrak{D}$-automorphisms of any $\mathfrak{C}$-object is finite, and $\mathcal{R}$ a ring such that, for any $\mathfrak{C}$-object C, $|\mathfrak{D}(C)|$ is invertible in $\mathcal{R}$. Then, for any contravariant functor $\mathfrak{m}^* : \mathfrak{ch}^*(\mathfrak{C}) \to \mathcal{R}$-$\mathfrak{mod}$ and any $n\in\mathbb{N}$, the inclusion $(\mathfrak{l}^{\mathfrak{D}}_{\mathfrak{C}})_*(\mathfrak{m}^{\mathfrak{D}}) \subset (\mathfrak{l}^{\mathfrak{o}}_{\mathfrak{C}})_*(\mathfrak{m}^{\mathfrak{o}})$ induces a group isomorphism*

$$\mathbb{H}^n_{\mathfrak{D}}(\mathfrak{C},\mathfrak{m}^*) \cong \mathbb{H}^n_{\mathfrak{o}}(\mathfrak{C},\mathfrak{m}^*) \qquad \text{A4.13.1.}$$

Proof: According to Corollary A4.9, we may assume that two $\mathfrak{C}$-isomorphic $\mathfrak{C}$-objects always coincide; for any $n\in\mathbb{N}$, let us denote by

$$\iota(\mathfrak{m}^{\mathfrak{D}})_n : \big((\mathfrak{l}^{\mathfrak{D}}_{\mathfrak{C}})_*(\mathfrak{m}^{\mathfrak{D}})\big)(\Delta_n) \longrightarrow \big((\mathfrak{l}^{\mathfrak{o}}_{\mathfrak{C}})_*(\mathfrak{m}^{\mathfrak{o}})\big)(\Delta_n) \qquad \text{A4.13.2}$$

the corresponding inclusion map; in our situation, it admits a section $\sigma(\mathfrak{m}^{\mathfrak{D}})_n$ mapping any $v = (v_\mathfrak{q})_{\mathfrak{q}\in\mathfrak{Fct}(\boldsymbol{\Delta}_n,\mathfrak{C})}$, where $v_\mathfrak{q} \in \mathfrak{m}(\mathfrak{q},\Delta_n)$, on

$$\sigma(\mathfrak{m}^{\mathfrak{D}})_n(v) = \Big(\frac{1}{z_\mathfrak{q}}\cdot\sum_{\mathfrak{q}',\nu} \mathfrak{b}(\nu,\mathrm{id}_{\Delta_n})(v_{\mathfrak{q}'})\Big)_{\mathfrak{q}\in\mathfrak{Fct}(\boldsymbol{\Delta}_n,\mathfrak{C})} \qquad \text{A4.13.3,}$$

where $\mathfrak{q}'$ runs over the set of functors from Δ_n to $\mathfrak{C}$ which are $\mathfrak{D}$-isomorphic to $\mathfrak{q}$, ν runs over the set of natural $\mathfrak{D}$-isomorphisms $\mathfrak{q}\cong\mathfrak{q}'$, and we set

$$z_\mathfrak{q} = \prod_{i\in\Delta_n} \big|\mathfrak{D}\big(\mathfrak{q}(i)\big)\big| \qquad \text{A4.13.4,}$$

which coincides with the cardinal of the set of such pairs $(\mathfrak{q}',\nu)$. Indeed, it is easily checked that $\sigma(\mathfrak{m}^{\mathfrak{D}})_n(v)$ belongs to $\big((\mathfrak{l}^{\mathfrak{D}}_{\mathfrak{C}})_*(\mathfrak{m}^{\mathfrak{D}})\big)(\Delta_n)$ and that we have

$$\sigma(\mathfrak{m}^{\mathfrak{D}})_n \circ \iota(\mathfrak{m}^{\mathfrak{D}})_n = \mathrm{id}_{((\mathfrak{l}^{\mathfrak{D}}_{\mathfrak{C}})_*(\mathfrak{m}^{\mathfrak{D}}))(\Delta_n)} \qquad \text{A4.13.5.}$$

Moreover, for any $i \le n+1$, it is straightforward to check the commutativity of the diagram

$$\begin{array}{ccc}
\big((\mathfrak{l}^{\mathfrak{D}}_{\mathfrak{C}})_*(\mathfrak{m}^{\mathfrak{D}})\big)(\Delta_{n+1}) & \xleftarrow{\sigma(\mathfrak{m}^{\mathfrak{D}})_{n+1}} & \big((\mathfrak{l}^{\mathfrak{o}}_{\mathfrak{C}})_*(\mathfrak{m}^{\mathfrak{o}})\big)(\Delta_{n+1}) \\
{\scriptstyle ((\mathfrak{l}^{\mathfrak{D}}_{\mathfrak{C}})_*(\mathfrak{m}^{\mathfrak{D}}))(\delta^n_i)}\uparrow & & \uparrow{\scriptstyle ((\mathfrak{l}^{\mathfrak{o}}_{\mathfrak{C}})_*(\mathfrak{m}^{\mathfrak{o}}))(\delta^n_i)} \\
\big((\mathfrak{l}^{\mathfrak{D}}_{\mathfrak{C}})_*(\mathfrak{m}^{\mathfrak{D}})\big)(\Delta_n) & \xleftarrow{\sigma(\mathfrak{m}^{\mathfrak{D}})_n} & \big((\mathfrak{l}^{\mathfrak{o}}_{\mathfrak{C}})_*(\mathfrak{m}^{\mathfrak{o}})\big)(\Delta_n)
\end{array} \qquad \text{A4.13.6.}$$

In particular, the family of sections $\sigma(\mathfrak{m}^{\mathfrak{D}}) = \{\sigma(\mathfrak{m}^{\mathfrak{D}})_n\}_{n\in\mathbb{N}}$ still determines a homomorphism between the differential complexes associated with the functors (cf. A3.2)

$$(\mathfrak{l}^{\mathfrak{D}}_{\mathfrak{C}})_*(\mathfrak{m}^{\mathfrak{D}}) : \boldsymbol{\Delta} \longrightarrow \mathcal{R}\text{-}\mathfrak{mod} \quad \text{and} \quad (\mathfrak{l}^{\mathfrak{o}}_{\mathfrak{C}})_*(\mathfrak{m}^{\mathfrak{o}}) : \boldsymbol{\Delta} \longrightarrow \mathcal{R}\text{-}\mathfrak{mod} \qquad \text{A4.13.7,}$$

so that, for any $n\in\mathbb{N}$, we get a group homomorphism

$$\begin{array}{cccc}
\mathbb{H}_n\big(\sigma(\mathfrak{m}^{\mathfrak{D}})\big) : & \mathbb{H}_n\big((\mathfrak{l}^{\mathfrak{o}}_{\mathfrak{C}})_*(\mathfrak{m}^{\mathfrak{o}})\big) & \longrightarrow & \mathbb{H}_n\big((\mathfrak{l}^{\mathfrak{D}}_{\mathfrak{C}})_*(\mathfrak{m}^{\mathfrak{D}})\big) \\
& \| & & \| \\
& \mathbb{H}^n_{\mathfrak{o}}(\mathfrak{C},\mathfrak{m}^*) & & \mathbb{H}^n_{\mathfrak{D}}(\mathfrak{C},\mathfrak{m}^*)
\end{array} \qquad \text{A4.13.8;}$$

moreover, it follows from equalities A4.13.5 that we have

$$\mathbb{H}_n\big(\sigma(\mathfrak{m}^{\mathfrak{D}})\big) \circ \mathbb{H}_n\big(\iota(\mathfrak{m}^{\mathfrak{D}})\big) = \mathrm{id}_{\mathbb{H}^n_{\mathfrak{D}}(\mathfrak{C},\mathfrak{m}^*)} \qquad \text{A4.13.9}$$

and we claim that we also have $\mathbb{H}_n\big(\iota(\mathfrak{m}^{\mathfrak{D}})\big) \circ \mathbb{H}_n\big(\sigma(\mathfrak{m}^{\mathfrak{D}})\big) = \mathrm{id}_{\mathbb{H}^n_{\mathfrak{o}}(\mathfrak{C},\mathfrak{m}^*)}$.

Once again using Lemma A4.2, for any $n\in\mathbb{N}$ we consider the group homomorphism

$$h_n : (\mathfrak{l}^{\mathfrak{o}}_{\mathfrak{C}})_*(\Delta_{n+1}) \longrightarrow (\mathfrak{l}^{\mathfrak{o}}_{\mathfrak{C}})_*(\Delta_n) \qquad \text{A4.13.10}$$

mapping $v = (v_{\mathfrak{r}})_{\mathfrak{r}\in\mathfrak{Fct}(\Delta_{n+1},\mathfrak{C})} \in (\mathfrak{l}^{\mathfrak{o}}_{\mathfrak{C}})_*(\Delta_{n+1})$ on

$$h_n(v) = \Big(\sum_{i=0}^{n} \frac{(-1)^i}{z_{\mathfrak{r}}}\cdot \sum_{\mathfrak{q}',\nu} \mathfrak{m}^*\big(\kappa_i^n(\nu), \varepsilon_i^{n+1}\big)\big(v_{\mathfrak{h}_i^n(\nu)}\big)\Big)_{\mathfrak{q}\in\mathfrak{Fct}(\boldsymbol{\Delta}_n,\mathfrak{C})} \qquad \text{A4.13.11}$$

where $\mathfrak{q}'$ runs over the set of functors $\Delta_n \to \mathfrak{C}$ which are $\mathfrak{D}$-isomorphic to $\mathfrak{q}$, and ν over the set of natural $\mathfrak{D}$-isomorphisms $\mathfrak{q} \cong \mathfrak{q}'$. Now, we claim that

$$\mathrm{id}_{(\mathfrak{l}^{\mathfrak{o}}_{\mathfrak{C}})_*(\Delta_n)} - \iota(\mathfrak{m}^{\mathfrak{D}})_n \circ \sigma(\mathfrak{m}^{\mathfrak{D}})_n = d_{n-1} \circ h_{n-1} + h_n \circ d_n \qquad \text{A4.13.12,}$$

where $d_n : (\mathfrak{l}^{\mathfrak{o}}_{\mathfrak{C}})_*(\Delta_n) \to (\mathfrak{l}^{\mathfrak{o}}_{\mathfrak{C}})_*(\Delta_{n+1})$ denotes the corresponding differential map (cf. A3.11).

As in the proof of Proposition A4.5, the first term of the right-hand member maps $v = (v_{\mathfrak{q}})_{\mathfrak{q}\in\mathfrak{Fct}(\Delta_n,\mathfrak{C})}$ on

$$\begin{aligned} &d_{n-1}\big(h_{n-1}(v)\big) \\ &= \sum_{i=0}^{n-1} (-1)^i d_{n-1}\Big(\sum_{\mathfrak{t}',\lambda} \frac{1}{z_{\mathfrak{t}}}\cdot \mathfrak{m}^*\big(\kappa_i^{n-1}(\lambda), \varepsilon_i^n\big)\big(v_{\mathfrak{h}_i^{n-1}(\lambda)}\big)\Big)_{\mathfrak{t}} \\ &= \Big(\sum_{i=0}^{n-1}\sum_{j=0}^{n} \frac{(-1)^{i+j}}{z_{\mathfrak{q}}}\cdot \sum_{\mathfrak{q}',\nu} w_{i,j}^{n-1}(\nu * \delta_j^{n-1})\Big)_{\mathfrak{q}} \end{aligned} \qquad \text{A4.13.13}$$

where $(\mathfrak{t},\mathfrak{t}')$ runs over the set of pairs of $\mathfrak{D}$-isomorphic functors from Δ_{n-1} to $\mathfrak{C}$, λ runs over the set of natural $\mathfrak{D}$-isomorphisms $\mathfrak{t} \cong \mathfrak{t}'$, $(\mathfrak{q},\mathfrak{q}')$ runs over the set of pairs of $\mathfrak{D}$-isomorphic functors from Δ_n to $\mathfrak{C}$, ν runs over the set of natural $\mathfrak{D}$-isomorphisms $\mathfrak{q} \cong \mathfrak{q}'$, and we set

$$\begin{aligned} &w_{i,j}^{n-1}(\nu * \delta_j^{n-1}) \\ &\qquad = \big(\mathfrak{m}^*\big(\kappa_i^{n-1}(\nu * \delta_j^{n-1}), \delta_j^{n-1} \circ \varepsilon_i^n\big)\big)\big(v_{h_i^{n-1}(\nu*\delta_j^{n-1})}\big) \end{aligned} \qquad \text{A4.13.14.}$$

Similarly, the second term of the right-hand member maps the same v on

$$\begin{aligned} h_n\big(d_n(v)\big) &= \sum_{j=0}^{n+1} (-1)^j h_n\big(\mathfrak{m}^*(\mathrm{id}_{\mathfrak{r}\circ\delta_j^{n-1}}, \delta_j^n)(v_{\mathfrak{r}\circ\delta_j^n})\big)_{\mathfrak{r}} \\ &= \Big(\sum_{j=0}^{n+1}\sum_{i=0}^{n} \frac{(-1)^{i+j}}{z_{\mathfrak{q}}}\cdot \sum_{\mathfrak{q}',\nu} u_{i,j}^n(\nu)\Big)_{\mathfrak{q}} \end{aligned} \qquad \text{A4.13.15}$$

where $\mathfrak{r}$ runs over the set of functors from Δ_{n+1} to $\mathfrak{C}$, $(\mathfrak{q},\mathfrak{q}')$ runs over the set of $\mathfrak{D}$-isomorphic pairs of functors from Δ_n to $\mathfrak{C}$, ν runs over the set of natural $\mathfrak{D}$-isomorphisms $\mathfrak{q} \cong \mathfrak{q}'$, and we set

$$u_{i,j}^n(\nu) = \big(\mathfrak{m}^*\big(\kappa_i^n(\nu) * \delta_j^{n-1}, \varepsilon_i^{n+1} \circ \delta_j^n\big)\big)\big(v_{h_i^n(\nu)\circ\delta_j^{n-1}}\big) \qquad \text{A4.13.16.}$$

Then, it follows from Lemma A4.2 that, in the last members of equalities A4.13.13 and A4.13.15, comparing the components corresponding to a functor $\mathfrak{q}:\Delta_n \to \mathfrak{C}$, the term (i,j) in the first sum cancels either with the term $(i+1,j)$ in the second sum if $j \le i$, or with the term $(i,j+1)$ if $i < j$, the terms $(i,i+1)$ and $(i+1,i+1)$ of the second sum cancel with each other, and it remains the last and the first terms of the second sum, which respectively coincide with $v_{\mathfrak{q}}$ and with $-\frac{1}{z_{\mathfrak{q}}}\cdot\sum_{\mathfrak{q}',\nu}\mathfrak{m}^*(\nu,\mathrm{id}_{\Delta_n})(v_{\mathfrak{q}'})$, where $\mathfrak{q}'$ runs over the set of functors $\mathbf{\Delta}_n \to \mathfrak{C}$ which are $\mathfrak{D}$-isomorphic to $\mathfrak{q}$ and ν over the set of natural $\mathfrak{D}$-isomorphisms $\mathfrak{q}\cong\mathfrak{q}'$. We are done.

A5 Regular cohomology in ordered categories

A5.1 Let us call *ordered category* any small category $\mathfrak{C}$ such that

45.1.1 *Any pair of $\mathfrak{C}$-objects C and C' which admit $\mathfrak{C}$-morphisms $C \to C'$ and $C' \to C$ are $\mathfrak{C}$-isomorphic and then all the $\mathfrak{C}$-morphisms $C \to C'$ are $\mathfrak{C}$-isomorphisms.*

Clearly, for any finite p-group P, any *Frobenius P-category* $\mathcal{F}$ is an ordered category. The point is that, for any ordered category $\mathfrak{C}$ and any *contravariant* functor $\mathfrak{a}^*:\mathfrak{ch}^*(\mathfrak{C}) \to \mathfrak{Ab}$ which factorizes *via* $\overline{\mathfrak{ch}^*(\mathfrak{C})}$ (cf. A1.8), on the one hand the *stable* cohomology groups $\mathbb{H}^n_*(\mathfrak{C},\mathfrak{a}^*)$ can be computed from the *regular $\mathfrak{C}$-chains* defined below, and on the other hand if $\mathfrak{C}$ is *finite* then there are only finitely many such *$\mathfrak{C}$-chains*.

A5.2 Let $\mathfrak{C}$ be an *ordered category*; we say that a *$\mathfrak{C}$-chain* $\mathfrak{q}:\Delta_n \to \mathfrak{C}$ is *regular* if, for any $0 \le i < j \le n$, the $\mathfrak{C}$-morphism $\mathfrak{q}(i\bullet j)$ is *not* a $\mathfrak{C}$-isomorphism; actually, $\mathfrak{q}$ is regular if and only if $\mathfrak{q}(i-1)\not\cong\mathfrak{q}(i)$ for any $1 \le i \le n$; moreover, for any *injective* order-preserving map $\delta:\Delta_m \to \Delta_n$, if $\mathfrak{q}$ is regular then $\mathfrak{q}\circ\delta$ is regular too. This last fact leads us to consider the 2-subcategory $\mathbf{\Delta}^{\mathrm{r}}$ of $\mathbf{\Delta}$ formed by the same objects, the *injective* order-preserving maps and the corresponding natural maps. Actually, since the definition of the *differential map* (cf. A3.2.2) only needs injective order-preserving maps, we consider that the functors $\mathbb{H}_n$ (cf. A3.2), where n runs over $\mathbb{N}$, have been defined over the category of functors $\mathfrak{Fct}(\mathbf{\Delta}^{\mathrm{r}},\mathfrak{Ab})$.

A5.3 Coherently, we denote by $\mathfrak{st}^{\mathrm{r}}:\mathbf{\Delta}^{\mathrm{r}} \to \mathfrak{CC}$ the restriction to $\mathbf{\Delta}^{\mathrm{r}}$ of the *standard representation* of $\mathbf{\Delta}$ (cf. A2.2), by

$$\mathfrak{Fct}_{\mathrm{r}}(\mathfrak{st}^{\mathrm{r}},\mathfrak{C}):(\mathbf{\Delta}^{\mathrm{r}})^\circ \longrightarrow \mathfrak{CC} \qquad \text{A5.3.1}$$

the *subrepresentation* of $\mathfrak{Fct}(\mathfrak{st}^{\mathrm{r}},\mathfrak{C})$ mapping Δ_n on the full subcategory $\mathfrak{Fct}_{\mathrm{r}}(\Delta_n,\mathfrak{C})$ of $\mathfrak{Fct}(\Delta_n,\mathfrak{C})$ over the *regular $\mathfrak{C}$-chains* $\mathfrak{q}:\Delta_n \to \mathfrak{C}$, and by

$$\mathfrak{ch}_{\mathrm{r}}(\mathfrak{C}) = \mathfrak{Fct}_{\mathrm{r}}(\mathfrak{st}^{\mathrm{r}},\mathfrak{C})\rtimes(\mathbf{\Delta}^{\mathrm{r}})^\circ \qquad \text{A5.3.2}$$

the corresponding *category of regular $\mathfrak{C}$-chains*, which is obviously a 2-subcategory of $\mathfrak{ch}(\mathfrak{C})$; moreover, we call *regular* any $\mathfrak{ch}(\mathfrak{C})$-object $(\mathfrak{q},\Delta_n)$ such that $\mathfrak{q}$ is regular. Similarly, we still consider the *proper category of regular $\mathfrak{C}$-chains* $\mathfrak{ch}^*_{\mathrm{r}}(\mathfrak{C})$ which is the full 2-subcategory of $\mathfrak{ch}^*(\mathfrak{C})$ over the regular $\mathfrak{ch}(\mathfrak{C})$-objects. To some extent, the next result mutually justifies the introduction of the *category of connected components* (cf. A1.8) and of the *regularity*.

Proposition A5.4 *Let $\mathfrak{C}$ be an ordered category. The inclusion 2-functor $\mathfrak{ch}^*_{\mathrm{r}}(\mathfrak{C}) \to \mathfrak{ch}^*(\mathfrak{C})$ induces an equivalence of categories $\mathfrak{ch}^*_{\mathrm{r}}(\mathfrak{C}) \cong \overline{\mathfrak{ch}^*(\mathfrak{C})}$.*

Proof: If $(\mathfrak{q}, \Delta_n)$ and $(\mathfrak{r}, \Delta_m)$ are regular $\mathfrak{ch}^*(\mathfrak{C})$-objects, if

$$(\nu, \delta) : (\mathfrak{r}, \Delta_m) \to (\mathfrak{q}, \Delta_n) \quad \text{and} \quad (\nu', \delta') : (\mathfrak{r}, \Delta_m) \to (\mathfrak{q}, \Delta_n) \qquad \text{A5.4.1}$$

are $\mathfrak{ch}^*(\mathfrak{C})$-morphisms and if we have $\mathrm{Nat}_{\mathfrak{ch}^*(\mathfrak{C})}\big((\nu', \delta'), (\nu, \delta)\big) \neq \emptyset$, then we have $\delta(i) \le \delta'(i)$ for any $i \in \Delta_n$ and, since $\mathfrak{C}$ is an ordered category, the isomorphisms

$$\nu_i : \mathfrak{r}\big(\delta(i)\big) \cong \mathfrak{q}(i) \quad \text{and} \quad \nu'_i : \mathfrak{r}\big(\delta'(i)\big) \cong \mathfrak{q}(i) \qquad \text{A5.4.2}$$

force $\mathfrak{r}\big(\delta(i) \bullet \delta'(i)\big)$ to be an isomorphism (cf. A5.1); hence, since $\mathfrak{r}$ is regular, we have $\delta(i) = \delta'(i)$ and then, according to definition A2.6.4, we necessarily have $\nu_i = \nu'_i$; thus, if two $\mathfrak{ch}^*_{\mathrm{r}}(\mathfrak{C})$-morphisms (ν, δ) and $\big(\nu', \delta'\big)$ have the same class in $\overline{\mathfrak{ch}^*(\mathfrak{C})}$, arguing by induction on the length of a sequence of natural maps connecting them, we conclude that they coincide, proving the faithfulness of the functor $\mathfrak{ch}^*_{\mathrm{r}}(\mathfrak{C}) \to \overline{\mathfrak{ch}^*(\mathfrak{C})}$.

On the other hand, if $\mathfrak{q} : \Delta_n \to \mathfrak{C}$ and $\mathfrak{r} : \Delta_m \to \mathfrak{C}$ are two functors and $(\nu, \delta) : (\mathfrak{r}, \Delta_m) \to (\mathfrak{q}, \Delta_n)$ is a $\mathfrak{ch}^*(\mathfrak{C})$-morphism, it follows from Lemma A5.5 below that it suffices to replace $(\mathfrak{r}, \Delta_m)$ and $(\mathfrak{q}, \Delta_n)$ by $\overline{\mathfrak{ch}^*(\mathfrak{C})}$-isomorphic $\overline{\mathfrak{ch}^*(\mathfrak{C})}$-objects to obtain that $\mathfrak{q}(i \bullet i{+}1)$ and $\mathfrak{r}(j \bullet j{+}1)$ are not isomorphisms for any $i \in \Delta_n$ and any $j \in \Delta_m$, and then $\mathfrak{q}$ and $\mathfrak{r}$ are regular. In this case, if $i' < i$ and we assume that $\delta(i') = \delta(i)$ then the isomorphisms

$$\nu_{i'} : \mathfrak{r}\big(\delta(i')\big) \cong \mathfrak{q}(i') \quad \text{and} \quad \nu_i : \mathfrak{r}\big(\delta(i)\big) \cong \mathfrak{q}(i) \qquad \text{A5.4.3}$$

force $\mathfrak{q}(i') \cong \mathfrak{q}(i)$ and therefore $\mathfrak{q}(i' \bullet i)$ is an isomorphism (cf. A5.1), a contradiction. We are done.

Lemma A5.5 *Let $\mathfrak{C}$ be a small category. For any $\mathfrak{C}$-chain $\mathfrak{q} : \Delta_n \to \mathfrak{C}$ and any $i \in \Delta_n$ such that $\mathfrak{q}(i \bullet i{+}1)$ is a $\mathfrak{C}$-isomorphism, the $\mathfrak{ch}^*(\mathfrak{C})$-morphisms*

$$\begin{aligned} (\mathrm{id}_{\mathfrak{q}\circ\delta_i^{n-1}}, \delta_i^{n-1}) &: (\mathfrak{q}, \Delta_n) \longrightarrow (\mathfrak{q} \circ \delta_i^{n-1}, \Delta_{n-1}) \\ (\mathrm{id}_{\mathfrak{q}\circ\delta_{i+1}^{n-1}}, \delta_{i+1}^{n-1}) &: (\mathfrak{q}, \Delta_n) \longrightarrow (\mathfrak{q} \circ \delta_{i+1}^{n-1}, \Delta_{n-1}) \end{aligned} \qquad \text{A5.5.1}$$

determine isomorphisms in $\overline{\mathfrak{ch}^*(\mathfrak{C})}$.

Proof: Set $\mathfrak{r} = \mathfrak{q}\circ\delta_i^{n-1}$ and $\mathfrak{t} = \mathfrak{q}\circ\delta_{i+1}^{n-1}$; on the one hand, it follows from A3.1.1 that

$$\begin{aligned}(\mathrm{id}_{\mathfrak{r}}, \delta_i^{n-1}) \circ (\mu, \varepsilon_i^n) &= (\mathrm{id}_{\mathfrak{r}}, \mathrm{id}_{\Delta_{n-1}})\\ (\mathrm{id}_{\mathfrak{t}}, \delta_{i+1}^{n-1}) \circ (\nu, \varepsilon_i^n) &= (\mathrm{id}_{\mathfrak{t}}, \mathrm{id}_{\Delta_{n-1}})\end{aligned} \qquad \text{A5.5.2.}$$

On the other hand, we have evident natural $\boldsymbol{\Delta}^\circ$-maps

$$\alpha_n : \mathrm{id}_{\Delta_n} \longrightarrow \delta_i^{n-1} \circ \varepsilon_i^n \quad \text{and} \quad \beta_n : \delta_{i+1}^{n-1} \circ \varepsilon_i^n \longrightarrow \mathrm{id}_{\Delta_n} \qquad \text{A5.5.3}$$

which determine natural maps (cf. definition A2.5.3)

$$\begin{aligned}\mathfrak{Fct}(\alpha_n, \mathfrak{C}) &: \mathfrak{Fct}(\mathrm{id}_{\Delta_n}, \mathfrak{C}) \longrightarrow \mathfrak{Fct}(\delta_i^{n-1} \circ \varepsilon_i^n, \mathfrak{C})\\ \mathfrak{Fct}(\beta_n, \mathfrak{C}) &: \mathfrak{Fct}(\delta_{i+1}^{n-1} \circ \varepsilon_i^n, \mathfrak{C}) \longrightarrow \mathfrak{Fct}(\mathrm{id}_{\Delta_n}, \mathfrak{C})\end{aligned} \qquad \text{A5.5.4;}$$

thus, setting $\nu = \mathfrak{Fct}(\beta_n, \mathfrak{C})_{\mathfrak{q}}$ and $\mu = \big(\mathfrak{Fct}(\alpha_n, \mathfrak{C})_{\mathfrak{q}}\big)^{-1}$ which makes sense since $\mathfrak{q}(i \bullet i+1)$ is a $\mathfrak{C}$-isomorphism, we get the $\mathfrak{ch}^*(\mathfrak{C})$-natural maps (cf. A2.6.4)

$$\begin{array}{ccccc}(\nu, \delta_{i+1}^{n-1} \circ \varepsilon_i^n) & \xrightarrow{\beta_n} & (\mathrm{id}_{\mathfrak{q}}, \mathrm{id}_{\Delta_n}) & \xrightarrow{\alpha_n} & (\mu, \delta_i^{n-1} \circ \varepsilon_i^n)\\ \| & & & & \|\\ (\nu, \varepsilon_i^n) \circ (\mathrm{id}_{\mathfrak{t}}, \delta_{i+1}^{n-1}) & & & & (\mu, \varepsilon_i^n) \circ (\mathrm{id}_{\mathfrak{r}}, \delta_i^{n-1})\end{array} \qquad \text{A5.5.5,}$$

so that the classes of $(\mu, \varepsilon_i^n) \circ (\mathrm{id}_{\mathfrak{r}}, \delta_i^{n-1})$ and $(\nu, \varepsilon_i^n) \circ (\mathrm{id}_{\mathfrak{t}}, \delta_{i+1}^{n-1})$ in $\overline{\mathfrak{ch}^*(\mathfrak{C})}$ coincide with the class of $(\mathrm{id}_{\mathfrak{q}}, \mathrm{id}_{\Delta_n})$. We are done.

A5.6 Let $\mathfrak{C}$ be an *ordered category* and $\mathfrak{a}^* : \mathfrak{ch}^*(\mathfrak{C}) \to \mathfrak{Ab}$ a *contravariant* functor which factorizes *via* $\overline{\mathfrak{ch}^*(\mathfrak{C})}$; for any $n \in \mathbb{N}$, we say that an element $a = (a_{\mathfrak{q}})_{\mathfrak{q}\in\mathfrak{Fct}(\boldsymbol{\Delta}_n,\mathfrak{C})}$ in $\big((\mathfrak{l}_{\mathfrak{C}}^*)_*(\mathfrak{a}^*)\big)(\Delta_n)$ is *regular* or *singular* according to $a_{\mathfrak{q}} = 0$ for any *nonregular* or for any *regular* $\mathfrak{C}$-chain $\mathfrak{q} : \Delta_n \to \mathfrak{C}$ respectively. Then, denoting by

$$\mathfrak{l}_{\mathfrak{C}}^{\mathrm{r}} : \mathfrak{ch}_{\mathrm{r}}^*(\mathfrak{C}) \longrightarrow (\boldsymbol{\Delta}^{\mathrm{r}})^\circ \qquad \text{A5.6.1}$$

the structural functor and by $\mathfrak{a}_{\mathrm{r}}^*$ the restriction of $\mathfrak{a}^*$ to $\mathfrak{ch}_{\mathrm{r}}^*(\mathfrak{C})$, we can identify $\big((\mathfrak{l}_{\mathfrak{C}}^{\mathrm{r}})_*(\mathfrak{a}_{\mathrm{r}}^*)\big)(\Delta_n)$ with the quotient of $\big((\mathfrak{l}_{\mathfrak{C}}^*)_*(\mathfrak{a}^*)\big)(\Delta_n)$ by the subgroup of all the singular elements; actually, denoting by $\mathfrak{i}_{\mathrm{r}} : \boldsymbol{\Delta}^{\mathrm{r}} \to \boldsymbol{\Delta}$ the inclusion 2-functor, we have an evident natural map

$$(\mathfrak{l}_{\mathfrak{C}}^*)_*(\mathfrak{a}^*) \circ \mathfrak{i}_{\mathrm{r}} \longrightarrow (\mathfrak{l}_{\mathfrak{C}}^{\mathrm{r}})_*(\mathfrak{a}_{\mathrm{r}}^*) \qquad \text{A5.6.2.}$$

Finally, for any $n \in \mathbb{N}$, we define the *n-th regular cohomology group of* $\mathfrak{C}$ *over* $\mathfrak{a}^*$ by

$$\mathbb{H}_{\mathrm{r}}^n(\mathfrak{C}, \mathfrak{a}^*) = \mathbb{H}_n\big((\mathfrak{l}_{\mathfrak{C}}^{\mathrm{r}})_*(\mathfrak{a}_{\mathrm{r}}^*)\big) \qquad \text{A5.6.3.}$$

Proposition A5.7 *Let $\mathfrak{C}$ be an ordered category and $\mathfrak{a}^* : \mathfrak{ch}^*(\mathfrak{C}) \to \mathfrak{Ab}$ a contravariant functor which factorizes via $\overline{\mathfrak{ch}^*(\mathfrak{C})}$; denote by $\mathfrak{a}^*_{\mathrm{r}}$ the restriction of $\mathfrak{a}^*$ to $\mathfrak{ch}^*_{\mathrm{r}}(\mathfrak{C})$. For any $n \in \mathbb{N}$, the natural map*

$$(\mathfrak{l}^*_{\mathfrak{C}})_*(\mathfrak{a}^*) \circ \mathfrak{i}_{\mathrm{r}} \longrightarrow (\mathfrak{l}^{\mathrm{r}}_{\mathfrak{C}})_*(\mathfrak{a}^*_{\mathrm{r}}) \qquad \text{A5.7.1}$$

induces a group isomorphism

$$\mathbb{H}^n_*(\mathfrak{C}, \mathfrak{a}^*) \cong \mathbb{H}^n_{\mathrm{r}}(\mathfrak{C}, \mathfrak{a}^*) \qquad \text{A5.7.2.}$$

Proof: It is clear that, for any $n \in \mathbb{N}$, the subgroup of $(\mathfrak{l}^*_{\mathfrak{C}})_*(\mathfrak{a}^*)$ of all the regular elements maps bijectively onto $(\mathfrak{l}^{\mathrm{r}}_{\mathfrak{C}})_*(\mathfrak{a}^*_{\mathrm{r}})$; set $d_n = d_n^{(\mathfrak{l}^*_{\mathfrak{C}})_*(\mathfrak{a}^*)}$ and $d^{\mathrm{r}}_n = d_n^{(\mathfrak{l}^{\mathrm{r}}_{\mathfrak{C}})_*(\mathfrak{a}^*)}$ for short. For $n = 0$, we actually have the equality

$$\big((\mathfrak{l}^*_{\mathfrak{C}})_*(\mathfrak{a}^*)\big)(\Delta_0) = \big((\mathfrak{l}^{\mathrm{r}}_{\mathfrak{C}})_*(\mathfrak{a}^*_{\mathrm{r}})\big)(\Delta_0) \qquad \text{A5.7.3}$$

and moreover, for any $a = (a_{\mathfrak{q}})_{\mathfrak{q} \in \mathfrak{Fct}(\Delta_0, \mathfrak{C})}$ in $\big((\mathfrak{l}^*_{\mathfrak{C}})_*(\mathfrak{a}^*)\big)(\Delta_0)$ and *any nonregular* $\mathfrak{C}$-chain $\mathfrak{t} : \Delta_1 \to \mathfrak{C}$, the $\mathfrak{t}$-component $d_0(a)_{\mathfrak{t}}$ of the element $d_0(a)$ vanishes since $\mathfrak{t}(0\bullet 1)$ is an isomorphism; consequently, we get

$$\mathbb{H}^0_*(\mathfrak{C}, \mathfrak{a}^*) = \mathrm{Ker}(d_0) = \mathrm{Ker}(d^{\mathrm{r}}_0) = \mathbb{H}^0_{\mathrm{r}}(\mathfrak{C}, \mathfrak{a}^*) \qquad \text{A5.7.4.}$$

From now on, we assume that $n \geq 1$; it is clear that, for any a^{r} in $\mathrm{Ker}(d^{\mathrm{r}}_n)$, there is a unique regular element $a \in \big((\mathfrak{l}^{\mathfrak{o}}_{\mathfrak{C}})_*(\mathfrak{a}^*)\big)(\Delta_n)$ lifting a^{r} (cf. A3.11.1) and it is easily checked that a belongs to $\big((\mathfrak{l}^*_{\mathfrak{C}})_*(\mathfrak{a}^*)\big)(\Delta_n)$ (cf. A3.18); moreover, for any $\mathfrak{C}$-chain $\mathfrak{t}: \Delta_{n+1} \to \mathfrak{C}$ we have

$$d_n(a)_{\mathfrak{t}} = \sum_{i=0}^{n+1} \big(\mathfrak{a}^*(\mathrm{id}_{\mathfrak{t}\circ\delta^n_i}, \delta^n_i)\big)(a_{\mathfrak{t}\circ\delta^n_i}) \qquad \text{A5.7.5}$$

and, if $\mathfrak{t}$ is nonregular, there is $i \leq n$ such that, for a suitable $\mathfrak{C}$-chain $\mathfrak{t}'$ isomorphic to $\mathfrak{t}$, we have $\mathfrak{t}' \circ \delta^n_i = \mathfrak{t}' \circ \delta^n_{i+1}$ and $\mathfrak{t} \circ \delta^n_j$ remains nonregular for any $j \in \Delta_n - \{i, i{+}1\}$; thus, since a belongs to $\big((\mathfrak{l}^*_{\mathfrak{C}})_*(\mathfrak{a}^*)\big)(\Delta_n)$, we get

$$\big(\mathfrak{a}^*(\mathrm{id}_{\mathfrak{t}\circ\delta^n_i}, \delta^n_i)\big)(a_{\mathfrak{t}\circ\delta^n_i}) = \big(\mathfrak{a}^*(\mathrm{id}_{\mathfrak{t}\circ\delta^n_{i+1}}, \delta^n_{i+1})\big)(a_{\mathfrak{t}\circ\delta^n_{i+1}}) \qquad \text{A5.7.6}$$

and, since a is regular, we still get $a_{\mathfrak{t}\circ\delta^n_j} = 0$ for any $j \in \Delta_n - \{i, i{+}1\}$, so that we finally get $d_n(a)_{\mathfrak{t}} = 0$; but, if $\mathfrak{t}$ is regular, it is clear that

$$d_n(a)_{\mathfrak{t}} = d^{\mathrm{r}}_n(a^{\mathrm{r}})_{\mathfrak{t}} = 0 \qquad \text{A5.7.7;}$$

consequently, we have $d_n(a) = 0$ which shows the surjectivity of homomorphism A5.7.2.

On the other hand, it follows from Lemma A5.9 below that, for any element a in $\mathrm{Ker}(d_n)$ there exists $b \in \big((\mathfrak{l}^*_{\mathfrak{C}})_*(\mathfrak{a}^*)\big)(\Delta_{n-1})$ such that the difference $a - d_{n-1}(b)$ is *regular*; hence, denoting by a^{r} and b^{r} the respective images of a in $\big((\mathfrak{l}^{\mathrm{r}}_{\mathfrak{C}})_*(\mathfrak{a}^*_{\mathrm{r}})\big)(\Delta_n)$ and of b in $\big((\mathfrak{l}^{\mathrm{r}}_{\mathfrak{C}})_*(\mathfrak{a}^*_{\mathrm{r}})\big)(\Delta_{n-1})$, if we assume that for some $c^{\mathrm{r}} \in \big((\mathfrak{l}^{\mathrm{r}}_{\mathfrak{C}})_*(\mathfrak{a}^*_{\mathrm{r}})\big)(\Delta_{n-1})$ we have

$$a^{\mathrm{r}} - d^{\mathrm{r}}_{n-1}(b^{\mathrm{r}}) = d^{\mathrm{r}}_{n-1}(c^{\mathrm{r}}) \qquad \text{A5.7.8}$$

then, for the unique regular element $c \in \big((\mathfrak{l}^*_{\mathfrak{C}})_*(\mathfrak{a}^*)\big)(\Delta_{n-1})$ lifting c^{r} we still have

$$a - d_{n-1}(b) = d_{n-1}(c) \qquad \text{A5.7.9.}$$

We are done.

A5.8 Note that an element $a = (a_{\mathfrak{q}})_{\mathfrak{q}\in\mathfrak{Fct}(\Delta_n,\mathfrak{C})}$ of $\big((\mathfrak{l}^*_{\mathfrak{C}})_*(\mathfrak{a}^*)\big)(\Delta_n)$ (cf. A3.18) is regular if and only if we have $a_{\mathfrak{q}} = 0$ for any $\mathfrak{C}$-chain $\mathfrak{q}\colon \Delta_n \to \mathfrak{C}$ such that for some $i \in \Delta_{n-1}$ we have

$$\mathfrak{q}(i) = \mathfrak{q}(i{+}1) \quad \text{and} \quad \mathfrak{q}(i \bullet (i{+}1)) = \mathrm{id}_{\mathfrak{q}(i)} \qquad \text{A5.8.1}$$

and, as a matter of fact, in the next lemma we can deal with any element of $\big((\mathfrak{l}^{\circ}_{\mathfrak{C}})_*(\mathfrak{a}^*)\big)(\Delta_n)$ (cf. A3.7) fulfilling this condition, without any hypothesis on $\mathfrak{C}$. Thus, let us say that a $\mathfrak{C}$-chain $\mathfrak{q}\colon \Delta_n \to \mathfrak{C}$ is *redundant* whenever it fulfills condition A5.8.1 for some $i \in \Delta_{n-1}$, and then, that an element $a = (a_{\mathfrak{q}})_{\mathfrak{q}\in\mathfrak{Fct}(\Delta_n,\mathfrak{C})}$ in $\big((\mathfrak{l}^{\circ}_{\mathfrak{C}})_*(\mathfrak{a}^*)\big)(\Delta_n)$ is *regular* if $a_{\mathfrak{q}} = 0$ for any *redundant* $\mathfrak{C}$-chain $\mathfrak{q} : \Delta_n \to \mathfrak{C}$.

Lemma A5.9 *Let $\mathfrak{a}^{\circ} : \mathfrak{ch}^{\circ}(\mathfrak{C}) \to \mathfrak{Ab}$ be a contravariant functor which factorizes via $\overline{\mathfrak{ch}^{\circ}(\mathfrak{C})}$ and set $d_n = d_n^{(\mathfrak{l}^{\circ}_{\mathfrak{C}})_*(\mathfrak{a}^{\circ})}$. For any $a \in \big((\mathfrak{l}^{\circ}_{\mathfrak{C}})_*(\mathfrak{a}^{\circ})\big)(\Delta_n)$ such that the element $d_n(a)$ is regular, there is $b \in \big((\mathfrak{l}^{\circ}_{\mathfrak{C}})_*(\mathfrak{a}^{\circ})\big)(\Delta_{n-1})$ such that the difference $a - d_{n-1}(b)$ is regular too. Moreover, if $\mathfrak{a}^{\circ}$ can be extended to a contravariant functor $\mathfrak{a}^* : \mathfrak{ch}^*(\mathfrak{C}) \to \mathfrak{Ab}$ and the element a belongs to $\big((\mathfrak{l}^*_{\mathfrak{C}})_*(\mathfrak{a}^*)\big)(\Delta_n)$ then b can be chosen in $\big((\mathfrak{l}^*_{\mathfrak{C}})_*(\mathfrak{a}^*)\big)(\Delta_{n-1})$.*

Proof: If a is not regular, then there is $\mathfrak{q} \in \mathfrak{Fct}(\Delta_n, \mathfrak{C})$ such that $a_{\mathfrak{q}} \neq 0$ while, for some $i < n$, it fulfills condition A5.8.1; that is to say, we have $\mathfrak{q} = \mathfrak{r}\circ\varepsilon^n_i$ for some $\mathfrak{r} \in \mathfrak{Fct}(\Delta_{n-1}, \mathfrak{C})$. We argue by induction on the minimal i fulfilling condition A5.8.1 and follow Eilenberg–MacLane's construction (cf. Theorem 4.1 in [23]); in particular, we are assuming that $a_{\mathfrak{r}\circ\varepsilon^n_\ell} = 0$ for any $\ell < i$ and any $\mathfrak{r} \in \mathfrak{Fct}(\Delta_{n-1}, \mathfrak{C})$ and that the element

$$d_n(a) = \Big(\sum_{j=0}^{n+1} (-1)^j \big(\mathfrak{a}^{\circ}(\mathrm{id}_{\mathfrak{t}\circ\delta^n_j}, \delta^n_j)\big)(a_{\mathfrak{t}\circ\delta^n_j})\Big)_{\mathfrak{t}\in\mathfrak{Fct}(\Delta_{n+1},\mathfrak{C})} \qquad \text{A5.9.1}$$

is regular.

But, for any $j < i$ we have $\varepsilon_i^{n+1} \circ \delta_j^n = \delta_j^{n-1} \circ \varepsilon_{i-1}^n$ (cf. A3.1.1) and therefore, according to the minimality of i, for any $\mathfrak{q} \in \mathfrak{Fct}(\Delta_n, \mathfrak{C})$, we still have $a_{\mathfrak{q}\circ(\varepsilon_i^{n+1}\circ\delta_j^n)} = 0$, so that we get

$$0 = \sum_{j=i}^{n+1} (-1)^j \big(\mathfrak{a}^{\mathrm{o}}(\mathrm{id}_{\mathfrak{q}\circ(\varepsilon_i^{n+1}\circ\delta_j^n)}, \delta_j^n)\big)(a_{\mathfrak{q}\circ(\varepsilon_i^{n+1}\circ\delta_j^n)}) \qquad \text{A5.9.2;}$$

moreover, we have $\varepsilon_i^{n+1} \circ \delta_i^n = \mathrm{id}_{\Delta_n} = \varepsilon_i^{n+1} \circ \delta_{i+1}^n$ (cf. A3.1.1), so that equality A5.9.2 becomes

$$\begin{aligned} &(-1)^i \big(\mathfrak{a}^{\mathrm{o}}(\mathrm{id}_{\mathfrak{q}}, \delta_i^n) - \mathfrak{a}^{\mathrm{o}}(\mathrm{id}_{\mathfrak{q}}, \delta_{i+1}^n)\big)(a_{\mathfrak{q}}) \\ &\quad = \sum_{j=i+1}^{n} (-1)^j \big(\mathfrak{a}^{\mathrm{o}}(\mathrm{id}_{\mathfrak{q}\circ(\varepsilon_i^{n+1}\circ\delta_{j+1}^n)}, \delta_{j+1}^n)\big)(a_{\mathfrak{q}\circ(\varepsilon_i^{n+1}\circ\delta_{j+1}^n)}) \end{aligned} \qquad \text{A5.9.3.}$$

Now, in $\big((\mathfrak{l}_{\mathfrak{C}}^{\mathrm{o}})_*(\mathfrak{a}^{\mathrm{o}})\big)(\Delta_{n-1})$ consider the element

$$b = \big(\mathfrak{a}^{\mathrm{o}}(\mathrm{id}_{\mathfrak{r}\circ\varepsilon_i^n}, \varepsilon_i^n)(a_{\mathfrak{r}\circ\varepsilon_i^n})\big)_{\mathfrak{r}\in\mathfrak{Fct}(\Delta_{n-1},\mathfrak{C})} \qquad \text{A5.9.4;}$$

note that, if we assume that $\mathfrak{a}^{\mathrm{o}}$ can be extended to a *contravariant* functor $\mathfrak{a}^* : \mathfrak{ch}^*(\mathfrak{C}) \to \mathfrak{Ab}$ and then that a belongs to $\big((\mathfrak{l}_{\mathfrak{C}}^*)_*(\mathfrak{a}^*)\big)(\Delta_n)$ (cf. 3.18), for any $\mathfrak{r}' \in \mathfrak{Fct}(\Delta_{n-1}, \mathfrak{C})$ isomorphic to $\mathfrak{r}$, and any natural isomorphism $\mu : \mathfrak{r} \cong \mathfrak{r}'$, we have the natural isomorphism

$$\mu * \varepsilon_i^n : \mathfrak{r} \circ \varepsilon_i^n \cong \mathfrak{r}' \circ \varepsilon_i^n \qquad \text{A5.9.5}$$

and therefore we get

$$\begin{aligned} \mathfrak{a}^*(\mu, \mathrm{id}_{\Delta_{n-1}})\Big(\big(\mathfrak{a}^*(\mathrm{id}_{\mathfrak{r}'\circ\varepsilon_i^n}, \varepsilon_i^n)\big)(a_{\mathfrak{r}'\circ\varepsilon_i^n})\Big) &= \big(\mathfrak{a}^*(\mu * \varepsilon_i^n, \varepsilon_i^n)\big)(a_{\mathfrak{r}'\circ\varepsilon_i^n}) \\ &= \mathfrak{a}^*(\mathrm{id}_{\mathfrak{r}\circ\varepsilon_i^n}, \varepsilon_i^n)\Big(\big(\mathfrak{a}^*(\mu * \varepsilon_i^n, \mathrm{id}_{\Delta_n})\big)(a_{\mathfrak{r}'\circ\varepsilon_i^n})\Big) \\ &= \big(\mathfrak{a}^*(\mathrm{id}_{\mathfrak{r}\circ\varepsilon_i^n}, \varepsilon_i^n)\big)(a_{\mathfrak{r}\circ\varepsilon_i^n}), \end{aligned} \qquad \text{A5.9.6,}$$

so that b belongs to $\big((\mathfrak{l}_{\mathfrak{C}}^*)_*(\mathfrak{a}^*)\big)(\Delta_{n-1})$.

In any case, for any $\mathfrak{q} \in \mathfrak{Fct}(\Delta_n, \mathfrak{C})$, the $\mathfrak{q}$-component of $d_{n-1}(b)$ is

$$d_{n-1}(b)_{\mathfrak{q}} = \sum_{j=0}^{n} (-1)^j \big(\mathfrak{a}^{\mathrm{o}}(\mathrm{id}_{\mathfrak{q}\circ(\delta_j^{n-1}\circ\varepsilon_i^n)}, \delta_j^{n-1} \circ \varepsilon_i^n)\big)(a_{\mathfrak{q}\circ(\delta_j^{n-1}\circ\varepsilon_i^n)}) \qquad \text{A5.9.7.}$$

First of all, let us consider the $(\mathfrak{r} \circ \varepsilon_\ell^n)$-component of $d_{n-1}(b)$ for any $\ell < i$ and any $\mathfrak{r} \in \mathfrak{Fct}(\Delta_{n-1}, \mathfrak{C})$; for any $\ell < i$, it is easily checked that

$$\varepsilon_\ell^n \circ \delta_j^{n-1} \circ \varepsilon_i^n = \begin{cases} \delta_j^{n-2} \circ \varepsilon_{i-1}^{n-1} \circ \varepsilon_{\ell-1}^n & \text{if } j < \ell \\ \delta_{j-1}^{n-2} \circ \varepsilon_{i-1}^{n-1} \circ \varepsilon_\ell^n & \text{if } j > \ell + 1 \end{cases} \qquad \text{A5.9.8,}$$

and therefore, according to the minimality of i, for any $j \in \Delta_n - \{\ell, \ell+1\}$ we get $a_{\mathfrak{r}\circ(\varepsilon_\ell^n \circ \delta_j^{n-1} \circ \varepsilon_i^n)} = 0$; hence, in this case this component, up to a sign, becomes (cf. A5.9.7)

$$\big(\mathfrak{a}^\circ(\mathrm{id}_{\mathfrak{r}\circ\varepsilon_i^n}, \delta_\ell^{n-1} \circ \varepsilon_i^n)\big)(a_{\mathfrak{r}\circ\varepsilon_i^n}) - \big(\mathfrak{a}^\circ(\mathrm{id}_{\mathfrak{r}\circ\varepsilon_i^n}, \delta_{\ell+1}^{n-1} \circ \varepsilon_i^n)\big)(a_{\mathfrak{r}\circ\varepsilon_i^n}) \qquad \text{A5.9.9.}$$

Moreover, since $\varepsilon_\ell^n \circ \delta_\ell^{n-1} = \mathrm{id}_{\Delta_{n-1}} = \varepsilon_\ell^n \circ \delta_{\ell+1}^{n-1}$ (cf. A3.1.1), we have the following $\mathfrak{ch}^\circ(\mathfrak{C})$-morphisms

$$\begin{aligned} (\mathrm{id}_\mathfrak{r}, \delta_\ell^{n-1}) &: (\mathfrak{r} \circ \varepsilon_\ell^n, \Delta_n) \longrightarrow (\mathfrak{r}, \Delta_{n-1}) \\ (\mathrm{id}_\mathfrak{r}, \delta_{\ell+1}^{n-1}) &: (\mathfrak{r} \circ \varepsilon_\ell^n, \Delta_n) \longrightarrow (\mathfrak{r}, \Delta_{n-1}) \end{aligned} \qquad \text{A5.9.10}$$

which have the same class in $\overline{\mathfrak{ch}^\circ(\mathfrak{C})}$; indeed, the $\boldsymbol{\Delta}$-natural map $\delta_{\ell+1}^{n-1} \bullet \delta_\ell^{n-1}$ from $\delta_{\ell+1}^{n-1}$ to δ_ℓ^{n-1} determines a natural map (cf. A2.5.3)

$$\mathfrak{Fct}(\delta_{\ell+1}^{n-1} \bullet \delta_\ell^{n-1}, \mathfrak{C}) : \mathfrak{Fct}(\delta_{\ell+1}^{n-1}, \mathfrak{C}) \longrightarrow \mathfrak{Fct}(\delta_\ell^{n-1}, \mathfrak{C}) \qquad \text{A5.9.11}$$

which sends $\mathfrak{r} \circ \varepsilon_\ell^n$ to $\mathrm{id}_\mathfrak{r}$ and therefore, by definition A2.6.4, $\delta_{\ell+1}^{n-1} \bullet \delta_\ell^{n-1}$ is also a $\mathfrak{ch}^\circ(\mathfrak{C})$-natural map between the morphisms A5.9.10. A first consequence of this fact is that, since we are assuming that $\mathfrak{a}^\circ$ factorizes *via* $\overline{\mathfrak{ch}^\circ(\mathfrak{C})}$, the difference A5.9.9 is equal to zero and therefore the $(\mathfrak{r} \circ \varepsilon_\ell^n)$-component of $d_n(b)$ vanishes for any $\ell < i$. A second consequence is that, setting $\varepsilon_{i,i}^{n+1} = \varepsilon_i^n \circ \varepsilon_i^{n+1}$, equality A5.9.3 applied to $\mathfrak{q} = \mathfrak{r} \circ \varepsilon_i^n$ becomes

$$0 = \sum_{j=i+1}^{n} (-1)^j \big(\mathfrak{a}^\circ(\mathrm{id}_{\mathfrak{r}\circ(\varepsilon_{i,i}^{n+1} \circ \delta_{j+1}^n)}, \delta_{j+1}^n)\big)(a_{\mathfrak{r}\circ(\varepsilon_{i,i}^{n+1} \circ \delta_{j+1}^n)}) \qquad \text{A5.9.12.}$$

Secondly, let us consider the $(\mathfrak{r} \circ \varepsilon_i^n)$-component of $d_{n-1}(b)$; it is easily checked that (cf. A3.1.1)

$$\varepsilon_i^n \circ \delta_j^{n-1} \circ \varepsilon_i^n = \begin{cases} \delta_j^{n-1} \circ \varepsilon_{i-1,i-1}^n & \text{if } j < i \\ \varepsilon_{i,i}^{n+1} \circ \delta_{j+1}^n & \text{if } j > i+1 \end{cases} \qquad \text{A5.9.13}$$

and recall again that $\varepsilon_i^n \circ \delta_i^{n-1} = \mathrm{id}_{\Delta_{n-1}} = \varepsilon_i^n \circ \delta_{i+1}^{n-1}$; thus, according to equality A5.9.7, this component becomes

$$\begin{aligned} &\sum_{j=i+2}^{n} (-1)^j \big(\mathfrak{a}^\circ(\mathrm{id}_{\mathfrak{r}\circ(\varepsilon_i^n \circ \delta_j^{n-1} \circ \varepsilon_i^n)}, \delta_j^{n-1} \circ \varepsilon_i^n)\big)(a_{\mathfrak{r}\circ(\varepsilon_i^n \circ \delta_j^{n-1} \circ \varepsilon_i^n)}) \\ &= \sum_{j=i+2}^{n} (-1)^j \big(\mathfrak{a}^\circ(\mathrm{id}_{\mathfrak{r}\circ(\varepsilon_{i,i}^{n+1} \circ \delta_{j+1}^n)}, \varepsilon_i^{n+1} \circ \delta_{j+1}^n)\big)(a_{\mathfrak{r}\circ(\varepsilon_{i,i}^{n+1} \circ \delta_{j+1}^n)}) \qquad \text{A5.9.14,} \\ &= (-1)^i \big(\mathfrak{a}^\circ(\mathrm{id}_{\mathfrak{r}\circ\varepsilon_i^n}, \delta_{i+1}^{n-1} \circ \varepsilon_i^n)\big)(a_{\mathfrak{r}\circ\varepsilon_i^n}) \end{aligned}$$

since the second member of the these equalities coincides with the image *via* $\mathfrak{a}^\circ(\mathrm{id}_{\mathfrak{r}\circ\varepsilon_{i,i}^{n+1}}, \varepsilon_i^{n+1})$ of the sum in A5.9.12 without the first summand.

Finally, we claim that the class of the $\mathfrak{ch}^o(\mathfrak{C})$-morphism

$$(\mathrm{id}_{\mathfrak{r}\circ\varepsilon_i^n}, \delta_{i+1}^{n-1}\circ\varepsilon_i^n) : \mathfrak{r}\circ\varepsilon_i^n \longrightarrow \mathfrak{r}\circ\varepsilon_i^n \qquad \text{A5.9.15}$$

in the category $\overline{\mathfrak{ch}^o(\mathfrak{C})}$ coincides with the corresponding identity morphism; indeed, we have a unique natural map $\nu : \delta_{i+1}^{n-1}\circ\varepsilon_i^n \rightarrow \mathrm{id}_{\Delta_n}$ sending $i+1$ to $i \bullet (i+1)$, which determines a natural map (cf. A2.5.3)

$$\mathfrak{Fct}(\nu, \mathfrak{C}) : \mathfrak{Fct}(\delta_{i+1}^{n-1}\circ\varepsilon_i^n, \mathfrak{C}) \longrightarrow \mathfrak{id}_{\mathfrak{Fct}(\Delta_n,\mathfrak{C})} \qquad \text{A5.9.16}$$

sending $\mathfrak{r}\circ\varepsilon_i^n$ to $\mathrm{id}_{\mathfrak{r}\circ\varepsilon_i^n}$, and therefore, according to definition A2.6.4, ν is also a $\mathfrak{ch}^o(\mathfrak{C})$-natural map from $(\mathrm{id}_{\mathfrak{r}\circ\varepsilon_i^n}, \delta_{i+1}^{n-1}\circ\varepsilon_i^n)$ to $(\mathrm{id}_{\mathfrak{r}\circ\varepsilon_i^n}, \mathrm{id}_{\Delta_n})$.

In conclusion, for any $\ell \leq i$ and any $\mathfrak{r} \in \mathfrak{Fct}(\Delta_{n-1}, \mathfrak{C})$ the $(\mathfrak{r}\circ\varepsilon_\ell^n)$-component of the element $a - (-1)^i d_{n-1}(b)$ vanishes and, since

$$d_n\big(a - (-1)^i d_{n-1}(b)\big) = d_n(a) \qquad \text{A5.9.17}$$

is regular, it follows from our induction hypothesis that there is an element $c = (c_\mathfrak{r})_{\mathfrak{r}\in\mathfrak{Fct}(\Delta_{n-1},\mathfrak{C})}$ in $\big((\mathfrak{l}_\mathfrak{C}^o)_*(\mathfrak{a}^o)\big)(\Delta_{n-1})$ such that the difference $a - d_{n-1}\big((-1)^i\, b + c\big)$ is regular. Moreover, if $\mathfrak{a}^o$ can be extended to a *contravariant* functor $\mathfrak{a}^* : \mathfrak{ch}^*(\mathfrak{C}) \rightarrow \mathfrak{Ab}$ and a belongs to $\big((\mathfrak{l}_\mathfrak{C}^*)_*(\mathfrak{a}^*)\big)(\Delta_n)$, we already know that b belongs to $\big((\mathfrak{l}_\mathfrak{C}^*)_*(\mathfrak{a}^*)\big)(\Delta_{n-1})$ and then it follows from our induction hypothesis that c can be chosen in $\big((\mathfrak{l}_\mathfrak{C}^*)_*(\mathfrak{a}^*)\big)(\Delta_{n-1})$ too. We are done.

References

[1] Jon Alperin, *Sylow intersections and fusion*, Journal of Algebra, 1 (1964), 110-113.

[2] Jon Alperin, *Weights for finite groups*, in Proc. Symp. Pure Math., 47 (1987), 369-379, Amer. Math. Soc., Providence.

[3] Jon Alperin and Michel Broué, *Local methods in block theory*, Annals of Math., 110 (1979), 143-157.

[4] Michael Aschbacher, *Normal subsystems of fusion systems*, preprint.

[5] Dave Benson, *Cohomology of sporadic groups, finite loop spaces and the Dickson invariants*, Proc. 1994 Durham Conference, London Math. Soc., Springer Lecture Notes, 252 (1998), 10-23.

[6] Serge Bouc, *The Dade group of a p-group,* Inventiones mathematicæ 164 (2006), 189-231.

[7] Serge Bouc and Jacques Thévenaz, *Gluing torsion endo-permutation modules*, preprint, 2006.

[8] Richard Brauer, *Investigations on group characters,* Annals of Math., 42 (1941), 937-958.

[9] Richard Brauer, *On the structure of blocks of characters of finite groups,* Proc. Second International Conference on Groups, Camberra, 1973, 103-130.

[10] Michel Broué, *Les ℓ-blocs des groupes $GL(n,q)$ et $U(n,q^2)$, et leurs structures locales,* Astérisque, 133/134 (1986), 159-188.

[11] Michel Broué and Lluís Puig, *Characters and Local Structure in G-algebras,* Journal of Algebra, 63 (1980), 306-317.

[12] Michel Broué and Lluís Puig, *A Frobenius theorem for blocks,* Inventiones mathematicæ, 56 (1980), 117-128.

[13] Carles Broto, Ran Levi and Bob Oliver, *The homotopy theory of fusion systems*, Journal of Amer. Math. Soc. 16 (2003), 779-856.

[14] Carles Broto, Ran Levi and Bob Oliver, *The theory of p-local groups, a survey*, Contemp. Math., 346 (2004), Amer. Math. Soc., Providence.

[15] Carles Broto, Natàlia Castellana, Jesper Grodal, Ran Levi and Bob Oliver, *Subgroup families controlling p-local finite groups*, Proc. Lond. Math. Soc., 91 (2005), 325-254.

[16] Jon Carlson, *The dimension of modules and their restriction over modular group algebras*, Journal of Algebra 69 (1981), 95-105.

[17] Jon Carlson and Jacques Thévenaz, *The classification of torsion endo-trivial modules*, Annals of Math., 162 (2005), 823-883.

[18] Henri Cartan and Samuel Eilenberg, *"Homological Algebra"*, Princeton Math. 19, 1956, Princeton University Press.

[19] Everett Dade, *Endopermutation modules over p-groups, I and II,* Annals of Math. 107 (1978), 459-494, and 108 (1978), 317-346.

[20] Everett Dade, *A correspondence of characters*, in Proc. Symp. Pure Math. 37 (1980), 401-403, Amer. Math. Soc., Providence.

[21] Everett Dade, *Counting characters in Blocks, I*, Inventiones mathematicæ 109 (1992), 187-210.

[22] Andreas Dress, *Contributions to the theory of induced representations*, in "*Algebraic K-theory II*", Lecture Notes in Math. 342 (1973), 183-240.

[23] Samuel Eilenberg and Saunders Mac Lane, *On the groups* $\mathbb{H}(\Pi, n)$*, I*, Annals of Math., 58 (1953), 55-106.

[24] Yun Fan and Lluís Puig, *On blocks with nilpotent coefficient extensions*, Algebras and Representation Theory, 1 (1998), 27-73, and Publisher revised form, 2 (1999), 209.

[25] Walter Feit and John Thompson, *Solvability of groups of odd order*, Pacific Journal of Math., 13 (1963), 775-1029.

[26] Georg Frobenius, *Über aufläsbare Gruppen, V*, Sitzungs. der König. Preu. Akad. der Wissen. zu Berlin, 1324-1329 (1901).

[27] George Glauberman, *A characteristic subgroup of a p-stable group*, Can. J. Math., 20 (1968), 1101-1135.

[28] Daniel Gorenstein, *"Finite groups"* Harper's Series, 1968, Harper and Row.

[29] James Green, *Some remarks on defect groups*, Math. Zeit., 107 (1968), 133-150.

[30] James Green, *Functors on categories of finite group representations*, Journal of Pure and Applied Algebra, 37 (1985), 265-298.

[31] Bertram Huppert, *"Endliche Gruppen I"*, Die Grundlehren der math. Wissenschaften, 134 (1967), Springer-Verlag, Berlin.

[32] Stefan Jackowski and James McClure, *Homotopy decomposition of classifying spaces via elementary abelian subgroups*, Topology, 31 (1992), 113-132.

[33] Reinhard Knörr and Geoffrey Robinson, *Some remarks on a conjecture of Alperin*, Journal of London Math. Soc. 39 (1989), 48-60.

[34] Burkhard Külshammer and Lluís Puig, *Extensions of nilpotent blocks*, Inventiones mathematicæ, 102 (1990), 17-71.

[35] Lluís Puig, *Structure locale dans les groupes finis*, Bulletin Soc. Math. France, Mémoire 47 (1976).

[36] Lluís Puig, *The source algebra of a nilpotent block*, preprint, Mathematisches Forschungsinstitut Oberwolfach, June 1981.

[37] Lluís Puig, *Pointed groups and construction of characters*, Math. Zeit. 176 (1981), 265-292.

[38] Lluís Puig, *La classification des groupes finis simples: bref apperçu et quelques conséquences*, Astérisque, 92/93 (1982), 101-128.

[39] Lluís Puig, *Local fusions in block source algebras*, Journal of Algebra, 104 (1986), 358-369.

[40] Lluís Puig, *Local extensions in endopermutation modules split: a proof of Dade's theorem*, in *"Seminaire sur les Groupes Finis III"*, Publications Math. de Paris VII, 25 (1986).

[41] Lluís Puig, *Nilpotent blocks and their source algebras*, Inventiones mathematicæ, 93 (1988), 77-116.

[42] Lluís Puig, *Pointed groups and construction of modules*, Journal of Algebra, 116 (1988), 7-129.

[43] Lluís Puig, *Affirmative answer to a question of Feit*, Journal of Algebra, 131 (1990), 513-526.

[44] Lluís Puig, *On the general structure of a block*, manuscript, MSRI, 1990.

[45] Lluís Puig, *Une correspondance de modules pour les blocks à groupes de défaut abéliens*, Geometriæ Dedicata, 37 (1991), 9-43.

[46] Lluís Puig, *Brauer-Frobenius categories*, manuscript, Paris, 1992.

[47] Lluís Puig, *On Joanna Scopes' Criterion of Equivalence for Blocks of Symmetric Groups*, Algebra Colloquium 1 (1994), 25-55.

[48] Lluís Puig, *"On the Morita and Rickard equivalences between Brauer blocks"*, Progress in Math., 1999, Birkhäuser, Basel.

[49] Lluís Puig, *The hyperfocal subalgebra of a block*, Inventiones mathematicæ, 141 (2000), 365-397.

[50] Lluís Puig, *Source algebras of p-central group extensions*, Journal of Algebra, 235 (2001), 359-398.

[51] Lluís Puig, *"Blocks of Finite Groups"*, Springer Monographs in Mathematics, 2002, Springer-Verlag, Berlin, Barcelona.

[52] Lluís Puig, *Frobenius categories*, Journal of Algebra, 303 (2006), 309-357.

[53] Lluís Puig, *Block Source Algebras in p-Solvable Groups*, to appear in Michigan Journal of Mathematics, 2009.

[54] Jean-Pierre Serre, *"Linear Representations of Finite Groups"*, Graduate Texts in Mathematics 42, 1977, Springer-Verlag, Berlin.

[55] Jean-Pierre Serre, *"Local Fields"*, Graduate Texts in Mathematics 67, 1979, Springer-Verlag, Berlin, Barcelona.

[56] Ronald Solomon, *Finite groups with Sylow* 2*-subgroup of type .3*, Journal of Algebra 28 (1974), 182-198.

[57] John Thompson , *Finite groups with fixed-point-free automorphisms of prime order*, Proc. Nat. Acad. Sci., 45 (1959), 578-581.

[58] John Thompson , *Quadratic pairs*, in "Actes du Congrés International de Math., 1970", Gauthier-Villars, Paris, 1971.

Index

d

e

f

g

h

i

j

k